P9-BHU-281

STUDENT'S SOLUTIONS MANUAL

JUDITH A. PENNA

Indiana University Purdue University Indianapolis

PREALGEBRA

FIFTH EDITION

Marvin L. Bittinger

Indiana University Purdue University Indianapolis

David J. Ellenbogen

Community College of Vermont

Barbara L. Johnson

Indiana University Purdue University Indianapolis

PEARSON
Addison
Wesley

Boston San Francisco New York
London Toronto Sydney Tokyo Singapore Madrid
Mexico City Munich Paris Cape Town Hong Kong Montreal

Reproduced by Pearson Addison-Wesley from electronic files supplied by the author.

Copyright © 2008 Pearson Education, Inc.
Publishing as Pearson Addison-Wesley, 75 Arlington Street, Boston, MA 02116.

ISBN-13   978-0-321-33707-8
ISBN-10   0-321-33707-7

2 3 4 5 6 BB 09 08 07

Contents

Chapter 1

Whole Numbers

Exercise Set 1.1

1. 2 3 $\boxed{5}$, 8 8 8

The digit 5 means 5 thousands.

3. 1, 4 8 8, $\boxed{5}$ 2 6

The digit 5 means 5 hundreds.

5. 1, 5 8 $\boxed{2}$, 3 7 0

The digit 2 names the numbers of thousands.

7. $\boxed{1}$, 5 8 2, 3 7 0

The digit 1 names the number of millions.

9. 5702 = 5 thousands + 7 hundreds + 0 tens + 2 ones, or 5 thousands + 7 hundreds + 2 ones

11. 93,986 = 9 ten thousands + 3 thousands + 9 hundreds + 8 tens + 6 ones

13. 2058 = 2 thousands + 0 hundreds + 5 tens + 8 ones, or 2 thousands + 5 tens + 8 ones

15. 1268 = 1 thousand + 2 hundreds + 6 tens + 8 ones

17. 519,955 = 5 hundred thousands + 1 ten thousand + 9 thousands + 9 hundreds + 5 tens + 5 ones

19. 308,845 = 3 hundred thousands + 0 ten thousands + 8 thousands + 8 hundreds + 4 tens + 5 ones, or 3 hundred thousands + 8 thousands + 8 hundreds + 4 tens + 5 ones

21. 4,302,737 = 4 millions + 3 hundred thousands + 0 ten thousands + 2 thousands + 7 hundreds + 3 tens + 7 ones, or 4 millions + 3 hundred thousands + 2 thousands + 7 hundreds + 3 tens + 7 ones

23. A word name for 85 is eighty-five.

25.

88,000

Eighty-eight thousand

27.

123,765

One hundred twenty-three thousand,
seven hundred sixty-five

29.

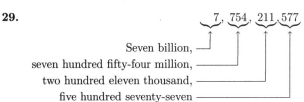

7, 754, 211,577

Seven billion,
seven hundred fifty-four million,
two hundred eleven thousand,
five hundred seventy-seven

31.

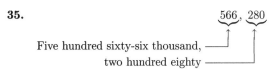

Two million,
two hundred thirty-three thousand,
eight hundred twelve

Standard notation is   2, 233, 812.

33.

Eight billion

Standard notation is   8,000,000,000.

35.

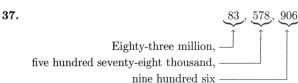

566, 280

Five hundred sixty-six thousand,
two hundred eighty

37.

83, 578, 906

Eighty-three million,
five hundred seventy-eight thousand,
nine hundred six

39.

Nine trillion,
four hundred sixty billion,

Standard notation is   9,460,000,000,000.

41.

Sixty-four million,
one hundred eighty-six thousand,

Standard notation is   64, 186, 000.

43. Discussion and Writing Exercise

45. First consider the whole numbers from 100 through 199. The 10 numbers 102, 112, 122, . . . , 192 contain the digit 2. In addition, the 10 numbers 120, 121, 122, . . . , 129 contain the digit 2. However, we do not count the number 122 in this group because it was counted in the first group of ten numbers. Thus, 19 numbers from 100 through 199 contain the digit 2. Using the same type of reasoning for the whole numbers from 300 to 400, we see that there are also 19 numbers in this group that contain the digit 2.

Finally, consider the 100 whole numbers 200 through 299. Each contains the digit 2.

Thus, there are 19 + 19 + 100, or 138 whole numbers between 100 and 400 that contain the digit 2 in their standard notation.

Exercise Set 1.2

1.

| One truck hauls 6 cu yd. | The other truck hauls 8 cu yd. | Altogether, they haul 14 cu yd. |

6 cu yd + 8 cu yd = 14 cu yd

3.

| One parcel contains 500 acres. | The other parcel contains 300 acres. | The total purchase is 800 acres. |

500 acres + 300 acres = 800 acres

5.
$$\begin{array}{r} 3\,6\,4 \\ +\ \ \ 2\,3 \\ \hline 3\,8\,7 \end{array}$$
Add ones, add tens, then add hundreds.

7.
$$\begin{array}{r} \overset{1}{} \\ 1\,7\,1\,6 \\ +\,3\,4\,8\,2 \\ \hline 5\,1\,9\,8 \end{array}$$
Add ones: We get 8. Add tens: We get 9 tens. Add hundreds: We get 11 hundreds, or 1 thousand + 1 hundred. Write 1 in the hundreds column and 1 above the thousands. Add thousands: We get 5 thousands.

9.
$$\begin{array}{r} \overset{1}{} \\ 8\,6 \\ +\,7\,8 \\ \hline 1\,6\,4 \end{array}$$
Add ones: We get 14 ones, or 1 ten + 4 ones. Write 4 in the ones column and 1 above the tens. Add tens: We get 16 tens.

11.
$$\begin{array}{r} \overset{1}{} \\ 9\,9 \\ +\ \ 1 \\ \hline 1\,0\,0 \end{array}$$
Add ones: We get 10 ones, or 1 ten + 0 ones. Write 0 in the ones column and 1 above the tens. Add tens: We get 10 tens.

13.
$$\begin{array}{r} \overset{1}{} \\ 8\,1\,1\,3 \\ +\ \ \ 3\,9\,0 \\ \hline 8\,5\,0\,3 \end{array}$$
Add ones: We get 3. Add tens: We get 10 tens, or 1 hundred + 0 tens. Write 0 in the tens column and 1 above the hundreds. Add hundreds: We get 5. Add thousands: We get 8.

15.
$$\begin{array}{r} \overset{1}{} \\ 3\,5\,6 \\ +\,4\,9\,1\,0 \\ \hline 5\,2\,6\,6 \end{array}$$
Add ones: We get 6. Add tens: We get 6. Add hundreds: We get 12 hundreds, or 1 thousand + 2 hundreds. Write 2 in the hundreds column and 1 above the thousands. Add thousands: We get 5.

17.
$$\begin{array}{r} \overset{1\,2\,1}{} \\ 3\,8\,7\,0 \\ 9\,2 \\ 7 \\ +\ \ \ 4\,9\,7 \\ \hline 4\,4\,6\,6 \end{array}$$
Add ones: We get 16 ones, or 1 ten + 6 ones. Write 6 in the ones column and 1 above the tens. Add tens: We get 26 tens, or 2 hundreds + 6 tens. Write 6 in the tens column and 2 above the hundreds. Add hundreds: We get 14 hundreds, or 1 thousand + 4 hundreds. Write 4 in the hundreds column and 1 above the thousands. Add thousands: We get 4.

19.
$$\begin{array}{r} \overset{1\ \ 1}{} \\ 4\,8\,2\,5 \\ +\,1\,7\,8\,3 \\ \hline 6\,6\,0\,8 \end{array}$$
Add ones: We get 8. Add tens: We get 10 tens. Write 0 in the tens column and 1 above the hundreds. Add hundreds: We get 16 hundreds. Write 6 in the hundreds column and 1 above the thousands. Add thousands: We get 6 thousands.

21.
$$\begin{array}{r} \overset{1\ \ 1\ \ 1}{} \\ 2\,3,4\,4\,3 \\ +\,1\,0,9\,8\,9 \\ \hline 3\,4,4\,3\,2 \end{array}$$
Add ones: We get 12 ones, or 1 ten + 2 ones. Write 2 in the ones column and 1 above the tens. Add tens: We get 13 tens. Write 3 in the tens column and 1 above the hundreds. Add hundreds: We get 14 hundreds. Write 4 in the hundreds column and 1 above the thousands. Add thousands: We get 4 thousands. Add ten thousands: We get 3 ten thousands.

23.
$$\begin{array}{r} \overset{1\ 1\ 1\ 1}{} \\ 7\,7,5\,4\,3 \\ +\,2\,3,7\,6\,7 \\ \hline 1\,0\,1,3\,1\,0 \end{array}$$
Add ones: We get 10 ones, or 1 ten + 0 ones. Write 0 in the ones column and 1 above the tens. Add tens: We get 11 tens. Write 1 in the tens column and 1 above the hundreds. Add hundreds: We get 13 hundreds. Write 3 in the hundreds column and 1 above the thousands. Add thousands: We get 11 thousands. Write 1 in the thousands column and 1 above the ten thousands. Add ten thousands: We get 10 ten thousands.

25. We look for pairs of numbers whose sums are 10, 20, 30, and so on.

$$\begin{array}{r} 45 \longrightarrow \quad 70 \\ 25 \nearrow \\ 36 \longrightarrow \quad 80 \\ 44 \nearrow \\ +80 \longrightarrow \quad 80 \\ \hline 230 \qquad\quad 230 \end{array}$$

27.
$$\begin{array}{r} \overset{1\ \ 1}{} \\ 1\,2,0\,7\,0 \\ 2\,9\,5\,4 \\ +\ \ \ 3\,4\,0\,0 \\ \hline 1\,8,4\,2\,4 \end{array}$$
Add ones: We get 4. Add tens: We get 12 tens, or 1 hundred + 2 tens. Write 2 in the tens column and 1 above the hundreds. Add hundreds: We get 14 hundreds, or 1 thousand + 4 hundreds. Write 4 in the hundreds column and 1 above the hundreds. Add thousands: We get 8 thousands. Add ten thousands: We get 1 ten thousand.

29.
$$\begin{array}{r} \overset{3\ 1\ 2}{} \\ 4\,8\,3\,5 \\ 7\,2\,9 \\ 9\,2\,0\,4 \\ 8\,9\,8\,6 \\ +\,7\,9\,3\,1 \\ \hline 3\,1,6\,8\,5 \end{array}$$
Add ones: We get 25. Write 5 in the ones column and 2 above the tens. Add tens: We get 18 tens. Write 8 in the tens column and 1 above the hundreds. Add hundreds: We get 36 hundreds. Write 6 in the hundreds column and 3 above the thousands. Add thousands: We get 31 thousands.

31. We regroup.

$$(2+5)+4 = 2+(5+4)$$

33. We regroup:

$6 + (3 + 2) = (6 + 3) + 2$

35. We reverse the order of the addends.

$2 + 7 = 7 + 2$

37. We reverse the order of the addends.

$6 + 1 = 1 + 6$

39. We reverse the order of the addends.

$2 + 9 = 9 + 2$

41. Add from the top.

We first add 7 and 9, getting 16; then 16 and 4, getting 20; then 20 and 8, getting 28.

Check by adding from the bottom.

We first add 8 and 4, getting 12; then 12 and 9, getting 21; then 21 and 7, getting 28.

43. Add from the top.

Check:

45. Perimeter = 14 mi + 13 mi + 8 mi + 10 mi + 47 mi + 22 mi

We carry out the addition.

$$\begin{array}{r} {\scriptstyle 2} \\ 1\,4 \\ 1\,3 \\ 8 \\ 1\,0 \\ 4\,7 \\ +\;2\,2 \\ \hline 1\,1\,4 \end{array}$$

The perimeter of the figure is 114 mi.

47. Perimeter = 200 ft + 85 ft + 200 ft + 85 ft

We carry out the addition.

$$\begin{array}{r} {\scriptstyle 1}\;{\scriptstyle 1} \\ 2\,0\,0 \\ 8\,5 \\ 2\,0\,0 \\ +\;\;8\,5 \\ \hline 5\,7\,0 \end{array}$$

The perimeter of the hockey rink is 570 ft.

49. Discussion and Writing Exercise

51. 4 ⟨8⟩ 6, 2 0 5

The digit 8 tells the number of ten thousands.

53. Discussion and Writing Exercise

55. $5,987,943 + 328,959 + 49,738,765$

Using a calculator to carry out the addition, we find that the sum is 56,055,667.

57. One method is described in the answer section in the text. Another method is: $1 + 100 = 101, \; 2 + 99 = 101, \ldots, 50 + 51 = 101$. Then $50 \cdot 101 = 5050$.

Exercise Set 1.3

1.
Number to begin with		Number spent		Number left
20	−	4	=	16

3.
Amount to begin with		Amount sold		Amount left
126 oz	−	13 oz	=	113 oz

5. $7 - 4 = 3$

This number gets added (after 3).

$7 = 3 + 4$

(By the commutative law of addition, $7 = 4 + 3$ is also correct.)

7. $13 - 8 = 5$

This number gets added (after 5).

$13 = 5 + 8$

(By the commutative law of addition, $13 = 8 + 5$ is also correct.)

9. $23 - 9 = 14$

This number gets added (after 14).

$23 = 14 + 9$

(By the commutative law of addition, $23 = 9 + 14$ is also correct.)

11. $43 - 16 = 27$

This number gets added (after 27).

$43 = 27 + 16$

(By the commutative law of addition, $43 = 16 + 27$ is also correct.)

13. $6 + 9 = 15$ $6 + 9 = 15$
    ↑   ↑
This addend gets   This addend gets
subtracted from   subtracted from
the sum.   ↓   the sum.   ↓
    $6 = 15 - 9$ $9 = 15 - 6$

15. $8 + 7 = 15$ $8 + 7 = 15$
    ↑   ↑
This addend gets   This addend gets
subtracted from   subtracted from
the sum.   ↓   the sum.   ↓
    $8 = 15 - 7$ $7 = 15 - 8$

17. $17 + 6 = 23$ $17 + 6 = 23$
    ↑   ↑
This addend gets   This addend gets
subtracted from   subtracted from
the sum.   ↓   the sum.   ↓
    $17 = 23 - 6$ $6 = 23 - 17$

19. $23 + 9 = 32$ $23 + 9 = 32$
    ↑   ↑
This addend gets   This addend gets
subtracted from   subtracted from
the sum.   ↓   the sum.   ↓
    $23 = 32 - 9$ $9 = 32 - 23$

21. We first write an addition sentence. Keep in mind that all numbers are in millions.

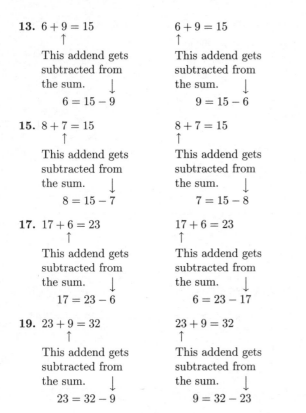

Now we write a related subtraction sentence.

$17 + \boxed{} = 32$

$\boxed{} = 32 - 17$ The addend 17 gets subtracted.

We have $32 - 17 = 15$.

23. We first write an addition sentence.

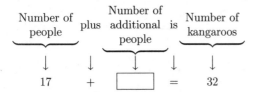

Now we write a related subtraction sentence.

$10 + \boxed{} = 23$

$\boxed{} = 23 - 10$ The addend 10 gets subtracted.

We have $23 - 10 = 13$.

25.
$$\begin{array}{r} 6\,5 \\ -\ 2\,1 \\ \hline 4\,4 \end{array}$$
Subtract ones, then subtract tens.

27.
$$\begin{array}{r} 8\,6\,6 \\ -\ 3\,3\,3 \\ \hline 5\,3\,3 \end{array}$$
Subtract ones, subtract tens, then subtract hundreds.

29.
$$\begin{array}{r} {}^{7}\!\!\!\not{8}\ {}^{16}\!\!\!\not{0} \\ -\ 4\,7 \\ \hline 3\,9 \end{array}$$
We cannot subtract 7 ones from 6 ones. Borrow 1 ten to get 16 ones. Subtract ones, then subtract tens.

31.
$$\begin{array}{r} 9\ {}^{7}\!\!\!\not{8}\ {}^{11}\!\!\!\not{1} \\ -\ 7\,4\,7 \\ \hline 2\,3\,4 \end{array}$$
We cannot subtract 7 ones from 1 one. Borrow 1 ten to get 11 ones. Subtract ones, subtract tens, then subtract hundreds.

33.
$$\begin{array}{r} 7\ {}^{6}\!\!\!\not{7}\ {}^{16}\!\!\!\not{6}\ 9 \\ -\ 2\,3\,8\,7 \\ \hline 5\,3\,8\,2 \end{array}$$
Subtract ones. We cannot subtract 8 tens from 6 tens. Borrow 1 hundred to get 16 tens. Subtract tens, subtract hundreds, then subtract thousands.

35.
$$\begin{array}{r} 7\ {}^{6}\!\!\!\not{6}\ {}^{16}\!\!\!\not{4}\ {}^{3}\!\!\!\not{4}\ {}^{10}\!\!\!\not{0} \\ -\ 3\,8\,0\,9 \\ \hline 3\,8\,3\,1 \end{array}$$
We cannot subtract 9 ones from 0 ones. Borrow 1 ten to get 10 ones. Subtract ones, then tens. We cannot subtract 8 hundreds from 6 hundreds. Borrow 1 thousand to get 16 hundreds. Subtract hundreds, then thousands.

37.
$$\begin{array}{r} {}^{11}\!\!\!\not{1}\ {}^{15}\!\!\!\not{2},\ {}^{13}\!\!\!\not{6}\ {}^{3}\!\!\!\not{4}\ {}^{17}\!\!\!\not{7} \\ -\quad 4\,8\,9\,9 \\ \hline 7\,7\,4\,8 \end{array}$$

39.
$$\begin{array}{r} {}^{8}\!\!\!\not{9}\ {}^{10}\!\!\!\not{0},\ 2\ {}^{2}\!\!\!\not{3}\ {}^{17}\!\!\!\not{7} \\ -\ 4\,7,\,2\,0\,9 \\ \hline 4\,3,\,0\,2\,8 \end{array}$$

41.
$$\begin{array}{r} {}^{7}\!\!\!\not{8}\ {}^{10}\!\!\!\not{0} \\ -\ 2\,4 \\ \hline 5\,6 \end{array}$$

43.
$$\begin{array}{r} 6\ {}^{8}\!\!\!\not{9}\ {}^{10}\!\!\!\not{0} \\ -\ 2\,3\,6 \\ \hline 4\,5\,4 \end{array}$$

45.
$$\begin{array}{r} 6\ {}^{7}\!\!\!\not{8}\ {}^{9}\!\!\!\not{0}\ {}^{18}\!\!\!\not{8} \\ -\ 3\,0\,5\,9 \\ \hline 3\,7\,4\,9 \end{array}$$
We have 8 hundreds or 80 tens. We borrow 1 ten to get 18 ones. We then have 79 tens. Subtract ones, then tens, then hundreds, then thousands.

47.
$$\begin{array}{r} {\scriptstyle 2\ 9\ 10} \\ 2\,\cancel{3\,0\,0} \\ -\quad 1\,0\,9 \\ \hline 2\,1\,9\,1 \end{array}$$

We have 3 hundreds or 30 tens. We borrow 1 ten to get 10 ones. We then have 29 tens. Subtract ones, then tens, then hundreds, then thousands.

49.
$$\begin{array}{r} {\scriptstyle 15} \\ {\scriptstyle \cancel{5}\ 10} \\ \cancel{1}\ \cancel{6}\ \cancel{0} \\ -\quad 7\,4 \\ \hline 8\,6 \end{array}$$

51.
$$\begin{array}{r} {\scriptstyle 3\ 10} \\ 7\,8\,\cancel{4}\,\cancel{0} \\ -3\,0\,2\,7 \\ \hline 4\,8\,1\,3 \end{array}$$

53.
$$\begin{array}{r} {\scriptstyle 13} \\ {\scriptstyle 7\ \cancel{3}\ 13} \\ 5\,\cancel{8}\,\cancel{4}\,\cancel{3} \\ -\quad 9\,8 \\ \hline 5\,7\,4\,5 \end{array}$$

55.
$$\begin{array}{r} {\scriptstyle 10\ 16} \\ {\scriptstyle 9\ \cancel{0}\ \cancel{6}\ 13} \\ \cancel{1\,0}\,1,\cancel{7}\,\cancel{3}\,4 \\ -\quad\quad 5\,7\,6\,0 \\ \hline 9\,5,9\,7\,4 \end{array}$$

57.
$$\begin{array}{r} {\scriptstyle 9\ 9\ 9\ 14} \\ \cancel{1\,0,0\,0}\,4 \\ -\quad\quad 2\,9 \\ \hline 9\,9\,7\,5 \end{array}$$

We have 1 ten thousand, or 1000 tens. We borrow 1 ten to get 10 ones. We then have 999 tens. Subtract ones, then tens, then hundreds, then thousands.

59.
$$\begin{array}{r} {\scriptstyle 8\ 9\ 17} \\ 8\,3,\cancel{9\,0}\,7 \\ -\quad\quad 8\,9 \\ \hline 8\,3,8\,1\,8 \end{array}$$

61.
$$\begin{array}{r} {\scriptstyle 6\ 9\ 9\ 10} \\ \cancel{7\,0\,0\,0} \\ -2\,7\,9\,4 \\ \hline 4\,2\,0\,6 \end{array}$$

We have 7 thousands or 700 tens. We borrow 1 ten to get 10 ones. We then have 699 tens. Subtract ones, then tens, then hundreds, then thousands.

63.
$$\begin{array}{r} {\scriptstyle 7\ 9\ 9\ 10} \\ 4\,\cancel{8,0\,0\,0} \\ -3\,7,6\,9\,5 \\ \hline 1\,0,3\,0\,5 \end{array}$$

We have 8 thousands or 800 tens. We borrow 1 ten to get 10 ones. We then have 799 tens. Subtract ones, then tens, then hundreds, then thousands, then ten thousands.

65. Discussion and Writing Exercise

67.
$$\begin{array}{r} {\scriptstyle 1\ 1} \\ 9\,4\,6 \\ +\quad 7\,8 \\ \hline 1\,0\,2\,4 \end{array}$$

Add ones: We get 14. Write 4 in the ones column and 1 above the tens. Add tens: We get 12. Write 2 in the tens column and 1 above the hundreds. Add hundreds: We get 10 hundreds.

69.
$$\begin{array}{r} {\scriptstyle 1\ 1\quad 1} \\ 5\,7,8\,7\,7 \\ +3\,2,4\,0\,6 \\ \hline 9\,0,2\,8\,3 \end{array}$$

Add ones: We get 13. Write 3 in the ones column and 1 above the tens. Add tens: We get 8. Add hundreds: We get 12. Write 2 in the hundreds column and 1 above the thousands. Add thousands: We get 10. Write 0 in the thousands column and 1 above the ten thousands. Add ten thousands: We get 9 ten thousands.

71.
$$\begin{array}{r} {\scriptstyle 1\ 1} \\ 5\,6\,7 \\ +7\,7\,8 \\ \hline 1\,3\,4\,5 \end{array}$$

Add ones: We get 15. Write 5 in the ones column and 1 above the tens. Add tens: We get 14. Write 4 in the tens column and 1 above the hundreds. Add hundreds: We get 13 hundreds.

73.
$$\begin{array}{r} {\scriptstyle 1\ 1\quad 1} \\ 1\,2,8\,8\,5 \\ +\quad 9\,8\,0\,7 \\ \hline 2\,2,6\,9\,2 \end{array}$$

Add ones: We get 12. Write 2 in the ones column and 1 above the tens. Add tens: We get 9. Add hundreds: We get 16. Write 6 in the hundreds column and 1 above the thousands. Add thousands: We get 12. Write 2 in the thousands column and 1 above the ten thousands. Add ten thousands. We get 2 ten thousands.

75.

$$6, \overbrace{375}, \overbrace{602}$$

Six million, ⎯

three hundred seventy-five thousand, ⎯

six hundred two ⎯

77. Discussion and Writing Exercise

79. $3,928,124 - 1,098,947$

Using a calculator to carry out the subtraction, we find that the difference is 2,829,177.

81.
$$\begin{array}{r} 9,\_\,4\,8,6\,2\,1 \\ -2,0\,9\,7,\_\,8\,1 \\ \hline 7,2\,5\,1,1\,4\,0 \end{array}$$

To subtract tens, we borrow 1 hundred to get 12 tens.

$$\begin{array}{r} {\scriptstyle 5\ 12} \\ 9,\_\,4\,8,\cancel{6}\,\cancel{2}\,1 \\ -2,0\,9\,7,\_\,8\,1 \\ \hline 7,2\,5\,1,1\,4\,0 \end{array}$$

In order to have 1 hundred in the difference, the missing digit in the subtrahend must be 4 $(5-4=1)$.

$$\begin{array}{r} {\scriptstyle 5\ 12} \\ 9,\_\,4\,8,\cancel{6}\,\cancel{2}\,1 \\ -2,0\,9\,7,4\,8\,1 \\ \hline 7,2\,5\,1,1\,4\,0 \end{array}$$

In order to subtract ten thousands, we must borrow 1 hundred thousand to get 14 ten thousands. The number of hundred thousands left must be 2 since the hundred thousands place in the difference is 2 $(2-0=2)$. Thus, the missing digit in the minuend must be $2+1$, or 3.

$$\begin{array}{r} {\scriptstyle 2\ 14\quad 5\ 12} \\ 9,\cancel{3}\,\cancel{4}\,8,\cancel{6}\,\cancel{2}\,1 \\ -2,0\,9\,7,4\,8\,1 \\ \hline 7,2\,5\,1,1\,4\,0 \end{array}$$

Exercise Set 1.4

1. Round 48 to the nearest ten.

4 $\boxed{8}$
↑

The digit 4 is in the tens place. Consider the next digit to the right. Since the digit, 8, is 5 or higher, round 4 tens up to 5 tens. Then change the digit to the right of the tens digit to zero.

The answer is 50.

3. Round 467 to the nearest ten.

4 6 $\boxed{7}$
  ↑

The digit 6 is in the tens place. Consider the next digit to the right. Since the digit, 7, is 5 or higher, round 6 tens up to 7 tens. Then change the digit to the right of the tens digit to zero.

The answer is 470.

5. Round 731 to the nearest ten.

7 3 $\boxed{1}$
  ↑

The digit 3 is in the tens place. Consider the next digit to the right. Since the digit, 1, is 4 or lower, round down, meaning that 3 tens stays as 3 tens. Then change the digit to the right of the tens digit to zero.

The answer is 730.

7. Round 895 to the nearest ten.

8 9 $\boxed{5}$
  ↑

The digit 9 is in the tens place. Consider the next digit to the right. Since the digit, 5, is 5 or higher, we round up. The 89 tens become 90 tens. Then change the digit to the right of the tens digit to zero.

The answer is 900.

9. Round 146 to the nearest hundred.

1 $\boxed{4}$ 6
↑

The digit 1 is in the hundreds place. Consider the next digit to the right. Since the digit, 4, is 4 or lower, round down, meaning that 1 hundred stays as 1 hundred. Then change all digits to the right of the hundreds digit to zeros.

The answer is 100.

11. Round 957 to the nearest hundred.

9 $\boxed{5}$ 7
↑

The digit 9 is in the hundreds place. Consider the next digit to the right. Since the digit, 5, is 5 or higher, round up. The 9 hundreds become 10 hundreds. Then change all digits to the right of the hundreds digit to zeros.

The answer is 1000.

13. Round 9079 to the nearest hundred.

9 0 $\boxed{7}$ 9
    ↑

The digit 0 is in the hundreds place. Consider the next digit to the right. Since the digit, 7, is 5 or higher, round 0 hundreds up to 1 hundred. Then change all digits to the right of the hundreds digit to zeros.

The answer is 9100.

15. Round 32,850 to the nearest hundred.

3 2, 8 $\boxed{5}$ 0
       ↑

The digit 8 is in the hundreds place. Consider the next digit to the right. Since the digit, 5, is 5 or higher, round 8 hundreds up to 9 hundreds. Then change all digits to the right of the hundreds digit to zero.

The answer is 32,900.

17. Round 5876 to the nearest thousand.

5 $\boxed{8}$ 7 6
  ↑

The digit 5 is in the thousands place. Consider the next digit to the right. Since the digit, 8, is 5 or higher, round 5 thousands up to 6 thousands. Then change all digits to the right of the thousands digit to zeros.

The answer is 6000.

19. Round 7500 to the nearest thousand.

7 $\boxed{5}$ 0 0
  ↑

The digit 7 is in the thousands place. Consider the next digit to the right. Since the digit, 5, is 5 or higher, round 7 thousands up to 8 thousands. Then change all the digits to the right of the thousands digit to zeros.

The answer is 8000.

21. Round 45,340 to the nearest thousand.

4 5, $\boxed{3}$ 4 0
     ↑

The digit 5 is in the thousands place. Consider the next digit to the right. Since the digit, 3, is 4 or lower, round down, meaning that 5 thousands stays as 5 thousands. Then change all the digits to the right of the thousands digit to zeros.

The answer is 45,000.

23. Round 373,405 to the nearest thousand.

3 7 3, $\boxed{4}$ 0 5
       ↑

The digit 3 is in the thousands place. Consider the next digit to the right. Since the digit, 4, is 4 or lower, round down, meaning that 3 thousands stays as 3 thousands. Then change all the digits to the right of the thousands digit to zeros.

The answer is 373,000.

25.
	Rounded to the nearest ten

```
    7 8        8 0
  + 9 7      + 1 0 0
  -----      -------
             1 8 0  ← Estimated answer
```

27.
	Rounded to the nearest ten

```
  8 0 7 4      8 0 7 0
 − 2 3 4 7   − 2 3 5 0
 --------    --------
             5 7 2 0  ← Estimated answer
```

29.
	Rounded to the nearest ten

```
    4 5        5 0
    7 7        8 0
    2 5        3 0
  + 5 6      + 6 0
  -----      -----
  3 4 3      2 2 0  ← Estimated answer
```

The sum 343 seems to be incorrect since 220 is not close to 343.

31.
	Rounded to the nearest ten

```
  6 2 2      6 2 0
    7 8        8 0
    8 1        8 0
 + 1 1 1    + 1 1 0
 ------     -------
  9 3 2      8 9 0  ← Estimated answer
```

The sum 932 seems to be incorrect since 890 is not close to 932.

33.
	Rounded to the nearest hundred

```
  7 3 4 8        7 3 0 0
 + 9 2 4 7      + 9 2 0 0
 --------       --------
              1 6, 5 0 0  ← Estimated answer
```

35.
	Rounded to the nearest hundred

```
  6 8 5 2      6 9 0 0
 − 1 7 4 8    − 1 7 0 0
 --------     --------
              5 2 0 0  ← Estimated answer
```

37. We round the cost of each option to the nearest hundred and add.

```
  7 4 5 0      7 5 0 0
  1 5 9 5      1 6 0 0
  1 5 4 0      1 5 0 0
 +  6 2 5     +  6 0 0
 --------     --------
            1 1, 2 0 0
```

The estimated cost is $11,200.

39. We round the cost of each option to the nearest hundred and add.

```
    8 8 2 0      8 8 0 0
    2 8 7 0      2 9 0 0
    6 2 4 5      6 2 0 0
  +  9 8 5     + 1 0 0 0
  --------     --------
              1 8, 9 0 0
```

The estimated cost is $18,900. Since this is more than Sara and Ben's budget of $17,700, they cannot afford their choices.

41. Answers will vary depending on the options chosen.

43.
	Rounded to the nearest hundred

```
  2 1 6      2 0 0
    8 4      1 0 0
  7 4 5      7 0 0
 + 5 9 5    + 6 0 0
 ------     -------
  1 6 4 0    1 6 0 0  ← Estimated answer
```

The sum 1640 seems to be correct since 1600 is close to 1640.

45.
	Rounded to the nearest hundred

```
  7 5 0      8 0 0
  4 2 8      4 0 0
    6 3      1 0 0
 + 2 0 5    + 2 0 0
 ------     -------
  1 4 4 6    1 5 0 0  ← Estimated answer
```

The sum 1446 seems to be correct since 1500 is close to 1446.

47.
	Rounded to the nearest thousand

```
  9 6 4 3      1 0, 0 0 0
  4 8 2 1       5 0 0 0
  8 9 4 3       9 0 0 0
 + 7 0 0 4     + 7 0 0 0
 --------      ---------
             3 1, 0 0 0  ← Estimated answer
```

49.
	Rounded to the nearest thousand

```
  9 2, 1 4 9      9 2, 0 0 0
 − 2 2, 5 5 5   − 2 3, 0 0 0
 ----------     ----------
              6 9, 0 0 0  ← Estimated answer
```

51.

Since 0 is to the left of 17, 0 < 17.

53.

Since 34 is to the right of 12, 34 > 12.

55.

Since 1000 is to the left of 1001, 1000 < 1001.

57.

Since 133 is to the right of 132, 133 > 132.

59.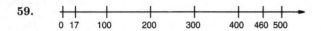

Since 460 is to the right of 17, $460 > 17$.

61.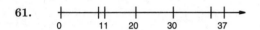

Since 37 is to the right of 11, $37 > 11$.

63. Since 2,083,660 lies to the left of 2,296,335 on the number line, we can write $2,083,660 < 2,296,335$.

Conversely, since 2,296,335 lies to the right of 2,083,660 on the number line, we could also write $2,296,335 > 2,083,660$.

65. Since 6482 lies to the right of 4641 on the number line, we can write $6482 > 4641$.

Conversely, since 4641 lies to the left of 6482 on the number line, we can also write $4641 < 6482$.

67. Discussion and Writing Exercise

69. $7992 = 7$ thousands $+ 9$ hundreds $+ 9$ tens $+ 2$ ones

71.
$$246,\ 605,\ 004,032$$

Two hundred forty-six billion, ⎯

six hundred five million, ⎯

four thousand, ⎯

thirty-two ⎯

73.
```
  1  1  1 1
  6 7, 7 8 9
+ 1 8, 9 6 5
-----------
  8 6, 7 5 4
```
Add ones. We get 14. Write 4 in the ones column and 1 above the tens. Add tens: We get 15 tens. Write 5 in the tens column and 1 above the hundreds. Add hundreds: We get 17 hundreds. Write 7 in the hundreds column and 1 above the thousands. Add thousands: We get 16 thousands. Write 6 in the thousands column and 1 above the ten thousands. Add ten thousands: We get 8 ten thousands.

75.
```
       16
   5  6 17
   6 7, 7 8 9
 - 1 8, 9 6 5
 -----------
   4 8, 8 2 4
```
Subtract ones: We get 4. Subtract tens: We get 2. We cannot subtract 9 hundreds from 7 hundreds. We borrow 1 thousand to get 17 hundreds. Subtract hundreds. We cannot subtract 8 thousands from 6 thousands. We borrow 1 ten thousand to get 16 thousands. Subtract thousands, then ten thousands.

77. Discussion and Writing Exercise

79. Using a calculator, we find that the sum is 30,411. This is close to the estimated sum found in Exercise 47.

81. Using a calculator, we find that the difference is 69,594. This is close to the estimated difference found in Exercise 49.

Exercise Set 1.5

1. Think of a rectangular array consisting of 21 rows with 21 objects in each row.

$21 \cdot 21 = 441$

3. Repeated addition fits best in this case.

$$\underbrace{\boxed{12\ \text{oz}} + \boxed{12\ \text{oz}} + \boxed{12\ \text{oz}} + \cdots + \boxed{12\ \text{oz}}}_{8\ \text{addends}}$$

$8 \cdot 12\ \text{oz} = 96\ \text{oz}$

5. Think of a rectangular array consisting of 4800 rows with 1200 objects in each row.

$4800 \cdot 1200 = 5,760,000$

7.
```
    8 7
  × 1 0
  -----
  8 7 0
```
Multiplying by 1 ten (We write 0 and then multiply 87 by 1.)

9.
```
      2 3 4 0
    × 1 0 0 0
    ---------
  2, 3 4 0, 0 0 0
```
Multiplying by 1 thousand (We write 000 and then multiply 2340 by 1.)

11.
```
      4
      6 5
  ×     8
  -------
    5 2 0
```
Multiplying by 8

13.
```
      2
      9 4
  ×     6
  -------
    5 6 4
```
Multiplying by 6

15.
```
      2
      5 0 9
  ×       3
  ---------
    1 5 2 7
```
Multiplying by 3

17.
```
    1 2 6
    9 2 2 9
  ×       7
  ---------
  6 4, 6 0 3
```
Multiplying by 7

19.
```
      2
      5 3
  ×   9 0
  -------
  4 7 7 0
```
Multiplying by 9 tens (We write 0 and then multiply 53 by 9.)

21.
```
      2
      3
      8 5
  ×   4 7
  -------
    5 9 5
  3 4 0 0
  -------
  3 9 9 5
```
Multiplying by 7
Multiplying by 40
Adding

23.
```
        2
      6 4 0
    ×   7 2
    1 2 8 0      Multiplying by 2
  4 4 8 0 0      Multiplying by 70
  4 6, 0 8 0     Adding
```

25.
```
      1 1
      1 1
      4 4 4
    ×   3 3
    1 3 3 2      Multiplying by 3
  1 3 3 2 0      Multiplying by 30
  1 4, 6 5 2     Adding
```

27.
```
          3
          7
        5 0 9
    ×   4 0 8
      4 0 7 2    Multiplying by 8
  2 0 3 6 0 0    Multiplying by 4 hundreds (We write 00
  2 0 7, 6 7 2   and then multiply 509 by 4.)
```

29.
```
        4 2
        1
        3 1
        8 5 3
    ×   9 3 6
      5 1 1 8    Multiplying by 6
    2 5 5 9 0    Multiplying by 30
  7 6 7 7 0 0    Multiplying by 900
  7 9 8, 4 0 8   Adding
```

31.
```
        1     2
        1
              1
        1 1 3
      6 4 2 8
    ×   3 2 2 4
    2 5 7 1 2      Multiplying by 4
  1 2 8 5 6 0      Multiplying by 20
  1 2 8 5 6 0 0    Multiplying by 200
  1 9 2 8 4 0 0 0  Multiplying by 3000
  2 0, 7 2 3, 8 7 2  Adding
```

33.
```
        1 3
      3 4 8 2
    ×   1 0 4
    1 3 9 2 8    Multiplying by 4
  3 4 8 2 0 0    Multiplying by 1 hundred  (We write 00
  3 6 2, 1 2 8   and then multiply 3482 by 1.)
```

35.
```
          2
          4
        5 0 0 6
    ×   4 0 0 8
      4 0 0 4 8    Multiplying by 8
  2 0 0 2 4 0 0 0  Multiplying by 4 thousands  (We
  2 0, 0 6 4, 0 4 8  write 000 and then multiply 5006 by
                     4.)
```

37.
```
        2       3
        3       4
        5 6 0 8
    ×   4 5 0 0
    2 8 0 4 0 0 0      Multiplying by 5 hundreds (We write
    2 2 4 3 2 0 0 0    00 and then multiply 5608 by 5.)
    2 5, 2 3 6, 0 0 0  Multiplying by 4000
                       Adding
```

39.
```
        2 1
        3 2
        3 3
        8 7 6
    ×   3 4 5
      4 3 8 0      Multiplying by 5
    3 5 0 4 0      Multiplying by 40
  2 6 2 8 0 0      Multiplying by 300
  3 0 2, 2 2 0     Adding
```

41.
```
        5 5 5
        1 1 1
        1 1 1
        3 3 3
        7 8 8 9
    ×   6 2 2 4
      3 1 5 5 6        Multiplying by 4
    1 5 7 7 8 0        Multiplying by 20
    1 5 7 7 8 0 0      Multiplying by 200
  4 7 3 3 4 0 0 0      Multiplying by 6000
  4 9, 1 0 1, 1 3 6·   Adding
```

43.
```
                    Rounded to
                    the nearest ten
      4 5             5 0
    × 6 7           × 7 0
    ───────        ─────────
                   3 5 0 0 ← Estimated answer
```

45.
```
                    Rounded to
                    the nearest ten
      3 4             3 0
    × 2 9           × 3 0
    ───────        ─────────
                   9 0 0 ← Estimated answer
```

47.
```
                      Rounded to
                    the nearest hundred
      8 7 6             9 0 0
    × 3 4 5           × 3 0 0
    ─────────        ───────────
                     2 7 0, 0 0 0 ← Estimated answer
```

49.
```
                      Rounded to
                    the nearest hundred
      4 3 2             4 0 0
    × 1 9 9           × 2 0 0
    ─────────        ───────────
                     8 0, 0 0 0 ← Estimated answer
```

51. a) First we round the cost of the car and the destination charges to the nearest hundred and add.
```
    2 7, 8 9 6        2 7, 9 0 0
  +      5 4 0      +      5 0 0
  ───────────       ─────────────
                      2 8, 4 0 0
```

The number of sales representatives, 112, rounded to the nearest hundred is 100. Now we multiply the rounded total cost of a car and the rounded number of representatives.

$$
\begin{array}{r}
2\,8,4\,0\,0 \\
\times \qquad 1\,0\,0 \\
\hline
2,8\,4\,0,0\,0\,0
\end{array}
$$

The cost of the purchase is approximately $2,840,000.

b) First we round the cost of the car to the nearest thousand and the destination charges to the nearest hundred and add.

$$
\begin{array}{r}
2\,7,8\,9\,6 \\
+ \qquad 5\,4\,0 \\
\hline
\end{array}
\qquad
\begin{array}{r}
2\,8,0\,0\,0 \\
+ \qquad 5\,0\,0 \\
\hline
2\,8,5\,0\,0
\end{array}
$$

From part (a) we know that the number of sales representatives, rounded to the nearest hundred, is 100. We multiply the rounded total cost of a car and the rounded number of representatives.

$$
\begin{array}{r}
2\,8,5\,0\,0 \\
\times \qquad 1\,0\,0 \\
\hline
2,8\,5\,0,0\,0\,0
\end{array}
$$

The cost of the purchase is approximately $2,850,000.

53. $A = 728 \text{ mi} \times 728 \text{ mi} = 529,984$ square miles

55. $A = l \times w = 6 \times 3 = 18$ square feet

57. $A = l \times w = 11 \text{ yd} \times 11 \text{ yd} = 121$ square yards

59. $A = l \times w = 48 \text{ mm} \times 3 \text{ mm} = 144$ square millimeters

61. $A = l \times w = 90 \text{ ft} \times 90 \text{ ft} = 8100$ square feet

63. Discussion and Writing Exercise

65.
$$
\begin{array}{r}
\overset{1 \quad\ \ 1}{}\\
4\,9\,0\,8 \\
5\,6\,6\,7 \\
+\ 2\,1\,1\,0 \\
\hline
1\,2,6\,8\,5
\end{array}
$$
Add ones: We get 15. Write 5 in the ones column and 1 above the tens. Add tens: We get 8. Add hundreds: We get 16. Write 6 in the hundreds column and 1 above the thousands. Add thousands: We get 12 thousands.

67.
$$
\begin{array}{r}
\overset{1 \quad\ \ 1\ 1}{}\\
3\,4\,0,7\,9\,8 \\
+\ \ \ 8\,6,6\,7\,9 \\
\hline
4\,2\,7,4\,7\,7
\end{array}
$$
Add ones: We get 17. Write 7 in the ones column and 1 above the tens. Add tens: We get 17. Write 7 in the tens column and 1 above the hundreds. Add hundreds: We get 14. Write 4 in the hundreds column and 1 above the thousands. Add thousands: We get 7. Add ten thousands: We get 12. Write 2 in the ten thousands column and 1 above the hundred thousands. Add hundred thousands: We get 4 hundred thousands.

69.
$$
\begin{array}{r}
\overset{8\ \ 10}{}\\
4\,\cancel{9}\,\cancel{0}\,8 \\
-\ 3\,6\,6\,7 \\
\hline
1\,2\,4\,1
\end{array}
$$
Subtract ones. We cannot subtract 6 tens from 0 tens. We have 9 hundreds or 90 tens. We borrow 1 hundred to get 10 tens. We have 8 hundreds. Subtract tens, hundreds, and thousands.

71.
$$
\begin{array}{r}
\overset{\quad 13}{}\\
\overset{2\ \ \cancel{8}\ \ 10\ \ \ 8\ \ 18}{}\\
\cancel{3}\,\cancel{4}\,\cancel{0},\cancel{7}\,\cancel{9}\,\cancel{8} \\
-\ \ \ \ \ 8\,6,6\,7\,9 \\
\hline
2\,5\,4,1\,1\,9
\end{array}
$$
We cannot subtract 9 ones from 8 ones. Borrow 1 ten to get 18 ones. Subtract ones. Then subtract tens and hundreds. We cannot subtract 6 thousands from 0 thousands. We have 4 ten thousands or 40 thousands. We borrow 1 ten thousand to get 10 thousands. Subtract thousands. We cannot subtract 8 ten thousands from 3 ten thousands. We borrow 1 hundred thousand to get 13 ten thousands. Subtract ten thousands and then hundred thousands.

73. Round $6,3\,7\,5,\boxed{6}\,0\,2$ to the nearest thousand.

The digit 5 is in the thousands place. Consider the next digit to the right. Since the digit 6 is 5 or higher, round 5 thousands to 6 thousands. Then change all digits to the right of the thousands digit to zero.

The answer is 6,376,000.

75. Discussion and Writing Exercise

77. Use a calculator to perform the computations in this exercise.

First find the total area of each floor:

$A = l \times w = 172 \times 84 = 14,448$ square feet

Find the area lost to the elevator and the stairwell:

$A = l \times w = 35 \times 20 = 700$ square feet

Subtract to find the area available as office space on each floor:

$14,448 - 700 = 13,748$ square feet

Finally, multiply by the number of floors, 18, to find the total area available as office space:

$18 \times 13,748 = 247,464$ square feet

79. First, find the area of the photo.

$A = l \times w = 8 \times 10 = 80$ square inches

From Exercise 5 we know that the printer prints 5,760,000 dots per square inch. We multiply to find the total number of dots.

$80 \times 5,760,000 = 460,800,000$

The printer will print 460,800,000 dots.

Exercise Set 1.6

1. Think of an array with 4 rows. The number of pounds in each row will go to a mule.

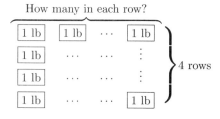

$$760 \div 4 = 190$$

3. Think of an array with 5 rows. The number of mL in each row will go in a beaker.

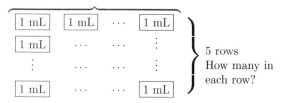

$$455 \div 5 = 91$$

5. $18 \div 3 = 6$ The 3 moves to the right. A related multiplication sentence is $18 = 6 \cdot 3$. (By the commutative law of multiplication, there is also another multiplication sentence: $18 = 3 \cdot 6$.)

7. $22 \div 22 = 1$ The 22 on the right of the $\div$ symbol moves to the right. A related multiplication sentence is $22 = 1 \cdot 22$. (By the commutative law of multiplication, there is also another multiplication sentence: $22 = 22 \cdot 1$.)

9. $54 \div 6 = 9$ The 6 moves to the right. A related multiplication sentence is $54 = 9 \cdot 6$. (By the commutative law of multiplication, there is also another multiplication sentence: $54 = 6 \cdot 9$.)

11. $37 \div 1 = 37$ The 1 moves to the right. A related multiplication sentence is $37 = 37 \cdot 1$. (By the commutative law of multiplication, there is also another multiplication sentence: $37 = 1 \cdot 37$.)

13. $9 \times 5 = 45$

Move a factor to the other side and then write a division.

$$9 \times 5 = 45 \qquad\qquad 9 \times 5 = 45$$

$$9 = 45 \div 5 \qquad\qquad 5 = 45 \div 9$$

15. Two related division sentences for $37 \cdot 1 = 37$ are:

$$37 = 37 \div 1 \qquad (\ 37 \cdot 1 = 37 \ \ )$$

and

$$1 = 37 \div 37 \qquad (\ 37 \cdot 1 = 37 \ \ )$$

17. $8 \times 8 = 64$

Since the factors are both 8, moving either one to the other side gives the related division sentence $8 = 64 \div 8$.

19. Two related division sentences for $11 \cdot 6 = 66$ are:

$$11 = 66 \div 6 \qquad (\ 11 \cdot 6 = 66 \ \ )$$

and

$$6 = 66 \div 11 \qquad (\ 11 \cdot 6 = 66 \ \ )$$

21.
$$\begin{array}{r} 12 \\ 6\,\overline{)72} \\ \underline{60} \\ 12 \\ \underline{12} \\ 0 \end{array}$$
Think: 7 tens $\div$ 6. Estimate 1 ten.
Think: 12 ones $\div$ 6. Estimate 2 ones.

The answer is 12.

23. $\dfrac{23}{23} = 1$ Any nonzero number divided by itself is 1.

25. $22 \div 1 = 22$ Any number divided by 1 is that same number.

27. $\dfrac{16}{0}$ is not defined, because division by 0 is not defined.

29.
$$\begin{array}{r} 55 \\ 5\,\overline{)277} \\ \underline{250} \\ 27 \\ \underline{25} \\ 2 \end{array}$$
Think: 2 hundreds $\div$ 5. There are no hundreds in the quotient.
Think: 27 tens $\div$ 5. Estimate 5 tens.
Think: 27 ones $\div$ 5. Estimate 5 ones.

The answer is 55 R 2.

31.
$$\begin{array}{r} 108 \\ 8\,\overline{)864} \\ \underline{800} \\ 64 \\ \underline{64} \\ 0 \end{array}$$
Think: 8 hundreds $\div$ 8. Estimate 1 hundred.
Think: 6 tens $\div$ 8. There are no tens in the quotient (other than the tens in 100). Write a 0 to show this.
Think: 64 ones $\div$ 8. Estimate 8 ones.

The answer is 108.

33.
$$\begin{array}{r} 307 \\ 4\,\overline{)1228} \\ \underline{1200} \\ 28 \\ \underline{28} \\ 0 \end{array}$$
Think: 12 hundreds $\div$ 4. Estimate 3 hundreds.
Think: 2 tens $\div$ 4. There are no tens in the quotient (other than the tens in 300). Write a 0 to show this.
Think: 28 ones $\div$ 4. Estimate 7 ones.

The answer is 307.

35.
$$\begin{array}{r} 753 \\ 6\,\overline{)4521} \\ \underline{4200} \\ 321 \\ \underline{300} \\ 21 \\ \underline{18} \\ 3 \end{array}$$
Think: 45 hundreds $\div$ 6. Estimate 7 hundreds.
Think: 32 tens $\div$ 6. Estimate 5 tens.
Think: 21 ones $\div$ 6. Estimate 3 ones.

The answer is 753 R 3.

37.
$$5\overline{)8515}$$
```
   1 7 0 3
5 )8 5 1 5
  5 0 0 0
  -------
  3 5 1 5
  3 5 0 0
  -------
      1 5
      1 5
    -----
        0
```
Think: 8 thousands ÷ 5. Estimate 1 thousand.

Think: 35 hundreds ÷ 5. Estimate 7 hundreds.

Think: 1 ten ÷ 5. There are no tens in the quotient (other than the tens in 1700). Write a 0 to show this.

Think: 15 ones ÷ 5. Estimate 3 ones.

The answer is 1703.

39.
```
     9 8 7
9 )8 8 8 8
  8 1 0 0
  -------
    7 8 8
    7 2 0
    -----
      6 8
      6 3
    -----
        5
```
Think: 88 hundreds ÷ 9. Estimate 9 hundreds.

Think: 78 tens ÷ 9. Estimate 8 tens.

Think: 68 ones ÷ 9. Estimate 7 ones.

The answer is 987 R 5.

41.
```
      1 2,7 0 0
1 0 )1 2 7,0 0 0
     1 0 0,0 0 0
     ----------
       2 7,0 0 0
       2 0,0 0 0
       --------
          7 0 0 0
          7 0 0 0
          ------
                0
```
Think: 12 ten thousands ÷ 10. Estimate 1 ten thousand.

Think: 27 thousands ÷ 10. Estimate 2 thousands.

Think: 70 hundreds ÷ 10. Estimate 7 hundreds.

Since the difference is 0, there are no tens or ones in the quotient (other than the tens and ones in 12,700). We write zeros to show this.

The answer is 12,700.

43.
```
            1 2 7
1 0 0 0 )1 2 7,0 0 0
        1 0 0,0 0 0
        ----------
          2 7,0 0 0
          2 0,0 0 0
          --------
            7 0 0 0
            7 0 0 0
            ------
                  0
```
Think: 1270 hundreds ÷ 1000. Estimate 1 hundred.

Think: 2700 tens ÷ 1000. Estimate 2 tens.

Think: 7000 ones ÷ 1000. Estimate 7 ones.

The answer is 127.

45.
```
       5 2
7 0 )3 6 9 2
    3 5 0 0
    -------
      1 9 2
      1 4 0
      -----
        5 2
```
Think: 369 tens ÷ 70. Estimate 5 tens.

Think: 192 ones ÷ 70. Estimate 2 ones.

The answer is 52 R 52.

47.
```
       2 9
3 0 )8 7 5
    6 0 0
    -----
    2 7 5
    2 7 0
    -----
        5
```
Think: 87 tens ÷ 30. Estimate 2 tens.

Think: 275 ones ÷ 30. Estimate 9 ones.

The answer is 29 R 5.

49.
```
          3
1 1 1 )3 2 1 9
      3 3 3 0
```
Round 111 to 100.

Think: 321 tens ÷ 100. Estimate 3 tens.

Since we cannot subtract 3330 from 3219, the estimate is too high.

```
         2 9
1 1 1 )3 2 1 9
      2 2 2 0
      -------
        9 9 9
        9 9 9
        -----
            0
```
Think: 321 tens ÷ 100. Estimate 2 tens.

Think: 999 ones ÷ 100. Estimate 9 ones.

The answer is 29.

51.
```
     1 0 5
8 )8 4 3
  8 0 0
  -----
    4 3
    4 0
  -----
      3
```
Think: 8 hundreds ÷ 8. Estimate 1 hundred.

Think: 4 tens ÷ 8. There are no tens in the quotient (other than the tens in 100). Write a 0 to show this.

Think: 43 ones ÷ 8. Estimate 5 ones.

The answer is 105 R 3.

53.
```
     1 6 0 9
5 )8 0 4 7
  5 0 0 0
  -------
  3 0 4 7
  3 0 0 0
  -------
      4 7
      4 5
    -----
        2
```
Think: 8 thousands ÷ 5. Estimate 1 thousand.

Think: 30 hundreds ÷ 5. Estimate 6 hundreds.

Think: 4 tens ÷ 5. There are no tens in the quotient (other than the tens in 1600). Write a 0 to show this.

Think: 47 ones ÷ 5. Estimate 9 ones.

The answer is 1609 R 2.

55.
```
     1 0 0 7
5 )5 0 3 6
  5 0 0 0
  -------
      3 6
      3 5
    -----
        1
```
Think: 5 thousands ÷ 5. Estimate 1 thousand.

Think: 0 hundreds ÷ 5. There are no hundreds in the quotient (other than the hundreds in 1000). Write a 0 to show this.

Think: 3 tens ÷ 5. There are no tens in the quotient (other than the tens in 1000). Write a 0 to show this.

Think: 36 ones ÷ 5. Estimate 7 ones.

The answer is 1007 R 1.

57.
```
         2 2
4 6 )1 0 5 8
    9 2 0
    -----
    1 3 8
      9 2
    -----
      4 6
```
Round 46 to 50.

Think: 105 tens ÷ 50. Estimate 2 tens.

Think: 138 ones ÷ 50. Estimate 2 ones.

Since 46 is not smaller than the divisor, 46, the estimate is too low.

```
          2 3
    4 6 ⟌1 0 5 8
          9 2 0
          1 3 8    Think: 138 ones ÷ 50. Estimate 3 ones.
          1 3 8
              0
```

The answer is 23.

59.
```
          1 0 7      Round 32 to 30.
    3 2 ⟌3 4 2 5     Think: 34 hundreds ÷ 30. Estimate
          3 2 0 0    1 hundred.
            2 2 5    Think: 22 tens ÷ 30. There are no
            2 2 4    tens in the quotient (other than the
                1    tens in 100). Write 0 to show this.
                     Think: 225 ones ÷ 30. Estimate 7
                     ones.
```

The answer is 107 R 1.

61.
```
            4        Round 24 to 20.
    2 4 ⟌8 8 8 0     Think: 88 hundreds ÷ 20. Estimate
        9 6 0 0      4 hundreds.
```

Since we cannot subtract 9600 from 8880, the estimate
is too high.

```
           3 8       Think: 88 hundreds ÷ 20. Estimate
    2 4 ⟌8 8 8 0     3 hundreds.
        7 2 0 0
        1 6 8 0      Think: 168 tens ÷ 20. Estimate 8
        1 9 2 0      tens.
```

Since we cannot subtract 1920 from 1680, the estimate
is too high.

```
          3 7 0      Think: 168 tens ÷ 20. Estimate 7
    2 4 ⟌8 8 8 0     tens.
        7 2 0 0
        1 6 8 0      Think: 0 ones ÷ 20. There are no
        1 6 8 0      ones in the quotient (other than the
              0      ones in 370). Write a 0 to show this.
```

The answer is 370.

63.
```
            5        Round 28 to 30.
    2 8 ⟌1 7, 0 6 7  Think: 170 hundreds ÷ 30. Esti-
        1 4 0 0 0    mate 5 hundreds.
         ⌐3 0⌐ 6 7
```

Since 30 is larger than the divisor, 28, the estimate is
too low.

```
          6 0 8      Think: 170 hundreds ÷ 30. Esti-
    2 8 ⟌1 7, 0 6 7  mate 6 hundreds.
        1 6 8 0 0
            2 6 7    Think: 26 tens ÷ 30. There are no
            2 2 4    tens in the quotient (other than the
             ⌐4 3⌐   tens in 600.) Write a zero to show
                     this.

                     Think: 267 ones ÷ 30. Estimate 8
                     ones.
```

Since 43 is larger than the divisor, 28, the estimate is
too low.

```
            6 0 9
    2 8 ⟌1 7, 0 6 7
        1 6 8 0 0
            2 6 7    Think: 267 ones ÷ 30. Estimate 9
            2 5 2    ones.
             1 5
```

The answer is 609 R 15.

65.
```
          3 0 4      Think: 243 hundreds ÷ 80. Esti-
    8 0 ⟌2 4, 3 2 0  mate 3 hundreds.
        2 4 0 0 0    Think: 32 tens ÷ 80. There are no
            3 2 0    tens in the quotient (other than the
            3 2 0    tens in 300). Write a 0 to show this.
                0    Think: 320 ones ÷ 80. Estimate 4
                     ones.
```

The answer is 304.

67.
```
            3 5 0 8
    2 8 5 ⟌9 9 9, 9 9 9
          8 5 5 0 0 0
          1 4 4 9 9 9
          1 4 2 5 0 0
              2 4 9 9
              2 2 8 0
                2 1 9
```

The answer is 3508 R 219.

69.
```
              8 0 7 0
    4 5 6 ⟌3, 6 7 9, 9 2 0
          3 6 4 8 0 0 0
            3 1 9 2 0
            3 1 9 2 0
                  0
```

The answer is 8070.

71. Discussion and Writing Exercise

73. The distance around an object is its <u>perimeter</u>.

75. For large numbers, <u>digits</u> are separated by commas into
groups of three, called <u>periods</u>.

77. In the sentence $28 \div 7 = 4$, the <u>dividend</u> is 28.

79. The <u>minuend</u> is the number from which another number
is being subtracted.

81. Discussion and Writing Exercise

83.

a	b	$a \cdot b$	$a+b$
	68	3672	
84			117
		32	12

To find a in the first row we divide $a \cdot b$ by b:

$$3672 \div 68 = 54$$

Then we add to find $a + b$:

$$54 + 68 = 122$$

To find b in the second row we subtract a from $a + b$:

$$117 - 84 = 33$$

Then we multiply to find $a \cdot b$:

$$84 \cdot 33 = 2772$$

To find a and b in the last row we find a pair of numbers whose product is 32 and whose sum is 12. Pairs of numbers whose product is 32 are 1 and 32, 2 and 16, 4 and 8. Since $4 + 8 = 12$, the numbers we want are 4 and 8. We will let $a = 4$ and $b = 8$. (We could also let $a = 8$ and $b = 4$).

The completed table is shown below.

a	b	$a \cdot b$	$a+b$
54	68	3672	122
84	33	2772	117
4	8	32	12

85. We divide 1231 by 42:

```
        2 9
  4 2 ⟌1 2 3 1
        8 4 0
        ─────
        3 9 1
        3 7 8
        ─────
          1 3
```

The answer is 29 R 13. Since 13 students will be left after 29 buses are filled, then 30 buses are needed.

Exercise Set 1.7

1. $x + 0 = 14$

We replace x by different numbers until we get a true equation. If we replace x by 14, we get a true equation: $14 + 0 = 14$. No other replacement makes the equation true, so the solution is 14.

3. $y \cdot 17 = 0$

We replace y by different numbers until we get a true equation. If we replace y by 0, we get a true equation: $0 \cdot 17 = 0$. No other replacement makes the equation true, so the solution is 0.

5.

$$
\begin{aligned}
13 + x &= 42 \\
13 + x - 13 &= 42 - 13 \quad &\text{Subtracting 13 on both sides} \\
0 + x &= 29 \quad &\text{13 plus } x \text{ minus 13 is } 0 + x. \\
x &= 29
\end{aligned}
$$

Check:

$$
\begin{array}{c}
13 + x = 42 \\ \hline
13 + 29 \;?\; 42 \\
42 \quad \big| \quad \text{TRUE}
\end{array}
$$

The solution is 29.

7.

$$
\begin{aligned}
12 &= 12 + m \\
12 - 12 &= 12 + m - 12 \quad &\text{Subtracting 12 on both sides} \\
0 &= 0 + m \quad &\text{12 plus } m \text{ minus 12 is } 0 + m. \\
0 &= m
\end{aligned}
$$

Check:

$$
\begin{array}{c}
12 = 12 + m \\ \hline
12 \;?\; 12 + 0 \\
\big| \quad 12 \quad \text{TRUE}
\end{array}
$$

The solution is 0.

9.

$$
\begin{aligned}
3 \cdot x &= 24 \\
\frac{3 \cdot x}{3} &= \frac{24}{3} \quad &\text{Dividing by 3 on both sides} \\
x &= 8 \quad &\text{3 times } x \text{ divided by 3 is } x.
\end{aligned}
$$

Check:

$$
\begin{array}{c}
3 \cdot x = 24 \\ \hline
3 \cdot 8 \;?\; 24 \\
24 \quad \big| \quad \text{TRUE}
\end{array}
$$

The solution is 8.

11.

$$
\begin{aligned}
112 &= n \cdot 8 \\
\frac{112}{8} &= \frac{n \cdot 8}{8} \quad &\text{Dividing by 8 on both sides} \\
14 &= n
\end{aligned}
$$

Check:

$$
\begin{array}{c}
112 = n \cdot 8 \\ \hline
112 \;?\; 14 \cdot 8 \\
\big| \quad 112 \quad \text{TRUE}
\end{array}
$$

The solution is 14.

13. $45 \times 23 = x$

To solve the equation we carry out the calculation.

```
      4 5
    × 2 3
    ─────
    1 3 5
    9 0 0
    ─────
  1 0 3 5
```

We can check by repeating the calculation. The solution is 1035.

15. $t = 125 \div 5$

To solve the equation we carry out the calculation.

```
        2 5
    5 ⟌1 2 5
      1 0 0
      ─────
        2 5
        2 5
        ───
          0
```

We can check by repeating the calculation. The solution is 25.

17. $p = 908 - 458$

To solve the equation we carry out the calculation.

$$\begin{array}{r} 9\,0\,8 \\ -\,4\,5\,8 \\ \hline 4\,5\,0 \end{array}$$

We can check by repeating the calculation. The solution is 450.

19. $x = 12,345 + 78,555$

To solve the equation we carry out the calculation.

$$\begin{array}{r} 1\,2,3\,4\,5 \\ +\,7\,8,5\,5\,5 \\ \hline 9\,0,9\,0\,0 \end{array}$$

We can check by repeating the calculation. The solution is 90,900.

21. $3 \cdot m = 96$

$$\frac{3 \cdot m}{3} = \frac{96}{3} \quad \text{Dividing by 3 on both sides}$$

$$m = 32$$

Check: $\dfrac{3 \cdot m = 96}{3 \cdot 32 \ ? \ 96}$

$\qquad\qquad 96 \ \Big| \quad$ TRUE

The solution is 32.

23. $715 = 5 \cdot z$

$$\frac{715}{5} = \frac{5 \cdot z}{5} \quad \text{Dividing by 5 on both sides}$$

$$143 - z$$

Check: $\dfrac{715 = 5 \cdot x}{715 \ ? \ 5 \cdot 143}$

$\qquad\qquad \Big| \ 715 \quad$ TRUE

The solution is 143.

25. $10 + x = 89$

$$10 + x - 10 = 89 - 10$$

$$x = 79$$

Check: $\dfrac{10 + x = 89}{10 + 79 \ ? \ 89}$

$\qquad\qquad 89 \ \Big| \quad$ TRUE

The solution is 79.

27. $61 = 16 + y$

$$61 - 16 = 16 + y - 16$$

$$45 = y$$

Check: $\dfrac{61 = 16 + y}{61 \ ? \ 16 + 45}$

$\qquad\qquad \Big| \ 61 \quad$ TRUE

The solution is 45.

29. $6 \cdot p = 1944$

$$\frac{6 \cdot p}{6} = \frac{1944}{6}$$

$$p = 324$$

Check: $\dfrac{6 \cdot p = 1944}{6 \cdot 324 \ ? \ 1944}$

$\qquad\qquad 1944 \ \Big| \quad$ TRUE

The solution is 324.

31. $5 \cdot x = 3715$

$$\frac{5 \cdot x}{5} = \frac{3715}{5}$$

$$x = 743$$

The number 743 checks. It is the solution.

33. $47 + n = 84$

$$47 + n - 47 = 84 - 47$$

$$n = 37$$

The number 37 checks. It is the solution.

35. $x + 78 = 144$

$$x + 78 - 78 = 144 - 78$$

$$x = 66$$

The number 66 checks. It is the solution.

37. $165 = 11 \cdot n$

$$\frac{165}{11} = \frac{11 \cdot n}{11}$$

$$15 = n$$

The number 15 checks. It is the solution.

39. $624 = t \cdot 13$

$$\frac{624}{13} = \frac{t \cdot 13}{13}$$

$$48 = t$$

The number 48 checks. It is the solution.

41. $x + 214 = 389$

$$x + 214 - 214 = 389 - 214$$

$$x = 175$$

The number 175 checks. It is the solution.

43. $567 + x = 902$

$$567 + x - 567 = 902 - 567$$

$$x = 335$$

The number 335 checks. It is the solution.

45. $18 \cdot x = 1872$

$$\frac{18 \cdot x}{18} = \frac{1872}{18}$$

$$x = 104$$

The number 104 checks. It is the solution.

47. $40 \cdot x = 1800$

$$\frac{40 \cdot x}{40} = \frac{1800}{40}$$

$$x = 45$$

The number 45 checks. It is the solution.

49. $2344 + y = 6400$

$$2344 + y - 2344 = 6400 - 2344$$

$$y = 4056$$

The number 4056 checks. It is the solution.

51. $8322 + 9281 = x$
       $17,603 = x$ Doing the addition

The number 17,603 checks. It is the solution.

53. $234 \cdot 78 = y$
     $18,252 = y$ Doing the multiplication

The number 18,252 checks. It is the solution.

55. $58 \cdot m = 11,890$
    $\dfrac{58 \cdot m}{58} = \dfrac{11,890}{58}$
        $m = 205$

The number 205 checks. It is the solution.

57. Discussion and Writing Exercise

59. $7 + 8 = 15$ $7 + 8 = 15$
     ↑   ↑

This number gets   This number gets
subtracted from the   subtracted from the
sum.   ↓   sum.   ↓
    $7 = 15 - 8$ $8 = 15 - 7$

61. Since 123 is to the left of 789 on the number line, $123 < 789$.

63. Since 688 is to the right of 0 on the number line, $688 > 0$.

65.
$$\begin{array}{r} 1\,4\,2 \\ 9\,\overline{\smash{)}1\,2\,8\,3} \\ 9\,0\,0 \\ \hline 3\,8\,3 \\ 3\,6\,0 \\ \hline 2\,3 \\ 1\,8 \\ \hline 5 \end{array}$$
Think: 12 hundreds ÷ 9. Estimate 1 hundred.
Think: 38 tens ÷ 9. Estimate 4 tens.
Think: 23 ones ÷ 9. Estimate 2 ones.

The answer is 142 R 5.

67.
$$\begin{array}{r} 3\,3\,4 \\ 1\,7\,\overline{\smash{)}5\,6\,7\,8} \\ 5\,1\,0\,0 \\ \hline 5\,7\,8 \\ 5\,1\,0 \\ \hline 6\,8 \\ 6\,8 \\ \hline 0 \end{array}$$
Think 56 hundreds ÷ 17. Estimate 3 hundreds.
Think 57 tens ÷ 17. Estimate 3 tens.
Think 68 ones ÷ 17. Estimate 4 ones.

The answer is 334.

69. Discussion and Writing Exercise

71. $23,465 \cdot x = 8,142,355$
   $\dfrac{23,465 \cdot x}{23,465} = \dfrac{8,142,355}{23,465}$
           $x = 347$ Using a calculator to divide

The number 347 checks. It is the solution.

Exercise Set 1.8

1. *Familiarize*. We visualize the situation. We are combining quantities, so addition can be used.

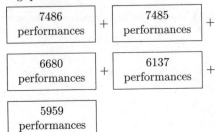

Let p = the total number of performances of all five shows.

***Translate*.** We translate to an equation.

$7486 + 7485 + 6680 + 6137 + 5959 = p$

***Solve*.** We carry out the addition.

$$\begin{array}{r} {\scriptstyle 2\ 3\ 2} \\ 7\,4\,8\,6 \\ 7\,4\,8\,5 \\ 6\,6\,8\,0 \\ 6\,1\,3\,7 \\ +\ 5\,9\,5\,9 \\ \hline 3\,3,7\,4\,7 \end{array}$$

Thus, $33,747 = p$.

***Check*.** We can repeat the calculation. We can also estimate by rounding, say to the nearest thousand.

$7486 + 7485 + 6680 + 6137 + 5959$

$\approx 7000 + 7000 + 7000 + 6000 + 6000$

$\approx 33,000 \approx 33,747$

Since the estimated answer is close to the calculated answer, our result is probably correct.

***State*.** The five longest-running Broadway shows had a total of 33,747 performances.

3. *Familiarize*. We visualize the situation. Let c = the number by which the performances of *The Phantom of the Opera* exceeded the performances of *A Chorus Line*.

Chorus Line performances	Excess *Phantom* performances
6137	c
Number of *Phantom* performances	
7486	

***Translate*.** We see this as a "how many more" situation.

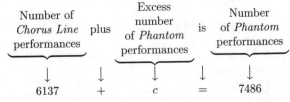

***Solve*.** We subtract 6137 on both sides of the equation.

$6137 + c = 7486$

$6137 + c - 6137 = 7486 - 6137$

$c = 1349$

Check. We can add the difference, 1349, to the subtrahend, 6137: $6137 + 1349 = 7486$. We can also estimate:

$$7486 - 6137 \approx 7500 - 6100$$
$$\approx 1400 \approx 1349$$

The answer checks.

State. There were 1349 more performances of *The Phantom of the Opera* than *A Chorus Line*.

5. Familiarize. We visualize the situation. Let m = the number of miles by which the Canadian border exceeds the Mexican border.

Mexican border 1933 mi	Excess miles in Canadian border m
Canadian border 3987 mi	

Translate. We see this as a "how many more" situation.

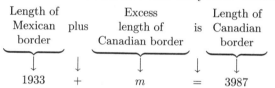

Length of Mexican border	plus	Excess length of Canadian border	is	Length of Canadian border
1933	+	m	=	3987

Solve. We subtract 1933 on both sides of the equation.

$$1933 + m = 3987$$
$$1933 + m - 1933 = 3987 - 1933$$
$$m = 2054$$

Check. We can add the difference, 2054, to the subtrahend, 1933: $1933 + 2054 = 3987$. We can also estimate:

$$3987 - 1933 \approx 4000 - 2000$$
$$\approx 2000 \approx 2054$$

The answer checks.

State. The Canadian border is 2054 mi longer than the Mexican border.

7. Familiarize. We first make a drawing. Let r = the number of rows.

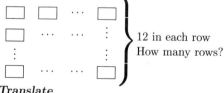

12 in each row
How many rows?

Translate.

Number of holes	divided by	Number per row	is	Number of rows
216	÷	12	=	r

Solve. We carry out the division.

```
        1 8
  1 2 ) 2 1 6
        1 2 0
        -----
          9 6
          9 6
        -----
            0
```

Thus, $18 = r$, or $r = 18$.

Check. We can check by multiplying: $12 \cdot 18 = 216$. Our answer checks.

State. There are 18 rows.

9. Familiarize. We visualize each situation. We are combining quantities, so addition can be used.

451,097	+	341,219
degrees awarded to men in 1970		degrees awarded to women in 1970

Let x = the total number of bachelor's degrees awarded in 1970.

775,424	+	573,079
degrees awarded to women in 2003		degrees awarded to men in 2003

Let y = the total number of bachelor's degrees awarded in 2003.

Translate. We translate each situation to an equation.

For 1970: $451,097 + 341,219 = x$

For 2003: $775,424 + 573,079 = y$

Solve. We carry out the additions.

```
      1 1
    4 5 1, 0 9 7
  + 3 4 1, 2 1 9
  -------------
    7 9 2, 3 1 6
```

Thus, $792,316 = x$.

```
    1       1 1
    7 7 5, 4 2 4
  + 5 7 3, 0 7 9
  -------------
  1, 3 4 8, 5 0 3
```

Thus, $1,348,503 = y$.

Check. We will estimate.

For 1970: $451,097 + 341,219$
$$\approx 450,000 + 340,000$$
$$\approx 790,000 \approx 792,316$$

For 2003: $775,424 + 573,079$
$$\approx 780,000 + 570,000$$
$$\approx 1,350,000 \approx 1,348,503$$

The answer checks.

State. In 1970 a total of 792,316 bachelor's degrees were awarded; the total in 2003 was 1,348,503.

11. Familiarize. We visualize the situation. Let w = the number by which the degrees awarded to women in 2003 exceeded those awarded to men.

Men's degrees 573,079	Excess women's degrees w
Women's degrees 775,424	

Translate. We see this as a "how much more" situation.

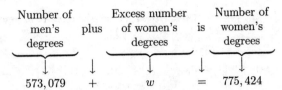

$$573,079 \quad + \quad w \quad = \quad 775,424$$

Solve. We subtract 573,079 on both sides of the equation.

$$573,079 + w = 775,424$$

$$573,079 + w - 573,079 = 775,424 - 573,079$$

$$w = 202,345$$

Check. We will estimate.

$$775,424 - 573,079$$

$$\approx 780,000 - 570,000$$

$$\approx 210,000 \approx 202,345$$

The answer checks.

State.   There were 202,345 more bachelor's degrees awarded to women than to men in 2003.

13. **Familiarize.** We visualize the situation. Let $m =$ the median mortgage debt in 2004.

Debt in 1989 $39,802	Excess debt in 2004 $48,388
Debt in 2004 m	

Translate. We translate to an equation.

$$\underbrace{\text{Debt in 1989}}_{39,802} + \underbrace{\text{Excess debt in 2004}}_{48,388} \text{ is } \underbrace{\text{Debt in 2004}}_{m}$$

Solve. We carry out the addition.

$$\begin{array}{r} {\scriptstyle 1\ 1\quad 1} \\ 3\,9,8\,0\,2 \\ +\,4\,8,3\,8\,8 \\ \hline 8\,8,1\,9\,0 \end{array}$$

Thus, $88,190 = m$.

Check. We can estimate.

$$39,802 + 48,388$$

$$\approx 40,000 + 48,000$$

$$\approx 88,000 \approx 88,190$$

The answer checks.

State. In 2004 the median mortgage debt was $88,190.

15. **Familiarize.** We visualize the situation. Let $l =$ the excess length of the Nile River, in miles.

Length of Missouri-Mississippi 3860 miles	Excess length of Nile l
Length of Nile 4100 miles	

Translate. This is a "how much more" situation. We translate to an equation.

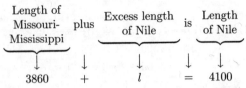

$$3860 \quad + \quad l \quad = \quad 4100$$

Solve. We subtract 3860 on both sides of the equation.

$$3860 + l = 4100$$

$$3860 + l - 3860 = 4100 - 3860$$

$$l = 240$$

Check. We can check by adding the difference, 240, to the subtrahend, 3860: $3860 + 240 = 4100$. Our answer checks.

State. The Nile River is 240 mi longer than the Missouri-Mississippi River.

17. **Familiarize.** We first draw a picture. Let $h =$ the number of hours in a week. Repeated addition works well here.

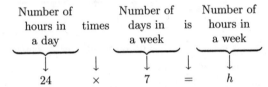

7 addends

Translate. We translate to an equation.

$$\underbrace{\text{Number of hours in a day}}_{24} \quad \text{times} \quad \underbrace{\text{Number of days in a week}}_{7} \quad \text{is} \quad \underbrace{\text{Number of hours in a week}}_{h}$$

$$24 \qquad \times \qquad 7 \qquad = \qquad h$$

Solve. We carry out the multiplication.

$$\begin{array}{r} 2\,4 \\ \times\ \ 7 \\ \hline 1\,6\,8 \end{array}$$

Thus, $168 = h$, or $h = 168$.

Check. We can repeat the calculation. We an also estimate:

$$24 \times 7 \approx 20 \times 10 = 200 \approx 168$$

Our answer checks.

State. There are 168 hours in a week.

19. **Familiarize.** We first draw a picture. Let $s =$ the number of squares in the puzzle. Repeated addition works well here.

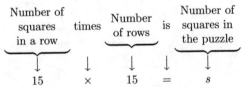

15 addends

Translate. We translate to an equation.

$$\underbrace{\text{Number of squares in a row}}_{15} \quad \text{times} \quad \underbrace{\text{Number of rows}}_{15} \quad \text{is} \quad \underbrace{\text{Number of squares in the puzzle}}_{s}$$

$$15 \qquad \times \qquad 15 \qquad = \qquad s$$

Solve. We carry out the multiplication.

$$\begin{array}{r} 1\,5 \\ \times\,1\,5 \\ \hline 7\,5 \\ 1\,5\,0 \\ \hline 2\,2\,5 \end{array}$$

Thus, $225 = s$.

Check. We can repeat the calculation. The answer checks.

State. There are 225 squares in the crossword puzzle.

21. Familiarize. We draw a picture of the situation. Let $c =$ the total cost of the purchase. Repeated addition works well here.

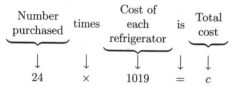

24 addends

Translate. We translate to an equation.

Number purchased	times	Cost of each refrigerator	is	Total cost
↓	↓	↓	↓	↓
24	×	1019	=	c

Solve. We carry out the multiplication.

$$
\begin{array}{r}
{\scriptstyle 1} \\
{\scriptstyle 3} \\
1\,0\,1\,9 \\
\times 2\,4 \\
\hline
4\,0\,7\,6 \\
2\,0\,3\,8\,0 \\
\hline
2\,4,4\,5\,6
\end{array}
$$

Thus, $24,456 = c$.

Check. We can repeat the calculation. We can also estimate: $24 \times 1019 \approx 24 \times 1000 \approx 24,000 \approx 24,456$. The answer checks.

State. The total cost of the purchase is $24,456.

23. Familiarize. We first draw a picture. Let $w =$ the number of full weeks the episodes can run.

5 in each row

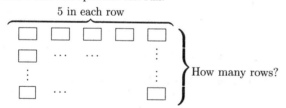

How many rows?

Translate. We translate to an equation.

Number of episodes	divided by	Number shown per week	is	Number of weeks
↓	↓	↓	↓	↓
177	÷	5	=	w

Solve. We carry out the division.

$$
\begin{array}{r}
3\,5 \\
5\,\overline{)1\,7\,7} \\
1\,5\,0 \\
\hline
2\,7 \\
2\,5 \\
\hline
2
\end{array}
$$

Check. We can check by multiplying the number of weeks by 5 and adding the remainder, 2:

$$5 \cdot 35 = 175, \qquad 175 + 2 = 177$$

State. 35 full weeks will pass before the station must start over. There will be 2 episodes left over.

25. Familiarize. We first draw a picture of the situation. Let $g =$ the number of gallons that will be used in 6136 mi of city driving.

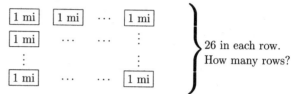

26 in each row. How many rows?

Translate. We translate to an equation.

Number of miles	divided by	Number of mpg	is	Number of gallons
↓	↓	↓	↓	↓
6136	÷	26	=	g

Solve. We carry out the division.

$$
\begin{array}{r}
2\,3\,6 \\
2\,6\,\overline{)6\,1\,3\,6} \\
5\,2\,0\,0 \\
\hline
9\,3\,6 \\
7\,8\,0 \\
\hline
1\,5\,6 \\
1\,5\,6 \\
\hline
0
\end{array}
$$

Thus, $236 = g$.

Check. We can check by multiplying the number of gallons by the number of miles per gallon: $26 \cdot 236 = 6136$. The answer checks.

State. The Hyundai Tucson GLS will use 236 gal of gasoline in 6136 mi of city driving.

27. Familiarize. We visualize the situation. Let $d =$ the number of miles by which the nonstop flight distance of the Boeing 747 exceeds the nonstop flight distance of the Boeing 777.

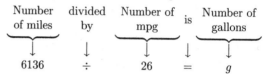

777 distance 5210 mi	Excess 747 distance d
747 distance 8826 mi	

Translate. This is a "how much more" situation.

777 distance	plus	Excess 747 distance	is	747 distance
↓	↓	↓	↓	↓
5210	+	d	=	8826

Solve. We subtract 5210 on both sides of the equation.

$$5210 + d = 8826$$
$$5210 + d - 5210 = 8826 - 5210$$
$$d = 3616$$

Check. We can estimate.

$8826 - 5210 \approx 8800 - 5200 \approx 3600 \approx 3616$

The answer checks.

State. The Boeing 747's nonstop flight distance is 3616 mi greater than that of the Boeing 777.

29. *Familiarize*. We draw a picture. Let g = the number of gallons of fuel needed for a 4-hr flight of the Boeing 747. Repeated addition works well here.

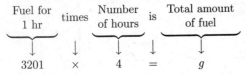

Translate. We translate to an equation.

Fuel for 1 hr	times	Number of hours	is	Total amount of fuel
↓	↓	↓	↓	↓
3201	×	4	=	g

Solve. We carry out the multiplication.

$$\begin{array}{r} 3\,2\,0\,1 \\ \times \qquad 4 \\ \hline 1\,2,8\,0\,4 \end{array}$$

Thus, $12,804 = g$.

Check. We can repeat the calculation. The answer checks.

State. For a 4-hr flight of the Boeing 747, 12,804 gal of fuel are needed.

31. *Familiarize*. This is a multistep problem. First we will find the cost for the crew. Then we will find the cost for the fuel. Finally, we will find the total cost for the crew and the fuel. Let c = the cost of the crew, f = the cost of the fuel, and t = the total cost.

Translate.

Crew cost for 1 hr	times	Number of hours	is	Total cost
↓	↓	↓	↓	↓
1948	×	3	=	c

Fuel cost for 1 hr	times	Number of hours	is	Total cost
↓	↓	↓	↓	↓
2867	×	3	=	f

Crew cost	plus	Fuel cost	is	Total cost
↓	↓	↓	↓	↓
c	+	f	=	t

Solve. First we carry out the multiplications to solve the first two equations.

$$\begin{array}{r} {\scriptstyle 2\ 1\ 2} \\ 1\,9\,4\,8 \\ \times \qquad 3 \\ \hline 5\,8\,4\,4 \end{array}$$

Thus, $5844 = c$.

$$\begin{array}{r} {\scriptstyle 2\ 2\ 2} \\ 2\,8\,6\,7 \\ \times \qquad 3 \\ \hline 8\,6\,0\,1 \end{array}$$

Thus, $8601 = f$.

Now we substitute 5844 for c and 8601 for f in the third equation and carry out the addition.

$$c + f = t$$
$$5844 + 8601 = t$$
$$14,445 = t$$

Check. We repeat the calculations. The answer checks.

State. The total cost for the crew and the fuel for a 3-hr flight of the Boeing 747 is \$14,445.

33. *Familiarize*. We first draw a picture. We let x = the amount of each payment.

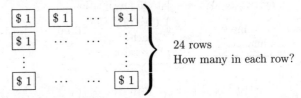

Translate. We translate to an equation.

Amount of loan	divided by	Number of payments	is	Amount of each payment
↓	↓	↓	↓	↓
5928	÷	24	=	x

Solve. We carry out the division.

$$\begin{array}{r} 2\,4\,7 \\ 24\,\overline{)\,5\,9\,2\,8} \\ 4\,8\,0\,0 \\ \hline 1\,1\,2\,8 \\ 9\,6\,0 \\ \hline 1\,6\,8 \\ 1\,6\,8 \\ \hline 0 \end{array}$$

Thus, $247 = x$, or $x = 247$.

Check. We can check by multiplying 247 by 24: $24 \cdot 247 = 5928$. The answer checks.

State. Each payment is \$247.

35. *Familiarize*. We first draw a picture. Let A = the area and P = the perimeter of the court, in feet.

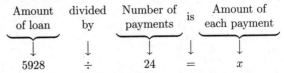

Translate. We write one equation to find the area and another to find the perimeter.

a) Using the formula for the area of a rectangle, we have
$$A = l \cdot w = 84 \cdot 50$$

b) Recall that the perimeter is the distance around the court.
$$P = 84 + 50 + 84 + 50$$

Solve. We carry out the calculations.

a)
$$
\begin{array}{r}
5\,0 \\
\times\; 8\,4 \\
\hline
2\,0\,0 \\
4\,0\,0\,0 \\
\hline
4\,2\,0\,0
\end{array}
$$

Thus, $A = 4200$.

b) $P = 84 + 50 + 84 + 50 = 268$

Check. We can repeat the calculation. The answers check.

State. a) The area of the court is 4200 square feet.

b) The perimeter of the court is 268 ft.

37. Familiarize. We visualize the situation. Let $a =$ the number of dollars by which the imports exceeded the exports.

Exports	Excess amount of imports
\$2,596,000,000	a
Imports	
\$31,701,000,000	

Translate. This as a "how much more" situation.

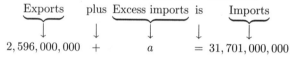

$$2{,}596{,}000{,}000 \;+\; a \;=\; 31{,}701{,}000{,}000$$

Solve. We subtract 2,596,000,000 on both sides of the equation.

$$2{,}596{,}000{,}000 + a = 31{,}701{,}000{,}000$$
$$2{,}596{,}000{,}000 + a - 2{,}596{,}000{,}000 = 31{,}701{,}000{,}000 - $$
$$2{,}596{,}000{,}000$$
$$a = 29{,}105{,}000{,}000$$

Check. We can estimate.

$$31{,}701{,}000{,}000 - 2{,}596{,}000{,}000$$
$$\approx 32{,}000{,}000{,}000 - 3{,}000{,}000{,}000$$
$$\approx 29{,}000{,}000{,}000 \approx 29{,}105{,}000{,}000$$

The answer checks.

State. Imports exceeded exports by \$29,105,000,000.

39. Familiarize. We visualize the situation. Let $p =$ the Colonial population in 1680.

Population in 1680	Increase in population
p	2,628,900
Population in 1780	
2,780,400	

Translate. This is a "how much more" situation.

Population in 1680 plus Increase in population is Population in 1780

$$p \;+\; 2{,}628{,}900 \;=\; 2{,}780{,}400$$

Solve. We subtract 2,628,900 on both sides of the equation.

$$p + 2{,}628{,}900 = 2{,}780{,}400$$
$$p + 2{,}628{,}900 - 2{,}628{,}900 = 2{,}780{,}400 - 2{,}628{,}900$$
$$p = 151{,}500$$

Check. Since $2{,}628{,}900 + 151{,}500 = 2{,}780{,}400$, the answer checks.

State. In 1680 the Colonial population was 151,500.

41. Familiarize. We draw a picture of the situation. Let $n =$ the number of 20-bar packages that can be filled.

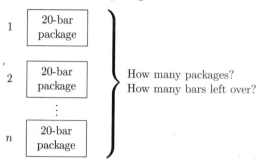

Translate. We translate to an equation.

Number of bars divided by Number per package is Number of packages

$$11{,}267 \;\div\; 20 \;=\; n$$

Solve. We carry out the division.

$$
\begin{array}{r}
5\,6\,3 \\
20\,\overline{)\,1\,1{,}2\,6\,7} \\
1\,0{,}0\,0\,0 \\
\hline
1\,2\,6\,7 \\
1\,2\,0\,0 \\
\hline
6\,7 \\
6\,0 \\
\hline
7
\end{array}
$$

Thus, $n = 563$ R 7.

Check. We can check by multiplying the number of packages by 20 and then adding the remainder, 7:

$$20 \cdot 563 = 11{,}260 \qquad 11{,}260 + 7 = 11{,}267$$

The answer checks.

State. 563 packages can be filled. There will be 7 bars left over.

43. Familiarize. First we find the distance in reality between two cities that are 6 in. apart on the map. We make a drawing. Let $d =$ the distance between the cities, in miles. Repeated addition works well here.

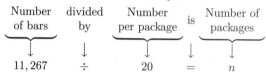

6 addends

Translate.

Number of miles per inch times Number of inches is Distance, in miles

$$64 \;\times\; 6 \;=\; d$$

Solve. We carry out the multiplication.

$$
\begin{array}{r}
6\,4 \\
\times\; 6 \\
\hline
3\,8\,4
\end{array}
$$

Thus, $384 = d$.

Check. We can repeat the calculation or estimate the product. Our answer checks.

State. Two cities that are 6 in. apart on the map are 384 miles apart in reality.

Next we find distance on the map between two cities that, in reality, are 1728 mi apart.

Familiarize. We visualize the situation. Let $m =$ the distance between the cities on the map.

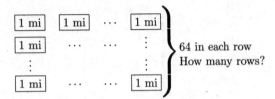

64 in each row
How many rows?

Translate.

Number of miles	divided by	Number of miles per inch	is	Distance, in inches.
1728	÷	64	=	m

Solve. We carry out the division.

$$
\begin{array}{r}
27 \\
64\overline{)1728} \\
1280 \\
\hline
448 \\
448 \\
\hline
0
\end{array}
$$

Thus, $27 = m$, or $m = 27$.

Check. We can check by multiplying: $64 \cdot 27 = 1728$.

Our answer checks.

State. The cities are 27 in. apart on the map.

45. Familiarize. First we draw a picture. Let $c =$ the number of columns. The number of columns is the same as the number of squares in each row.

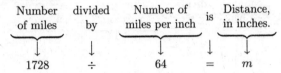

21 rows
How many in each row?

Translate. We translate to an equation.

Number of squares	divided by	Number of rows	is	Number of columns.
441	÷	21	=	c

Solve. We carry out the division.

$$
\begin{array}{r}
21 \\
21\overline{)441} \\
420 \\
\hline
21 \\
21 \\
\hline
0
\end{array}
$$

Thus, $21 = c$.

Check. We can check by multiplying the number of rows by the number of columns: $24 \cdot 21 = 441$. The answer checks.

State. The puzzle has 21 columns.

47. Familiarize. We visualize the situation as we did in Exercise 39. Let $c =$ the number of cartons that can be filled.

Translate.

Number of books	divided by	Number per carton	is	Number of full cartons.
1355	÷	24	=	c

Solve. We carry out the division.

$$
\begin{array}{r}
56 \\
24\overline{)1355} \\
1200 \\
\hline
155 \\
144 \\
\hline
11
\end{array}
$$

Check. We can check by multiplying the number of cartons by 24 and adding the remainder, 11:

$$24 \cdot 56 = 1344, \qquad 1344 + 11 = 1355$$

Our answer checks.

State. 56 cartons can be filled. There will be 11 books left over. If 1355 books are to be shipped, it will take 57 cartons.

49. Familiarize. This is a multistep problem.

We must find the total price of the 5 video games. Then we must find how many 10's there are in the total price. Let $p =$ the total price of the games.

To find the total price of the 5 video games we can use repeated addition.

$$\underbrace{\boxed{\$64} + \boxed{\$64} + \boxed{\$64} + \boxed{\$64} + \boxed{\$64}}_{5\ \text{addends}}$$

Translate.

Price per game	times	Number of games	is	Total price of games
64	·	5	=	p

Solve. First we carry out the multiplication.

$$64 \cdot 5 = p$$
$$320 = p$$

The total price of the 5 video games is $320. Repeated addition can be used again to find how many 10's there are in $320. We let $x =$ the number of $10 bills required.

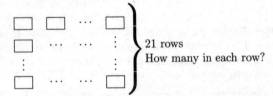

Translate to an equation and solve.

$$10 \cdot x = 320$$
$$\frac{10 \cdot x}{10} = \frac{320}{10}$$
$$x = 32$$

Check. We repeat the calculations. The answer checks.

State. It takes 32 ten dollar bills.

51. Familiarize. This is a multistep problem. We must find the total amount of the debits. Then we subtract this amount from the original balance and add the amount of the deposit. Let a = the total amount of the debits. To find this we can add.

Translate.

$$\underbrace{\text{First debit}}_{46} \overset{\text{plus}}{+} \underbrace{\text{Second debit}}_{87} \overset{\text{plus}}{+} \underbrace{\text{Third debit}}_{129} \overset{\text{is}}{=} \underbrace{\text{Total amount}}_{a}$$

Solve. First we carry out the addition.

```
  1 2
    4 6
    8 7
+ 1 2 9
-------
  2 6 2
```

Thus, $262 = a$.

Now let b = the amount left in the account after the debits.

$$\underbrace{\text{Amount left}}_{b} \overset{\text{is}}{=} \underbrace{\text{Original amount}}_{568} \overset{\text{minus}}{-} \underbrace{\text{Amount of debits}}_{262}$$

We solve this equation by carrying out the subtraction.

```
  5 6 8
- 2 6 2
-------
  3 0 6
```

Thus, $b = 306$.

Finally, let f = the final amount in the account after the deposit is made.

$$\underbrace{\text{Final amount}}_{f} \overset{\text{is}}{=} \underbrace{\text{Amount after debits}}_{306} \overset{\text{plus}}{+} \underbrace{\text{Amount of deposit}}_{94}$$

We solve this equation by carrying out the addition.

```
  1 1
  3 0 6
+   9 4
-------
  4 0 0
```

Thus, $f = 400$.

Check. We repeat the calculations. The answer checks.

State. There is $400 left in the account.

53. Familiarize. This is a multistep problem. We begin by visualizing the situation.

One pound 3500 calories			
100 cal	100 cal	...	100 cal
8 min	8 min		8 min

Let x = the number of hundreds in 3500. Repeated addition applies here.

Translate. We translate to an equation.

$$\underbrace{\text{100 calories}}_{100} \overset{\text{times}}{\cdot} \underbrace{\text{How many 100's}}_{x} \overset{\text{is 3500?}}{= 3500}$$

Solve. We divide by 100 on both sides of the equation.

$$100 \cdot x = 3500$$
$$\frac{100 \cdot x}{100} = \frac{3500}{100}$$
$$x = 35$$

We know that running for 8 min will burn 100 calories. This must be done 35 times in order to lose one pound. Let t = the time it takes to lose one pound. We have:

$$t = 35 \times 8$$
$$t = 280$$

Check. $280 \div 8 = 35$, so there are 35 8's in 280 min, and $35 \cdot 100 = 3500$, the number of calories that must be burned in order to lose one pound. The answer checks.

State. You must run for 280 min, or 4 hr, 40 min, at a brisk pace in order to lose one pound.

55. Familiarize. This is a multistep problem. We begin by visualizing the situation.

One pound 3500 calories			
100 cal	100 cal	...	100 cal
15 min	15 min		15 min

From Exercise 53 we know that there are 35 100's in 3500. From the chart we know that doing aerobic exercise for 15 min burns 100 calories. Thus we must do 15 min of exercise 35 times in order to lose one pound. Let t = the number of minutes of aerobic exercise required to lose one pound.

Translate. We translate to an equation.

$$\underbrace{\text{Number of times}}_{35} \overset{\text{times}}{\times} \underbrace{\text{Number of minutes}}_{15} \overset{\text{is}}{=} \underbrace{\text{Total time}}_{t}$$

```
    1 5
  × 3 5
-------
    7 5
  4 5 0
-------
  5 2 5
```

Thus, $525 = t$.

Check. $525 \div 15 = 35$, so there are 35 15's in 525 min, and $35 \cdot 100 = 3500$, the number of calories that must be burned in order to lose one pound. The answer checks.

State. You must do aerobic exercise for 525 min, or 8 hr, 45 min, in order to lose one pound.

57. *Familiarize*. This is a multistep problem. We will find the number of bones in both hands and the number in both feet and then the total of these two numbers. Let $h =$ the number of bones in two human hands, $f =$ the number of bones in two human feet, and $t =$ the total number of bones in two hands and two feet.

Translate. We translate to three equations.

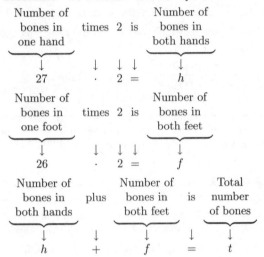

Solve. We solve each equation.
$$27 \cdot 2 = h \qquad 26 \cdot 2 = f$$
$$54 = h \qquad 52 = f$$

$$h + f = t$$
$$54 + 52 = t$$
$$106 = t$$

Check. We repeat the calculations. The answer checks.

State. In all, a human has 106 bones in both hands and both feet.

59. *Familiarize*. This is a multistep problem. First we find the writing area on one card and then the total writing area on 100 cards. Let $a =$ the writing area on one card, in square inches, and $t =$ the total writing area on 100 cards. Keep in mind that we can write on both sides of each card. Recall that the formula for the area of a rectangle is length × width.

Translate. We translate to two equations.

$$\underbrace{\text{Writing area on one card}}_{a} \text{ is } \underbrace{\text{Two}}_{=} \text{ times } \underbrace{\text{Length}}_{2} \text{ times } \underbrace{\text{Width}}_{5} \cdot 3$$

$$\underbrace{\text{Writing area on 100 cards}}_{t} \text{ is } \underbrace{100}_{= 100} \text{ times } \underbrace{\text{Writing area on one card}}_{a}$$

Solve. First we carry out the multiplication in the first equation.
$$a = 2 \cdot 5 \cdot 3$$
$$a = 30$$

Now substitute 30 for a in the second equation and carry out the multiplication.
$$t = 100 \cdot a$$
$$t = 100 \cdot 30$$
$$t = 3000$$

Check. We can repeat the calculations. The answer checks.

State. The total writing area on 100 cards is 3000 square inches.

61. Discussion and Writing Exercise

63. Round 234,562 to the nearest hundred.

$$2\ 3\ 4,\ 5\ \boxed{6}\ 2$$
$$\uparrow$$

The digit 5 is in the hundreds place. Consider the next digit to the right. Since the digit, 6, is 5 or higher, round 5 hundreds up to 6 hundreds. Then change all digits to the right of the hundreds place to zeros.

The answer is 234,600.

65. Round 234,562 to the nearest thousand.

$$2\ 3\ 4,\ \boxed{5}\ 6\ 2$$
$$\uparrow$$

The digit 4 is in the thousands place. Consider the next digit to the right. Since the digit, 5, is 5 or higher, round 4 thousands up to 5 thousands. Then change all digits to the right of the thousands place to zeros.

The answer is 235,000.

67.

	Rounded to the nearest thousand
$28,430$	$28,000$
$-\ 11,977$	$-\ 12,000$
	$16,000 \leftarrow$ Estimated answer

69.

	Rounded to the nearest thousand
5800	6000
$-\ 2100$	$-\ 2000$
	$4000 \leftarrow$ Estimated answer

71.

	Rounded to the nearest hundred
799	800
$\times\ 887$	$\times\ \ \ 900$
	$720,000 \leftarrow$ Estimated answer

73. Discussion and Writing Exercise

75. *Familiarize*. This is a multistep problem. First we will find the differences in the distances traveled in 1 second. Then we will find the differences for 18 seconds. Let $d =$ the difference in the number of miles light would travel per second in a vacuum and in ice. Let $g =$ the difference in the number of miles light would travel per second in a vacuum and in glass.

Translate. Each is a "how much more" situation.

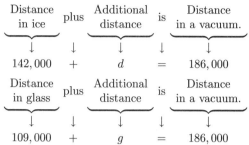

$$142,000 + d = 186,000$$

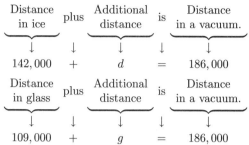

$$109,000 + g = 186,000$$

Solve. We begin by solving each equation.

$$142,000 + d = 186,000$$
$$142,000 + d - 142,000 = 186,000 - 142,000$$
$$d = 44,000$$

$$109,000 + g = 186,000$$
$$109,000 + g - 109,000 = 186,000 - 109,000$$
$$g = 77,000$$

Now to find the differences in the distances in 18 seconds, we multiply each solution by 18.

For ice: $18 \cdot 44,000 = 792,000$

For glass: $18 \cdot 77,000 = 1,386,000$

Check. We repeat the calculations. Our answers check.

State. In 18 seconds light travels 792,000 miles farther in ice and 1,386,000 miles farther in glass than in a vacuum.

Exercise Set 1.9

1. Exponential notation for $3 \cdot 3 \cdot 3 \cdot 3$ is 3^4.

3. Exponential notation for $5 \cdot 5$ is 5^2.

5. Exponential notation for $7 \cdot 7 \cdot 7 \cdot 7 \cdot 7$ is 7^5.

7. Exponential notation for $10 \cdot 10 \cdot 10$ is 10^3.

9. $7^2 = 7 \cdot 7 = 49$

11. $9^3 = 9 \cdot 9 \cdot 9 = 729$

13. $12^4 = 12 \cdot 12 \cdot 12 \cdot 12 = 20,736$

15. $11^2 = 11 \cdot 11 = 121$

17. $12 + (6 + 4) = 12 + 10$ Doing the calculation inside the parentheses
$= 22$ Adding

19. $52 - (40 - 8) = 52 - 32$ Doing the calculation inside the parentheses
$= 20$ Subtracting

21. $1000 \div (100 \div 10)$
$= 1000 \div 10$ Doing the calculation inside the parentheses
$= 100$ Dividing

23. $(256 \div 64) \div 4 = 4 \div 4$ Doing the calculation inside the parentheses
$= 1$ Dividing

25. $(2 + 5)^2 = 7^2$ Doing the calculation inside the parentheses
$= 49$ Evaluating the exponential expression

27. $(11 - 8)^2 - (18 - 16)^2$
$= 3^2 - 2^2$ Doing the calculations inside the parentheses
$= 9 - 4$ Evaluating the exponential expressions
$= 5$ Subtracting

29. $16 \cdot 24 + 50 = 384 + 50$ Doing all multiplications and divisions in order from left to right
$= 434$ Doing all additions and subtractions in order from left to right

31. $83 - 7 \cdot 6 = 83 - 42$ Doing all multiplications and divisions in order from left to right
$= 41$ Doing all additions and subtractions in order from left to right

33. $10 \cdot 10 - 3 \times 4$
$= 100 - 12$ Doing all multiplications and divisions in order from left to right
$= 88$ Doing all additions and subtractions in order from left to right

35. $4^3 \div 8 - 4$
$= 64 \div 8 - 4$ Evaluating the exponential expression
$= 8 - 4$ Doing all multiplications and divisions in order from left to right
$= 4$ Doing all additions and subtractions in order from left to right

37. $17 \cdot 20 - (17 + 20)$
$= 17 \cdot 20 - 37$ Carrying out the operation inside parentheses
$= 340 - 37$ Doing all multiplications and divisions in order from left to right
$= 303$ Doing all additions and subtractions in order from left to right

39. $6 \cdot 10 - 4 \cdot 10$
$= 60 - 40$ Doing all multiplications and divisions in order from left to right
$= 20$ Doing all additions and subtractions in order from left to right

41. $300 \div 5 + 10$
$= 60 + 10$ Doing all multiplications and divisions in order from left to right
$= 70$ Doing all additions and subtractions in order from left to right

43. $3 \cdot (2+8)^2 - 5 \cdot (4-3)^2$

$= 3 \cdot 10^2 - 5 \cdot 1^2$ Carrying out operations inside parentheses

$= 3 \cdot 100 - 5 \cdot 1$ Evaluating the exponential expressions

$= 300 - 5$ Doing all multiplications and divisions in order from left to right

$= 295$ Doing all additions and subtractions in order from left to right

45. $4^2 + 8^2 \div 2^2 = 16 + 64 \div 4$
$$= 16 + 16$$
$$= 32$$

47. $10^3 - 10 \cdot 6 - (4 + 5 \cdot 6) = 10^3 - 10 \cdot 6 - (4 + 30)$
$$= 10^3 - 10 \cdot 6 - 34$$
$$= 1000 - 10 \cdot 6 - 34$$
$$= 1000 - 60 - 34$$
$$= 940 - 34$$
$$= 906$$

49. $6 \times 11 - (7+3) \div 5 - (6-4) = 6 \times 11 - 10 \div 5 - 2$
$$= 66 - 2 - 2$$
$$= 64 - 2$$
$$= 62$$

51. $120 - 3^3 \cdot 4 \div (5 \cdot 6 - 6 \cdot 4)$
$$= 120 - 3^3 \cdot 4 \div (30 - 24)$$
$$= 120 - 3^3 \cdot 4 \div 6$$
$$= 120 - 27 \cdot 4 \div 6$$
$$= 120 - 108 \div 6$$
$$= 120 - 18$$
$$= 102$$

53. $2^3 \cdot 2^8 \div 2^6 = 8 \cdot 256 \div 64$
$$= 2048 \div 64$$
$$= 32$$

55. We add the numbers and then divide by the number of addends.
$$\frac{\$64 + \$97 + \$121}{3} = \frac{\$282}{3} = \$94$$

57. We add the numbers and then divide by the number of addends.
$$\frac{320 + 128 + 276 + 880}{4} = \frac{1604}{4} = 401$$

59. $8 \times 13 + \{42 \div [18 - (6+5)]\}$

$= 8 \times 13 + \{42 \div [18 - 11]\}$

$= 8 \times 13 + \{42 \div 7\}$

$= 8 \times 13 + 6$

$= 104 + 6$

$= 110$

61. $[14 - (3+5) \div 2] - [18 \div (8-2)]$

$= [14 - 8 \div 2] - [18 \div 6]$

$= [14 - 4] - 3$

$= 10 - 3$

$= 7$

63. $(82 - 14) \times [(10 + 45 \div 5) - (6 \cdot 6 - 5 \cdot 5)]$

$= (82 - 14) \times [(10 + 9) - (36 - 25)]$

$= (82 - 14) \times [19 - 11]$

$= 68 \times 8$

$= 544$

65. $4 \times \{(200 - 50 \div 5) - [(35 \div 7) \cdot (35 \div 7) - 4 \times 3]\}$

$= 4 \times \{(200 - 10) - [5 \cdot 5 - 4 \times 3]\}$

$= 4 \times \{190 - [25 - 12]\}$

$= 4 \times \{190 - 13\}$

$= 4 \times 177$

$= 708$

67. $\{[18 - 2 \cdot 6] - [40 \div (17 - 9)]\} +$
$$\{48 - 13 \times 3 + [(50 - 7 \cdot 5) + 2]\}$$

$= \{[18 - 12] - [40 \div 8]\} +$
$$\{48 - 13 \times 3 + [(50 - 35) + 2]\}$$

$= \{6 - 5\} + \{48 - 13 \times 3 + [15 + 2]\}$

$= 1 + \{48 - 13 \times 3 + 17\}$

$= 1 + \{48 - 39 + 17\}$

$= 1 + 26$

$= 27$

69. Discussion and Writing Exercise

71. $$x + 341 = 793$$
$$x + 341 - 341 = 793 - 341$$
$$x = 452$$

The solution is 452.

73. $$7 \cdot x = 91$$
$$\frac{7 \cdot x}{7} = \frac{91}{7}$$
$$x = 13$$

The solution is 13.

75. $$3240 = y + 898$$
$$3240 - 898 = y + 898 - 898$$
$$2342 = y$$

The solution is 2342.

77. $$25 \cdot t = 625$$
$$\frac{25 \cdot t}{25} = \frac{625}{25}$$
$$t = 25$$

The solution is 25.

79. *Familiarize.* We first make a drawing.

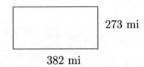

273 mi

382 mi

Translate. We use the formula for the area of a rectangle.
$$A = l \cdot w = 382 \cdot 273$$

Solve. We carry out the multiplication.
$$A = 382 \cdot 273 = 104,286$$

Check. We repeat the calculation. The answer checks.

State. The area is 104,286 square miles.

81. Discussion and Writing Exercise

83. $1 + 5 \cdot 4 + 3 = 1 + 20 + 3$
$$= 24 \qquad \text{Correct answer}$$

To make the incorrect answer correct we add parentheses:
$$1 + 5 \cdot (4 + 3) = 36$$

85. $12 \div 4 + 2 \cdot 3 - 2 = 3 + 6 - 2$
$$= 7 \qquad \text{Correct answer}$$

To make the incorrect answer correct we add parentheses:
$$12 \div (4 + 2) \cdot 3 - 2 = 4$$

Chapter 1 Review Exercises

1. $4,67\boxed{8},952$

The digit 8 means 8 thousands.

2. $1\boxed{3},768,940$

The digit 3 names the number of millions.

3. $2793 = 2$ thousands $+ 7$ hundreds $+ 9$ tens $+ 3$ ones

4. $56,078 = 5$ ten thousands $+ 6$ thousands $+ 0$ hundreds $+ 7$ tens $+ 8$ ones, or 5 ten thousands $+ 6$ thousands $+ 7$ tens $+ 8$ ones

5. $4,007,101 = 4$ millions $+ 0$ hundred thousands $+ 0$ ten thousands $+ 7$ thousands $+ 1$ hundred $+ 0$ tens $+ 1$ one, or 4 millions $+ 7$ thousands $+ 1$ hundred $+ 1$ one

6.

$$\overbrace{67},\overbrace{819}$$

Sixty-seven thousand, ——
eight hundred nineteen ——

7.

$$\overbrace{2},\overbrace{781},\overbrace{427}$$

Two million, ——
seven hundred eighty-one thousand, ——
four hundred twenty-seven ——

8.

$$\overbrace{1},\overbrace{065},\overbrace{070},\overbrace{607}$$

One billion, ——
sixty-five million, ——
seventy thousand, ——
six hundred seven ——

9. Four hundred seventy-six thousand, ——
five hundred eighty-eight ——

Standard notation is $\overbrace{476},\overbrace{588}$.

10. Two billion, ——
four hundred thousand, ——

Standard notation is $\overbrace{2},000,\overbrace{400},000$.

11.
$$\begin{array}{r} \overset{\,1\,1}{7\,3\,0\,4} \\ +\ 6\,9\,6\,8 \\ \hline 1\,4,2\,7\,2 \end{array}$$

12.
$$\begin{array}{r} \overset{1\ 1\ \ 1}{2\,7,6\,0\,9} \\ +\,3\,8,4\,1\,5 \\ \hline 6\,6,0\,2\,4 \end{array}$$

13.
$$\begin{array}{r} \overset{1\ \ 1}{2\,7\,0\,3} \\ 4\,1\,2\,5 \\ 6\,0\,0\,4 \\ +\ 8\,9\,5\,6 \\ \hline 2\,1,7\,8\,8 \end{array}$$

14.
$$\begin{array}{r} \overset{1\ 1}{9\,1,4\,2\,6} \\ +\ \ 7,4\,9\,5 \\ \hline 9\,8,9\,2\,1 \end{array}$$

15. $10 - 6 = 4$

This number gets added (after 4).
$$10 = 6 + 4$$

(By the commutative law of addition, $10 = 4 + 6$ is also correct.)

16. $8 + 3 = 11$

This addend gets subtracted from the sum.
$$8 = 11 - 3$$

$8 + 3 = 11$

This addend gets subtracted from the sum.
$$3 = 11 - 8$$

17.
$$\begin{array}{r} \overset{\quad\ 13}{\overset{7\ \ 9\ \ \cancel{3}\ 15}{\cancel{8}\,\cancel{0}\,\cancel{4}\,\cancel{5}}} \\ -\ 2\,8\,9\,7 \\ \hline 5\,1\,4\,8 \end{array}$$

18.
$$\begin{array}{r} \overset{8\ \ 9\ \ 9\ 11}{\cancel{9}\,\cancel{0}\,\cancel{0}\,\cancel{1}} \\ -\ 7\,3\,1\,2 \\ \hline 1\,6\,8\,9 \end{array}$$

19.
$$\begin{array}{r} \overset{5\ \ 9\ \ 9\ 13}{\cancel{6}\,\cancel{0}\,\cancel{0}\,\cancel{3}} \\ -\ 3\,7\,2\,9 \\ \hline 2\,2\,7\,4 \end{array}$$

20.
$$\begin{array}{r} \overset{\qquad 16\ 13}{\overset{2\ \ \cancel{6}\ \ \cancel{3}\ \ 9\ 15}{\cancel{3}\,7,4\,\cancel{0}\,\cancel{5}}} \\ -\ 1\,9,6\,4\,8 \\ \hline 1\,7,7\,5\,7 \end{array}$$

21. Round 345,759 to the nearest hundred.

$$3\,4\,5,7\,\boxed{5}\,9$$

The digit 7 is in the hundreds place. Consider the next digit to the right. Since the digit, 5, is 5 or higher, round 7 hundreds up to 8 hundreds. Then change the digits to the right of the hundreds digit to zero.

The answer is 345,800.

22. Round 345,759 to the nearest ten.

$$3\,4\,5,7\,5\,\boxed{9}$$
$$\uparrow$$

The digit 5 is in the tens place. Consider the next digit to the right. Since the digit, 9, is 5 or higher, round 5 tens up to 6 tens. Then change the digit to the right of the tens digit to zero.

The answer is 345,760.

23. Round 345,759 to the nearest thousand.

$$3\,4\,5,\,\boxed{7}\,5\,9$$
$$\uparrow$$

The digit 5 is in the thousands place. Consider the next digit to the right. Since the digit, 7, is 5 or higher, round 5 thousands up to 6 thousands. Then change the digits to the right of the thousands digit to zero.

The answer is 346,000.

24. Round 345,759 to the nearest hundred thousand.

$$3\,\boxed{4}\,5,7\,5\,9$$
$$\uparrow$$

The digit 3 is in the hundred thousands place. Consider the next digit to the right. Since the digit, 4, is 4 or lower, round down, meaning that 3 hundred thousands stays as 3 hundred thousands. Then change the digits to the right of the hundred thousands digit to zero.

The answer is 300,000.

25.

	Rounded to the nearest hundred
$\begin{array}{r} 4\,1,3\,4\,8 \\ +\,1\,9,7\,4\,9 \\ \hline \end{array}$	$\begin{array}{r} 4\,1,3\,0\,0 \\ +\,1\,9,7\,0\,0 \\ \hline 6\,1,0\,0\,0 \end{array}$ ← Estimated answer

26.

	Rounded to the nearest hundred
$\begin{array}{r} 3\,8,6\,5\,2 \\ -\,2\,4,5\,4\,9 \\ \hline \end{array}$	$\begin{array}{r} 3\,8,7\,0\,0 \\ -\,2\,4,5\,0\,0 \\ \hline 1\,4,2\,0\,0 \end{array}$ ← Estimated answer

27.

	Rounded to the nearest hundred
$\begin{array}{r} 3\,9\,6 \\ \times\,7\,4\,8 \\ \hline \end{array}$	$\begin{array}{r} 4\,0\,0 \\ \times\,7\,0\,0 \\ \hline 2\,8\,0,0\,0\,0 \end{array}$ ← Estimated answer

28. Since 67 is to the right of 56 on the number line, $67 > 56$.

29. Since 1 is to the left of 23 on the number line, $1 < 23$.

30.
$$\begin{array}{r} \overset{2}{1\,7,0\,0\,0} \\ \times\quad\;\;3\,0\,0 \\ \hline 5,1\,0\,0,0\,0\,0 \end{array}$$

Multiplying by 300
(Write 00 and then multiply 17,000 by 3.)

31.
$$\begin{array}{r} \overset{6\;3\;4}{7\,8\,4\,6} \\ \times\quad\;\;8\,0\,0 \\ \hline 6,2\,7\,6,8\,0\,0 \end{array}$$

Multiplying by 800
(Write 00 and then multiply 7846 by 8.)

32.
$$\begin{array}{r} \overset{1\;3}{}\,\overset{2\;5}{}\,\overset{2\;4}{} \\ 7\,2\,6 \\ \times\,6\,9\,8 \\ \hline 5\,8\,0\,8 \\ 6\,5\,3\,4\,0 \\ 4\,3\,5\,6\,0\,0 \\ \hline 5\,0\,6,7\,4\,8 \end{array}$$

Multiplying by 8
Multiplying by 9
Multiplying by 6

33.
$$\begin{array}{r} \overset{3\;2}{}\,\overset{6\;4}{} \\ 5\,8\,7 \\ \times\quad4\,7 \\ \hline 4\,1\,0\,9 \\ 2\,3\,4\,8\,0 \\ \hline 2\,7,5\,8\,9 \end{array}$$

Multiplying by 7
Multiplying by 4

34.
$$\begin{array}{r} 8\,3\,0\,5 \\ \times\quad\;\;6\,4\,2 \\ \hline 1\,6\,6\,1\,0 \\ 3\,3\,2\,2\,0\,0 \\ 4\,9\,8\,3\,0\,0\,0 \\ \hline 5,3\,3\,1,8\,1\,0 \end{array}$$

35. $56 \div 7 = 8$ The 7 moves to the right. A related multiplication sentence is $56 = 8 \cdot 7$. (By the commutative law of multiplication, there is also another multiplication sentence: $56 = 7 \cdot 8$.)

36. $13 \cdot 4 = 52$

Move a factor to the other side and then write a division.

$13 \cdot 4 = 52$ $13 \cdot 4 = 52$

$13 = 52 \div 4$ $4 = 52 \div 13$

37.
$$\begin{array}{r} 1\,2 \\ 5\overline{)6\,3} \\ 5\,0 \\ \hline 1\,3 \\ 1\,0 \\ \hline 3 \end{array}$$

The answer is 12 R 3.

38.
$$\begin{array}{r} 5 \\ 1\,6\overline{)8\,0} \\ 8\,0 \\ \hline 0 \end{array}$$

The answer is 5.

39.
$$\begin{array}{r} 9\,1\,3 \\ 7\overline{)6\,3\,9\,4} \\ 6\,3\,0\,0 \\ \hline 9\,4 \\ 7\,0 \\ \hline 2\,4 \\ 2\,1 \\ \hline 3 \end{array}$$

The answer is 913 R 3.

40.

$$
\begin{array}{r}
384 \\
8\overline{)3073} \\
2400 \\
\hline
673 \\
640 \\
\hline
33 \\
32 \\
\hline
1
\end{array}
$$

The answer is 384 R 1.

41.

$$
\begin{array}{r}
4 \\
60\overline{)286} \\
240 \\
\hline
46
\end{array}
$$

The answer is 4 R 46.

42.

$$
\begin{array}{r}
54 \\
79\overline{)4266} \\
3950 \\
\hline
316 \\
316 \\
\hline
0
\end{array}
$$

The answer is 54.

43.

$$
\begin{array}{r}
452 \\
38\overline{)17,176} \\
15200 \\
\hline
1976 \\
1900 \\
\hline
76 \\
76 \\
\hline
0
\end{array}
$$

The answer is 452.

44.

$$
\begin{array}{r}
5008 \\
14\overline{)70,112} \\
70000 \\
\hline
112 \\
112 \\
\hline
0
\end{array}
$$

The answer is 5008.

45.

$$
\begin{array}{r}
4389 \\
12\overline{)52,668} \\
48000 \\
\hline
4668 \\
3600 \\
\hline
1068 \\
960 \\
\hline
108 \\
108 \\
\hline
0
\end{array}
$$

The answer is 4389.

46. $46 \cdot n = 368$

$\dfrac{46 \cdot n}{46} = \dfrac{368}{46}$

$n = 8$

Check: $46 \cdot n = 368$

$46 \cdot 8 \ ? \ 368$

$368 \ \big| \ $ TRUE

The solution is 8.

47. $47 + x = 92$

$47 + x - 47 = 92 - 47$

$x = 45$

Check: $47 + x = 92$

$47 + 45 \ ? \ 92$

$92 \ \big| \ $ TRUE

The solution is 45.

48. $1 \cdot y = 58$

$y = 58 \qquad (1 \cdot y = y)$

The number 58 checks. It is the solution.

49. $24 = x + 24$

$24 - 24 = x + 24 - 24$

$0 = x$

The number 0 checks. It is the solution.

50. Exponential notation for $4 \cdot 4 \cdot 4$ is 4^3.

51. $10^4 = 10 \cdot 10 \cdot 10 \cdot 10 = 10{,}000$

52. $6^2 = 6 \cdot 6 = 36$

53. $8 \cdot 6 + 17 = 48 + 17$ Multiplying

$\qquad\qquad\quad = 65$ Adding

54. $10 \cdot 24 - (18 + 2) \div 4 - (9 - 7)$

$= 10 \cdot 24 - 20 \div 4 - 2$ Doing the calculations inside the parentheses

$= 240 - 5 - 2$ Multiplying and dividing

$= 235 - 2$ Subtracting from

$= 233$ left to right

55. $7 + (4 + 3)^2 = 7 + 7^2$

$\qquad\qquad\qquad = 7 + 49$

$\qquad\qquad\qquad = 56$

56. $7 + 4^2 + 3^2 = 7 + 16 + 9$

$\qquad\qquad\qquad = 23 + 9$

$\qquad\qquad\qquad = 32$

57. $(80 \div 16) \times [(20 - 56 \div 8) + (8 \cdot 8 - 5 \cdot 5)]$

$= 5 \times [(20 - 7) + (64 - 25)]$

$= 5 \times [13 + 39]$

$= 5 \times 52$

$= 260$

58. We add the numbers and divide by the number of addends.

$$\dfrac{157 + 170 + 168}{3} = \dfrac{495}{3} = 165$$

59. *Familiarize*. Let $x =$ the additional amount of money, in dollars, Natasha needs to buy the desk.

Translate. This is a "how much more" situation.

Money available	plus	Additional amount	is	Price of desk
↓	↓	↓	↓	↓
196	+	x	=	698

Solve. We subtract 196 on both sides of the equation.

$$196 + x = 698$$
$$196 + x - 196 = 698 - 196$$
$$x = 502$$

Check. We can estimate.

$$196 + 502 \approx 200 + 500 \approx 700 \approx 698$$

The answer checks.

State. Natasha needs $502 dollars.

60. Familiarize. Let b = the balance in Taneesha's account after the deposit.

Translate.

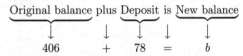

Solve. We add on the left side.

$$406 + 78 = b$$
$$484 = b$$

Check. We can repeat the calculation. The answer checks.

State. The new balance is $484.

61. Familiarize. Let y = the first year in which the copper content of pennies was reduced.

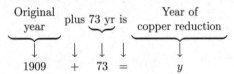

Solve. We add on the left side.

$$1909 + 73 = y$$
$$1982 = y$$

Check. We can estimate.

$$1909 + 73 \approx 1910 + 70 \approx 1980 \approx 1982$$

The answer checks.

State. The copper content of pennies was first reduced in 1982.

62. Familiarize. We first make a drawing. Let c = the number of cartons filled.

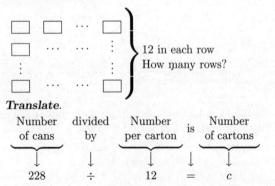

Translate.

Solve. We carry out the division.

$$
\begin{array}{r}
1\,9 \\
1\,2\,\overline{)\,2\,2\,8} \\
1\,2\,0 \\
\hline
1\,0\,8 \\
1\,0\,8 \\
\hline
0
\end{array}
$$

Thus, $19 = c$, or $c = 19$.

Check. We can check by multiplying: $12 \cdot 19 = 228$. Our answer checks.

State. 19 cartons were filled.

63. Familiarize. Let b = the number of beehives the farmer needs.

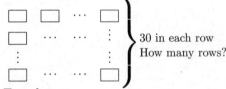

Translate.

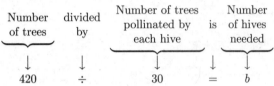

Solve. We carry out the division.

$$
\begin{array}{r}
1\,4 \\
3\,0\,\overline{)\,4\,2\,0} \\
3\,0\,0 \\
\hline
1\,2\,0 \\
1\,2\,0 \\
\hline
0
\end{array}
$$

Thus, $14 = b$, or $b = 14$.

Check. We can check by multiplying: $30 \cdot 14 = 420$. The answer checks.

State. The farmer needs 14 beehives.

64. Familiarize. This is a multistep problem. Let s = the cost of 13 stoves, r = the cost of 13 refrigerators, and t = the total cost of the stoves and refrigerators.

Translate.

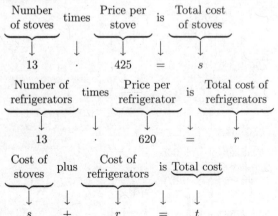

Solve. We first carry out the multiplications in the first two equations.

$$13 \cdot 425 = s \qquad 13 \cdot 620 = r$$
$$5525 = s \qquad 8060 = r$$

Now we substitute 5525 for s and 8060 for r in the third equation and then add on the left side.

$$s + r = t$$
$$5525 + 8060 = t$$
$$13,585 = t$$

Check. We repeat the calculations. The answer checks.

State. The total cost was $13,585.

65. *Familiarize*. This is a multistep problem. Let $b =$ the total amount budgeted for food, clothing, and entertainment and let $r =$ the income remaining after these allotments.

Translate.

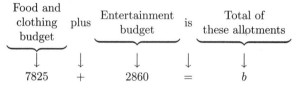

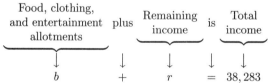

Solve. We add on the left side to solve the first equation.

$$7825 + 2860 = b$$
$$10,685 = b$$

Now we substitute 10,685 for b in the second equation and solve for r.

$$b + r = 38,283$$
$$10,685 + r = 38,283$$
$$10,685 + r - 10,685 = 38,283 - 10,685$$
$$r = 27,598$$

Check. We repeat the calculations. The answer checks.

State. After the allotments for food, clothing, and entertainment, $27,598 remains.

66. *Familiarize*. We make a drawing. Let $b =$ the number of beakers that will be filled.

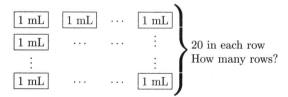

Translate.

Amount of alcohol	divided by	Amount per beaker	is	Number of beakers filled
↓	↓	↓	↓	↓
2753	÷	20	=	b

Solve. We carry out the division.

```
        1 3 7
20 ) 2 7 5 3
     2 0 0 0
     -------
       7 5 3
       6 0 0
     -------
       1 5 3
       1 4 0
     -------
         1 3
```

Thus, 137 R 13 = b.

Check. We can check by multiplying the number of beakers by 137 and then adding the remainder, 13.

$$137 \cdot 20 = 2740 \text{ and } 2740 + 13 = 2753$$

The answer checks.

State. 137 beakers can be filled; 13 mL will be left over.

67. $A = l \cdot w = 14 \text{ ft} \cdot 7 \text{ ft} = 98$ square ft

Perimeter $= 14 \text{ ft} + 7 \text{ ft} + 14 \text{ ft} + 7 \text{ ft} = 42 \text{ ft}$

68. *Discussion and Writing Exercise*. A vat contains 1152 oz of hot sauce. If 144 bottles are to be filled equally, how much will each bottle contain? Answers may vary.

69. *Discussion and Writing Exercise*. No; if subtraction were associative, then $a - (b - c) = (a - b) - c$ for any a, b, and c. But, for example,

$$12 - (8 - 4) = 12 - 4 = 8,$$

whereas

$$(12 - 8) - 4 = 4 - 4 = 0.$$

Since $8 \neq 0$, this examples shows that subtraction is not associative.

70.
$$\begin{array}{r} 9\,d \\ \times \quad d\,2 \\ \hline 8\,0\,3\,6 \end{array}$$

By using rough estimates, we see that the factor $d2 \approx 8100 \div 90 = 90$ or $d2 \approx 8000 \div 100 = 80$. Since $99 \times 92 = 9108$ and $98 \times 82 = 8036$, we have $d = 8$.

71.
$$\begin{array}{r} 9\,a\,1 \\ 2\,b\,1\,\overline{)\,2\,3\,6{,}4\,2\,1} \end{array}$$

Since $250 \times 1000 = 250,000 \approx 236,421$ we deduce that $2b1 \approx 250$ and $9a1 \approx 1000$. By trial we find that $a = 8$ and $b = 4$.

72. At the beginning of each day the tunnel reaches 500 ft − 200 ft, or 300 ft, farther into the mountain than it did the day before. We calculate how far the tunnel reaches into the mountain at the beginning of each day, starting with Day 2.

Day 2: 300 ft

Day 3: 300 ft + 300 ft = 600 ft

Day 4: 600 ft + 300 ft = 900 ft

Day 5: 900 ft + 300 ft = 1200 ft

Day 6: 1200 ft + 300 ft = 1500 ft

We see that the tunnel reaches 1500 ft into the mountain at the beginning of Day 6. On Day 6 the crew tunnels an additional 500 ft, so the tunnel reaches 1500 ft + 500 ft, or 2000 ft, into the mountain. Thus, it takes 6 days to reach the copper deposit.

Chapter 1 Test

1. $\boxed{5}$ 4 6, 7 8 9

The digit 5 tells the number of hundred thousands.

2. 8843 = 8 thousands + 8 hundreds + 4 tens + 3 ones

3.

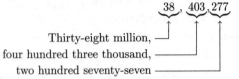

$$38 , 403 , 277$$

Thirty-eight million, —
four hundred three thousand, —
two hundred seventy-seven —

4.
```
    6 8 1 1     Add ones, add tens, add hundreds,
 +  3 1 7 8     and then add thousands.
 ─────────
    9 9 8 9
```

5.
```
      1 1    1
     4 5, 8 8 9
  + 1 7, 9 0 2
  ───────────
     6 3, 7 9 1
```

6. We look for pairs of numbers whose sums are 10, 20, 30, and so on.

```
 12  ──→  20
  8
  3  ──→  10
  7
 +4  ──→   4
 ──       ──
 34       34
```

7.
```
      6 2 0 3
  +   4 3 1 2
  ───────────
   1 0, 5 1 5
```

8.
```
    7 9 8 3       Subtract ones, subtract tens, subtract
 -  4 3 5 3       hundreds, and then subtract thousands.
 ─────────
    3 6 3 0
```

9.
```
         6 14
    2 9 7 4̸
 -  1 9 3 5
 ─────────
    1 0 3 9
```

10.
```
      8 9 17
    8 9̸ 0̸ 7
 -  2 0 5 9
 ─────────
    6 8 4 8
```

11.
```
          12
      1 2̸ 9 16
    2̸ 3, 0̸ 6̸ 7
 -  1 7, 8 9 2
 ───────────
       5 1 7 5
```

12.
```
     5 6 7
    4 5 6 8
 ×         9
 ─────────
  4 1, 1 1 2
```

13.
```
      5 4 3
     8 8 7 6
 ×     6 0 0      Multiply by 6 hundreds (We write 00
 ───────────      and then multiply 8876 by 6.)
 5, 3 2 5, 6 0 0
```

14.
```
        6 5
 ×      3 7
 ───────
    4 5 5        Multiplying by 7
  1 9 5 0        Multiplying by 30
 ───────
  2 4 0 5        Adding
```

15.
```
         6 7 8
 ×       7 8 8
 ─────────
     5 4 2 4
   5 4 2 4 0
 4 7 4 6 0 0
 ───────────
 5 3 4, 2 6 4
```

16.
```
        3
  4 ⟌ 1 5
      1 2
      ───
        3
```
The answer is 3 R 3.

17.
```
        7 0
  6 ⟌ 4 2 0
      4 2 0
      ─────
          0
          0
          ─
          0
```
The answer is 70.

18.
```
          9 7
  8 9 ⟌ 8 6 3 3
       8 0 1 0
       ───────
         6 2 3
         6 2 3
         ─────
             0
```
The answer is 97.

19.
```
            8 0 5
  4 4 ⟌ 3 5, 4 2 8
       3 5 2 0 0
       ─────────
           2 2 8
           2 2 0
           ─────
               8
```
The answer is 805 R 8.

20. *Familiarize*. Let $n =$ the number of 12-packs that can be filled. We can think of this as repeated subtraction, taking successive sets of 12 snack cakes and putting them into n packages.

Translate.

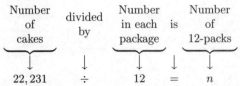

Number of cakes	divided by	Number in each package	is	Number of 12-packs
22, 231	÷	12	=	n

Solve. We carry out the division.

```
          1 8 5 2
  1 2 ) 2 2 , 2 3 1
        1 2 0 0 0
        ‾‾‾‾‾‾‾‾‾
        1 0 2 3 1
          9 6 0 0
          ‾‾‾‾‾‾‾
            6 3 1
            6 0 0
            ‾‾‾‾‾
              3 1
              2 4
              ‾‾‾
                7
```

Then 1852 R 7 = n.

Check. We multiply the number of cartons by 12 and then add the remainder, 7.

$$12 \cdot 1852 = 22,224$$
$$22,224 + 7 = 22,231$$

The answer checks.

State. 1852 twelve-packs can be filled. There will be 7 cakes left over.

21. **Familiarize**. Let $a =$ the total land area of the five largest states, in square meters. Since we are combining the areas of the states, we can add.

Translate.

$$571,951 + 261,797 + 155,959 + 145,552 + 121,356 = a$$

Solve. We carry out the addition.

```
  2 1  3 3 2
  5 7 1 , 9 5 1
  2 6 1 , 7 9 7
  1 5 5 , 9 5 9
  1 4 5 , 5 5 2
+ 1 2 1 , 3 5 6
‾‾‾‾‾‾‾‾‾‾‾‾‾‾‾
1 , 2 5 6 , 6 1 5
```

Then $1,256,615 = a$.

Check. We can repeat the calculation. We can also estimate the result by rounding. We will round to the nearest ten thousand.

$$571,951 + 261,797 + 155,959 + 145,552 + 121,356$$
$$\approx 570,000 + 260,000 + 160,000 + 150,000 + 120,000$$
$$= 1,260,000$$

Since $1,260,000 \approx 1,256,615$, we have a partial check.

State. The total land area of Alaska, Texas, California, Montana, and New Mexico is $1,256,615$ m².

22. a) We will use the formula Perimeter $= 2 \cdot$ length $+ 2 \cdot$ width to find the perimeter of each pool table in inches. We will use the formula Area $=$ length $\cdot$ width to find the area of each pool table, in in².

For the 50 in. by 100 in. table:

$$\text{Perimeter} = 2 \cdot 100 \text{ in.} + 2 \cdot 50 \text{ in.}$$
$$= 200 \text{ in.} + 100 \text{ in.}$$
$$= 300 \text{ in.}$$
$$\text{Area} = 100 \text{ in.} \cdot 50 \text{ in.} = 5000 \text{ in}^2$$

For the 44 in. by 88 in. table:

$$\text{Perimeter} = 2 \cdot 88 \text{ in.} + 2 \cdot 44 \text{ in.}$$
$$= 176 \text{ in.} + 88 \text{ in.}$$
$$= 264 \text{ in.}$$
$$\text{Area} = 88 \text{ in.} \cdot 44 \text{ in.} = 3872 \text{ in}^2$$

For the 38 in. by 76 in. table:

$$\text{Perimeter} = 2 \cdot 76 \text{ in.} + 2 \cdot 38 \text{ in.}$$
$$= 152 \text{ in.} + 76 \text{ in.}$$
$$= 228 \text{ in.}$$
$$\text{Area} = 76 \text{ in.} \cdot 38 \text{ in.} = 2888 \text{ in}^2$$

b) Let $a =$ the number of square inches by which the area of the largest table exceeds the area of the smallest table. We subtract to find a.

$$a = 5000 \text{ in}^2 - 2888 \text{ in}^2 = 2112 \text{ in}^2$$

23. **Familiarize**. Let $v =$ the number of Nevada voters who voted early in the 2000 presidential election.

Translate.

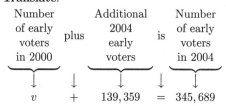

Solve. We subtract 139,359 on both sides of the equation.

$$v + 139,359 = 345,689$$
$$v + 139,359 - 139,359 = 345,689 - 139,359$$
$$v = 206,330$$

Check. We can add the difference, 206,330, to the subtrahend, 139,359: $139,359 + 206,330 = 345,689$. The answer checks.

State. 206,330 voters voted early in Nevada in 2000.

24. **Familiarize**. There are three parts to this problem. First we find the total weight of each type of fruit and then we add. Let $x =$ the total weight of the oranges, $y =$ the total weight of the apples, and $t =$ the total weight of both fruits together.

Translate.

For the oranges:

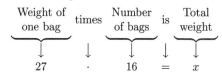

For the apples:

```
Weight of   times   Number   is   Total
one bag             of bags       weight
   ↓          ↓        ↓       ↓      ↓
   32         ·        43      =      y
```

For the total weight of both fruits:

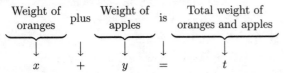

$$x \quad + \quad y \quad = \quad t$$

Solve. We solve the first two equations and then add the solutions.

$$27 \cdot 16 = x$$
$$432 = x$$

$$32 \cdot 43 = y$$
$$1376 = y$$

$$x + y = t$$
$$432 + 1376 = t$$
$$1808 = t$$

Check. We repeat the calculations. The answer checks.

State. The total weight of 16 bags of oranges and 43 bags of apples is 1808 lb.

25. **Familiarize.** Let s = the number of staplers that can be filled. We can think of this as repeated subtraction, taking successive sets of 250 staples and putting them into s staplers.

Translate.

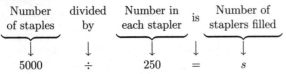

$$5000 \quad \div \quad 250 \quad = \quad s$$

Solve. We carry out the division.

$$\begin{array}{r} 20 \\ 250\overline{)5000} \\ \underline{5000} \\ 0 \\ \underline{0} \\ \overline{0} \end{array}$$

Then $20 = s$.

Check. We can multiply the number of staplers filled by the number of staples in each one.

$$20 \cdot 250 = 5000$$

The answer checks.

State. 20 staplers can be filled from a box of 5000 staples.

26. $$28 + x = 74$$
$$28 + x - 28 = 74 - 28 \quad \text{Subtracting 28 on both sides}$$
$$x = 46$$

Check: $$28 + x = 74$$
$$\overline{28 + 46 \; ? \; 74}$$
$$74 \; \big| \quad \text{TRUE}$$

The solution is 46.

27. $169 \div 13 = n$

We carry out the division.

$$\begin{array}{r} 13 \\ 13\overline{)169} \\ \underline{130} \\ 39 \\ \underline{39} \\ 0 \end{array}$$

The solution is 13.

28. $$38 \cdot y = 532$$
$$\frac{38 \cdot y}{38} = \frac{532}{38} \quad \text{Dividing by 38 on both sides}$$
$$y = 14$$

Check: $$38 \cdot y = 532$$
$$\overline{38 \cdot 14 \; ? \; 532}$$
$$532 \; \big| \quad \text{TRUE}$$

The solution is 14.

29. $$381 = 0 + a$$
$$381 = a \quad \text{Adding on the right side}$$
The solution is 381.

30. Round 34,578 to the nearest thousand.

$$3\,4,\boxed{5}\,7\,8$$
$$\uparrow$$

The digit 4 is in the thousands place. Consider the next digit to the right, 5. Since 5 is 5 or higher, round 4 thousands up to 5 thousands. Then change all the digits to the right of thousands to zeros.

The answer is 35,000.

31. Round 34,578 to the nearest ten.

$$3\,4,5\,7\,\boxed{8}$$
$$\uparrow$$

The digit 7 is in the tens place. Consider the next digit to the right, 8. Since 8 is 5 or higher, round 7 tens up to 8 tens. Then change the digit to the right of tens to zero.

The answer is 34,580.

32. Round 34,578 to the nearest hundred.

$$3\,4,5\,\boxed{7}\,8$$
$$\uparrow$$

The digit 5 is in the hundreds place. Consider the next digit to the right, 7. Since 7 is 5 or higher, round 5 hundreds up to 6 hundreds. Then change all the digits to the right of hundreds to zeros.

The answer is 34,600.

33.   Rounded to
                                  the nearest hundred

$$\begin{array}{r} 2\,3,6\,4\,9 \\ +\,5\,4,7\,4\,6 \\ \hline \end{array} \qquad \begin{array}{r} 2\,3,6\,0\,0 \\ +\,5\,4,7\,0\,0 \\ \hline 7\,8,3\,0\,0 \end{array} \leftarrow \text{Estimated answer}$$

34.
$$\begin{array}{c} \text{Rounded to} \\ \text{the nearest hundred} \end{array}$$

$$
\begin{array}{rr}
5\,4,7\,5\,1 & 5\,4,8\,0\,0 \\
-\,2\,3,6\,4\,9 & -\,2\,3,6\,0\,0 \\
\hline
& 3\,1,2\,0\,0 \leftarrow \text{Estimated answer}
\end{array}
$$

35.
$$\begin{array}{c} \text{Rounded to} \\ \text{the nearest hundred} \end{array}$$

$$
\begin{array}{rr}
8\,2\,4 & 8\,0\,0 \\
\times\,4\,8\,9 & \times\,5\,0\,0 \\
\hline
& 4\,0\,0,0\,0\,0 \leftarrow \text{Estimated answer}
\end{array}
$$

36. Since 34 is to the right of 17 on the number line, $34 > 17$.

37. Since 117 is to the left of 157 on the number line, $117 < 157$.

38. Exponential notation for $12 \cdot 12 \cdot 12 \cdot 12$ is 12^4.

39. $7^3 = 7 \cdot 7 \cdot 7 = 343$

40. $10^5 = 10 \cdot 10 \cdot 10 \cdot 10 \cdot 10 = 100,000$

41. $25^2 = 25 \cdot 25 = 625$

42.
$$
\begin{aligned}
& 35 - 1 \cdot 28 \div 4 + 3 \\
&= 35 - 28 \div 4 + 3 \quad \text{Doing all multiplications and} \\
&= 35 - 7 + 3 \quad\quad\;\; \text{divisions in order from left to right} \\
&= 28 + 3 \quad\quad\quad\;\; \text{Doing all additions and subtractions} \\
&= 31 \quad\quad\quad\quad\quad \text{in order from left to right}
\end{aligned}
$$

43.
$$
\begin{aligned}
& 10^2 - 2^2 \div 2 \\
&= 100 - 4 \div 2 \quad \text{Evaluating the exponential} \\
&\quad\quad\quad\quad\quad\quad \text{expressions} \\
&= 100 - 2 \quad\quad\;\; \text{Dividing} \\
&= 98 \quad\quad\quad\quad\; \text{Subtracting}
\end{aligned}
$$

44.
$$
\begin{aligned}
& (25 - 15) \div 5 \\
&= 10 \div 5 \quad \text{Doing the calculation inside the parentheses} \\
&= 2 \quad\quad\;\; \text{Dividing}
\end{aligned}
$$

45.
$$
\begin{aligned}
& 8 \times \{(20 - 11) \cdot [(12 + 48) \div 6 - (9 - 2)]\} \\
&= 8 \times \{9 \cdot [60 \div 6 - 7]\} \\
&= 8 \times \{9 \cdot [10 - 7]\} \\
&= 8 \times \{9 \cdot 3\} \\
&= 8 \times 27 \\
&= 216
\end{aligned}
$$

46.
$$
\begin{aligned}
& 2^4 + 24 \div 12 \\
&= 16 + 24 \div 12 \quad \text{Evaluating the exponential} \\
&\quad\quad\quad\quad\quad\quad \text{expression} \\
&= 16 + 2 \quad\quad\quad\; \text{Dividing} \\
&= 18 \quad\quad\quad\quad\;\; \text{Adding}
\end{aligned}
$$

47. We add the numbers and then divide by the number of addends.
$$
\frac{97 + 98 + 87 + 86}{4} = \frac{368}{4} = 92
$$

48. *Familiarize*. We make a drawing.

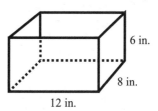

Observe that the dimensions of two sides of the container are 8 in. by 6 in. The area of each is 8 in. $\cdot$ 6 in. and their total area is $2 \cdot 8$ in. $\cdot 6$ in. The dimensions of the other two sides are 12 in. by 6 in. The area of each is 12 in. $\cdot 6$ in. and their total area is $2 \cdot 12$ in. $\cdot 6$ in. The dimensions of the bottom of the box are 12 in. by 8 in. and its area is 12 in. $\cdot 8$ in. Let $c =$ the number of square inches of cardboard that are used for the container.

***Translate*.** We add the areas of the sides and the bottom of the container.
$$
2 \cdot 8 \text{ in.} \cdot 6 \text{ in.} + 2 \cdot 12 \text{ in.} \cdot 6 \text{ in.} + 12 \text{ in.} \cdot 8 \text{ in.} = c
$$

***Solve*.** We carry out the calculation.
$$
\begin{aligned}
2 \cdot 8 \text{ in.} \cdot 6 \text{ in.} + 2 \cdot 12 \text{ in.} \cdot 6 \text{ in.} + 12 \text{ in.} \cdot 8 \text{ in.} &= c \\
96 \text{ in}^2 + 144 \text{ in}^2 + 96 \text{ in}^2 &= c \\
336 \text{ in}^2 &= c
\end{aligned}
$$

***Check*.** We can repeat the calculations. The answer checks.

***State*.** 336 in^2 of cardboard are used for the container.

49. We can reduce the number of trials required by simplifying the expression on the left side of the equation and then using the addition principle.
$$
\begin{aligned}
359 - 46 + a \div 3 \times 25 - 7^2 &= 339 \\
359 - 46 + a \div 3 \times 25 - 49 &= 339 \\
359 - 46 + \frac{a}{3} \times 25 - 49 &= 339 \\
359 - 46 + \frac{25 \cdot a}{3} - 49 &= 339 \\
313 + \frac{25 \cdot a}{3} - 49 &= 339 \\
264 + \frac{25 \cdot a}{3} &= 339 \\
264 + \frac{25 \cdot a}{3} - 264 &= 339 - 264 \\
\frac{25 \cdot a}{3} &= 75
\end{aligned}
$$

We see that when we multiply a by 25 and divide by 3, the result is 75. By trial, we find that $\dfrac{25 \cdot 9}{3} = \dfrac{225}{3} = 75$, so $a = 9$. We could also reason that since $75 = 25 \cdot 3$ and $9/3 = 3$, we have $a = 9$.

50. *Familiarize*. First observe that a 10-yr loan with monthly payments has a total of $10 \cdot 12$, or 120, payments. Let $m =$ the number of monthly payments represented by \$9160 and let $p =$ the number of payments remaining after \$9160 has been repaid.

***Translate*.** First we will translate to an equation that can be used to find m. Then we will write an equation that can be used to find p.

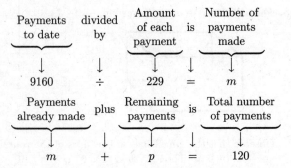

Solve. To solve the first equation we carry out the division.

$$229\overline{\smash{\big)}9160} \quad \begin{array}{r} 40 \\ \underline{9160} \\ 0 \\ 0 \\ \hline 0 \end{array}$$

Thus, $m = 40$.

Now we solve the second equation.

$$m + p = 120$$

$$40 + p = 120 \qquad \text{Substituting 40 for } m$$

$$40 + p - 40 = 120 - 40$$

$$p = 80$$

Check. We can approach the problem in a different way to check the answer. In 10 years, Cara's loan payments will total $120 \cdot \$229$, or \$27,480. If \$9160 has already been paid, then $\$27,480 - \$9160$, or \$18,320, remains to be paid. Since $80 \cdot \$229 = \$18,320$, the answer checks.

State. 80 payments remain on the loan.

51. $\dfrac{90 + 90 + 90 + 80 + 80 + 80 + 80 + 74}{8} = \dfrac{664}{8} = 83$

Chapter 2

Introduction to Integers and Algebraic Expressions

Exercise Set 2.1

1. The integer $-34,000,000$ corresponds to a \$34 million fine.

3. The integer 40 corresponds to receiving \$40 for a ton of paper; the integer -15 corresponds to paying \$15 to get rid of a ton of paper.

5. The integer 820 corresponds to receiving an \$820 refund; the integer -541 corresponds to owing \$541.

7. The integer -280 corresponds to 280 ft below sea level; the integer 14,491 corresponds to an elevation of 14,491 ft.

9. Since -8 is to the left of 0, we have $-8 < 0$.

11. Since 9 is to the right of 0, we have $9 > 0$.

13. Since 8 is to the right of -8, we have $8 > -8$.

15. Since -6 is to the left of -4, we have $-6 < -4$.

17. Since -8 is to the left of -5, we have $-8 < -5$.

19. Since -13 is to the left of -9, we have $-13 < -9$.

21. Since -3 is to the right of -4, we have $-3 > -4$.

23. The distance from 57 to 0 is 57, so $|57| = 57$.

25. The distance from 0 to 0 is 0, so $|0| = 0$.

27. The distance from -24 to 0 is 24, so $|-24| = 24$.

29. The distance from 53 to 0 is 53, so $|53| = 53$.

31. This distance from -8 to 0 is 8, so $|-8| = 8$.

33. To find the opposite of x when x is -7, we reflect -7 to the other side of 0. We have $-(-7) = 7$. The opposite of -7 is 7.

35. To find the opposite of x when x is 7, we reflect 7 to the other side of 0. We have $-(7) = -7$. The opposite of 7 is -7.

37. When we try to reflect 0 to the other side of 0, we go nowhere. The opposite of 0 is 0.

39. To find the opposite of x when x is -19, we reflect -19 to the other side of 0. We have $-(-19) = 19$.

41. To find the opposite of x when x is 42, we reflect 42 to the other side of 0. We have $-(42) = -42$.

43. The opposite of -8 is 8. $\quad -(-8) = 8$

45. The opposite of 7 is -7. $\quad -(7) = -7$

47. The opposite of -29 is 29. $\quad -(-29) = 29$

49. The opposite of -22 is 22. $\quad -(-22) = 22$

51. The opposite of 1 is -1. $\quad -(1) = -1$

53. We replace x by 7. We wish to find $-(-7)$. Reflecting 7 to the other side of 0 gives us -7 and then reflecting back gives us 7. Thus, $-(-x) = 7$ when x is 7.

55. We replace x by -9. We wish to find $-(-(-9))$. Reflecting -9 to the other side of 0 gives us 9 and then reflecting back gives us -9. Thus, $-(-x) = -9$ when x is -9.

57. We replace x by -17. We wish to find $-(-(-17))$. Reflecting -17 to the other side of 0 gives us 17 and then reflecting back gives us -17. Thus $-(-x) = -17$ when x is -17.

59. We replace x by 23. We wish to find $-(-23)$. Reflecting 23 to the other side of 0 gives us -23 and then reflecting back gives us 23. Thus, $-(-x) = 23$ when x is 23.

61. We replace x by -1. We wish to find $-(-(-1))$. Reflecting -1 to the other side of 0 gives us 1 and then reflecting back gives us -1. Thus, $-(-x) = -1$ when x is -1.

63. We replace x by 85. We wish to find $-(-85)$. Reflecting 85 to the other side of 0 gives us -85 and then reflecting back gives us 85. Thus, $-(-x) = 85$ when x is 85.

65. We have $-|-x| = -|-47|$. Since $|-47| = 47$, it follows that $-|-47| = -47$.

67. We have $-|-x| = -|-345|$. Since $|-345| = 345$, it follows that $-|-345| = -345$.

69. We have $-|-x| = -|-0|$. Since $|-0| = |0| = 0$, it follows that $-|-0| = -|0| = -0 = 0$.

71. We have $-|-x| = -|-(-8)| = -|8|$. Since $|8| = 8$, it follows that $-|-(-8)| = -8$.

73. Discussion and Writing Exercise

75.
$$\begin{array}{r} \overset{1\ 1}{3\ 2\ 7} \\ +\ 4\ 9\ 8 \\ \hline 8\ 2\ 5 \end{array}$$

77.
$$\begin{array}{r} \overset{\ \ 2}{\overset{\ \ 3}{2\ 0\ 9}} \\ \times\ \ \ 3\ 4 \\ \hline 8\ 3\ 6 \quad \text{Multiplying 209 by 4} \\ 6\ 2\ 7\ 0 \quad \text{Multiplying 209 by 30} \\ \hline 7\ 1\ 0\ 6 \quad \text{Adding} \end{array}$$

79. $9^2 = 9 \cdot 9 = 81$

81. $5(8 - 6) = 5(2) = 10$

83. Discussion and Writing Exercise

85. Answers may vary. On many scientific calculators we would add 549 and 387 and then take the opposite of the sum.

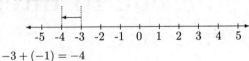

87. $|-5| = 5$ and $|-2| = 2$. Since 5 is to the right of 2, we have $|-5| > |-2|$.

89. $|-8| = 8$ and $|8| = 8$, so $|-8| = |8|$.

91. $|-8| = 8$, so $-|-8| = -(8) = -8$.

93. $|7| = 7$, so $-|7| = -(7) = -7$.

95. The integers whose distance from 0 is less than 2 are -1, 0, and 1. These are the solutions.

97. First note that $2^{10} = 1024$, $|-6| = 6$, $|3| = 3$, $2^7 = 128$, $7^2 = 49$, and $10^2 = 100$. Listing the entire set of integers in order from least to greatest, we have -100, -5, 0, $|3|$, 4, $|-6|$, 7^2, 10^2, 2^7, 2^{10}.

Exercise Set 2.2

1. Add: $-7 + 2$

$-7 + 2 = -5$

3. Add: $-9 + 5$

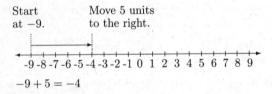

$-9 + 5 = -4$

5. Add: $-3 + 9$

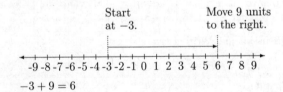

$-3 + 9 = 6$

7. $-7 + 7$

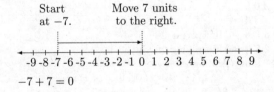

$-7 + 7 = 0$

9. $-3 + (-1)$

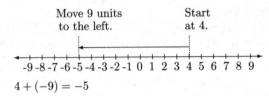

$-3 + (-1) = -4$

11. $4 + (-9)$

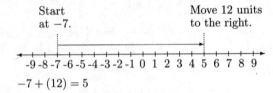

$4 + (-9) = -5$

13. $-7 + (12)$

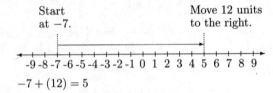

$-7 + (12) = 5$

15. $-3 + (-9)$ Two negative integers

Add the absolute values: $3 + 9 = 12$
Make the answer negative: $-3 + (-9) = -12$

17. $-6 + (-5)$ Two negative integers

Add the absolute values: $6 + 5 = 11$
Make the answer negative: $-6 + (-5) = -11$

19. $5 + (-5) = 0$

For any integer a, $a + (-a) = 0$.

21. $-2 + 2 = 0$

For any integer a, $-a + a = 0$.

23. $0 + 6 = 6$

For any integer a, $0 + a = a$.

25. $13 + (-13) = 0$

For any integer a, $a + (-a) = 0$.

27. $-25 + 0 = -25$

For any integer a, $a + 0 = a$.

29. $0 + (-27) = -27$

For any integer a, $0 + a = a$.

31. $-31 + 31 = 0$

For any integer a, $-a + a = 0$.

33. $-8 + 0 = -8$

When 0 is added to any number, that number remains unchanged.

35. $9 + (-4)$ The absolute values are 9 and 4. The difference is 5. The positive number has the larger absolute value, so the answer is positive. $9 + (-4) = 5$

37. $-4 + (-5)$ Two negative integers

Add the absolute values: $4 + 5 = 9$

Make the answer negative: $-4 + (-5) = -9$

39. $0 + (-5) = -5$

For any integer a, $0 + a = a$.

41. $14 + (-5)$ The absolute values are 14 and 5. The difference is 9. The positive number has the larger absolute value, so the answer is positive. $14 + (-5) = 9$

43. $-11 + 8$ The absolute values are 11 and 8. The difference is 3. Since the negative number has the larger absolute value, the answer is negative. $-11 + 8 = -3$

45. $-19 + 19 = 0$

For any integer a, $-a + a = 0$.

47. $-17 + 7$ The absolute values are 17 and 7. The difference is 10. Since the negative number has the larger absolute value, the answer is negative. $-17 + 7 = -10$

49. $-17 + (-7)$ Two negative integers

Add the absolute values: $17 + 7 = 24$

Make the answer negative: $-17 + (-7) = -24$

51. $11 + (-16)$ The absolute values are 11 and 16. The difference is 5. Since the negative number has the larger absolute value, the answer is negative. $11 + (-16) = -5$

53. $-15 + (-6)$ Two negative integers

Add the absolute values: $15 + 6 = 21$

Make the answer negative: $-15 + (-6) = -21$

55. $11 + (-9)$ The absolute values are 11 and 9. The difference is 2. The positive number has the larger absolute value, so the answer is positive. $11 + (-9) = 2$

57. $-11 + 17$ The absolute values are 11 and 17. The difference is 6. The positive number has the larger absolute value, so the answer is positive. $-11 + 17 = 6$

59. We will add from left to right.

$$-15 + (-7) + 1 = -22 + 1$$
$$= -21$$

61. We will add from left to right.

$$30 + (-10) + 5 = 20 + 5$$
$$= 25$$

63. We will add from left to right.

$$-23 + (-9) + 15 = -32 + 15$$
$$= -17$$

65. We will add from left to right.

$$40 + (-40) + 6 = 0 + 6$$
$$= 6$$

67. $12 + (-65) + (-12)$

Note that $12 + (-12) = 0$. Then we have $0 + (-65)$, so the sum is -65.

69. We will add from left to right.

$$
\begin{aligned}
&-24 + (-37) + (-19) + (-45) + (-35) \\
&= \quad -61 + (-19) + (-45) + (-35) \\
&= \quad\quad\quad -80 + (-45) + (-35) \\
&= \quad\quad\quad\quad\quad -125 + (-35) \\
&= \quad\quad\quad\quad\quad\quad\quad -160
\end{aligned}
$$

71. $28 + (-44) + 17 + 31 + (-94)$

a) $28 + 17 + 31 = 76$ Adding the positive numbers

b) $-44 + (-94) = -138$ Adding the negative numbers

c) $76 + (-138) = -62$ Adding the results

73. $-19 + 73 + (-23) + 19 + (-73)$

a) $-19 + 19 = 0$ Adding one pair of opposites

b) $73 + (-73) = 0$ Adding the other pair of opposites

c) We have $0 + (-23) = -23$

75. Discussion and Writing Exercise

77.
$$
\begin{array}{r}
\overset{3\ \ 13}{5\,\cancel{4}\,\cancel{3}} \\
-\ 2\ 1\ 9 \\
\hline
3\ 2\ 4
\end{array}
$$

79.
$$
\begin{array}{r}
\overset{8\ \ 11}{2\ 8\,\cancel{9}\,\cancel{1}} \\
-\ 1\ 4\ 0\ 7 \\
\hline
1\ 4\ 8\ 4
\end{array}
$$

81. 3 ten thousands + 9 thousands + 4 hundreds + 1 ten + 7 ones

83. a) Locate the digit in the thousands place.

$$3\,2,\,\boxed{8}\,3\,1$$
$$\uparrow$$

b) Then consider the next digit to the right.

c) Since the digit is 5 or higher, round 2 thousands up to 3 thousands.

d) Change all digits to the right of thousands to zeros.

The answer is 33,000.

85.
$$
\begin{array}{r}
3\ 2 \\
9\overline{)2\ 8\ 8} \\
\underline{2\ 7\ 0} \\
1\ 8 \\
\underline{1\ 8} \\
0
\end{array}
$$

The answer is 32.

87. Discussion and Writing Exercise

89. $-|27| + (-|-13|) = -27 + (-13) = -40$

91. We use a calculator.

$-3496 + (-2987) = -6483$

93. We use a calculator.

$-7846 + 5978 = -1868$

95. If $-x$ is positive, it is the reflection of a negative number x across 0 on the number line. Thus, $-x$ is positive for all negative numbers x.

97. If n is positive, $-n$ is negative. Then $-n + m$, the sum of two negative numbers, is negative.

99. If n is negative and m is less than n, then m is also negative. Then $n + m$, the sum of two negative numbers, is negative.

Exercise Set 2.3

1. $2 - 7 = 2 + (-7) = -5$

3. $0 - 8 = 0 + (-8) = -8$

5. $-7 - (-4) = -7 + 4 = -3$

7. $-11 - (-11) = -11 + 11 = 0$

9. $13 - 17 = 13 + (-17) = -4$

11. $20 - 27 = 20 + (-27) = -7$

13. $-9 - (-4) = -9 + 4 = -5$

15. $-40 - (-40) = -40 + 40 = 0$

17. $7 - 7 = 7 + (-7) = 0$

19. $7 - (-7) = 7 + 7 = 14$

21. $8 - (-3) = 8 + 3 = 11$

23. $-6 - 8 = -6 + (-8) = -14$

25. $-3 - (-9) = -3 + 9 = 6$

27. $1 - 9 = 1 + (-9) = -8$

29. $-6 - (-5) = -6 + 5 = -1$

31. $8 - (-10) = 8 + 10 = 18$

33. $0 - 10 = 0 + (-10) = -10$

35. $-5 - (-2) = -5 + 2 = -3$

37. $-7 - 14 = -7 + (-14) = -21$

39. $0 - (-5) = 0 + 5 = 5$

41. $-8 - 0 = -8 + 0 = -8$

43. $7 - (-5) = 7 + 5 = 12$

45. $6 - 25 = 6 + (-25) = -19$

47. $-42 - 26 = -42 + (-26) = -68$

49. $-72 - 9 = -72 + (-9) = -81$

51. $24 - (-92) = 24 + 92 = 116$

53. $-50 - (-50) = -50 + 50 = 0$

55. $-30 - (-85) = -30 + 85 = 55$

57. $7 - (-5) + 4 - (-3) = 7 + 5 + 4 + 3 = 19$

59.
$$-31 + (-28) - (-14) - 17$$
$$= -31 + (-28) + 14 + (-17)$$
$$= -31 + (-28) + (-17) + 14 \quad \text{Using a commutative law}$$
$$= -76 + 14 \quad \text{Adding the negative numbers}$$
$$= -62$$

61. $-34 - 28 + (-33) - 44 = (-34) + (-28) + (-33) + (-44) = -139$

63.
$$-93 - (-84) - 41 - (-56)$$
$$= -93 + 84 + (-41) + 56$$
$$= -93 + (-41) + 84 + 56 \quad \text{Using a commutative law}$$
$$= -134 + 140 \quad \text{Adding negatives and adding positives}$$
$$= 6$$

65.
$$-5 - (-30) + 30 + 40 - (-12)$$
$$= -5 + 30 + 30 + 40 + 12$$
$$= -5 + 112 \quad \text{Adding the positive numbers}$$
$$= 107$$

67. $132 - (-21) + 45 - (-21) = 132 + 21 + 45 + 21 = 219$

69. We subtract the beginning page number from the final page number.
$$62 - 37 = 25$$
Alicia read 25 pages.

71. The integer 8 corresponds to 8 lb above the ideal weight, and -9 corresponds to 9 lb below it. We subtract the lower weight from the higher weight:
$$8 - (-9) = 8 + 9 = 17$$
Rod lost 17 lb.

73. We subtract the lower temperature from the higher temperature.
$$27 - (-128) = 27 + 128 = 155$$
The difference in temperatures is $155°$C.

75. We subtract the initial reading from the final reading.
$$29 - (-21) = 29 + 21 = 50$$
50 minutes were recorded. Thus, the entire 60-minute show was not recorded.

77. We start with the original temperature, add the rise in temperature, and subtract the drop in temperature.
$$32 + 15 - 50 = 32 + 15 + (-50) = 47 + (-50) = -3$$
The final temperature was $-3°$.

79. The integer -5000 represents a loss of $5000, and the integer 8000 represents a profit of $8000. We subtract the amount of the loss from the profit.
$$8000 - (-5000) = 8000 + 5000 = 13,000$$
The store made $13,000 more in 2005 than in 2004.

81. To find the elevation that is 2293 ft deeper than -7718 ft, we subtract the additional depth from the original depth.
$$-7718 - 2293 = -7718 + (-2293) = -10,011$$
In 2005 the elevation of the deepwater drilling record was $-10,011$ ft.

83. First we subtract the cost of the tolls from the original balance in the account.

$$13 - 20 = 13 + (-20) = -7$$

Then we subtract the cost of the fines and fees from the new balance in the account.

$$-7 - 80 = -7 + (-80) = -87$$

The Murrays were $87 in debt as a result of their travel on toll roads.

85. Discussion and Writing Exercise

87. $4^3 = 4 \cdot 4 \cdot 4 = 64$

89. $1^7 = 1 \cdot 1 \cdot 1 \cdot 1 \cdot 1 \cdot 1 \cdot 1 = 1$

91. *Familiarize.* Let n = the number of 12-oz cans that can be filled. We think of an array consisting of 96 oz with 12 oz in each row.

The number n corresponds to the number of rows in the array.

Translate and Solve. We translate to an equation and solve it.

$$96 \div 12 = n \qquad 12 \overline{\smash{\big)}\,96} \begin{array}{r} 8 \\ \underline{9\,6} \\ 0 \end{array}$$

Check. We multiply the number of cans by 12: $8 \cdot 12 = 96$. The result checks.

State. Eight 12-oz cans can be filled.

93.
$$5 + 4^2 + 2 \cdot 7$$
$$= 5 + 16 + 2 \cdot 7$$
$$= 5 + 16 + 14$$
$$= 21 + 14$$
$$= 35$$

95. $(9 + 7)(9 - 7) = (16)(2) = 32$

97. Discussion and Writing Exercise

99. Use a calculator to do this exercise.
$$123,907 - 433,789 = -309,882$$

101. False; $3 - 0 \neq 0 - 3$.

103. True

105. True

107. a is the number we add to -57 to get -34. If we think of starting at -57 on the number line and moving to -34, we move 17 units to the right, so $a = 17$.

109. The changes during weeks 1 to 5 are represented by the integers -13, -16, 36, -11, and 19, respectively. We add to find the total rise or fall:

$$-13 + (-16) + 36 + (-11) + 19 = 15$$

The market rose 15 points during the 5 week period.

Exercise Set 2.4

1. $-2 \cdot 8 = -16$

3. $-9 \cdot 2 = -18$

5. $8 \cdot (-6) = -48$

7. $-10 \cdot 3 = -30$

9. $-3 \cdot (-5) = 15$

11. $-9 \cdot (-2) = 18$

13. $(-6)(-7) = 42$

15. $-10(-2) = 20$

17. $12(-10) = -120$

19. $-6(-50) = 300$

21. $(-72)(-1) = 72$

23. $(-20)17 = -340$

25. $-47 \cdot 0 = 0$

27. $0(-14) = 0$

29.
$$3 \cdot (-8) \cdot (-1)$$
$$= -24 \cdot (-1) \quad \text{Multiplying the first two numbers}$$
$$= 24$$

31.
$$7(-4)(-3)5$$
$$= 7 \cdot 12 \cdot 5 \quad \text{Multiplying the negative numbers}$$
$$= 84 \cdot 5$$
$$= 420$$

33.
$$-2(-5)(-7)$$
$$= 10 \cdot (-7) \quad \text{Multiplying the first two numbers}$$
$$= -70$$

35.
$$(-5)(-2)(-3)(-1)$$
$$= 10 \cdot 3 \qquad \text{Multiplying the first two numbers and the last two numbers}$$
$$= 30$$

37.
$$(-15)(-29)0 \cdot 8$$
$$= 435 \cdot 0 \qquad \text{Multiplying the first two numbers and the last two numbers}$$
$$= 0$$

(We might have noted at the outset that the product would be 0 since one of the numbers in the product is 0.)

39.
$$(-7)(-1)(7)(-6)$$
$$= 7(-42) \qquad \text{Multiplying the first two numbers and the last two numbers}$$
$$= -294$$

41. $(-6)^2 = (-6)(-6) = 36$

43.
$$(-5)^3 = (-5)(-5)(-5)$$
$$= 25(-5)$$
$$= -125$$

45. $(-10)^4 = (-10)(-10)(-10)(-10)$
$$= 100 \cdot 100$$
$$= 10,000$$

47. $-2^4 = -1 \cdot 2^4$
$$= -1 \cdot 2 \cdot 2 \cdot 2 \cdot 2$$
$$= -1 \cdot 4 \cdot 4$$
$$= -1 \cdot 16$$
$$= -16$$

49. $(-3)^5 = (-3)(-3)(-3)(-3)(-3)$
$$= 9 \cdot 9 \cdot (-3)$$
$$= 81(-3)$$
$$= -243$$

51. $(-1)^{12}$
$$= (-1) \cdot (-1) \cdot (-1) \cdot (-1) \cdot (-1) \cdot (-1) \cdot (-1) \cdot (-1) \cdot$$
$$(-1) \cdot (-1) \cdot (-1) \cdot (-1)$$
$$= 1 \cdot 1 \cdot 1 \cdot 1 \cdot 1 \cdot 1$$
$$= 1 \cdot 1 \cdot 1$$
$$= 1 \cdot 1$$
$$= 1$$

53. -3^6
$$= -1 \cdot 3^6$$
$$= -1 \cdot 3 \cdot 3 \cdot 3 \cdot 3 \cdot 3 \cdot 3$$
$$= -1 \cdot 9 \cdot 9 \cdot 9$$
$$= -9 \cdot 81$$
$$= -729$$

55. $-4^3 = -1 \cdot 4^3$
$$= -1 \cdot 4 \cdot 4 \cdot 4$$
$$= -4 \cdot 16$$
$$= -64$$

57. -8^4 is read "the opposite of eight to the fourth power."

59. $(-9)^{10}$ is read "negative nine to the tenth power."

61. Discussion and Writing Exercise

63. a) Locate the digit in the hundreds place.

5 3 2, 4 $\boxed{5}$ 1
        ↑

b) Then consider the next digit to the right.

c) Since that digit is 5 or higher, round 4 hundreds up to 5 hundreds.

d) Change all digits to the right of hundreds to zeros. The answer is 532,500.

65.
```
        8 0
  3 6 ⟌ 2 8 8 0
        2 8 8 0
        ───────
              0
              0
            ───
              0
```
The answer is 80.

67. $10 - 2^3 + 6 \div 2$
$$= 10 - 8 + 6 \div 2 \quad \text{Evaluating the exponential expression}$$
$$= 10 - 8 + 3 \quad \text{Dividing}$$
$$= 2 + 3 \quad \text{Adding and subtracting in}$$
$$= 5 \quad \text{order from left to right}$$

69. *Familiarize*. We first make a drawing.

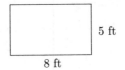

5 ft

8 ft

Let A = the area.

***Translate*.** Using the formula for area, we have $A = l \cdot w = 8 \cdot 5$.

***Solve*.** We carry out the multiplication.
$$A = 8 \cdot 5 = 40$$

***Check*.** We repeat our calculation.

***State*.** The area of the rug is 40 ft^2.

71. *Familiarize*. Let n = the number of trips the ferry will make carrying 12 cars. We can think of this problem as repeated subtraction.

***Translate*.**

Number of cars	divided by	Number per trip	is	Number of 12-car trips
↓	↓	↓	↓	↓
53	÷	12	=	n

***Solve*.** We carry out the division.
```
        4
  1 2 ⟌ 5 3
        4 8
        ───
          5
```
$$53 \div 12 = n$$
$$4 \text{ R } 5 = n$$

After the ferry makes 4 trips, carrying 12 cars on each trip, 5 cars remained to be ferried. Thus, a fifth trip will be required to ferry the remaining cars.

***Check*.** We can check by multiplying the number of full trips by 12 and then adding the remainder, 5.
$$4 \cdot 12 = 48$$
$$48 + 5 = 53$$
Since 53 cars were to be ferried, the answer checks.

***State*.** A total of 5 trips will be required.

73. Discussion and Writing Exercise

75. $(-3)^5(-1)^{379} = -243(-1) = 243$

77. $-9^4 + (-9)^4 = -6561 + 6561 = 0$

79. $|(-2)^5 + 3^2| - (3-7)^2$
$= |-32 + 9| - (-4)^2$
$= |-23| - 16$
$= 23 - 16$
$= 7$

81. Use a calculator. On many scientific calculators the keystrokes are $\boxed{4}\,\boxed{7}\,\boxed{x^2}\,\boxed{+/-}$. We get -2209.

83. Use a calculator. On many scientific calculators the keystrokes are $\boxed{1}\,\boxed{9}\,\boxed{+/-}\,\boxed{x^y}\,\boxed{4}\,\boxed{=}$. We get $130{,}321$. (Some calculators have a $\boxed{y^x}$ key rather than an $\boxed{x^y}$ key.)

85. Use a calculator. On many scientific calculators the keystrokes are $\boxed{(}\,\boxed{7}\,\boxed{3}\,\boxed{-}\,\boxed{8}\,\boxed{6}\,\boxed{)}\,\boxed{x^y}\,\boxed{3}\,\boxed{=}$. We get -2197. (Some calculators have a $\boxed{y^x}$ key rather than an $\boxed{x^y}$ key.)

87. Use a calculator. On many scientific calculators the keystrokes are $\boxed{9}\,\boxed{3}\,\boxed{5}\,\boxed{+/-}\,\boxed{\times}\,\boxed{5}\,\boxed{+/-}\,\boxed{x^y}\,\boxed{3}\,\boxed{=}$. We get $116{,}875$. (Some calculators have a $\boxed{y^x}$ key rather than an $\boxed{x^y}$ key.)

89. The new balance will be
$\$68 - 7(\$13) = \$68 - \$91 = -\$23$.

91. a) If $[(-5)^m]^n$ is to be negative, first m must be an odd number so that $(-5)^m$ is negative. Similarly, n must also be odd in order for $[(-5)^m]^n$ to be negative. Thus, both m and n must be odd numbers.

b) If $[(-5)^m]^n$ is to be positive, at least one of m and n must be an even number. For example, if m is even then $(-5)^m$ is positive and so is $[(-5)^m]^n$ regardless of whether n is even or odd. If m is odd, then $(-5)^m$ is negative and n must be even in order for $[(-5)^m]^n$ to be positive.

Exercise Set 2.5

1. $28 \div (-4) = -7$ Check: $-7(-4) = 28$

3. $\dfrac{28}{-2} = -14$ Check: $-14(-2) = 28$

5. $\dfrac{18}{-2} = -9$ Check: $-9(-2) = 18$

7. $\dfrac{-48}{-12} = 4$ Check: $4(-12) = -48$

9. $\dfrac{-72}{8} = -9$ Check: $-9 \cdot 8 = -72$

11. $-100 \div (-50) = 2$ Check: $2(-50) = -100$

13. $-344 \div 8 = -43$ Check: $-43 \cdot 8 = -344$

15. $\dfrac{200}{-25} = -8$ Check: $-8(-25) = 200$

17. $\dfrac{-56}{0}$ is undefined.

19. $\dfrac{88}{-11} = -8$ Check: $-8(-11) = 88$

21. $-\dfrac{276}{12} = \dfrac{-276}{12} = -23$ Check: $-23 \cdot 12 = -276$

23. $\dfrac{0}{-2} = 0$ Check: $0 \cdot (-2) = 0$

25. $\dfrac{19}{-1} = -19$ Check: $-19(-1) = 19$

27. $-41 \div 1 = -41$ Check: $-41 \cdot 1 = -41$

29. $5 - 2 \cdot 3 - 6 = 5 - 6 - 6$ Multiplying
$\qquad = -1 - 6$ Doing all additions and subtractions in order
$\qquad = -7$ from left to right

31. $9 - 2(3-8) = 9 - 2(-5)$ Subtracting inside parentheses
$\qquad = 9 + 10$ Multiplying
$\qquad = 19$ Adding

33. $16 \cdot (-24) + 50 = -384 + 50$ Multiplying
$\qquad = -334$ Adding

35. $40 - 3^2 - 2^3$
$= 40 - 9 - 8$ Evaluating the exponential expressions
$= 31 - 8$ Doing all additions and subtractions
$= 23$ in order from left to right

37. $4 \cdot (6+8)/(4+3)$
$= 4 \cdot 14/7$ Adding inside parentheses
$= 56/7$ Doing all multiplications and divisions
$= 8$ in order from left to right

39. $4 \cdot 5 - 2 \cdot 6 + 4 = 20 - 12 + 4$ Multiplying
$\qquad = 8 + 4$
$\qquad = 12$

41. $\dfrac{9^2 - 1}{1 - 3^2}$
$= \dfrac{81 - 1}{1 - 9}$ Evaluating the exponential expressions
$= \dfrac{80}{-8}$ Subtracting in the numerator and in the denominator
$= -10$

43. $8(-7) + 6(-5) = -56 - 30$ Multiplying
$\qquad = -86$

45. $20 \div 5(-3) + 3 = 4(-3) + 3$ Dividing
$\qquad = -12 + 3$ Multiplying
$\qquad = -9$ Adding

47. $18 - 0(3^2 - 5^2 \cdot 7 - 4)$
Observe that $a \cdot 0 = 0$ for an integer a. Then $0(3^2 - 5^2 \cdot 7 - 4) = 0$, so the result is $18 - 0$, or 18.

49. $4 \cdot 5^2 \div 10$
$= 4 \cdot 25 \div 10$ Evaluating the exponential expression
$= 100 \div 10$ Multiplying
$= 10$ Dividing

51. $(3-8)^2 \div (-1)$
$= (-5)^2 \div (-1)$ Subtracting inside parentheses
$= 25 \div (-1)$ Evaluating the exponential
                          expression
$= -25$

53. $17 - 10^3$
$= 17 - 1000$ Evaluating the exponential
                          expression
$= -983$ Subtracting

55. $2 + 10^2 \div 5 \cdot 2^2$
$= 2 + 100 \div 5 \cdot 4$ Evaluating the exponential
                                  expressions
$= 2 + 20 \cdot 4$ Doing all multiplications and di-
$= 2 + 80$ visions in order from left to right
$= 82$ Adding

57. $12 - 20^3 = 12 - 8000$
$= -7988$

59. $2 \times 10^3 - 5000 = 2 \times 1000 - 5000$
$= 2000 - 5000$
$= -3000$

61. $6[9 - (3-4)] = 6[9 - (-1)]$ Subtracting inside the
                                              innermost parentheses
$= 6[9+1]$
$= 6[10]$
$= 60$

63. $-1000 \div (-100) \div 10 = 10 \div 10$ Doing the divi-
                                                          sions in order
$= 1$ from left to right

65. $8 - |7-9| \cdot 3 = 8 - |-2| \cdot 3$
$= 8 - 2 \cdot 3$
$= 8 - 6$
$= 2$

67. $9 - |7 - 3^2| = 9 - |7 - 9|$
$= 9 - |-2|$
$= 9 - 2$
$= 7$

69. $\dfrac{6^3 - 7 \cdot 3^4 - 2^5 \cdot 9}{(1 - 2^3)^3 + 7^3}$

$= \dfrac{216 - 7 \cdot 81 - 32 \cdot 9}{(1 - 8)^3 + 343}$

$= \dfrac{216 - 567 - 288}{(-7)^3 + 343}$

$= \dfrac{-351 - 288}{-343 + 343}$

$= -\dfrac{639}{0}$

Since division by 0 is not defined, this expression is not defined.

71. $\dfrac{2 \cdot 3^2 \div (3^2 - (2+1))}{5^2 - 6^2 - 2^2(-3)}$

$= \dfrac{2 \cdot 3^2 \div (3^2 - 3)}{25 - 36 - 4(-3)}$

$= \dfrac{2 \cdot 3^2 \div (9 - 3)}{25 - 36 + 12}$

$= \dfrac{2 \cdot 3^2 \div 6}{-11 + 12}$

$= \dfrac{2 \cdot 9 \div 6}{1}$

$= \dfrac{18 \div 6}{1}$

$= \dfrac{3}{1}$

$= 3$

73. $\dfrac{(-5)^3 + 17}{10(2-6) - 2(5+2)}$

$= \dfrac{-125 + 17}{10(2-6) - 2(5+2)}$ Evaluating the exponential
                                              expression

$= \dfrac{-125 + 17}{10(-4) - 2 \cdot 7}$ Doing the calculations within
                                              parentheses

$= \dfrac{-125 + 17}{-40 - 14}$ Multiplying

$= \dfrac{-108}{-54}$ Adding and subtracting

$= 2$

75. $\dfrac{2 \cdot 4^3 - 4 \cdot 32}{19^3 - 17^4}$

$= \dfrac{2 \cdot 64 - 4 \cdot 32}{6859 - 83,521}$ Evaluating the exponential ex-
                                                          pressions

$= \dfrac{128 - 128}{6859 - 83,521}$ Multiplying

$= \dfrac{0}{-76,662}$ Subtracting

$= 0$ Dividing

77. Discussion and Writing Exercise

79. *Familiarize.* We first make a drawing.

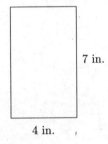

7 in.

4 in.

Let A = the area.

Translate. Using the formula for area, we have $A = l \cdot w = 7 \cdot 4$.

Solve. We carry out the multiplication.

$A = 7 \cdot 4 = 28$

Check. We repeat the calculation.

State. The area of the ad was 28 in².

81. *Familiarize*. We let g = the number of gallons needed to travel 384 miles. Think of a rectangular array consisting of 384 miles with 32 miles in each row. The number g is the number of rows.

Translate.

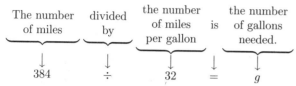

Solve. We carry out the division.

$$
\begin{array}{r}
1\,2 \\
3\,2\,\overline{)3\,8\,4} \\
3\,2\,0 \\
\hline
6\,4 \\
6\,4 \\
\hline
0
\end{array}
$$

Check. We multiply the number of gallons by the number of miles per gallon:

$$12 \cdot 32 = 384$$

We get the number of miles to be traveled, so the answer checks.

State. It will take 12 gallons of gasoline to travel 384 miles.

83. *Familiarize*. We let c = the number of calories in a 1-oz serving. Think of a rectangular array consisting of 1050 calories arranged in 7 rows. The number c is the number of calories in each row.

Translate.

Total calories	divided by	number of ounces	is	number of calories in 1 oz.
↓	↓	↓	↓	↓
1050	÷	7	=	c

Solve. We carry out the division.

$$
\begin{array}{r}
1\,5\,0 \\
7\,\overline{)1\,0\,5\,0} \\
7\,0\,0 \\
\hline
3\,5\,0 \\
3\,5\,0 \\
\hline
0 \\
0 \\
\hline
0
\end{array}
$$

Check. We multiply the number of calories in 1 oz by 7:

$$150 \cdot 7 = 1050$$

The result checks

State. There are 150 calories in a 1-oz serving.

85. *Familiarize*. Let p = the number of whole pieces of gum each person will receive. We can think of this problem as repeated subtraction.

Translate.

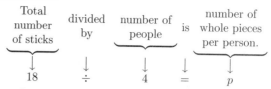

Solve. We carry out the division.

$$
\begin{array}{r}
4 \\
4\,\overline{)1\,8} \\
1\,6 \\
\hline
2
\end{array}
$$

$$18 \div 4 = p$$
$$4 \text{ R } 2 = p$$

Check. We multiply the number of whole pieces of gum per person by the number of people and then add the number of remaining pieces.

$$4 \cdot 4 = 16$$
$$16 + 2 = 18$$

We got the number of sticks of gum in a package, 18, so the answer checks.

State. Each person will receive 4 pieces of gum, and there will be 2 extra pieces remaining.

87. Discussion and Writing Exercise

89.
$$
\frac{9 - 3^2}{2 \cdot 4^2 - 5^2 \cdot 9 + 8^2 \cdot 7}
$$
$$
= \frac{9 - 9}{2 \cdot 16 - 25 \cdot 9 + 64 \cdot 7}
$$
$$
= \frac{0}{32 - 225 + 448}
$$
$$
= \frac{0}{-193 + 448}
$$
$$
= \frac{0}{255}
$$
$$
= 0
$$

91.
$$
\frac{(25 - 4^2)^3}{17^2 - 16^2} \cdot ((-6)^2 - 6^2) = \frac{(25 - 16)^3}{289 - 256} \cdot (36 - 36)
$$
$$
= \frac{9^3}{289 - 256} \cdot 0
$$
$$
= 0
$$

93. Use a calculator.
$$
\frac{19 - 17^2}{13^2 - 34} = -2
$$

95. Use a calculator.
$$
28^2 - 36^2 / 4^2 + 17^2 = 992
$$

97. (1 5 x^2 − 5 x^y 3) ÷ (3 x^2 + 4 x^2) = (Some calculators have a y^x key rather than an x^y key.)

99. Entering the given keystrokes and then pressing = , we get 5.

101. $-n$ and m are both negative, so $\dfrac{-n}{m}$ is the quotient of two negative numbers and, thus, is positive.

103. $\dfrac{-n}{m}$ is positive (see Exercise 101), so $-\left(\dfrac{-n}{m}\right)$ is the opposite of a positive number and, thus, is negative.

105. $-n$ is negative and $-m$ is positive, so $\dfrac{-n}{-m}$ is the quotient of a negative and a positive number and, thus, is negative. Then $-\left(\dfrac{-n}{-m}\right)$ is the opposite of a negative number and, thus, is positive.

Exercise Set 2.6

1. $12n = 12 \cdot 2 = 24\cancel{c}$

3. $\dfrac{x}{y} = \dfrac{6}{-3} = -2$

5. $\dfrac{2q}{p} = \dfrac{2 \cdot 3}{6} = \dfrac{6}{6} = 1$

7. $\dfrac{72}{r} = \dfrac{72}{4} = 18$ yr

9. $3 + 5 \cdot x = 3 + 5 \cdot 2 = 3 + 10 = 13$

11. $2l + 2w = 2 \cdot 3 + 2 \cdot 4 = 6 + 8 = 14$ ft

13. $2(l + w) = 2(3 + 4) = 2 \cdot 7 = 14$ ft

15. $7a - 7b = 7 \cdot 5 - 7 \cdot 2 = 35 - 14 = 21$

17. $7(a - b) = 7(5 - 2) = 7 \cdot 3 = 21$

19. $16t^2 = 16 \cdot 5^2 = 16 \cdot 25 = 400$ ft

21. $\begin{aligned} a + (b - a)^2 &= 6 + (4 - 6)^2 \\ &= 6 + (-2)^2 \\ &= 6 + 4 \\ &= 10 \end{aligned}$

23. $\begin{aligned} 9a + 9b &= 9 \cdot 13 + 9(-13) \\ &= 117 - 117 \\ &= 0 \end{aligned}$

25. $\begin{aligned} \dfrac{n^2 - n}{2} &= \dfrac{9^2 - 9}{2} \\ &= \dfrac{81 - 9}{2} \\ &= \dfrac{72}{2} \\ &= 36 \end{aligned}$

27. $\begin{aligned} m^3 - m^2 &= 5^3 - 5^2 \\ &= 125 - 25 \\ &= 100 \end{aligned}$

29. $-\dfrac{a}{b}$, $\dfrac{-a}{b}$, and $\dfrac{a}{-b}$ all represent the same number. Thus we can also write $\dfrac{-5}{t}$ as $\dfrac{-5}{t}$ and $\dfrac{5}{-t}$.

31. $-\dfrac{a}{b}$, $\dfrac{-a}{b}$, and $\dfrac{a}{-b}$ all represent the same number. Thus we can also write $\dfrac{-n}{b}$ as $-\dfrac{n}{b}$ and $\dfrac{n}{-b}$.

33. $-\dfrac{a}{b}$, $\dfrac{-a}{b}$, and $\dfrac{a}{-b}$ all represent the same number. Thus we can also write $\dfrac{9}{-p}$ as $-\dfrac{9}{p}$ and $\dfrac{-9}{p}$.

35. $-\dfrac{a}{b}$, $\dfrac{-a}{b}$, and $\dfrac{a}{-b}$ all represent the same number. Thus we can also write $\dfrac{-14}{w}$ as $-\dfrac{14}{w}$ and $\dfrac{14}{-w}$.

37. $\dfrac{-a}{b} = \dfrac{-45}{9} = -5$;

$\dfrac{a}{-b} = \dfrac{45}{-9} = -5$;

$-\dfrac{a}{b} = -\dfrac{45}{9} = -5$

39. $\dfrac{-a}{b} = \dfrac{-81}{3} = -27$;

$\dfrac{a}{-b} = \dfrac{81}{-3} = -27$;

$-\dfrac{a}{b} = -\dfrac{81}{3} = -27$

41. $(-3x)^2 = (-3 \cdot 2)^2 = (-6)^2 = 36$;
$-3x^2 = -3(2)^2 = -3 \cdot 4 = -12$

43. $5x^2 = 5(3)^2 = 5 \cdot 9 = 45$;
$5x^2 = 5(-3)^2 = 5 \cdot 9 = 45$

45. $x^3 = 6^3 = 6 \cdot 6 \cdot 6 = 216$;
$x^3 = (-6)^3 = (-6) \cdot (-6) \cdot (-6) = -216$

47. $x^8 = 1^8 = 1 \cdot 1 \cdot 1 \cdot 1 \cdot 1 \cdot 1 \cdot 1 \cdot 1 = 1$;
$x^8 = (-1)^8 =$
$(-1) \cdot (-1) \cdot (-1) \cdot (-1) \cdot (-1) \cdot (-1) \cdot (-1) \cdot (-1) = 1$

49. $a^5 = 2^5 = 2 \cdot 2 \cdot 2 \cdot 2 \cdot 2 = 32$;
$a^5 = (-2)^5 = (-2)(-2)(-2)(-2)(-2) = -32$

51. $5(a + b) = 5 \cdot a + 5 \cdot b = 5a + 5b$

53. $4(x + 1) = 4 \cdot x + 4 \cdot 1 = 4x + 4$

55. $2(b + 5) = 2 \cdot b + 2 \cdot 5 = 2b + 10$

57. $7(1 - t) = 7 \cdot 1 - 7 \cdot t = 7 - 7t$

59. $6(5x - 2) = 6 \cdot 5x - 6 \cdot 2 = 30x - 12$

61. $8(x + 7 + 6y) = 8 \cdot x + 8 \cdot 7 + 8 \cdot 6y = 8x + 56 + 48y$

63. $-7(y - 2) = -7 \cdot y - (-7) \cdot 2 = -7y - (-14) = -7y + 14$

65. $(x + 2)3 = x \cdot 3 + 2 \cdot 3 = 3x + 6$

67. $-4(x - 3y - 2z) = -4 \cdot x - (-4)3y - (-4)2z =$
$-4x - (-12y) - (-8z) = -4x + 12y + 8z$

69. $8(a - 3b + c) = 8 \cdot a - 8 \cdot 3b + 8 \cdot c =$
$8a - 24b + 8c$

71. $4(x - 3y - 7z) = 4 \cdot x - 4 \cdot 3y - 4 \cdot 7z =$
$4x - 12y - 28z$

73. $(4a - 5b + c - 2d)5 = 4a \cdot 5 - 5b \cdot 5 + c \cdot 5 - 2d \cdot 5 =$
$20a - 25b + 5c - 10d$

75. Discussion and Writing Exercise

77. Twenty-three million, forty-three thousand, nine hundred twenty-one

79.
$$\begin{array}{r} 5\,2\,8\,3 \\ -\,2\,4\,7\,5 \\ \hline \end{array} \qquad \begin{array}{r} 5\,2\,8\,0 \\ -\,2\,4\,8\,0 \\ \hline 2\,8\,0\,0 \end{array}$$

81. *Familiarize.* Since we are combining snowfall amounts, addition can be used. We let $s =$ the total snowfall.

Translate. We translate to an equation.
$$9 + 8 = s$$

Solve. We carry out the addition.
$$9 + 8 = 17$$
Thus, $17 = s$, or $s = 17$.

Check. We repeat the calculation.

State. It snowed 17 in. altogether.

83. *Familiarize.* This is a multistep problem. We let $x =$ the total price of the plain pizzas, $y =$ the total price of the pepperoni pizzas, and $t =$ the total price of the pizza.

Translate.

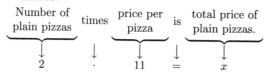

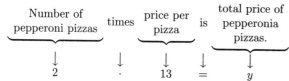

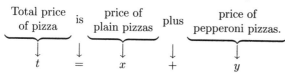

Solve. We solve each equation.
$$2 \cdot 11 = x \qquad 2 \cdot 13 = y \qquad t = x + y$$
$$22 = x \qquad\quad 26 = y \qquad t = 22 + 26$$
$$t = 48$$

Check. We can repeat the calculations. We can also check by rounding, multiplying, and adding. The answer checks.

State. Brett's wife paid $48 for the pizza.

85. Discussion and Writing Exercise

87. We substitute 370 for C in the formula in Example 5.
$$\frac{9C}{5} + 32 = \frac{9 \cdot 370}{5} + 32 = \frac{3330}{5} + 32 = 666 + 32 = 698$$
The temperature $370°$Celsius corresponds to a Fahrenheit temperature of $698°$.

89. Use a calculator.
$$a - b^3 + 17a = 19 - (-16)^3 + 17 \cdot 19 = 4438$$

91. Use a calculator.
$$r^3 + r^2t - rt^2 = (-9)^3 + (-9)^2 \cdot 7 - (-9) \cdot 7^2 = 279$$

93. $a^{1996} - a^{1997} = (-1)^{1996} - (-1)^{1997}$
$$= 1 - (-1)$$
$$= 1 + 1$$
$$= 2$$

95. $(m^3 - mn)^m = (4^3 - 4 \cdot 6)^4$
$$= (64 - 4 \cdot 6)^4$$
$$= (64 - 24)^4$$
$$= 40^4$$
$$= 2,560,000$$

97. $-32 \boxed{\times} (88 \boxed{-} 29) = -1888$

99. True

101. True

103. Discussion and Writing Exercise

Exercise Set 2.7

1. $2a + 5b - 7c = 2a + 5b + (-7c)$
The terms are $2a$, $5b$, and $-7c$.

3. $9mn - 6n + 8 = 9mn + (-6n) + 8$
The terms are $9mn$, $-6n$, and 8.

5. $3x^2y - 4y^2 - 2z^3 = 3x^2y + (-4y^2) + (-2z^3)$
The terms are $3x^2y$, $-4y^2$, and $-2z^3$.

7. $5x + 9x = (5+9)x = 14x$

9. $10a - 13a = (10 - 13)a = -3a$

11. $2x + 6z + 9x = 2x + 9x + 6z$
$$= (2+9)x + 6z$$
$$= 11x + 6z$$

13. $27a + 70 - 40a - 8 = 27a - 40a + 70 - 8$
$$= (27 - 40)a + 70 - 8$$
$$= -13a + 62$$

15. $9 + 5t + 7y - t - y - 13$
$$= 9 - 13 + 5t - 1 \cdot t + 7y - 1 \cdot y$$
$$= (9 - 13) + (5 - 1)t + (7 - 1)y$$
$$= -4 + 4t + 6y$$

17. $a + 3b + 5a - 2 + b$
$$= a + 5a + 3b + b - 2$$
$$= (1 + 5)a + (3 + 1)b - 2$$
$$= 6a + 4b - 2$$

19. $-8 + 11a - 5b + 6a - 7b + 7$
$$= 11a + 6a - 5b - 7b - 8 + 7$$
$$= (11 + 6)a + (-5 - 7)b + (-8 + 7)$$
$$= 17a - 12b - 1$$

21. $8x^2 + 3y - x^2 = 8x^2 - x^2 + 3y$
$\qquad\qquad = (8-1)x^2 + 3y$
$\qquad\qquad = 7x^2 + 3y$

23. $11x^4 + 2y^3 - 4x^4 - y^3 = 11x^4 - 4x^4 + 2y^3 - y^3$
$\qquad\qquad\qquad\qquad = (11-4)x^4 + (2-1)y^3$
$\qquad\qquad\qquad\qquad = 7x^4 + y^3$

25. $9a^2 - 4a + a - 3a^2 = 9a^2 - 3a^2 - 4a + a$
$\qquad\qquad\qquad = (9-3)a^2 + (-4+1)a$
$\qquad\qquad\qquad = 6a^2 - 3a$

27. $\quad x^3 - 5x^2 + 2x^3 - 3x^2 + 4$
$= x^3 + 2x^3 - 5x^2 - 3x^2 + 4$
$= (1+2)x^3 + (-5-3)x^2 + 4$
$= 3x^3 - 8x^2 + 4$

29. $\quad 7a^3 + 4ab - 5 - 7ab + 8$
$= 7a^3 + 4ab - 7ab - 5 + 8$
$= 7a^3 + (4-7)ab + (-5+8)$
$= 7a^3 - 3ab + 3$

31. $\quad 9x^3y + 4xy^3 - 6xy^3 + 3xy$
$= 9x^3y + (4-6)xy^3 + 3xy$
$= 9x^3y - 2xy^3 + 3xy$

33. $\quad 3a^6 - 9b^4 + 2a^6b^4 - 7a^6 - 2b^4$
$= 3a^6 - 7a^6 - 9b^4 - 2b^4 + 2a^6b^4$
$= (3-7)a^6 + (-9-2)b^4 + 2a^6b^4$
$= -4a^6 - 11b^4 + 2a^6b^4$

35. $P = 2 \cdot (l+w)$ $\qquad$ Perimeter of a rectangle
$P = 2 \cdot (3 \text{ ft} + 2 \text{ ft})$
$P = 2 \cdot (3+2) \text{ ft}$
$P = 2 \cdot 5 \text{ ft}$
$P = 10 \text{ ft}$

37. $\quad$ Perimeter
$= 7 \text{ km} + 7 \text{ km} + 7 \text{ km} + 7 \text{ km} + 7 \text{ km} + 7 \text{ km}$
$= (7+7+7+7+7+7) \text{ km}$
$= 42 \text{ km}$

39. Perimeter $= 3 \text{ m} + 1 \text{ m} + 3 \text{ m} + 1 \text{ m}$
$\qquad\qquad = (3+1+3+1) \text{ m}$
$\qquad\qquad = 8 \text{ m}$

41. A singles court is 78 ft by 27 ft.
$P = 2l + 2w = 2 \cdot 78 \text{ ft} + 2 \cdot 27 \text{ ft}$
$\qquad\qquad = 156 \text{ ft} + 54 \text{ ft}$
$\qquad\qquad = 210 \text{ ft}$

43. The rectangle formed by the services lines and the singles sideline is 42 ft by 27 ft.
$P = 2l + 2w = 2 \cdot 42 \text{ ft} + 2 \cdot 27 \text{ ft}$
$\qquad\qquad = 84 \text{ ft} + 54 \text{ ft}$
$\qquad\qquad = 138 \text{ ft}$

45. $P = 2(l+w) = 2(10 \text{ ft} + 8 \text{ ft})$
$\qquad\qquad = 2 \cdot 18 \text{ ft} = 36 \text{ ft}$

47. $P = 4s$
$\qquad = 4 \cdot 14 \text{ in.} = 56 \text{ in.}$

49. $P = 4s$
$\qquad = 4 \cdot 65 \text{ cm} = 260 \text{ cm}$

51. $P = 2(l+w) = 2(20 \text{ ft} + 12 \text{ ft})$
$\qquad\qquad = 2 \cdot 32 \text{ ft} = 64 \text{ ft}$

53. Discussion and Writing Exercise

55. **_Familiarize_.** Let $s =$ the number of servings of Shaw's Corn Flakes in one box. Visualize a rectangular array consisting of 510 grams with 30 grams in each row. Then s is the number of rows.

Translate.

Total weight	divided by	weight of one serving	is	number of servings.
↓	↓	↓	↓	↓
510	÷	30	=	s

Solve. We carry out the division.

$$
\begin{array}{r}
1\,7 \\
30\overline{)5\,1\,0} \\
3\,0\,0 \\
\hline
2\,1\,0 \\
2\,1\,0 \\
\hline
0
\end{array}
$$

We have $17 = s$, or $s = 17$.

Check. We multiply the number of servings by the weight of a serving.

$$17 \cdot 30 = 510$$

We get 510 oz, the total weight of the corn flakes, so the answer checks.

State. There are 17 servings in a box of Shaw's Corn Flakes.

57. $\quad 5 + 3 \cdot 2^3$
$= 5 + 3 \cdot 8$ $\quad$ Evaluating the exponential expression
$= 5 + 24$ $\quad$ Multiplying
$= 29$ $\quad$ Adding

59. $\quad 12 \div 3 \cdot 2$
$= 4 \cdot 2$ $\quad$ Dividing and multiplying in order
$= 8$ $\qquad$ from left to right

61. $\quad 15 - 3 \cdot 2 + 7$
$= 15 - 6 + 7$ $\quad$ Multiplying
$= 9 + 7$ $\qquad$ Subtracting and adding in order
$= 16$ $\qquad\quad$ from left to right

63. $\quad 25 = t + 9$
$\quad 25 - 9 = t + 9 - 9$
$\qquad 16 = t$
The solution is 16.

65.
$$45 = 3x$$
$$\frac{45}{3} = \frac{3x}{3}$$
$$15 = x$$

The solution is 15.

67. Discussion and Writing Exercise

69. $5(x + 3) + 2(x - 7) = 5x + 15 + 2x - 14 = 7x + 1$

71. $2(3 - 4a) + 5(a - 7) = 6 - 8a + 5a - 35 = -3a - 29$

73. $-5(2 + 3x + 4y) + 7(2x - y) =$
$-10 - 15x - 20y + 14x - 7y = -10 - x - 27y$

75. *Familiarize.* First we will find the amount of sealant needed to caulk each door and each window, keeping in mind that the bottom of each door requires no caulk. Thus, for each door we add the lengths of the other three sides and for each window we find the perimeter. Then we will find the amount of caulk required for all the doors and windows. Next we will determine how many sealant cartridges are needed and, finally, we will find the cost of the sealant.

Translate.

The amount of caulk required for each door is given by

7 ft + 3 ft + 7 ft.

The perimeter of each window is given by

$P = 2(l + w) = 2(3 \text{ ft} + 4 \text{ ft}).$

Solve. First we do the calculations in the Translate step.

For each door: 7 ft + 3 ft + 7 ft = 17 ft

For each window: $P = 2(3 \text{ ft} + 4 \text{ ft}) = 2 \cdot 7 \text{ ft} = 14 \text{ ft}$

We multiply to find the amount of caulk required for 3 doors and of 13 windows.

Doors: $3 \cdot 17 \text{ ft} = 51 \text{ ft}$

Windows: $13 \cdot 14 \text{ ft} = 182 \text{ ft}$

We add to find the total of the perimeters:

51 ft + 182 ft = 233 ft

Next we divide to determine how many sealant cartridges are needed:

$$\begin{array}{r} 4 \\ 5\,6\,\overline{)2\,3\,3} \\ \underline{2\,2\,4} \\ 9 \end{array}$$

The answer is 4 R 9. Since 9 ft will be left unsealed after 4 cartridges are used, Andrea should buy 5 sealant cartridges.

Finally, we multiply to find the cost of 5 sealant cartridges.

$5 \cdot \$5.95 = \$29.75$

Check. We repeat the calculations. The result checks.

State. It will cost Andrea $29.75 to seal the windows and doors.

77. *Familiarize.* The inside of the rack is a square whose side has a length that is the total diameter of 4 balls, or $4 \cdot 57$ mm, or 228 mm. We find the perimeter of a square with side 228 mm.

Translate.

$P = 4s = 4 \cdot 228$ mm

Solve. We calculate the perimeter.

$P = 4 \cdot 228$ mm $= 912$ mm

Check. We repeat the calculation. The answer checks.

State. The inside perimeter of the storage rack is 912 mm.

Exercise Set 2.8

1.
$$2x = 10 \qquad\qquad 5x = 25$$
$$\frac{2x}{2} = \frac{10}{2} \qquad\qquad \frac{5x}{5} = \frac{25}{5}$$
$$x = 5 \qquad\qquad x = 5$$

We see that $2x = 10$ and $5x = 25$ are equivalent equations.

3. Combining like terms in $4a - 3 + 3a$, we have
$$\begin{aligned} 4a - 3 + 3a &= (4 + 3)a - 3 \\ &= 7a - 3. \end{aligned}$$
We see that $7a - 3$ and $4a - 3 + 3a$ are equivalent expressions.

5. Combining like terms in $8 + 4r - 5$, we have
$$\begin{aligned} 8 + 4r - 5 &= 4r + (8 - 5) \\ &= 4r + 3. \end{aligned}$$
We see that $4r + 3$ and $8 + 4r - 5$ are equivalent expressions.

7.
$$x - 9 = 8 \qquad\qquad x + 3 = 20$$
$$x - 9 + 9 = 8 + 9 \qquad x + 3 - 3 = 20 - 3$$
$$x = 17 \qquad\qquad x = 17$$

We see that $x - 9 = 8$ and $x + 3 = 20$ are equivalent equations.

9.
$$3(t + 2) = 3t + 6$$
$$5 + 3t + 1 = 3t + (5 + 1) = 3t + 6$$

We see that $3(t + 2)$ and $5 + 3t + 1$ are equivalent expressions.

11.
$$x + 4 = -8 \qquad\qquad 2x = -24$$
$$x + 4 - 4 = -8 - 4 \qquad \frac{2x}{2} = -\frac{24}{2}$$
$$x = -12 \qquad\qquad x = -12$$

We see that $x + 4 = -8$ and $2x = -24$ are equivalent equations.

13.
$$x - 6 = -9$$
$$x - 6 + 6 = -9 + 6 \quad \text{Adding 6 to both sides}$$
$$x + 0 = -3$$
$$x = -3$$

Check:
$$\begin{array}{c|c} \multicolumn{2}{c}{x - 6 = -9} \\ \hline -3 - 6 \; ? \; -9 \\ -9 & -9 \quad \text{TRUE} \end{array}$$

The solution is -3.

15. $x - 4 = -12$

$x - 4 + 4 = -12 + 4$ Adding 4 to both sides

$x + 0 = -8$

$x = -8$

Check: $\dfrac{x - 4 = -12}{}$

$\dfrac{-8 - 4 \; ? \; -12}{}$

$-12 \;\Big|\; -12$ TRUE

The solution is -8.

17. $a + 7 = 25$

$a + 7 - 7 = 25 - 7$ Subtracting 7 from both sides

$a + 0 = 18$

$a = 18$

The solution is 18.

19. $x + 8 = -6$

$x + 8 - 8 = -6 - 8$ Subtracting 8 from both sides

$x + 0 = -14$

$x = -14$

The solution is -14.

21. $24 = t - 8$

$24 + 8 = t - 8 + 8$ Adding 8 to both sides

$32 = t + 0$

$32 = t$

The solution is 32.

23. $-12 = x + 5$

$-12 - 5 = x + 5 - 5$ Subtracting 5 from both sides

$-17 + 0 = x$

$-17 = x$

The solution is -17.

25. $-5 + a = 12$

$5 - 5 + a = 5 + 12$ Adding 5 to both sides

$0 + a = 12$

$a = 17$

The solution is 17.

27. $-8 = -8 + t$

$8 - 8 = 8 - 8 + t$ Adding 8 to both sides

$0 = 0 + t$

$0 = t$

The solution is 0.

29. $6x = -24$

$\dfrac{6x}{6} = \dfrac{-24}{6}$ Dividing both sides by 6

$x = -4$

The solution is -4.

31. $-3t = 42$

$\dfrac{-3t}{-3} = \dfrac{42}{-3}$ Dividing both sides by -3

$t = -14$

The solution is -14.

33. $-7n = -35$

$\dfrac{-7n}{-7} = \dfrac{-35}{-7}$ Dividing both sides by -7

$n = 5$

The solution is 5.

35. $0 = 6x$

$\dfrac{0}{6} = \dfrac{6x}{6}$ Dividing both sides by 6

$0 = x$

The solution is 0.

37. $55 = -5t$

$\dfrac{55}{-5} = \dfrac{-5t}{-5}$ Dividing both sides by -5

$-11 = t$

The solution is -11.

39. $-x = 56$

$\dfrac{-x}{-1} = \dfrac{56}{-1}$ Dividing both sides by -1

$x = -56$

The solution is -56.

41. $n(-4) = -48$

$\dfrac{n(-4)}{-4} = \dfrac{-48}{-4}$ Dividing both sides by -4

$n = 12$

The solution is 12.

43. $-x = -390$

$\dfrac{-x}{-1} = \dfrac{-390}{-1}$ Dividing both sides by -1

$x = 390$

The solution is 390.

45. $t - 6 = -2$

To undo the addition of -6, or the subtraction of 6, we subtract -6, or simply add 6, to both sides.

$t - 6 = -2$

$t - 6 + 6 = -2 + 6$

$t + 0 = 4$

$t = 4$

The solution is 4.

47. $6x = -54$

To undo multiplication by 6, we divide both sides by 6.

$6x = -54$

$\dfrac{6x}{6} = \dfrac{-54}{6}$

$x = -9$

The solution is -9.

49. $15 = -x$

$-1 \cdot 15 = -1 \cdot (-x)$ Multiplying both sides by -1

$-15 = x$

The solution is -15.

51. $-21 = x + 5$

$-21 - 5 = x + 5 - 5$ Subtracting 5 from both sides

$-26 = x$

The solution is -26.

53. $35 = -7t$

To undo multiplication by -7, we divide both sides by -7.

$35 = -7t$

$\dfrac{35}{-7} = \dfrac{-7t}{-7}$

$-5 = t$

The solution is -5.

55. $-17x = 68$

To undo multiplication by -17, we divide both sides by -17.

$-17x = 68$

$\dfrac{-17x}{-17} = \dfrac{68}{-17}$

$x = -4$

The solution is -4.

57. $18 + t = -160$

To undo the addition of 18, we subtract 18 from both sides.

$18 + t = -160$

$18 + t - 18 = -160 - 18$

$t + 0 = -178$

$t = -178$

The solution is -178.

59. $-27 = x + 23$

To undo the addition of 23, we subtract 23 from both sides.

$-27 = x + 23$

$-27 - 23 = x + 23 - 23$

$-50 = x + 0$

$-50 = x$

The solution is -50.

61. $5x - 1 = 34$

$5x - 1 + 1 = 34 + 1$ Adding 1 to both sides

$5x + 0 = 35$

$5x = 35$

$\dfrac{5x}{5} = \dfrac{35}{5}$ Dividing both sides by 5

$x = 7$

Check: $\dfrac{5x - 1 = 34}{\begin{array}{c|c} 5 \cdot 7 - 1 \ ? \ 34 & \\ 35 - 1 & \\ 34 & 34 \ \text{TRUE} \end{array}}$

The solution is 7.

63. $4t + 2 = 14$

$4t + 2 - 2 = 14 - 2$ Subtracting 2 from both sides

$4t + 0 = 12$

$4t = 12$

$\dfrac{4t}{4} = \dfrac{12}{4}$ Dividing both sides by 4

$t = 3$

Check: $\dfrac{4t + 2 = 14}{\begin{array}{c|c} 4 \cdot 3 + 2 \ ? \ 14 & \\ 12 + 2 & \\ 14 & 14 \ \text{TRUE} \end{array}}$

The solution is 3.

65. $6a + 1 = -17$

$6a + 1 - 1 = -17 - 1$ Subtracting 1 from both sides

$6a + 0 = -18$

$6a = -18$

$\dfrac{6a}{6} = \dfrac{-18}{6}$ Dividing both sides by 6

$a = -3$

The solution is -3.

67. $2x - 9 = -23$

$2x - 9 + 9 = -23 + 9$ Adding 9 to both sides

$2x + 0 = -14$

$2x = -14$

$\dfrac{2x}{2} = \dfrac{-14}{2}$ Dividing both sides by 2

$x = -7$

The solution is -7.

69. $-2x + 1 = 17$

$-2x + 1 - 1 = 17 - 1$ Subtracting 1 from both sides

$-2x + 0 = 16$

$-2x = 16$

$\dfrac{-2x}{-2} = \dfrac{16}{-2}$ Dividing both sides by -2

$x = -8$

The solution is -8.

71. $-8t - 3 = -67$

$-8t - 3 + 3 = -67 + 3$ Adding 3 to both sides

$-8t + 0 = -64$

$-8t = -64$

$\dfrac{-8t}{-8} = \dfrac{-64}{-8}$ Dividing both sides by -8

$t = 8$

The solution is 8.

73.
$$-x + 9 = -15$$
$$-x + 9 - 9 = -15 - 9 \quad \text{Subtracting 9 from both sides}$$
$$-x + 0 = -24$$
$$-x = -24$$
$$-1(-x) = -1(-24) \quad \text{Multiplying both sides by } -1$$
$$x = 24$$
The solution is 24.

75.
$$7 = 2x - 5$$
$$7 + 5 = 2x - 5 + 5 \quad \text{Adding 5 to both sides}$$
$$12 = 2x + 0$$
$$12 = 2x$$
$$\frac{12}{2} = \frac{2x}{2} \quad \text{Dividing both sides by 2}$$
$$6 = x$$
The solution is 6.

77.
$$13 = 3 + 2x$$
$$13 - 3 = 3 + 2x - 3 \quad \text{Subtracting 3 from both sides}$$
$$10 = 2x + 0$$
$$10 = 2x$$
$$\frac{10}{2} = \frac{2x}{2} \quad \text{Dividing both sides by 2}$$
$$5 = x$$
The solution is 5.

79.
$$13 = 5 - x$$
$$13 - 5 = 5 - x - 5 \quad \text{Subtracting 5 from both sides}$$
$$8 = 0 - x$$
$$8 = -x$$
$$-1 \cdot 8 = -1(-x) \quad \text{Multiplying both sides by } -1$$
$$-8 = x$$
The solution is -8.

81. Discussion and Writing Exercise

83. A <u>polygon</u> is a closed geometric figure.

85. Numbers we multiply together are called <u>factors</u>.

87. The result of an addition is a <u>sum</u>.

89. The <u>absolute value</u> of a number is its distance from zero on a number line.

91. Discussion and Writing Exercise

93.
$$2x - 7x = -40$$
$$-5x = -40 \quad \text{Collecting like terms}$$
$$\frac{-5x}{-5} = \frac{-40}{-5}$$
$$x = 8$$
The solution is 8.

95.
$$17 - 3^2 = 4 + t - 5^2$$
$$17 - 9 = 4 + t - 25$$
$$8 = t - 21 \quad \text{Collecting like terms}$$
$$8 + 21 = t - 21 + 21 \quad \text{Adding 21 to both sides}$$
$$29 = t$$
The solution is 29.

97.
$$(-7)^2 - 5 = t + 4^3$$
$$49 - 5 = t + 64$$
$$44 = t + 64$$
$$44 - 64 = t + 64 - 64 \quad \text{Subtracting 64 from both sides}$$
$$-20 = t$$
The solution is -20.

99.
$$x - (19)^3 = -18^3$$
$$x - 6859 = -5832$$
$$x - 6859 + 6859 = -5832 + 6859$$
$$x = 1027$$
The solution is 1027.

101.
$$35^3 = = -125t$$
$$42,875 = -125t$$
$$\frac{42,875}{-125} = \frac{-125t}{-125}$$
$$-343 = t$$
The solution is -343.

103.
$$529 - 143x = -1902$$
$$529 - 143x - 529 = -1902 - 529$$
$$-143x = -2431$$
$$\frac{-143x}{-143} = \frac{-2431}{-143}$$
$$x = 17$$
The solution is 17.

Chapter 2 Review Exercises

1. The integer 527 corresponds to having $527 in an account; the integer -53 corresponds to a $53 debt.

2. Since 0 is to the right of -5, we have $0 > -5$.

3. Since -7 is to the left of 6, we have $-7 < 6$.

4. Since -4 is to the right of -19, we have $-4 > -19$.

5. The distance from -39 to 0 is 39, so $|-39| = 39$.

6. The distance from 23 to 0 is 23, so $|23| = 23$.

7. The distance from 0 to 0 is 0, so $|0| = 0$.

8. When $x = -72$, $-x = -(-72) = 72$.

9. When $x = 59$, $-(-x) = -(-59) = 59$.

10. $-14 + 5$ The absolute values are 14 and 5. The difference is 9. The negative number has the larger absolute value, so the answer is negative. $-14 + 5 = -9$

11. $-5 + (-6)$

Add the absolute values: $5 + 6 = 11$

Make the answer negative: $-5 + (-6) = -11$

12. $14 + (-8)$ The absolute values are 14 and 8. The difference is 6. The positive number has the larger absolute value, so the answer is positive. $14 + (-8) = 6$

13. $0 + (-24) = -24$

When 0 is added to any number, that number remains unchanged.

14. $17 - 29 = 17 + (-29) = -12$

15. $9 - (-14) = 9 + 14 = 23$

16. $-8 - (-7) = -8 + 7 = -1$

17. $-3 - (-10) = -3 + 10 = 7$

18. $-3 + 7 + (-8)$

$= -3 + (-8) + 7$ Using a commutative law

$= -11 + 7$

$= -4$

19. $8 - (-9) - 7 + 2$

$= 8 + 9 + (-7) + 2$

$= 19 + (-7)$ Adding the positive numbers

$= 12$

20. $-23 \cdot (-4) = 92$

21. $7(-12) = -84$

22. $2(-4)(-5)(-1)$

$= -8 \cdot 5$ Multiplying the first two numbers and the last two numbers

$= -40$

23. $15 \div (-5) = -3$ Check: $-3(-5) = 15$

24. $\dfrac{-55}{11} = -5$ Check: $-5 \cdot 11 = -55$

25. $\dfrac{0}{7} = 0$ Check: $0 \cdot 7 = 0$

26. $7 \div 1^2 \cdot (-3) - 4$

$= 7 \div 1 \cdot (-3) - 4$ Evaluating the exponential expression

$= 7 \cdot (-3) - 4$ Dividing

$= -21 - 4$ Multiplying

$= -25$ Subtracting

27. $(-3)|4 - 3^2| - 5$

$= (-3)|4 - 9| - 5$

$= (-3)|-5| - 5$

$= -3 \cdot 5 - 5$

$= -15 - 5$

$= -20$

28. $3a + b = 3 \cdot 4 + (-5) = 12 + (-5) = 7$

29. $\dfrac{-x}{y} = \dfrac{-30}{5} = -6$

$\dfrac{x}{-y} = \dfrac{30}{-5} = -6$

$-\dfrac{x}{y} = -\dfrac{30}{5} = -6$

30. $4(5x + 9) = 4 \cdot 5x + 4 \cdot 9 = 20x + 36$

31. $3(2a - 4b + 5) = 3 \cdot 2a - 3 \cdot 4b + 3 \cdot 5 = 6a - 12b + 15$

32. $5a + 12a = (5 + 12)a = 17a$

33. $-7x + 13x = (-7 + 13)x = 6x$

34. $9m + 14 - 12m - 8$

$= 9m - 12m + 14 - 8$

$= (9 - 12)m + (14 - 8)$

$= -3m + 6$

35. $P = 2l + 2w = 2 \cdot 10 \text{ in.} + 2 \cdot 8 \text{ in.}$

$= 20 \text{ in.} + 16 \text{ in.} = 36 \text{ in.}$

36. $P = 4s = 4 \cdot 25 \text{ cm} = 100 \text{ cm}$

37. $x - 9 = -17$

$x - 9 + 9 = -17 + 9$

$x = -8$

The solution is -8.

38. $-4t = 36$

$\dfrac{-4t}{-4} = \dfrac{36}{-4}$

$t = -9$

The solution is -9.

39. $13 = -x$

$-1 \cdot 13 = -1 \cdot (-x)$

$-13 = x$

The solution is -13.

40. $56 = 6x - 10$

$56 + 10 = 6x - 10 + 10$

$66 = 6x$

$\dfrac{66}{6} = \dfrac{6x}{6}$

$11 = x$

The solution is 11.

41. $-x + 3 = -12$

$-x + 3 - 3 = -12 - 3$

$-x = -15$

$\dfrac{-x}{-1} = \dfrac{-15}{-1}$

$x = 15$

The solution is 15.

42.
$$18 = 4 - 2x$$
$$18 - 4 = 4 - 2x - 4$$
$$14 = -2x$$
$$\frac{14}{-2} = \frac{-2x}{-2}$$
$$-7 = x$$

The solution is -7.

43. *Discussion and Writing Exercise.* Equivalent expressions are expressions that have the same value when evaluated for various replacements of the variable(s). Equivalent equations are equations that have the same solution(s).

44. *Discussion and Writing Exercise.* A number's absolute value is the number itself if the number is nonnegative, and the opposite of the number if the number is negative. In neither case is the result less than the number itself, so "no," a number's absolute value is never less than the number itself.

45. *Discussion and Writing Exercise.* The notation "$-x$" means "the opposite of x." If x is a negative number, then $-x$ is a positive number. For example, if $x = -2$, then $-x = 2$.

46. *Discussion and Writing Exercise.* The expressions $(a-b)^2$ and $(b-a)^2$ are equivalent for all choices of a and b because $a - b$ and $b - a$ are opposites. When opposites are raised to an even power, the results are the same.

47.
$$87 \div 3 \cdot 29^3 - (-6)^6 + 1957$$
$$= 87 \div 3 \cdot 24,389 - 46,656 + 1957$$
$$= 29 \cdot 24,389 - 46,656 + 1957$$
$$= 707,281 - 46,656 + 1957$$
$$= 660,625 + 1957$$
$$= 662,582$$

48.
$$1969 + (-8)^5 - 17 \cdot 15^3$$
$$= 1969 + (-32,768) - 17 \cdot 3375$$
$$= 1969 + (-32,768) - 57,375$$
$$= -30,799 - 57,375$$
$$= -88,174$$

49.
$$\frac{113 - 17^3}{15 + 8^3 - 507} = \frac{113 - 4913}{15 + 512 - 507}$$
$$= \frac{-4800}{527 - 507}$$
$$= \frac{-4800}{20}$$
$$= -240$$

50. $8 + x^3$ will be negative for all values of x for which x^3 is less than -8. Thus, $8 + x^3$ will be negative for $x < -2$.

51. $|x| > x$ for all negative values of x, or for $x < 0$.

Chapter 2 Test

1. The integer -542 corresponds to selling 542 fewer shirts than expected; the integer 307 corresponds to selling 307 more shirts than expected.

2. Since -14 is to the right of -21, we have $-14 > -21$.

3. The distance from -739 to 0 is 739, so $|-739| = 739$.

4. When $x = -19$, $-(-x) = -(-(-19)) = -(19) = -19$.

5. $6 + (-17)$ The absolute values are 6 and 17. The difference is 11. The negative number has the larger absolute value, so the answer is negative. $6 + (-17) = -11$

6. $-9 + (-12)$

Add the absolute values: $9 + 12 = 21$

Make the answer negative: $-9 + (-12) = -21$

7. $-8 + 17$ The absolute values are 8 and 17. The difference is 9. The positive number has the larger absolute value, so the answer is positive. $-8 + 17 = 9$

8. $0 - 12 = 0 + (-12) = -12$

When 0 is added to any number, that number remains unchanged.

9. $7 - 22 = 7 + (-22) = -15$

10. $-5 - 19 = -5 + (-19) = -24$

11. $-8 - (-27) = -8 + 27 = 19$

12.
$$31 - (-3) - 5 + 9$$
$$= 31 + 3 + (-5) + 9$$
$$= 43 + (-5) \quad \text{Adding the positive numbers}$$
$$= 38$$

13. $(-4)^3 = -4(-4)(-4) = 16(-4) = -64$

14. $27(-10) = -270$

15. $-9 \cdot 0 = 0$

16. $-72 \div (-9) = 8$ Check: $8(-9) = -72$

17. $\dfrac{-56}{7} = -8$ Check: $-8 \cdot 7 = -56$

18.
$$8 \div 2 \cdot 2 - 3^2 = 8 \div 2 \cdot 2 - 9$$
$$= 4 \cdot 2 - 9$$
$$= 8 - 9$$
$$= -1$$

19.
$$29 - (3 - 5)^2 = 29 - (-2)^2$$
$$= 29 - 4$$
$$= 25$$

20. We subtract the lower temperature from the higher temperature.
$$-67 - (-81) = -67 + 81 = 14$$

The average high temperature is $14°$F higher than the average low temperature.

21. We subtract the final mark from the first mark.

$$8 - (-15) = 8 + 15 = 23$$

Thus, 23 min of tape were rewound.

22. $\dfrac{a-b}{6} = \dfrac{-8-10}{6} = \dfrac{-18}{6} = -3$

23. $7(2x + 3y - 1) = 7 \cdot 2x + 7 \cdot 3y - 7 \cdot 1 = 14x + 21y - 7$

24. $9x - 14 - 5x - 3 = 9x - 5x - 14 - 3$
$$= (9 - 5)x + (-14 - 3)$$
$$= 4x - 17$$

25. $-7x = -35$
$$\dfrac{-7x}{-7} = \dfrac{-35}{-7}$$
$$x = 5$$

The solution is 5.

26. $a + 9 = -3$
$$a + 9 - 9 = -3 - 9$$
$$a = -12$$

The solution is -12.

27. The amount of trim needed is given by the perimeter of the room, less the 3 ft width of the door, plus the lengths of the three sides of the door that will get trim.

Perimeter of room: $P = 2(l + w)$
$$= 2(14 \text{ ft} + 12 \text{ ft})$$
$$= 2(26 \text{ ft})$$
$$= 52 \text{ ft}$$

Subtract the width of the door: $52 \text{ ft} - 3 \text{ ft} = 49 \text{ ft}$

Trim on door: $7 \text{ ft} + 3 \text{ ft} + 7 \text{ ft} = 17 \text{ ft}$

Total length of trim: $49 \text{ ft} + 17 \text{ ft} = 66 \text{ ft}$

28. $9 - 5[x + 2(3 - 4x)] + 14$
$$= 9 - 5[x + 6 - 8x] + 14$$
$$= 9 - 5(-7x + 6) + 14$$
$$= 9 + 35x - 30 + 14$$
$$= 35x - 7$$

29. $15x + 3(2x - 7) - 9(4 + 5x)$
$$= 15x + 6x - 21 - 36 - 45x$$
$$= -24x - 57$$

30. $49 \cdot 14^3 \div 7^4 + 1926^2 \div 6^2$
$$= 49 \cdot 2744 \div 2401 + 3,709,476 \div 36$$
$$= 134,456 \div 2401 + 3,709,476 \div 36$$
$$= 56 + 3,709,476 \div 36$$
$$= 56 + 103,041$$
$$= 103,097$$

31. $3487 - 16 \div 4 \cdot 4 \div 2^8 \cdot 14^4$
$$= 3487 - 16 \div 4 \cdot 4 \div 256 \cdot 38,416$$
$$= 3487 - 4 \cdot 4 \div 256 \cdot 38,416$$
$$= 3487 - 16 \div 256 \cdot 38,416$$
$$= 3487 - 2401 \quad \text{Dividing and then multiplying}$$
$$= 1086$$

Cumulative Review Chapters 1 - 2

1. Standard notation is 181,599,900.

2. A word name for 5,380,001,437 is five billion, three hundred eighty million, one thousand, four hundred thirty-seven.

3.
```
   1 1
  1 5, 8 9 2
+    2, 9 3 5
-----------
  1 8, 8 2 7
```

4.
```
    1 2 2
  7 9 8 9
    7 8 9
+      7 9
---------
  8 8 5 7
```

5.
```
    7 12
  8̶ 2̶ 7 6
-     4 3 0
---------
  7 8 4 6
```

6.
```
  2 9  9 16
  3̶0̶-0̶-0̶̶
-     5 7 8
---------
    2 4 2 8
```

7.
```
       1
     6 2 1
  ×    2 7
  --------
   4 3 4 7
  1 2 4 2 0
  --------
  1 6, 7 6 7
```

8.
```
        1      1
      1        1
      2 5 0 5
  ×   3 3 0 0
  ------------
    7 5 1 5 0 0
  7 5 1 5 0 0 0
  ------------
  8, 2 6 6, 5 0 0
```

9. $43 \cdot (-8) = -344$

10. $-12(-6) = 72$

11.
```
        1 0 4
  6 3 ⟌6 5 5 2
       6 3 0 0
       ------
         2 5 2
         2 5 2
         -----
             0
```

The answer is 104.

12.
$$\begin{array}{r} 6\,2 \\ 6\,2\,\overline{\big)\,3\,8\,4\,4} \\ 3\,7\,2 \\ \hline 1\,2\,4 \\ 1\,2\,4 \\ \hline 0 \end{array}$$

The answer is 62.

13. $0 \div (-67) = 0$ Check: $-67 \cdot 0 = 0$

14. $60 \div (-12) = -5$ Check: $-5(-12) = 60$

15. Round 427,931 to the nearest thousand.

$$4\,2\,7,\;\boxed{9}\,3\,1$$
$$\uparrow$$

The digit 7 is in the thousands place. Consider the next digit to the right. Since the digit, 9, is 5 or higher, we round up meaning that 7 thousands become 8 thousands. Then change all digits to the right of the thousands digit to zeros.

The answer is 428,000.

16. Round 5309 to the nearest hundred.

$$5\,3\;\boxed{0}\,9$$
$$\uparrow$$

The digit 3 is in the hundreds place. Consider the next digit to the right. Since the digit, 0, is 4 or lower, we round down, meaning that 3 hundreds stays as 3 hundreds. Then change all digits to the right of the hundreds place to zeros.

The answer is 5300.

17.

	Rounded to the nearest hundred
$7\,4\,9,5\,5\,9$	$7\,4\,9,6\,0\,0$
$+\,3\,0\,1,3\,6\,2$	$+\,3\,0\,1,4\,0\,0$
	$\overline{1,0\,5\,1,0\,0\,0}$

18.

	Rounded to the nearest hundred
$7\,4\,9$	$7\,0\,0$
$\times\,5\,3\,1$	$\times\,5\,0\,0$
	$\overline{3\,5\,0,0\,0\,0}$

19. Since -26 is to the left of 19, we have $-26 < 19$.

20. The distance from -279 to 0 is 279, so $|-279| = 279$.

21.
$$35 - 25 \div 5 + 2 \times 3$$
$$= 35 - 5 + 2 \times 3$$
$$= 35 - 5 + 6$$
$$= 30 + 6$$
$$= 36$$

22.
$$\{17 - [8 - (5 - 2 \times 2)]\} \div (3 + 12 \div 6)$$
$$= \{17 - [8 - (5 - 4)]\} \div (3 + 2)$$
$$= \{17 - [8 - 1]\} \div 5$$
$$= \{17 - 7\} \div 5$$
$$= 10 \div 5$$
$$= 2$$

23.
$$10 \div 1(-5) - 6^2$$
$$= 10 \div 1(-5) - 36$$
$$= 10(-5) - 36$$
$$= -50 - 36$$
$$= -86$$

24. $5^3 = 5 \cdot 5 \cdot 5 = 25 \cdot 5 = 125$

25. $\dfrac{x + y}{5} = \dfrac{11 + 4}{5} = \dfrac{15}{5} = 3$

26. $7x^2 = 7(-2)^2 = 7(4) = 28$

27. $-2(x + 5) = -2 \cdot x + (-2) \cdot 5 = -2x + (-10) = -2x - 10$

28. $6(3x - 2y + 4) = 6 \cdot 3x - 6 \cdot 2y + 6 \cdot 4 = 18x - 12y + 24$

29. $-12 + (-14)$

Add the absolute values: $12 + 14 = 26$

Make the answer negative: $-12 + (-14) = -26$

30. $(-3)(-10) = 30$

31. $23 - 38 = 23 + (-38) = -15$

32. $64 \div (-2) = -32$ Check: $-32(-2) = 64$

33. $-12 - (-25) = -12 + 25 = 13$

34. $(-2)(-3)(-5) = 6(-5) = -30$

35. $3 - (-8) + 2 - (-3) = 3 + 8 + 2 + 3 = 16$

36.
$$16 \div 2(-8) + 7 = 8(-8) + 7$$
$$= -64 + 7$$
$$= -57$$

37.
$$x + 8 = 35$$
$$x + 8 - 8 = 35 - 8$$
$$x = 27$$
The solution is 27.

38.
$$-12t = 36$$
$$\frac{-12t}{-12} = \frac{36}{-12}$$
$$t = -3$$
The solution is -3.

39.
$$6 - x = -9$$
$$6 - x - 6 = -9 - 6$$
$$-x = -15$$
$$\frac{-x}{-1} = \frac{-15}{-1}$$
$$x = 15$$
The solution is 15.

40.
$$-39 = 4x - 7$$
$$-39 + 7 = 4x - 7 + 7$$
$$-32 = 4x$$
$$\frac{-32}{4} = \frac{4x}{4}$$
$$-8 = x$$
The solution is -8.

41. *Familiarize*. Let y = the number of years the character aged.

***Translate*.** We can think of this as a "how many more" situation.

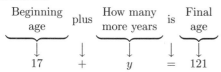

$$
\underbrace{\text{Beginning age}}_{17} \ \underset{+}{\text{plus}} \ \underbrace{\text{How many more years}}_{y} \ \underset{=}{\text{is}} \ \underbrace{\text{Final age}}_{121}
$$

***Solve*.** We subtract 17 on both sides of the equation.
$$17 + y = 121$$
$$17 + y - 17 = 121 - 17$$
$$y = 104$$

***Check*.** We can add the difference to the subtrahend: $17 + 104 = 121$. The answer checks.

***State*.** The character aged 104 yr.

42. *Familiarize*. We are combining quantities, so addition can be used. Let r = the total number of rooms in the four hotels.

***Translate*.** We translate to an equation.
$$5034 + 4408 + 4008 + 3770 = r$$

***Solve*.** We carry out the addition.

$$
\begin{array}{r}
\scriptstyle 1\ 1\ 2 \\
5\ 0\ 3\ 4 \\
4\ 4\ 0\ 8 \\
4\ 0\ 0\ 8 \\
+\ \ 3\ 7\ 7\ 0 \\
\hline
1\ 7,2\ 2\ 0
\end{array}
$$

***Check*.** We can repeat the calculation. We can also estimate by rounding, say to the nearest thousand.
$$5034 + 4408 + 4008 + 3770$$
$$\approx 5000 + 4000 + 4000 + 4000$$
$$\approx 17,000 \approx 17,220$$

Since the estimated answer is close to the calculated answer, our result is probably correct.

***State*.** The four hotels have a total of 17,220 rooms.

43. *Familiarize*. Let p = the amount of each weekly paycheck.

***Translate*.** We translate to an equation.

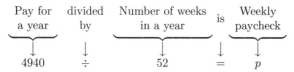

$$
\underbrace{\text{Pay for a year}}_{4940} \ \underset{\div}{\text{divided by}} \ \underbrace{\text{Number of weeks in a year}}_{52} \ \underset{=}{\text{is}} \ \underbrace{\text{Weekly paycheck}}_{p}
$$

***Solve*.** We carry out the division.
$$4940 \div 52 = p$$
$$95 = p$$

***Check*.** We can check by multiplying: $95 \cdot 52 = 4940$ The answer checks.

***State*.** Amanda's weekly paycheck is $95.

44. *Familiarize*. We will find and compare the total costs of the refrigerators. Let x = the total cost of the refrigerator from Eastside Appliance and let y = the total cost of the refrigerator from Westside Appliance.

***Translate*.** We translate to two equations.

For Eastside Appliance:

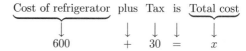

$$
\underbrace{\text{Cost of refrigerator}}_{600} \ \underset{+}{\text{plus}} \ \underset{30}{\text{Tax}} \ \underset{=}{\text{is}} \ \underbrace{\text{Total cost}}_{x}
$$

For Westside Appliance:

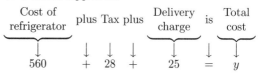

$$
\underbrace{\text{Cost of refrigerator}}_{560} \ \underset{+}{\text{plus}} \ \underset{28}{\text{Tax}} \ \underset{+}{\text{plus}} \ \underbrace{\text{Delivery charge}}_{25} \ \underset{=}{\text{is}} \ \underbrace{\text{Total cost}}_{y}
$$

***Solve*.** We carry out the additions.
$$600 + 30 = x \qquad 560 + 28 + 25 = y$$
$$630 = x \qquad\qquad 613 = y$$

Since $630 > 613$, the refrigerator from Westside Appliance costs less than the refrigerator from Eastside Appliance and is thus the better buy.

***Check*.** We repeat the calculations. The answer checks.

***State*.** The refrigerator from Westside Appliance is the better buy.

45. $7x - 9 + 3x - 5 = 7x + 3x - 9 - 5$
$$= (7 + 3)x + (-9 - 5)$$
$$= 10x - 14$$

46. *Familiarize*. This is a multistep problem. We begin by finding the number of 24-can cases that can be filled. Let c = the number of cases.

***Translate*.** We translate to an equation.

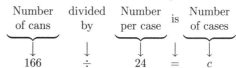

$$
\underbrace{\text{Number of cans}}_{166} \ \underset{\div}{\text{divided by}} \ \underbrace{\text{Number per case}}_{24} \ \underset{=}{\text{is}} \ \underbrace{\text{Number of cases}}_{c}
$$

***Solve*.** We carry out the division.

$$
\begin{array}{r}
6 \\
24\overline{\smash{\big)}\,166} \\
\underline{1\ 4\ 4} \\
2\ 2
\end{array}
$$

Then $c = 6$ R 22. This means that 6 cases can be filled and 22 cans will be left over.

Now let s = the number of six-packs that can be filled with the 22 left-over cans.

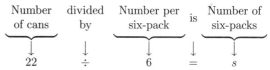

$$
\underbrace{\text{Number of cans}}_{22} \ \underset{\div}{\text{divided by}} \ \underbrace{\text{Number per six-pack}}_{6} \ \underset{=}{\text{is}} \ \underbrace{\text{Number of six-packs}}_{s}
$$

We carry out the division.

$$
\begin{array}{r}
3 \\
6\overline{\smash{\big)}\,22} \\
\underline{1\ 8} \\
4
\end{array}
$$

Then $s = 3$ R 4. This means that 3 six-packs can be filled and there will be 4 cans left over.

Check. In 6 cases there are $6 \cdot 24$, or 144 cans; in 3 six-packs there are $3 \cdot 6$, or 18 cans. With the 4 cans left over the total number of cans is $144 + 18 + 4$, or 166. The answer checks.

State. The distributor can fill 6 cases and 3 six-packs and there will be 4 loose cans.

47.
$$a - \{3a - [4a - (2a - 4a)]\}$$
$$= a - \{3a - [4a - (-2a)]\}$$
$$= a - \{3a - [4a + 2a]\}$$
$$= a - \{3a - 6a\}$$
$$= a - \{-3a\}$$
$$= a + 3a$$
$$= 4a$$

48.
$$37 \cdot 64 \div 4^2 \cdot 2 - (7^3 - (-4)^5)$$
$$= 37 \cdot 64 \div 4^2 \cdot 2 - (343 - (-1024))$$
$$= 37 \cdot 64 \div 4^2 \cdot 2 - (343 + 1024)$$
$$= 37 \cdot 64 \div 4^2 \cdot 2 - 1367$$
$$= 37 \cdot 64 \div 16 \cdot 2 - 1367$$
$$= 2368 \div 16 \cdot 2 - 1367$$
$$= 148 \cdot 2 - 1367$$
$$= 296 - 1367$$
$$= -1071$$

49.
$$5|x| - 2 = 13$$
$$5|x| - 2 + 2 = 13 + 2$$
$$5|x| = 15$$
$$\frac{5|x|}{5} = \frac{15}{5}$$
$$|x| = 3$$

The solutions are the number whose distance from 0 is 3. Thus, the solutions are 3 and -3.

Chapter 3

Fraction Notation: Multiplication and Division

Exercise Set 3.1

1. $1 \cdot 7 = 7$ $6 \cdot 7 = 42$
$2 \cdot 7 = 14$ $7 \cdot 7 = 49$
$3 \cdot 7 = 21$ $8 \cdot 7 = 56$
$4 \cdot 7 = 28$ $9 \cdot 7 = 63$
$5 \cdot 7 = 35$ $10 \cdot 7 = 70$

3. $1 \cdot 20 = 20$ $6 \cdot 20 = 120$
$2 \cdot 20 = 40$ $7 \cdot 20 = 140$
$3 \cdot 20 = 60$ $8 \cdot 20 = 160$
$4 \cdot 20 = 80$ $9 \cdot 20 = 180$
$5 \cdot 20 = 100$ $10 \cdot 20 = 200$

5. $1 \cdot 3 = 3$ $6 \cdot 3 = 18$
$2 \cdot 3 = 6$ $7 \cdot 3 = 21$
$3 \cdot 3 = 9$ $8 \cdot 3 = 24$
$4 \cdot 3 = 12$ $9 \cdot 3 = 27$
$5 \cdot 3 = 15$ $10 \cdot 3 = 30$

7. $1 \cdot 12 = 12$ $6 \cdot 12 = 72$
$2 \cdot 12 = 24$ $7 \cdot 12 = 84$
$3 \cdot 12 = 36$ $8 \cdot 12 = 96$
$4 \cdot 12 = 48$ $9 \cdot 12 = 108$
$5 \cdot 12 = 60$ $10 \cdot 12 = 120$

9. $1 \cdot 10 = 10$ $6 \cdot 10 = 60$
$2 \cdot 10 = 20$ $7 \cdot 10 = 70$
$3 \cdot 10 = 30$ $8 \cdot 10 = 80$
$4 \cdot 10 = 40$ $9 \cdot 10 = 90$
$5 \cdot 10 = 50$ $10 \cdot 10 = 100$

11. $1 \cdot 25 = 25$ $6 \cdot 25 = 150$
$2 \cdot 25 = 50$ $7 \cdot 25 = 175$
$3 \cdot 25 = 75$ $8 \cdot 25 = 200$
$4 \cdot 25 = 100$ $9 \cdot 25 = 225$
$5 \cdot 25 = 125$ $10 \cdot 25 = 250$

13. We divide 61 by 3.

$$
\begin{array}{r}
2\,0 \\
3\overline{\smash{)}6\,1} \\
6\,0 \\
\hline
1 \\
0 \\
\hline
1
\end{array}
$$

Since the remainder is not 0 we know that 61 is not divisible by 3.

15. We divide 527 by 7.

$$
\begin{array}{r}
7\,5 \\
7\overline{\smash{)}5\,2\,7} \\
4\,9\,0 \\
\hline
3\,7 \\
3\,5 \\
\hline
2
\end{array}
$$

Since the remainder is not 0 we know that 527 is not divisible by 7.

17. We divide 8127 by 9.

$$
\begin{array}{r}
9\,0\,3 \\
9\overline{\smash{)}8\,1\,2\,7} \\
8\,1\,0\,0 \\
\hline
2\,7 \\
2\,7 \\
\hline
0
\end{array}
$$

The remainder of 0 indicates that 8127 is divisible by 9.

19. Because $8 + 4 = 12$ and 12 is divisible by 3, 84 is divisible by 3.

21. 5553 is not divisible by 5 because the ones digit is neither 0 nor 5.

23. 671,500 is divisible by 10 because the ones digit is 0.

25. Because $1 + 7 + 7 + 3 = 18$ and 18 is divisible by 9, 1773 is divisible by 9.

27. 21,687 is not divisible by 2 because the ones digit is not even.

29. 32,109 is not divisible by 6 because it is not even.

31. 6825 is not divisible by 2 because the ones digit is not even.

Because $6 + 8 + 2 + 5 = 21$ and 21 is divisible by 3, 6825 is divisible by 3.

6825 is divisible by 5 because the ones digit is 5.

6825 is not divisible by 6 because it is not even.

Because $6 + 8 + 2 + 5 = 21$ and 21 is not divisible by 9, 6825 is not divisible by 9.

6825 is not divisible by 10 because the ones digit is not 0.

33. 119,117 is not divisible by 2 because the ones digit is not even.

Because $1 + 1 + 9 + 1 + 1 + 7 = 20$ and 20 is not divisible by 3, then 119,117 is not divisible by 3.

119,117 is not divisible by 5 because the ones digit is neither 0 nor 5.

119,117 is not divisible by 6 because it is not even.

Because $1 + 1 + 9 + 1 + 1 + 7 = 20$ and 20 is not divisible by 9, then 119,117 is not divisible by 9.

119,117 is not divisible by 10 because the ones digit is not 0.

35. 127,575 is not divisible by 2 because the ones digit is not even.

Because $1 + 2 + 7 + 5 + 7 + 5 = 27$ and 27 is divisible by 3, then 127,575 is divisible by 3.

127,575 is divisible by 5 because the ones digit is 5.

127,575 is not divisible by 6 because it is not even.

Because $1 + 2 + 7 + 5 + 7 + 5 = 27$ and 27 is divisible by 9, then 127,575 is divisible by 9.

127,575 is not divisible by 10 because the ones digit is not 0.

37. 9360 is divisible by 2 because the ones digit is even.

Because $9 + 3 + 6 + 0 = 18$ and 18 is divisible by 3, 9360 is divisible by 3.

9360 is divisible by 5 because the ones digit is 0.

We saw above that 9360 is even and that it is divisible by 3, so it is divisible by 6.

Because $9 + 3 + 6 + 0 = 18$ and 18 is divisible by 9, 9360 is divisible by 9.

9360 is divisible by 10 because the ones digit is 0.

39. A number is divisible by 3 if the sum of the digits is divisible by 3.

46 is not divisible by 3 because $4 + 6 = 10$ and 10 is not divisible by 3.

224 is not divisible by 3 because $2 + 2 + 4 = 8$ and 8 is not divisible by 3.

19 is not divisible by 3 because $1 + 9 = 10$ and 10 is not divisible by 3.

555 is divisible by 3 because $5 + 5 + 5 = 15$ and 15 is divisible by 3.

300 is divisible by 3 because $3 + 0 + 0 = 3$ and 3 is divisible by 3.

36 is divisible by 3 because $3 + 6 = 9$ and 9 is divisible by 3.

45,270 is divisible by 3 because $4 + 5 + 2 + 7 + 0 = 18$ and 18 is divisible by 3.

4444 is not divisible by 3 because $4 + 4 + 4 + 4 = 16$ and 16 is not divisible by 3.

85 is not divisible by 3 because $8 + 5 = 13$ and 13 is not divisible by 3.

711 is divisible by 3 because $7 + 1 + 1 = 9$ and 9 is divisible by 3.

13,251 is divisible by 3 because $1 + 3 + 2 + 5 + 1 = 12$ and 12 is divisible by 3.

254,765 is not divisible by 3 because $2+5+4+7+6+5 = 29$ and 29 is not divisible by 3.

256 is not divisible by 3 because $2 + 5 + 6 = 13$ and 13 is not divisible by 3.

8064 is divisible by 3 because $8 + 0 + 6 + 4 = 18$ and 18 is divisible by 3.

1867 is not divisible by 3 because $1 + 8 + 6 + 7 = 22$ and 22 is not divisible by 3.

21,568 is not divisible by 3 because $2 + 1 + 5 + 6 + 8 = 22$ and 22 is not divisible by 3.

41. A number is divisible by 10 if its ones digit is 0.

Of the numbers under consideration, only 300 and 45,270 have one digits of 0. Therefore, only 300 and 45,270 are divisible by 10.

43. For a number to be divisible by 6, the sum of the digits must be divisible by 3 and the ones digit must be 0, 2, 4, 6 or 8 (even). It is most efficient to determine if the ones digit is even first and then, if so, to determine if the sum of the digits is divisible by 3.

46 is not divisible by 6 because 46 is not divisible by 3.

$$4 + 6 = 10$$
$$\uparrow$$
Not divisible by 3

224 is not divisible by 6 because 224 is not divisible by 3.

$$2 + 2 + 4 = 8$$
$$\uparrow$$
Not divisible by 3

19 is not divisible by 6 because 19 is not even.

$$19$$
$$\uparrow$$
Not even

555 is not divisible by 6 because 555 is not even.

$$555$$
$$\uparrow$$
Not even

300 is divisible by 6.

$$300 \qquad 3 + 0 + 0 = 3$$
$$\uparrow \qquad\qquad\quad \uparrow$$
Even   Divisible by 3

36 is divisible by 6.

$$36 \qquad 3 + 6 = 9$$
$$\uparrow \qquad\qquad \uparrow$$
Even   Divisible by 3

45,270 is divisible by 6.

$$45,270 \qquad 4 + 5 + 2 + 7 + 0 = 18$$
$$\uparrow \qquad\qquad\qquad \uparrow$$
Even   Divisible by 3

4444 is not divisible by 6 because 4444 is not divisible by 3.

$$4 + 4 + 4 + 4 = 16$$
$$\uparrow$$

Not divisible by 3

85 is not divisible by 6 because 85 is not even.

$$85$$
$$\uparrow$$
Not even

711 is not divisible by 6 because 711 is not even.

$$711$$
$$\uparrow$$
Not even

13,251 is not divisible by 6 because 13,251 is not even.

$$13,251$$
$$\uparrow$$
Not even

254,765 is not divisible by 6 because 254,765 is not even.

$$254{,}765$$
$$\uparrow$$
Not even

256 is not divisible by 6 because 256 is not divisible by 3.

$$2 + 5 + 6 = 13$$
$$\uparrow$$
Not divisible by 3

8064 is divisible by 6.

$$8064 \qquad 8 + 0 + 6 + 4 = 18$$
$$\uparrow \qquad\qquad\qquad \uparrow$$
Even $\qquad\qquad$ Divisible by 3

1867 is not divisible by 6 because 1867 is not even.

$$1867$$
$$\uparrow$$
Not even

21,568 is not divisible by 6 because 21,568 is not divisible by 3.

$$2 + 1 + 5 + 6 + 8 = 22$$
$$\uparrow$$
Not divisible by 3

45. A number is divisible by 2 if its <u>ones digit</u> is even.

5<u>6</u> is divisible by 2 because <u>6</u> is even.
32<u>4</u> is divisible by 2 because <u>4</u> is even.
78<u>4</u> is divisible by 2 because <u>4</u> is even.
55,55<u>5</u> is not divisible by 2 because <u>5</u> is not even.
20<u>0</u> is divisible by 2 because <u>0</u> is even.
4<u>2</u> is divisible by 2 because <u>2</u> is even.
50<u>1</u> is not divisible by 2 because <u>1</u> is not even.
300<u>9</u> is not divisible by 2 because <u>9</u> is not even.

7<u>5</u> is not divisible by 2 because <u>5</u> is not even.
81<u>2</u> is divisible by 2 because <u>2</u> is even.
234<u>5</u> is not divisible by 2 because <u>5</u> is not even.
200<u>1</u> is not divisible by 2 because <u>1</u> is not even.
3<u>5</u> is not divisible by 2 because <u>5</u> is not even.
40<u>2</u> is divisible by 2 because <u>2</u> is even.
111,11<u>1</u> is not divisible by 2 because <u>1</u> is not even.
100<u>5</u> is not divisible by 2 because <u>5</u> is not even.

47. A number is divisible by 5 if the ones digit is 0 or 5.

5<u>6</u> is not divisible by 5 because the ones digit (6) is not 0 or 5.

32<u>4</u> is not divisible by 5 because the ones digit (4) is not 0 or 5.

78<u>4</u> is not divisible by 5 because the ones digit (4) is not 0 or 5.

55,55<u>5</u> is divisible by 5 because the ones digit is 5.

20<u>0</u> is divisible by 5 because the ones digit is 0.

4<u>2</u> is not divisible by 5 because the ones digit (2) is not 0 or 5.

50<u>1</u> is not divisible by 5 because the ones digit (1) is not 0 or 5.

300<u>9</u> is not divisible by 5 because the ones digit (9) is not 0 or 5.

7<u>5</u> is divisible by 5 because the ones digit is 5.

81<u>2</u> is not divisible by 5 because the ones digit (2) is not 0 or 5.

234<u>5</u> is divisible by 5 because the ones digit is 5.

200<u>1</u> is not divisible by 5 because the ones digit (1) is not 0 or 5.

3<u>5</u> is divisible by 5 because the ones digit is 5.

40<u>2</u> is not divisible by 5 because the ones digit (2) is not 0 or 5.

111,11<u>1</u> is not divisible by 5 because the ones digit (1) is not 0 or 5.

100<u>5</u> is divisible by 5 because the ones digit is 5.

49. A number is divisible by 9 if the sum of the digits is divisible by 9.

56 is not divisible by 9 because $5 + 6 = 11$ and 11 is not divisible by 9.

324 is divisible by 9 because $3 + 2 + 4 = 9$ and 9 is divisible by 9.

784 is not divisible by 9 because $7 + 8 + 4 = 19$ and 19 is not divisible by 9.

55,555 is not divisible by 9 because $5 + 5 + 5 + 5 + 5 = 25$ and 25 is not divisible by 9.

200 is not divisible by 9 because $2 + 0 + 0 = 2$ and 2 is not divisible by 9.

42 is not divisible by 9 because $4 + 2 = 6$ and 6 is not divisible by 9.

501 is not divisible by 9 because $5 + 0 + 1 = 6$ and 6 is not divisible by 9.

3009 is not divisible by 9 because $3 + 0 + 0 + 9 = 12$ and 12 is not divisible by 9.

75 is not divisible by 9 because $7 + 5 = 12$ and 12 is not divisible by 9.

812 is not divisible by 9 because $8 + 1 + 2 = 11$ and 11 is not divisible by 9.

2345 is not divisible by 9 because $2 + 3 + 4 + 5 = 14$ and 14 is not divisible by 9.

2001 is not divisible by 9 because $2 + 0 + 0 + 1 = 3$ and 3 is not divisible by 9.

35 is not divisible by 9 because $3 + 5 = 8$ and 8 is not divisible by 9.

402 is not divisible by 9 because $4 + 0 + 2 = 6$ and is not divisible by 9.

111,111 is not divisible by 9 because $1 + 1 + 1 + 1 + 1 + 1 = 6$ and 6 is not divisible by 9.

1005 is not divisible by 9 because $1 + 0 + 0 + 5 = 6$ and 6 is not divisible by 9.

51. Discussion and Writing Exercise

53. $\qquad 16 \cdot t = 848$

$\qquad \dfrac{16 \cdot t}{16} = \dfrac{848}{16} \quad$ Dividing by 16 on both sides

$\qquad\qquad t = 53$

The solution is 53.

55.
$$23 + x = 15$$
$$23 + x - 23 = 15 - 23 \quad \text{Subtracting 23 on both sides}$$
$$x = -8$$

The solution is -8.

57. *Familiarize*. This is a multistep problem. Find the total cost of the sweaters and the total cost of the jackets and then find the sum of the two.

We let s = the total cost of the sweaters and t = the total cost of the jackets.

Translate. We write two equations.

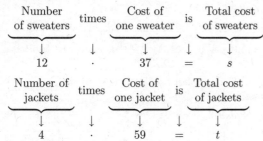

Solve. We carry out the multiplication.

$$12 \cdot 37 = s$$
$$444 = s \quad \text{Doing the multiplication}$$

The total cost of the 12 sweaters is \$444.

$$4 \cdot 59 = t$$
$$236 = t \quad \text{Doing the multiplication}$$

The total cost of the 4 jackets is \$236.

Now we find the total amount spent. We let a = this amount.

$$
\underbrace{\text{Total cost}}_{444} \underbrace{\text{plus}}_{+} \underbrace{\text{Total cost}}_{236} \underbrace{\text{is}}_{=} \underbrace{\text{Total amount}}_{a}
$$

To solve the equation, carry out the addition.

$$
\begin{array}{r}
4\,4\,4 \\
+\,2\,3\,6 \\
\hline
6\,8\,0
\end{array}
$$

Check. We can repeat the calculations. The answer checks.

State. The total cost is \$680.

59. $5^3 = 5 \cdot 5 \cdot 5 = 125$

61. $4^5 = 4 \cdot 4 \cdot 4 \cdot 4 \cdot 4 = 1024$

63. $\underbrace{9 \cdot 9 \cdot 9 \cdot 9 \cdot 9}_{5 \text{ factors}} = 9^5$

65. Discussion and Writing Exercise

67. When we use a calculator to divide the largest five-digit number, 99,999, by 47 we get 2127.638298. This tells us that 99,999 is not divisible by 47 but that 2127×47, or 99,969, is divisible by 47 and that it is the largest such five-digit number.

69. We list multiples of 2, 3, and 5 and find the smallest number that is on all 3 lists.

Multiples of 2: 2, 4, 6, 8, 10, 12, 14, 16, 18, 20, 22, 24, 26, 28, <u>30</u>, 32, $\cdots$

Multiples of 3: 3, 6, 9, 12, 15, 18, 21, 24, 27, <u>30</u>, 33, $\cdots$

Multiples of 5: 5, 10, 15, 20, 25, <u>30</u>, 35, $\cdots$

The smallest number that is simultaneously a multiple of 2, 3, and 5 is 30.

71. We list multiples of 4, 6, and 10 and find the smallest number that is on all 3 lists.

Multiples of 4: 4, 8, 12, 16, 20, 24, 28, 32, 36, 40, 44, 48, 52, 56, <u>60</u>, 64, $\cdots$

Multiples of 6: 6, 12, 18, 24, 30, 36, 42, 48, 54, <u>60</u>, 66, $\cdots$

Multiples of 10: 10, 20, 30, 40, 50, <u>60</u>, 70, $\cdots$

The smallest number that is simultaneously a multiple of 4, 6, and 10 is 60.

73. First note that 85 is a multiple of 17 ($85 = 5 \cdot 17$). Thus, any multiple of 85 will also be a multiple of 17. Then, using a calculator, we list multiples of 43 and 85 and find the smallest number that is on both lists. We find that this number is 3655.

75. First note that 120 is a multiple of 30 ($120 = 4 \cdot 30$). Thus, any multiple of 120 is also a multiple of 30. Now we list multiples of 70 and 120 and find the smallest number that is on both lists.

Multiples of 70: 70, 140, 210, 280, 350, 420, 490, 560, 630, 700, 770, <u>840</u>, 910, $\ldots$

Multiples of 120: 120, 240, 360, 480, 600, 720, <u>840</u>, 960, $\ldots$

The smallest number that is simultaneously a multiple of 30, 70, and 120 is 840.

77. Discussion and Writing Exercise

79. The sum of the given digits is $9 + 5 + 8$, or 22. If the number is divisible by 99, it is also divisible by 9 since 99 is divisible by 9. The smallest number that is divisible by 9 and also greater than 22 is 27. Then the sum of the two missing digits must be at least $27 - 22$, or 5. We try various combinations of two digits whose sum is 5, using a calculator to divide the resulting number by 99:

95,058 is not divisible by 99.

95,148 is not divisible by 99.

95,238 is divisible by 99.

Thus, the missing digits are 2 and 3 and the number is 95,238.

Exercise Set 3.2

1. Since 18 is even, we know that 2 is a factor. Since the sum of the digits is 9 and 9 is divisible by both 3 and 9, we know that 3 and 9 are both factors. Also, since 2 and 3 are factors, 6 is a factor as well. We write a list of factorizations.

$$18 = 1 \cdot 18 \qquad 18 = 3 \cdot 6$$
$$18 = 2 \cdot 9$$

Factors: 1, 2, 3, 6, 9, 18

3. Since 54 is even, we know that 2 is a factor. Since the sum of the digits is 9 and 9 is divisible by both 3 and 9, we know that 3 and 9 are both factors. Also, since 2 and 3 are factors, 6 is a factor as well. We write a list of factorizations.

$54 = 1 \cdot 54$ $54 = 3 \cdot 18$
$54 = 2 \cdot 27$ $54 = 6 \cdot 9$

Factors: 1, 2, 3, 6, 9, 18, 27, 54

5. Since 9 is divisible by 3, we know that 3 is a factor. We write a list of factorizations.

$9 = 1 \cdot 9$
$9 = 3 \cdot 3$

Factors: 1, 3, 9

7. The number 13 is prime. It has only 1 and 13 as factors.

9. The number 17 is prime. It has only the factors 1 and 17.

11. The number 22 has factors 1, 2, 11, and 22. Since it has at least one factor other than itself and 1, it is composite.

13. The number 48 has factors 1, 2, 3, 4, 6, 8, 12, 16, 24, and 48. Since it has at least one factor other than itself and 1, it is composite.

15. The number 53 is prime. It has only the factors 1 and 53.

17. 1 is neither prime nor composite.

19. The number 81 has factors 1, 3, 9, 27, and 81.

Since it has at least one factor other than itself and 1, it is composite.

21. The number 47 is prime. It has only the factors 1 and 47.

23. The number 29 is prime. It has only the factors 1 and 29.

25.
$$\begin{array}{r} 3 \\ 3\,\overline{\smash{)}\,9} \\ 3\,\overline{\smash{)}\,27} \end{array}$$ ←   3 is prime.

$27 = 3 \cdot 3 \cdot 3$

27.
$$\begin{array}{r} 7 \\ 2\,\overline{\smash{)}\,14} \end{array}$$ ←   7 is prime.

$14 = 2 \cdot 7$

29.
$$\begin{array}{r} 5 \\ 2\,\overline{\smash{)}\,10} \\ 2\,\overline{\smash{)}\,20} \\ 2\,\overline{\smash{)}\,40} \\ 2\,\overline{\smash{)}\,80} \end{array}$$ ←   5 is prime.

$80 = 2 \cdot 2 \cdot 2 \cdot 2 \cdot 5$

We can also use a factor tree.

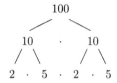

31.
$$\begin{array}{r} 5 \\ 5\,\overline{\smash{)}\,25} \end{array}$$ ←   5 is prime. (25 is not divisible by 2 or 3. We move to 5.)

$25 = 5 \cdot 5$

33.
$$\begin{array}{r} 31 \\ 2\,\overline{\smash{)}\,62} \end{array}$$ ←   31 is prime.

$62 = 2 \cdot 31$

35.
$$\begin{array}{r} 5 \\ 5\,\overline{\smash{)}\,25} \\ 2\,\overline{\smash{)}\,50} \\ 2\,\overline{\smash{)}\,100} \end{array}$$ ←   5 is prime.

$100 = 2 \cdot 2 \cdot 5 \cdot 5$

We can also use a factor tree.

```
          100
         /    \
       10   ·   10
      / \      / \
     2 · 5 · 2 · 5
```

37.
$$\begin{array}{r} 13 \\ 11\,\overline{\smash{)}\,143} \end{array}$$ ←   13 is prime. (143 is not divisible by 2, 3, 5, or 7. We move to 11.)

$143 = 11 \cdot 13$

39.
$$\begin{array}{r} 11 \\ 11\,\overline{\smash{)}\,121} \end{array}$$ ←   11 is prime. (121 is not divisible by 2, 3, 5, or 7. We move to 11.)

$121 = 11 \cdot 11$

41.
$$\begin{array}{r} 13 \\ 7\,\overline{\smash{)}\,91} \\ 3\,\overline{\smash{)}\,273} \end{array}$$ ←   13 is prime. (273 is not divisible by 2. We move to 3.)

$273 = 3 \cdot 7 \cdot 13$

43.
$$\begin{array}{r} 7 \\ 5\,\overline{\smash{)}\,35} \\ 5\,\overline{\smash{)}\,175} \end{array}$$ ←   7 is prime. (175 is not divisible by 2 or 3. We move to 5.)

$175 = 5 \cdot 5 \cdot 7$

45.
$$\begin{array}{r} 19 \\ 11\,\overline{\smash{)}\,209} \end{array}$$ ←   19 is prime. (209 is not divisible by 2, 3, 5, or 7. We move to 11.)

$209 = 11 \cdot 19$

47.
$$\begin{array}{r} 43 \\ 2\,\overline{\smash{)}\,86} \end{array}$$ ←   43 is prime.

$86 = 2 \cdot 43$

49.
$$\begin{array}{r} 31 \\ 7\,\overline{\smash{)}\,217} \end{array}$$ ←   31 is prime. (217 is not divisible by 2, 3, or 5. We move to 7.)

$217 = 7 \cdot 31$

Chapter 3: Fraction Notation: Multiplication and Division

64

51.

$$\begin{array}{r} 7 \quad \leftarrow \ 7 \text{ is prime.} \\ 5\,\overline{)\,3\,5} \\ 5\,\overline{)\,1\,7\,5} \\ 2\,\overline{)\,8\,7\,5} \\ 2\,\overline{)\,1\,7\,5\,0} \\ 2\,\overline{)\,3\,5\,0\,0} \\ 2\,\overline{)\,7\,0\,0\,0} \end{array}$$

$7000 = 2 \cdot 2 \cdot 2 \cdot 5 \cdot 5 \cdot 5 \cdot 7$

53.

$$\begin{array}{r} 1\,7 \quad \leftarrow \ 17 \text{ is prime.} \\ 1\,1\,\overline{)\,1\,8\,7} \\ 3\,\overline{)\,5\,6\,1} \\ 2\,\overline{)\,1\,1\,2\,2} \end{array}$$

$1122 = 2 \cdot 3 \cdot 11 \cdot 17$

55. Since 100 is even we know that 2 is a factor. Using other tests for divisibility, we determine that 5 and 10 are also factors. We write a list of factorizations.

$100 = 1 \cdot 100 \qquad 100 = 5 \cdot 20$
$100 = 2 \cdot 50 \qquad 100 = 10 \cdot 10$
$100 = 4 \cdot 25$

Factors: 1, 2, 4, 5, 10, 20, 25, 50, 100

57. Using tests for divisibility we determine that 5 is a factor. We write a list of factorizations.

$385 = 1 \cdot 385 \qquad 385 = 7 \cdot 55$
$385 = 5 \cdot 77 \qquad 385 = 11 \cdot 35$

Factors: 1, 5, 7, 11, 35, 55, 77, 385

59. Using tests for divisibility we determine that 3 and 9 are factors. We write a list of factorizations.

$81 = 1 \cdot 81 \qquad 81 = 9 \cdot 9$
$81 = 3 \cdot 27$

Factors: 1, 3, 9, 27, 81

61. Using tests for divisibility we determine that 3, 5, and 9 are factors. We write a list of factorizations.

$225 = 1 \cdot 225 \qquad 225 = 9 \cdot 25$
$225 = 3 \cdot 75 \qquad 225 = 15 \cdot 15$
$225 = 5 \cdot 45$

Factors: 1, 3, 5, 9, 15, 25, 45, 75, 225

63. Discussion and Writing Exercise

65. $-2 \cdot 13 = -26$ (The signs are different, so the answer is negative.)

67. $-17 + 25$ The absolute values are 17 and 25. The difference is 8. The positive number has the larger absolute value, so the answer is positive. $-17 + 25 = 8$

69. $53 \div 53 = 1$

71. $0 \div 22 = 0$ (0 divided by a nonzero number is 0.)

73. $-42 \div 1 = -42$ (The signs are different, so the answer is negative.)

75. Discussion and Writing Exercise

77. Discussion and Writing Exercise

79. Using a calculator to perform successive divisions by prime numbers, we find that $102,971 = 11 \cdot 11 \cdot 23 \cdot 37$.

81. Using a calculator to perform successive divisions by prime numbers, we find that $168,840 = 2 \cdot 2 \cdot 2 \cdot 3 \cdot 3 \cdot 5 \cdot 7 \cdot 67$

83. Answers may vary. One arrangement is a 3-dimensional rectangular array consisting of 2 tiers of 12 objects each where each tier consists of a rectangular array of 4 rows with 3 objects each.

85. The factors of 63 whose sum is 16 are 7 and 9.

The factors of 36 whose sum is 20 are 2 and 18.

The factors of 72 whose sum is 38 are 2 and 36.

The factors of 140 whose sum is 24 are 10 and 14.

The factors of 96 whose sum is 20 are 8 and 12.

The factors of 48 whose sum is 14 are 6 and 8.

The factors of 168 whose sum is 29 are 8 and 21.

The factors of 110 whose sum is 21 are 10 and 11.

The factors of 90 whose sum is 19 are 9 and 10.

The factors of 432 whose sum is 42 are 18 and 24.

The factors of 63 whose sum is 24 are 3 and 21.

Exercise Set 3.3

1. The top number is the numerator, and the bottom number is the denominator.

$$\frac{3}{4} \quad \begin{array}{l} \leftarrow \text{Numerator} \\ \leftarrow \text{Denominator} \end{array}$$

3. $\dfrac{7}{-9} \quad \begin{array}{l} \leftarrow \text{Numerator} \\ \leftarrow \text{Denominator} \end{array}$

5. $\dfrac{2x}{3z} \quad \begin{array}{l} \leftarrow \text{Numerator} \\ \leftarrow \text{Denominator} \end{array}$

7. The dollar is divided into 4 parts of the same size, and 2 of them are shaded. This is $2 \cdot \dfrac{1}{4}$ or $\dfrac{2}{4}$. Thus, $\dfrac{2}{4}$ (two-fourths) of the dollar is shaded.

9. The yard is divided into 8 parts of the same size, and 1 of them is shaded. Thus, $\dfrac{1}{8}$ (one-eighth) of the yard is shaded.

11. The window is divided into 9 parts of the same size, and 4 of them are shaded. Thus, $\dfrac{4}{9}$ (four-ninths) of the window is shaded.

13. The acre is divided into 4 parts of the same size, and 3 of them are shaded. This is $3 \cdot \dfrac{1}{4}$ or $\dfrac{3}{4}$ of the acre.

15. The pie is divided into 8 equal parts. The unit is $\dfrac{1}{8}$. The denominator is 8. We have 4 parts shaded. This tells us that the numerator is 4. Thus, $\dfrac{4}{8}$ is shaded.

17. The square mile is divided into 12 equal parts. The unit is $\dfrac{1}{12}$. The denominator is 12. We have 6 parts shaded. This tells us that the numerator is 6. Thus, $\dfrac{6}{12}$ is shaded.

19. Each inch on the ruler is divided into 16 equal parts. The shading extends to the 12th mark, so $\frac{12}{16}$ is shaded.

21. Each inch on the ruler is divided into 16 equal parts. The shading extends to the 38th mark, so $\frac{7}{16}$ is shaded.

23. There are 8 circles, and 5 are shaded. Thus, $\frac{5}{8}$ of the circles are shaded.

25. There are 7 objects in the set, and 4 of the objects are shaded. Thus, $\frac{4}{7}$ of the set is shaded.

27. The gas gauge is divided into 8 equal parts.

   a) The needle is 2 marks from the E (empty) mark, so the amount of gas in the tank is $\frac{2}{8}$ of a full tank.

   b) The needle is 6 marks from the F (full) mark, so $\frac{6}{8}$ of a full tank of gas has been burned.

29. The gas gauge is divided into 8 equal parts.

   a) The needle is 3 marks from the E (empty) mark, so the amount of gas in the tank is $\frac{3}{8}$ of a full tank.

   b) The needle is 5 marks from the F (full) mark, so $\frac{5}{8}$ of a full tank of gas has been burned.

31. We have 2 gold bars, each divided into 8 parts. We take 9 of those parts. This is $9 \cdot \frac{1}{8}$, or $\frac{9}{8}$. Thus, $\frac{9}{8}$ of a gold bar is shaded.

33. We have 3 feet, each divided into 6 equal parts. We take 7 of those parts. This is $7 \cdot \frac{1}{6}$, or $\frac{7}{6}$. Thus, $\frac{7}{6}$ of a foot is shaded.

35. We have 2 spools, each divided into 5 parts. We take 7 of those parts. This is $7 \cdot \frac{1}{5}$, or $\frac{7}{5}$. Thus, $\frac{7}{5}$ of a spool is shaded.

37. a) The ratio is $\frac{390}{13}$.

   b) The ratio is $\frac{13}{390}$.

39. The ratio is $\frac{850}{1000}$.

41. a) There are 7 people in the set and 3 are women, so the desired ratio is $\frac{3}{7}$.

   b) There are 3 women and 4 men, so the ratio of women to men is $\frac{3}{4}$.

   c) There are 7 people in the set and 4 are men, so the desired ratio is $\frac{4}{7}$.

   d) There are 4 men and 3 women, so the ratio of men to women is $\frac{4}{3}$.

43. a) In Orlando there are 35 police officers per 10,000 residents, so the ratio is $\frac{35}{10,000}$.

   b) In New York there are 50 police officers per 10,000 residents, so the ratio is $\frac{50}{10,000}$.

   c) In Detroit there are 44 police officers per 10,000 residents, so the ratio is $\frac{44}{10,000}$.

   d) In Washington there are 63 police officers per 10,000 residents, so the ratio is $\frac{63}{10,000}$.

   e) In St. Louis there are 43 police officers per 10,000 residents, so the ratio is $\frac{43}{10,000}$.

   f) In Santa Fe there are 21 police officers per 10,000 residents, so the ratio is $\frac{21}{10,000}$.

45. Remember: $\frac{0}{n} = 0$, for any integer n that is not 0.

$$\frac{0}{17} = 0$$

Think of dividing an object into 17 parts and taking none of them. We get 0.

47. Remember: $\frac{n}{1} = n$.

$$\frac{15}{1} = 15$$

Think of taking 15 objects and dividing them into 1 part. (We do not divide them.) We have 15 objects.

49. Remember: $\frac{n}{n} = 1$, for any integer n that is not 0.

$$\frac{20}{20} = 1$$

If we divide an object into 20 parts and take 20 of them, we get all of the object (1 whole object).

51. Remember: $\frac{n}{n} = 1$, for any integer n that is not 0.

$$\frac{-14}{-14} = 1$$

53. Remember: $\frac{0}{n} = 0$, for any integer n that is not 0.

$$\frac{0}{-234} = 0$$

55. Remember: $\frac{n}{n} = 1$, for any integer n that is not 0.

$$\frac{3n}{3n} = 1$$

57. Remember: $\frac{n}{n} = 1$, for any integer n that is not 0.

$$\frac{9x}{9x} = 1$$

59. Remember: $\frac{n}{1} = n$

$$\frac{-63}{1} = -63$$

61. Remember: $\dfrac{0}{n} = 0$, for any integer n that is not 0.

$$\frac{0}{2a} = 0$$

63. Remember: $\dfrac{n}{0}$ is not defined.

$$\frac{52}{0} \text{ is undefined.}$$

65. Remember: $\dfrac{n}{1} = n$

$$\frac{7n}{1} = 7n$$

67. $\dfrac{6}{7-7} = \dfrac{6}{0}$

Remember: $\dfrac{n}{0}$ is not defined. Thus, $\dfrac{6}{7-7}$ is undefined.

69. Discussion and Writing Exercise

71. $-7(30) = -210$
(The signs are different, so the answer is negative.)

73. $(-71)(-12)0 = -71 \cdot 0 = 0$
(We might have observed at the outset that the answer is 0 since one of the factors is 0.)

75. *Familiarize.* Let $c =$ the number of excess calories in the Burger King meal.

Translate. We can think of this as a "how much more" situation.

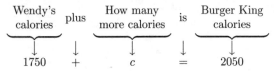

Wendy's calories	plus	How many more calories	is	Burger King calories
↓	↓	↓	↓	↓
1750	+	c	=	2050

Solve. We subtract 1750 on both sides of the equation.

$$1750 + c = 2050$$
$$1750 + c - 1750 = 2050 - 1750$$
$$c = 300$$

Check. Since $1750 + 300 = 2050$, the answer checks.

State. The Burger King meal has 300 more calories than the Wendy's meal.

77. Discussion and Writing Exercise

79. $365 = 52 \cdot 7 + 1$, so in one year there are 52 full weeks plus one additional day. Since 2006 began on a Sunday, the additional day is not a Monday. (It is a Sunday.) Thus, of the 365 days in 2006, 52 were Mondays, so $\dfrac{52}{365}$ were Mondays.

81. The surface of the earth is divided into $3 + 1$, or 4 parts. Three of them are taken up by water, so $\dfrac{3}{4}$ is water. One of them is land, so $\dfrac{1}{4}$ is land.

83. Since the denominators are all the same, the numerators tell us the relative sizes of the fractions. The smallest fraction is the one with the smallest numerator and so on. Accordingly, we label the smallest sector "No working

TV." Then the next smallest sector is labeled "One TV," the next smallest "Two TVs," and the largest sector is labeled "Three or more TVs."

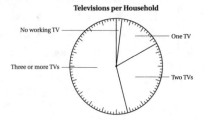

Televisions per Household

85. We can think of the object as being divided into 6 sections, each the size of one of the sections shaded. Since 2 sections are shaded, $\dfrac{2}{6}$ of the object is shaded. We could also express this as $\dfrac{1}{3}$.

87. We can think of the object as being divided into 16 sections, each the size of one of the sections shaded. Since a portion of the object that is the equivalent of 6 sections is shaded, $\dfrac{6}{16}$ of the object is shaded. We could also express this as $\dfrac{3}{8}$.

Exercise Set 3.4

1. $3 \cdot \dfrac{1}{8} = \dfrac{3 \cdot 1}{8} = \dfrac{3}{8}$

3. $(-5) \times \dfrac{1}{6} = \dfrac{-5 \times 1}{6} = \dfrac{-5}{6}$, or $-\dfrac{5}{6}$

5. $\dfrac{2}{3} \cdot 7 = \dfrac{2 \cdot 7}{3} = \dfrac{14}{3}$

7. $(-1)\dfrac{7}{9} = \dfrac{(-1)7}{9} = \dfrac{-7}{9}$, or $-\dfrac{7}{9}$

9. $\dfrac{5}{6} \cdot x = \dfrac{5 \cdot x}{6} = \dfrac{5x}{6}$

11. $\dfrac{2}{5}(-3) = \dfrac{2(-3)}{5} = \dfrac{-6}{5}$, or $-\dfrac{6}{5}$

13. $a \cdot \dfrac{2}{7} = \dfrac{a \cdot 2}{7} = \dfrac{2a}{7}$

15. $17 \times \dfrac{m}{6} = \dfrac{17 \times m}{6} = \dfrac{17m}{6}$

17. $-3 \cdot \dfrac{-2}{5} = \dfrac{-3}{1} \cdot \dfrac{-2}{5} = \dfrac{-3(-2)}{1 \cdot 5} = \dfrac{6}{5}$

19. $-\dfrac{2}{7}(-x) = \dfrac{-2}{7} \cdot \dfrac{-x}{1} = \dfrac{-2(-x)}{7 \cdot 1} = \dfrac{2x}{7}$

21. $\dfrac{1}{3} \cdot \dfrac{1}{5} = \dfrac{1 \cdot 1}{3 \cdot 5} = \dfrac{1}{15}$

23. $\left(-\dfrac{1}{4}\right) \times \dfrac{1}{10} = -\dfrac{1 \times 1}{4 \times 10} = -\dfrac{1}{40}$, or $\dfrac{-1}{40}$

25. $\dfrac{2}{3} \times \dfrac{1}{5} = \dfrac{2 \times 1}{3 \times 5} = \dfrac{2}{15}$

27. $\dfrac{2}{y} \cdot \dfrac{x}{9} = \dfrac{2 \cdot x}{y \cdot 9} = \dfrac{2x}{9y}$

29. $\left(-\dfrac{3}{4}\right)\left(-\dfrac{3}{4}\right) = \dfrac{(-3)(-3)}{4 \cdot 4} = \dfrac{9}{16}$

31. $\dfrac{2}{3} \cdot \dfrac{7}{13} = \dfrac{2 \cdot 7}{3 \cdot 13} = \dfrac{14}{39}$

33. $\dfrac{1}{10}\left(\dfrac{-3}{5}\right) = \dfrac{1(-3)}{10 \cdot 5} = \dfrac{-3}{50}$, or $-\dfrac{3}{50}$

35. $\dfrac{7}{8} \cdot \dfrac{a}{8} = \dfrac{7 \cdot a}{8 \cdot 8} = \dfrac{7a}{64}$

37. $\dfrac{1}{y} \cdot \dfrac{1}{100} = \dfrac{1 \cdot 1}{y \cdot 100} = \dfrac{1}{100y}$

39. $\dfrac{-14}{15} \cdot \dfrac{13}{19} = \dfrac{-14 \cdot 13}{15 \cdot 19} = \dfrac{-182}{285}$, or $-\dfrac{182}{285}$

41. *Familiarize*. We draw a picture. We let $h =$ the amount of sliced almonds needed.

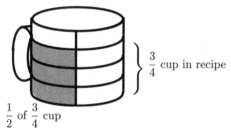

$\dfrac{3}{4}$ cup in recipe

$\dfrac{1}{2}$ of $\dfrac{3}{4}$ cup

Translate. We are finding $\dfrac{1}{2}$ of $\dfrac{3}{4}$, so the multiplication sentence $\dfrac{1}{2} \cdot \dfrac{3}{4} = h$ corresponds to the situation.

Solve. We multiply:

$$\dfrac{1}{2} \cdot \dfrac{3}{4} = \dfrac{1 \cdot 3}{2 \cdot 4} = \dfrac{3}{8}$$

Check. We repeat the calculation. The answer checks.

State. $\dfrac{3}{8}$ cup of sliced almonds is needed.

43. *Familiarize*. Recall that area is length times width. We draw a picture. We will let A = the area of the table top.

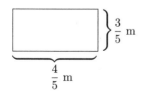

$\dfrac{3}{5}$ m

$\dfrac{4}{5}$ m

Translate. Then we translate.

Area is length times width

$\downarrow$ $\quad$ $\downarrow$ $\quad$ $\downarrow$ $\quad$ $\downarrow$ $\quad$ $\downarrow$

A $\quad = \quad \dfrac{4}{5} \quad \times \quad \dfrac{3}{5}$

Solve. The sentence tells us what to do. We multiply.

$$\dfrac{4}{5} \times \dfrac{3}{5} = \dfrac{4 \times 3}{5 \times 5} = \dfrac{12}{25}$$

Check. We repeat the calculation. The answer checks.

State. The area is $\dfrac{12}{25}$ m^2.

45. *Familiarize*. We know that $\dfrac{4}{5}$ of all municipal waste is dumped in landfills. We also know that $\dfrac{1}{10}$ of the waste in landfills is landscape trimmings. We let $y =$ the fractional part of municipal waste that is landscape trimmings that are landfilled.

Translate. The multiplication sentence $\dfrac{1}{10} \cdot \dfrac{4}{5} = y$ corresponds to this situation.

Solve. We carry out the multiplication.

$$\dfrac{1}{10} \cdot \dfrac{4}{5} = \dfrac{1 \cdot 4}{10 \cdot 5} = \dfrac{4}{50}$$

Check. We repeat the calculation. The answer checks.

State. $\dfrac{4}{50}$ of municipal waste is landscape trimmings that are landfilled.

47. *Familiarize*. A picture of the situation appears in the text. Let $f =$ the fraction of the floor that has been tiled.

Translate. The multiplication sentence $\dfrac{3}{5} \cdot \dfrac{3}{4} = f$ corresponds to the situation.

Solve. We multiply.

$$\dfrac{3}{5} \cdot \dfrac{3}{4} = \dfrac{3 \cdot 3}{5 \cdot 4} = \dfrac{9}{20}$$

Check. We repeat the calculation. The answer checks.

State. $\dfrac{9}{20}$ of the floor has been tiled.

49. *Familiarize*. We know that 1 of 35 high school basketball players plays college basketball. That is, $\dfrac{1}{35}$ of high school basketball players play college basketball. In addition, we know that 1 of 75 college players plays professional basketball. That is, $\dfrac{1}{75}$ of college basketball players play professional basketball or $\dfrac{1}{75}$ of the $\dfrac{1}{35}$ of high school players who play college basketball also play professionally. We let $f =$ the fractional part of high school basketball players that plays professional basketball.

Translate. The multiplication sentence $f = \dfrac{1}{75} \cdot \dfrac{1}{35}$ corresponds to this situation.

Solve. We carry out the multiplication.

$$\dfrac{1}{75} \cdot \dfrac{1}{35} = \dfrac{1 \cdot 1}{75 \cdot 35} = \dfrac{1}{2625}$$

Check. We repeat the calculation. The answer checks.

State. The fractional part of high school players that plays professional basketball is $\dfrac{1}{2625}$.

51. Discussion and Writing Exercise

53.
$$\begin{array}{r} 9 \\ 2\,0\,\overline{)1\,8\,0} \\ \underline{1\,8\,0} \\ 0 \end{array}$$

The answer is 9.

55. $450 \div (-9) = -50$

(The signs are different, so the answer is negative.)

57. $\dfrac{-35}{5} = -7$

(The signs are different, so the answer is negative.)

59. $\dfrac{-65}{-5} = 13$

(The signs are the same, so the answer is positive.)

61. $4,6\,7\,\boxed{8}\,,9\,5\,2$

The digit 8 means 8 thousands.

63. $7\,1\,4\,\boxed{8}$

The digit 8 means 8 ones.

65. Discussion and Writing Exercise

67. **Familiarize**. Let $t =$ the number of gallons of two-cycle oil in a freshly filled chainsaw.

Translate. The multiplication sentence $\dfrac{1}{16} \cdot \dfrac{1}{5} = t$ corresponds to this situation.

Solve. We carry out the multiplication.
$$\frac{1}{16} \cdot \frac{1}{5} = \frac{1 \cdot 1}{16 \cdot 5} = \frac{1}{80}$$

Check. We repeat the calculation. The answer checks.

State. There is $\dfrac{1}{80}$ gal of two-cycle oil in a freshly filled chainsaw.

69. Use a calculator.
$$\left(-\frac{57}{61}\right)^3 = -\frac{185,193}{226,981}, \text{ or } \frac{-185,193}{226,981}$$

71. $\left(-\dfrac{1}{2}\right)^5\left(\dfrac{3}{5}\right) = -\dfrac{1}{32}\left(\dfrac{3}{5}\right) = -\dfrac{3}{160}, \text{ or } \dfrac{-3}{160}$

73. $-\dfrac{2}{3}xy = -\dfrac{2}{3} \cdot \dfrac{2}{5}\left(-\dfrac{1}{7}\right)$

$= -\dfrac{4}{15}\left(-\dfrac{1}{7}\right)$

$= \dfrac{4}{105}$

75. Use a calculator.
$$-\frac{4}{7}ab = -\frac{4}{7} \cdot \frac{93}{107} \cdot \frac{13}{41} = -\frac{4836}{30,709}$$

Exercise Set 3.5

1. Since $10 \div 2 = 5$, we multiply by $\dfrac{5}{5}$.
$$\frac{1}{2} = \frac{1}{2} \cdot \frac{5}{5} = \frac{1 \cdot 5}{2 \cdot 5} = \frac{5}{10}$$

3. Since $-48 \div 4 = -12$, we multiply by $\dfrac{-12}{-12}$.
$$\frac{3}{4} = \frac{3}{4}\left(\frac{-12}{-12}\right) = \frac{3(-12)}{4(-12)} == \frac{-36}{-48}$$

5. Since $50 \div 10 = 5$, we multiply by $\dfrac{5}{5}$.
$$\frac{7}{10} = \frac{7}{10} \cdot \frac{5}{5} = \frac{7 \cdot 5}{10 \cdot 5} = \frac{35}{50}$$

7. Since $5t \div 5 = t$, we multiply by $\dfrac{t}{t}$.
$$\frac{11}{5} = \frac{11}{5} \cdot \frac{t}{t} = \frac{11 \cdot t}{5 \cdot t} = \frac{11t}{5t}$$

9. Since $48 \div 12 = 4$, we multiply by $\dfrac{4}{4}$.
$$\frac{5}{12} = \frac{5}{12} \cdot \frac{4}{4} = \frac{5 \cdot 4}{12 \cdot 4} = \frac{20}{48}$$

11. Since $54 \div 18 = 3$, we multiply by $\dfrac{3}{3}$.
$$-\frac{17}{18} = -\frac{17}{18} \cdot \frac{3}{3} = -\frac{17 \cdot 3}{18 \cdot 3} = -\frac{51}{54}$$

13. Since $-40 \div -8 = 5$, we multiply by $\dfrac{5}{5}$.
$$\frac{3}{-8} = \frac{3}{-8} \cdot \frac{5}{5} = \frac{3 \cdot 5}{-8 \cdot 5} = \frac{15}{-40}$$

15. Since $132 \div 22 = 6$, we multiply by $\dfrac{6}{6}$.
$$\frac{-7}{22} = \frac{-7}{22} \cdot \frac{6}{6} = \frac{-7 \cdot 6}{22 \cdot 6} = \frac{-42}{132}$$

17. Since $8x \div 8 = x$, we multiply by $\dfrac{x}{x}$.
$$\frac{5}{8} = \frac{5}{8} \cdot \frac{x}{x} = \frac{5x}{8x}$$

19. Since $7a \div 7 = a$, we multiply by $\dfrac{a}{a}$.
$$\frac{10}{7} = \frac{10}{7} \cdot \frac{a}{a} = \frac{10a}{7a}$$

21. Since $9ab \div 9 = ab$, we multiply by $\dfrac{ab}{ab}$.
$$\frac{4}{9} \cdot \frac{ab}{ab} = \frac{4ab}{9ab}$$

23. Since $27b \div 9 = 3b$, we multiply by $\dfrac{3b}{3b}$.
$$\frac{4}{9} = \frac{4}{9} \cdot \frac{3b}{3b} = \frac{12b}{27b}$$

25. $\dfrac{2}{4} = \dfrac{1\cdot 2}{2\cdot 2}$ ⟵ Factor the numerator
⟵ Factor the denominator

$= \dfrac{1}{2}\cdot\dfrac{2}{2}$ ⟵ Factor the fraction

$= \dfrac{1}{2}\cdot 1$ ⟵ $\dfrac{2}{2}=1$

$= \dfrac{1}{2}$ ⟵ Removing a factor of 1

27. $-\dfrac{6}{9} = -\dfrac{2\cdot 3}{3\cdot 3}$ ⟵ Factor the numerator
⟵ Factor the denominator

$= -\dfrac{2}{3}\cdot\dfrac{3}{3}$ ⟵ Factor the fraction

$= -\dfrac{2}{3}\cdot 1$ ⟵ $\dfrac{3}{3}=1$

$= -\dfrac{2}{3}$ ⟵ Removing a factor of 1

29. $\dfrac{10}{25} = \dfrac{2\cdot 5}{5\cdot 5}$ ⟵ Factor the numerator
⟵ Factor the denominator

$= \dfrac{2}{5}\cdot\dfrac{5}{5}$ ⟵ Factor the fraction

$= \dfrac{2}{5}\cdot 1$ ⟵ $\dfrac{5}{5}=1$

$= \dfrac{2}{5}$ ⟵ Removing a factor of 1

31. $\dfrac{27}{-3} = \dfrac{9\cdot 3}{-1\cdot 3} = \dfrac{9}{-1}\cdot\dfrac{3}{3} = \dfrac{9}{-1}\cdot 1 = -9$

33. $\dfrac{27}{36} = \dfrac{9\cdot 3}{9\cdot 4} = \dfrac{9}{9}\cdot\dfrac{3}{4} = 1\cdot\dfrac{3}{4} = \dfrac{3}{4}$

35. $-\dfrac{24}{14} = -\dfrac{12\cdot 2}{7\cdot 2} = -\dfrac{12}{7}\cdot\dfrac{2}{2} = -\dfrac{12}{7}$

37. $\dfrac{16n}{48n} = \dfrac{1\cdot 16n}{3\cdot 16n} = \dfrac{1}{3}\cdot\dfrac{16n}{16n} = \dfrac{1}{3}$

39. $\dfrac{-17}{51} = \dfrac{-1\cdot 17}{3\cdot 17} = \dfrac{-1}{3}\cdot\dfrac{17}{17} = \dfrac{-1}{3}$

41. $\dfrac{420}{480} = \dfrac{2\cdot 2\cdot 3\cdot 5\cdot 7}{2\cdot 2\cdot 2\cdot 2\cdot 2\cdot 3\cdot 5}$

$= \dfrac{2}{2}\cdot\dfrac{2}{2}\cdot\dfrac{3}{3}\cdot\dfrac{5}{5}\cdot\dfrac{7}{2\cdot 2\cdot 2}$

$= \dfrac{7}{2\cdot 2\cdot 2}$

$= \dfrac{7}{8}$

43. $\dfrac{153}{136} = \dfrac{3\cdot 3\cdot 17}{2\cdot 2\cdot 2\cdot 17}$

$= \dfrac{3\cdot 3}{2\cdot 2\cdot 2}\cdot\dfrac{17}{17}$

$= \dfrac{3\cdot 3}{2\cdot 2\cdot 2}$

$= \dfrac{9}{8}$

45. $\dfrac{132}{143} = \dfrac{11\cdot 12}{11\cdot 13} = \dfrac{11}{11}\cdot\dfrac{12}{13} = \dfrac{12}{13}$

47. $\dfrac{221}{247} = \dfrac{13\cdot 17}{13\cdot 19} = \dfrac{13}{13}\cdot\dfrac{17}{19} = \dfrac{17}{19}$

49. $\dfrac{3ab}{8ab} = \dfrac{3\cdot a\cdot b}{8\cdot a\cdot b} = \dfrac{3}{8}\cdot\dfrac{a}{a}\cdot\dfrac{b}{b} = \dfrac{3}{8}$

51. $\dfrac{9xy}{6x} = \dfrac{3\cdot 3\cdot x\cdot y}{2\cdot 3\cdot x} = \dfrac{3\cdot y}{2}\cdot\dfrac{3}{3}\cdot\dfrac{x}{x} = \dfrac{3y}{2}$

53. $\dfrac{-18a}{20ab} = \dfrac{-9\cdot 2\cdot a}{10\cdot 2\cdot a\cdot b} = \dfrac{-9}{10\cdot b}\cdot\dfrac{2}{2}\cdot\dfrac{a}{a} = \dfrac{-9}{10b}$

55. We multiply these two numbers: We multiply these two numbers:

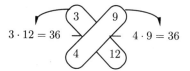

$3\cdot 12 = 36$ $4\cdot 9 = 36$

Since $36 = 36$, $\dfrac{3}{4} = \dfrac{9}{12}$.

57. We multiply these two numbers: We multiply these two numbers:

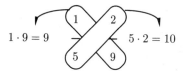

$1\cdot 9 = 9$ $5\cdot 2 = 10$

Since $9 \neq 10$, $\dfrac{1}{5}$ and $\dfrac{2}{9}$ do not name the same number. Thus, $\dfrac{1}{5} \neq \dfrac{2}{9}$.

59. We multiply these two numbers: We multiply these two numbers:

$3\cdot 16 = 48$ $8\cdot 6 = 48$

Since $48 = 48$, $\dfrac{3}{8} = \dfrac{6}{16}$.

61. We multiply these two numbers: We multiply these two numbers:

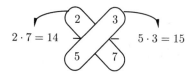

$2\cdot 7 = 14$ $5\cdot 3 = 15$

Since $14 \neq 15$, $\dfrac{2}{5}$ and $\dfrac{3}{7}$ do not name the same number. Thus, $\dfrac{2}{5} \neq \dfrac{3}{7}$.

63.

We multiply these two numbers: We multiply these two numbers:

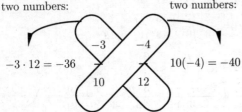

$-3 \cdot 12 = -36$ $10(-4) = -40$

Since $-36 \neq -40$, $\dfrac{-3}{10}$ and $\dfrac{-4}{12}$ do not name the same number. Thus, $\dfrac{-3}{10} \neq \dfrac{-4}{12}$.

65. We rewrite $-\dfrac{12}{9}$ as $\dfrac{-12}{9}$ and check cross products.

We multiply these two numbers: We multiply these two numbers:

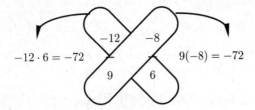

$-12 \cdot 6 = -72$ $9(-8) = -72$

Since $-72 = -72$, $-\dfrac{12}{9} = \dfrac{-8}{6}$.

67. We rewrite $-\dfrac{17}{7}$ as $\dfrac{17}{-7}$ and check cross products.

We multiply these two numbers: We multiply these two numbers:

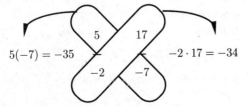

$5(-7) = -35$ $-2 \cdot 17 = -34$

Since $-35 \neq -34$, $\dfrac{5}{-2}$ and $\dfrac{17}{-7}$ do not name the same number. Thus, $\dfrac{5}{-2} \neq \dfrac{17}{-7}$, or $\dfrac{5}{-2} \neq -\dfrac{17}{7}$.

69.

We multiply these two numbers: We multiply these two numbers:

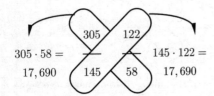

$305 \cdot 58 =$ $145 \cdot 122 =$
$17,690$ $17,690$

Since $17,690 = 17,690$, $\dfrac{305}{145} = \dfrac{122}{58}$.

71. Discussion and Writing Exercise

73. *Familiarize.* We make a drawing. We let A = the area.

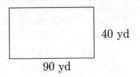

40 yd

90 yd

Translate. Using the formula for area, we have
$$A = l \cdot w = 90 \cdot 40.$$
Solve. We carry out the multiplication.
$$\begin{array}{r} 4\,0 \\ \times \quad 9\,0 \\ \hline 3\,6\,0\,0 \end{array}$$
Thus, $A = 3600$.

Check. We repeat the calculation. The answer checks.

State. The area is 3600 yd².

75. $-12(-5) = 60$

(The signs are the same, so the product is positive.)

77. $-9 \cdot 7 = -63$

(The signs are different, so the product is negative.)

79. $\quad 30 \cdot x = 150$

$\quad \dfrac{30 \cdot x}{30} = \dfrac{150}{30}$ Dividing both sides by 30

$\quad\quad\quad x = 5$

The solution is 5.

81. $\quad\quad 5280 = 1760 + t$

$\quad 5280 - 1760 = 1760 + t - 1760$ Subtracting 1760 from both sides

$\quad\quad\quad 3520 = t$

The solution is 3520.

83. Discussion and Writing Exercise

85. $\dfrac{391}{667} = \dfrac{17 \cdot 23}{23 \cdot 29} = \dfrac{17}{29} \cdot \dfrac{23}{23} = \dfrac{17}{29}$

87. $-\dfrac{1073x}{555y} = -\dfrac{29 \cdot 37 \cdot x}{15 \cdot 37 \cdot y} = -\dfrac{29 \cdot x}{15 \cdot y} \cdot \dfrac{37}{37} = -\dfrac{29x}{15y}$

89. $\dfrac{4247}{4619} = \dfrac{31 \cdot 137}{31 \cdot 149} = \dfrac{31}{31} \cdot \dfrac{137}{149} = \dfrac{137}{149}$

91. The part of the population that is shy is $\dfrac{4}{10}$. We simplify:
$$\dfrac{4}{10} = \dfrac{2 \cdot 2}{5 \cdot 2} = \dfrac{2}{5} \cdot \dfrac{2}{2} = \dfrac{2}{5}$$
Since 4 out of 10 people are shy, then $10 - 4$, or 6, are not shy. The part of the population that is not shy is $\dfrac{6}{10}$. We simplify:
$$\dfrac{6}{10} = \dfrac{3 \cdot 2}{5 \cdot 2} = \dfrac{3}{5} \cdot \dfrac{2}{2} = \dfrac{3}{5}$$

93. Derrek Lee's batting average was $\dfrac{199}{594}$; Michael Young's batting average was $\dfrac{221}{668}$. We test these fractions for equality:

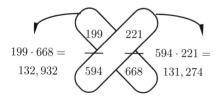

$199 \cdot 668 =$
$132,932$ $594 \cdot 221 =$
$131,274$

Since $132,932 \neq 131,274$, $\dfrac{199}{594}$ and $\dfrac{221}{668}$ do not name the same number. Thus, $\dfrac{199}{594} \neq \dfrac{221}{668}$ and the batting averages are not the same.

Exercise Set 3.6

1. $\dfrac{3}{8} \cdot \dfrac{7}{3} = \dfrac{3 \cdot 7}{8 \cdot 3} = \dfrac{3}{3} \cdot \dfrac{7}{8} = 1 \cdot \dfrac{7}{8} = \dfrac{7}{8}$

3. $\dfrac{7}{8} \cdot \dfrac{-1}{7} = \dfrac{7(-1)}{8 \cdot 7} = \dfrac{7}{7} \cdot \dfrac{-1}{8} = \dfrac{-1}{8}$, or $-\dfrac{1}{8}$

5. $\dfrac{1}{8} \cdot \dfrac{6}{7} = \dfrac{1 \cdot 6}{8 \cdot 7} = \dfrac{1 \cdot 2 \cdot 3}{2 \cdot 4 \cdot 7} = \dfrac{2}{2} \cdot \dfrac{1 \cdot 3}{4 \cdot 7} = \dfrac{3}{28}$

7. $\dfrac{1}{6} \cdot \dfrac{4}{3} = \dfrac{1 \cdot 4}{6 \cdot 3} = \dfrac{1 \cdot 2 \cdot 2}{2 \cdot 3 \cdot 3} = \dfrac{2}{2} \cdot \dfrac{1 \cdot 2}{3 \cdot 3} = \dfrac{2}{9}$

9. $\dfrac{12}{-5} \cdot \dfrac{9}{8} = \dfrac{12 \cdot 9}{-5 \cdot 8} = \dfrac{4 \cdot 3 \cdot 9}{-5 \cdot 2 \cdot 4} = \dfrac{4}{4} \cdot \dfrac{3 \cdot 9}{-5 \cdot 2} = \dfrac{3 \cdot 9}{-5 \cdot 2} =$
$\dfrac{27}{-10}$, or $-\dfrac{27}{10}$

11. $\dfrac{5x}{9} \cdot \dfrac{4}{5} = \dfrac{5x \cdot 4}{9 \cdot 5} = \dfrac{5 \cdot x \cdot 4}{9 \cdot 5} = \dfrac{5}{5} \cdot \dfrac{x \cdot 4}{9} = \dfrac{4x}{9}$

13. $\dfrac{1}{4} \cdot 12 = \dfrac{1 \cdot 12}{4} = \dfrac{12}{4} = \dfrac{4 \cdot 3}{4 \cdot 1} = \dfrac{4}{4} \cdot \dfrac{3}{1} = \dfrac{3}{1} = 3$

15. $21 \cdot \dfrac{1}{3} = \dfrac{21 \cdot 1}{3} = \dfrac{21}{3} = \dfrac{3 \cdot 7}{3 \cdot 1} = \dfrac{3}{3} \cdot \dfrac{7}{1} = \dfrac{7}{1} = 7$

17. $-16\left(-\dfrac{3}{4}\right) = \dfrac{16 \cdot 3}{4} = \dfrac{4 \cdot 4 \cdot 3}{4 \cdot 1} = \dfrac{4}{4} \cdot \dfrac{4 \cdot 3}{1} =$
$\dfrac{4 \cdot 3}{1} = \dfrac{12}{1} = 12$

19. $\dfrac{3}{8} \cdot 8a = \dfrac{3 \cdot 8a}{8} = \dfrac{3 \cdot 8 \cdot a}{8 \cdot 1} = \dfrac{8}{8} \cdot \dfrac{3 \cdot a}{1} = \dfrac{3a}{1} = 3a$

21. $\left(-\dfrac{3}{8}\right)\left(-\dfrac{8}{3}\right) = \dfrac{3 \cdot 8}{8 \cdot 3} = \dfrac{3 \cdot 8}{3 \cdot 8} = 1$

23. $\dfrac{a}{b} \cdot \dfrac{b}{a} = \dfrac{a \cdot b}{b \cdot a} = \dfrac{a \cdot b}{a \cdot b} = 1$

25. $\dfrac{1}{26} \cdot 143a = \dfrac{1 \cdot 143a}{26} = \dfrac{1 \cdot 11 \cdot 13 \cdot a}{2 \cdot 13} =$
$\dfrac{13}{13} \cdot \dfrac{1 \cdot 11a}{2} = \dfrac{11a}{2}$

27. $176\left(\dfrac{1}{-6}\right) = \dfrac{176 \cdot 1}{-6} = \dfrac{2 \cdot 88 \cdot 1}{-3 \cdot 2} = \dfrac{2}{2} \cdot \dfrac{88 \cdot 1}{-3} =$
$\dfrac{88}{-3}$, or $-\dfrac{88}{3}$

29. $-8x \cdot \dfrac{1}{-8x} = \dfrac{8x \cdot 1}{8x \cdot 1} = 1$

31. $\dfrac{2x}{9} \cdot \dfrac{27}{2x} = \dfrac{2x \cdot 27}{9 \cdot 2x} = \dfrac{2x \cdot 9 \cdot 3}{9 \cdot 2x \cdot 1} =$
$\dfrac{2x \cdot 9}{2x \cdot 9} \cdot \dfrac{3}{1} = 3$

33. $\dfrac{7}{10} \cdot \dfrac{34}{150} = \dfrac{7 \cdot 34}{10 \cdot 150} = \dfrac{7 \cdot 2 \cdot 17}{2 \cdot 5 \cdot 150} = \dfrac{2}{2} \cdot \dfrac{7 \cdot 17}{5 \cdot 150} = \dfrac{119}{750}$

35. $\dfrac{36}{85} \cdot \dfrac{25}{-99} = \dfrac{9 \cdot 4 \cdot 5 \cdot 5}{5 \cdot 17 \cdot 9(-11)} = \dfrac{9 \cdot 5}{9 \cdot 5} \cdot \dfrac{4 \cdot 5}{17(-11)} =$
$\dfrac{20}{-187}$, or $-\dfrac{20}{187}$

37. $\dfrac{-98}{99} \cdot \dfrac{27a}{175a} = \dfrac{7(-14) \cdot 9 \cdot 3 \cdot a}{9 \cdot 11 \cdot 7 \cdot 25 \cdot a} =$
$\dfrac{7 \cdot 9 \cdot a}{7 \cdot 9 \cdot a} \cdot \dfrac{-14 \cdot 3}{11 \cdot 25} = \dfrac{-42}{275}$, or $-\dfrac{42}{275}$

39. $\dfrac{110}{33} \cdot \dfrac{-24}{25x} = -\dfrac{2 \cdot 5 \cdot 11 \cdot 3 \cdot 8}{3 \cdot 11 \cdot 5 \cdot 5 \cdot x} = -\dfrac{2 \cdot 8}{5 \cdot x} \cdot \dfrac{3 \cdot 5 \cdot 11}{3 \cdot 5 \cdot 11} =$
$\dfrac{-16}{5x}$, or $-\dfrac{16}{5x}$

41. $\left(-\dfrac{11}{24}\right)\dfrac{3}{5} = -\dfrac{11 \cdot 3}{24 \cdot 5} = -\dfrac{11 \cdot 3}{3 \cdot 8 \cdot 5} = \dfrac{3}{3}\left(-\dfrac{11}{8 \cdot 5}\right) =$
$-\dfrac{11}{40}$

43. $\dfrac{10a}{21} \cdot \dfrac{3}{8b} = \dfrac{10a \cdot 3}{21 \cdot 8b} = \dfrac{2 \cdot 5 \cdot a \cdot 3}{3 \cdot 7 \cdot 2 \cdot 4 \cdot b} = \dfrac{2 \cdot 3}{2 \cdot 3} \cdot \dfrac{5 \cdot a}{7 \cdot 4 \cdot b} = \dfrac{5a}{28b}$

45. *Familiarize.* Let $n =$ the number of inches the screw will go into the piece of oak when it is turned 10 complete rotations.

Translate. We write an equation.

Total distance	is	Distance for one revolution	times	Number of revolutions
↓	↓	↓	↓	↓
n	$=$	$\dfrac{1}{16}$	$\cdot$	10

Solve. We carry out the multiplication.

$n = \dfrac{1}{16} \cdot 10 = \dfrac{1 \cdot 10}{16}$
$= \dfrac{1 \cdot 2 \cdot 5}{2 \cdot 8} = \dfrac{2}{2} \cdot \dfrac{1 \cdot 5}{8}$
$= \dfrac{5}{8}$

Check. We can repeat the calculation. We can also determine that the answer seems reasonable since we multiplied 10 by a number less than 10 and the result is less than 10. The answer checks.

State. The screw will go $\dfrac{5}{8}$ in. into the piece of oak when it is turned 10 complete rotations.

47. *Familiarize.* Let $s =$ the swimming speed of a dolphin, in mph.

Translate. We write an equation.

$$\underbrace{\text{Dolphin's speed}} \text{ is } \frac{3}{5} \text{ of } \underbrace{\text{Whale's speed}}$$

$$s = \frac{3}{5} \cdot 30$$

Solve. We carry out the multiplication.

$$s = \frac{3}{5} \cdot 30 = \frac{3 \cdot 30}{5}$$
$$= \frac{3 \cdot 5 \cdot 6}{5 \cdot 1} = \frac{5}{5} \cdot \frac{3 \cdot 6}{1}$$
$$= 18$$

Check. We can repeat the calculation. We can also determine that the answer seems reasonable since we multiplied 30 by a number less than 30 and the result is less than 30. The answer checks.

State. The swimming speed of a dolphin is 18 mph.

49. *Familiarize*. We visualize the situation. We let n = the number of addresses that will be incorrect after one year.

Mailing list 2500 addresses			
1/4 of the addresses n			

Translate.

$$\underbrace{\text{Number incorrect}} \text{ is } \frac{1}{4} \text{ of } \underbrace{\text{Number of addresses}}$$

$$n = \frac{1}{4} \cdot 2500$$

Solve. We carry out the multiplication.

$$n = \frac{1}{4} \cdot 2500 = \frac{1 \cdot 2500}{4} = \frac{2500}{4}$$
$$= \frac{4 \cdot 625}{4 \cdot 1} = \frac{4}{4} \cdot \frac{625}{1}$$
$$= 625$$

Check. We can repeat the calculation. We can also determine that the answer seems reasonable since we multiplied 2500 by a number less than 1 and the result is less than 2500. The answer checks.

State. After one year 625 addresses will be incorrect.

51. *Familiarize*. Let f = the fraction of Americans who eat breakfast at home. This fraction is $\frac{3}{4}$ of $\frac{2}{3}$.

Translate. We translate to a multiplication sentence.

$$\frac{3}{4} \cdot \frac{2}{3} = f$$

Solve. We multiply and simplify.

$$f = \frac{3}{4} \cdot \frac{2}{3} = \frac{3 \cdot 2}{4 \cdot 3} = \frac{3 \cdot 2 \cdot 1}{2 \cdot 2 \cdot 3} = \frac{2 \cdot 3}{2 \cdot 3} \cdot \frac{1}{2} = \frac{1}{2}$$

Check. We can repeat the calculation. The answer checks.

State. $\frac{1}{2}$ of Americans eat breakfast at home.

53. *Familiarize*. We draw a picture.

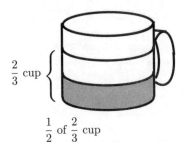

$\frac{2}{3}$ cup

$\frac{1}{2}$ of $\frac{2}{3}$ cup

We let n = the amount of flour the chef should use.

Translate. The multiplication sentence

$$\frac{1}{2} \cdot \frac{2}{3} = n$$

corresponds to the situation.

Solve. We multiply and simplify:

$$n = \frac{1}{2} \cdot \frac{2}{3} = \frac{1 \cdot 2}{2 \cdot 3} = \frac{2}{2} \cdot \frac{1}{3} = \frac{1}{3}$$

Check. We can repeat the calculation. We can also determine that the answer seems reasonable since we multiplied $\frac{2}{3}$ by a number less than 1 and the result is less than $\frac{2}{3}$. The answer checks.

State. The chef should use $\frac{1}{3}$ cup of flour.

55. *Familiarize*. We visualize the situation. Let a = the assessed value of the house.

Value of house $154,000	
3/4 of the value $a	

Translate. We write an equation.

$$\underbrace{\text{Assessed value}} \text{ is } \frac{3}{4} \text{ of } \underbrace{\text{the value of the house}}$$

$$a = \frac{3}{4} \cdot 154{,}000$$

Solve. We carry out the multiplication.

$$a = \frac{3}{4} \cdot 154{,}000 = \frac{3 \cdot 154{,}000}{4}$$
$$= \frac{3 \cdot 4 \cdot 38{,}500}{4 \cdot 1} = \frac{4}{4} \cdot \frac{3 \cdot 38{,}500}{1}$$
$$= 115{,}500$$

Check. We can repeat the calculation. We can also determine that the answer seems reasonable since we multiplied 154,000 by a number less than 1 and the result is less than 154,000. The answer checks.

State. The assessed value of the house is $115,500.

57. *Familiarize*. We draw a picture.

$\dfrac{2}{3}$ in.

1 in.
240 miles

We let $n =$ the number of miles represented by $\dfrac{2}{3}$ in.

Translate. The multiplication sentence

$$n = \frac{2}{3} \cdot 240$$

corresponds to the situation.

Solve. We multiply and simplify:

$$n = \frac{2}{3} \cdot 240 = \frac{2 \cdot 240}{3} = \frac{2 \cdot 3 \cdot 80}{1 \cdot 3}$$
$$= \frac{3}{3} \cdot \frac{2 \cdot 80}{1} = \frac{2 \cdot 80}{1}$$
$$= 160$$

Check. We can repeat the calculation. We can also determine that the answer seems reasonable since we multiplied 240 by a number less than 1 and the result is less than 240.

State. $\dfrac{2}{3}$ in. on the map represents 160 miles.

59. *Familiarize*. This is a multistep problem. First we find the amount of each of the given expenses. Then we find the total of these expenses and take it away from the annual income to find how much is spent for other expenses.

$39,600 $39,600 $39,600

$\dfrac{1}{4}$ for food $\dfrac{1}{5}$ for housing $\dfrac{1}{10}$ for clothing

$39,600 $39,600

$\dfrac{1}{9}$ for savings $\dfrac{1}{4}$ for taxes

We let f, h, c, s, and t represent the amounts spent on food, housing, clothing, savings, and taxes, respectively.

Translate. The following multiplication sentences correspond to the situation.

$$\frac{1}{4} \cdot 39{,}600 = f \qquad \frac{1}{9} \cdot 39{,}600 = s$$

$$\frac{1}{5} \cdot 39{,}600 = h \qquad \frac{1}{4} \cdot 39{,}600 = t$$

$$\frac{1}{10} \cdot 39{,}600 = c$$

Solve. We multiply and simplify.

$$f = \frac{1}{4} \cdot 39{,}600 = \frac{39{,}600}{4} = \frac{4 \cdot 9900}{4 \cdot 1} = \frac{4}{4} \cdot \frac{9900}{1} =$$
9900

$$h = \frac{1}{5} \cdot 39{,}600 = \frac{39{,}600}{5} = \frac{5 \cdot 7920}{5 \cdot 1} = \frac{5}{5} \cdot \frac{7920}{1} =$$
7920

$$c = \frac{1}{10} \cdot 39{,}600 = \frac{39{,}600}{10} = \frac{10 \cdot 3960}{10 \cdot 1} = \frac{10}{10} \cdot \frac{3960}{1} =$$
3960

$$s = \frac{1}{9} \cdot 39{,}600 = \frac{39{,}600}{9} = \frac{9 \cdot 4400}{9 \cdot 1} = \frac{9}{9} \cdot \frac{4400}{1} =$$
4400

$$t = \frac{1}{4} \cdot 39{,}600 = \frac{39{,}600}{4} = \frac{4 \cdot 9900}{4 \cdot 1} = \frac{4}{4} \cdot \frac{9900}{1} =$$
9900

We add to find the total of these expenses.

$$\begin{array}{r} \$ 9\,900 \\ 7\,920 \\ 3\,960 \\ 4\,400 \\ 9\,900 \\ \hline \$ 36{,}080 \end{array}$$

We let $m =$ the amount spent on other expenses and subtract to find this amount.

Annual income	minus	Total of itemized expenses	is	Total spent on other expenses
↓	↓	↓	↓	↓
$39,600	−	$36,080	=	m
		$3520	=	m Subtracting

Check. We repeat the calculations. The results check.

State. $9900 is spent for food, $7920 for housing, $3960 for clothing, $4400 for savings, $9900 for taxes, and $3520 for other expenses.

61. $A = \dfrac{1}{2} \cdot b \cdot h$ Area of a triangle

$A = \dfrac{1}{2} \cdot 15 \text{ in.} \cdot 8 \text{ in.}$ Substituting 15 in. for b and 8 in. for h

$A = \dfrac{15 \cdot 8}{2} \text{ in}^2$

$A = 60 \text{ in}^2$

63. $A = \dfrac{1}{2} \cdot b \cdot h$ Area of a triangle

$A = \dfrac{1}{2} \cdot 5 \text{ mm} \cdot \dfrac{7}{2} \text{ mm}$ Substituting 5 mm for b and $\dfrac{7}{2}$ mm for h

$A = \dfrac{5 \cdot 7}{2 \cdot 2} \text{ mm}^2$

$A = \dfrac{35}{4} \text{ mm}^2$

65. $A = \dfrac{1}{2} \cdot b \cdot h$ Area of a triangle

$A = \dfrac{1}{2} \cdot \dfrac{9}{2} \text{ m} \cdot \dfrac{7}{2} \text{ m}$ Substituting $\dfrac{9}{2}$ m for b and

$\dfrac{7}{2}$ m for h

$A = \dfrac{9 \cdot 7}{2 \cdot 2 \cdot 2} \text{ m}^2$

$A = \dfrac{63}{8} \text{ m}^2$

67. *Familiarize*. We look for figures whose areas we can calculate using area formulas we already know.

Translate. The figure consists of a rectangle with a length of 10 mi and a width of 8 mi and of a triangle with a base of $13 - 10$, or 3 mi, and a height of 8 mi. We use the formula $A = l \cdot w$ for the area of a rectangle and the formula $A = \dfrac{1}{2} \cdot b \cdot h$ for the area of a triangle and add the two areas.

Solve. For the rectangle: $A = l \cdot w = 10 \text{ mi} \cdot 8 \text{ mi} = 80 \text{ mi}^2$

For the triangle: $A = \dfrac{1}{2} \cdot b \cdot h = \dfrac{1}{2} \cdot 3 \text{ mi} \cdot 8 \text{ mi} = 12 \text{ mi}^2$

Then we add: $80 \text{ mi}^2 + 12 \text{ mi}^2 = 92 \text{ mi}^2$

Check. We repeat the calculations.

State. The area of the figure is 92 mi^2.

69. *Familiarize*. We look for figures with areas we can calculate using area formulas that we already know. We let $T = $ the area of the front of the tie.

Translate. The front of the tie can be thought of as two triangles, each with base 4 cm + 1 cm, or 5 cm, and height $\dfrac{3}{2}$ cm. We can use the formula $A = \dfrac{1}{2} \cdot b \cdot h$ for the area of a triangle and then multiply by 2.

$\underbrace{\text{Area of tie}}$ is twice $\underbrace{\text{Area of triangle}}$.

$\quad\quad\downarrow \quad\quad\quad \downarrow \quad\ \downarrow \quad\quad\quad\quad\quad \downarrow$

$\quad\ T \quad\ = \quad 2\cdot \quad \dfrac{1}{2} \cdot (5 \text{ cm}) \cdot \left(\dfrac{3}{2} \text{ cm}\right)$

Solve. We carry out the calculation.

$T = 2 \cdot \dfrac{1}{2} \cdot (5 \text{ cm}) \cdot \left(\dfrac{3}{2} \text{ cm}\right)$

$\quad = 1 \cdot 5 \text{ cm} \cdot \dfrac{3}{2} \text{ cm}$

$\quad = \dfrac{5 \cdot 3}{2} \text{ cm}^2$

$\quad = \dfrac{15}{2} \text{ cm}^2$

Check. We can repeat the calculations. The answer checks.

State. The area of the front of the tie is $\dfrac{15}{2} \text{ cm}^2$.

71. Discussion and Writing Exercise

73. $48 \cdot i = 1680$

$\dfrac{48 \cdot t}{48} = \dfrac{1680}{48}$

$\quad\quad t = 35$

The solution is 35.

75. $3125 = 25 \cdot t$

$\dfrac{3125}{25} = \dfrac{25 \cdot t}{25}$ Dividing by 25 on both sides

$\quad 125 = t$

The solution is 125.

77. $t + 28 = 5017$

$\quad\quad t = 5017 - 28$

$\quad\quad t = 4989$

The solution is 4989.

79. $\quad\quad\quad 8797 = y + 2299$

$8797 - 2299 = y + 2299 - 2299$ Subtracting 2299 on both sides

$\quad\quad 6498 = y$

The solution is 6498.

81. Discussion and Writing Exercise

83. $\dfrac{201}{535} \cdot \dfrac{4601}{6499} = \dfrac{201 \cdot 4601}{535 \cdot 6499}$

$\quad\quad\quad\quad\quad = \dfrac{3 \cdot 67 \cdot 43 \cdot 107}{5 \cdot 107 \cdot 67 \cdot 97}$

$\quad\quad\quad\quad\quad = \dfrac{67 \cdot 107}{67 \cdot 107} \cdot \dfrac{3 \cdot 43}{5 \cdot 97}$

$\quad\quad\quad\quad\quad = \dfrac{129}{485}$

85. $\dfrac{667}{899} \cdot \dfrac{558}{621} = \dfrac{667 \cdot 558}{899 \cdot 621}$

$\quad\quad\quad\quad\quad = \dfrac{23 \cdot 29 \cdot 2 \cdot 3 \cdot 3 \cdot 31}{29 \cdot 31 \cdot 3 \cdot 3 \cdot 3 \cdot 23}$

$\quad\quad\quad\quad\quad = \dfrac{3 \cdot 3 \cdot 23 \cdot 29 \cdot 31}{3 \cdot 3 \cdot 23 \cdot 29 \cdot 31} \cdot \dfrac{2}{3}$

$\quad\quad\quad\quad\quad = \dfrac{2}{3}$

87. *Familiarize*. We know that $\dfrac{7}{8}$ of the students are high school graduates and $\dfrac{1}{7}$ of all students are left-handed. Also, if we divide the group of students into 3 equal parts and take 2 of them, we have the fractional part of the students who are over the age of 20. Then the 1 part remaining, or $\dfrac{1}{3}$ of the students, are 20 yr old or younger. Thus, we want to find $\dfrac{7}{8}$ of $\dfrac{1}{7}$ of $\dfrac{1}{3}$ of 480 students. We let $s = $ this number.

Translate. The multiplication sentence

$s = \dfrac{7}{8} \cdot \dfrac{1}{7} \cdot \dfrac{1}{3} \cdot 480$

corresponds to this situation.

Solve. We carry out the multiplication.

$s = \dfrac{7}{8} \cdot \dfrac{1}{7} \cdot \dfrac{1}{3} \cdot 480 = \dfrac{7 \cdot 1 \cdot 1 \cdot 480}{8 \cdot 7 \cdot 3}$

$\quad = \dfrac{7 \cdot 1 \cdot 1 \cdot 3 \cdot 8 \cdot 20}{8 \cdot 7 \cdot 3 \cdot 1}$

$\quad = \dfrac{7 \cdot 3 \cdot 8}{7 \cdot 3 \cdot 8} \cdot \dfrac{1 \cdot 1 \cdot 20}{1}$

$\quad = 20$

Check. We can repeat the calculations. The answer checks.

State. 20 students are left-handed high school graduates 20 yr old or younger.

89. Area of each triangular end:

$A = \frac{1}{2} \cdot b \cdot h = \frac{1}{2} \cdot 30 \text{ mm} \cdot 26 \text{ mm} = 390 \text{ mm}^2$

Area of each rectangular side:

$A = l \cdot w = 140 \text{ mm} \cdot 30 \text{ mm} = 4200 \text{ mm}^2$

Total area: $2 \cdot 390 \text{ mm}^2 + 3 \cdot 4200 \text{ mm}^2 =$
$780 \text{ mm}^2 + 12{,}600 \text{ mm}^2 = 13{,}380 \text{ mm}^2$

91. From Exercise 70 we know that the total area of the sides and entrances of the building, including the area of the windows and doors, is 6800 ft². The area of each window is

$A = l \cdot w = 4 \text{ ft} \cdot 3 \text{ ft} = 12 \text{ ft}^2.$

The area of each entrance is

$A = l \cdot w = 8 \text{ ft} \cdot 6 \text{ ft} = 48 \text{ ft}^2.$

From the drawing we see that there are 26 windows and 2 entrances. Then the total area of the windows and entrances is

$26 \cdot 12 \text{ ft}^2 + 2 \cdot 48 \text{ ft}^2 = 312 \text{ ft}^2 + 96 \text{ ft}^2 = 408 \text{ ft}^2.$

We subtract to find the area of the sides and ends of the building excluding the area of the windows and entrances.

$6800 \text{ ft}^2 - 408 \text{ ft}^2 = 6392 \text{ ft}^2$

The building requires 6392 ft² of siding.

Exercise Set 3.7

1. $\frac{7}{3}$ Interchange the numerator and denominator.

The reciprocal of $\frac{7}{3}$ is $\frac{3}{7}$. $\left(\frac{7}{3} \cdot \frac{3}{7} = \frac{21}{21} = 1\right)$

3. Think of 9 as $\frac{9}{1}$.

$\frac{9}{1}$ Interchange the numerator and denominator.

The reciprocal of 9 is $\frac{1}{9}$. $\left(\frac{9}{1} \cdot \frac{1}{9} = \frac{9}{9} = 1\right)$

5. $\frac{1}{7}$ Interchange the numerator and denominator.

The reciprocal of $\frac{1}{7}$ is 7. $\left(\frac{7}{1} = 7; \frac{1}{7} \cdot \frac{7}{1} = \frac{7}{7} = 1\right)$

7. $-\frac{10}{3}$ Interchange the numerator and denominator.

The reciprocal of $-\frac{10}{3}$ is $-\frac{3}{10}$. $\left(-\frac{10}{3}\left(-\frac{3}{10}\right) = \frac{30}{30} = 1\right)$

9. $\frac{3}{17}$ Interchange the numerator and denominator.

The reciprocal of $\frac{3}{17}$ is $\frac{17}{3}$. $\left(\frac{3}{17} \cdot \frac{17}{3} = \frac{51}{51} = 1\right)$

11. $\frac{-3n}{m}$ Interchange the numerator and denominator.

The reciprocal of $\frac{-3n}{m}$ is $\frac{m}{-3n}$.

$\left(\frac{-3n}{m} \cdot \frac{m}{-3n} = \frac{-3mn}{-3mn} = 1\right)$

13. $\frac{8}{-15}$ Interchange the numerator and denominator.

The reciprocal of $\frac{8}{-15}$ is $\frac{-15}{8}$. $\left(\frac{8}{-15}\left(\frac{-15}{8}\right) = \frac{-120}{-120} = 1\right)$

15. Think of $7m$ as $\frac{7m}{1}$.

$\frac{7m}{1}$ Interchange the numerator and denominator.

The reciprocal of $7m$ is $\frac{1}{7m}$. $\left(\frac{7m}{1} \cdot \frac{1}{7m} = \frac{7m}{7m} = 1\right)$

17. $\frac{1}{4a}$ Interchange the numerator and denominator.

The reciprocal of $\frac{1}{4a}$ is $\frac{4a}{1}$, or $4a$.

$\left(\frac{1}{4a} \cdot \frac{4a}{1} = \frac{4a}{4a} = 1\right)$

19. The reciprocal of $-\frac{1}{3z}$ is $-\frac{3z}{1}$, or $-3z$.

$\left(-\frac{1}{3z} \cdot (-3z) = \frac{3z}{3z} = 1\right)$

21. $\frac{3}{7} \div \frac{3}{4} = \frac{3}{7} \cdot \frac{4}{3}$ Multiplying by the reciprocal of the divisor

$= \frac{3 \cdot 4}{7 \cdot 3}$ Multiplying numerators and denominators

$= \frac{3}{3} \cdot \frac{4}{7} = \frac{4}{7}$ Removing a factor equal to 1

23. $\frac{7}{6} \div \frac{5}{-3} = \frac{7}{6} \cdot \frac{-3}{5}$ Multiplying by the reciprocal of the divisor

$= \frac{7(-1)(3)}{2 \cdot 3 \cdot 5}$ Factoring

$= \frac{3}{3} \cdot \frac{7(-1)}{2 \cdot 5}$

$= \frac{-7}{10}$, or $-\frac{7}{10}$

25. $\frac{4}{3} \div \frac{1}{3} = \frac{4}{3} \cdot 3 = \frac{4 \cdot 3}{3} = \frac{3}{3} \cdot 4 = 4$

27. $\left(-\frac{1}{3}\right) \div \frac{1}{6} = -\frac{1}{3} \cdot 6 = -\frac{1 \cdot 2 \cdot 3}{1 \cdot 3} = -\frac{1 \cdot 3}{1 \cdot 3} \cdot 2 = -2$

29. $\left(-\dfrac{10}{21}\right) \div \left(-\dfrac{2}{15}\right) = \left(-\dfrac{10}{21}\right) \cdot \left(-\dfrac{15}{2}\right) = \dfrac{10 \cdot 15}{21 \cdot 2} =$

$\dfrac{2 \cdot 5 \cdot 3 \cdot 5}{3 \cdot 7 \cdot 2} = \dfrac{2 \cdot 3}{2 \cdot 3} \cdot \dfrac{5 \cdot 5}{7} = \dfrac{25}{7}$

31. $\dfrac{3}{8} \div 24 = \dfrac{3}{8} \cdot \dfrac{1}{24} = \dfrac{3 \cdot 1}{8 \cdot 3 \cdot 8} = \dfrac{3}{3} \cdot \dfrac{1}{8 \cdot 8} = \dfrac{1}{64}$

33. $\dfrac{12}{7} \div (4x) = \dfrac{12}{7} \cdot \dfrac{1}{4x} = \dfrac{4 \cdot 3 \cdot 1}{7 \cdot 4 \cdot x} = \dfrac{4}{4} \cdot \dfrac{3 \cdot 1}{7 \cdot x} = \dfrac{3}{7x}$

35. $(-12) \div \dfrac{3}{2} = -12 \cdot \dfrac{2}{3} = -\dfrac{3 \cdot 4 \cdot 2}{3 \cdot 1}$

$\qquad = -\dfrac{3}{3} \cdot \dfrac{4 \cdot 2}{1} = -\dfrac{8}{1} = -8$

37. $28 \div \dfrac{4}{5a} = 28 \cdot \dfrac{5a}{4} = \dfrac{28 \cdot 5a}{4} = \dfrac{4 \cdot 7 \cdot 5 \cdot a}{4 \cdot 1} = \dfrac{4}{4} \cdot \dfrac{7 \cdot 5 \cdot a}{1}$

$\qquad = 35a$

39. $\left(-\dfrac{5}{8}\right) \div \left(-\dfrac{5}{8}\right) = -\dfrac{5}{8}\left(-\dfrac{8}{5}\right) = \dfrac{5 \cdot 8}{8 \cdot 5} = \dfrac{5 \cdot 8}{5 \cdot 8} = 1$

41. $\dfrac{-8}{15} \div \dfrac{4}{5} = \dfrac{-8}{15} \cdot \dfrac{5}{4} = \dfrac{-8 \cdot 5}{15 \cdot 4} = \dfrac{-2 \cdot 4 \cdot 5}{3 \cdot 5 \cdot 4} =$

$\dfrac{4 \cdot 5}{4 \cdot 5} \cdot \dfrac{-2}{3} = \dfrac{-2}{3},$ or $-\dfrac{2}{3}$

43. $\dfrac{77}{64} \div \dfrac{49}{18} = \dfrac{77}{64} \cdot \dfrac{18}{49} = \dfrac{7 \cdot 11 \cdot 2 \cdot 9}{2 \cdot 32 \cdot 7 \cdot 7} =$

$\dfrac{2 \cdot 7}{2 \cdot 7} \cdot \dfrac{11 \cdot 9}{32 \cdot 7} = \dfrac{99}{224}$

45. $120a \div \dfrac{45}{14} = 120a \cdot \dfrac{14}{45} = \dfrac{8 \cdot 15 \cdot a \cdot 14}{3 \cdot 15} =$

$\dfrac{15}{15} \cdot \dfrac{8 \cdot a \cdot 14}{3} = \dfrac{112a}{3}$

47. $\dfrac{\frac{2}{5}}{\frac{3}{7}} = \dfrac{2}{5} \div \dfrac{3}{7} = \dfrac{2}{5} \cdot \dfrac{7}{3} = \dfrac{2 \cdot 7}{5 \cdot 3} = \dfrac{14}{15}$

49. $\dfrac{\frac{7}{20}}{\frac{8}{5}} = \dfrac{7}{20} \div \dfrac{8}{5} = \dfrac{7}{20} \cdot \dfrac{5}{8} = \dfrac{7 \cdot 5}{20 \cdot 8} = \dfrac{7 \cdot 5}{4 \cdot 5 \cdot 8} =$

$\dfrac{5}{5} \cdot \dfrac{7}{4 \cdot 8} = \dfrac{7}{32}$

51. $\dfrac{-\frac{15}{8}}{\frac{9}{10}} = -\dfrac{15}{8} \div \dfrac{9}{10} = -\dfrac{15}{8} \cdot \dfrac{10}{9} = -\dfrac{15 \cdot 10}{8 \cdot 9} =$

$-\dfrac{3 \cdot 5 \cdot 2 \cdot 5}{2 \cdot 4 \cdot 3 \cdot 3} = -\dfrac{5 \cdot 5}{4 \cdot 3} \cdot \dfrac{3 \cdot 2}{3 \cdot 2} = -\dfrac{25}{12}$

53. $\dfrac{-\frac{9}{16}}{-\frac{6}{5}} = -\dfrac{9}{16} \div \left(-\dfrac{6}{5}\right) = -\dfrac{9}{16} \cdot \left(-\dfrac{5}{6}\right) =$

$\dfrac{9 \cdot 5}{16 \cdot 6} = \dfrac{3 \cdot 3 \cdot 5}{16 \cdot 2 \cdot 3} = \dfrac{3}{3} \cdot \dfrac{3 \cdot 5}{16 \cdot 2} = \dfrac{15}{32}$

55. Discussion and Writing Exercise

57. The equation $14 + (2 + 30) = (14 + 2) + 30$ illustrates the <u>associative</u> law of addition.

59. A natural number that has exactly two different factors, only itself and 1, is called a <u>prime</u> number.

61. Since $a + 0 = a$ for any number a, the number 0 is the <u>additive</u> identity.

63. The sum of 6 and -6 is 0; we say that 6 and -6 are <u>opposites</u> of each other.

65. Discussion and Writing Exercise

67. $\left(\dfrac{4}{15} \div \dfrac{2}{25}\right)^2 = \left(\dfrac{4}{15} \cdot \dfrac{25}{2}\right)^2$

$\qquad = \left(\dfrac{4 \cdot 25}{15 \cdot 2}\right)^2$

$\qquad = \left(\dfrac{2 \cdot 2 \cdot 5 \cdot 5}{3 \cdot 5 \cdot 2}\right)^2$

$\qquad = \left(\dfrac{2 \cdot 5}{2 \cdot 5} \cdot \dfrac{2 \cdot 5}{3}\right)$

$\qquad = \left(\dfrac{10}{3}\right)^2$

$\qquad = \dfrac{100}{9}$

69. $\left(\dfrac{9}{10} \div \dfrac{2}{5} \div \dfrac{3}{8}\right)^2 = \left(\dfrac{9}{10} \cdot \dfrac{5}{2} \div \dfrac{3}{8}\right)^2$

$\qquad = \left(\dfrac{9 \cdot 5}{10 \cdot 2} \div \dfrac{3}{8}\right)^2$

$\qquad = \left(\dfrac{9 \cdot 5}{2 \cdot 5 \cdot 2} \div \dfrac{3}{8}\right)^2$

$\qquad = \left(\dfrac{9}{2 \cdot 2} \div \dfrac{3}{8}\right)^2$

$\qquad = \left(\dfrac{9}{2 \cdot 2} \cdot \dfrac{8}{3}\right)^2$

$\qquad = \left(\dfrac{9 \cdot 8}{2 \cdot 2 \cdot 3}\right)^2$

$\qquad = \left(\dfrac{3 \cdot 3 \cdot 2 \cdot 2 \cdot 2}{2 \cdot 2 \cdot 3 \cdot 1}\right)^2$

$\qquad = \left(\dfrac{3 \cdot 2}{1}\right)^2$

$\qquad = 6^2$

$\qquad = 36$

71. $\left(\dfrac{14}{15} \div \dfrac{49}{65} \cdot \dfrac{77}{260}\right)^2 = \left(\dfrac{14}{15} \cdot \dfrac{65}{49} \cdot \dfrac{77}{260}\right)^2$

$\qquad = \left(\dfrac{2 \cdot 7 \cdot 5 \cdot 13 \cdot 7 \cdot 11}{3 \cdot 5 \cdot 7 \cdot 7 \cdot 2 \cdot 2 \cdot 5 \cdot 13}\right)^2$

$\qquad = \left(\dfrac{2 \cdot 5 \cdot 7 \cdot 7 \cdot 13}{2 \cdot 5 \cdot 7 \cdot 7 \cdot 13} \cdot \dfrac{11}{2 \cdot 3 \cdot 5}\right)^2$

$\qquad = \left(\dfrac{11}{30}\right)^2$

$\qquad = \dfrac{121}{900}$

73. Use a calculator.

$\dfrac{711}{1957} \div \dfrac{10,033}{13,081} = \dfrac{711}{1957} \cdot \dfrac{13,081}{10,033}$

$\qquad = \dfrac{711 \cdot 13,081}{1957 \cdot 10,033}$

$\qquad = \dfrac{3 \cdot 3 \cdot 79 \cdot 103 \cdot 127}{19 \cdot 103 \cdot 79 \cdot 127}$

$\qquad = \dfrac{79 \cdot 103 \cdot 127}{79 \cdot 103 \cdot 127} \cdot \dfrac{3 \cdot 3}{19}$

$\qquad = \dfrac{9}{19}$

75. Use a calculator.

$\dfrac{451}{289} \div \dfrac{123}{340} = \dfrac{451}{289} \cdot \dfrac{340}{123}$

$\qquad = \dfrac{451 \cdot 340}{289 \cdot 123}$

$\qquad = \dfrac{11 \cdot 41 \cdot 17 \cdot 20}{17 \cdot 17 \cdot 3 \cdot 41}$

$\qquad = \dfrac{41 \cdot 17}{41 \cdot 17} \cdot \dfrac{11 \cdot 20}{17 \cdot 3}$

$\qquad = \dfrac{220}{51}$

Exercise Set 3.8

1. $\qquad \dfrac{4}{5}x = 12$

$\dfrac{5}{4} \cdot \dfrac{4}{5}x = \dfrac{5}{4} \cdot 12 \qquad$ The reciprocal of $\dfrac{4}{5}$ is $\dfrac{5}{4}$.

$\qquad 1x = \dfrac{5 \cdot 4 \cdot 3}{4}$

$\qquad x = 15 \qquad$ Removing the factor $\dfrac{4}{4}$

Check: $\qquad \dfrac{4}{5}x = 12$

$\qquad\dfrac{4}{5} \cdot 15 \ ? \ 12$

$\qquad\dfrac{4 \cdot 3 \cdot 5}{5 \cdot 1}$

$\qquad\qquad 12 \ \bigg| \ 12 \qquad$ TRUE

The solution is 15.

3. $\qquad \dfrac{7}{3}a = 21$

$\dfrac{3}{7} \cdot \dfrac{7}{3}a = \dfrac{3}{7} \cdot 21 \qquad$ The reciprocal of $\dfrac{7}{3}$ is $\dfrac{3}{7}$.

$\qquad 1a = \dfrac{3 \cdot 3 \cdot 7}{7}$

$\qquad a = 9 \qquad$ Removing the factor $\dfrac{7}{7}$

Check: $\qquad \dfrac{7}{3}a = 21$

$\qquad\dfrac{7}{3} \cdot 9 \ ? \ 21$

$\qquad\dfrac{7 \cdot 3 \cdot 3}{3 \cdot 1}$

$\qquad\qquad 21 \ \bigg| \ 21 \qquad$ TRUE

The solution is 9.

5. $\qquad \dfrac{2}{9}x = -10$

$\dfrac{9}{2} \cdot \dfrac{2}{9}x = \dfrac{9}{2}(-10) \qquad$ The reciprocal of $\dfrac{2}{9}$ is $\dfrac{9}{2}$.

$\qquad 1x = -\dfrac{9 \cdot 2 \cdot 5}{2 \cdot 1}$

$\qquad x = -45 \qquad$ Removing the factor $\dfrac{2}{2}$.

Check: $\qquad \dfrac{2}{9}x = -10$

$\qquad\dfrac{2}{9}(-45) \ ? \ -10$

$\qquad -\dfrac{2 \cdot 5 \cdot 9}{9 \cdot 1}$

$\qquad\qquad -10 \ \bigg| \ -10 \qquad$ TRUE

The solution is -45.

7. $\qquad 6a = \dfrac{12}{17}$

$\dfrac{1}{6} \cdot 6a = \dfrac{1}{6} \cdot \dfrac{12}{17} \qquad$ The reciprocal of 6 is $\dfrac{1}{6}$.

$\qquad 1a = \dfrac{2 \cdot 6}{6 \cdot 17}$

$\qquad a = \dfrac{2}{17} \qquad$ Removing the factor $\dfrac{6}{6}$

Check: $\qquad 6a = \dfrac{12}{17}$

$\qquad 6 \cdot \dfrac{2}{17} \ ? \ \dfrac{12}{17}$

$\qquad\dfrac{12}{17} \ \bigg| \ \dfrac{12}{17} \qquad$ TRUE

The solution is $\dfrac{2}{17}$.

9.
$$\frac{1}{4}x = \frac{3}{5}$$
$$\frac{4}{1} \cdot \frac{1}{4}x = \frac{4}{1} \cdot \frac{3}{5}$$
$$x = \frac{12}{5}$$

$\frac{12}{5}$ checks and is the solution.

11.
$$\frac{3}{2}t = -\frac{8}{7}$$
$$\frac{2}{3} \cdot \frac{3}{2}t = \frac{2}{3}\left(-\frac{8}{7}\right)$$
$$t = -\frac{16}{21}$$

$-\frac{16}{21}$ checks and is the solution.

13.
$$\frac{4}{5} = -10a$$
$$-\frac{1}{10} \cdot \frac{4}{5} = -\frac{1}{10}(-10a)$$
$$-\frac{2 \cdot 2}{2 \cdot 5 \cdot 5} = a$$
$$-\frac{2}{25} = a$$

$-\frac{2}{25}$ checks and is the solution.

15.
$$\frac{9}{5}x = \frac{3}{10}$$
$$\frac{5}{9} \cdot \frac{9}{5}x = \frac{5}{9} \cdot \frac{3}{10}$$
$$x = \frac{5 \cdot 3 \cdot 1}{3 \cdot 3 \cdot 2 \cdot 5}$$
$$x = \frac{1}{6}$$

$\frac{1}{6}$ checks and is the solution.

17.
$$-\frac{9}{10}x = 8$$
$$-\frac{10}{9}\left(-\frac{9}{10}x\right) = -\frac{10}{9} \cdot 8$$
$$x = -\frac{10 \cdot 8}{9}$$
$$x = -\frac{80}{9}$$

$-\frac{80}{9}$ checks and is the solution.

19.
$$a \cdot \frac{9}{7} = -\frac{3}{14}$$
$$a \cdot \frac{9}{7} \cdot \frac{7}{9} = -\frac{3}{14} \cdot \frac{7}{9}$$
$$a \cdot 1 = -\frac{3 \cdot 7 \cdot 1}{2 \cdot 7 \cdot 3 \cdot 3}$$
$$a = -\frac{1}{6}$$

$-\frac{1}{6}$ checks and is the solution.

21.
$$-x = \frac{7}{13}$$
$$-1(-x) = -1 \cdot \frac{7}{13}$$
$$x = -\frac{7}{13}$$

$-\frac{7}{13}$ checks and is the solution.

23.
$$-x = -\frac{27}{31}$$
$$-1(-x) = -1\left(-\frac{27}{31}\right)$$
$$x = \frac{27}{31}$$

$\frac{27}{31}$ checks and is the solution.

25.
$$7t = 6$$
$$\frac{1}{7} \cdot 7t = \frac{1}{7} \cdot 6$$
$$t = \frac{6}{7}$$

$\frac{6}{7}$ checks and is the solution.

27.
$$-24 = -10a$$
$$-\frac{1}{10}(-24) = -\frac{1}{10}(-10a)$$
$$\frac{2 \cdot 12}{2 \cdot 5} = a$$
$$\frac{12}{5} = a$$

$\frac{12}{5}$ checks and is the solution.

29.
$$-\frac{14}{9} = \frac{10}{3}t$$
$$\frac{3}{10}\left(-\frac{14}{9}\right) = \frac{3}{10} \cdot \frac{10}{3}t$$
$$-\frac{3 \cdot 2 \cdot 7}{2 \cdot 5 \cdot 3 \cdot 3} = t$$
$$-\frac{7}{15} = t$$

$-\frac{7}{15}$ checks and is the solution.

31.
$$n \cdot \frac{4}{15} = \frac{12}{25}$$
$$n \cdot \frac{4}{15} \cdot \frac{15}{4} = \frac{12}{25} \cdot \frac{15}{4}$$
$$n = \frac{4 \cdot 3 \cdot 3 \cdot 5}{5 \cdot 5 \cdot 4}$$
$$n = \frac{9}{5}$$

$\frac{9}{5}$ checks and is the solution.

33.
$$-\frac{7}{20}x = -\frac{21}{10}$$

$$-\frac{20}{7}\left(-\frac{7}{20}x\right) = -\frac{20}{7}\left(-\frac{21}{10}\right)$$

$$x = \frac{2\cdot 10\cdot 3\cdot 7}{7\cdot 10}$$

$$x = 6$$

6 checks and is the solution.

35.
$$-\frac{25}{17} = -\frac{35}{34}a$$

$$-\frac{34}{35}\left(-\frac{25}{17}\right) = -\frac{34}{35}\left(-\frac{35}{34}a\right)$$

$$\frac{2\cdot 17\cdot 5\cdot 5}{5\cdot 7\cdot 17} = a$$

$$\frac{10}{7} = a$$

$\frac{10}{7}$ checks and is the solution.

37. Familiarize. We draw a picture. Let $t =$ the number of times Benny will be able to brush his teeth.

t brushings

Translate. The multiplication that corresponds to the situation is
$$\frac{2}{5}\cdot t = 30.$$

Solve. We solve the equation by dividing on both sides by $\frac{2}{5}$ and carrying out the division:

$$t = 30 \div \frac{2}{5} = 30\cdot\frac{5}{2} = \frac{2\cdot 15\cdot 5}{2\cdot 1} = \frac{2}{2}\cdot\frac{15\cdot 5}{1} = 75$$

Check. We repeat the calculation. The answer checks.

State. Benny can brush his teeth 75 times with a 30-g tube of toothpaste.

39. Familiarize. Let $g =$ the number of gallons of gasoline the tanker holds when it is fully loaded.

Translate. We translate to an equation.

$$\underbrace{1400\text{ gal}}\quad\text{is}\quad\underbrace{\frac{7}{9}}\quad\text{of}\quad\underbrace{\text{a full load}}$$
$$\downarrow\qquad\downarrow\quad\downarrow\quad\downarrow\qquad\downarrow$$
$$1400\quad = \quad\frac{7}{9}\quad\cdot\qquad g$$

Solve. We solve the equation.

$$1400 = \frac{7}{9}\cdot g$$

$$1400 \div \frac{7}{9} = g$$

$$1400\cdot\frac{9}{7} = g$$

$$\frac{1400\cdot 9}{7} = g$$

$$\frac{7\cdot 200\cdot 9}{7\cdot 1} = g$$

$$\frac{7}{7}\cdot\frac{200\cdot 9}{1} = g$$

$$1800 = g$$

Check. $\frac{7}{9}$ of 1800 gal is $\frac{7}{9}\cdot 1800 = \frac{7\cdot 1800}{9} = \frac{7\cdot 9\cdot 200}{9\cdot 1} = \frac{9}{9}\cdot\frac{7\cdot 200}{1} = 1400$ gal. The answer checks.

State. The tanker holds 1800 gal of gasoline when it is full.

41. Familiarize. Let $w =$ the number of worker bees it takes to produce $\frac{3}{4}$ tsp of honey.

Translate.

$$\underbrace{\begin{array}{c}\text{Amount produced}\\\text{by one bee}\end{array}}\quad\text{times}\quad\underbrace{\begin{array}{c}\text{Number}\\\text{of bees}\end{array}}\quad\text{is}\quad\underbrace{\frac{3}{4}\text{ tsp}}$$
$$\downarrow\qquad\qquad\downarrow\qquad\downarrow\qquad\downarrow\quad\downarrow$$
$$\frac{1}{12}\qquad\cdot\qquad w\qquad = \quad\frac{3}{4}$$

Solve. We solve the equation.

$$\frac{1}{12}\cdot w = \frac{3}{4}$$

$$w = \frac{3}{4}\div\frac{1}{12} = \frac{3}{4}\cdot\frac{12}{1} = \frac{3\cdot 12}{4\cdot 1}$$

$$= \frac{3\cdot 3\cdot 4}{4\cdot 1} = \frac{3\cdot 3}{1}\cdot\frac{4}{4} = \frac{3\cdot 3}{1}$$

$$= 9$$

Check. Since $\frac{1}{12}\cdot 9 = \frac{9}{12} = \frac{3}{4}$, the answer checks.

State. It takes 9 worker bees to produce $\frac{3}{4}$ tsp of honey.

43. Familiarize. We make a drawing. Let $p =$ the number of packages that can be made from 15 lb of cheese.

p packages

Translate. The problem translates to the following equation:
$$p = 15 \div \frac{3}{4}.$$

Solve. We carry out the division.

$$p = 15 \div \frac{3}{4}$$
$$= 15 \cdot \frac{4}{3}$$
$$= \frac{3 \cdot 5 \cdot 4}{3 \cdot 1} = \frac{3}{3} \cdot \frac{5 \cdot 4}{1}$$
$$= 20$$

Check. If 20 packages, each containing $\frac{3}{4}$ lb of cheese, are made, a total of

$$20 \cdot \frac{3}{4} = \frac{5 \cdot 4 \cdot 3}{4 \cdot 1} = \frac{4}{4} \cdot \frac{5 \cdot 3}{1} = 15,$$

or 15 lb of cheese is used. The answer checks.

State. 20 packages can be made.

45. *Familiarize*. Let c = the amount of clay each art department will receive, in tons.

Translate. The problem translates to the following equation:

$$c = \frac{3}{4} \div 6.$$

Solve. We carry out the division.

$$c = \frac{3}{4} \div 6$$
$$= \frac{3}{4} \cdot \frac{1}{6}$$
$$= \frac{3 \cdot 1}{4 \cdot 2 \cdot 3} = \frac{3}{3} \cdot \frac{1}{4 \cdot 2}$$
$$= \frac{1}{8}$$

Check. If each of 6 art departments get $\frac{1}{8}$ T of clay, the total amount of clay is

$$6 \cdot \frac{1}{8} = \frac{6 \cdot 1}{8} = \frac{2 \cdot 3 \cdot 1}{2 \cdot 4} = \frac{3}{4} \text{ T.}$$

The answer checks.

State. Each art department will receive $\frac{1}{8}$ T of clay.

47. *Familiarize*. We make a drawing. Let w = the number of walkways that can be covered with 6 yd of gravel.

w walkways

Translate. The problem translates to the following situation:

$$w = 6 \div \frac{3}{4}.$$

Solve. We carry out the division.

$$w = 6 \div \frac{3}{4}$$
$$= 6 \cdot \frac{4}{3}$$
$$= \frac{2 \cdot 3 \cdot 4}{3 \cdot 1} = \frac{3}{3} \cdot \frac{2 \cdot 4}{1}$$
$$= 8$$

Check. If each of 8 walkways is covered with $\frac{3}{4}$ yd of gravel, a total of

$$8 \cdot \frac{3}{4} = \frac{2 \cdot 4 \cdot 3}{4 \cdot 1} = \frac{4}{4} \cdot \frac{2 \cdot 3}{1} = 6,$$

or 6 yd of gravel is used. The answer checks.

State. 8 walkways can be covered with one dump truck load of gravel.

49. *Familiarize*. We make a drawing. Let c = the number of customers that can be accommodated with a 30 yd batch of mulch.

c customers

Translate. The problem translates to the following situation.

$$c = 30 \div \frac{2}{3}.$$

Solve. We carry out the division.

$$c = 30 \div \frac{2}{3}$$
$$= 30 \cdot \frac{3}{2}$$
$$= \frac{2 \cdot 15 \cdot 3}{2 \cdot 1} = \frac{2}{2} \cdot \frac{15 \cdot 3}{1}$$
$$= 45$$

Check. If each of 45 customers gets $\frac{2}{3}$ yd of mulch, a total of

$$45 \cdot \frac{2}{3} = \frac{3 \cdot 15 \cdot 2}{3 \cdot 1} = \frac{3}{3} \cdot \frac{15 \cdot 2}{1} = 30,$$

or 30 yd of mulch is used. The answer checks.

State. 45 customers can be accommodated with 30 yd of mulch.

51. *Familiarize*. We draw a picture.

$\frac{3}{4}$ yd per pair

We let s = the number of pairs of basketball shorts that can be made.

Translate. The problem translates to the following equation:

$$s = 24 \div \frac{3}{4}$$

Solve. We carry out the division.

$$s = 24 \div \frac{3}{4}$$
$$= 24 \cdot \frac{4}{3}$$
$$= \frac{3 \cdot 8 \cdot 4}{1 \cdot 3} = \frac{3}{3} \cdot \frac{8 \cdot 4}{1}$$
$$= 32$$

Check. If each of 32 pairs of shorts requires $\frac{3}{4}$ yd of nylon, a total of

$$32 \cdot \frac{3}{4} = \frac{32 \cdot 3}{4} = \frac{4 \cdot 8 \cdot 3}{4} = 8 \cdot 3,$$

or 24 yd of nylon is needed. Our answer checks.

State. 32 pairs of basketball shorts can be made from 24 yd of nylon.

53. Familiarize. Let p = the pitch of the screw, in inches. The distance the screw has traveled into the wallboard is found by multiplying the pitch by the number of complete rotations.

Translate. We translate to an equation.

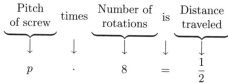

Solve. We divide on both sides of the equation by 8 and carry out the division.

$$p = \frac{1}{2} \div 8 = \frac{1}{2} \cdot \frac{1}{8} = \frac{1 \cdot 1}{2 \cdot 8} = \frac{1}{16}$$

Check. We repeat the calculation. The answer checks.

State. The pitch of the screw is $\frac{1}{16}$ in.

55. Discussion and Writing Exercise

57. $-23 + 49 = 26$

(Find the difference of the absolute values. The positive integer has the larger absolute value, so the answer is positive.)

59. $-38 - 29 = -67$

(Add the absolute values. The answer is negative.)

61. $36 \div (-3)^2 \times (7 - 2) = 36 \div (-3)^2 \times 5$
$= 36 \div 9 \times 5$
$= 4 \times 5$
$= 20$

63. $13x + 4x = (13 + 4)x = 17x$

65. $2a + 3 + 5a = 2a + 5a + 3$
$= (2 + 5)a + 3$
$= 7a + 3$

67. Discussion and Writing Exercise

69. $2x - 7x = -\frac{10}{9}$

$-5x = -\frac{10}{9}$

$-\frac{1}{5}(-5x) = -\frac{1}{5}\left(-\frac{10}{9}\right)$

$x = \frac{2 \cdot 5}{5 \cdot 9}$

$x = \frac{2}{9}$

71. Familiarize. Let w = the weight of the package when it is completely filled.

Translate.

$\frac{3}{4}$ of total weight is $\frac{21}{32}$ lb

$\frac{3}{4} \cdot w = \frac{21}{32}$

Solve. We solve the equation.

$$\frac{3}{4} \cdot w = \frac{21}{32}$$

$$\frac{4}{3} \cdot \frac{3}{4} \cdot w = \frac{4}{3} \cdot \frac{21}{32}$$

$$w = \frac{4 \cdot 3 \cdot 7}{3 \cdot 4 \cdot 8} = \frac{4 \cdot 3}{4 \cdot 3} \cdot \frac{7}{8}$$

$$w = \frac{7}{8}$$

Check. We find $\frac{3}{4}$ of $\frac{7}{8}$ lb.

$$\frac{3}{4} \cdot \frac{7}{8} = \frac{3 \cdot 7}{4 \cdot 8} = \frac{21}{32} \text{ lb}$$

The answer checks.

State. The package could hold $\frac{7}{8}$ lb of coffee beans when it is completely filled.

73. Familiarize. Let x = the number of slices yielded by the $\frac{3}{32}$-in. cuts and y = the number of slices yielded by the $\frac{5}{32}$-in. cuts. Half the block is $\frac{1}{2} \cdot 12$ in., or 6 in.

Translate. The problem translates to the following situations.

$$x = 6 \div \frac{3}{32} \text{ and } y = 6 \div \frac{5}{32}$$

Solve. We carry out the division.

$$x = 6 \div \frac{3}{32}$$

$$= 6 \cdot \frac{32}{3}$$

$$= \frac{2 \cdot 3 \cdot 32}{3 \cdot 1} = \frac{3}{3} \cdot \frac{2 \cdot 32}{1}$$

$$= 64$$

$$y = 6 \div \frac{5}{32}$$

$$= 6 \cdot \frac{32}{5}$$

$$= \frac{192}{5}$$

$$= 38 \text{ R } 2$$

The $\frac{3}{32}$-in. cuts yield 64 slices. The $\frac{5}{32}$-in. cuts yield 38 slices that are $\frac{5}{32}$ in. thick and an additional slice (indicated by the remainder) that is less than $\frac{5}{32}$ in., so this cutting yields 39 slices. Then the total number of slices is $64 + 39$, or 103.

Check. We repeat the calculations. The answer checks.

State. The cutting will yield 103 slices of cheese.

75. **Familiarize.** Let $w =$ the number of walkways that can be covered with 6 yd of gravel. Let $c =$ the cost of covering each walkway. We will find w and c and then multiply to find how much Eric will receive.

Translate. The problem translates to the following situations.

$$w = 6 \div \frac{3}{5} \quad \text{and} \quad c = 85 \cdot \frac{3}{5}$$

Solve. We carry out the calculations.

$$w = 6 \div \frac{3}{5}$$
$$= 6 \cdot \frac{5}{3}$$
$$= \frac{2 \cdot 3 \cdot 5}{3 \cdot 1} = \frac{3}{3} \cdot \frac{2 \cdot 5}{1}$$
$$= 10$$

Eric will use his full load and will cover 10 walkways.

$$c = 85 \cdot \frac{3}{5}$$
$$= \frac{5 \cdot 17 \cdot 3}{5 \cdot 1} = \frac{5}{5} \cdot \frac{17 \cdot 3}{1}$$
$$= 51$$

If Eric is paid $51 for each of 10 walkways, then he receives $51 \cdot 10 = \$510$. (Note that, since we know Eric will use the full load of 6 yd of gravel, we could also have found this result by multiplying $85 \cdot 6$ to get $510.)

Check. We repeat the calculations. The answer checks.

State. Eric will receive $510 for a full load.

77. **Familiarize.** First we find the number n of customers who can be provided with $\frac{3}{4}$ yd of mulch from a 25-yd batch. Then we will multiply to find the amount a that Green Season Gardening would receive for the mulch.

Translate. We have
$$n = 25 \div \frac{3}{4}.$$

Solve.
$$n = 25 \div \frac{3}{4}$$
$$= 25 \cdot \frac{4}{3}$$
$$= \frac{100}{3}$$
$$= 33 \text{ R } 1$$

Thus, 33 customers can receive $\frac{3}{4}$ yd of mulch. Now we multiply to find the amount Green Season Gardening would receive for the mulch.

$$a = 33 \cdot \frac{3}{4} \cdot 65$$
$$= \frac{6435}{4}$$
$$= 1608.75$$

Check. We repeat the calculations. The answer checks.

State. Green Season Gardening would receive $1608.75 for the mulch. (Note that when each of 33 customers receives $\frac{3}{4}$ yd of mulch, then $33 \cdot \frac{3}{4}$, or $\frac{99}{4}$ yd is used. Since $25 = 25 \cdot \frac{4}{4} = \frac{100}{4}$, this means that $\frac{1}{4}$ yd is left unused after the $\frac{99}{4}$ yd are distributed.)

Chapter 3 Review Exercises

1. $1 \cdot 8 = 8$ $6 \cdot 8 = 48$
    $2 \cdot 8 = 16$ $7 \cdot 8 = 56$
    $3 \cdot 8 = 24$ $8 \cdot 8 = 64$
    $4 \cdot 8 = 32$ $9 \cdot 8 = 72$
    $5 \cdot 8 = 40$ $10 \cdot 8 = 80$

2. 3920 is even because the ones digit is even; $3+9+2+0 = 14$ and 14 is not divisible by 3, so 3920 is not divisible by 3. Since 3920 is not divisible by 3, it is not divisible by 6.

3. Because $6 + 8 + 5 + 3 + 7 = 29$ and 29 is not divisible by 3, then 68,537 is not divisible by 3.

4. 673 is not divisible by 5 because the ones digit is neither 0 nor 5.

5. 4936 is divisible by 2 because the ones digit is even.

6. Because $5 + 2 + 3 + 8 = 18$ and 18 is divisible by 9, then 5238 is divisible by 9.

7. Since the ones digit of 60 is 0 we know that 2, 5, and 10 are factors. Since the sum of the digits is 6 and 6 is divisible by 3, then 3 is a factor. Since 2 and 3 are factors, 6 is also a factor. We write a list of factorizations.
    $60 = 1 \cdot 60$ $60 = 4 \cdot 15$
    $60 = 2 \cdot 30$ $60 = 5 \cdot 12$
    $60 = 3 \cdot 20$ $60 = 6 \cdot 10$
    Factors: 1, 2, 3, 4, 5, 6, 10, 12, 15, 20, 30, 60

8. 176 is even so 2 is a factor. None of the other tests for divisibility yields additional factors, so we find as many two-factor factorizations as we can:
    $176 = 1 \cdot 176$ $176 = 8 \cdot 22$
    $176 = 2 \cdot 88$ $176 = 11 \cdot 16$
    $176 = 4 \cdot 44$
    Factors: 1, 2, 4, 8, 11, 16, 22, 44, 88, 176

9. The only factors of 37 are 1 and 37, so 37 is prime.

10. 1 is neither prime nor composite.

11. The number 91 has factors 1, 7, 13, and 91, so it is composite.

12.
$$\begin{array}{r} 7 \quad \leftarrow \text{ 7 is prime.} \\ 5 \overline{\smash{\big)}\, 3\,5} \\ 2 \overline{\smash{\big)}\, 7\,0} \end{array}$$
$70 = 2 \cdot 5 \cdot 7$

13.
$$
\begin{array}{r}
3 \\
3\,\overline{\smash{)}\,9} \\
2\,\overline{\smash{)}\,18} \\
2\,\overline{\smash{)}\,36} \\
2\,\overline{\smash{)}\,72}
\end{array}
\qquad \leftarrow \text{3 is prime.}
$$
$72 = 2 \cdot 2 \cdot 2 \cdot 3 \cdot 3$

14.
$$
\begin{array}{r}
5 \\
3\,\overline{\smash{)}\,15} \\
3\,\overline{\smash{)}\,45}
\end{array}
\qquad \leftarrow \text{5 is prime.}
$$
$45 = 3 \cdot 3 \cdot 5$

15.
$$
\begin{array}{r}
5 \\
5\,\overline{\smash{)}\,25} \\
3\,\overline{\smash{)}\,75} \\
2\,\overline{\smash{)}\,150}
\end{array}
\qquad \leftarrow \text{5 is prime.}
$$
$150 = 2 \cdot 3 \cdot 5 \cdot 5$

16.
$$
\begin{array}{r}
3 \\
3\,\overline{\smash{)}\,9} \\
3\,\overline{\smash{)}\,27} \\
3\,\overline{\smash{)}\,81} \\
2\,\overline{\smash{)}\,162} \\
2\,\overline{\smash{)}\,324} \\
2\,\overline{\smash{)}\,648}
\end{array}
\qquad \leftarrow \text{3 is prime.}
$$
$648 = 2 \cdot 2 \cdot 2 \cdot 3 \cdot 3 \cdot 3 \cdot 3$

17.
$$
\begin{array}{r}
5 \\
5\,\overline{\smash{)}\,25} \\
3\,\overline{\smash{)}\,75} \\
2\,\overline{\smash{)}\,150} \\
2\,\overline{\smash{)}\,300} \\
2\,\overline{\smash{)}\,600} \\
2\,\overline{\smash{)}\,1200}
\end{array}
\qquad \leftarrow \text{5 is prime.}
$$
$1200 = 2 \cdot 2 \cdot 2 \cdot 2 \cdot 3 \cdot 5 \cdot 5$

18. The top number is the numerator, and the bottom number is the denominator.
$$
\begin{array}{l}
9 \quad \leftarrow \text{Numerator} \\
\overline{7} \quad \leftarrow \text{Denominator}
\end{array}
$$

19. The object is divided into 8 equal parts. The unit is $\frac{1}{8}$. The denominator is 8. We have 3 parts shaded. This tells us that the numerator is 3. Thus, $\frac{3}{8}$ is shaded.

20. We can regard this as 2 bars of 6 parts each and take 7 of those parts. The unit is $\frac{1}{6}$. The denominator is 6 and the numerator is 7. Thus, $\frac{7}{6}$ is shaded.

21. a) The ratio is $\frac{3}{5}$.

 b) The ratio is $\frac{5}{3}$.

 c) There are $3 + 5$, or 8, members of the committee. The desired ratio is $\frac{3}{8}$.

22. $\frac{0}{n} = 0$, for any number n that is not 0.
$$\frac{0}{6} = 0$$

23. $\frac{n}{n} = 1$, for any number n that is not 0.
$$\frac{74}{74} = 1$$

24. $\frac{n}{1} = n$, for any number n.
$$\frac{48}{1} = 48$$

25. Remember: $\frac{n}{n} = 1$ for any number n that is not 0.
$$\frac{7x}{7x} = 1$$

26. $-\frac{10}{15} = -\frac{2 \cdot 5}{3 \cdot 5} = -\frac{2}{3} \cdot \frac{5}{5} = -\frac{2}{3}$

27. $\frac{7}{28} = \frac{7 \cdot 1}{7 \cdot 4} = \frac{7}{7} \cdot \frac{1}{4} = \frac{1}{4}$

28. $\frac{-42}{42} = \frac{-1 \cdot 42}{1 \cdot 42} = \frac{-1}{1} \cdot \frac{42}{42} = \frac{-1}{1} = -1$

29. $\frac{9m}{12m} = \frac{3 \cdot 3 \cdot m}{3 \cdot 4 \cdot m} = \frac{3 \cdot m}{3 \cdot m} \cdot \frac{3}{4} = \frac{3}{4}$

30. $\frac{12}{30} = \frac{2 \cdot 6}{5 \cdot 6} = \frac{2}{5} \cdot \frac{6}{6} = \frac{2}{5}$

31. Remember: $\frac{n}{0}$ is not defined.

 $\frac{-27}{0}$ is undefined.

32. Remember: $\frac{n}{1} = n$
$$\frac{6x}{1} = 6x$$

33. $\frac{-9}{-27} = \frac{-9 \cdot 1}{-9 \cdot 3} = \frac{-9}{-9} \cdot \frac{1}{3} = \frac{1}{3}$

34. Since $21 \div 7 = 3$, we multiply by $\frac{3}{3}$.
$$\frac{5}{7} = \frac{5}{7} \cdot \frac{3}{3} = \frac{5 \cdot 3}{7 \cdot 3} = \frac{15}{21}$$

35. Since $55 \div 11 = 5$, we multiply by $\frac{5}{5}$.
$$\frac{-6}{11} = \frac{-6}{11} \cdot \frac{5}{5} = \frac{-6 \cdot 5}{11 \cdot 5} = \frac{-30}{55}$$

36. 3 and 100 have no prime factors in common, so $\frac{3}{100}$ cannot be simplified.
$$\frac{8}{100} = \frac{2 \cdot 4}{25 \cdot 4} = \frac{2}{25} \cdot \frac{4}{4} = \frac{2}{25}$$
$$\frac{10}{100} = \frac{10 \cdot 1}{10 \cdot 10} = \frac{10}{10} \cdot \frac{1}{10} = \frac{1}{10}$$
$$\frac{15}{100} = \frac{3 \cdot 5}{20 \cdot 5} = \frac{3}{20} \cdot \frac{5}{5} = \frac{3}{20}$$

21 and 100 have no prime factors in common, so $\frac{21}{100}$ cannot be simplified.

43 and 100 have no prime factors in common, so $\frac{43}{100}$ cannot be simplified.

37. We multiply these   We multiply these
      two numbers: two numbers:

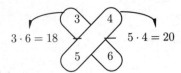

$3 \cdot 6 = 18$ $5 \cdot 4 = 20$

Since $18 \neq 20$, $\dfrac{3}{5}$ and $\dfrac{4}{6}$ do not name the same number.
Thus, $\dfrac{3}{5} \neq \dfrac{4}{6}$.

38. We multiply these   We multiply these
      two numbers: two numbers:

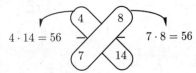

$4 \cdot 14 = 56$ $7 \cdot 8 = 56$

Since $56 = 56$, $\dfrac{4}{7} = \dfrac{8}{14}$.

39. We multiply these   We multiply these
      two numbers: two numbers:

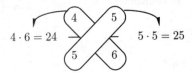

$4 \cdot 6 = 24$ $5 \cdot 5 = 25$

Since $24 \neq 25$, $\dfrac{4}{5}$ and $\dfrac{5}{6}$ do not name the same number.
Thus, $\dfrac{4}{5} \neq \dfrac{5}{6}$.

40. We multiply these   We multiply these
      two numbers: two numbers:

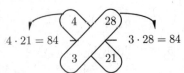

$4 \cdot 21 = 84$ $3 \cdot 28 = 84$

Since $84 = 84$, $\dfrac{4}{3} = \dfrac{28}{21}$.

41. Interchange the numerator and denominator.

The reciprocal of $\dfrac{2}{13}$ is $\dfrac{13}{2}$. $\left(\dfrac{2}{13} \cdot \dfrac{13}{2} = \dfrac{26}{26} = 1 \right)$

42. Think of -7 as $\dfrac{-7}{1}$. Interchange the numerator and denominator.

The reciprocal of -7 is $\dfrac{1}{-7}$, or $\dfrac{-1}{7}$, or $-\dfrac{1}{7}$.

$\left(-7 \cdot \left(-\dfrac{1}{7} \right) = \dfrac{7}{7} = 1 \right)$

43. Interchange the numerator and denominator.

The reciprocal of $\dfrac{1}{8}$ is $\dfrac{8}{1}$, or 8. $\left(\dfrac{1}{8} \cdot 8 = \dfrac{8}{8} = 1 \right)$

44. Interchange the numerator and denominator.

The reciprocal of $\dfrac{3x}{5y}$ is $\dfrac{5y}{3x}$. $\left(\dfrac{3x}{5y} \cdot \dfrac{5y}{3x} = \dfrac{15xy}{15xy} = 1 \right)$

45. $\dfrac{2}{9} \cdot \dfrac{7}{5} = \dfrac{2 \cdot 7}{9 \cdot 5} = \dfrac{14}{45}$

46. $\dfrac{3}{x} \cdot \dfrac{y}{7} = \dfrac{3 \cdot y}{x \cdot 7} = \dfrac{3y}{7x}$

47. $\dfrac{3}{4} \cdot \dfrac{8}{9} = \dfrac{3 \cdot 8}{4 \cdot 9} = \dfrac{3 \cdot 2 \cdot 4}{4 \cdot 3 \cdot 3} = \dfrac{3 \cdot 4}{3 \cdot 4} \cdot \dfrac{2}{3} = \dfrac{2}{3}$

48. $-\dfrac{5}{7} \cdot \dfrac{1}{10} = -\dfrac{5 \cdot 1}{7 \cdot 10} = -\dfrac{5 \cdot 1}{7 \cdot 2 \cdot 5} = -\dfrac{1}{7 \cdot 2} \cdot \dfrac{5}{5} = -\dfrac{1}{14}$

49. $\dfrac{3a}{10} \cdot \dfrac{2}{15a} = \dfrac{3a \cdot 2}{10 \cdot 15a} = \dfrac{3 \cdot a \cdot 2 \cdot 1}{2 \cdot 5 \cdot 3 \cdot 5 \cdot a} = \dfrac{2 \cdot 3 \cdot a}{2 \cdot 3 \cdot a} \cdot \dfrac{1}{5 \cdot 5} = \dfrac{1}{25}$

50. $\dfrac{4a}{7} \cdot \dfrac{7}{4a} = \dfrac{4a \cdot 7}{7 \cdot 4a} = 1$

51. $9 \div \dfrac{5}{3} = \dfrac{9}{1} \cdot \dfrac{3}{5} = \dfrac{9 \cdot 3}{1 \cdot 5} = \dfrac{27}{5}$

52. $\dfrac{3}{14} \div \dfrac{6}{7} = \dfrac{3}{14} \cdot \dfrac{7}{6} = \dfrac{3 \cdot 7}{14 \cdot 6} = \dfrac{3 \cdot 7 \cdot 1}{2 \cdot 7 \cdot 2 \cdot 3} = \dfrac{3 \cdot 7}{3 \cdot 7} \cdot \dfrac{1}{2 \cdot 2} =$

$\dfrac{1}{2 \cdot 2} = \dfrac{1}{4}$

53. $120 \div \dfrac{3}{5} = 120 \cdot \dfrac{5}{3} = \dfrac{120 \cdot 5}{3} = \dfrac{3 \cdot 40 \cdot 5}{3 \cdot 1} = \dfrac{3}{3} \cdot \dfrac{40 \cdot 5}{1} =$

$\dfrac{40 \cdot 5}{1} = 200$

54. $-\dfrac{5}{36} \div \left(-\dfrac{25}{12} \right) = -\dfrac{5}{36} \cdot \left(-\dfrac{12}{25} \right) = \dfrac{5 \cdot 12}{36 \cdot 25} =$

$\dfrac{5 \cdot 12 \cdot 1}{3 \cdot 12 \cdot 5 \cdot 5} = \dfrac{5 \cdot 12}{5 \cdot 12} \cdot \dfrac{1}{3 \cdot 5} = \dfrac{1}{15}$

55. $21 \div \dfrac{7}{2a} = \dfrac{21}{1} \cdot \dfrac{2a}{7} = \dfrac{3 \cdot 7 \cdot 2a}{1 \cdot 7} = \dfrac{7}{7} \cdot \dfrac{3 \cdot 2a}{1} = 6a$

56. $-\dfrac{23}{25} \div \dfrac{23}{25} = -\dfrac{23}{25} \cdot \dfrac{25}{23} = -\dfrac{23 \cdot 25}{25 \cdot 23} = -1$

57. $\dfrac{\frac{21}{30}}{\frac{14}{15}} = \dfrac{21}{30} \cdot \dfrac{15}{14} = \dfrac{21 \cdot 15}{30 \cdot 14} = \dfrac{3 \cdot 7 \cdot 3 \cdot 5}{2 \cdot 3 \cdot 5 \cdot 2 \cdot 7} =$

$\dfrac{3 \cdot 5 \cdot 7}{3 \cdot 5 \cdot 7} \cdot \dfrac{3}{2 \cdot 2} = \dfrac{3}{4}$

58. $\dfrac{-\frac{3}{40}}{-\frac{35}{54}} = -\dfrac{3}{40} \cdot \left(-\dfrac{35}{54} \right) = \dfrac{3 \cdot 35}{40 \cdot 54} = \dfrac{3 \cdot 5 \cdot 7}{5 \cdot 8 \cdot 3 \cdot 18} =$

$\dfrac{3 \cdot 5}{3 \cdot 5} \cdot \dfrac{7}{8 \cdot 18} = \dfrac{7}{144}$

59. $A = \dfrac{1}{2} \cdot b \cdot h$

$A = \dfrac{1}{2} \cdot 14 \text{ m} \cdot 6 \text{ m}$

$A = \dfrac{14 \cdot 6}{2} \text{ m}^2$

$A = 42 \text{ m}^2$

60. $A = \dfrac{1}{2} \cdot b \cdot h$

$A = \dfrac{1}{2} \cdot \dfrac{7}{2} \text{ ft} \cdot 10 \text{ ft}$

$A = \dfrac{7 \cdot 10}{2 \cdot 2} \text{ ft}^2$

$A = \dfrac{35}{2} \text{ ft}^2$

61. $\dfrac{2}{3}x = 160$

$\dfrac{3}{2} \cdot \dfrac{2}{3}x = \dfrac{3}{2} \cdot 160$

$1x = \dfrac{3 \cdot 2 \cdot 80}{2}$

$x = 240$

The solution is 240.

62. $\dfrac{3}{8} = -\dfrac{5}{4}t$

$-\dfrac{4}{5} \cdot \dfrac{3}{8} = -\dfrac{4}{5}\left(-\dfrac{5}{4}t\right)$

$-\dfrac{4 \cdot 3}{5 \cdot 2 \cdot 4} = 1t$

$-\dfrac{3}{10} = t$

The solution is $-\dfrac{3}{10}$.

63. $-\dfrac{1}{7}n = -4$

$-7\left(-\dfrac{1}{7}n\right) = -7(-4)$

$n = 28$

The solution is 28.

64. *Familiarize*. Let d = the number of days it will take to repave the road.

***Translate*.**

Number of miles repaved each day	times	Number of days	is	Total number of miles repaved
$\downarrow$	$\downarrow$	$\downarrow$	$\downarrow$	$\downarrow$
$\dfrac{1}{12}$	$\cdot$	d	$=$	$\dfrac{3}{4}$

***Solve*.** We divide by $\dfrac{1}{12}$ on both sides of the equation.

$d = \dfrac{3}{4} \div \dfrac{1}{12}$

$d = \dfrac{3}{4} \cdot \dfrac{12}{1} = \dfrac{3 \cdot 12}{4 \cdot 1} = \dfrac{3 \cdot 3 \cdot 4}{4 \cdot 1}$

$= \dfrac{4}{4} \cdot \dfrac{3 \cdot 3}{1} = \dfrac{3 \cdot 3}{1} = 9$

***Check*.** We repeat the calculation. The answer checks.

***State*.** It will take 9 days to repave the road.

65. *Familiarize*. Let s = the amount of sugar required for $\dfrac{1}{2}$ of the recipe, in cups. We want to find $\dfrac{1}{2}$ of $\dfrac{3}{4}$ cup.

***Translate*.** We write a multiplication sentence.

$\dfrac{1}{2} \cdot \dfrac{3}{4} = s$

***Solve*.** We carry out the multiplication.

$s = \dfrac{1}{2} \cdot \dfrac{3}{4} = \dfrac{1 \cdot 3}{2 \cdot 4} = \dfrac{3}{8}$

***Check*.** We repeat the calculation. The answer checks.

***State*.** $\dfrac{3}{8}$ cup of sugar should be used for $\dfrac{1}{2}$ of the recipe.

66. *Familiarize*. This is a multistep problem. First we find the length of the total trip. Then we find how many kilometers were left to drive. We draw a picture. We let n = the length of the total trip.

***Translate*.** We translate to an equation.

Fraction of trip completed	times	Total length of trip	is	Amount already traveled
$\downarrow$	$\downarrow$	$\downarrow$	$\downarrow$	$\downarrow$
$\dfrac{5}{8}$	$\cdot$	n	$=$	180

***Solve*.** We solve the equation as follows:

$\dfrac{5}{8} \cdot n = 180$

$n = 180 \div \dfrac{5}{8} = 180 \cdot \dfrac{8}{5} = \dfrac{5 \cdot 36 \cdot 8}{5 \cdot 1}$

$= \dfrac{5}{5} \cdot \dfrac{36 \cdot 8}{1} = \dfrac{36 \cdot 8}{1} = 288$

The total trip was 288 km.

Now we find how many kilometers were left to travel. Let t = this number.

Length of total trip	minus	Distance traveled	is	Distance left to travel
$\downarrow$	$\downarrow$	$\downarrow$	$\downarrow$	$\downarrow$
288	$-$	180	$=$	t

We carry out the subtraction:

$288 - 180 = t$

$108 = t$

***Check*.** We repeat the calculation. The results check.

***State*.** The total trip was 288 km. There were 108 km left to travel.

67. *Familiarize*. Let d = the distance each person will swim, in miles.

***Translate*.**

Number of swimmers	times	Distance each swims	is	Total distance
$\downarrow$	$\downarrow$	$\downarrow$	$\downarrow$	$\downarrow$
4	$\cdot$	d	$=$	$\dfrac{2}{3}$

Solve. We solve the equation.

$$4 \cdot d = \frac{2}{3}$$

$$\frac{1}{4} \cdot 4 \cdot d = \frac{1}{4} \cdot \frac{2}{3}$$

$$d = \frac{1 \cdot 2}{4 \cdot 3} = \frac{1 \cdot 2}{2 \cdot 2 \cdot 3}$$

$$= \frac{2}{2} \cdot \frac{1}{2 \cdot 3} = \frac{1}{6}$$

Check. Since $4 \cdot \frac{1}{6} = \frac{4}{6} = \frac{2}{3}$, the answer checks.

State. Each person will swim $\frac{1}{6}$ mi.

68. **Familiarize**. Let $c =$ the number of metric tons of corn produced in the U.S. in 2003.

Translate.

U.S. corn production	is	$\frac{2}{5}$	of	Total world corn production

$$c = \frac{2}{5} \cdot 640,000,000$$

Solve. We carry out the multiplication.

$$c = \frac{2}{5} \cdot 640,00,000 = \frac{2 \cdot 640,000,000}{5}$$

$$= \frac{2 \cdot 5 \cdot 128,000,000}{5 \cdot 1} = \frac{5}{5} \cdot \frac{2 \cdot 128,000,000}{1}$$

$$= \frac{2 \cdot 128,000,000}{1} = 256,000,000$$

Check. We repeat the calculation. The answer checks.

State. The U.S. produced 256,000,000 metric tons of corn in 2003.

69. *Discussion and Writing Exercise*. Taking $\frac{1}{2}$ of a number is equivalent to multiplying the number by $\frac{1}{2}$. Dividing by $\frac{1}{2}$ is equivalent to multiplying by the reciprocal of $\frac{1}{2}$, or 2. Thus taking $\frac{1}{2}$ of a number is not the same as dividing by $\frac{1}{2}$.

70. *Discussion and Writing Exercise*. Because $\frac{2}{8}$ simplifies to $\frac{1}{4}$, it is incorrect to suggest that $\frac{2}{8}$ is the simplified form of $\frac{20}{80}$.

71.

$$\frac{15x}{14z} \cdot \frac{17yz}{35xy} \div \left(-\frac{3}{7}\right)^2$$

$$= \frac{15x}{14z} \cdot \frac{17yz}{35xy} \div \frac{9}{49}$$

$$= \frac{15x \cdot 17yz}{14z \cdot 35xy} \div \frac{9}{49}$$

$$= \frac{15x \cdot 17yz}{14z \cdot 35xy} \cdot \frac{49}{9}$$

$$= \frac{15x \cdot 17yz \cdot 49}{14z \cdot 35xy \cdot 9}$$

$$= \frac{3 \cdot 5 \cdot x \cdot 17 \cdot y \cdot z \cdot 7 \cdot 7}{2 \cdot 7 \cdot z \cdot 5 \cdot 7 \cdot x \cdot y \cdot 3 \cdot 3}$$

$$= \frac{3 \cdot 5 \cdot 7 \cdot 7 \cdot x \cdot y \cdot z}{3 \cdot 5 \cdot 7 \cdot 7 \cdot x \cdot y \cdot z} \cdot \frac{17}{2 \cdot 3}$$

$$= \frac{17}{6}$$

72. The digit must be even and the sum of the digits must be divisible by 3. Let $d =$ the digit to be inserted. Then $5 + 7 + 4 + d$, or $16 + d$, must be divisible by 3. The only even digits for which $16 + d$ is divisible by 3 are 2 and 8.

73. 13 and 31 are both prime numbers, so 13 is a palindrome prime.

19 is prime but 91 is not ($91 = 7 \cdot 13$), so 19 is not a palindrome prime.

16 is not prime ($16 = 2 \cdot 8 = 4 \cdot 4$), so it is not a palindrome prime.

11 is prime and when its digits are reversed we have 11 again, so 11 is a palindrome prime.

15 is not prime ($15 = 3 \cdot 5$), so it is not a palindrome prime.

24 is not prime ($24 = 2 \cdot 12 = 3 \cdot 8 = 4 \cdot 6$), so it is not a palindrome prime.

29 is prime but 92 is not ($92 = 2 \cdot 46 = 4 \cdot 23$), so 29 is not a palindrome prime.

101 is prime and when its digits are reversed we get 101 again, so 101 is a palindrome prime.

201 is not prime ($201 = 3 \cdot 67$), so it is not a palindrome prime.

37 and 73 are both prime numbers, so 37 is a palindrome prime.

74. $\dfrac{19}{24} \div \dfrac{a}{b} = \dfrac{19}{24} \cdot \dfrac{b}{a} = \dfrac{19 \cdot b}{24 \cdot a} = \dfrac{187,853}{268,224}$

Then, assuming the quotient has not been simplified, we have

$$19 \cdot b = 187,853 \quad \text{and} \quad 24 \cdot a = 268,224$$

$$b = \frac{187,853}{19} \quad \text{and} \quad a = \frac{268,224}{24}$$

$$b = 9887 \quad \text{and} \quad a = 11,176.$$

75.
$$\frac{1751}{267}x = \frac{3193}{2759}$$

$$\frac{267}{1751} \cdot \frac{1751}{267}x = \frac{267}{1751} \cdot \frac{3193}{2759}$$

$$x = \frac{267 \cdot 3193}{1751 \cdot 2759}$$

$$x = \frac{3 \cdot 89 \cdot 31 \cdot 103}{17 \cdot 103 \cdot 31 \cdot 89}$$

$$x = \frac{89 \cdot 31 \cdot 103}{89 \cdot 31 \cdot 103} \cdot \frac{3}{17}$$

$$x = \frac{3}{17}$$

The solution is $\frac{3}{17}$.

Chapter 3 Test

1. Because $5 + 6 + 8 + 2 = 21$ and 21 is divisible by 3, then 5682 is divisible by 3.

2. 7018 is not divisible by 5 because the ones digit is neither 0 nor 5.

3. Since the ones digit of 90 is 0 we know that 2, 5, and 10 are factors. Since the sum of the digits is 9 and 9 is divisible by both 3 and 9, we know that 3 and 9 are factors. Since 2 and 3 are factors, 6 is also a factor. We write a list of factorizations.

$90 = 1 \cdot 90$ $90 = 5 \cdot 18$
$90 = 2 \cdot 45$ $90 = 6 \cdot 15$
$90 = 3 \cdot 30$ $90 = 9 \cdot 10$

Factors: 1, 2, 3, 5, 6, 9, 10, 15, 18, 30, 45, 90

4. The number 93 has factors 1, 3, 31, and 93. Since it has at least one factor other than itself and 1, it is composite.

5.
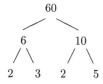
$36 = 2 \cdot 2 \cdot 3 \cdot 3$

6. We use a factor tree.

```
        60
       /  \
      6    10
     / \   / \
    2   3 2   5
```

$60 = 2 \cdot 3 \cdot 2 \cdot 5$, or $2 \cdot 2 \cdot 3 \cdot 5$

7. $\dfrac{4}{9}$ ← Numerator
         ← Denominator

8. The figure is divided into 4 equal parts, so the unit is $\frac{1}{4}$ and the denominator is 4. Three of the units are shaded, so the numerator is 3. Thus, $\frac{3}{4}$ is shaded.

9. There are 7 objects in the set, so the denominator is 7. Three of the objects are shaded, so the numerator is 3. Thus, $\frac{3}{7}$ of the set is shaded.

10. a) The ratio is $\frac{1112}{1202}$.

   b) The ratio is $\frac{90}{1202}$.

11. Remember: $\frac{n}{1} = n$.

$$\frac{32}{1} = 32$$

12. Remember: $\frac{n}{n} = 1$ for any integer n that is not 0.

$$\frac{-12}{-12} = 1$$

13. Remember: $\frac{0}{n} = 0$ for any integer n that is not 0.

$$\frac{0}{16} = 0$$

14. $\dfrac{-8}{24} = \dfrac{-1 \cdot 8}{3 \cdot 8} = \dfrac{-1}{3} \cdot \dfrac{8}{8} = \dfrac{-1}{3}$

15. $\dfrac{9x}{45x} = \dfrac{9 \cdot x \cdot 1}{5 \cdot 9 \cdot x} = \dfrac{9 \cdot x}{9 \cdot x} \cdot \dfrac{1}{5} = \dfrac{1}{5}$

16. $\dfrac{7}{63} = \dfrac{7 \cdot 1}{7 \cdot 9} = \dfrac{7}{7} \cdot \dfrac{1}{9} = \dfrac{1}{9}$

17. We multiply these   We multiply these
        two numbers: two numbers:

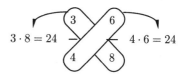

$3 \cdot 8 = 24$ $4 \cdot 6 = 24$

Since $24 = 24$, $\dfrac{3}{4} \leftarrow \dfrac{6}{8}$.

18. We multiply these   We multiply these
        two numbers: two numbers:

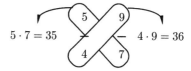

$5 \cdot 7 = 35$ $4 \cdot 9 = 36$

Since $35 \neq 36$, $\dfrac{5}{4}$ and $\dfrac{9}{7}$ do not name the same number. Thus, $\dfrac{5}{4} \neq \dfrac{9}{7}$.

19. Since $40 \div 8 = 5$, we multiply by $\dfrac{5}{5}$.

$$\frac{3}{8} = \frac{3}{8} \cdot \frac{5}{5} = \frac{3 \cdot 5}{8 \cdot 5} = \frac{15}{40}$$

20. Interchange the numerator and denominator.

The reciprocal of $\dfrac{a}{42}$ is $\dfrac{42}{a}$. $\left(\dfrac{a}{42} \cdot \dfrac{42}{a} = \dfrac{42a}{42a} = 1 \right)$

21. Think of -9 as $\dfrac{-9}{1}$. Interchange the numerator and denominator.

The reciprocal of -9 is $\dfrac{1}{-9}$, or $-\dfrac{1}{9}$, or $\dfrac{-1}{9}$.

$$\left(-9\cdot\left(-\dfrac{1}{9}\right)=\dfrac{9\cdot1}{9}=1\right)$$

22. $\dfrac{5}{7}\cdot\dfrac{7}{2}=\dfrac{5\cdot7}{7\cdot2}=\dfrac{7}{7}\cdot\dfrac{5}{2}=\dfrac{5}{2}$

23. $\dfrac{2}{11}\div\dfrac{3}{4}=\dfrac{2}{11}\cdot\dfrac{4}{3}=\dfrac{2\cdot4}{11\cdot3}=\dfrac{8}{33}$

24. $3\cdot\dfrac{x}{8}=\dfrac{3\cdot x}{8}=\dfrac{3x}{8}$

25. $\dfrac{\frac{4}{7}}{-\frac{8}{3}}=\dfrac{4}{7}\cdot\left(-\dfrac{3}{8}\right)=-\dfrac{4\cdot3}{7\cdot8}=-\dfrac{4\cdot3}{7\cdot2\cdot4}=$

$-\dfrac{3}{7\cdot2}\cdot\dfrac{4}{4}=-\dfrac{3}{14}$

26. $12\div\dfrac{2}{3}=12\cdot\dfrac{3}{2}=\dfrac{12\cdot3}{2}=\dfrac{2\cdot6\cdot3}{2\cdot1}=\dfrac{2}{2}\cdot\dfrac{6\cdot3}{1}=\dfrac{6\cdot3}{1}=18$

27. $\dfrac{4a}{13}\cdot\dfrac{9b}{30ab}=\dfrac{4a\cdot9b}{13\cdot30ab}=\dfrac{2\cdot2\cdot a\cdot3\cdot3\cdot b}{13\cdot2\cdot3\cdot5\cdot a\cdot b}=$

$\dfrac{2\cdot3\cdot a\cdot b}{2\cdot3\cdot a\cdot b}\cdot\dfrac{2\cdot3}{13\cdot5}=\dfrac{6}{65}$

28. *Familiarize*. Let $c=$ the number of pounds of cheese each person receives.

Translate. We write a division sentence.

$$c=\dfrac{3}{4}\div5$$

Solve. We carry out the division.

$$c=\dfrac{3}{4}\div5=\dfrac{3}{4}\cdot\dfrac{1}{5}=\dfrac{3\cdot1}{4\cdot5}=\dfrac{3}{20}$$

Check. Since $\dfrac{3}{20}\cdot5=\dfrac{3\cdot5}{20}=\dfrac{3\cdot5}{4\cdot5}=\dfrac{3}{4}$, the answer checks.

State. Each person receives $\dfrac{3}{20}$ lb of cheese.

29. *Familiarize*. Let $w=$ Monroe's weight, in pounds. We want to find $\dfrac{5}{7}$ of 175 lb.

Translate. We write a multiplication sentence.

$$w=\dfrac{5}{7}\cdot175$$

Solve. We carry out the multiplication.

$$w=\dfrac{5}{7}\cdot175=\dfrac{5\cdot175}{7}=\dfrac{5\cdot7\cdot25}{7\cdot1}$$

$$=\dfrac{7}{7}\cdot\dfrac{5\cdot25}{1}$$

$$=125$$

Check. We can repeat the calculation. The answer checks.

State. Monroe weighs 125 lb.

30. $\dfrac{7}{8}\cdot x=56$

$\quad x=56\div\dfrac{7}{8}\qquad$ Dividing by $\dfrac{7}{8}$ on both sides

$\quad x=56\cdot\dfrac{8}{7}$

$\quad=\dfrac{56\cdot8}{7}=\dfrac{7\cdot8\cdot8}{7\cdot1}=\dfrac{7}{7}\cdot\dfrac{8\cdot8}{1}=\dfrac{8\cdot8}{1}=64$

The solution is 64.

31.
$$\dfrac{7}{10}=\dfrac{-2}{5}\cdot t$$
$$\dfrac{5}{-2}\cdot\dfrac{7}{10}=\dfrac{5}{-2}\cdot\dfrac{-2}{5}\cdot t$$
$$\dfrac{5\cdot7}{-2\cdot10}=1t$$
$$\dfrac{5\cdot7}{-2\cdot2\cdot5}=t$$
$$\dfrac{7}{-2\cdot2}\cdot\dfrac{5}{5}=t$$
$$\dfrac{7}{-4}=t$$

The solution is $\dfrac{7}{-4}$, or $-\dfrac{7}{4}$, or $\dfrac{-7}{4}$.

32. $A=\dfrac{1}{2}\cdot b\cdot h$

$A=\dfrac{1}{2}\cdot13\text{ m}\cdot7\text{ m}$

$A=\dfrac{13\cdot7}{2}\text{ m}^2=\dfrac{91}{2}\text{ m}^2$

33. *Familiarize*. This is a multistep problem. First we will find half the amount of salt for one batch of pancakes. Then we will find 5 times this amount. Let $s=$ half the amount of salt in a single batch, in teaspoons.

Translate. We translate to an equation.

$$s=\dfrac{1}{2}\cdot\dfrac{3}{4}$$

Solve. We carry out the multiplication.

$$s=\dfrac{1}{2}\cdot\dfrac{3}{4}=\dfrac{1\cdot3}{2\cdot4}=\dfrac{3}{8}$$

Half the amount of salt in a single batch of pancakes is $\dfrac{3}{8}$ tsp. Let $p=$ the number of teaspoons of salt in 5 batches.

The equation that corresponds to this situation is

$$p=5\cdot\dfrac{3}{8}.$$

We solve the equation by carrying out the multiplication.

$$p=5\cdot\dfrac{3}{8}=\dfrac{5\cdot3}{8}=\dfrac{15}{8}$$

Check. We repeat the calculations. The answer checks.

State. Jacqueline will need $\dfrac{15}{8}$ tsp of salt.

34. *Familiarize*. This is a multistep problem. First we will find the number of acres Karl received. Then we will find how much of that land Irene received. Let $k=$ the number of acres of land Karl received.

Translate. We translate to an equation.

$$k = \frac{7}{8} \cdot \frac{2}{3}$$

Solve. We carry out the multiplication.

$$k = \frac{7}{8} \cdot \frac{2}{3} = \frac{7 \cdot 2}{8 \cdot 3} = \frac{7 \cdot 2}{2 \cdot 4 \cdot 3} = \frac{2}{2} \cdot \frac{7}{4 \cdot 3} = \frac{7}{12}$$

Karl received $\frac{7}{12}$ acre of land. Let $a =$ the number of acres Irene received. An equation that corresponds to this situation is

$$a = \frac{1}{4} \cdot \frac{7}{12}.$$

We solve the equation by carrying out the multiplication.

$$a = \frac{1}{4} \cdot \frac{7}{12} = \frac{1 \cdot 7}{4 \cdot 12} = \frac{7}{48}$$

Check. We repeat the calculations. The answer checks.

State. Irene received $\frac{7}{48}$ acre of land.

35. First we will evaluate the exponential expression; then we will multiply and divide in order from left to right.

$$\left(-\frac{3}{8}\right)^2 \div \frac{6}{7} \cdot \frac{2}{9} \div (-5) = \frac{9}{64} \div \frac{6}{7} \cdot \frac{2}{9} \div (-5)$$

$$= \frac{9}{64} \cdot \frac{7}{6} \cdot \frac{2}{9} \div (-5)$$

$$= \frac{9 \cdot 7}{64 \cdot 6} \cdot \frac{2}{9} \div (-5)$$

$$= \frac{9 \cdot 7 \cdot 2}{64 \cdot 6 \cdot 9} \div (-5)$$

$$= \frac{9 \cdot 7 \cdot 2}{64 \cdot 6 \cdot 9} \cdot \left(-\frac{1}{5}\right)$$

$$= -\frac{9 \cdot 7 \cdot 2 \cdot 1}{64 \cdot 6 \cdot 9 \cdot 5}$$

$$= -\frac{9 \cdot 7 \cdot 2 \cdot 1}{64 \cdot 2 \cdot 3 \cdot 9 \cdot 5}$$

$$= -\frac{9 \cdot 2}{9 \cdot 2} \cdot \frac{7 \cdot 1}{64 \cdot 3 \cdot 5}$$

$$= -\frac{7}{960}, \text{ or } \frac{-7}{960}$$

36.

$$\frac{33}{38} \cdot \frac{34}{55} = \frac{17}{35} \cdot \frac{15}{19}x$$

$$\frac{33 \cdot 34}{38 \cdot 55} = \frac{17 \cdot 15}{35 \cdot 19}x$$

$$\frac{3 \cdot 11 \cdot 2 \cdot 17}{2 \cdot 19 \cdot 5 \cdot 11} = \frac{17 \cdot 3 \cdot 5}{5 \cdot 7 \cdot 19}x$$

$$\frac{2 \cdot 11}{2 \cdot 11} \cdot \frac{3 \cdot 17}{19 \cdot 5} = \frac{5}{5} \cdot \frac{17 \cdot 3}{7 \cdot 19}x$$

$$\frac{3 \cdot 17}{19 \cdot 5} = \frac{17 \cdot 3}{7 \cdot 19}x$$

$$\frac{3 \cdot 17}{19 \cdot 5} \div \frac{17 \cdot 3}{7 \cdot 19} = x \quad \text{Dividing by } \frac{17 \cdot 3}{7 \cdot 19} \text{ on both sides}$$

$$\frac{3 \cdot 17}{19 \cdot 5} \cdot \frac{7 \cdot 19}{17 \cdot 3} = x$$

$$\frac{3 \cdot 17 \cdot 7 \cdot 19}{19 \cdot 5 \cdot 17 \cdot 3} = x$$

$$\frac{3 \cdot 17 \cdot 19}{3 \cdot 17 \cdot 19} \cdot \frac{7}{5} = x$$

$$\frac{7}{5} = x$$

The solution is $\frac{7}{5}$.

Cumulative Review Chapters 1 - 3

1. A word name for 2,056,783 is two million, fifty-six thousand, seven hundred eighty-three.

2.
$$
\begin{array}{r}
\overset{1}{} \\
2\,7\,4\,3 \\
+\,8\,2\,3\,9 \\
\hline
1\,0,9\,8\,2
\end{array}
$$

3. $-29 + (-14)$

Add the absolute values: $29 + 14 = 43$

Make the answer negative: $-29 + (-14) = -43$

4. $-45 + 12$ The absolute values are 45 and 12. The difference is 33. The negative number has the larger absolute value, so the answer is negative. $-45 + 12 = -33$

5.
$$
\begin{array}{r}
\overset{11}{} \\
\overset{2}{}\,\overset{1}{}\,\overset{14}{} \\
6\,3\,2\,4 \\
-\,4\,1\,9\,5 \\
\hline
2\,1\,2\,9
\end{array}
$$

6. $27 - 50 = 27 + (-50) = -23$

7. $-12 - (-4) = -12 + 4 = -8$

8.
$$
\begin{array}{r}
\overset{1}{} \\
\overset{1}{}\,\overset{1}{} \\
7\,3\,5 \\
\times2\,3 \\
\hline
2\,2\,0\,5 \\
1\,4\,7\,0\,0 \\
\hline
1\,6,9\,0\,5
\end{array}
$$

9. $-52 \cdot 6 = -312$

10. $\frac{6}{7} \cdot (-35x) = -\frac{6 \cdot 35x}{7} = -\frac{6 \cdot 5 \cdot 7 \cdot x}{7 \cdot 1} =$

$-\frac{6 \cdot 5 \cdot x}{1} \cdot \frac{7}{7} = -30x$

11. $\frac{2}{9} \cdot \frac{21}{10} = \frac{2 \cdot 21}{9 \cdot 10} = \frac{2 \cdot 3 \cdot 7}{3 \cdot 3 \cdot 2 \cdot 5} = \frac{2 \cdot 3}{2 \cdot 3} \cdot \frac{7}{3 \cdot 5} = \frac{7}{15}$

12.
$$\begin{array}{r} 2\,3\,5 \\ 1\,3\,\overline{)3\,0\,5\,8} \\ \underline{2\,6\,0\,0} \\ 4\,5\,8 \\ \underline{3\,9\,0} \\ 6\,8 \\ \underline{6\,5} \\ 3 \end{array}$$

The answer is 235 R 3.

13. $-85 \div 5 = -17$ Check: $-17 \cdot 5 = -85$

14. $-16 \div \frac{4}{7} = -16 \cdot \frac{7}{4} = -\frac{16 \cdot 7}{4} = -\frac{4 \cdot 4 \cdot 7}{4 \cdot 1} =$

$-\frac{4 \cdot 7}{1} \cdot \frac{4}{4} = -28$

15. $\frac{3}{7} \div \frac{9}{14} = \frac{3}{7} \cdot \frac{14}{9} = \frac{3 \cdot 14}{7 \cdot 9} = \frac{3 \cdot 2 \cdot 7}{7 \cdot 3 \cdot 3} = \frac{3 \cdot 7}{3 \cdot 7} \cdot \frac{2}{3} = \frac{2}{3}$

16. $4\,5\,0\,\boxed{9}$
$\uparrow$

Locate the number in the tens place, 0. Consider the next digit to the right, 9. Since 9 is 5 or higher, round 0 tens up to 1 ten. Then change the digit to the right of the tens digit to zeros.

The answer is 4510.

17.

	Rounded to the nearest hundred
$\begin{array}{r} 9\,2\,1 \\ \times\,4\,5\,3 \\ \hline \end{array}$	$\begin{array}{r} 9\,0\,0 \\ \times\,5\,0\,0 \\ \hline 4\,5\,0,0\,0\,0 \end{array}$

18. The distance of -479 from 0 is 479, so $|-479| = 479$.

19. $10^2 \div 5(-2) - 8(2 - 8)$
$= 10^2 \div 5(-2) - 8(-6)$
$= 100 \div 5(-2) - 8(-6)$
$= 20(-2) - 8(-6)$
$= -40 - 8(-6)$
$= -40 + 48$
$= 8$

20. The number 98 has factors 1, 2, 7, and 98. Since it has at least one factor other than itself and 1, it is composite.

21. $a - b^2 = -5 - 4^2 = -5 - 16 = -21$

22. $a + 24 = 49$
$a + 24 - 24 = 49 - 24$
$a = 25$
The solution is 25.

23. $7x = 49$
$\frac{7x}{7} = \frac{49}{7}$
$x = 7$
The solution is 7.

24. $\frac{2}{9} \cdot a = -10$
$\frac{9}{2} \cdot \frac{2}{9} \cdot a = \frac{9}{2} \cdot (-10)$
$a = -\frac{9 \cdot 10}{2} = -\frac{9 \cdot 2 \cdot 5}{2 \cdot 1}$
$= -\frac{9 \cdot 5}{1} \cdot \frac{2}{2} = -45$
The solution is -45.

25. *Familiarize.* Let m = the number of miles per gallon by which the van's gas mileage exceeded the truck's.

Translate. We can think of this as a "how much more" situation.

Truck's miles	plus	How many more miles	is	Van's miles
$\downarrow$	$\downarrow$	$\downarrow$	$\downarrow$	$\downarrow$
17	$+$	m	$=$	25

Solve. We solve the equation.
$17 + m = 25$
$17 + m - 17 = 25 - 17$
$m = 8$

Check. Since $17 + 8 = 25$, the answer checks.

State. The van got 8 more miles per gallon than the truck.

26. *Familiarize.* Let m = the number of ounces of coffee in each mug.

Translate. We write a division sentence.
$m = 48 \div 6$

Solve. We carry out the division.
$m = 48 \div 6 = 8$

Check. Since $6 \cdot 8 = 48$, the answer checks.

State. Each mug will contain 8 oz of coffee.

27. $8 - 4x - 13 + 9x = -4x + 9x + 8 - 13$
$= (-4 + 9)x + (8 - 13)$
$= 5x - 5$

28. $-12x + 7y + 15x = -12x + 15x + 7y$
$= (-12 + 15)x + 7y$
$= 3x + 7y$

29. Remember: $\frac{n}{n} = 1$ for any integer n that is not 0.
$\frac{97}{97} = 1$

30. Remember: $\frac{0}{n} = 0$ for any integer n that is not 0.
$\frac{0}{81} = 0$

31. Remember: $\dfrac{n}{1} = -n$.

$\dfrac{63}{1} = 63$

32. $\dfrac{-10}{54} = -\dfrac{10}{54} = -\dfrac{2 \cdot 5}{2 \cdot 27} = -\dfrac{5}{27} \cdot \dfrac{2}{2} = -\dfrac{5}{27}$

33. Interchange the numerator and denominator.

The reciprocal of $\dfrac{2}{5}$ is $\dfrac{5}{2}$. $\left(\dfrac{2}{5} \cdot \dfrac{5}{2} = \dfrac{10}{10} = 1 \right)$

34. Interchange the numerator and denominator.

The reciprocal of $\dfrac{1}{57}$ is $\dfrac{57}{1}$, or 57. $\left(\dfrac{1}{57} \cdot \dfrac{57}{1} = \dfrac{57}{57} = 1 \right)$

35. Since $70 \div 10 = 7$, we multiply by $\dfrac{7}{7}$.

$\dfrac{3}{10} = \dfrac{3}{10} \cdot \dfrac{7}{7} = \dfrac{21}{70}$

36. *Familiarize.* Let $d =$ the number of students who live in dorms. We want to find $\dfrac{5}{8}$ of 7000.

Translate. We write a multiplication sentence.

$d = \dfrac{5}{8} \cdot 7000$

Solve. We carry out the multiplication.

$\begin{aligned} d &= \dfrac{5}{8} \cdot 7000 = \dfrac{5 \cdot 7000}{8} \\ &= \dfrac{5 \cdot 8 \cdot 875}{8 \cdot 1} = \dfrac{8}{8} \cdot \dfrac{5 \cdot 875}{1} \\ &= 4375 \end{aligned}$

Check. We repeat the calculation. The answer checks.

State. 4375 students live in dorms.

37. *Familiarize.* Let $t =$ the number of quarts of tea the thermos holds when it is full.

Translate.

$\underset{\downarrow}{\dfrac{3}{5}} \ \underset{\downarrow}{\text{of}} \ \underbrace{\text{a full thermos}}_{\downarrow} \ \underset{\downarrow}{\text{is}} \ \underbrace{\text{3 qt}}_{\downarrow}$

$\dfrac{3}{5} \qquad \cdot \qquad t \qquad = \qquad 3$

Solve. We solve the equation.

$\begin{aligned} \dfrac{3}{5} \cdot t &= 3 \\ \dfrac{5}{3} \cdot \dfrac{3}{5} \cdot t &= \dfrac{5}{3} \cdot 3 \\ t &= \dfrac{5 \cdot 3}{3} = \dfrac{3}{3} \cdot \dfrac{5}{1} \\ t &= 5 \end{aligned}$

Check. Since $\dfrac{3}{5}$ of 5 qt is $\dfrac{3}{5} \cdot 5 = \dfrac{3 \cdot 5}{3} = 3$ qt, the answer checks.

State. The thermos holds 5 qt when it is full.

38. *Familiarize.* Let $d =$ the number of miles Tony has jogged.

Translate.

$\underbrace{\text{Distance Tony has jogged}} \ \text{is} \ \dfrac{2}{3} \ \text{of} \ \underbrace{\dfrac{9}{10}}_{} \text{mi}$

$\qquad\qquad\downarrow \qquad\qquad\quad \downarrow \ \downarrow \ \downarrow \qquad \downarrow$

$\qquad\qquad d \qquad\qquad\quad = \dfrac{2}{3} \ \cdot \quad \dfrac{9}{10}$

Solve. We carry out the multiplication.

$\begin{aligned} d &= \dfrac{2}{3} \cdot \dfrac{9}{10} = \dfrac{2 \cdot 9}{3 \cdot 10} = \dfrac{2 \cdot 3 \cdot 3}{3 \cdot 2 \cdot 5} \\ &= \dfrac{2 \cdot 3}{2 \cdot 3} \cdot \dfrac{3}{5} = \dfrac{3}{5} \end{aligned}$

Check. We repeat the calculation. The answer checks.

State. Tony has jogged $\dfrac{3}{5}$ mi.

39.

$\begin{aligned} \dfrac{ab}{c} &= \dfrac{-\dfrac{2}{5} \cdot \dfrac{10}{13}}{\dfrac{26}{27}} = -\dfrac{2}{5} \cdot \dfrac{10}{13} \cdot \dfrac{27}{26} \\ &= -\dfrac{2 \cdot 10}{5 \cdot 13} \cdot \dfrac{27}{26} = -\dfrac{2 \cdot 10 \cdot 27}{5 \cdot 13 \cdot 26} \\ &= -\dfrac{2 \cdot 2 \cdot 5 \cdot 27}{5 \cdot 13 \cdot 2 \cdot 13} = -\dfrac{2 \cdot 27}{13 \cdot 13} \cdot \dfrac{2 \cdot 5}{2 \cdot 5} \\ &= -\dfrac{54}{169} \end{aligned}$

40.

$\begin{aligned} -|xy|^2 &= -\left| -\dfrac{3}{5} \cdot \dfrac{1}{2} \right|^2 = -\left| -\dfrac{3 \cdot 1}{5 \cdot 2} \right|^2 \\ &= -\left| -\dfrac{3}{10} \right|^2 = -\left(\dfrac{3}{10} \right)^2 \\ &= -\dfrac{9}{100} \end{aligned}$

41. *Familiarize.* This is a multistep problem. First we will find the total earnings for three days. In three days Wayne and Patty each earn $3 \cdot \$85$ and Janet earns $3 \cdot \$90$. Let $t =$ the total earnings of the three people for three days.

Translate. We can add.

$t = 3 \cdot 85 + 3 \cdot 85 + 3 \cdot 90$

Solve. We carry out the computation.

$\begin{aligned} t &= 3 \cdot 85 + 3 \cdot 85 + 3 \cdot 90 \\ &= 255 + 255 + 270 \\ &= 780 \end{aligned}$

Now let $x =$ the amount spent on entertainment. We write a multiplication sentence.

$x = \dfrac{2}{5} \cdot 780$

We solve the equation.

$\begin{aligned} x &= \dfrac{2}{5} \cdot 780 = \dfrac{2 \cdot 780}{5} = \dfrac{2 \cdot 5 \cdot 156}{5 \cdot 1} \\ &= \dfrac{5}{5} \cdot \dfrac{2 \cdot 156}{1} \\ &= 312 \end{aligned}$

Now let $s =$ the amount that is saved.

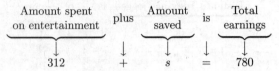

Amount spent on entertainment	plus	Amount saved	is	Total earnings
312	+	s	=	780

We solve this equation.

$$312 + s = 780$$
$$312 + s - 312 = 780 - 312$$
$$s = 468$$

Check. We repeat the calculation. The answer checks.

State. Wayne, Patty, and Janet will save \$468.

Chapter 4

Fraction Notation: Addition and Subtraction

Exercise Set 4.1

In this section we will find the LCM using the multiples method in Exercises 1 - 19 and the prime factorization method in Exercises 21 - 43.

1. 1. 10 is the larger number and is a multiple of 5, so it is the LCM.

The LCM = 10.

3. 1. 25 is the larger number, but it is not a multiple of 10.

2. Check multiples of 25:

$$2 \cdot 25 = 50 \qquad \text{A multiple of 10}$$

The LCM = 50.

5. 1. 40 is the larger number and is a multiple of 20, so it is the LCM.

The LCM = 40.

7. 1. 27 is the larger number, but it is not a multiple of 18.

2. Check multiples of 27:

$$2 \cdot 27 = 54 \qquad \text{A multiple of 18}$$

The LCM = 54.

9. 1. 50 is the larger number, but it is not a multiple of 30.

2. Check multiples of 50:

$$2 \cdot 50 = 100 \qquad \text{Not a multiple of 30}$$
$$3 \cdot 50 = 150 \qquad \text{A multiple of 30}$$

The LCM = 150.

11. 1. 40 is the larger number, but it is not a multiple of 30.

2. Check multiples of 40:

$$2 \cdot 40 = 80 \qquad \text{Not a multiple of 30}$$
$$3 \cdot 40 = 120 \qquad \text{A multiple of 30}$$

The LCM = 120.

13. 1. 24 is the larger number, but it is not a multiple of 18.

2. Check multiples of 24:

$$2 \cdot 24 = 48 \qquad \text{Not a multiple of 18}$$
$$3 \cdot 24 = 72 \qquad \text{A multiple of 18}$$

The LCM = 72.

15. 1. 70 is the larger number, but it is not a multiple of 60.

2. Check multiples of 70:

$$2 \cdot 70 = 140 \qquad \text{Not a multiple of 60}$$
$$3 \cdot 70 = 210 \qquad \text{Not a multiple of 60}$$
$$4 \cdot 70 = 280 \qquad \text{Not a multiple of 60}$$
$$5 \cdot 70 = 350 \qquad \text{Not a multiple of 60}$$
$$6 \cdot 70 = 420 \qquad \text{A multiple of 60}$$

The LCM = 420.

17. 1. 36 is the larger number, but it is not a multiple of 16.

2. Check multiples of 36:

$$2 \cdot 36 = 72 \qquad \text{Not a multiple of 16}$$
$$3 \cdot 36 = 108 \qquad \text{Not a multiple of 16}$$
$$4 \cdot 36 = 144 \qquad \text{A multiple of 16}$$

The LCM = 144.

19. 1. 20 is the larger number, but it is not a multiple of 18.

2. Check multiples of 20:

$$2 \cdot 20 = 40 \qquad \text{Not a multiple of 18}$$
$$3 \cdot 20 = 60 \qquad \text{Not a multiple of 18}$$
$$4 \cdot 20 = 80 \qquad \text{Not a multiple of 18}$$
$$5 \cdot 20 = 100 \qquad \text{Not a multiple of 18}$$
$$6 \cdot 20 = 120 \qquad \text{Not a multiple of 18}$$
$$7 \cdot 20 = 140 \qquad \text{Not a multiple of 18}$$
$$8 \cdot 20 = 160 \qquad \text{Not a multiple of 18}$$
$$9 \cdot 20 = 180 \qquad \text{A multiple of 18}$$

The LCM = 180.

21. 1. Write the prime factorization of each number. Because 2, 3, and 7 are all prime we write $2 = 2$, $3 = 3$, and $7 = 7$.

2. a) None of the factorizations contains the other two.

b) We begin with 2. Since 3 contains a factor of 3, we multiply by 3:

$$2 \cdot 3$$

Next we multiply $2 \cdot 3$ by 7, the factor of 7 that is missing:

$$2 \cdot 3 \cdot 7$$

The LCM is $2 \cdot 3 \cdot 7$, or 42.

3. To check, note that 2, 3, and 7 appear in the LCM the greatest number of times that each appears as a factor of 2, 3, or 7. The LCM is $2 \cdot 3 \cdot 7$, or 42.

23. 1. Write the prime factorization of each number.

$$3 = 3$$
$$6 = 2 \cdot 3$$
$$15 = 3 \cdot 5$$

2. a) None of the factorizations contains the other two.

b) We first consider 3 and 6. Since the factorization of 6 contains 3, we next multiply $2 \cdot 3$ by the factor of 15 that is missing, 5. The LCM is $2 \cdot 3 \cdot 5$, or 30.

3. To check, note that 2, 3, and 5 appear in the LCM the greatest number of times that each appears as a factor of 3, 6, or 15. The LCM is $2 \cdot 3 \cdot 5$, or 30.

25. 1. Write the prime factorization of each number.

$$24 = 2 \cdot 2 \cdot 2 \cdot 3$$
$$36 = 2 \cdot 2 \cdot 3 \cdot 3$$
$$12 = 2 \cdot 2 \cdot 3$$

2. a) None of the factorizations contains the other two.

   b) We begin with the factorization of 24, $2 \cdot 2 \cdot 2 \cdot 3$. Since 36 contains a second factor of 3, we multiply by another factor of 3:

   $$2 \cdot 2 \cdot 2 \cdot 3 \cdot 3$$

   Next we look for factors of 12 that are still missing. There are none. The LCM is

   $$2 \cdot 2 \cdot 2 \cdot 3 \cdot 3, \text{ or } 72.$$

3. To check, note that 2 and 3 appear in the LCM the greatest number of times that each appears as a factor of 24, 36, or 12. The LCM is $2 \cdot 2 \cdot 2 \cdot 3 \cdot 3$, or 72.

27. 1. Write the prime factorization of each number.

   $$5 = 5$$
   $$12 = 2 \cdot 2 \cdot 3$$
   $$15 = 3 \cdot 5$$

2. a) None of the factorizations contains the other two.

   b) We begin with the factorization of 12, $2 \cdot 2 \cdot 3$. Since 5 contains a factor of 5, we multiply by 5:

   $$2 \cdot 2 \cdot 3 \cdot 5$$

   Next we look for factors of 15 that are still missing. There are none. The LCM is $2 \cdot 2 \cdot 3 \cdot 5$, or 60.

3. The result checks.

29. 1. Write the prime factorization of each number.

   $$9 = 3 \cdot 3$$
   $$12 = 2 \cdot 2 \cdot 3$$
   $$6 = 2 \cdot 3$$

2. a) None of the factorizations contains the other two.

   b) We begin with the factorization of 12, $2 \cdot 2 \cdot 3$. Since 9 contains a second factor of 3, we multiply by another factor of 3:

   $$2 \cdot 2 \cdot 3 \cdot 3$$

   Next we look for factors of 6 that are still missing. There are none. The LCM is $2 \cdot 2 \cdot 3 \cdot 3$, or 36.

3. The result checks.

31. 1. Write the prime factorization of each number.

   $$180 = 2 \cdot 2 \cdot 3 \cdot 3 \cdot 5$$
   $$100 = 2 \cdot 2 \cdot 5 \cdot 5$$
   $$450 = 2 \cdot 3 \cdot 3 \cdot 5 \cdot 5$$

2. a) None of the factorizations contains the other two.

   b) We begin with the factorization of 450, $2 \cdot 3 \cdot 3 \cdot 5 \cdot 5$. Since 180 contains another factor of 2, we multiply by 2:

   $$2 \cdot 3 \cdot 3 \cdot 5 \cdot 5 \cdot 2$$

   Next we look for factors of 100 that are still missing. There are none. The LCM is $2 \cdot 3 \cdot 3 \cdot 5 \cdot 5 \cdot 2$, or 900.

3. The result checks.

33. 1. Write the prime factorization of each number.

   $$75 = 3 \cdot 5 \cdot 5$$
   $$100 = 2 \cdot 2 \cdot 5 \cdot 5$$

2. a) Neither factorization contains the other.

   b) We begin with the factorization of 100, $2 \cdot 2 \cdot 5 \cdot 5$. Since 75 contains a factor of 3, we multiply by 3:

   $$2 \cdot 2 \cdot 5 \cdot 5 \cdot 3$$

   The LCM is $2 \cdot 2 \cdot 5 \cdot 5 \cdot 3$, or 300.

3. The result checks.

35. 1. We have the following factorizations:

   $$ab = a \cdot b$$
   $$bc = b \cdot c$$

2. a) Neither factorization contains the other.

   b) Consider the factorization of ab, $a \cdot b$. Since bc contains a factor of c, we multiply by c.

   $$a \cdot b \cdot c$$

   The LCM is $a \cdot b \cdot c$, or abc.

3. The result checks.

37. 1. We have the following factorizations:

   $$3x = 3 \cdot x$$
   $$9x^2 = 3 \cdot 3 \cdot x \cdot x$$

2. a) One factorization, $3 \cdot 3 \cdot x \cdot x$, contains the other. Thus the LCM is $3 \cdot 3 \cdot x \cdot x$, or $9x^2$.

39. 1. We have the following factorizations:

   $$4x^3 = 2 \cdot 2 \cdot x \cdot x \cdot x$$
   $$x^2y = x \cdot x \cdot y$$

2. a) Neither factorization contains the other.

   b) Consider the factorization of $4x^3$, $2 \cdot 2 \cdot x \cdot x \cdot x$. Since x^2y contains a factor of y, we multiply by y.

   $$2 \cdot 2 \cdot x \cdot x \cdot x \cdot y$$

   The LCM is $2 \cdot 2 \cdot x \cdot x \cdot x \cdot y$, or $4x^3y$.

3. The result checks.

41. 1. We have the following factorizations:

   $$6r^3st^4 = 2 \cdot 3 \cdot r \cdot r \cdot r \cdot s \cdot t \cdot t \cdot t \cdot t$$
   $$8rs^2t = 2 \cdot 2 \cdot 2 \cdot r \cdot s \cdot s \cdot t$$

2. a) Neither factorization contains the other.

   b) Consider the factorization of $6r^3st^4$, $2 \cdot 3 \cdot r \cdot r \cdot r \cdot s \cdot t \cdot t \cdot t \cdot t$. Since $8rs^2t$ contains two more factors of 2 and one more factor of s, we multiply by $2 \cdot 2 \cdot s$.

   $$2 \cdot 3 \cdot r \cdot r \cdot r \cdot s \cdot t \cdot t \cdot t \cdot t \cdot 2 \cdot 2 \cdot s$$

   The LCM is $2 \cdot 3 \cdot 2 \cdot 2 \cdot r \cdot r \cdot r \cdot s \cdot s \cdot t \cdot t \cdot t \cdot t$, or $24r^3s^2t^4$.

3. The result checks.

43. 1. We have the following factorizations:

   $$a^3b = a \cdot a \cdot a \cdot b$$
   $$b^2c = b \cdot b \cdot c$$
   $$ac^2 = a \cdot c \cdot c$$

2. a) No one factorization contains the others.

  b) Consider the factorization of a^3b,
$a \cdot a \cdot a \cdot b$. Since b^2c contains another factor of b and a factor of c, we multiply by $b \cdot c$.

$$a \cdot a \cdot a \cdot b \cdot b \cdot c$$

Now consider ac^2. Since ac^2 contains another factor of c, we multiply by c.

$$a \cdot a \cdot a \cdot b \cdot b \cdot c \cdot c$$

The LCM is $a \cdot a \cdot a \cdot b \cdot b \cdot c \cdot c$, or $a^3b^2c^2$.

3. The result checks.

45. We find the LCM of the number of years it takes Jupiter and Saturn to make a complete revolution around the sun.

    Jupiter: $12 = 2 \cdot 2 \cdot 3$

    Saturn: $30 = 2 \cdot 3 \cdot 5$

The LCM $= 2 \cdot 2 \cdot 3 \cdot 5$, or 60. Thus, Jupiter and Saturn will appear in the exact same direction in the night sky as seen from Earth tonight once every 60 years.

47. Discussion and Writing Exercise

49. $-38 + 52$

The absolute values are 38 and 52. The difference is 14. The positive number has the larger absolute value, so the answer is positive.

$$-38 + 52 = 14$$

51.
$$\begin{array}{r} {\scriptstyle 1} \\ {\scriptstyle 1\ 1} \\ 3\ 4\ 5 \\ \times\ \ \ 2\ 3 \\ \hline 1\ 0\ 3\ 5 \\ 6\ 9\ 0\ 0 \\ \hline 7\ 9\ 3\ 5 \end{array}$$

53. $\dfrac{4}{5} \div \left(-\dfrac{7}{10}\right) = \dfrac{4}{5} \cdot \left(-\dfrac{10}{7}\right) = -\dfrac{4 \cdot 10}{5 \cdot 7} = -\dfrac{4 \cdot 2 \cdot 5}{5 \cdot 7} =$

$-\dfrac{4 \cdot 2}{7} \cdot \dfrac{5}{5} = -\dfrac{8}{7}$

55. Discussion and Writing Exercise

57. Discussion and Writing Exercise

59. a) 7800 is the larger number, but it is not a multiple of 2700.

  b) Check multiples using a calculator:

$2 \cdot 7800 = 15,600$	Not a multiple of 2700
$3 \cdot 7800 = 23,400$	Not a multiple of 2700
$4 \cdot 7800 = 31,200$	Not a multiple of 2700
$5 \cdot 7800 = 39,000$	Not a multiple of 2700
$6 \cdot 7800 = 46,800$	Not a multiple of 2700
$7 \cdot 7800 = 54,600$	Not a multiple of 2700
$8 \cdot 7800 = 62,400$	Not a multiple of 2700
$9 \cdot 7800 = 70,200$	A multiple of 2700

  c) The LCM is 70,200.

61. a) 24,339 is the larger number, but it is not a multiple of 17,385.

  b) Check multiples using a calculator:

$2 \cdot 24,339 = 48,678$	Not a multiple of 17,385
$3 \cdot 24,339 = 73,017$	Not a multiple of 17,385
$4 \cdot 24,339 = 97,356$	Not a multiple of 17,385
$5 \cdot 24,339 = 121,695$	A multiple of 17,385

  c) The LCM is 121,695.

63. The smallest number of strands that can be used is the LCM of 10 and 3. The prime factorizations are $10 = 2 \cdot 5$ and $3 = 3$. Thus, the LCM is $2 \cdot 5 \cdot 3$, or 30, so the smallest number of strands that can be used is 30.

65. We find the LCM of 30 and 14.

    $30 = 2 \cdot 3 \cdot 5$

    $14 = 2 \cdot 7$

We form the LCM using the greatest power of each factor.

The LCM is $2 \cdot 3 \cdot 5 \cdot 7$, or 210, so the prescriptions will both be refilled on the same day in 210 days.

67. 1. From Example 9 we know that the LCM of 27, 90, and 84 is $2 \cdot 3 \cdot 3 \cdot 5 \cdot 3 \cdot 2 \cdot 7$, so the LCM of 27, 90, 84, 210, 108, and 50 must contain at least these factors. We write the prime factorizations of 210, 108, and 50:

    $210 = 2 \cdot 3 \cdot 5 \cdot 7$

    $108 = 2 \cdot 2 \cdot 3 \cdot 3 \cdot 3$

    $50 = 2 \cdot 5 \cdot 5$

  2. a) Neither of the four factorizations above contains the other three.

    b) Begin with the LCM of 27, 90, and 84, $2 \cdot 3 \cdot 3 \cdot 5 \cdot 3 \cdot 2 \cdot 7$. Neither 210 nor 108 contains any factors that are missing in this factorization. Next we look for factors of 50 that are missing. Since 50 contains a second factor of 5, we multiply by 5:

$$2 \cdot 3 \cdot 3 \cdot 5 \cdot 3 \cdot 2 \cdot 7 \cdot 5$$

The LCM is $2 \cdot 3 \cdot 3 \cdot 5 \cdot 3 \cdot 2 \cdot 7 \cdot 5$, or 18,900.

  3. The result checks.

69. Answers may vary.

$$56 = 2 \cdot 2 \cdot 2 \cdot 7$$

Three pairs are $2 \cdot 2 \cdot 2$ and 7, or 8 and 7; $2 \cdot 2 \cdot 2$ and $2 \cdot 7$, or 8 and 14; and $2 \cdot 2 \cdot 2$ and $2 \cdot 2 \cdot 7$, or 8 and 28.

Exercise Set 4.2

1. $\dfrac{4}{9} + \dfrac{1}{9} = \dfrac{4+1}{9} = \dfrac{5}{9}$

3. $\dfrac{4}{7} + \dfrac{3}{7} = \dfrac{4+3}{7} = \dfrac{7}{7} = 1$

5. $\dfrac{7}{10} + \dfrac{3}{-10} = \dfrac{7}{10} + \dfrac{-3}{10} = \dfrac{7+(-3)}{10} = \dfrac{4}{10} = \dfrac{2 \cdot 2}{2 \cdot 5} =$

$\dfrac{2}{2} \cdot \dfrac{2}{5} = 1 \cdot \dfrac{2}{5} = \dfrac{2}{5}$

7. $\dfrac{9}{a} + \dfrac{4}{a} = \dfrac{9+4}{a} = \dfrac{13}{a}$

9. $\dfrac{-7}{11} + \dfrac{3}{11} = \dfrac{-7+3}{11} = \dfrac{-4}{11}$, or $-\dfrac{4}{11}$

11. $\dfrac{2}{9}x + \dfrac{5}{9}x = \left(\dfrac{2}{9} + \dfrac{5}{9}\right)x = \dfrac{7}{9}x$

13. $\quad \dfrac{3}{32}t + \dfrac{13}{32}t$

$\qquad = \left(\dfrac{3}{32} + \dfrac{13}{32}\right)t$

$\qquad = \dfrac{16}{32}t$

$\qquad = \dfrac{16 \cdot 1}{16 \cdot 2}t$

$\qquad = \dfrac{16}{16} \cdot \dfrac{1}{2}t$

$\qquad = \dfrac{1}{2}t$

15. $-\dfrac{2}{x} + \left(-\dfrac{7}{x}\right) = \dfrac{-2}{x} + \dfrac{-7}{x} = \dfrac{-2+(-7)}{x} = \dfrac{-9}{x}$,

$\qquad$ or $-\dfrac{9}{x}$

17. $\quad \dfrac{1}{8} + \dfrac{1}{6}$ $\qquad$ $8 = 2 \cdot 2 \cdot 2$ and $6 = 2 \cdot 3$, so the LCD is $2 \cdot 2 \cdot 2 \cdot 3$, or 24

$\quad = \dfrac{1}{8} \cdot \dfrac{3}{3} + \dfrac{1}{6} \cdot \dfrac{4}{4}$

$\qquad\qquad$ Think: $6 \times \square = 24$. The answer is 4, so we multiply by 1, using $\dfrac{4}{4}$.

$\qquad\qquad$ Think: $8 \times \square = 24$. The answer is 3, so we multiply by 1, using $\dfrac{3}{3}$.

$\quad = \dfrac{3}{24} + \dfrac{4}{24}$

$\quad = \dfrac{7}{24}$

19. $\quad \dfrac{-4}{5} + \dfrac{7}{10}$ $\qquad$ 5 is a factor of 10, so the LCD is 10.

$\quad = \dfrac{-4}{5} \cdot \dfrac{2}{2} + \dfrac{7}{10}$ $\leftarrow$ This fraction already has the LCD as denominator.

$\qquad\qquad$ Think: $5 \times \square = 10$. The answer is 2, so we multiply by 1, using $\dfrac{2}{2}$.

$\quad = \dfrac{-8}{10} + \dfrac{7}{10}$

$\quad = \dfrac{-1}{10}$, or $-\dfrac{1}{10}$

21. $\quad \dfrac{7}{12} + \dfrac{3}{8}$ $\qquad$ $12 = 2 \cdot 2 \cdot 3$ and $8 = 2 \cdot 2 \cdot 2$, so the LCD is $2 \cdot 2 \cdot 2 \cdot 3$, or 24.

$\quad = \dfrac{7}{12} \cdot \dfrac{2}{2} + \dfrac{3}{8} \cdot \dfrac{3}{3}$

$\qquad\qquad$ Think: $8 \times \square = 24$. The answer is 3, so we multiply by 1, using $\dfrac{3}{3}$.

$\qquad\qquad$ Think: $12 \times \square = 24$. The answer is 2, so we multiply by 1, using $\dfrac{2}{2}$.

$\quad = \dfrac{14}{24} + \dfrac{9}{24} = \dfrac{23}{24}$

23. $\quad \dfrac{3}{20} + 4$

$\quad = \dfrac{3}{20} + \dfrac{4}{1}$ $\qquad$ Rewriting 4 in fractional notation

$\quad = \dfrac{3}{20} + \dfrac{4}{1} \cdot \dfrac{20}{20}$ $\qquad$ The LCD is 20.

$\quad = \dfrac{3}{20} + \dfrac{80}{20}$

$\quad = \dfrac{83}{20}$

25. $\quad \dfrac{5}{-8} + \dfrac{5}{6}$

$\quad = \dfrac{-5}{8} + \dfrac{5}{6}$ $\qquad$ Recall that $\dfrac{m}{-n} = \dfrac{-m}{n}$. The LCD is 24. (See Exercise 17.)

$\quad = \dfrac{-5}{8} \cdot \dfrac{3}{3} + \dfrac{5}{6} \cdot \dfrac{4}{4}$

$\quad = \dfrac{-15}{24} + \dfrac{20}{24}$

$\quad = \dfrac{5}{24}$

27. $\quad \dfrac{3}{10}x + \dfrac{7}{100}x$

$\quad = \dfrac{3}{10} \cdot \dfrac{10}{10} \cdot x + \dfrac{7}{100}x$ $\qquad$ 10 is a factor of 100, so the LCD is 100.

$\quad = \dfrac{30}{100}x + \dfrac{7}{100}x$

$\quad = \dfrac{37}{100}x$

29. $\quad \dfrac{5}{12} + \dfrac{8}{15}$ $\qquad$ $12 = 2 \cdot 2 \cdot 3$ and $15 = 3 \cdot 5$, so the LCM is $2 \cdot 2 \cdot 3 \cdot 5$, or 60.

$\quad = \dfrac{5}{12} \cdot \dfrac{5}{5} + \dfrac{8}{15} \cdot \dfrac{4}{4}$

$\quad = \dfrac{25}{60} + \dfrac{32}{60} = \dfrac{57}{60}$

$\quad = \dfrac{3 \cdot 19}{3 \cdot 20} = \dfrac{3}{3} \cdot \dfrac{19}{20}$

$\quad = \dfrac{19}{20}$

31. $\dfrac{-7}{10} + \dfrac{-29}{100}$ 10 is a factor of 100, so the LCD is 100.

$= \dfrac{-7}{10} \cdot \dfrac{10}{10} + \dfrac{-29}{100}$

$= \dfrac{-70}{100} + \dfrac{-29}{100} = \dfrac{-99}{100}$, or $-\dfrac{99}{100}$

33. $-\dfrac{1}{10}x + \dfrac{1}{15}x$

$= -\dfrac{1}{2 \cdot 5}x + \dfrac{1}{3 \cdot 5}x$ The LCD is $2 \cdot 5 \cdot 3$.

$= -\dfrac{1}{2 \cdot 5} \cdot \dfrac{3}{3}x + \dfrac{1}{3 \cdot 5} \cdot \dfrac{2}{2}x$

$= -\dfrac{3}{30}x + \dfrac{2}{30}x$

$= -\dfrac{1}{30}x$

35. $-5t + \dfrac{2}{7}t$

$= \dfrac{-5}{1}t + \dfrac{2}{7}t$ The LCD is 7.

$= \dfrac{-5}{1} \cdot \dfrac{7}{7} \cdot t + \dfrac{2}{7}t$

$= \dfrac{-35}{7}t + \dfrac{2}{7}t$

$= \dfrac{-33}{7}t$, or $-\dfrac{33}{7}t$

37. $-\dfrac{5}{12} + \dfrac{7}{-24}$

$\dfrac{-5}{12} + \dfrac{-7}{24}$ 12 is a factor of 24, so the LCD is 24.

$= \dfrac{-5}{12} \cdot \dfrac{2}{2} + \dfrac{-7}{24}$

$= \dfrac{-10}{24} + \dfrac{-7}{24}$

$= \dfrac{-17}{24}$, or $-\dfrac{17}{24}$

39. $\dfrac{4}{10} + \dfrac{3}{100} + \dfrac{7}{1000}$ 10 and 100 are factors of 1000, so the LCD is 1000.

$= \dfrac{4}{10} \cdot \dfrac{100}{100} + \dfrac{3}{100} \cdot \dfrac{10}{10} + \dfrac{7}{1000}$

$= \dfrac{400}{1000} + \dfrac{30}{1000} + \dfrac{7}{1000}$

$= \dfrac{437}{1000}$

41. $\dfrac{3}{10} + \dfrac{5}{12} + \dfrac{8}{15}$

$= \dfrac{3}{2 \cdot 5} + \dfrac{5}{2 \cdot 2 \cdot 3} + \dfrac{8}{3 \cdot 5}$ Factoring the denominators
The LCD is $2 \cdot 5 \cdot 2 \cdot 3$.

$= \dfrac{3}{2 \cdot 5} \cdot \dfrac{2 \cdot 3}{2 \cdot 3} + \dfrac{5}{2 \cdot 2 \cdot 3} \cdot \dfrac{5}{5} + \dfrac{8}{3 \cdot 5} \cdot \dfrac{2 \cdot 2}{2 \cdot 2}$

In each case we multiply by 1 to obtain the LCD.

$= \dfrac{3 \cdot 2 \cdot 3}{2 \cdot 5 \cdot 2 \cdot 3} + \dfrac{5 \cdot 5}{2 \cdot 2 \cdot 3 \cdot 5} + \dfrac{8 \cdot 2 \cdot 2}{3 \cdot 5 \cdot 2 \cdot 2}$

$= \dfrac{18}{2 \cdot 5 \cdot 2 \cdot 3} + \dfrac{25}{2 \cdot 5 \cdot 2 \cdot 3} + \dfrac{32}{2 \cdot 5 \cdot 2 \cdot 3}$

$= \dfrac{75}{2 \cdot 5 \cdot 2 \cdot 3}$

$= \dfrac{3 \cdot 5 \cdot 5}{2 \cdot 5 \cdot 2 \cdot 3} = \dfrac{3 \cdot 5}{3 \cdot 5} \cdot \dfrac{5}{2 \cdot 2}$

$= \dfrac{5}{4}$

43. $\dfrac{5}{6} + \dfrac{25}{52} + \dfrac{7}{4}$

$= \dfrac{5}{2 \cdot 3} + \dfrac{25}{2 \cdot 2 \cdot 13} + \dfrac{7}{2 \cdot 2}$ LCD is $2 \cdot 3 \cdot 2 \cdot 13$.

$= \dfrac{5}{2 \cdot 3} \cdot \dfrac{2 \cdot 13}{2 \cdot 13} + \dfrac{25}{2 \cdot 2 \cdot 13} \cdot \dfrac{3}{3} + \dfrac{7}{2 \cdot 2} \cdot \dfrac{3 \cdot 13}{3 \cdot 13}$

$= \dfrac{5 \cdot 2 \cdot 13}{2 \cdot 3 \cdot 2 \cdot 13} + \dfrac{25 \cdot 3}{2 \cdot 2 \cdot 13 \cdot 3} + \dfrac{7 \cdot 3 \cdot 13}{2 \cdot 2 \cdot 3 \cdot 13}$

$= \dfrac{130}{2 \cdot 3 \cdot 2 \cdot 13} + \dfrac{75}{2 \cdot 3 \cdot 2 \cdot 13} + \dfrac{273}{2 \cdot 3 \cdot 2 \cdot 13}$

$= \dfrac{478}{2 \cdot 3 \cdot 2 \cdot 13}$

$= \dfrac{2 \cdot 239}{2 \cdot 3 \cdot 2 \cdot 13} = \dfrac{2}{2} \cdot \dfrac{239}{3 \cdot 2 \cdot 13}$

$= \dfrac{239}{78}$

45. $\dfrac{2}{9} + \dfrac{7}{10} + \dfrac{-4}{15}$

$= \dfrac{2}{3 \cdot 3} + \dfrac{7}{2 \cdot 5} + \dfrac{-4}{3 \cdot 5}$ LCD is $3 \cdot 3 \cdot 2 \cdot 5$.

$= \dfrac{2}{3 \cdot 3} \cdot \dfrac{2 \cdot 5}{2 \cdot 5} + \dfrac{7}{2 \cdot 5} \cdot \dfrac{3 \cdot 3}{3 \cdot 3} + \dfrac{-4}{3 \cdot 5} \cdot \dfrac{3 \cdot 2}{3 \cdot 2}$

$= \dfrac{2 \cdot 2 \cdot 5}{3 \cdot 3 \cdot 2 \cdot 5} + \dfrac{7 \cdot 3 \cdot 3}{2 \cdot 5 \cdot 3 \cdot 3} + \dfrac{-4 \cdot 3 \cdot 2}{3 \cdot 5 \cdot 3 \cdot 2}$

$= \dfrac{20}{3 \cdot 3 \cdot 2 \cdot 5} + \dfrac{63}{3 \cdot 3 \cdot 2 \cdot 5} + \dfrac{-24}{3 \cdot 3 \cdot 2 \cdot 5}$

$= \dfrac{59}{3 \cdot 3 \cdot 2 \cdot 5}$

$= \dfrac{59}{90}$

47.
$$-\frac{3}{4} + \frac{1}{5} + \frac{-7}{10}$$
$$= \frac{-3}{4} + \frac{1}{5} + \frac{-7}{10}$$
$$= \frac{-3}{2 \cdot 2} + \frac{1}{5} + \frac{-7}{2 \cdot 5} \qquad \text{The LCD is } 2 \cdot 2 \cdot 5.$$
$$= \frac{-3}{2 \cdot 2} \cdot \frac{5}{5} + \frac{1}{5} \cdot \frac{2 \cdot 2}{2 \cdot 2} + \frac{-7}{2 \cdot 5} \cdot \frac{2}{2}$$
$$= \frac{-15}{2 \cdot 2 \cdot 5} + \frac{4}{5 \cdot 2 \cdot 2} + \frac{-14}{2 \cdot 5 \cdot 2}$$
$$= \frac{-25}{2 \cdot 2 \cdot 5} = \frac{-5 \cdot 5}{2 \cdot 2 \cdot 5} = \frac{-5}{2 \cdot 2} \cdot \frac{5}{5}$$
$$= \frac{-5}{4}, \text{ or } -\frac{5}{4}$$

49. Since there is a common denominator, compare the numerators.
$$3 > 2, \text{ so } \frac{3}{8} > \frac{2}{8}.$$

51. The LCD is 6. We multiply $\frac{2}{3}$ by 1 to make the denominators the same.
$$\frac{2}{3} \cdot \frac{2}{2} = \frac{4}{6}$$
The denominator of $\frac{5}{6}$ is the LCD.
Since $4 < 5$, it follows that $\frac{4}{6} < \frac{5}{6}$, so $\frac{2}{3} < \frac{5}{6}$.

53. The LCD is 21. We multiply by 1 to make the denominators the same.
$$\frac{-2}{3} \cdot \frac{7}{7} = \frac{-14}{21}$$
$$\frac{-5}{7} \cdot \frac{3}{3} = \frac{-15}{21}$$
Since $-14 > -15$, it follows that $\frac{-14}{21} > \frac{-15}{21}$, so
$$\frac{-2}{3} > \frac{-5}{7}.$$

55. The LCD is 30. We multiply by 1 to make the denominators the same.
$$\frac{9}{15} \cdot \frac{2}{2} = \frac{18}{30}$$
$$\frac{7}{10} \cdot \frac{3}{3} = \frac{21}{30}$$
Since $18 < 21$, it follows that $\frac{18}{30} < \frac{21}{30}$, so $\frac{9}{15} < \frac{7}{10}$.

57. Express $-\frac{1}{5}$ as $\frac{-1}{5}$. The LCD is 20. Multiply by 1 to make the denominators the same.
$$\frac{3}{4} \cdot \frac{5}{5} = \frac{15}{20}$$
$$\frac{-1}{5} \cdot \frac{4}{4} = \frac{-4}{20}$$
Since $15 > -4$, it follows that $\frac{15}{20} > \frac{-4}{20}$, so $\frac{3}{4} > -\frac{1}{5}$. We might have observed at the outset that one number is positive and the other is negative and it follows that the positive number is greater than the negative number.

59. The LCD is 60. We multiply by 1 to make the denominators the same.
$$\frac{-7}{20} \cdot \frac{3}{3} = \frac{-21}{60}$$
$$\frac{-6}{15} \cdot \frac{4}{4} = \frac{-24}{60}$$
Since $-21 > -24$, it follows that $\frac{-21}{60} > \frac{-24}{60}$, so
$$\frac{-7}{20} > \frac{-6}{15}.$$

61. The LCD is 60. We multiply by 1 to make the denominators the same.
$$\frac{3}{10} \cdot \frac{6}{6} = \frac{18}{60}$$
$$\frac{5}{12} \cdot \frac{5}{5} = \frac{25}{60}$$
$$\frac{4}{15} \cdot \frac{4}{4} = \frac{16}{60}$$
Since $16 < 18$ and $18 < 25$, when we arrange $\frac{18}{60}, \frac{25}{60}$, and $\frac{16}{60}$ from smallest to largest we have $\frac{16}{60}, \frac{18}{60}, \frac{25}{60}$. Then it follows that when we arrange the original fractions from smallest to largest we have $\frac{4}{15}, \frac{3}{10}, \frac{5}{12}$.

63. *Familiarize.* We draw a picture. We let $p =$ the number of pounds of candy Todd bought.

$\frac{1}{4}$ lb	$\frac{1}{2}$ lb
\multicolumn{2}{c}{p}	

Translate. The problem can be translated to an equation as follows:

Pounds of gumdrops	plus	Pounds of caramels	is	Total pounds of candy
$\downarrow$	$\downarrow$	$\downarrow$	$\downarrow$	$\downarrow$
$\frac{1}{4}$	$+$	$\frac{1}{2}$	$=$	p

Solve. We carry out the addition. Since 4 is a multiple of 2, the LCM of the denominators is 4.
$$\frac{1}{4} + \frac{1}{2} = p$$
$$\frac{1}{4} + \frac{1}{2} \cdot \frac{2}{2} = p$$
$$\frac{1}{4} + \frac{2}{4} = p$$
$$\frac{3}{4} = p$$

Check. We check by repeating the calculation. We also note that the sum is larger than either of the individual weights, so the answer seems reasonable.

State. Todd bought $\frac{3}{4}$ lb of candy.

65. *Familiarize*. We draw a picture. We let D = the total distance walked.

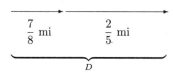

$\dfrac{7}{8}$ mi $\dfrac{2}{5}$ mi

D

Translate. The problem can be translated to an equation as follows:

Distance to student union	plus	Distance to class	is	Total distance
↓	↓	↓	↓	↓
$\dfrac{7}{8}$	$+$	$\dfrac{2}{5}$	$=$	D

Solve. To solve the equation, carry out the addition. Since $8 = 2 \cdot 2 \cdot 2$ and $5 = 5$, the LCM of the denominators is $2 \cdot 2 \cdot 2 \cdot 5$, or 40.

$$\frac{7}{8} + \frac{2}{5} = D$$
$$\frac{7}{8} \cdot \frac{5}{5} + \frac{2}{5} \cdot \frac{8}{8} = D$$
$$\frac{35}{40} + \frac{16}{40} = D$$
$$\frac{51}{40} = D$$

Check. We repeat the calculation. We also note that the sum is larger than either of the original distances, so the answer seems reasonable.

State. Kate walked $\dfrac{51}{40}$ mi.

67. *Familiarize*. We draw a picture and let f = the total amount of flour used.

$\dfrac{1}{2}$ lb	$\dfrac{1}{4}$ lb	$\dfrac{1}{3}$ lb
	f	

Translate. The problem can be translated to an equation as follows:

Amount for rolls	plus	Amount for donuts	plus	Amount for cookies	is	Total amount
↓	↓	↓	↓	↓	↓	↓
$\dfrac{1}{2}$	$+$	$\dfrac{1}{4}$	$+$	$\dfrac{1}{3}$	$=$	f

Solve. We carry out the addition. Since $2 = 2$, $4 = 2 \cdot 2$, and $3 = 3$, the LCM of the denominators is $2 \cdot 2 \cdot 3$, or 12.

$$\frac{1}{2} + \frac{1}{4} + \frac{1}{3} = f$$
$$\frac{1}{2} \cdot \frac{6}{6} + \frac{1}{4} \cdot \frac{3}{3} + \frac{1}{3} \cdot \frac{4}{4} = f$$
$$\frac{6}{12} + \frac{3}{12} + \frac{4}{12} = f$$
$$\frac{13}{12} = f$$

Check. We repeat the calculation. We also note that the sum is larger than any of the individual amounts, as expected.

State. $\dfrac{13}{12}$ lb of flour was used.

69. *Familiarize*. We draw a picture and let r = the total rainfall, in inches.

$\dfrac{1}{2}$ in.	$\dfrac{3}{8}$ in.
	r

Translate. The problem can be translated to an equation as follows:

Morning rain	plus	Afternoon rain	is	Total rainfall
↓	↓	↓	↓	↓
$\dfrac{1}{2}$	$+$	$\dfrac{3}{8}$	$=$	r

Solve. We carry out the addition. Since 8 is a multiple of 2, the LCM of the denominators is 8.

$$\frac{1}{2} + \frac{3}{8} = r$$
$$\frac{1}{2} \cdot \frac{4}{4} + \frac{3}{8} = r$$
$$\frac{4}{8} + \frac{3}{8} = r$$
$$\frac{7}{8} = r$$

Check. We repeat the calculations. We also note that the sum is larger than either of the individual amounts, so the answer seems reasonable.

State. Altogether it rained $\dfrac{7}{8}$ in.

71. *Familiarize*. We draw a picture and let d = the number of miles the naturalist hikes.

Lookout   Nest Campsite

$\dfrac{3}{5}$ mi $\dfrac{3}{10}$ mi $\dfrac{3}{4}$ mi

d

Translate. We translate to an equation as follows:

$$
\begin{array}{ccccccc}
\underbrace{\text{Miles to lookout}} & \text{plus} & \underbrace{\text{Miles to nest}} & \text{plus} & \underbrace{\text{Miles to campsite}} & \text{is} & \underbrace{\text{Total miles}}\\
\downarrow & \downarrow & \downarrow & \downarrow & \downarrow & \downarrow & \downarrow\\
\frac{3}{5} & + & \frac{3}{10} & + & \frac{3}{4} & = & d
\end{array}
$$

Solve. We carry out the addition. Since $5 = 5$, $10 = 2 \cdot 5$, and $4 = 2 \cdot 2$, the LCM of the denominators is $2 \cdot 2 \cdot 5$, or 20.

$$\frac{3}{5} + \frac{3}{10} + \frac{3}{4} = d$$

$$\frac{3}{5} \cdot \frac{4}{4} + \frac{3}{10} \cdot \frac{2}{2} + \frac{3}{4} \cdot \frac{5}{5} = d$$

$$\frac{12}{20} + \frac{6}{20} + \frac{15}{20} = d$$

$$\frac{33}{20} = d$$

Check. We repeat the calculation. We also note that the sum is larger than any of the individual distances, as expected.

State. The naturalist hiked a total of $\frac{33}{20}$ mi.

73. Familiarize. First we will find the amount of liquid needed. Let $l =$ this amount, in quarts.

Translate.

$$
\begin{array}{ccccc}
\underbrace{\substack{\text{Amount of}\\\text{ginger ale}}} & \text{plus} & \underbrace{\substack{\text{Amount of}\\\text{strawberry}\\\text{soda}}} & \text{is} & \underbrace{\substack{\text{Total}\\\text{amount of}\\\text{liquid}}}\\
\downarrow & \downarrow & \downarrow & \downarrow & \downarrow\\
\frac{1}{5} & + & \frac{3}{5} & = & l
\end{array}
$$

Solve. We carry out the addition.

$$\frac{1}{5} + \frac{3}{5} = l$$

$$\frac{4}{5} = l$$

Let $d =$ the number of quarts of liquid needed if the recipe is doubled. We write a multiplication sentence and carry out the multiplication.

$$d = 2 \cdot \frac{4}{5}$$

$$d = \frac{2 \cdot 4}{5} = \frac{8}{5}$$

Let $h =$ the number of quarts of liquid needed if the recipe is halved. We write another multiplication sentence and carry out the multiplication.

$$h = \frac{1}{2} \cdot \frac{4}{5}$$

$$h = \frac{1 \cdot 4}{2 \cdot 5} = \frac{1 \cdot 2 \cdot 2}{2 \cdot 5} = \frac{2}{2} \cdot \frac{1 \cdot 2}{5} = \frac{2}{5}$$

Check. We repeat the calculations. The answers check.

State. The recipe requires $\frac{4}{5}$ qt of liquid. If the recipe is doubled, $\frac{8}{5}$ qt is required; $\frac{2}{5}$ qt is required if the recipe is halved.

75. Familiarize. We draw a picture. We let $t =$ the total thickness.

Translate. We translate to an equation.

$$
\begin{array}{cccc}
\underbrace{\substack{\text{Thickness of}\\\text{tile}}} & \text{plus} & \underbrace{\substack{\text{Thickness}\\\text{of cement}}} & \text{plus}\\
\downarrow & \downarrow & \downarrow & \downarrow\\
\frac{5}{8} & + & \frac{3}{32} & +
\end{array}
$$

$$
\begin{array}{ccc}
\underbrace{\substack{\text{Thickness of}\\\text{subflooring}}} & \text{is} & \underbrace{\substack{\text{Total}\\\text{thickness}}}\\
\downarrow & \downarrow & \downarrow\\
\frac{7}{8} & = & t
\end{array}
$$

Solve. We carry out the addition. The LCD is 32 since 8 is a factor of 32.

$$\frac{5}{8} + \frac{3}{32} + \frac{7}{8} = t$$

$$\frac{5}{8} \cdot \frac{4}{4} + \frac{3}{32} + \frac{7}{8} \cdot \frac{4}{4} = t$$

$$\frac{20}{32} + \frac{3}{32} + \frac{28}{32} = t$$

$$\frac{51}{32} = t$$

Check. We repeat the calculation. We also note that the sum is larger than any of the individual thicknesses, as expected.

State. The result is $\frac{51}{32}$ in. thick.

77. Discussion and Writing Exercise

79. $-7 - 6 = -7 + (-6) = -13$

81. $9 - 17 = 9 + (-17) = -8$

83. $\dfrac{x-y}{3} = \dfrac{7-(-3)}{3} = \dfrac{7+3}{3} = \dfrac{10}{3}$

85. Familiarize. Let $x =$ the amount by which spending on books in 2005 exceeded spending on books in 2004.

Translate. We consider this to be a "missing addend" situation.

$$
\begin{array}{ccccc}
\underbrace{\substack{\text{Amount spent}\\\text{in 2004}}} & \text{plus} & \underbrace{\substack{\text{How much}\\\text{more}}} & \text{is} & \underbrace{\substack{\text{Amount spent}\\\text{in 2005?}}}\\
\downarrow & \downarrow & \downarrow & \downarrow & \downarrow\\
238 & + & x & = & 304
\end{array}
$$

Solve. We subtract 238 on both sides of the equation.

$$238 + x = 304$$
$$238 + x - 238 = 304 - 238$$
$$x = 66$$

Check. We can repeat the calculation. We can also round and estimate: $238 + 66 \approx 240 + 70 \approx 310 \approx 304$. The answer checks.

State. Freshmen planned to spend $66 more on textbooks in 2005 than in 2004.

87. *Familiarize*. Let $s =$ the amount by which spending on shoes in 2004 exceeded spending on shoes in 2005.

Translate. We consider this to be a "missing addend" situation.

Amount spent in 2005	plus	How much more	is	Amount spent in 2004?
↓	↓	↓	↓	↓
36	+	s	=	93

Solve. We subtract 36 on both sides of the equation.

$$36 + s = 93$$
$$36 + s - 36 = 93 - 36$$
$$s = 57$$

Check. We can repeat the calculation. The answer checks.

State. Freshmen planned to spent $57 more on shoes in 2004 than in 2005.

89. *Familiarize*. Let $t =$ the total planned expenditure in 2004. We are combining amounts, so we add.

Translate. We translate to an equation.

$$238 + 760 + 58 + 83 + 32 + 93 = t$$

Solve. We carry out the addition.

$$\begin{array}{r} {\scriptstyle 3\ 2} \\ 2\ 3\ 8 \\ 7\ 6\ 0 \\ 5\ 8 \\ 8\ 3 \\ 3\ 2 \\ +\ \ \ 9\ 3 \\ \hline 1\ 2\ 6\ 4 \end{array}$$

Check. We can repeat the calculation. The answer checks.

State. The total planned expenditure in 2004 was $1264.

91. Discussion and Writing Exercise

93.
$$\frac{3}{10}t + \frac{2}{7} + \frac{2}{15}t + \frac{3}{5}$$
$$= \left(\frac{3}{10} + \frac{2}{15}\right)t + \left(\frac{2}{7} + \frac{3}{5}\right)$$
$$= \left(\frac{3}{10} \cdot \frac{3}{3} + \frac{2}{15} \cdot \frac{2}{2}\right)t + \left(\frac{2}{7} \cdot \frac{5}{5} + \frac{3}{5} \cdot \frac{7}{7}\right)$$
$$= \left(\frac{9}{30} + \frac{4}{30}\right)t + \left(\frac{10}{35} + \frac{21}{35}\right)$$
$$= \frac{13}{30}t + \frac{31}{35}$$

95.
$$5t^2 + \frac{6}{a}t + 2t^2 + \frac{3}{a}t$$
$$= (5 + 2)t^2 + \left(\frac{6}{a} + \frac{3}{a}\right)t$$
$$= 7t^2 + \frac{9}{a}t$$

97. Use a calculator to do this exercise. First, add on the left.

$$\frac{12}{169} + \frac{53}{103} = \frac{10,193}{17,407}$$

Now compare $\frac{10,193}{17,407}$ and $\frac{10,192}{17,407}$. The denominators are the same. Since $10,193 > 10,192$, it follows that $\frac{10,193}{17,407} > \frac{10,192}{17,407}$, so $\frac{12}{169} + \frac{53}{103} > \frac{10,192}{17,407}$.

99. *Familiarize*. First we find the fractional part of the band's pay that the guitarist received. We let $f =$ this fraction.

Translate. We translate to an equation.

One-third	of	one-half	plus	one-fifth	of	one-half	is	fractional part
↓	↓	↓	↓	↓	↓	↓	↓	↓
$\frac{1}{3}$	$\cdot$	$\frac{1}{2}$	+	$\frac{1}{5}$	$\cdot$	$\frac{1}{2}$	=	f

Solve. We carry out the calculation.

$$\frac{1}{3} \cdot \frac{1}{2} + \frac{1}{5} \cdot \frac{1}{2} = f$$
$$\frac{1}{6} + \frac{1}{10} = f \quad \text{LCD is 30.}$$
$$\frac{1}{6} \cdot \frac{5}{5} + \frac{1}{10} \cdot \frac{3}{3} = f$$
$$\frac{5}{30} + \frac{3}{30} = f$$
$$\frac{8}{30} = f$$
$$\frac{4}{15} = f$$

Now we find how much of the $1200 received by the band was paid to the guitarist. We let $p =$ the amount.

Four-fifteenths	of	$1200	=	guitarist's pay
↓	↓	↓	↓	↓
$\frac{4}{15}$	$\cdot$	1200	=	p

We solve the equation.

$$\frac{4}{15} \cdot 1200 = p$$
$$\frac{4 \cdot 1200}{15} = p$$
$$\frac{4 \cdot 3 \cdot 5 \cdot 80}{3 \cdot 5} = p$$
$$320 = p$$

Check. We repeat the calculations.

State. The guitarist received $\frac{4}{15}$ of the band's pay. This was $320.

101. Trying various placements and using a calculator to evaluate them yields $4 + \dfrac{6}{3} \cdot 5$, or $4 + \dfrac{5}{3} \cdot 6$. Each is equal to 14.

103. First we find the LCM of the denominators.

$$4 = 2 \cdot 2$$
$$21 = 3 \cdot 7$$
$$15 = 3 \cdot 5$$
$$9 = 3 \cdot 3$$
$$17 = 17$$
$$12 = 2 \cdot 2 \cdot 3$$
$$22 = 2 \cdot 11$$

The LCM is $2 \cdot 2 \cdot 3 \cdot 3 \cdot 5 \cdot 7 \cdot 11 \cdot 17 = 235{,}620$.

Now write each fraction with this denominator.

$$\frac{3}{4} = \frac{176{,}715}{235{,}620}$$
$$\frac{17}{21} = \frac{190{,}740}{235{,}620}$$
$$\frac{13}{15} = \frac{204{,}204}{235{,}620}$$
$$\frac{7}{9} = \frac{183{,}260}{235{,}620}$$
$$\frac{15}{17} = \frac{207{,}900}{235{,}620}$$
$$\frac{13}{12} = \frac{255{,}255}{235{,}620}$$
$$\frac{19}{22} = \frac{203{,}490}{235{,}620}$$

Now arrange the numerators in order from smallest to largest. This yields the following arrangement of the original fractions.

$$\frac{3}{4}, \ \frac{7}{9}, \ \frac{17}{21}, \ \frac{19}{22}, \ \frac{13}{15}, \ \frac{15}{17}, \ \frac{13}{12}$$

Exercise Set 4.3

1. When denominators are the same, subtract the numerators and keep the denominator.

$$\frac{5}{6} - \frac{1}{6} = \frac{5-1}{6} = \frac{4}{6} = \frac{2 \cdot 2}{2 \cdot 3} = \frac{2}{2} \cdot \frac{2}{3} = \frac{2}{3}$$

3. When denominators are the same, subtract the numerators and keep the denominator.

$$\frac{9}{16} - \frac{13}{16} = \frac{9-13}{16} = \frac{-4}{16} = \frac{-1 \cdot 4}{4 \cdot 4} = \frac{-1}{4} \cdot \frac{4}{4} = \frac{-1}{4},$$
or $-\dfrac{1}{4}$

5. $\dfrac{8}{a} - \dfrac{6}{a} = \dfrac{8-6}{a} = \dfrac{2}{a}$

7. $-\dfrac{2}{9} - \dfrac{5}{9} = \dfrac{-2-5}{9} = \dfrac{-7}{9}$, or $-\dfrac{7}{9}$

9. $-\dfrac{3}{8} - \dfrac{1}{8} = \dfrac{-3-1}{8} = \dfrac{-4}{8} = \dfrac{-1 \cdot 4}{2 \cdot 4} = \dfrac{-1}{2} \cdot \dfrac{4}{4} = \dfrac{-1}{2}$, or $-\dfrac{1}{2}$

11. $\dfrac{10}{3t} - \dfrac{4}{3t} = \dfrac{10-4}{3t} = \dfrac{6}{3t} = \dfrac{3}{3} \cdot \dfrac{2}{t} = \dfrac{2}{t}$

13. $\dfrac{3}{5a} - \dfrac{7}{5a} = \dfrac{3-7}{5a} = \dfrac{-4}{5a}$, or $-\dfrac{4}{5a}$

15. The LCM of 8 and 16 is 16.

$$\frac{7}{8} - \frac{1}{16} = \frac{7}{8} \cdot \frac{2}{2} \ - \frac{1}{16} \quad \longleftarrow \text{This fraction already has the}$$
LCM as the denominator.

Think: $8 \times \square = 16$. The answer is 2, so we multiply by 1, using $\dfrac{2}{2}$.

$$= \frac{14}{16} - \frac{1}{16} = \frac{13}{16}$$

17. The LCM of 15 and 5 is 15.

$$\frac{7}{15} - \frac{4}{5} = \quad \frac{7}{15} \quad - \frac{4}{5} \cdot \frac{3}{2}$$

Think: $5 \times \square = 15$. The answer is 3, so we multiply by 1, using $\dfrac{3}{3}$.

This fraction already has the LCM as the denominator.

$$= \frac{7}{15} - \frac{12}{15}$$
$$= \frac{-5}{15} = \frac{5}{5} \cdot \frac{-1}{3}$$
$$= \frac{-1}{3}, \text{ or } -\frac{1}{3}$$

19. The LCM of 4 and 20 is 20.

$$\frac{3}{4} - \frac{1}{20} = \frac{3}{4} \cdot \frac{5}{5} - \frac{1}{20}$$
$$= \frac{15}{20} - \frac{1}{20} = \frac{14}{20}$$
$$= \frac{2 \cdot 7}{2 \cdot 10} = \frac{2}{2} \cdot \frac{7}{10}$$
$$= \frac{7}{10}$$

21. The LCM of 15 and 12 is 60.

$$\frac{2}{15} - \frac{5}{12} = \frac{2}{15} \cdot \frac{4}{4} - \frac{5}{12} \cdot \frac{5}{5}$$
$$= \frac{8}{60} - \frac{25}{60} = \frac{8-25}{60}$$
$$= \frac{-17}{60}, \text{ or } -\frac{17}{60}$$

23. The LCM of 10 and 100 is 100.

$$\frac{7}{10} - \frac{23}{100} = \frac{7}{10} \cdot \frac{10}{10} - \frac{23}{100}$$
$$= \frac{70}{100} - \frac{23}{100} = \frac{47}{100}$$

25. The LCM of 15 and 25 is 75.

$$\frac{7}{15} - \frac{3}{25} = \frac{7}{15} \cdot \frac{5}{5} - \frac{3}{25} \cdot \frac{3}{3}$$
$$= \frac{35}{75} - \frac{9}{75} = \frac{26}{75}$$

27. The LCM of 10 and 100 is 100.

$$\frac{69}{100} - \frac{9}{10} = \frac{69}{100} - \frac{9}{10} \cdot \frac{10}{10}$$
$$= \frac{69}{100} - \frac{90}{100} = \frac{69 - 90}{100}$$
$$= \frac{-21}{100}, \text{ or } -\frac{21}{100}$$

29. The LCM of 8 and 3 is 24.

$$\frac{1}{8} - \frac{2}{3} = \frac{1}{8} \cdot \frac{3}{3} - \frac{2}{3} \cdot \frac{8}{8}$$
$$= \frac{3}{24} - \frac{16}{24}$$
$$= \frac{-13}{24}, \text{ or } -\frac{13}{24}$$

31. The LCM of 10 and 25 is 50.

$$-\frac{3}{10} - \frac{7}{25} = -\frac{3}{10} \cdot \frac{5}{5} - \frac{7}{25} \cdot \frac{2}{2}$$
$$= -\frac{15}{50} - \frac{14}{50}$$
$$= -\frac{29}{50}, \text{ or } -\frac{29}{50}$$

33. The LCM of 3 and 5 is 15.

$$\frac{2}{3} - \frac{4}{5} = \frac{2}{3} \cdot \frac{5}{5} - \frac{4}{5} \cdot \frac{3}{3}$$
$$= \frac{10}{15} - \frac{12}{15}$$
$$= \frac{-2}{15}, \text{ or } -\frac{2}{15}$$

35. The LCM of 18 and 24 is 72.

$$\frac{-5}{18} - \frac{7}{24} = \frac{-5}{18} \cdot \frac{4}{4} - \frac{7}{24} \cdot \frac{3}{3}$$
$$= \frac{-20}{72} - \frac{21}{72}$$
$$= \frac{-41}{72}, \text{ or } -\frac{41}{72}$$

37. The LCM of 90 and 120 is 360.

$$\frac{13}{90} - \frac{17}{120} = \frac{13}{90} \cdot \frac{4}{4} - \frac{17}{120} \cdot \frac{3}{3}$$
$$= \frac{52}{360} - \frac{51}{360}$$
$$= \frac{1}{360}$$

39. The LCM of 3 and 9 is 9.

$$\frac{2}{3}x - \frac{4}{9}x = \frac{2}{3} \cdot \frac{3}{3} \cdot x - \frac{4}{9}x$$
$$= \frac{6}{9}x - \frac{4}{9}x$$
$$= \frac{2}{9}x$$

41. The LCM of 5 and 4 is 20.

$$\frac{2}{5}a - \frac{3}{4}a = \frac{2}{5} \cdot \frac{4}{4} \cdot a - \frac{3}{4} \cdot \frac{5}{5}a$$
$$= \frac{8}{20}a - \frac{15}{20}a$$
$$= \frac{-7}{20}a, \text{ or } -\frac{7}{20}a$$

43.
$$x - \frac{4}{9} = \frac{3}{9}$$
$$x - \frac{4}{9} + \frac{4}{9} = \frac{3}{9} + \frac{4}{9} \quad \text{Adding } \frac{4}{9} \text{ to both sides}$$
$$x + 0 = \frac{7}{9}$$
$$x = \frac{7}{9}$$

The solution is $\frac{7}{9}$.

45.
$$a + \frac{2}{11} = \frac{6}{11}$$
$$a + \frac{2}{11} - \frac{2}{11} = \frac{6}{11} - \frac{2}{11} \quad \text{Subtracting } \frac{2}{11} \text{ from both sides}$$
$$a + 0 = \frac{4}{11}$$
$$a = \frac{4}{11}$$

The solution is $\frac{4}{11}$.

47.
$$x + \frac{1}{3} = \frac{7}{9}$$
$$x + \frac{1}{3} - \frac{1}{3} = \frac{7}{9} - \frac{1}{3} \quad \text{Subtracting } \frac{1}{3} \text{ from both sides}$$
$$x + 0 = \frac{7}{9} - \frac{1}{3} \cdot \frac{3}{3} \quad \text{The LCD is 9. We multiply by 1 to get the LCD.}$$
$$x = \frac{7}{9} - \frac{3}{9} = \frac{4}{9}$$

The solution is $\frac{4}{9}$.

49.
$$a - \frac{3}{8} = \frac{3}{4}$$

$$a - \frac{3}{8} + \frac{3}{8} = \frac{3}{4} + \frac{3}{8} \quad \text{Adding } \frac{3}{8} \text{ on both sides}$$

$$a + 0 = \frac{3}{4} \cdot \frac{2}{2} + \frac{3}{8} \quad \begin{array}{l}\text{The LCD is 8. We mul-}\\ \text{tiply by 1 to get the}\\ \text{LCD.}\end{array}$$

$$a = \frac{6}{8} + \frac{3}{8} = \frac{9}{8}$$

The solution is $\frac{9}{8}$.

51.
$$\frac{2}{3} + x = \frac{4}{5}$$

$$\frac{2}{3} + x - \frac{2}{3} = \frac{4}{5} - \frac{2}{3} \quad \text{Subtracting } \frac{2}{3} \text{ on both sides}$$

$$x + 0 = \frac{4}{5} \cdot \frac{3}{3} - \frac{2}{3} \cdot \frac{5}{5} \quad \begin{array}{l}\text{The LCD is 15. We}\\ \text{multiply by 1 to get}\\ \text{the LCD.}\end{array}$$

$$x = \frac{12}{15} - \frac{10}{15} = \frac{2}{15}$$

The solution is $\frac{2}{15}$.

53.
$$\frac{3}{8} + a = \frac{1}{12}$$

$$\frac{3}{8} + a - \frac{3}{8} = \frac{1}{12} - \frac{3}{8} \quad \begin{array}{l}\text{Subtracting } \frac{3}{8} \text{ on both}\\ \text{sides}\end{array}$$

$$a + 0 = \frac{1}{12} \cdot \frac{2}{2} - \frac{3}{8} \cdot \frac{3}{3} \quad \begin{array}{l}\text{The LCD is 24. We}\\ \text{multiply by 1 to get}\\ \text{the LCD.}\end{array}$$

$$a = \frac{2}{24} - \frac{9}{24} = \frac{2-9}{24}$$

$$a = \frac{-7}{24}, \text{ or } -\frac{7}{24}$$

The solution is $-\frac{7}{24}$.

55.
$$n - \frac{3}{10} = -\frac{1}{6}$$

$$n - \frac{3}{10} + \frac{3}{10} = -\frac{1}{6} + \frac{3}{10} \quad \begin{array}{l}\text{Adding } \frac{3}{10} \text{ to both}\\ \text{sides}\end{array}$$

$$n + 0 = -\frac{1}{6} \cdot \frac{5}{5} + \frac{3}{10} \cdot \frac{3}{3} \quad \begin{array}{l}\text{The LCD is 30. We}\\ \text{multiply by 1 to get}\\ \text{the LCD.}\end{array}$$

$$n = -\frac{5}{30} + \frac{9}{30}$$

$$n = \frac{4}{30}$$

$$n = \frac{2 \cdot 2}{2 \cdot 15} = \frac{2}{2} \cdot \frac{2}{15}$$

$$n = \frac{2}{15}$$

The solution is $\frac{2}{15}$.

57.
$$x + \frac{3}{4} = -\frac{1}{2}$$

$$x + \frac{3}{4} - \frac{3}{4} = -\frac{1}{2} - \frac{3}{4} \quad \begin{array}{l}\text{Subtracting } \frac{3}{4} \text{ on}\\ \text{both sides}\end{array}$$

$$x + 0 = -\frac{1}{2} \cdot \frac{2}{2} - \frac{3}{4} \quad \begin{array}{l}\text{The LCD is 4. We}\\ \text{multiply by 1 to get}\\ \text{the LCD.}\end{array}$$

$$x = -\frac{2}{4} - \frac{3}{4} = \frac{-2}{4} - \frac{3}{4}$$

$$x = \frac{-2-3}{4}$$

$$x = \frac{-5}{4}, \text{ or } -\frac{5}{4}$$

The solution is $-\frac{5}{4}$.

59. *Familiarize*. We visualize the situation. Let $d =$ the distance that remains to be swum.

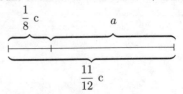

Translate. This is a "missing addend" situation that can be translate as follows:

Distance already swum	plus	Distance remaining to be swum	is	Total distance
↓	↓	↓	↓	↓
$\frac{1}{5}$	$+$	d	$=$	$\frac{1}{2}$

Solve. We subtract $\frac{1}{5}$ from both sides of the equation.

$$\frac{1}{5} + d - \frac{1}{5} = \frac{1}{2} - \frac{1}{5}$$

$$d + 0 = \frac{1}{2} \cdot \frac{5}{5} - \frac{1}{5} \cdot \frac{2}{2} \quad \begin{array}{l}\text{The LCD is 10. We}\\ \text{multiply by 1 to get the}\\ \text{LCD.}\end{array}$$

$$d = \frac{5}{10} - \frac{2}{10} = \frac{3}{10}$$

Check. We return to the original problem and add.

$$\frac{1}{5} + \frac{3}{10} = \frac{1}{5} \cdot \frac{2}{2} + \frac{3}{10} = \frac{2}{10} + \frac{3}{10} = \frac{5}{10} = \frac{5}{5} \cdot \frac{1}{2} = \frac{1}{2}$$

The answer checks.

State. Deb should swim $\frac{3}{10}$ mi farther.

61. *Familiarize*. We visualize the situation. Let $a =$ the amount of cheese left in the bowl, in cups.

Translate. This is a "how much more" situation.

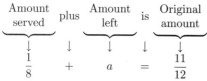

$$\frac{1}{8} + a = \frac{11}{12}$$

Solve. We subtract $\frac{1}{8}$ on both sides of the equation.

$$\frac{1}{8} + a - \frac{1}{8} = \frac{11}{12} - \frac{1}{8}$$

$$a + 0 = \frac{11}{12} \cdot \frac{2}{2} - \frac{1}{8} \cdot \frac{3}{3}$$
The LCD is 24. We multiply by 1 to get the LCD.

$$a = \frac{22}{24} - \frac{3}{24}$$

$$a = \frac{19}{24}$$

Check. We return to the original problem and add.

$$\frac{1}{8} + \frac{19}{24} = \frac{1}{8} \cdot \frac{3}{3} + \frac{19}{24} = \frac{3}{24} + \frac{19}{24} = \frac{22}{24} = \frac{2 \cdot 11}{2 \cdot 12} = \frac{2}{2} \cdot \frac{11}{12} = \frac{11}{12}$$

The answer checks.

State. There is $\frac{19}{24}$ cup of cheese left in the bowl.

63. *Familiarize.* Using the label on the drawing in the text, we let r = the amount by which the board should be planed down, in inches.

Translate. This is a "missing addend" situation.

Desired thickness plus Excess amount is Original thickness

$$\frac{3}{4} + r = \frac{15}{16}$$

Solve. We subtract $\frac{3}{4}$ from both sides of the equation.

$$\frac{3}{4} + r - \frac{3}{4} = \frac{15}{16} - \frac{3}{4}$$

$$r + 0 = \frac{15}{16} - \frac{3}{4} \cdot \frac{4}{4}$$
The LCD is 16. We multiply by 1 to get the LCD.

$$r = \frac{15}{16} - \frac{12}{16}$$

$$r = \frac{3}{16}$$

Check. We return to the original problem and add.

$$\frac{3}{4} + \frac{3}{16} = \frac{3}{4} \cdot \frac{4}{4} + \frac{3}{16} = \frac{12}{16} + \frac{3}{16} = \frac{15}{16}$$

The answer checks.

State. The board should be planed down $\frac{3}{16}$ in.

65. *Familiarize.* We visualize the situation. Let c = the amount of cheese remaining, in pounds.

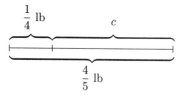

Translate. This is a "how much more" situation.

Amount served plus Amount remaining is Original amount

$$\frac{1}{4} + c = \frac{4}{5}$$

Solve. We subtract $\frac{1}{4}$ on both sides of the equation.

$$\frac{1}{4} + c - \frac{1}{4} = \frac{4}{5} - \frac{1}{4}$$

$$c + 0 = \frac{4}{5} \cdot \frac{4}{4} - \frac{1}{4} \cdot \frac{5}{5}$$
The LCD is 20.

$$c = \frac{16}{20} - \frac{5}{20}$$

$$c = \frac{11}{20}$$

Check. Since $\frac{1}{4} + \frac{11}{20} = \frac{1}{4} \cdot \frac{5}{5} + \frac{11}{20} = \frac{5}{20} + \frac{11}{20} = \frac{16}{20} = \frac{4 \cdot 4}{4 \cdot 5} = \frac{4}{4} \cdot \frac{4}{5} = \frac{4}{5}$, the answer checks.

State. $\frac{11}{20}$ lb of cheese remains on the wheel.

67. *Familiarize.* We visualize the situation. Let t = the time spent on country driving, in hours.

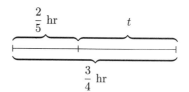

Translate. This is a "how much more" situation.

City driving time plus Country driving time is Total time

$$\frac{2}{5} + t = \frac{3}{4}$$

Solve. We subtract $\frac{2}{5}$ from both sides of the equation.

$$\frac{2}{5} + t - \frac{2}{5} = \frac{3}{4} - \frac{2}{5}$$

$$t + 0 = \frac{3}{4} \cdot \frac{5}{5} - \frac{2}{5} \cdot \frac{4}{4}$$
The LCD is 20.

$$t = \frac{15}{20} - \frac{8}{20} = \frac{7}{20}$$

Check. We return to the original problem and add.

$$\frac{2}{5} + \frac{7}{20} = \frac{2}{5} \cdot \frac{4}{4} + \frac{7}{20} = \frac{8}{20} + \frac{7}{20} = \frac{15}{20} = \frac{3 \cdot 5}{4 \cdot 5} = \frac{3}{4}$$

The answer checks.

State. Jorge spent $\frac{7}{20}$ hr on country driving.

69. _Familiarize_. We visualize the situation. Let s = the amount of syrup that should be added, in cups.

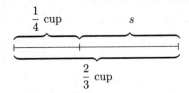

Translate. This is a "missing addend" situation.

$$\underbrace{\text{Original}\atop\text{amount}}\ \ \underset{\text{plus}}{} \ \ \underbrace{\text{Additional}\atop\text{amount}}\ \ \underset{\text{is}}{} \ \ \underbrace{\text{Total}\atop\text{amount}}$$

$$\frac{1}{4}\ \ \ +\ \ \ \ s\ \ \ \ =\ \ \ \frac{2}{3}$$

Solve. We subtract $\frac{1}{4}$ from both sides of the equation.

$$\frac{1}{4}+s-\frac{1}{4}=\frac{2}{3}-\frac{1}{4}$$

$$s+0=\frac{2}{3}\cdot\frac{4}{4}-\frac{1}{4}\cdot\frac{3}{3}\qquad\text{The LCD is 12. We mul-}$$
$$\text{tiply by 1 to get the LCD.}$$

$$s=\frac{8}{12}-\frac{3}{12}=\frac{5}{12}$$

Check. We return to the original problem and add.

$$\frac{1}{4}+\frac{5}{12}=\frac{1}{4}\cdot\frac{3}{3}+\frac{5}{12}=\frac{3}{12}+\frac{5}{12}=\frac{8}{12}=\frac{4}{4}\cdot\frac{2}{3}=\frac{2}{3}$$

The answer checks.

State. Blake should add $\frac{5}{12}$ cup of syrup to the batter.

71. Discussion and Writing Exercise

73. $\dfrac{3}{7}\div\dfrac{9}{4}=\dfrac{3}{7}\cdot\dfrac{4}{9}=\dfrac{3\cdot4}{7\cdot9}$

$$=\frac{3\cdot4}{7\cdot3\cdot3}=\frac{3}{3}\cdot\frac{4}{7\cdot3}$$

$$=\frac{4}{21}$$

75. $7\div\dfrac{1}{3}=7\cdot\dfrac{3}{1}=\dfrac{7\cdot3}{1}$

$$=\frac{21}{1}=21$$

77. _Familiarize_. Let d = the number of days it will take to fill an order for 66,045 tire gauges.

Translate.

$$\underbrace{\text{Gauges produced}\atop\text{per day}}\ \underset{\text{times}}{}\ \underbrace{\text{Number}\atop\text{of days}}\ \underset{\text{is}}{}\ \underbrace{\text{Total}\atop\text{order}}$$

$$3885\qquad\cdot\qquad d\qquad=\ 66,045$$

Solve. We divide by 3885 on both sides of the equation.

$$\frac{3885\cdot d}{3885}=\frac{66,045}{3885}$$
$$d=17$$

Check. Since $3885\cdot17=66,045$, the answer checks.

State. It will take 17 days to produce 66,045 tire gauges.

79.
$$3x-8=25$$
$$3x-8+8=25+8\qquad\text{Adding 8 to both sides}$$
$$3x+0=33$$
$$3x=33$$
$$\frac{3x}{3}=\frac{33}{3}\qquad\text{Dividing both sides by 3}$$
$$x=11$$

The solution is 11.

81. Discussion and Writing Exercise

83. The LCM of 8, 4, and 16 is 16.

$$\frac{7}{8}-\frac{3}{4}-\frac{1}{16}$$
$$=\frac{7}{8}\cdot\frac{2}{2}-\frac{3}{4}\cdot\frac{4}{4}-\frac{1}{16}$$
$$=\frac{14}{16}-\frac{12}{16}-\frac{1}{16}$$
$$=\frac{14-12-1}{16}$$
$$=\frac{1}{16}$$

85. $\dfrac{2}{5}-\dfrac{1}{6}(-3)^2=\dfrac{2}{5}-\dfrac{1}{6}\cdot9$

$$=\frac{2}{5}-\frac{9}{6}$$
$$=\frac{2}{5}\cdot\frac{6}{6}-\frac{9}{6}\cdot\frac{5}{5}\qquad\text{The LCD is 30.}$$
$$=\frac{12}{30}-\frac{45}{30}$$
$$=\frac{-33}{30}=\frac{3}{3}\cdot\frac{-11}{10}$$
$$=\frac{-11}{10},\text{ or }-\frac{11}{10}$$

87. $-4\cdot\dfrac{3}{7}-\dfrac{1}{7}\cdot\dfrac{4}{5}=\dfrac{-12}{7}-\dfrac{4}{35}$

$$=\frac{-12}{7}\cdot\frac{5}{5}-\frac{4}{35}\qquad\text{The LCD is 35.}$$
$$=\frac{-60}{35}-\frac{4}{35}$$
$$=\frac{-64}{35},\text{ or }-\frac{64}{35}$$

89. $\left(-\dfrac{2}{5}\right)^3-\left(-\dfrac{3}{10}\right)^3$

$$=-\frac{8}{125}-\left(-\frac{27}{1000}\right)$$
$$=-\frac{8}{125}+\frac{27}{1000}$$
$$=-\frac{8}{125}\cdot\frac{8}{8}+\frac{27}{1000}\qquad\text{The LCD is 1000.}$$
$$=-\frac{64}{1000}+\frac{27}{1000}$$
$$=-\frac{37}{1000}$$

91. Familiarize. Let t = the fractional portion of the business Trey owns.

Translate. This is a "how much more" situation.

Becky's portion	plus	Clay's portion	plus	Trey's portion	is	One entire business
$\downarrow$	$\downarrow$	$\downarrow$	$\downarrow$	$\downarrow$	$\downarrow$	$\downarrow$
$\frac{1}{3}$	$+$	$\frac{1}{2}$	$+$	t	$=$	1

Solve. First we collect like terms on the left side.

$$\frac{1}{3} + \frac{1}{2} + t = 1$$
$$\frac{1}{3} \cdot \frac{2}{2} + \frac{1}{2} \cdot \frac{3}{3} + t = 1 \quad \text{The LCD is 6.}$$
$$\frac{2}{6} + \frac{3}{6} + t = 1$$
$$\frac{5}{6} + t = 1$$
$$\frac{5}{6} + t - \frac{5}{6} = 1 - \frac{5}{6}$$
$$t + 0 = \frac{6}{6} - \frac{5}{6}$$
$$t = \frac{1}{6}$$

Check. We return to the original problem and add.
$$\frac{1}{3} + \frac{1}{2} + \frac{1}{6} = \frac{1}{3} \cdot \frac{2}{2} + \frac{1}{2} \cdot \frac{3}{3} + \frac{1}{6} = \frac{2}{6} + \frac{3}{6} + \frac{1}{6} = \frac{6}{6} = 1$$
The answer checks.

State. Trey owns $\frac{1}{6}$ of the business.

93. Familiarize. This is a multistep problem. First we find the portion of shoppers who stay for 1-2 hr. Let s = this portion of the shoppers.

Translate. This is a "missing addend" situation.

Portions shown in graph	plus	Remaining portion	is	One entire group
$\downarrow$	$\downarrow$	$\downarrow$	$\downarrow$	$\downarrow$
$\frac{25}{50} + \frac{5}{50} + \frac{2}{50}$	$+$	s	$=$	1

Solve. First we collect like terms on the left.
$$\frac{26}{50} + \frac{5}{50} + \frac{2}{50} + s = 1$$
$$\frac{33}{50} + s = 1$$
$$\frac{33}{50} + s - \frac{33}{50} = 1 - \frac{33}{50} \quad \text{Subtracting } \frac{33}{50} \text{ from both sides}$$
$$s + 0 = 1 \cdot \frac{50}{50} - \frac{33}{50} \quad \text{The LCD is 50.}$$
$$s = \frac{50}{50} - \frac{33}{50}$$
$$s = \frac{17}{50}$$

Now we add the portion of shoppers who stay less than one hour and the portion who stay 1-2 hr to find the portion of shoppers who stay 0-2 hr.

$$\frac{26}{50} + \frac{17}{50} = \frac{43}{50}$$

Check. We repeat the calculation.

State. $\frac{43}{50}$ of shoppers stay 0-2 hr when visiting a mall.

95. First, we solve an equation to find the portion of the tape used at the 4-hr speed.

Tape speed	times	Portion used	is	Time used
$\downarrow$	$\downarrow$	$\downarrow$	$\downarrow$	$\downarrow$
4	$\cdot$	x	$=$	$\frac{1}{2}$

$$x = \frac{1}{2} \div 4$$
$$x = \frac{1}{2} \cdot \frac{1}{4}$$
$$x = \frac{1}{8}$$

Thus, $\frac{1}{8}$ of the tape was used at the 4-hr speed.

Now we solve an equation to find the portion of the tape used at the 2-hr speed.

Tape speed	times	Portion used	is	Time used
$\downarrow$	$\downarrow$	$\downarrow$	$\downarrow$	$\downarrow$
2	$\cdot$	y	$=$	$\frac{3}{4}$

$$y = \frac{3}{4} \div 2$$
$$y = \frac{3}{4} \cdot \frac{1}{2}$$
$$y = \frac{3}{8}$$

Thus, $\frac{3}{8}$ of the tape is used at the 2-hr speed.

Next, we solve an equation to find the portion of the tape that is unused.

Portion used at 4-hr speed	+	Portion used at 2-hr speed	+	Unused portion	=	Entire tape
$\downarrow$	$\downarrow$	$\downarrow$	$\downarrow$	$\downarrow$	$\downarrow$	$\downarrow$
$\frac{1}{8}$	$+$	$\frac{3}{8}$	$+$	p	$=$	1

$$p = 1 - \frac{1}{8} - \frac{3}{8}$$
$$p = \frac{4}{8} = \frac{1}{2}$$

One-half of the tape is unused.

To find how much time is left on the tape at the 6-hr speed, we solve an equation.

$$\underbrace{\text{Unused portion}} \quad \text{of} \quad \underbrace{\text{Tape speed}} \quad \text{is} \quad \underbrace{\text{Time left}}$$

$$\downarrow \qquad \downarrow \qquad \downarrow \qquad \downarrow \qquad \downarrow$$

$$\frac{1}{2} \qquad \cdot \qquad 6 \qquad = \qquad t$$

$$\frac{6}{2} = t$$

$$3 = t$$

There are three hours left at the 6-hr speed.

97. Familiarize. Let d = the fractional portion of the dealership that Paul owns. Then Ella owns the same portion and together the Romanos and the Chrenkas own $\frac{7}{12} + \frac{1}{6} + d + d$.

Translate.

$$\underbrace{\text{Total portions}} \quad \text{are} \quad \underbrace{\text{1 dealership}}$$

$$\downarrow \qquad \qquad \downarrow \qquad \downarrow$$

$$\frac{7}{12} + \frac{1}{6} + d + d \quad = \qquad 1$$

Solve. First we collect like terms on the left.

$$\frac{7}{12} + \frac{1}{6} + d + d = 1$$

$$\frac{7}{12} + \frac{1}{6} \cdot \frac{2}{2} + 2d = 1 \quad \text{The LCD is 12.}$$

$$\frac{7}{12} + \frac{2}{12} + 2d = 1$$

$$\frac{9}{12} + 2d = 1$$

$$\frac{9}{12} + 2d - \frac{9}{12} = 1 - \frac{9}{12}$$

$$2d + 0 = 1 \cdot \frac{12}{12} - \frac{9}{12}$$

$$2d = \frac{3}{12}$$

$$\frac{1}{2} \cdot 2d = \frac{1}{2} \cdot \frac{3}{12}$$

$$d = \frac{3}{24} = \frac{3}{3} \cdot \frac{1}{8}$$

$$d = \frac{1}{8}$$

Check. We return to the original problem and add.

$$\frac{7}{12} + \frac{1}{6} + \frac{1}{8} + \frac{1}{8} = \frac{7}{12} \cdot \frac{2}{2} + \frac{1}{6} \cdot \frac{4}{4} + \frac{1}{8} \cdot \frac{3}{3} + \frac{1}{8} \cdot \frac{3}{3} =$$

$$\frac{14}{24} + \frac{4}{24} + \frac{3}{24} + \frac{3}{24} = \frac{24}{24} = 1$$

The answer checks.

State. Paul owns $\frac{1}{8}$ of the dealership.

99. Use a calculator.

$$x + \frac{16}{323} = \frac{10}{187}$$

$$x + \frac{16}{323} - \frac{16}{323} = \frac{10}{187} - \frac{16}{323}$$

$$x + 0 = \frac{10}{11 \cdot 17} - \frac{16}{17 \cdot 19}$$

$$x = \frac{10}{11 \cdot 17} \cdot \frac{19}{19} - \frac{16}{17 \cdot 19} \cdot \frac{11}{11} \quad \begin{array}{l}\text{The LCD is} \\ 11 \cdot 17 \cdot 19.\end{array}$$

$$x = \frac{190}{11 \cdot 17 \cdot 19} - \frac{176}{17 \cdot 19 \cdot 11}$$

$$x = \frac{14}{11 \cdot 17 \cdot 19}$$

$$x = \frac{14}{3553}$$

The solution is $\frac{14}{3553}$.

101.
$$\frac{10 + a}{23} = \frac{330}{391} - \frac{a}{17}$$

$$\frac{10 + a}{23} \cdot \frac{17}{17} = \frac{330}{391} - \frac{a}{17} \cdot \frac{23}{23}$$

$$\frac{17(10 + a)}{391} = \frac{330}{391} - \frac{23a}{391}$$

$$\frac{170 + 17a}{391} = \frac{330 - 23a}{391}$$

Since the denominators are the same, the numerators are the same.

$$170 + 17a = 330 - 23a$$

$$170 + 17a + 23a = 330 - 23a + 23a$$

$$170 + 40a = 330$$

$$170 + 40a - 170 = 330 - 170$$

$$40a = 160$$

$$\frac{40a}{40} = \frac{160}{40}$$

$$a = 4$$

103. Use the two cuts to cut the bar into three pieces as follows: one piece is $\frac{1}{7}$ of the bar, one is $\frac{2}{7}$ of the bar, and then the remaining piece is $\frac{4}{7}$ of the bar. On Day 1, give the contractor $\frac{1}{7}$ of the bar. On Day 2, have him/her return the $\frac{1}{7}$ and give him/her $\frac{2}{7}$ of the bar. On Day 3, add $\frac{1}{7}$ to what the contractor already has, making $\frac{3}{7}$ of the bar. On Day 4, have the contractor return the $\frac{1}{7}$ and $\frac{2}{7}$ pieces and give him/her the $\frac{4}{7}$ piece. On Day 5, add the $\frac{1}{7}$ piece to what the contractor already has, making $\frac{5}{7}$ of the bar. On Day 6, have the contractor return the $\frac{1}{7}$ piece and give him/her the $\frac{2}{7}$ to go with the $\frac{4}{7}$ piece he/she also has, making $\frac{6}{7}$ of the bar. On Day 7, give him/her the $\frac{1}{7}$

piece again. Now the contractor has all three pieces, or the entire bar. This assumes that he/she does not spend any part of the gold during the week.

Exercise Set 4.4

1.
$$6x - 3 = 15$$
$$6x - 3 + 3 = 15 + 3 \quad \text{Using the addition principle}$$
$$6x + 0 = 18$$
$$6x = 18$$
$$\frac{1}{6} \cdot 6x = \frac{1}{6} \cdot 18 \quad \text{Using the multiplication principle}$$
$$1x = \frac{18}{6}$$
$$x = 3$$

Check:
$$\frac{6x - 3 = 15}{6 \cdot 3 - 3 \;?\; 15}$$
$$\begin{array}{c|c} 18 - 3 & \\ 15 & 15 \quad \text{TRUE} \end{array}$$

The solution is 3.

3.
$$5x + 7 = -8$$
$$5x + 7 - 7 = -8 - 7 \quad \text{Using the addition principle}$$
$$5x = -15$$
$$\frac{1}{5} \cdot 5x = \frac{1}{5} \cdot (-15) \quad \text{Using the multiplication principle}$$
$$1x = \frac{-15}{5}$$
$$x = -3$$

Check:
$$\frac{5x + 7 = -8}{5(-3) + 7 \;?\; -8}$$
$$\begin{array}{c|c} -15 + 7 & \\ -8 & -8 \quad \text{TRUE} \end{array}$$

The solution is -3.

5.
$$31 = 3x - 5$$
$$31 + 5 = 3x - 5 + 5 \quad \text{Using the addition principle}$$
$$36 = 3x + 0$$
$$36 = 3x$$
$$\frac{1}{3} \cdot 36 = \frac{1}{3} \cdot 3x \quad \text{Using the multiplication principle}$$
$$\frac{36}{3} = 1x$$
$$12 = x$$

Check:
$$\frac{31 = 3x - 5}{31 \;?\; 3 \cdot 12 - 5}$$
$$\begin{array}{c|c} & 36 - 5 \\ 31 & 31 \quad \text{TRUE} \end{array}$$

The solution is 12.

7.
$$4x - 5 = \frac{1}{3}$$
$$4x - 5 + 5 = \frac{1}{3} + 5 \quad \text{Using the addition principle}$$
$$4x + 0 = \frac{1}{3} + 5 \cdot \frac{3}{3}$$
$$4x = \frac{1}{3} + \frac{15}{3}$$
$$4x = \frac{16}{3}$$
$$\frac{1}{4} \cdot 4x = \frac{1}{4} \cdot \frac{16}{3} \quad \text{Using the multiplication principle}$$
$$1x = \frac{1 \cdot 16}{4 \cdot 3} = \frac{1 \cdot 4 \cdot 4}{4 \cdot 3}$$
$$x = \frac{4}{4} \cdot \frac{1 \cdot 4}{3} = \frac{4}{3}$$

Check:
$$\frac{4x - 5 = \frac{1}{3}}{4 \cdot \frac{4}{3} - 5 \;?\; \frac{1}{3}}$$
$$\begin{array}{c|c} \frac{16}{3} - 5 & \\ \frac{16}{3} - \frac{15}{3} & \\ \frac{1}{3} & \frac{1}{3} \quad \text{TRUE} \end{array}$$

The solution is $\frac{4}{3}$.

9.
$$\frac{3}{2}t - \frac{1}{4} = \frac{1}{2}$$
$$\frac{3}{2}t - \frac{1}{4} + \frac{1}{4} = \frac{1}{2} + \frac{1}{4} \quad \text{Adding } \frac{1}{4} \text{ to each side}$$
$$\frac{3}{2}t + 0 = \frac{2}{4} + \frac{1}{4}$$
$$\frac{3}{2}t = \frac{3}{4}$$
$$\frac{2}{3} \cdot \frac{3}{2}t = \frac{2}{3} \cdot \frac{3}{4} \quad \text{Multiplying both sides by } \frac{2}{3}$$
$$1t = \frac{\cancel{2} \cdot \cancel{3}}{3 \cdot \cancel{2} \cdot 2}$$
$$t = \frac{1}{2}$$

Check:
$$\frac{\frac{3}{2}t - \frac{1}{4} = \frac{1}{2}}{\frac{3}{2} \cdot \frac{1}{2} - \frac{1}{4} \;?\; \frac{1}{2}}$$
$$\begin{array}{c|c} \frac{3}{4} - \frac{1}{4} & \\ \frac{2}{4} & \\ \frac{1}{2} & \frac{1}{2} \quad \text{TRUE} \end{array}$$

The solution is $\frac{1}{2}$.

11.
$$\frac{2}{5}x + \frac{3}{10} = \frac{3}{5}$$

$$\frac{2}{5}x + \frac{3}{10} - \frac{3}{10} = \frac{3}{5} - \frac{3}{10} \quad \text{Subtracting } \frac{3}{10} \text{ from both sides}$$

$$\frac{2}{5}x + 0 = \frac{6}{10} - \frac{3}{10}$$

$$\frac{2}{5}x = \frac{3}{10}$$

$$\frac{5}{2} \cdot \frac{2}{5}x = \frac{5}{2} \cdot \frac{3}{10} \quad \text{Multiplying both sides by } \frac{5}{2}$$

$$1x = \frac{\cancel{5} \cdot 3}{2 \cdot \cancel{5} \cdot 2}$$

$$x = \frac{3}{4}$$

Check:
$$\frac{\frac{2}{5}x + \frac{3}{10} = \frac{3}{5}}{}$$

$$\frac{2}{5} \cdot \frac{3}{4} + \frac{3}{10} \,?\, \frac{3}{5}$$

$$\frac{2 \cdot 3}{5 \cdot 2 \cdot 2} + \frac{3}{10}$$

$$\frac{3}{10} + \frac{3}{10}$$

$$\frac{6}{10}$$

$$\frac{3}{5} \quad \bigg| \quad \frac{3}{5} \qquad \text{TRUE}$$

The solution is $\frac{3}{4}$.

13.
$$5 - \frac{3}{4}x = 3$$

$$5 - \frac{3}{4}x - 5 = 3 - 5 \quad \text{Subtracting 5 from both sides}$$

$$-\frac{3}{4}x = -2$$

$$-\frac{4}{3}\left(-\frac{3}{4}x\right) = -\frac{4}{3}(-2) \quad \text{Multiplying both sides by } -\frac{4}{3}$$

$$1x = \frac{8}{3}$$

$$x = \frac{8}{3}$$

The number $\frac{8}{3}$ checks and is the solution.

15.
$$-1 + \frac{2}{5}t + 1 = -\frac{4}{5}$$

$$-1 + \frac{2}{5}t + 1 = -\frac{4}{5} + 1 \quad \text{Adding 1 to both sides}$$

$$\frac{2}{5}t = -\frac{4}{5} + \frac{5}{5}$$

$$\frac{2}{5}t = \frac{1}{5}$$

$$\frac{5}{2} \cdot \frac{2}{5}t = \frac{5}{2} \cdot \frac{1}{5}$$

$$1t = \frac{\cancel{5} \cdot 1}{2 \cdot \cancel{5}}$$

$$t = \frac{1}{2}$$

The number $\frac{1}{2}$ checks and is the solution.

17.
$$12 = 8 + \frac{7}{2}t$$

$$12 - 8 = 8 + \frac{7}{12}t - 8 \quad \text{Subtracting 8 from both sides}$$

$$4 = \frac{7}{2}t$$

$$\frac{2}{7} \cdot 4 = \frac{2}{7} \cdot \frac{7}{2}t \quad \text{Multiply both sides by } \frac{2}{7}$$

$$\frac{8}{7} = t$$

The number $\frac{8}{7}$ checks and is the solution.

19.
$$-4 = \frac{2}{3}x - 7$$

$$-4 + 7 = \frac{2}{3}x - 7 + 7 \quad \text{Adding 7 to both sides}$$

$$3 = \frac{2}{3}x$$

$$\frac{3}{2} \cdot 3 = \frac{3}{2} \cdot \frac{2}{3}x \quad \text{Multiplying both sides by } \frac{3}{2}$$

$$\frac{9}{2} = x$$

The number $\frac{9}{2}$ checks and is the solution.

21.
$$7 = a + \frac{14}{5}$$

$$7 - \frac{14}{5} = a + \frac{14}{5} - \frac{14}{5} \quad \text{Using the addition principle}$$

$$7 \cdot \frac{5}{5} - \frac{14}{5} = a + 0$$

$$\frac{35}{5} - \frac{14}{5} = a$$

$$\frac{21}{5} = a$$

The number $\frac{21}{5}$ checks and is the solution.

23.
$$\frac{2}{5}t - 1 = \frac{7}{5}$$

$$\frac{2}{5}t - 1 + 1 = \frac{7}{5} + 1 \qquad \text{Using the addition principle}$$

$$\frac{2}{5}t + 0 = \frac{7}{5} + 1 \cdot \frac{5}{5}$$

$$\frac{2}{5}t = \frac{7}{5} + \frac{5}{5}$$

$$\frac{2}{5}t = \frac{12}{5}$$

$$\frac{5}{2} \cdot \frac{2}{5}t = \frac{5}{2} \cdot \frac{12}{5} \qquad \text{Using the multiplication principle}$$

$$1t = \frac{5 \cdot 12}{2 \cdot 5}$$

$$t = \frac{5 \cdot 2 \cdot 6}{2 \cdot 5 \cdot 1} = \frac{5 \cdot 2}{5 \cdot 2} \cdot \frac{6}{1}$$

$$t = 6$$

The number 6 checks and is the solution.

25.
$$\frac{39}{8} = \frac{11}{4} + \frac{1}{2}x$$

$$\frac{39}{8} - \frac{11}{4} = \frac{11}{4} + \frac{1}{2}x - \frac{11}{4} \qquad \text{Using the addition principle}$$

$$\frac{39}{8} - \frac{11}{4} \cdot \frac{2}{2} = \frac{1}{2}x + 0$$

$$\frac{39}{8} - \frac{22}{8} = \frac{1}{2}x$$

$$\frac{17}{8} = \frac{1}{2}x$$

$$2 \cdot \frac{17}{8} = 2 \cdot \frac{1}{2}x \qquad \text{Using the multiplication principle}$$

$$\frac{2 \cdot 17}{8} = 1x$$

$$\frac{2 \cdot 17}{2 \cdot 4} = x$$

$$\frac{17}{4} = x$$

The number $\frac{17}{4}$ checks and is the solution.

27.
$$\frac{13}{3}x + \frac{11}{2} = \frac{35}{4}$$

$$\frac{13}{3}x + \frac{11}{2} - \frac{11}{2} = \frac{35}{4} - \frac{11}{2} \qquad \text{Using the addition principle}$$

$$\frac{13}{3}x + 0 = \frac{35}{4} - \frac{11}{2} \cdot \frac{2}{2}$$

$$\frac{13}{3}x = \frac{35}{4} - \frac{22}{4}$$

$$\frac{13}{3}x = \frac{13}{4}$$

$$\frac{3}{13} \cdot \frac{13}{3}x = \frac{3}{13} \cdot \frac{13}{4} \qquad \text{Using the multiplication principle}$$

$$1x = \frac{3 \cdot 13}{13 \cdot 4}$$

$$x = \frac{3}{4}$$

The number $\frac{3}{4}$ checks and is the solution.

29.
$$\frac{1}{2}x - \frac{1}{4} = \frac{1}{2}$$

$$4\left(\frac{1}{2}x - \frac{1}{4}\right) = 4 \cdot \frac{1}{2} \qquad \text{Multiplying both sides by the LCD, 4}$$

$$\frac{4 \cdot 1}{2}x - 4 \cdot \frac{1}{4} = \frac{4 \cdot 1}{2}$$

$$\frac{2 \cdot 2}{2}x - 1 = \frac{2 \cdot 2}{2}$$

$$2x - 1 = 2$$

$$2x - 1 + 1 = 2 + 1 \qquad \text{Adding 1 to both sides}$$

$$2x = 3$$

$$\frac{2x}{2} = \frac{3}{2} \qquad \text{Dividing both sides by 2}$$

$$x = \frac{3}{2}$$

The number $\frac{3}{2}$ checks and is the solution.

31.
$$7 = \frac{4}{9}t + 5$$

$$9 \cdot 7 = 9\left(\frac{4}{9}t + 5\right) \qquad \text{Multiplying both sides by the LCD, 9}$$

$$63 = \frac{9 \cdot 4}{9}t + 9 \cdot 5$$

$$63 = 4t + 45$$

$$63 - 45 = 4t + 45 - 45 \qquad \text{Subtracting 45 from both sides}$$

$$18 = 4t$$

$$\frac{18}{4} = \frac{4t}{4} \qquad \text{Dividing both sides by 4}$$

$$\frac{9 \cdot 2}{2 \cdot 2} = 1t$$

$$\frac{9}{2} = t$$

The number $\frac{9}{2}$ checks and is the solution.

33.
$$-3 = \frac{3}{4}t - \frac{1}{2}$$

$$4(-3) = 4\left(\frac{3}{4}t - \frac{1}{2}\right) \qquad \text{Multiplying both sides by the LCD, 4}$$

$$-12 = \frac{4 \cdot 3}{4}t - \frac{4 \cdot 1}{2}$$

$$-12 = \frac{4 \cdot 3}{4}t - \frac{2 \cdot 2}{2}$$

$$-12 = 3t - 2$$

$$-12 + 2 = 3t - 2 + 2 \qquad \text{Adding 2 to both sides}$$

$$-10 = 3t$$

$$\frac{-10}{3} = \frac{3t}{3} \qquad \text{Dividing both sides by 3}$$

$$-\frac{10}{3} = t$$

The number $-\frac{10}{3}$ checks and is the solution.

35.
$$\frac{4}{3} - \frac{5}{6}x = \frac{3}{2}$$

$$6\left(\frac{4}{3} - \frac{5}{6}x\right) = 6 \cdot \frac{3}{2} \quad \text{Multiplying both sides by the LCD, 6}$$

$$\frac{6 \cdot 4}{3} - \frac{6 \cdot 5}{6}x = \frac{6 \cdot 3}{2}$$

$$\frac{2 \cdot \cancel{3} \cdot 4}{\cancel{3}} - \frac{\cancel{6} \cdot 5}{\cancel{6}}x = \frac{\cancel{2} \cdot 3 \cdot 3}{\cancel{2}}$$

$$8 - 5x = 9$$

$$8 - 5x - 8 = 9 - 8 \quad \text{Subtracting 8 from both sides}$$

$$-5x = 1$$

$$\frac{-5x}{5} = \frac{1}{-5} \quad \text{Dividing both sides by } -5$$

$$x = -\frac{1}{5}$$

The number $-\frac{1}{5}$ checks and is the solution.

37.
$$-\frac{3}{4} = \frac{5}{6} - \frac{1}{2}x$$

$$12\left(-\frac{3}{4}\right) = 12\left(-\frac{5}{6} - \frac{1}{2}x\right) \quad \text{Multiplying both sides by the LCD, 12}$$

$$-\frac{12 \cdot 3}{4} = -\frac{12 \cdot 5}{6} - \frac{12 \cdot 1}{2}x$$

$$-\frac{\cancel{4} \cdot 3 \cdot 3}{\cancel{4}} = -\frac{2 \cdot \cancel{6} \cdot 5}{\cancel{6}} - \frac{\cancel{2} \cdot 6}{\cancel{2}}x$$

$$-9 = -10 - 6x$$

$$-9 + 10 = -10 - 6x + 10 \quad \text{Adding 10 to both sides}$$

$$1 = -6x$$

$$\frac{1}{-6} = \frac{-6x}{-6} \quad \text{Dividing both sides by } -6$$

$$-\frac{1}{6} = x$$

The number $-\frac{1}{6}$ checks and is the solution.

39.
$$\frac{4}{3} - \frac{1}{5}t = \frac{3}{4}$$

$$60\left(\frac{4}{3} - \frac{1}{5}t\right) = 60 \cdot \frac{3}{4} \quad \text{Multiplying both sides by the LCD, 60}$$

$$\frac{60 \cdot 4}{3} - \frac{60 \cdot 1}{5}t = \frac{60 \cdot 3}{4}$$

$$\frac{\cancel{3} \cdot 20 \cdot 4}{\cancel{3}} - \frac{\cancel{5} \cdot 12}{\cancel{5}}t = \frac{\cancel{4} \cdot 15 \cdot 3}{\cancel{4}}$$

$$80 - 12t = 45$$

$$80 - 12t - 80 = 45 - 80 \quad \text{Subtracting 80 from both sides}$$

$$-12t = -35$$

$$\frac{-12t}{-12} = \frac{-35}{-12}$$

$$t = \frac{35}{12}$$

The number $\frac{35}{12}$ checks and is the solution.

41. Discussion and Writing Exercise

43. $39 \div (-3) = -13$

Think: What number multiplied by -3 gives 39? The number is -13.

45. $(-72) \div (-4) = 18$

Think: What number multiplied by -4 gives -72? The number is 18.

47. -200 represents the \$200 withdrawal, 90 represents the \$90 deposit, and -40 represents the \$40 withdrawal. We add these numbers to find the change in the balance.

$$-200 + 90 + (-40) = -110 + (-40) = -150$$

The account balance decreased by \$150.

49. $\frac{10}{7} \div 2m = \frac{10}{7} \cdot \frac{1}{2m} = \frac{10 \cdot 1}{7 \cdot 2m} = \frac{2 \cdot 5 \cdot 1}{7 \cdot 2 \cdot m} = \frac{2}{2} \cdot \frac{5 \cdot 1}{7 \cdot m} = \frac{5}{7m}$

51. Discussion and Writing Exercise

53. Use a calculator.
$$\frac{553}{2451}a - \frac{13}{57} = \frac{29}{43}$$
$$\frac{553}{2451}a - \frac{13}{57} + \frac{13}{57} = \frac{29}{43} + \frac{13}{57}$$
$$\frac{553}{2451}a = \frac{29}{43} + \frac{13}{57}$$
$$\frac{553}{2451}a = \frac{29}{43} \cdot \frac{57}{57} + \frac{13}{57} \cdot \frac{43}{43}$$
$$\frac{553}{2451}a = \frac{1653}{43 \cdot 57} + \frac{559}{57 \cdot 43}$$
$$\frac{553}{2451}a = \frac{2212}{2451}$$
$$\frac{2451}{553} \cdot \frac{553}{2451}a = \frac{2451}{553} \cdot \frac{2212}{2451}$$
$$a = 4$$

The solution is 4.

55. Use a calculator.
$$\frac{71}{73} = \frac{19}{47} - \frac{53}{91}t$$
$$\frac{71}{73} - \frac{19}{47} = \frac{19}{47} - \frac{53}{91}t - \frac{19}{47}$$
$$\frac{71}{73} \cdot \frac{47}{47} - \frac{19}{47} \cdot \frac{73}{73} = -\frac{53}{91}t$$
$$\frac{3337}{3431} - \frac{1387}{3431} = -\frac{53}{91}t$$
$$\frac{1950}{3431} = -\frac{53}{91}t$$
$$-\frac{91}{53} \cdot \frac{1950}{3431} = -\frac{91}{53}\left(-\frac{53}{91}\right)t$$
$$-\frac{177,450}{181,843} = t$$

The solution is $-\frac{177,450}{181,843}$.

57.
$$-\frac{a}{5} + \frac{31}{4} = \frac{16}{3}$$
$$-\frac{1}{5}a + \frac{31}{4} = \frac{16}{3} \qquad \left(-\frac{1}{5} \cdot a = -\frac{a}{5}\right)$$
$$-\frac{1}{5}a = \frac{16}{3} - \frac{31}{4}$$
$$-\frac{1}{5}a = \frac{16}{3} \cdot \frac{4}{4} - \frac{31}{4} \cdot \frac{3}{3}$$
$$-\frac{1}{5}a = \frac{64}{12} - \frac{93}{12}$$
$$-\frac{1}{5}a = -\frac{29}{12}$$
$$-5\left(-\frac{1}{5}a\right) = -5\left(-\frac{29}{12}\right)$$
$$a = \frac{145}{12}$$

The solution is $\frac{145}{12}$.

59.
$$\frac{49}{8} + \frac{2x}{9} = 4$$
$$\frac{2x}{9} = 4 - \frac{49}{8}$$
$$\frac{2x}{9} = \frac{32}{8} - \frac{49}{8}$$
$$\frac{2x}{9} = -\frac{17}{8}$$
$$x = \frac{9}{2}\left(-\frac{17}{8}\right)$$
$$x = -\frac{153}{16}$$

The solution is $-\frac{153}{16}$.

61.
$$5 + 2 + 4 + 5 + 2 + 2x = 21$$
$$18 + 2x = 21$$
$$18 + 2x - 18 = 21 - 18$$
$$2x = 3$$
$$\frac{2x}{2} = \frac{3}{2}$$
$$x = \frac{3}{2}$$

The value of x is $\frac{3}{2}$ cm.

63.
$$7n + 6 + 5n = 15$$
$$12n + 6 = 15$$
$$12n + 6 - 6 = 15 - 6$$
$$12n = 9$$
$$\frac{12n}{12} = \frac{9}{12}$$
$$n = \frac{\cancel{3} \cdot 3}{\cancel{3} \cdot 4}$$
$$n = \frac{3}{4}$$

The value of n is $\frac{3}{4}$ cm.

Exercise Set 4.5

1.
$7\frac{2}{3} = \frac{23}{3}$

a. Multiply: $3 \cdot 7 = 21$.

b. Add: $21 + 2 = 23$.

c. Keep the denominator.

3.
$6\frac{1}{4} = \frac{25}{4}$

a. Multiply: $6 \cdot 4 = 24$.

b. Add: $24 + 1 = 25$.

c. Keep the denominator.

5. $-20\frac{1}{8} = -\frac{161}{8}$ $\quad(20 \cdot 8 = 160; 160 + 1 = 161;$ include the negative sign)

7. $5\frac{1}{10} = \frac{51}{10}$ $\quad(5 \cdot 10 = 50; 50 + 1 = 51)$

9. $20\frac{3}{5} = \frac{103}{5}$ $\quad(20 \cdot 5 = 100; 100 + 3 = 103)$

11. $-8\frac{2}{7} = -\frac{58}{7}$ $\quad(8 \cdot 7 = 56; 56 + 2 = 58;$ include the negative sign)

13. $6\frac{9}{10} = \frac{69}{10}$ $\quad(6 \cdot 10 = 60; 60 + 9 = 69)$

15. $-12\frac{3}{4} = -\frac{51}{4}$ $\quad(12 \cdot 4 = 48; 48 + 3 = 51;$ include the negative sign)

17. $5\frac{7}{10} = \frac{57}{10}$ $\quad(5 \cdot 10 = 50; 50 + 7 = 57)$

19. $-5\frac{7}{100} = -\frac{507}{100}$ $\quad(5 \cdot 100 = 500; 500 + 7 = 507;$ include the negative sign)

21. To convert $\frac{16}{3}$ to a mixed numeral, we divide.

$$\frac{16}{3} = 5\frac{1}{3}$$

23. To convert $\frac{45}{6}$ to a mixed numeral, we divide.

$$\frac{45}{6} = 7\frac{3}{6} = 7\frac{1}{2}$$

25.

$$\frac{57}{10} = 5\frac{7}{10}$$

27. $9\overline{\smash{\big)}\,65}$ with 7 above, 63 below, remainder 2
$$\frac{65}{9} = 7\frac{2}{9}$$

29. $6\overline{\smash{\big)}\,33}$ with 5 above, 30 below, remainder 3
$$\frac{33}{6} = 5\frac{3}{6} = 5\frac{1}{2}$$

Since $\dfrac{33}{6} = 5\dfrac{1}{2}$, we have $-\dfrac{33}{6} = -5\dfrac{1}{2}$.

31. $4\overline{\smash{\big)}\,46}$ with 11 above, 40, 6, 4, remainder 2
$$\frac{46}{4} = 11\frac{2}{4} = 11\frac{1}{2}$$

33. $8\overline{\smash{\big)}\,12}$ with 1 above, 8, remainder 4
$$\frac{12}{8} = 1\frac{4}{8} = 1\frac{1}{2}$$

Since $\dfrac{12}{8} = 1\dfrac{1}{2}$, we have $-\dfrac{12}{8} = -1\dfrac{1}{2}$.

35. $5\overline{\smash{\big)}\,307}$ with 61 above, 300, 7, 5, remainder 2
$$\frac{307}{5} = 61\frac{2}{5}$$

37. $50\overline{\smash{\big)}\,413}$ with 8 above, 400, remainder 13
$$\frac{413}{50} = 8\frac{13}{50}$$

Since $\dfrac{413}{50} = 8\dfrac{13}{50}$, we have $-\dfrac{413}{50} = -8\dfrac{13}{50}$.

39.   We first divide as usual.

$$8\overline{\smash{\big)}\,869}$$ with 108 above, 800, 69, 64, remainder 5

The answer is 108 R 5. We write a mixed numeral for the quotient as follows: $108\dfrac{5}{8}$.

41.   We first divide as usual.

$$7\overline{\smash{\big)}\,6345}$$ with 906 above, 6300, 45, 42, remainder 3

The answer is 906 R 3. We write a mixed numeral for the quotient as follows: $906\dfrac{3}{7}$.

43. $21\overline{\smash{\big)}\,852}$ with 40 above, 840, remainder 12

We get $40\dfrac{12}{21}$. This simplifies as $40\dfrac{4}{7}$.

45.   First we find $302 \div 15$.

$$15\overline{\smash{\big)}\,302}$$ with 20 above, 300, 2, 0, remainder 2
$$\frac{302}{15} = 20\frac{2}{15}$$

Since $302 \div 15 = 20\dfrac{2}{15}$, we have $-302 \div 15 = -20\dfrac{2}{15}$.

47.   First we find $471 \div 21$.

$$21\overline{\smash{\big)}\,471}$$ with 22 above, 420, 51, 42, remainder 9
$$\frac{471}{21} = 22\frac{9}{21} = 22\frac{3}{7}$$

Since $471 \div 21 = 22\dfrac{3}{7}$, we have $471 \div (-21) = -22\dfrac{3}{7}$.

49.   There are 5 health organizations in the list. We add the expenses for these organizations and then divide by 5.
$$\frac{\$4 + \$6 + \$7 + \$10 + \$14}{5} = \frac{\$41}{5} = \$8\frac{1}{5}$$

51.   We add the expenses for the first 6 organizations and then divide by 6.
$$\frac{\$1 + \$4 + \$5 + \$6 + \$7 + \$8}{6} = \frac{\$31}{6} = \$5\frac{1}{6}$$

53.   Discussion and Writing Exercise

55. $\dfrac{7}{9} \cdot \dfrac{24}{21} = \dfrac{7 \cdot 24}{9 \cdot 21}$
$$= \frac{7 \cdot 3 \cdot 8}{3 \cdot 3 \cdot 3 \cdot 7}$$
$$= \frac{3 \cdot 7}{3 \cdot 7} \cdot \frac{8}{3 \cdot 3}$$
$$= \frac{8}{9}$$

57. $\dfrac{7}{10} \cdot \dfrac{5}{14} = \dfrac{7 \cdot 5}{10 \cdot 14} = \dfrac{7 \cdot 5 \cdot 1}{2 \cdot 5 \cdot 2 \cdot 7} = \dfrac{7 \cdot 5}{7 \cdot 5} \cdot \dfrac{1}{2 \cdot 2} =$
$$\frac{1}{2 \cdot 2} = \frac{1}{4}$$

59. $-\dfrac{17}{25} \cdot \dfrac{15}{34} = -\dfrac{17 \cdot 15}{25 \cdot 34}$
$$= -\frac{17 \cdot 3 \cdot 5}{5 \cdot 5 \cdot 2 \cdot 17}$$
$$= -\frac{17 \cdot 5}{17 \cdot 5} \cdot \frac{3}{5 \cdot 2}$$
$$= -\frac{3}{10}$$

61.   Discussion and Writing Exercise

63. Use a calculator.

$$\frac{128,236}{541} = 237\frac{19}{541}$$

65. $\frac{56}{7} + \frac{2}{3} = 8 + \frac{2}{3}$ $(56 \div 7 = 8)$

$\qquad\qquad = 8\frac{2}{3}$

67. $\frac{12}{5} + \frac{19}{15} = \frac{36}{15} + \frac{19}{15} = \frac{55}{15}$

$$\begin{array}{r} 3 \\ 15\overline{\smash{\big)}\,5\,5} \\ 4\,5 \\ \hline 1\,0 \end{array} \qquad \frac{55}{15} = 3\frac{10}{15} = 3\frac{2}{3}$$

Thus, $\frac{12}{5} + \frac{19}{15} = 3\frac{2}{3}$.

69. There are 365 days in a year and 7 days in a week, so we write a mixed numeral for $\frac{365}{7}$.

$$\begin{array}{r} 5\,2 \\ 7\overline{\smash{\big)}\,3\,6\,5} \\ 3\,5\,0 \\ \hline 1\,5 \\ 1\,4 \\ \hline 1 \end{array} \qquad \frac{365}{7} = 52\frac{1}{7}$$

There are $52\frac{1}{7}$ weeks in a year.

Exercise Set 4.6

1.
$$\begin{array}{r} 6 \\ +5\frac{2}{5} \\ \hline 11\frac{2}{5} \end{array}$$

3.
$$\begin{array}{r} 2\frac{7}{8} \\ +6\frac{5}{8} \\ \hline 8\frac{12}{8} = 8 + \frac{12}{8} \\ = 8 + 1\frac{1}{2} \\ = 9\frac{1}{2} \end{array}$$

To find a mixed numeral for $\frac{12}{8}$ we divide:

$$\begin{array}{r} 1 \\ 8\overline{\smash{\big)}\,1\,2} \\ 8 \\ \hline 4 \end{array} \qquad \frac{12}{8} = 1\frac{4}{8} = 1\frac{1}{2}$$

5. The LCD is 12.

$$\begin{array}{r} 4\,\boxed{\frac{1}{4} \cdot \frac{3}{3}} = \quad 4\frac{3}{12} \\ +1\,\boxed{\frac{2}{3} \cdot \frac{4}{4}} = +1\frac{8}{12} \\ \hline 5\frac{11}{12} \end{array}$$

7. The LCD is 12.

$$\begin{array}{r} 7\,\boxed{\frac{3}{4} \cdot \frac{3}{3}} = \quad 7\frac{9}{12} \\ +5\,\boxed{\frac{5}{6} \cdot \frac{2}{2}} = +5\frac{10}{12} \\ \hline 12\frac{19}{12} = 12 + \frac{19}{12} \\ = 12 + 1\frac{7}{12} \\ = 13\frac{7}{12} \end{array}$$

9. The LCD is 10.

$$\begin{array}{r} 3\,\boxed{\frac{2}{5} \cdot \frac{2}{2}} = \quad 3\frac{4}{10} \\ +8\frac{7}{10} = +8\frac{7}{10} \\ \hline 11\frac{11}{10} = 11 + \frac{11}{10} \\ = 11 + 1\frac{1}{10} \\ = 12\frac{1}{10} \end{array}$$

11. The LCD is 24.

$$\begin{array}{r} 6\,\boxed{\frac{3}{8} \cdot \frac{3}{3}} = \quad 6\frac{9}{24} \\ +10\,\boxed{\frac{5}{6} \cdot \frac{4}{4}} = +10\frac{20}{24} \\ \hline 16\frac{29}{24} = 16 + \frac{29}{24} \\ = 16 + 1\frac{5}{24} \\ = 17\frac{5}{24} \end{array}$$

13. The LCD is 10.

$$\begin{array}{r} 18\,\boxed{\frac{4}{5} \cdot \frac{2}{2}} = 18\frac{8}{10} \\ +2\frac{7}{10} = +2\frac{7}{10} \\ \hline 20\frac{15}{10} = 20 + \frac{15}{10} \\ = 20 + 1\frac{5}{10} \\ = 21\frac{5}{10} \\ = 21\frac{1}{2} \end{array}$$

15. The LCD is 8.

$$14\frac{5}{8} \quad = \quad 14\frac{5}{8}$$

$$+13\boxed{\frac{1}{4}\cdot\frac{2}{2}} = +13\frac{2}{8}$$

$$\overline{\qquad\qquad\qquad 27\frac{7}{8}}$$

17.
$$4\frac{1}{5} = 3\frac{6}{5}$$
$$-2\frac{3}{5} = -2\frac{3}{5}$$
$$\overline{\qquad\quad 1\frac{3}{5}}$$

> Since $\frac{1}{5}$ is smaller than $\frac{3}{5}$, we cannot subtract until we borrow:
> $$4\frac{1}{5} = 3+\frac{5}{5}+\frac{1}{5} = 3+\frac{6}{5} = 3\frac{6}{5}$$

19. The LCD is 10.

$$9\boxed{\frac{3}{5}\cdot\frac{2}{2}} = \quad 9\frac{6}{10}$$

$$-3\boxed{\frac{1}{2}\cdot\frac{5}{5}} = -3\frac{5}{10}$$

$$\overline{\qquad\qquad\qquad 6\frac{1}{10}}$$

21. The LCD is 24.

$$34\boxed{\frac{1}{3}\cdot\frac{8}{8}} = \quad 34\frac{8}{24} = \quad 33\frac{32}{24}$$

$$-12\boxed{\frac{5}{8}\cdot\frac{3}{3}} = -12\frac{15}{24} = -12\frac{15}{24}$$

$$\overline{\qquad\qquad\qquad\qquad\qquad\qquad 21\frac{17}{24}}$$

$\left(\text{Since } \frac{8}{24} \text{ is smaller than } \frac{15}{24}, \text{ we cannot subtract until we}\right.$
borrow: $34\frac{8}{24} = 33+\frac{24}{24}+\frac{8}{24} = 33+\frac{32}{24} = 33\frac{32}{24}.\Big)$

23.
$$19 \quad = \quad 18\frac{4}{4} \quad \left(19 = 18+1 = 18+\frac{4}{4} = 18\frac{4}{4}\right)$$
$$- \ 5\frac{3}{4} = - \ 5\frac{3}{4}$$
$$\overline{\qquad\qquad\quad 13\frac{1}{4}}$$

25.
$$34 \quad = \quad 33\frac{8}{8} \quad \left(34 = 33+1 = 33+\frac{8}{8} = 33\frac{8}{8}\right)$$
$$- \ 18\frac{5}{8} = - \ 18\frac{5}{8}$$
$$\overline{\qquad\qquad\quad 15\frac{3}{8}}$$

27. The LCD is 12.

$$21\boxed{\frac{1}{6}\cdot\frac{2}{2}} = \quad 21\frac{2}{12} = \quad 20\frac{14}{12}$$

$$-13\boxed{\frac{3}{4}\cdot\frac{3}{3}} = -13\frac{9}{12} = -13\frac{9}{12}$$

$$\overline{\qquad\qquad\qquad\qquad\qquad\qquad 7\frac{5}{12}}$$

$\left(\text{Since } \frac{2}{12} \text{ is smaller than } \frac{9}{12}, \text{ we cannot subtract until we}\right.$
borrow: $21\frac{2}{12} = 20+\frac{12}{12}+\frac{2}{12} = 20+\frac{14}{12} = 20\frac{14}{12}.\Big)$

29. The LCD is 18.

$$25\boxed{\frac{1}{9}\cdot\frac{2}{2}} = \quad 25\frac{2}{18} = \quad 24\frac{20}{18}$$

$$-13\boxed{\frac{5}{6}\cdot\frac{3}{3}} = -13\frac{15}{18} = -13\frac{15}{18}$$

$$\overline{\qquad\qquad\qquad\qquad\qquad\qquad 11\frac{5}{18}}$$

$\left(\text{Since } \frac{2}{18} \text{ is smaller than } \frac{15}{18}, \text{ we cannot subtract until we}\right.$
borrow: $25\frac{2}{18} = 24+\frac{18}{18}+\frac{2}{18} = 24+\frac{20}{18} = 24\frac{20}{18}.\Big)$

31.
$$1\frac{3}{14}t + 7\frac{2}{21}t$$
$$= \left(1\frac{3}{14}+7\frac{2}{21}\right)t \quad \text{Using the distributive law}$$
$$= \left(1\frac{9}{42}+7\frac{4}{42}\right)t \quad \text{The LCD is 42.}$$
$$= 8\frac{13}{42}t \qquad\qquad \text{Adding}$$

33.
$$9\frac{1}{2}x - 7\frac{3}{8}x$$
$$= \left(9\frac{1}{2}-7\frac{3}{8}\right)x \quad \text{Using the distributive law}$$
$$= \left(9\frac{4}{8}-7\frac{3}{8}\right)x \quad \text{The LCD is 8.}$$
$$= 2\frac{1}{8}x \qquad\qquad \text{Subtracting}$$

35.
$$5\frac{9}{10}t + 2\frac{7}{8}t$$
$$= \left(5\frac{9}{10}+2\frac{7}{8}\right)t \quad \text{Using the distributive law}$$
$$= \left(5\frac{36}{40}+2\frac{35}{40}\right)t \quad \text{The LCD is 40.}$$
$$= 7\frac{71}{40}t = 8\frac{31}{40}t$$

37. $37\frac{5}{9}t - 25\frac{4}{5}t$

$= \left(37\frac{5}{9} - 25\frac{4}{5}\right)t$ Using the distributive law

$= \left(37\frac{25}{45} - 25\frac{36}{45}\right)t$ The LCD is 45.

$= \left(36\frac{70}{45} - 25\frac{36}{45}\right)t$

$= 11\frac{34}{45}t$

39. $2\frac{5}{6}x + 3\frac{1}{3}x$

$= \left(2\frac{5}{6} + 3\frac{1}{3}\right)x$ Using the distributive law

$= \left(2\frac{5}{6} + 3\frac{2}{6}\right)x$ The LCD is 6.

$= 5\frac{7}{6}x = 6\frac{1}{6}x$

41. $1\frac{3}{11}x + 8\frac{2}{3}x$

$= \left(1\frac{3}{11} + 8\frac{2}{3}\right)x$ Using the distributive law

$= \left(1\frac{9}{33} + 8\frac{22}{33}\right)x$ The LCD is 33.

$= 9\frac{31}{33}x$

43. *Familiarize*. Let f = the number of yards of fabric needed to make the outfit.

***Translate*.** We write an equation.

Fabric for dress	+	Fabric for band	+	Fabric for jacket	is	Total fabric
↓	↓	↓	↓	↓	↓	↓
$1\frac{3}{8}$	+	$\frac{5}{8}$	+	$3\frac{3}{8}$	=	f

***Solve*.** We add.

$$
\begin{array}{r}
1\frac{3}{8}\\
\frac{5}{8}\\
+3\frac{3}{8}\\
\hline
4\frac{11}{8} = 4 + \frac{11}{8}\\
= 4 + 1\frac{3}{8}\\
= 5\frac{3}{8}
\end{array}
$$

***Check*.** We can repeat the calculation. Also note that the answer is reasonable since it is larger than any of the individual amounts of fabric.

***State*.** The outfit requires $5\frac{3}{8}$ yd of fabric.

45. *Familiarize*. We let w = the total weight of the meat.

***Translate*.** We write an equation.

Weight of one package	plus	Weight of second package	is	Total weight
↓	↓	↓	↓	↓
$1\frac{2}{3}$	+	$5\frac{3}{4}$	=	w

***Solve*.** We carry out the addition. The LCD is 12.

$$
\begin{array}{r}
1\,\frac{2}{3} \cdot \frac{4}{4} = 1\frac{8}{12}\\
+5\,\frac{3}{4} \cdot \frac{3}{3} = +5\frac{9}{12}\\
\hline
6\frac{17}{12} = 6 + \frac{17}{12}\\
= 6 + 1\frac{5}{12}\\
= 7\frac{5}{12}
\end{array}
$$

***Check*.** We repeat the calculation. We also note that the answer is larger than either of the individual weights, so the answer seems reasonable.

***State*.** The total weight of the meat was $7\frac{5}{12}$ lb.

47. *Familiarize*. We let h = Juan's excess height.

***Translate*.** We have a missing addend situation.

Height of daughter	plus	How much more height	is	Juan's height
↓	↓	↓	↓	↓
$180\frac{3}{4}$	+	h	=	$187\frac{1}{10}$

***Solve*.** We solve the equation as follows:

$$h = 187\frac{1}{10} - 180\frac{3}{4}$$

$$187\,\frac{1}{10} \cdot \frac{2}{2} = 187\frac{2}{20}$$

$$180\,\frac{3}{4} \cdot \frac{5}{5} = 180\frac{15}{20}$$

$$
\begin{array}{r}
187\frac{1}{10} = 187\frac{2}{20} = 186\frac{22}{20}\\
-180\frac{3}{4} = -180\frac{15}{20} = -180\frac{15}{20}\\
\hline
6\frac{7}{20}
\end{array}
$$

Thus, $h = 6\frac{7}{20}$.

***Check*.** We add Juan's excess height to his daughter's height:

$$180\frac{3}{4} + 6\frac{7}{20} = 180\frac{15}{20} + 6\frac{7}{20} = 186\frac{22}{20} = 187\frac{2}{20} = 187\frac{1}{10}$$

The answer checks.

***State*.** Juan is $6\frac{7}{20}$ cm taller.

49. Familiarize. We draw a picture, letting $x =$ the amount of pipe that was used, in inches.

$$\vdash\!\!-10\tfrac{5}{16}\text{ in.}\!-\!\vdash\!-8\tfrac{3}{4}\text{ in.}\!-\!\dashv$$
$$\vdash\!\!-\!\!-\!\!-\!\!-\!\!-\!\!-\ x\ -\!\!-\!\!-\!\!-\!\!-\!\!-\!\dashv$$

Translate. We write an addition sentence.

First length plus Second length is Total length

$$10\tfrac{5}{16}\ +\ 8\tfrac{3}{4}\ =\ x$$

Solve. We carry out the addition. The LCD is 16.

$$10\tfrac{5}{16}\ =\ 10\tfrac{5}{16}$$
$$+\ 8\,\boxed{\tfrac{3}{4}\cdot\tfrac{4}{4}}\ =\ +\ 8\tfrac{12}{16}$$
$$18\tfrac{17}{16}=18+\tfrac{17}{16}$$
$$=18+1\tfrac{1}{16}$$
$$=19\tfrac{1}{16}$$

Check. We repeat the calculation. We also note that the total length is larger than either of the individual lengths, so the answer seems reasonable.

State. The plumber used $19\tfrac{1}{16}$ in. of pipe.

51. Familiarize. We let $f =$ the total number of yards of fabric Art bought.

Translate.

Length of first piece plus Length of second piece is Total length

$$9\tfrac{1}{4}\ +\ 10\tfrac{5}{6}\ =\ f$$

Solve. We carry out the addition.

$$9\,\boxed{\tfrac{1}{4}\cdot\tfrac{3}{3}}\ =\ 9\tfrac{3}{12}$$
$$+10\,\boxed{\tfrac{5}{6}\cdot\tfrac{2}{2}}\ =\ +10\tfrac{10}{12}$$
$$19\tfrac{13}{12}=19+\tfrac{13}{12}$$
$$=19+1\tfrac{1}{12}$$
$$=20\tfrac{1}{12}$$

Check. We can subtract one of the shorter lengths from the total and check to determine if we get the other length.
$$20\tfrac{1}{12}-9\tfrac{1}{4}=20\tfrac{1}{12}-9\tfrac{3}{12}=19\tfrac{13}{12}-9\tfrac{3}{12}=10\tfrac{10}{12}=10\tfrac{5}{6}$$
The answer checks.

State. Art bought a total of $20\tfrac{1}{12}$ yd of fabric.

53. Familiarize. This is a multistep problem. First we find the number of gallons of fertilizer left in the tank after the application. Let $x =$ this amount.

Translate.

Amount applied plus How much more fertilizer is Amount originally in tank

$$178\tfrac{2}{3}\ +\ x\ =\ 283\tfrac{5}{8}$$

Solve. We solve the equation.

$$x=283\tfrac{5}{8}-178\tfrac{2}{3}$$

We have:

$$283\,\boxed{\tfrac{5}{8}\cdot\tfrac{3}{3}}=283\tfrac{15}{24}$$
$$-178\,\boxed{\tfrac{2}{3}\cdot\tfrac{8}{8}}=-178\tfrac{16}{24}$$

Now we carry out the subtraction.

$$283\tfrac{15}{24}=282\tfrac{39}{24}$$
$$-178\tfrac{16}{24}=-178\tfrac{16}{2}$$
$$104\tfrac{23}{24}$$

Next we let $f =$ the number of gallons of fertilizer in the tank after the delivery.

Amount after application plus Amount delivered is Final amount in tank

$$104\tfrac{23}{24}\ +\ 250\ =\ f$$

We carry out the addition.

$$104\tfrac{23}{24}$$
$$+250$$
$$354\tfrac{23}{24}$$

Check. We repeat the calculations. The answer checks.

State. After the delivery the tank contains $354\tfrac{23}{24}$ gal of fertilizer.

55. Familiarize. Let $m =$ how much farther Angela will run in the marathon, in miles.

Translate.

Fun Run distance plus Additional marathon distance is Marathon distance

$$6\tfrac{1}{5}\ +\ m\ =\ 26\tfrac{7}{32}$$

Solve. We solve the equation as follows:

$$m = 26\frac{7}{32} - 6\frac{1}{5}$$

$$26 \boxed{\frac{7}{32} \cdot \frac{5}{5}} = \quad 26\frac{35}{160}$$

$$6 \boxed{\frac{1}{5} \cdot \frac{32}{32}} = - \;\; 6\frac{32}{160}$$

$$\overline{} \quad \overline{\quad 20\frac{3}{160}}$$

Check. We add the additional marathon distance to the Fun Run distance:

$$6\frac{1}{5} + 20\frac{3}{160} = 6\frac{32}{160} + 20\frac{3}{160} = 26\frac{35}{160} = 26\frac{7}{32}$$

The answer checks.

State. Angela will run $20\frac{3}{160}$ mi farther in the marathon.

57. Familiarize. Let x = the number of cups of ingredients. This is the sum of the measures listed.

Translate.

$$\underbrace{\text{Number of cups}}_{\downarrow} \;\; \text{is} \;\; \underbrace{\text{Sum of measures listed}}_{\downarrow}$$

$$x \qquad\quad = 1\frac{1}{2} + 1\frac{1}{2} + 2\frac{1}{2} + 1\frac{1}{2} + \frac{3}{4}$$

Solve.

$$x = 1\frac{1}{2} + 1\frac{1}{2} + 2\frac{1}{2} + 1\frac{1}{2} + \frac{3}{4}$$

$$x = 1\frac{2}{4} + 1\frac{2}{4} + 2\frac{2}{4} + 1\frac{2}{4} + \frac{3}{4}$$

$$x = 5\frac{11}{4} = 5 + \frac{11}{4} = 5 + 2\frac{3}{4}$$

$$x = 7\frac{3}{4}$$

Check. We repeat the calculations. The answer checks.

State. There are $7\frac{3}{4}$ cups of ingredients listed.

59. Familiarize. We make a drawing. We let t = the number of hours Sue worked on the third day.

$$\vdash\!\!-\; 2\frac{1}{2}\text{ hr }\; -\!\!\vdash\; -\; 4\frac{1}{5}\text{ hr }\; -\!\!\!\vdash\!-\; t\; -\!\!\dashv$$

$$\vdash\!\!-\!\!-\!\!-\!\!-\; 10\frac{1}{2}\text{ hr }\; -\!\!-\!\!-\!\!-\!\!\dashv$$

Translate. We write an addition sentence.

$$2\frac{1}{2} + 4\frac{1}{5} + t = 10\frac{1}{2}$$

Solve. This is a two-step problem.

First we add $2\frac{1}{2} + 4\frac{1}{5}$ to find the time worked on the first two days. The LCD is 10.

$$2 \boxed{\frac{1}{2} \cdot \frac{5}{5}} = \quad 2\frac{5}{10}$$

$$+\; 4 \boxed{\frac{1}{5} \cdot \frac{2}{2}} = +\; 4\frac{2}{10}$$

$$\overline{} \quad \overline{\quad 6\frac{7}{10}}$$

Then we subtract $6\frac{7}{10}$ from $10\frac{1}{2}$ to find the time worked on the third day. The LCD is 10.

$$6\frac{7}{10} + t = 10\frac{1}{2}$$

$$t = 10\frac{1}{2} - 6\frac{7}{10}$$

$$10\boxed{\frac{1}{2} \cdot \frac{5}{5}} = \quad 10\frac{5}{10} = \quad 9\frac{15}{10}$$

$$-\; 6\frac{7}{10} \quad = -\; 6\frac{7}{10} = -6\frac{7}{10}$$

$$\overline{} \qquad \overline{} \qquad \overline{\quad 3\frac{8}{10} = 3\frac{4}{5}}$$

Check. We repeat the calculations.

State. Sue worked $3\frac{4}{5}$ hr the third day.

61. The figure is equivalent to a rectangle with length $16\frac{1}{2}$ in. and width $9\frac{1}{4}$ in. We add to find the perimeter.

$$16\frac{1}{2} + 9\frac{1}{4} + 16\frac{1}{2} + 9\frac{1}{4}$$

$$= 16\frac{2}{4} + 9\frac{1}{4} + 16\frac{2}{4} + 9\frac{1}{4}$$

$$= 50\frac{6}{4} = 50 + \frac{6}{4}$$

$$= 50 + 1\frac{2}{4} = 50 + 1\frac{1}{2}$$

$$= 51\frac{1}{2}$$

The perimeter is $51\frac{1}{2}$ in.

63. We add to find the perimeter.

$$4 + 3\frac{3}{4} + 6\frac{3}{4} + 4 + 3\frac{3}{4} + 5\frac{1}{2}$$

$$= 4 + 3\frac{3}{4} + 6\frac{3}{4} + 4 + 3\frac{3}{4} + 5\frac{2}{4}$$

$$= 25\frac{11}{4} = 25 + \frac{11}{4}$$

$$= 25 + 2\frac{3}{4} = 27\frac{3}{4}$$

The perimeter is $27\frac{3}{4}$ ft.

65. We add to find the perimeter.

$$1\frac{5}{12} + \frac{17}{24} + 1 + 1 + \frac{17}{24}$$

$$= 1\frac{10}{24} + \frac{17}{24} + 1 + 1 + \frac{17}{24}$$

$$= 3\frac{44}{24} = 3 + \frac{44}{24}$$

$$= 3 + 1\frac{20}{24} = 3 + 1\frac{5}{6}$$

$$= 4\frac{5}{6}$$

The perimeter is $4\frac{5}{6}$ ft.

67. We see that d and the two smallest distances combined are the same as the largest distance. We translate and solve.

$$2\frac{3}{4} + d + 2\frac{3}{4} = 12\frac{7}{8}$$

$$d = 12\frac{7}{8} - 2\frac{3}{4} - 2\frac{3}{4}$$

$$= 10\frac{1}{8} - 2\frac{3}{4} \quad \text{Subtracting } 2\frac{3}{4} \text{ from } 12\frac{7}{8}$$

$$= 7\frac{3}{8} \quad \text{Subtracting } 2\frac{3}{4} \text{ from } 10\frac{1}{8}$$

The length of d is $7\frac{3}{8}$ ft.

69. Familiarize. We let $b = $ the length of the bolt.

Translate. From the drawing we see that the length of the small bolt is the sum of the diameters of the two tubes and the thicknesses of the two washers and the nut. Thus, we have

$$b = \frac{1}{2} + \frac{1}{16} + \frac{3}{4} + \frac{1}{16} + \frac{3}{16}.$$

Solve. We carry out the addition. The LCD is 16.

$$b = \frac{1}{2} + \frac{1}{16} + \frac{3}{4} + \frac{1}{16} + \frac{3}{16} =$$

$$\frac{1}{2} \cdot \frac{8}{8} + \frac{1}{16} + \frac{3}{4} \cdot \frac{4}{4} + \frac{1}{16} + \frac{3}{16} =$$

$$\frac{8}{16} + \frac{1}{16} + \frac{12}{16} + \frac{1}{16} + \frac{3}{16} = \frac{25}{16} = 1\frac{9}{16}$$

Check. We repeat the calculation.

State. The smallest bolt is $1\frac{9}{16}$ in. long.

71. $8\frac{3}{5} - 9\frac{2}{5} = 8\frac{3}{5} + \left(-9\frac{2}{5}\right)$

Since $9\frac{2}{5}$ is greater than $8\frac{3}{5}$, the answer will be negative. The difference in absolute values is

$$\begin{array}{rcr} 9\frac{2}{5} & = & 8\frac{7}{5} \\ -8\frac{3}{5} & = & -8\frac{3}{5} \\ \hline & & \frac{4}{5} \end{array}$$

so $8\frac{3}{5} - 9\frac{2}{5} = -\frac{4}{5}$.

73. $3\frac{1}{2} - 6\frac{3}{4} = 3\frac{1}{2} + \left(-6\frac{3}{4}\right)$

Since $6\frac{3}{4}$ is greater than $3\frac{1}{2}$, the answer will be negative. The difference in absolute values is

$$\begin{array}{rcccr} 6\frac{3}{4} & = & 6\frac{3}{4} & = & 6\frac{3}{4} \\ -3\frac{1}{2} & = & -3\boxed{\frac{1}{2} \cdot \frac{1}{2}} & = & -3\frac{2}{4} \\ \hline & & & & 3\frac{1}{4} \end{array}$$

so $3\frac{1}{2} - 6\frac{3}{4} = -3\frac{1}{4}$.

75. $3\frac{4}{5} - 7\frac{2}{3} = 3\frac{4}{5} + \left(-7\frac{2}{3}\right)$

Since $7\frac{2}{3}$ is greater than $3\frac{4}{5}$, the answer will be negative. The difference in absolute values is

$$\begin{array}{rcccccr} 7\frac{2}{3} & = & 7\boxed{\frac{2}{3} \cdot \frac{5}{5}} & = & 7\frac{10}{15} & = & 6\frac{25}{15} \\ -3\frac{4}{5} & = & -3\boxed{\frac{4}{5} \cdot \frac{3}{3}} & = & -3\frac{12}{15} & = & -3\frac{12}{15} \\ \hline & & & & & & 3\frac{13}{15} \end{array}$$

so $3\frac{4}{5} - 7\frac{2}{3} = -3\frac{13}{15}$.

77. $-3\frac{1}{5} - 4\frac{2}{5} = -3\frac{1}{5} + \left(-4\frac{2}{5}\right)$

We add the absolute values and make the answer negative.

$$\begin{array}{r} 3\frac{1}{5} \\ +4\frac{2}{5} \\ \hline 7\frac{3}{5} \end{array}$$

Thus, $-3\frac{1}{5} - 4\frac{2}{5} = -7\frac{3}{5}$.

79. $-4\frac{2}{5} - 6\frac{3}{7} = -4\frac{2}{5} + \left(-6\frac{3}{7}\right)$

We add the absolute values and make the answer negative.

$$\begin{array}{rcccr} 4\frac{2}{5} & = & 4\boxed{\frac{2}{5} \cdot \frac{7}{7}} & = & 4\frac{14}{35} \\ +6\frac{3}{7} & = & +6\boxed{\frac{3}{7} \cdot \frac{5}{5}} & = & +6\frac{15}{35} \\ \hline & & & & 10\frac{29}{35} \end{array}$$

Thus, $-4\frac{2}{5} - 6\frac{3}{7} = -10\frac{29}{35}$.

81. $-6\frac{1}{9} - \left(-4\frac{2}{9}\right) = -6\frac{1}{9} + 4\frac{2}{9}$

Since $-6\frac{1}{9}$ has the greater absolute value, the answer will be negative. The difference in absolute values is

$$6\frac{1}{9} = 5\frac{10}{9}$$
$$-4\frac{2}{9} = -4\frac{2}{9}$$
$$\underline{\phantom{-4\frac{2}{9}}}\qquad \underline{\phantom{-4\frac{2}{9}}}$$
$$1\frac{8}{9}$$

so $-6\frac{1}{9} - \left(-4\frac{2}{9}\right) = -1\frac{8}{9}$.

83. Discussion and Writing Exercise

85. *Familiarize*. We visualize the situation. Repeated subtraction, or division, works well here.

$$\underbrace{\boxed{\tfrac{3}{4}\text{ lb}}\ \boxed{\tfrac{3}{4}\text{ lb}}\cdots\boxed{\tfrac{3}{4}\text{ lb}}}$$

12 lb fills how many packages?

Let $n =$ the number of packages that can be made.

Translate. We translate to an equation.

$$n = 12 \div \frac{3}{4}$$

Solve. We carry out the division.

$$n = 12 \div \frac{3}{4} = 12 \cdot \frac{4}{3} = \frac{12 \cdot 4}{3}$$
$$= \frac{3 \cdot 4 \cdot 4}{3 \cdot 1} = \frac{3}{3} \cdot \frac{4 \cdot 4}{1}$$
$$= 16$$

Check. If each of 16 packages contains $\frac{3}{4}$ lb of cheese, a total of

$$16 \cdot \frac{3}{4} = \frac{16 \cdot 3}{4} = \frac{4 \cdot 4 \cdot 3}{4} = 4 \cdot 3,$$

or 12 lb of cheese is used. The answer checks.

State. 16 packages of cheese can be made from a 12-lb slab.

87. The sum of the digits is $9 + 9 + 9 + 3 = 30$. Since 30 is divisible by 3, then 9993 is divisible by 3.

89. The sum of the digits is $2 + 3 + 4 + 5 = 14$. Since 14 is not divisible by 9, then 2345 is not divisible by 9.

91. The ones digit of 2335 is not 0, so 2335 is not divisible by 10.

93. 18,888 is even because the ones digit is even. Because $1 + 8 + 8 + 8 + 8 = 33$ and 33 is divisible by 3, then 18,888 is divisible by 3. Thus, 18,888 is divisible by 6.

95. $\dfrac{15}{9} \cdot \dfrac{18}{39} = \dfrac{15 \cdot 18}{9 \cdot 39} = \dfrac{3 \cdot 5 \cdot 2 \cdot 3 \cdot 3}{3 \cdot 3 \cdot 3 \cdot 13}$

$\qquad = \dfrac{3 \cdot 3 \cdot 3}{3 \cdot 3 \cdot 3} \cdot \dfrac{5 \cdot 2}{13}$

$\qquad = \dfrac{10}{13}$

97. Discussion and Writing Exercise

99. Use a calculator.

$$3289\frac{1047}{1189} = \quad 3289\ \frac{1047}{1189} = \quad 3289\frac{1047}{1189}$$
$$+5278\frac{32}{41} \quad = +5278\ \boxed{\frac{32}{41} \cdot \frac{29}{29}} = +5278\frac{928}{1189}$$
$$\overline{\qquad\qquad\qquad\qquad\qquad\qquad\qquad} $$
$$8567\frac{1975}{1189} = 8568\frac{786}{1189}$$

101. Use a calculator.

$$5848\frac{17}{29} - 4230\frac{19}{73} = 5848\frac{1241}{2117} - 4230\frac{551}{2117} =$$
$$1618\frac{690}{2117}, \text{ so } 4230\frac{19}{73} - 5848\frac{17}{29} = -1618\frac{690}{2117}$$

103. $35\frac{2}{3} + n = 46\frac{1}{4}$

$$n = 46\frac{1}{4} - 35\frac{2}{3}$$
$$n = 46\frac{3}{12} - 35\frac{8}{12}$$
$$n = 45\frac{15}{12} - 35\frac{8}{12}$$
$$n = 10\frac{7}{12}$$

105. $\qquad -15\frac{7}{8} = 12\frac{1}{2} + t$

$$-15\frac{7}{8} - 12\frac{1}{2} = t$$
$$-28\frac{3}{8} = t$$

107. The resulting rectangle will have length $\left(8\frac{1}{2} + 1\frac{1}{8} + 8\frac{1}{2}\right)$ in. and width $9\frac{3}{4}$ in. We add to find the perimeter.

$$8\frac{1}{2} + 1\frac{1}{8} + 8\frac{1}{2} + 9\frac{3}{4} + 8\frac{1}{2} + 1\frac{1}{8} + 8\frac{1}{2} + 9\frac{3}{4}$$
$$= 8\frac{4}{8} + 1\frac{1}{8} + 8\frac{4}{8} + 9\frac{6}{8} + 8\frac{4}{8} + 1\frac{1}{8} + 8\frac{4}{8} + 9\frac{6}{8}$$
$$= 52\frac{30}{8} = 52 + \frac{30}{8}$$
$$= 52 + 3\frac{6}{8} = 52 + 3\frac{3}{4}$$
$$= 55\frac{3}{4}$$

The perimeter is $55\frac{3}{4}$ in.

Exercise Set 4.7

1. $\qquad 16 \cdot 1\frac{2}{5}$

$\qquad = \dfrac{16}{1} \cdot \dfrac{7}{5}$ \quad Writing fraction notation

$\qquad = \dfrac{16 \cdot 7}{1 \cdot 5} = \dfrac{112}{5}$

$\qquad = 22\dfrac{2}{5}$

3. $6\dfrac{2}{3} \cdot \dfrac{1}{4}$

$= \dfrac{20}{3} \cdot \dfrac{1}{4}$ Writing fraction notation

$= \dfrac{20 \cdot 1}{3 \cdot 4} = \dfrac{4 \cdot 5 \cdot 1}{3 \cdot 4} = \dfrac{4}{4} \cdot \dfrac{5 \cdot 1}{3} = \dfrac{5}{3} = 1\dfrac{2}{3}$

5. $20\left(-2\dfrac{5}{6}\right) = \dfrac{20}{1} \cdot \left(-\dfrac{17}{6}\right) = -\dfrac{20 \cdot 17}{1 \cdot 6} = -\dfrac{2 \cdot 10 \cdot 17}{2 \cdot 3} =$

$\dfrac{2}{2}\left(-\dfrac{10 \cdot 17}{3}\right) = -\dfrac{170}{3} = -56\dfrac{2}{3}$

7. $3\dfrac{1}{2} \cdot 4\dfrac{2}{3} = \dfrac{7}{2} \cdot \dfrac{14}{3} = \dfrac{7 \cdot 14}{2 \cdot 3} = \dfrac{7 \cdot 2 \cdot 7}{2 \cdot 3} = \dfrac{2}{2} \cdot \dfrac{7 \cdot 7}{3} =$

$\dfrac{49}{3} = 16\dfrac{1}{3}$

9. $-2\dfrac{3}{10} \cdot 4\dfrac{2}{5} = -\dfrac{23}{10} \cdot \dfrac{22}{5} = -\dfrac{23 \cdot 22}{10 \cdot 5} = -\dfrac{23 \cdot 2 \cdot 11}{2 \cdot 5 \cdot 5} =$

$\dfrac{2}{2}\left(-\dfrac{23 \cdot 11}{5 \cdot 5}\right) = -\dfrac{253}{25} = -10\dfrac{3}{25}$

11. $\left(-6\dfrac{3}{10}\right)\left(-5\dfrac{7}{10}\right) = \dfrac{63}{10} \cdot \dfrac{57}{10} = \dfrac{3591}{100} = 35\dfrac{91}{100}$

13. $30 \div 2\dfrac{3}{5}$

$= 30 \div \dfrac{13}{5}$ Writing fractional notation

$= 30 \cdot \dfrac{5}{13}$ Multiplying by the reciprocal

$= \dfrac{30 \cdot 5}{13} = \dfrac{150}{13} = 11\dfrac{7}{13}$

15. $8\dfrac{2}{5} \div 7$

$= \dfrac{42}{5} \div 7$ Writing fractional notation

$= \dfrac{42}{5} \cdot \dfrac{1}{7}$ Multiplying by the reciprocal

$= \dfrac{42 \cdot 1}{5 \cdot 7} = \dfrac{6 \cdot 7}{5 \cdot 7} = \dfrac{7}{7} \cdot \dfrac{6}{5} = \dfrac{6}{5} = 1\dfrac{1}{5}$

17. $5\dfrac{1}{4} \div 2\dfrac{3}{5} = \dfrac{21}{4} \div \dfrac{13}{5} = \dfrac{21}{4} \cdot \dfrac{5}{13} = \dfrac{21 \cdot 5}{4 \cdot 13} =$

$\dfrac{105}{52} = 2\dfrac{1}{52}$

19. $-5\dfrac{1}{4} \div 2\dfrac{3}{7} = -\dfrac{21}{4} \div \dfrac{17}{7} = -\dfrac{21}{4} \cdot \dfrac{7}{17} = -\dfrac{21 \cdot 7}{4 \cdot 17} =$

$-\dfrac{147}{68} = -2\dfrac{11}{68}$

21. $5\dfrac{1}{10} \div 4\dfrac{3}{10} = \dfrac{51}{10} \div \dfrac{43}{10} = \dfrac{51}{10} \cdot \dfrac{10}{43} = \dfrac{51 \cdot 10}{10 \cdot 43}$

$= \dfrac{10}{10} \cdot \dfrac{51}{43} = \dfrac{51}{43} = 1\dfrac{8}{43}$

23. $20\dfrac{1}{4} \div (-90) = \dfrac{81}{4} \div (-90) = \dfrac{81}{4}\left(-\dfrac{1}{90}\right) = -\dfrac{81 \cdot 1}{4 \cdot 90} =$

$-\dfrac{9 \cdot 9 \cdot 1}{4 \cdot 9 \cdot 10} = \dfrac{9}{9} \cdot \left(-\dfrac{9 \cdot 1}{4 \cdot 10}\right) = -\dfrac{9}{40}$

25. $lw = 2\dfrac{3}{5} \cdot 9$

$= \dfrac{13}{5} \cdot 9$

$= \dfrac{117}{5} = 23\dfrac{2}{5}$

27. $rs = 5 \cdot 3\dfrac{1}{7}$

$= 5 \cdot \dfrac{22}{7}$

$= \dfrac{110}{7} = 15\dfrac{5}{7}$

29. $mt = 6\dfrac{2}{9}\left(-4\dfrac{3}{5}\right)$

$= \dfrac{56}{9}\left(-\dfrac{23}{5}\right)$

$= -\dfrac{1288}{45} = -28\dfrac{28}{45}$

31. $R \cdot S \div T = 4\dfrac{2}{3} \cdot 1\dfrac{3}{7} \div (-5)$

$= \dfrac{14}{3} \cdot \dfrac{10}{7} \div (-5)$

$= \dfrac{14 \cdot 10}{3 \cdot 7} \div (-5)$

$= \dfrac{2 \cdot 7 \cdot 10}{3 \cdot 7} \div (-5)$

$= \dfrac{7}{7} \cdot \dfrac{2 \cdot 10}{3} \div (-5)$

$= \dfrac{20}{3} \div (-5)$

$= \dfrac{20}{3} \cdot \left(-\dfrac{1}{5}\right)$

$= -\dfrac{20 \cdot 1}{3 \cdot 5}$

$= -\dfrac{4 \cdot 5}{3 \cdot 5}$

$= -\dfrac{4}{3}$

$= -1\dfrac{1}{3}$

33. $r + ps = 5\dfrac{1}{2} + 3 \cdot 2\dfrac{1}{4}$

$= 5\dfrac{1}{2} + \dfrac{3}{1} \cdot \dfrac{9}{4}$

$= 5\dfrac{1}{2} + \dfrac{27}{4}$

$= 5\dfrac{1}{2} + 6\dfrac{3}{4}$

$= 5\dfrac{2}{4} + 6\dfrac{3}{4}$

$= 11\dfrac{5}{4}$

$= 12\dfrac{1}{4}$

35. $m + n \div p = 7\frac{2}{5} + 4\frac{1}{2} \div 6$

$$= 7\frac{2}{5} + \frac{9}{2} \div 6$$

$$= 7\frac{2}{5} + \frac{9}{2} \cdot \frac{1}{6}$$

$$= 7\frac{2}{5} + \frac{9 \cdot 1}{2 \cdot 6} = 7\frac{2}{5} + \frac{3 \cdot 3 \cdot 1}{2 \cdot 2 \cdot 3}$$

$$= 7\frac{2}{5} + \frac{3 \cdot 1}{2 \cdot 2} = 7\frac{2}{5} + \frac{3}{4}$$

$$= 7\frac{8}{20} + \frac{15}{20} = 7\frac{23}{20}$$

$$= 8\frac{3}{20}$$

37. Familiarize. Let b = the number of beagles registered with The American Kennel Club.

Translate.

$3\frac{4}{9}$	times	Number of beagles registered	is	Number of Labrador retrievers registered
↓	↓	↓	↓	↓
$3\frac{4}{9}$	$\cdot$	b	$=$	$155,000$

Solve. We divide by $3\frac{4}{9}$ on both sides of the equation.

$$b = 155,000 \div 3\frac{4}{9}$$

$$b = 155,000 \div \frac{31}{9}$$

$$b = 155,000 \div \frac{9}{31} = \frac{155,000 \cdot 9}{31}$$

$$b = \frac{31 \cdot 5000 \cdot 9}{31 \cdot 1} = \frac{31}{31} \cdot \frac{5000 \cdot 9}{1} = \frac{5000 \cdot 9}{1}$$

$$b = 45,000$$

Check. Since $3\frac{4}{9} \cdot 45,000 = \frac{31}{9} \cdot 45,000 = 155,000$, the answer checks.

State. There are 45,000 beagles registered with The American Kennel Club.

39. Familiarize. Let s = the number of teaspoons 10 average American women consume in one day.

Translate. A multiplication corresponds to this situation.

$$s = 10 \cdot 1\frac{1}{3}$$

Solve. We carry out the multiplication.

$$s = 10 \cdot 1\frac{1}{3} = 10 \cdot \frac{4}{3} = \frac{10 \cdot 4}{3} = \frac{40}{3} = 13\frac{1}{3}$$

Check. We repeat the calculation. The answer checks.

State. In one day 10 average American women consume $13\frac{1}{3}$ tsp of sodium.

41. Familiarize, Translate, and Solve. To find the ingredients for $\frac{1}{2}$ recipe, we multiply each ingredient by $\frac{1}{2}$.

$$1\frac{2}{3} \cdot \frac{1}{2} = \frac{5}{3} \cdot \frac{1}{2} = \frac{5 \cdot 1}{3 \cdot 2} = \frac{5}{6}$$

$$3 \cdot \frac{1}{2} = \frac{3 \cdot 1}{2} = \frac{3}{2} = 1\frac{1}{2}$$

$$4\frac{1}{2} \cdot \frac{1}{2} = \frac{9}{2} \cdot \frac{1}{2} = \frac{9 \cdot 1}{2 \cdot 2} = \frac{9}{4} = 2\frac{1}{4}$$

$$1 \cdot \frac{1}{2} = \frac{1 \cdot 1}{2} = \frac{1}{2}$$

$$3\frac{3}{4} \cdot \frac{1}{2} = \frac{15}{4} \cdot \frac{1}{2} = \frac{15 \cdot 1}{4 \cdot 2} = \frac{15}{8} = 1\frac{7}{8}$$

$$\frac{3}{4} \cdot \frac{1}{2} = \frac{3 \cdot 1}{4 \cdot 2} = \frac{3}{8}$$

$$1\frac{1}{2} \cdot \frac{1}{2} = \frac{3}{2} \cdot \frac{1}{2} = \frac{3 \cdot 1}{2 \cdot 2} = \frac{3}{4}$$

Check. We repeat the calculations.

State. The ingredients for $\frac{1}{2}$ recipe are $\frac{5}{6}$ cup water, $1\frac{1}{2}$ tablespoons canola oil, $2\frac{1}{4}$ teaspoons sugar, $\frac{1}{2}$ teaspoon salt, $1\frac{7}{8}$ cups bread flour, $\frac{3}{8}$ cup Grape-Nuts cereal, and $\frac{3}{4}$ teaspoon active dry yeast.

Familiarize, Translate and Solve. To find the ingredients for 3 recipes, we multiply each ingredient by 3.

$$1\frac{2}{3} \cdot 3 = \frac{5}{3} \cdot 3 = \frac{5 \cdot \cancel{3}}{\cancel{3} \cdot 1} = 5$$

$$3 \cdot 3 = 9$$

$$4\frac{1}{2} \cdot 3 = \frac{9}{2} \cdot 3 = \frac{9 \cdot 3}{2} = \frac{27}{2} = 13\frac{1}{2}$$

$$1 \cdot 3 = 3$$

$$3\frac{3}{4} \cdot 3 = \frac{15}{4} \cdot 3 = \frac{15 \cdot 3}{4} = \frac{45}{4} = 11\frac{1}{4}$$

$$\frac{3}{4} \cdot 3 = \frac{3 \cdot 3}{4} = \frac{9}{4} = 2\frac{1}{4}$$

$$1\frac{1}{2} \cdot 3 = \frac{3}{2} \cdot 3 = \frac{3 \cdot 3}{2} = \frac{9}{2} = 4\frac{1}{2}$$

Check. We can repeat the calculations.

State. The ingredients for 3 recipes are 5 cups water, 9 tablespoons canola oil, $13\frac{1}{2}$ teaspoons sugar, 3 teaspoons salt, $11\frac{1}{4}$ cups bread flour, $2\frac{1}{4}$ cups Grape-Nuts cereal, and $4\frac{1}{2}$ teaspoons active dry yeast.

43. Familiarize. Let h = the number of hours spent shopping on the Internet in 2005.

Translate.

$9\frac{1}{8}$	times	Internet shopping time	is	Television time
↓	↓	↓	↓	↓
$9\frac{1}{8}$	$\cdot$	h	$=$	1825

Solve. We solve the equation.

$$9\frac{1}{8} \cdot h = 1825$$

$$\frac{73}{8} \cdot h = 1825$$

$$\frac{8}{73} \cdot \frac{73}{8} \cdot h = \frac{8}{73} \cdot 1825$$

$$1 \cdot h = \frac{8 \cdot 1825}{73}$$

$$h = \frac{8 \cdot 25 \cdot \cancel{73}}{\cancel{73} \cdot 1}$$

$$h = 200$$

Check. Since $9\frac{1}{8} \cdot 200 = \frac{73}{8} \cdot 200 = \frac{73 \cdot 200}{8} = \frac{73 \cdot 8 \cdot 25}{8 \cdot 1} = 1825$, the answer checks.

State. The average person spent about 200 hr shopping on the Internet in 2005.

45. Familiarize. We let t = the Fahrenheit temperature.

Translate.

Celsius temperature	times $1\frac{4}{5}$	plus	$32°$	is	Fahrenheit temperature
↓	↓	↓	↓	↓	↓
20	$\cdot$ $1\frac{4}{5}$	$+$	32	$=$	t

Solve. We multiply and then add, according to the rules for order of operations.

$$t = 20 \cdot 1\frac{4}{5} + 32 = \frac{20}{1} \cdot \frac{9}{5} + 32 = \frac{20 \cdot 9}{1 \cdot 5} + 32 =$$

$$\frac{4 \cdot 5 \cdot 9}{1 \cdot 5} + 32 = \frac{5}{5} \cdot \frac{4 \cdot 9}{1} + 32 = 36 + 32 = 68$$

Check. We repeat the calculation.

State. 68° Fahrenheit corresponds to 20° Celsius.

47. Familiarize. First we will find the total width of the columns and determine if this is less than or greater than $8\frac{1}{2}$ in. Then, if the total is less than $8\frac{1}{2}$ in., we will find the difference between $8\frac{1}{2}$ in. and the total width and then divide by 2 to find the width of each margin. Let w = the total width of the columns.

Translate. First we write an equation for finding the total width of the columns.

$$w = 2 \cdot 1\frac{1}{2} + 5 \cdot \frac{3}{4}$$

Solve. We solve the equation.

$$w = 2 \cdot 1\frac{1}{2} + 5 \cdot \frac{3}{4}$$

$$w = 2 \cdot \frac{3}{2} + 5 \cdot \frac{3}{4}$$

$$w = \frac{2 \cdot 3}{2} + \frac{5 \cdot 3}{4}$$

$$w = 3 + \frac{15}{4} = 3 + 3\frac{3}{4}$$

$$w = 6\frac{3}{4}$$

Since the total width of the columns is less than $8\frac{1}{2}$ in., the table will fit on a piece of standard paper. Let l = the number of inches by which the width of the paper exceeds the width of the table. Then we have:

$$l = 8\frac{1}{2} - 6\frac{3}{4} = 8\frac{2}{4} - 6\frac{3}{4}$$

$$l = 7\frac{6}{4} - 6\frac{3}{4}$$

$$l = 1\frac{3}{4}$$

This tells us that the total width of the margins will be $1\frac{3}{4}$ in. Since the margins are of equal width, we divide by 2 to find the width of each margin. Let m = this width, in inches. We have:

$$m = 1\frac{3}{4} \div 2$$

$$m = \frac{7}{4} \div 2$$

$$m = \frac{7}{4} \cdot \frac{1}{2} = \frac{7}{8}$$

Check. We repeat the calculations.

State. The table will fit on a standard piece of paper. Each margin will be $\frac{7}{8}$ in. wide.

49. Familiarize. We draw a picture.

$\frac{1}{3}$ lb	$\frac{1}{3}$ lb	$\cdots\cdots$	$\frac{1}{3}$ lb

$$\longleftarrow \quad 5\frac{1}{2} \text{ lb} \quad \longrightarrow$$

We let s = the number of servings that can be prepared from $5\frac{1}{2}$ lb of salmon fillet.

Translate. The situation corresponds to a division sentence.

$$s = 5\frac{1}{2} \div \frac{1}{3}$$

Solve. We carry out the division.

$$s = 5\frac{1}{2} \div \frac{1}{3} = \frac{11}{2} \div \frac{1}{3}$$

$$= \frac{11}{2} \cdot \frac{3}{1} = \frac{33}{2}$$

$$= 16\frac{1}{2}$$

Check. We check by multiplying. If $16\frac{1}{2}$ servings are prepared, then

$$16\frac{1}{2} \cdot \frac{1}{3} = \frac{33}{2} \cdot \frac{1}{3} = \frac{3 \cdot 11 \cdot 1}{2 \cdot 3} = \frac{3}{3} \cdot \frac{11 \cdot 1}{2} = \frac{11}{2} = 5\frac{1}{2} \text{ lb}$$

of flounder is used. Our answer checks.

State. $16\frac{1}{2}$ servings can be prepared from $5\frac{1}{2}$ lb of salmon fillet.

51. _Familiarize_. We let w = the weight of $5\frac{1}{2}$ cubic feet of water.

Translate. We write an equation.

$$
\underbrace{\text{Weight per}}_{\text{cubic foot}} \cdot \underbrace{\text{Number of}}_{\text{cubic feet}} = \underbrace{\text{Total}}_{\text{weight}}
$$

$$
\begin{array}{ccccc}
\downarrow & \downarrow & \downarrow & \downarrow & \downarrow \\
62\frac{1}{2} & \cdot & 5\frac{1}{2} & = & w
\end{array}
$$

Solve. To solve the equation we carry out the multiplication.

$$
w = 62\frac{1}{2} \cdot 5\frac{1}{2}
$$
$$
= \frac{125}{2} \cdot \frac{11}{2} = \frac{125 \cdot 11}{2 \cdot 2}
$$
$$
= \frac{1375}{4} = 343\frac{3}{4}
$$

Check. We repeat the calculation. We also note that $62\frac{1}{2} \approx 60$ and $5\frac{1}{2} \approx 5$. Then the product is about 300. Our answer seems reasonable.

State. The weight of $5\frac{1}{2}$ cubic feet of water is $343\frac{3}{4}$ lb.

53. _Familiarize_. We let t = the number of inches of tape used in 60 sec of recording.

Translate. We write an equation.

$$
\underbrace{\text{Inches per}}_{\text{second}} \cdot \underbrace{\text{Number of}}_{\text{seconds}} = \underbrace{\text{Tape}}_{\text{used}}
$$

$$
\begin{array}{ccccc}
\downarrow & \downarrow & \downarrow & \downarrow & \downarrow \\
1\frac{3}{8} & \cdot & 60 & = & t
\end{array}
$$

Solve. We carry out the multiplication.

$$
t = 1\frac{3}{8} \cdot 60 = \frac{11}{8} \cdot 60
$$
$$
= \frac{11 \cdot 4 \cdot 15}{2 \cdot 4} = \frac{11 \cdot 15}{2} \cdot \frac{4}{4}
$$
$$
= \frac{165}{2} = 82\frac{1}{2}
$$

Check. We repeat the calculation.

State. $82\frac{1}{2}$ in. of tape are used in 60 sec of recording in short-play mode.

55. _Familiarize_. We let m = the number of miles per gallon the car got.

Translate. We write an equation.

$$
\underbrace{\begin{array}{c}\text{Total number}\\\text{of}\\\text{miles traveled}\end{array}}_{} \div \underbrace{\begin{array}{c}\text{Number of}\\\text{gallons of}\\\text{gas used}\end{array}}_{} = \underbrace{\begin{array}{c}\text{Miles}\\\text{per}\\\text{gallon}\end{array}}_{}
$$

$$
\begin{array}{ccccc}
\downarrow & \downarrow & \downarrow & \downarrow & \downarrow \\
213 & \div & 14\frac{2}{10} & = & m
\end{array}
$$

Solve. To solve the equation we carry out the division.

$$
m = 213 \div 14\frac{2}{10} = 213 \div \frac{142}{10}
$$
$$
= 213 \cdot \frac{10}{142} = \frac{3 \cdot 71 \cdot 2 \cdot 5}{2 \cdot 71 \cdot 1}
$$
$$
= \frac{2 \cdot 71}{2 \cdot 71} \cdot \frac{3 \cdot 5}{1} = 15
$$

Check. We repeat the calculation.

State. The car got 15 miles per gallon of gas.

57. _Familiarize_. To compute an average, we add the values and then divide the sum by the number of values. Let w = the average birthweight, in pounds.

Translate. We have

$$
w = \frac{2\frac{9}{16} + 2\frac{9}{32} + 2\frac{1}{8} + 2\frac{5}{16}}{4}.
$$

Solve. First we add.

$$
2\frac{9}{16} + 2\frac{9}{32} + 2\frac{1}{8} + 2\frac{5}{16}
$$
$$
= 2\frac{18}{32} + 2\frac{9}{32} + 2\frac{4}{32} + 2\frac{10}{32}
$$
$$
= 8\frac{41}{32} = 9\frac{9}{32}
$$

Then we divide:

$$
9\frac{9}{32} \div 4 = \frac{297}{32} \div 4
$$
$$
= \frac{297}{32} \cdot \frac{1}{4}
$$
$$
= \frac{297}{128} = 2\frac{41}{128}
$$

Check. As a partial check we note that the average weight is larger than the smallest weight and smaller than the largest weight. We could also repeat the calculations.

State. The average birthweight of the quadruplets was $2\frac{41}{128}$ lb.

59. We add the numbers and divide by the number of addends.

$$
\frac{7\frac{3}{5} + 9\frac{1}{10} + 6\frac{1}{2} + 6\frac{9}{10} + 7\frac{1}{5}}{5}
$$
$$
= \frac{7\frac{6}{10} + 9\frac{1}{10} + 6\frac{5}{10} + 6\frac{9}{10} + 7\frac{2}{10}}{5}
$$
$$
= \frac{35\frac{23}{10}}{5} = \frac{37\frac{3}{10}}{5}
$$
$$
= \frac{\frac{373}{10}}{5} = \frac{373}{10} \cdot \frac{1}{5}
$$
$$
= \frac{373}{50} = 7\frac{23}{50}
$$

The average time was $7\frac{23}{50}$ sec.

61. _Familiarize_. We can refer to the drawing in the text. Let a = the total area of the sod.

Translate. The total area is the sum of the areas of the two rectangles.

$$a = 20 \cdot 15\frac{1}{2} + 12\frac{1}{2} \cdot 10\frac{1}{2}$$

Solve. We perform the multiplication and then add.

$$a = 20 \cdot 15\frac{1}{2} + 12\frac{1}{2} \cdot 10\frac{1}{2}$$

$$= 20 \cdot \frac{31}{2} + \frac{25}{2} \cdot \frac{21}{2}$$

$$= \frac{20 \cdot 31}{2} + \frac{25 \cdot 21}{4}$$

$$= \frac{\cancel{2} \cdot 10 \cdot 31}{\cancel{2} \cdot 1} + \frac{525}{4}$$

$$= 310 + 131\frac{1}{4}$$

$$= 441\frac{1}{4}$$

Check. We can perform a partial check by estimating the total area as $20 \cdot 16 + 13 \cdot 11 = 320 + 143 = 463 \approx 441\frac{1}{4}$. Our answer seems reasonable.

State. The total area of the sod is $441\frac{1}{4}$ ft^2.

63. Familiarize. The figure contains a square with sides of $10\frac{1}{2}$ ft and a rectangle with dimensions of $8\frac{1}{2}$ ft by 4 ft. The area of the shaded region consists of the area of the square less the area of the rectangle. Let $A =$ the area of the shaded region, in square feet.

Translate. We write an equation.

$$A = 10\frac{1}{2} \cdot 10\frac{1}{2} - 8\frac{1}{2} \cdot 4$$

Solve. We multiply and then subtract.

$$A = 10\frac{1}{2} \cdot 10\frac{1}{2} - 8\frac{1}{2} \cdot 4$$

$$A = \frac{21}{2} \cdot \frac{21}{2} - \frac{17}{2} \cdot 4$$

$$A = \frac{441}{4} - \frac{68}{2}$$

$$A = \frac{441}{4} - \left(\frac{68}{2} \cdot \frac{2}{2}\right)$$

$$A = \frac{441}{4} - \frac{136}{4}$$

$$A = \frac{305}{4} = 76\frac{1}{4}$$

Check. We repeat the calculation.

State. The area of the shaded region is $76\frac{1}{4}$ ft^2.

65. Familiarize. The figure is a rectangle with dimensions $7\frac{1}{4}$ cm by $4\frac{1}{2}$ cm with a rectangular area cut out of it. One dimension of the cut-out area is $4\frac{1}{4}$ cm. We subtract to find the other dimension:

$$4\frac{1}{2} - \left(1\frac{1}{2} + 1\frac{3}{4}\right) = 4\frac{1}{2} - \left(1\frac{2}{4} + 1\frac{3}{4}\right) = 4\frac{1}{2} - 2\frac{5}{4} = 4\frac{1}{2} - 3\frac{1}{4} =$$

$$4\frac{2}{4} - 3\frac{1}{4} = 1\frac{1}{4}.$$

The area of the shaded region consists of the area of the larger rectangle less the area of the smaller rectangle. Let $A =$ the area of the shaded region, in square centimeters.

Translate.

$$A = 7\frac{1}{4} \cdot 4\frac{1}{2} - 4\frac{1}{4} \cdot 1\frac{1}{4}$$

Solve.

$$A = 7\frac{1}{4} \cdot 4\frac{1}{2} - 4\frac{1}{4} \cdot 1\frac{1}{4}$$

$$A = \frac{29}{4} \cdot \frac{9}{2} - \frac{17}{4} \cdot \frac{5}{4}$$

$$A = \frac{29 \cdot 9}{4 \cdot 2} - \frac{17 \cdot 5}{4 \cdot 4}$$

$$A = \frac{261}{8} - \frac{85}{16}$$

$$A = \frac{261}{8} \cdot \frac{2}{2} - \frac{85}{16}$$

$$A = \frac{522}{16} - \frac{85}{16}$$

$$A = \frac{437}{16} = 27\frac{5}{16}$$

Check. We can perform a partial check by estimating the area as $7 \cdot 5 - 4 \cdot 1 = 35 - 4 = 31$. Our answer seems reasonable.

State. The area of the shaded region is $27\frac{5}{16}$ cm^2.

67. Discussion and Writing Exercise

69. The set $\{\ldots, -3, -2, -1, 0, 1, 2, 3, \ldots\}$ is the set of <u>integers</u>.

71. The numbers 91, 95, and 111 are examples of <u>composite</u> numbers.

73. To add fractions with different denominators, we must first find the <u>least common multiple</u> of the denominators.

75. In the expression $\frac{c}{d}$, we call c the <u>numerator</u>.

77. Discussion and Writing Exercise

79. $-8 \div \frac{1}{2} + \frac{3}{4} + \left(-5 - \frac{5}{8}\right)^2 = -8 \div \frac{1}{2} + \frac{3}{4} + \left(-\frac{40}{8} - \frac{5}{8}\right)^2 =$

$-8 \div \frac{1}{2} + \frac{3}{4} + \left(-\frac{45}{8}\right)^2 = -8 \div \frac{1}{2} + \frac{3}{4} + \frac{2025}{64} =$

$-8 \cdot 2 + \frac{3}{4} + \frac{2025}{64} = -16 + \frac{3}{4} + \frac{2025}{64} =$

$-\frac{1024}{64} + \frac{48}{64} + \frac{2025}{64} = \frac{1049}{64} = 16\frac{25}{64}$

81. $\frac{1}{3} \div \left(\frac{1}{2} - \frac{1}{5}\right) \times \frac{1}{4} + \frac{1}{6}$

$= \frac{1}{3} \div \left(\frac{5}{10} - \frac{2}{10}\right) \times \frac{1}{4} + \frac{1}{6}$

$= \frac{1}{3} \div \frac{3}{10} \times \frac{1}{4} + \frac{1}{6}$

$= \frac{1}{3} \times \frac{10}{3} \times \frac{1}{4} + \frac{1}{6}$

$= \frac{10}{9} \times \frac{1}{4} + \frac{1}{6}$

$= \frac{2 \times 5 \times 1}{9 \times 2 \times 2} + \frac{1}{6} = \frac{2}{2} \cdot \frac{5 \times 1}{9 \times 2} + \frac{1}{6}$

$= \frac{5}{18} + \frac{1}{6} = \frac{5}{18} + \frac{3}{18} = \frac{8}{18} = \frac{4}{9}$

83. $\dfrac{1}{r} = \dfrac{1}{40} + \dfrac{1}{60} + \dfrac{1}{80}$

$\dfrac{1}{r} = \dfrac{1}{40} \cdot \dfrac{6}{6} + \dfrac{1}{60} \cdot \dfrac{4}{4} + \dfrac{1}{80} \cdot \dfrac{3}{3}$

$\dfrac{1}{r} = \dfrac{6}{240} + \dfrac{4}{240} + \dfrac{3}{240}$

$\dfrac{1}{r} = \dfrac{13}{240}$

Then r is the reciprocal of $\dfrac{13}{240}$, so $r = \dfrac{240}{13}$, or $18\dfrac{6}{13}$.

85. *Familiarize.* Let w = the amount of hot water required for two showers and two loads of wash. Note that washing one load of clothes requires $1\dfrac{3}{5} \cdot 20$ gallons of hot water.

Translate. We write an equation.

$$w = 2 \cdot 20 + 2 \cdot 1\dfrac{3}{5} \cdot 20$$

Solve. We perform the multiplications and then we add.

$w = 2 \cdot 20 + 2 \cdot 1\dfrac{3}{5} \cdot 20$

$w = 40 + 2 \cdot \dfrac{8}{5} \cdot 20$

$w = 40 + \dfrac{2 \cdot 8 \cdot 20}{5}$

$w = 40 + \dfrac{320}{5}$

$w = 40 + 64$

$w = 104$

Check. We repeat the calculations. The answer checks.

State. Two showers and two loads of wash require 104 gallons of hot water.

Chapter 4 Review Exercises

1. $16 = 2 \cdot 2 \cdot 2 \cdot 2 = 2^4$
$20 = 2 \cdot 2 \cdot 5 = 2^2 \cdot 5$

We form the LCM using the greatest power of each factor. The LCM is $2^4 \cdot 5$, or 80.

2. 1.) 45 is not a multiple of 18.

2.) Check multiples:

$2 \cdot 45 = 90$ A multiple of 18

The LCM is 90.

3. Note that 3 and 6 are factors of 30. Since the largest number, 30, has the other two numbers as factors, it is the LCM.

4. $\dfrac{2}{9} + \dfrac{5}{9} = \dfrac{2+5}{9} = \dfrac{7}{9}$

5. $\dfrac{7}{x} + \dfrac{2}{x} = \dfrac{7+2}{x} = \dfrac{9}{x}$

6. The LCM of 5 and 15 is 15.
$-\dfrac{6}{5} + \dfrac{11}{15} = -\dfrac{6}{5} \cdot \dfrac{3}{3} + \dfrac{11}{15} = -\dfrac{18}{15} + \dfrac{11}{15} = -\dfrac{7}{15}$

7. The LCM of 16 and 24 is 48.
$\dfrac{5}{16} + \dfrac{3}{24} = \dfrac{5}{16} \cdot \dfrac{3}{3} + \dfrac{3}{24} \cdot \dfrac{2}{2}$
$= \dfrac{15}{48} + \dfrac{6}{48} = \dfrac{21}{48}$
$= \dfrac{3 \cdot 7}{3 \cdot 16} = \dfrac{3}{3} \cdot \dfrac{7}{16} = \dfrac{7}{16}$

8. $\dfrac{7}{9} - \dfrac{5}{9} = \dfrac{7-5}{9} = \dfrac{2}{9}$

9. The LCM of 4 and 8 is 8.
$\dfrac{1}{4} - \dfrac{3}{8} = \dfrac{1}{4} \cdot \dfrac{2}{2} - \dfrac{3}{8} = \dfrac{2}{8} - \dfrac{3}{8} = -\dfrac{1}{8}$

10. The LCM of 27 and 9 is 27.
$\dfrac{10}{27} - \dfrac{2}{9} = \dfrac{10}{27} - \dfrac{2}{9} \cdot \dfrac{3}{3} = \dfrac{10}{27} - \dfrac{6}{27} = \dfrac{10-6}{27} = \dfrac{4}{27}$

11. The LCM of 6 and 9 is 18.
$\dfrac{5}{6} - \dfrac{2}{9} = \dfrac{5}{6} \cdot \dfrac{3}{3} - \dfrac{2}{9} \cdot \dfrac{2}{2}$
$= \dfrac{15}{18} - \dfrac{4}{18} = \dfrac{11}{18}$

12. The LCD is $7 \cdot 9$, or 63.
$\dfrac{4}{7} \cdot \dfrac{9}{9} = \dfrac{36}{63}$
$\dfrac{5}{9} \cdot \dfrac{7}{7} = \dfrac{35}{63}$
Since $36 > 35$, it follows that $\dfrac{36}{63} > \dfrac{35}{63}$, so $\dfrac{4}{7} > \dfrac{5}{9}$.

13. The LCD is $9 \cdot 13$, or 117.
$\dfrac{-8}{9} \cdot \dfrac{13}{13} = \dfrac{-104}{117}$
$\dfrac{-11}{13} \cdot \dfrac{9}{9} = \dfrac{-99}{117}$
Since $-104 < -99$, it follows that $\dfrac{-104}{117} < \dfrac{-99}{117}$, so $-\dfrac{8}{9} < -\dfrac{11}{13}$.

14. $x + \dfrac{2}{5} = \dfrac{7}{8}$

$x + \dfrac{2}{5} - \dfrac{2}{5} = \dfrac{7}{8} - \dfrac{2}{5}$

$x + 0 = \dfrac{7}{8} \cdot \dfrac{5}{5} - \dfrac{2}{5} \cdot \dfrac{8}{8}$

$x = \dfrac{35}{40} - \dfrac{16}{40}$

$x = \dfrac{19}{40}$

The solution is $\dfrac{19}{40}$.

15. $7a - 3 = 25$

$7a - 3 + 3 = 25 + 3$

$7a = 28$

$\dfrac{1}{7} \cdot 7a = \dfrac{1}{7} \cdot 28$

$a = 4$

The solution is 4.

16.
$$5 + \frac{16}{3}x = \frac{5}{9}$$
$$5 + \frac{16}{3}x - 5 = \frac{5}{9} - 5$$
$$\frac{16}{3}x = \frac{5}{9} - \frac{45}{9}$$
$$\frac{16}{3}x = -\frac{40}{9}$$
$$\frac{3}{16} \cdot \frac{16}{3}x = \frac{3}{16}\left(-\frac{40}{9}\right)$$
$$x = -\frac{3 \cdot 40}{16 \cdot 9} = -\frac{3 \cdot 5 \cdot 8}{2 \cdot 8 \cdot 3 \cdot 3}$$
$$x = -\frac{5}{2 \cdot 3} \cdot \frac{3 \cdot 8}{3 \cdot 8} = -\frac{5}{6}$$

The solution is $-\dfrac{5}{6}$.

17.
$$\frac{22}{5} = \frac{16}{5} + \frac{5}{2}x$$
$$\frac{22}{5} - \frac{16}{5} = \frac{16}{5} + \frac{5}{2}x - \frac{16}{5}$$
$$\frac{6}{5} = \frac{5}{2}x$$
$$\frac{2}{5} \cdot \frac{6}{5} = \frac{2}{5} \cdot \frac{5}{2}x$$
$$\frac{12}{25} = x$$

The solution is $\dfrac{12}{25}$.

18.
$$\frac{5}{3}x + \frac{5}{6} = \frac{3}{2}$$
$$6\left(\frac{5}{3}x + \frac{5}{6}\right) = 6 \cdot \frac{3}{2} \quad \text{The LCD is 6.}$$
$$\frac{6 \cdot 5}{3}x + 6 \cdot \frac{5}{6} = \frac{18}{2}$$
$$\frac{2 \cdot \cancel{3} \cdot 5}{\cancel{3} \cdot 1}x + 5 = 9$$
$$10x + 5 = 9$$
$$10x + 5 - 5 = 9 - 5$$
$$10x = 4$$
$$\frac{10x}{10} = \frac{4}{10}$$
$$x = \frac{2}{5}$$

The solution is $\dfrac{2}{5}$.

19. $7\dfrac{1}{2} = \dfrac{15}{2}$ $(7 \cdot 2 = 14,\ 14 + 1 = 15)$

20. $30\dfrac{4}{9} = \dfrac{274}{9}$ $(30 \cdot 9 = 270,\ 270 + 4 = 274)$

21. $-9\dfrac{2}{7} = -\left(9 + \dfrac{2}{7}\right) = -\dfrac{65}{7}$

$(9 \cdot 7 = 63;\ 63 + 2 = 65;\ \text{include the negative sign.})$

22.
$$\begin{array}{r} 2 \\ 5\overline{)1\,3} \\ 1\,0 \\ \hline 3 \end{array}$$
$$\frac{13}{5} = 2\frac{3}{5}$$

23. First consider $\dfrac{27}{4}$.
$$\begin{array}{r} 6 \\ 4\overline{)2\,7} \\ 2\,4 \\ \hline 3 \end{array}$$
$$\frac{27}{4} = 6\frac{3}{4}$$

Since $\dfrac{27}{4} = 6\dfrac{3}{4}$, we have $\dfrac{-27}{4} = -6\dfrac{3}{4}$.

24.
$$\begin{array}{r} 7 \\ 8\overline{)5\,7} \\ 5\,6 \\ \hline 1 \end{array}$$
$$\frac{57}{8} = 7\frac{1}{8}$$

25.
$$\begin{array}{r} 3 \\ 2\overline{)7} \\ 6 \\ \hline 1 \end{array}$$
$$\frac{7}{2} = 3\frac{1}{2}$$

26. First we find $7896 \div 9$.
$$\begin{array}{r} 8\,7\,7 \\ 9\overline{)7\,8\,9\,6} \\ 7\,2\,0\,0 \\ \hline 6\,9\,6 \\ 6\,3\,0 \\ \hline 6\,6 \\ 6\,3 \\ \hline 3 \end{array}$$

Since $877\dfrac{3}{9} = 877\dfrac{1}{3}$, we have $7896 \div (-9) = -877\dfrac{1}{3}$.

27. $\dfrac{80 + 82 + 85}{3} = \dfrac{247}{3} = 82\dfrac{1}{3}$

28.
$$\begin{array}{r} 7\dfrac{3}{5} \\ +2\dfrac{4}{5} \\ \hline 9\dfrac{7}{5} = 9 + \dfrac{7}{5} \\ = 9 + 1\dfrac{2}{5} \\ = 10\dfrac{2}{5} \end{array}$$

29.
$$\begin{array}{r} 6\ \boxed{\dfrac{1}{3} \cdot \dfrac{5}{5}} = \ \ 6\,\dfrac{5}{15} \\ +5\ \boxed{\dfrac{2}{5} \cdot \dfrac{3}{3}} = +5\,\dfrac{6}{15} \\ \hline 11\,\dfrac{11}{15} \end{array}$$

30. $-3\frac{5}{6} + \left(-5\frac{1}{6}\right)$

We add the absolute values and make the answer negative.

$$3\frac{5}{6}$$
$$+5\frac{1}{6}$$
$$\overline{}$$
$$8\frac{6}{6} = 8 + 1 = 9$$

Thus, $-3\frac{5}{6} + \left(-5\frac{1}{6}\right) = -9$.

31. $-2\frac{3}{4} + 4\frac{1}{2} = 4\frac{1}{2} - 2\frac{3}{4}$

$$4\,\boxed{\frac{1}{2}\cdot\frac{2}{2}} = \quad 4\frac{2}{4} = \quad 3\frac{6}{4}$$
$$-2\frac{3}{4} \quad = -2\frac{3}{4} = -2\frac{3}{4}$$
$$\overline{} \quad \overline{} \quad \overline{}$$
$$1\frac{3}{4}$$

32. $\quad 14 \quad = \quad 13\frac{9}{9}$

$$-\ 6\frac{2}{9} = -\ 6\frac{2}{9}$$
$$\overline{} \quad \overline{}$$
$$7\frac{7}{9}$$

33. $\quad 9\,\boxed{\frac{3}{5}\cdot\frac{3}{3}} = \quad 9\,\frac{9}{15} = \quad 8\frac{24}{15}$

$$-4\quad\frac{13}{15} \quad = -4\frac{13}{15} = -4\frac{13}{15}$$
$$\overline{} \quad \overline{} \quad \overline{}$$
$$4\frac{11}{15}$$

34. $4\frac{5}{8} - 9\frac{3}{4} = 4\frac{5}{8} + \left(-9\frac{3}{4}\right)$

Since $9\frac{3}{4}$ is greater than $4\frac{5}{8}$, the answer will be negative. The difference of the absolute values is

$$9\,\boxed{\frac{3}{4}\cdot\frac{2}{2}} = \quad 9\,\frac{6}{8}$$
$$-4\quad\frac{5}{8} \quad = -4\frac{5}{8}$$
$$\overline{} \quad \overline{}$$
$$5\frac{1}{8}$$

so $4\frac{5}{8} - 9\frac{3}{4} = -5\frac{1}{8}$.

35. $-7\frac{1}{2} - 6\frac{3}{4} = -7\frac{1}{2} + \left(-6\frac{3}{4}\right)$

We add the absolute values and make the answer negative.

$$7\,\boxed{\frac{1}{2}\cdot\frac{2}{2}} = \quad 7\frac{2}{4}$$
$$+6\quad\frac{3}{4} \quad = +6\frac{3}{4}$$
$$\overline{} \quad \overline{}$$
$$13\frac{5}{4} = 13 + \frac{5}{4}$$
$$= 13 + 1\frac{1}{4}$$
$$= 14\frac{1}{4}$$

Thus, $-7\frac{1}{2} - 6\frac{3}{4} = -14\frac{1}{4}$.

36. $\dfrac{4}{9}x + \dfrac{1}{3}x = \dfrac{4}{9}x + \dfrac{1}{3}\cdot\dfrac{3}{3}x$

$$= \frac{4}{9}x + \frac{3}{9}x$$
$$= \frac{7}{9}x$$

37. $8\frac{3}{10}a - 5\frac{1}{8}a = \left(8\frac{3}{10} - 5\frac{1}{8}\right)a$

$$= \left(8\frac{12}{40} - 5\frac{5}{40}\right)a$$
$$= 3\frac{7}{40}a$$

38. $6\cdot 2\frac{2}{3} = 6\cdot\dfrac{8}{3} = \dfrac{6\cdot 8}{3} = \dfrac{2\cdot 3\cdot 8}{3\cdot 1} = \dfrac{3}{3}\cdot\dfrac{2\cdot 8}{1} = 16$

39. $-5\frac{1}{4}\cdot\dfrac{2}{3} = -\dfrac{21}{4}\cdot\dfrac{2}{3} = -\dfrac{21\cdot 2}{4\cdot 3} = -\dfrac{3\cdot 7\cdot 2}{2\cdot 2\cdot 3} =$

$$\frac{2\cdot 3}{2\cdot 3}\cdot\left(-\frac{7}{2}\right) = -\frac{7}{2} = -3\frac{1}{2}$$

40. $2\frac{1}{5}\cdot 1\frac{1}{10} = \dfrac{11}{5}\cdot\dfrac{11}{10} = \dfrac{11\cdot 11}{5\cdot 10} = \dfrac{121}{50} = 2\frac{21}{50}$

41. $2\frac{2}{5}\cdot 2\frac{1}{2} = \dfrac{12}{5}\cdot\dfrac{5}{2} = \dfrac{12\cdot 5}{5\cdot 2} = \dfrac{2\cdot 6\cdot 5}{5\cdot 2\cdot 1} = \dfrac{2\cdot 5}{2\cdot 5}\cdot\dfrac{6}{1} = 6$

42. $-54 \div 2\frac{1}{4} = -54 \div \dfrac{9}{4} = -54\cdot\dfrac{4}{9} = \dfrac{-54\cdot 4}{9} =$

$$\frac{-2\cdot 3\cdot 9\cdot 4}{9\cdot 1} = \frac{9}{9}\cdot\frac{-2\cdot 3\cdot 4}{1} = -24$$

43. $2\frac{2}{5} \div \left(-1\frac{7}{10}\right) = \dfrac{12}{5} \div \left(-\dfrac{17}{10}\right) = \dfrac{12}{5}\cdot\left(-\dfrac{10}{17}\right) =$

$$-\frac{12\cdot 10}{5\cdot 17} = -\frac{12\cdot 2\cdot 5}{5\cdot 17} = \frac{5}{5}\cdot\left(-\frac{12\cdot 2}{17}\right) = -\frac{24}{17} = -1\frac{7}{17}$$

44. $3\frac{1}{4} \div 26 = \dfrac{13}{4} \div 26 = \dfrac{13}{4}\cdot\dfrac{1}{26} = \dfrac{13\cdot 1}{4\cdot 26} = \dfrac{13\cdot 1}{4\cdot 2\cdot 13} =$

$$\frac{13}{13}\cdot\frac{1}{4\cdot 2} = \frac{1}{8}$$

45. $4\frac{1}{5} \div 4\frac{2}{3} = \dfrac{21}{5} \div \dfrac{14}{3} = \dfrac{21}{5}\cdot\dfrac{3}{14} = \dfrac{21\cdot 3}{5\cdot 14} = \dfrac{3\cdot 7\cdot 3}{5\cdot 2\cdot 7} =$

$$\frac{7}{7}\cdot\frac{3\cdot 3}{5\cdot 2} = \frac{9}{10}$$

46. $5x - y = 5 \cdot 3\frac{1}{5} - 2\frac{2}{7}$

$\qquad = 5 \cdot \frac{16}{5} - 2\frac{2}{7}$

$\qquad = \frac{\cancel{5} \cdot 16}{1 \cdot \cancel{5}} - 2\frac{2}{7}$

$\qquad = 16 - 2\frac{2}{7}$

$\qquad = 15\frac{7}{7} - 2\frac{2}{7}$

$\qquad = 13\frac{5}{7}$

47. $2a \div b = 2 \cdot 5\frac{2}{11} \div 3\frac{4}{5}$

$\qquad = 2 \cdot \frac{57}{11} \div \frac{19}{5}$

$\qquad = \frac{2 \cdot 57}{11} \div \frac{19}{5}$

$\qquad = \frac{2 \cdot 57}{11} \cdot \frac{5}{19}$

$\qquad = \frac{2 \cdot 57 \cdot 5}{11 \cdot 19}$

$\qquad = \frac{2 \cdot 3 \cdot \cancel{19} \cdot 5}{11 \cdot \cancel{19}}$

$\qquad = \frac{30}{11}$

$\qquad = 2\frac{8}{11}$

48. ***Familiarize.*** Let $c =$ the number of cassettes that can be placed on each shelf.

Translate. We write a division sentence.

$$c = 27 \div 1\frac{1}{8}$$

Solve. We carry out the division.

$c = 27 \div 1\frac{1}{8} = 27 \div \frac{9}{8}$

$\qquad = 27 \cdot \frac{8}{9} = \frac{27 \cdot 8}{9}$

$\qquad = \frac{3 \cdot \cancel{9} \cdot 8}{\cancel{9} \cdot 1}$

$\qquad = 24$

Check. Since $24 \cdot 1\frac{1}{8} = 24 \cdot \frac{9}{8} = \frac{24 \cdot 9}{8} = \frac{3 \cdot 8 \cdot 9}{8} = 27$, the answer checks.

State. 24 cassettes can be placed on each shelf.

49. ***Familiarize.*** Let $p =$ the number of pizzas that remained.

Translate.

$$p = \frac{3}{8} + 1\frac{1}{2} + 1\frac{1}{4}$$

Solve. We carry out the addition.

$\qquad \frac{3}{8} \quad = \quad \frac{3}{8}$

$\quad 1 \boxed{\frac{1}{2} \cdot \frac{4}{4}} = \quad 1\frac{4}{8}$

$+1 \boxed{\frac{1}{4} \cdot \frac{2}{2}} = +1\frac{2}{8}$

$\qquad\qquad\qquad\qquad 2\frac{9}{8} = 2 + \frac{9}{8}$

$\qquad\qquad\qquad\qquad\qquad = 2 + 1\frac{1}{8}$

$\qquad\qquad\qquad\qquad\qquad = 3\frac{1}{8}$

Check. We repeat the calculation. The answer checks.

State. Altogether, $3\frac{1}{8}$ pizzas remained.

50. ***Familiarize.*** Let $d =$ the distance Mica traveled, in miles.

Translate.

$$d = \frac{1}{10} + \frac{1}{2}$$

Solve. We carry out the addition.

$$d = \frac{1}{10} + \frac{1}{2} = \frac{1}{10} + \frac{1}{2} \cdot \frac{5}{5} = \frac{1}{10} + \frac{5}{10} = \frac{6}{10} = \frac{\cancel{2} \cdot 3}{\cancel{2} \cdot 5} = \frac{3}{5}$$

Check. We repeat the calculation. The answer checks.

State. Mica traveled $\frac{3}{5}$ mi.

51. ***Familiarize.*** Let $c =$ how many fewer feet of cable Crew B can install per hour than Crew A.

Translate. This is a "how much more" situation.

Crew B's feet per hour	plus	How many more feet per hour	is	Crew A's feet per hour
↓	↓	↓	↓	↓
$31\frac{2}{3}$	$+$	c	$=$	$38\frac{1}{8}$

Solve.

$\qquad\qquad 31\frac{2}{3} + c = 38\frac{1}{8}$

$\quad 31\frac{2}{3} + c - 31\frac{2}{3} = 38\frac{1}{8} - 31\frac{2}{3}$

$\qquad\qquad\qquad\quad c = 38\frac{3}{24} - 31\frac{16}{24}$

$\qquad\qquad\qquad\quad c = 37\frac{27}{24} - 31\frac{16}{24}$

$\qquad\qquad\qquad\quad c = 6\frac{11}{24}$

Check. Since $31\frac{2}{3} + 6\frac{11}{24} = 31\frac{16}{24} + 6\frac{11}{24} = 37\frac{27}{24} = 37 + 1\frac{3}{24} = 38\frac{3}{24} = 38\frac{1}{8}$, the answer checks.

State. Crew B can install $6\frac{11}{24}$ fewer feet per hour than Crew A.

52. Familiarize. Let $s =$ the number of cups of shortening in the lower calorie cake.

Translate.

New amount of shortening	plus	Amount of prune puree	is	Original amount of shortening
$\downarrow$	$\downarrow$	$\downarrow$	$\downarrow$	$\downarrow$
s	$+$	$3\frac{5}{8}$	$=$	12

Solve. We subtract $3\frac{5}{8}$ on both sides of the equation.

$$12 \quad = 11\frac{8}{8}$$
$$-3\frac{5}{8} = -3\frac{5}{8}$$
$$\overline{\qquad\qquad 8\frac{3}{8}}$$

Thus, $s = 8\frac{3}{8}$.

Check. $8\frac{3}{8} + 3\frac{5}{8} = 11\frac{8}{8} = 12$, so the answer checks.

State. The lower calorie recipe uses $8\frac{3}{8}$ cups of shortening.

53. Familiarize. We draw a picture.

$$8\frac{1}{2} \text{ in.}$$

$9\frac{3}{4}$ in. $\qquad\qquad 9\frac{3}{4}$ in.

$$8\frac{1}{2} \text{ in.}$$

Translate. We let $D =$ the distance around the book.

Top distance	plus	Right-side distance	plus	Bottom distance	plus	Left-side distance	is	Total distance
$\downarrow$	$\downarrow$	$\downarrow$	$\downarrow$	$\downarrow$	$\downarrow$	$\downarrow$	$\downarrow$	$\downarrow$
$8\frac{1}{2}$	$+$	$9\frac{3}{4}$	$+$	$8\frac{1}{2}$	$+$	$9\frac{3}{4}$	$=$	D

Solve. To solve we carry out the addition. The LCD is 4.

$$8\boxed{\frac{1}{2}\cdot\frac{2}{2}} = 8\frac{2}{4}$$
$$9\frac{3}{4} = 9\frac{3}{4}$$
$$8\boxed{\frac{1}{2}\cdot\frac{2}{2}} = 8\frac{2}{4}$$
$$+9\frac{3}{4} = +9\frac{3}{4}$$
$$\overline{\qquad 34\frac{10}{4} = 36\frac{2}{4} = 36\frac{1}{2}}$$

Check. We repeat the calculation.

State. The distance around the book is $36\frac{1}{2}$ in.

54. Familiarize. Let $p =$ the population of Louisiana.

Translate.

Population of Louisiana	is	$2\frac{1}{2}$	times	Population of West Virginia
$\downarrow$	$\downarrow$	$\downarrow$	$\downarrow$	$\downarrow$
p	$=$	$2\frac{1}{2}$	$\cdot$	$1,800,000$

Solve. We carry out the multiplication.

$$p = 2\frac{1}{2}\cdot 1,800,000 = \frac{5}{2}\cdot 1,800,000$$
$$= \frac{5\cdot 1,800,000}{2} = \frac{5\cdot \not2\cdot 900,000}{\not2\cdot 1}$$
$$= 4,500,000$$

Check. We repeat the calculation. The answer checks.

State. The population of Louisiana is about 4,500,000.

55. We find the area of each rectangle and then add to find the total area. Recall that the area of a rectangle is length × width.

Area of rectangle A:
$$12\times 9\frac{1}{2} = 12\times\frac{19}{2} = \frac{12\times 19}{2} = \frac{2\cdot 6\cdot 19}{2\cdot 1} = \frac{2}{2}\cdot\frac{6\cdot 19}{1} =$$
114 in^2

Area of rectangle B:
$$8\frac{1}{2}\times 7\frac{1}{2} = \frac{17}{2}\times\frac{15}{2} = \frac{17\times 15}{2\times 2} = \frac{255}{4} = 63\frac{3}{4}\text{ in}^2$$

Sum of the areas:
$$114\text{ in}^2 + 63\frac{3}{4}\text{ in}^2 = 177\frac{3}{4}\text{ in}^2$$

56. We subtract the area of rectangle B from the area of rectangle A.

$$114 \quad = 113\frac{4}{4}$$
$$-63\frac{3}{4} = -63\frac{3}{4}$$
$$\overline{\qquad\qquad 50\frac{1}{4}}$$

The area of rectangle A is $50\frac{1}{4}$ in^2 greater than the area of rectangle B.

57. Discussion and Writing Exercise. The student multiplied the whole numbers and multiplied the fractions. The mixed numerals should be converted to fraction notation before multiplying.

58. Discussion and Writing Exercise. Yes. We may need to find a common denominator before adding or subtracting. To find the least common denominator, we use the last common multiple of the denominators.

59. We use a calculator to find the prime factorization of each number.

$$141 = 3 \cdot 47$$
$$2419 = 41 \cdot 59$$
$$1357 = 23 \cdot 59$$

The LCM is $3 \cdot 47 \cdot 41 \cdot 59 \cdot 23$, or $7{,}844{,}817$.

60.
$$\frac{1}{100} + \frac{1}{150} + \frac{1}{200} = \frac{1}{100} \cdot \frac{6}{6} + \frac{1}{150} \cdot \frac{4}{4} + \frac{1}{200} \cdot \frac{3}{3}$$
$$= \frac{6}{600} + \frac{4}{600} + \frac{3}{600}$$
$$= \frac{13}{600}$$

Thus, $\frac{1}{r} = \frac{13}{600}$, so r is the reciprocal of $\frac{13}{600}$, or $\frac{600}{13}$.

61. a) $\dfrac{\square}{11}$ is greater than $\dfrac{1}{2}$ when the numerator is greater than $\dfrac{1}{2}$ of the denominator. Since $\dfrac{1}{2} \cdot 11 = \dfrac{11}{2} = 5\dfrac{1}{2}$, the smallest integer numerator possible is 6.

b) $\dfrac{\square}{8}$ is greater than $\dfrac{1}{2}$ when the numerator is greater than $\dfrac{1}{2}$ of the denominator. Since $\dfrac{1}{2} \cdot 8 = 4$, the smallest integer numerator possible is 5.

c) $\dfrac{\square}{23}$ is greater than $\dfrac{1}{2}$ when the numerator is greater than $\dfrac{1}{2}$ of the denominator. Since $\dfrac{1}{2} \cdot 23 = \dfrac{23}{2} = 11\dfrac{1}{2}$, the smallest integer numerator possible is 12.

d) $\dfrac{\square}{35}$ is greater than $\dfrac{1}{2}$ when the numerator is greater than $\dfrac{1}{2}$ of the denominator. Since $\dfrac{1}{2} \cdot 35 = \dfrac{35}{2} = 17\dfrac{1}{2}$, the smallest integer numerator possible is 18.

e) $\dfrac{-51}{\square}$ is greater than $\dfrac{1}{2}$ when the denominator is greater than twice he numerator. Since $2(-51) = -102$, the smallest integer denominator possible is -101.

f) $\dfrac{-78}{\square}$ is greater than $\dfrac{1}{2}$ when the denominator is greater than twice he numerator. Since $2(-78) = -156$, the smallest integer denominator possible is -155.

g) $\dfrac{-2}{\square}$ is greater than $\dfrac{1}{2}$ when the denominator is greater than twice he numerator. Since $2(-2) = -4$, the smallest integer denominator possible is -3.

h) $\dfrac{-1}{\square}$ is greater than $\dfrac{1}{2}$ when the denominator is greater than twice he numerator. Since $2(-1) = -2$, the smallest integer denominator possible is -1.

62. a) $\dfrac{7}{\square}$ is greater than 1 when the denominator is less than the numerator. Thus, for this fraction the largest integer denominator possible is 6.

b) $\dfrac{11}{\square}$ is greater than 1 when the denominator is less than the numerator. Thus, for this fraction the largest integer denominator possible is 10.

c) $\dfrac{47}{\square}$ is greater than 1 when the denominator is less than the numerator. Thus, for this fraction the largest integer denominator possible is 46.

d) $\dfrac{\dfrac{9}{8}}{\square}$ is greater than 1 when the denominator is less than the numerator. Since $\dfrac{9}{8} = 1\dfrac{1}{8}$, the largest integer denominator possible is 1.

e) $\dfrac{\square}{-13}$ is greater than 1 when the numerator is less than the denominator. Thus, for this fraction the largest integer denominator possible is -14.

f) $\dfrac{\square}{-27}$ is greater than 1 when the numerator is less than the denominator. Thus, for this fraction the largest integer denominator possible is -28.

g) $\dfrac{\square}{-1}$ is greater than 1 when the numerator is less than the denominator. Thus, for this fraction the largest integer denominator possible is -2.

h) $\dfrac{\square}{-\dfrac{1}{2}}$ is greater than 1 when the numerator is less than the denominator. Thus, for this fraction the largest integer denominator possible is -1.

Chapter 4 Test

1. $12 = 2 \cdot 2 \cdot 3 = 2^2 \cdot 3$
$16 = 2 \cdot 2 \cdot 2 \cdot 2 = 2^4$

We form the LCM using the greatest power of each factor.

The LCM is $2^4 \cdot 3$, or 48.

2. $\dfrac{1}{2} + \dfrac{5}{2} = \dfrac{1+5}{2} = \dfrac{6}{2} = 3$

3.
$$-\frac{7}{8} + \frac{2}{3}$$
$$= \frac{-7}{8} + \frac{2}{3} \qquad \text{8 and 3 have no common factors,}$$
$$\qquad\qquad\qquad \text{so the LCD is } 8 \cdot 3, \text{ or } 24.$$
$$= \frac{-7}{8} \cdot \frac{3}{3} + \frac{2}{3} \cdot \frac{8}{8}$$
$$= \frac{-21}{24} + \frac{16}{24}$$
$$= \frac{-5}{24}$$

4. $\dfrac{5}{t} - \dfrac{3}{t} = \dfrac{5-3}{t} = \dfrac{2}{t}$

5. The LCM of 6 and 4 is 12.

$$\frac{5}{6} - \frac{3}{4} = \frac{5}{6} \cdot \frac{2}{2} - \frac{3}{4} \cdot \frac{3}{3}$$

$$= \frac{10}{12} - \frac{9}{12} = \frac{1}{12}$$

6. The LCM of 8 and 24 is 24.

$$\frac{5}{8} - \frac{17}{24} = \frac{5}{8} \cdot \frac{3}{3} - \frac{17}{24}$$

$$= \frac{15}{24} - \frac{17}{24} = \frac{15 - 17}{24}$$

$$= \frac{-2}{24} = \frac{-1 \cdot 2}{2 \cdot 12}$$

$$= \frac{-1}{12}, \text{ or } -\frac{1}{12}$$

7.
$$x + \frac{2}{3} = \frac{11}{12}$$

$$x + \frac{2}{3} - \frac{2}{3} = \frac{11}{12} - \frac{2}{3} \qquad \text{Subtracting } \frac{2}{3} \text{ on both sides}$$

$$x + 0 = \frac{11}{12} - \frac{2}{3} \cdot \frac{4}{4} \qquad \text{The LCD is 12.}$$

$$x = \frac{11}{12} - \frac{8}{12} = \frac{3}{12}$$

$$x = \frac{3 \cdot 1}{3 \cdot 4} = \frac{3}{3} \cdot \frac{1}{4}$$

$$x = \frac{1}{4}$$

8.
$$-5x - 3 = 9$$

$$-5x - 3 + 3 = 9 + 3$$

$$-5x = 12$$

$$\frac{-5x}{-5} = \frac{12}{-5}$$

$$x = \frac{12}{-5}$$

The solution is $\frac{12}{-5}$, or $\frac{-12}{5}$, or $-\frac{12}{5}$.

9. The LCM of the denominators is 12.

$$\frac{3}{4} = \frac{1}{2} + \frac{5}{3}x$$

$$12 \cdot \frac{3}{4} = 12\left(\frac{1}{2} + \frac{5}{3}x\right)$$

$$\frac{36}{4} = 12 \cdot \frac{1}{2} + \frac{12 \cdot 5}{3}x$$

$$9 = \frac{12}{2} + \frac{60}{3}x$$

$$9 = 6 + 20x$$

$$9 - 6 = 6 + 20x - 6$$

$$3 = 20x$$

$$\frac{3}{20} = \frac{20x}{20}$$

$$\frac{3}{20} = x$$

The solution is $\frac{3}{20}$.

10. The LCD is 175.

$$\frac{6}{7} \cdot \frac{25}{25} = \frac{150}{175}$$

$$\frac{21}{25} \cdot \frac{7}{7} = \frac{147}{175}$$

Since $150 > 147$, it follows that $\frac{150}{175} > \frac{147}{175}$, so $\frac{6}{7} > \frac{21}{25}$.

11. $3\frac{1}{2} = \frac{7}{2}$ $(3 \cdot 2 = 6, \ 6 + 1 = 7)$

12. $-9\frac{3}{8} = -\left(9 + \frac{3}{8}\right) = -\frac{75}{8}$

$(9 \cdot 8 = 72; \ 72 + 3 = 75;$ include the negative sign.$)$

13. First consider $\frac{74}{9}$.

$$9 \overline{\smash{\big)}\, 74} \quad \begin{array}{r} 8 \\ \hline 72 \\ \hline 2 \end{array} \qquad \frac{74}{9} = 8\frac{2}{9}$$

Since $\frac{74}{9} = 8\frac{2}{9}$, we have $-\frac{74}{9} = -8\frac{2}{9}$.

14.
$$11 \overline{\smash{\big)}\, 1789} \quad \begin{array}{r} 162 \\ \hline 1100 \\ \hline 689 \\ 660 \\ \hline 29 \\ 22 \\ \hline 7 \end{array}$$

The answer is $162\frac{7}{11}$.

15.
$$\begin{array}{r} 6\frac{2}{5} \\ +7\frac{4}{5} \\ \hline 13\frac{6}{5} = 13 + \frac{6}{5} \\ = 13 + 1\frac{1}{5} \\ = 14\frac{1}{5} \end{array}$$

16. The LCD is 12.

$$\begin{array}{r} 3\,\boxed{\dfrac{1}{4} \cdot \dfrac{3}{3}} = 3\dfrac{3}{12} \\ +9\,\boxed{\dfrac{1}{6} \cdot \dfrac{2}{2}} = +9\dfrac{2}{12} \\ \hline 12\dfrac{5}{12} \end{array}$$

17. The LCD is 24.

$$10\ \boxed{\frac{1}{6}\cdot\frac{4}{4}} = 10\ \frac{4}{24} = 9\ \frac{28}{24}$$

$$-5\ \boxed{\frac{7}{8}\cdot\frac{3}{3}} = -5\ \frac{21}{24} = -5\ \frac{21}{24}$$

$$4\ \frac{7}{24}$$

$$\left(\text{Since }\frac{4}{24}\text{ is smaller than }\frac{21}{24}\text{, we cannot subtract until we}\right.$$

$$\left.\text{borrow: }10\frac{4}{24} = 9 + \frac{24}{24} + \frac{4}{24} = 9 + \frac{28}{24} = 9\frac{28}{24}.\right)$$

18. $14 + \left(-5\frac{3}{7}\right) = 13\frac{7}{7} + \left(-5\frac{3}{7}\right) = 8\frac{4}{7}$

19. $3\frac{4}{5} - 9\frac{1}{2} = 3\frac{4}{5} + \left(-9\frac{1}{2}\right)$

Since $-9\frac{1}{2}$ has the larger absolute value, the answer will be negative. We find the difference in absolute values.

$$9\ \boxed{\frac{1}{2}\cdot\frac{5}{5}} = 9\ \frac{5}{10} = 8\ \frac{15}{10}$$

$$-3\ \boxed{\frac{4}{5}\cdot\frac{2}{2}} = -3\ \frac{8}{10} = -3\ \frac{8}{10}$$

$$5\ \frac{7}{10}$$

Thus, $3\frac{4}{5} - 9\frac{1}{2} = -5\frac{7}{10}$.

20. $\frac{3}{8}x - \frac{1}{2}x = \frac{3}{8}x - \frac{1}{2}\cdot\frac{4}{4}\cdot x$

$$= \frac{3}{8}x - \frac{4}{8}x$$

$$= -\frac{1}{8}x$$

21. $5\frac{2}{11}a - 3\frac{1}{5}a = \left(5\frac{2}{11} - 3\frac{1}{5}\right)a$

$$= \left(5\frac{10}{55} - 3\frac{11}{55}\right)a$$

$$= \left(4\frac{65}{55} - 3\frac{11}{55}\right)a$$

$$= 1\frac{54}{55}a$$

22. $9\cdot 4\frac{1}{3} = 9\cdot\frac{13}{3} = \frac{9\cdot 13}{3} = \frac{3\cdot 3\cdot 13}{3\cdot 1} = \frac{3}{3}\cdot\frac{3\cdot 13}{1} = 39$

23. $6\frac{3}{4}\cdot\left(-2\frac{2}{3}\right) = \frac{27}{4}\cdot\left(-\frac{8}{3}\right) = -\frac{27\cdot 8}{4\cdot 3} = -\frac{3\cdot 9\cdot 2\cdot 4}{4\cdot 3\cdot 1} =$

$$-\frac{9\cdot 2}{1}\cdot\frac{3\cdot 4}{3\cdot 4} = -18$$

24. $33\div 5\frac{1}{2} = 33\div\frac{11}{2} = 33\cdot\frac{2}{11} = \frac{33\cdot 2}{11} = \frac{3\cdot 11\cdot 2}{11\cdot 1} =$

$$\frac{11}{11}\cdot\frac{3\cdot 2}{1} = 6$$

25. $2\frac{1}{3}\div 1\frac{1}{6} = \frac{7}{3}\div\frac{7}{6} = \frac{7}{3}\cdot\frac{6}{7} = \frac{7\cdot 6}{3\cdot 7} = \frac{7\cdot 2\cdot 3}{3\cdot 7\cdot 1} =$

$$\frac{7\cdot 3}{7\cdot 3}\cdot\frac{2}{1} = 2$$

26. $\frac{2}{3}ab = \frac{2}{3}\cdot 7\cdot 4\frac{1}{5} = \frac{2\cdot 7}{3}\cdot 4\frac{1}{5} = \frac{2\cdot 7}{3}\cdot\frac{21}{5} = \frac{2\cdot 7\cdot 21}{3\cdot 5} =$

$$\frac{2\cdot 7\cdot 3\cdot 7}{3\cdot 5} = \frac{98}{5},\text{ or }19\frac{3}{5}$$

27. $4 + mn = 4 + 7\frac{2}{5}\cdot 3\frac{1}{4}$

$$= 4 + \frac{37}{5}\cdot\frac{13}{4}$$

$$= 4 + \frac{481}{20}$$

$$= 4 + 24\frac{1}{20}$$

$$= 28\frac{1}{20}$$

28. *Familiarize.* Let $t =$ the number of pounds of turkey required for 5 batches of chili.

Translate. We write a multiplication sentence.

$$t = 5\cdot 1\frac{1}{2}$$

Solve. We carry out the multiplication.

$$t = 5\cdot 1\frac{1}{2} = 5\cdot\frac{3}{2} = \frac{15}{2} = 7\frac{1}{2}$$

Check. We can repeat the calculation. The answer checks.

State. $7\frac{1}{2}$ lb of turkey is required for 5 batches of chili.

29. *Familiarize.* Let $b =$ the number of books in the order.

Translate.

Weight per book	times	Number of books	is	Total weight
↓	↓	↓	↓	↓
$2\frac{3}{4}$	$\cdot$	b	$=$	220

Solve

$$2\frac{3}{4}\cdot b = 220$$

$$\frac{11}{4}\cdot b = 220$$

$$\frac{4}{11}\cdot\frac{11}{4}\cdot b = \frac{4}{11}\cdot 220$$

$$b = \frac{4\cdot 220}{11} = \frac{4\cdot \cancel{11}\cdot 20}{\cancel{11}\cdot 1}$$

$$b = 80$$

Check. Since $2\frac{3}{4}\cdot 80 = \frac{11}{4}\cdot 80 = \frac{880}{4} = 220$, the answer checks.

State. There are 80 books in the order.

30. *Familiarize.* We add the three lengths across the top to find a and the three lengths across the bottom to find b.

Translate.

$$a = 1\frac{1}{8} + \frac{3}{4} + 1\frac{1}{8}$$

$$b = \frac{3}{4} + 3 + \frac{3}{4}$$

Solve. We carry out the additions.

$$a = 1\frac{1}{8} + \frac{6}{8} + 1\frac{1}{8} = 2\frac{8}{8} = 2 + 1 = 3$$

$$b = \frac{3}{4} + 3 + \frac{3}{4} = 3\frac{6}{4} = 3 + 1\frac{2}{4} = 3 + 1\frac{1}{2} = 4\frac{1}{2}$$

Check. We can repeat the calculations. The answer checks.

State. a) The short length a across the top is 3 in.

b) The length b across the bottom is $4\frac{1}{2}$ in.

31. *Familiarize.* Let t = the number of inches by which $\frac{3}{4}$ in. exceeds the actual thickness of the plywood.

Translate.

$$\underbrace{\text{Actual thickness}}_{} \quad \text{plus} \quad \underbrace{\text{Excess thickness}}_{} \quad \text{is} \quad \underbrace{\frac{3}{4} \text{ in.}}_{}$$

$$\downarrow \qquad\qquad \downarrow \qquad\qquad \downarrow \qquad\qquad \downarrow \quad \downarrow$$

$$\frac{11}{16} \qquad\quad + \qquad\qquad t \qquad\qquad = \quad \frac{3}{4}$$

Solve. We will subtract $\frac{11}{16}$ on both sides of the equation.

$$\frac{11}{16} + t = \frac{3}{4}$$

$$\frac{11}{16} + t - \frac{11}{16} = \frac{3}{4} - \frac{11}{16}$$

$$t = \frac{3}{4} \cdot \frac{4}{4} - \frac{11}{16}$$

$$t = \frac{12}{16} - \frac{11}{16}$$

$$t = \frac{1}{16}$$

Check. Since $\frac{11}{16} + \frac{1}{16} = \frac{12}{16} = \frac{3}{4}$, the answer checks.

State. A $\frac{3}{4}$-in. piece of plywood is actually $\frac{1}{16}$ in. thinner than its name indicates.

32. We add the heights and divide by the number of addends.

$$\frac{6\frac{5}{12} + 5\frac{11}{12} + 6\frac{7}{12}}{3} = \frac{17\frac{23}{12}}{3} = \frac{17 + 1\frac{11}{12}}{3} =$$

$$\frac{18\frac{11}{12}}{3} = \frac{227}{12} \div 3 = \frac{227}{12} \cdot \frac{1}{3} = \frac{227}{36} = 6\frac{11}{36}$$

The women's average height is $6\frac{11}{36}$ ft.

33. The length of the act can be expressed as a fraction with a denominator of 25 and a numerator that is the LCM of 6 and 8.

$$6 = 2 \cdot 3$$
$$8 = 2 \cdot 2 \cdot 2 = 2^3$$

The LCM is $2^3 \cdot 3$, or 24, so the act lasts $\frac{24}{25}$ min.

34. *Familiarize.* First compare $\frac{1}{7}$ mi and $\frac{1}{8}$ mi. The LCD is 56.

$$\frac{1}{7} = \frac{1}{7} \cdot \frac{8}{8} = \frac{8}{56}$$

$$\frac{1}{8} = \frac{1}{8} \cdot \frac{7}{7} = \frac{7}{56}$$

Since $8 > 7$, then $\frac{8}{56} > \frac{7}{56}$ so $\frac{1}{7} > \frac{1}{8}$.

This tells us that Cheri walks farther than Trent.

Next we will find how much farther Cheri walks on each lap and then multiply by 17 to find how much farther she walks in 17 laps. Let d represent how much farther Cheri walks on each lap, in miles.

Translate. An equation that fits this situation is

$$\frac{1}{8} + d = \frac{1}{7}, \text{ or } \frac{7}{56} + d = \frac{8}{56}$$

Solve.

$$\frac{7}{56} + d = \frac{8}{56}$$

$$\frac{7}{56} + d - \frac{7}{56} = \frac{8}{56} - \frac{7}{56}$$

$$d = \frac{1}{56}$$

Now we multiply: $17 \cdot \frac{1}{56} = \frac{17}{56}$.

Check. We can think of the problem in a different way.

In 17 laps Cheri walks $17 \cdot \frac{1}{7}$, or $\frac{17}{7}$ mi, and Trent walks $17 \cdot \frac{1}{8}$, or $\frac{17}{8}$ mi. Then $\frac{17}{7} - \frac{17}{8} = \frac{17}{7} \cdot \frac{8}{8} - \frac{17}{8} \cdot \frac{7}{7} = \frac{136}{56} - \frac{119}{56} = \frac{17}{56}$, so Cheri walks $\frac{17}{56}$ mi farther and our answer checks.

State. Cheri walks $\frac{17}{56}$ mi farther than Trent.

35. a) We find some common multiples of 8 and 6.

Multiples of 8: 8, 16, 24, 32, 40, 48, 56, 64, 72, ...
Multiples of 6: 6, 12, 18, 24, 30, 36, 42, 48, 54, 60, 66, 72, ...

Some common multiples are 24, 48, and 72. These are some class sizes for which study groups of 8 students or of 6 students can be organized with no students left out.

b) The smallest such class size is the least common multiple, 24.

36. a) $\dfrac{1}{1 \cdot 2} = \dfrac{1}{2}$

b) $\dfrac{1}{1 \cdot 2} + \dfrac{1}{2 \cdot 3} = \dfrac{1}{2} + \dfrac{1}{6} = \dfrac{1}{2} \cdot \dfrac{3}{3} + \dfrac{1}{6} = \dfrac{3}{6} + \dfrac{1}{6} = \dfrac{4}{6} = \dfrac{2 \cdot 2}{2 \cdot 3} = \dfrac{2}{2} \cdot \dfrac{2}{3} = \dfrac{2}{3}$

c) $\dfrac{1}{1 \cdot 2} + \dfrac{1}{2 \cdot 3} + \dfrac{1}{3 \cdot 4} = \left(\dfrac{1}{1 \cdot 2} + \dfrac{1}{2 \cdot 3}\right) + \dfrac{1}{3 \cdot 4} =$

$\dfrac{2}{3} + \dfrac{1}{12} = \dfrac{2}{3} \cdot \dfrac{4}{4} + \dfrac{1}{12} = \dfrac{8}{12} + \dfrac{1}{12} = \dfrac{9}{12} = \dfrac{3 \cdot 3}{3 \cdot 4} =$

$$\frac{3}{3}\cdot\frac{3}{4}=\frac{3}{4}$$

d) $\dfrac{1}{1\cdot2}+\dfrac{1}{2\cdot3}+\dfrac{1}{3\cdot4}+\dfrac{1}{4\cdot5}=$

$\left(\dfrac{1}{1\cdot2}+\dfrac{1}{2\cdot3}+\dfrac{1}{3\cdot4}\right)+\dfrac{1}{4\cdot5}=\dfrac{3}{4}+\dfrac{1}{20}=\dfrac{3}{4}\cdot\dfrac{5}{5}+\dfrac{1}{20}=$

$\dfrac{15}{20}+\dfrac{1}{20}=\dfrac{16}{20}=\dfrac{4\cdot4}{4\cdot5}=\dfrac{4}{4}\cdot\dfrac{4}{5}=\dfrac{4}{5}$

e) In each case, the sum is a fraction in which the numerator is the smaller factor in the denominator of the last addend and the denominator is the larger factor in the denominator of the last addend. Then the given sum is $\dfrac{9}{10}$.

Cumulative Review Chapters 1 - 4

1. a) **Familiarize.** Let $c=$ the number of inches by which the width of the $\frac{1}{16}$-in. craft excelsior exceeds that used for erosion control.

Translate.

Width of erosion control excelsior	plus	Excess width of craft excelsior	is	Width of craft excelsior
$\frac{1}{24}$	$+$	c	$=$	$\frac{1}{16}$

Solve.
$$\frac{1}{24}+c=\frac{1}{16}$$
$$c=\frac{1}{16}-\frac{1}{24}$$
$$c=\frac{1}{16}\cdot\frac{3}{3}-\frac{1}{24}\cdot\frac{2}{2}$$
$$c=\frac{3}{48}-\frac{2}{48}$$
$$c=\frac{1}{48}$$

Check. $\dfrac{1}{24}+\dfrac{1}{48}=\dfrac{1}{24}\cdot\dfrac{2}{2}+\dfrac{1}{48}=\dfrac{2}{48}+\dfrac{1}{48}=$

$\dfrac{3}{48}=\dfrac{3\cdot1}{3\cdot16}=\dfrac{3}{3}\cdot\dfrac{1}{16}=\dfrac{1}{16}$, so the answer checks.

State. The $\frac{1}{16}$-in. craft excelsior is $\frac{1}{48}$ in. wider than the erosion control excelsior.

b) **Familiarize.** Let $c=$ the number of inches by which the width of the $\frac{1}{8}$-in. craft excelsior exceeds that used for erosion control.

Translate.

Width of erosion control excelsior	plus	Excess width of craft excelsior	is	Width of craft excelsior
$\frac{1}{24}$	$+$	c	$=$	$\frac{1}{8}$

Solve.
$$\frac{1}{24}+c=\frac{1}{8}$$
$$c=\frac{1}{8}-\frac{1}{24}$$
$$c=\frac{1}{8}\cdot\frac{3}{3}-\frac{1}{24}$$
$$c=\frac{3}{24}-\frac{1}{24}=\frac{2}{24}=\frac{2\cdot1}{2\cdot12}=\frac{2}{2}\cdot\frac{1}{12}$$
$$c=\frac{1}{12}$$

Check. $\dfrac{1}{24}+\dfrac{1}{12}=\dfrac{1}{24}+\dfrac{1}{12}\cdot\dfrac{2}{2}=\dfrac{1}{24}+\dfrac{2}{24}=$

$\dfrac{3}{24}=\dfrac{3\cdot1}{3\cdot8}=\dfrac{3}{3}\cdot\dfrac{1}{8}=\dfrac{1}{8}$, so the answer checks.

State. The $\frac{1}{8}$-in. craft excelsior is $\frac{1}{12}$ in. wider than the erosion control excelsior.

2. **Familiarize.** Let $n=$ the number of DVDs the shelf will hold.

Translate.

Width of each DVD	times	Number of DVDs	is	Length of shelf
$\frac{7}{16}$	$\cdot$	n	$=$	27

Solve.
$$\frac{7}{16}\cdot n=27$$
$$n=27\div\frac{7}{16}$$
$$n=27\cdot\frac{16}{7}=\frac{27\cdot16}{7}$$
$$n=\frac{432}{7}=61\frac{5}{7}$$

Since a fractional part of a DVD cannot be placed on the shelf, we see that it will hold 61 DVDs.

Check. $\dfrac{7}{16}\cdot61=\dfrac{427}{16}=26\dfrac{11}{16}$ is less than 27 in. and

$\dfrac{7}{16}\cdot62=\dfrac{217}{8}=27\dfrac{1}{8}$ is greater than 27 in., so we know that the shelf will hold at most 61 DVDs.

State. The shelf will hold 61 DVDs.

3. a) **Familiarize.** Let $t=$ the total number of miles David and Sally Jean skied.

Translate.

First day's distance	plus	Second day's distance	plus	Third day's distance	is	Total distance skied
$3\frac{2}{3}$	$+$	$6\frac{1}{8}$	$+$	$4\frac{3}{4}$	$=$	t

Solve. We carry out the addition.

$$3 \boxed{\frac{2}{3} \cdot \frac{8}{8}} = 3\frac{16}{24}$$

$$6 \boxed{\frac{1}{8} \cdot \frac{3}{3}} = 6\frac{3}{24}$$

$$+4 \boxed{\frac{3}{4} \cdot \frac{6}{6}} = +4\frac{18}{24}$$

$$13\frac{37}{24} = 13 + \frac{37}{24}$$

$$= 13 + 1\frac{13}{24}$$

$$= 14\frac{13}{24}$$

Check. We repeat the calculation. The answer checks.

State. David and Sally Jean skied a total of $14\frac{13}{24}$ mi.

b) From part (a) we know that the sum of the three distances is $14\frac{13}{24}$. We divide this number by 3 to find the average number of miles skied per day.

$$\frac{14\frac{13}{24}}{3} = \frac{\frac{349}{24}}{3} = \frac{349}{24} \cdot \frac{1}{3} = \frac{349}{72} = 4\frac{61}{72}$$

An average of $4\frac{61}{72}$ mi was skied each day.

4. a) The total area is the sum of the areas of the two individual rectangles. We use the formula Area = length × width twice and add the results.

$$8\frac{1}{2} \cdot 11 + 6\frac{1}{2} \cdot 7\frac{1}{2}$$

$$= \frac{17}{2} \cdot 11 + \frac{13}{2} \cdot \frac{15}{2}$$

$$= \frac{187}{2} + \frac{195}{4} = \frac{187}{2} \cdot \frac{2}{2} + \frac{195}{4}$$

$$= \frac{374}{4} + \frac{195}{4} = \frac{569}{4}$$

$$= 142\frac{1}{4}$$

The area of the carpet is $142\frac{1}{4}$ ft^2.

b) The perimeter can be thought of as the sum of the perimeter of the larger rectangle and the two longer sides of the smaller rectangle. Thus, we have

$$
\begin{aligned}
& 8\frac{1}{2} \\
& 11 \\
& 8\frac{1}{2} \\
& 11 \\
& 7\frac{1}{2} \\
+ & 7\frac{1}{2} \\
\hline
& 52\frac{4}{2} = 52 + 2 = 54
\end{aligned}
$$

The perimeter is 54 ft.

5. *Familiarize*. Let p = the number of people who can get equal shares of the money.

Translate.

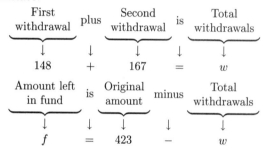

Amount of each share	times	Number of people	is	Total amount divided
↓	↓	↓	↓	↓
16	·	p	=	496

Solve.

$$16 \cdot p = 496$$

$$p = \frac{496}{16}$$

$$p = 31$$

Check. \$16 · 31 = \$496, so the answer checks.

State. 31 people can get equal \$16 shares from a total of \$496.

6. *Familiarize*. Let w = the total amount withdrawn for expenses and let f = the amount left in the fund after the withdrawals.

Translate.

First withdrawal	plus	Second withdrawal	is	Total withdrawals
↓	↓	↓	↓	↓
148	+	167	=	w

Amount left in fund	is	Original amount	minus	Total withdrawals
↓	↓	↓	↓	↓
f	=	423	−	w

Solve. We carry out the addition to solve the first equation.

$$148 + 167 = w$$

$$315 = w$$

Now we substitute 315 for w in the second equation and carry out the subtraction.

$$f = 423 - w$$

$$f = 423 - 315$$

$$f = 108$$

Check. We repeat the calculations. The answer checks.

State. \$108 remains in the fund.

7. *Familiarize*. Let x = the amount of salt required for $\frac{1}{2}$ recipe and y = the amount required for 5 recipes.

Translate. We multiply by $\frac{1}{2}$ to find the amount of salt required for $\frac{1}{2}$ recipe and by 5 to find the amount for 5 recipes.

$$x = \frac{1}{2} \cdot \frac{4}{5}, \quad y = 5 \cdot \frac{4}{5}$$

Solve. We carry out the multiplications.

$$x = \frac{1}{2} \cdot \frac{4}{5} = \frac{1 \cdot 4}{2 \cdot 5} = \frac{1 \cdot 2 \cdot 2}{2 \cdot 5} = \frac{2}{2} \cdot \frac{1 \cdot 2}{5} = \frac{2}{5}$$

$$y = 5 \cdot \frac{4}{5} = \frac{5 \cdot 4}{5} = \frac{5}{5} \cdot 4 = 4$$

Check. We repeat the calculations. The answer checks.

State. $\frac{2}{5}$ tsp of salt is required for $\frac{1}{2}$ recipe, and 4 tsp is required for 5 recipes.

8. Familiarize. Let w = the weight of 15 books, in pounds.

Translate. We write a multiplication sentence.

$$w = 15 \cdot 2\frac{3}{5}$$

Solve. We carry out the multiplication.

$$w = 15 \cdot 2\frac{3}{5} = 15 \cdot \frac{13}{5} = \frac{15 \cdot 13}{5}$$
$$= \frac{3 \cdot 5 \cdot 13}{5 \cdot 1} = \frac{5}{5} \cdot \frac{3 \cdot 13}{1} = 39$$

Check. We repeat the calculation. The answer checks.

State. The weight of 15 books is 39 lb.

9. Familiarize. Let n = the number of $2\frac{3}{8}$-ft pieces that can be cut from a 38-ft wire.

Translate. We write a division sentence.

$$n = 38 \div 2\frac{3}{8}$$

Solve. We carry out the division.

$$n = 38 \div 2\frac{3}{8} = 38 \div \frac{19}{8} = 38 \cdot \frac{8}{19}$$
$$= \frac{38 \cdot 8}{19} = \frac{2 \cdot 19 \cdot 8}{19 \cdot 1} = \frac{19}{19} \cdot \frac{2 \cdot 8}{1}$$
$$= 16$$

Check. $16 \cdot 2\frac{3}{8} = 16 \cdot \frac{19}{8} = \frac{16 \cdot 19}{8} = \frac{2 \cdot 8 \cdot 19}{8 \cdot 1} =$

$\frac{8}{8} \cdot \frac{2 \cdot 19}{1} = 38$, so the answer checks.

State. 16 pieces can be cut from the wire.

10. Familiarize. Let w = the total number of miles Jermaine and Oleta walked.

Translate.

Jermaine's distance	plus	Oleta's distance	is	Total distance
$\downarrow$	$\downarrow$	$\downarrow$	$\downarrow$	$\downarrow$
$\frac{9}{10}$	$+$	$\frac{75}{100}$	$=$	w

Solve. We carry out the addition. The LCD is 100.

$$\frac{9}{10} + \frac{75}{100} = \frac{9}{10} \cdot \frac{10}{10} + \frac{75}{100} = \frac{90}{100} + \frac{75}{100} = \frac{165}{100} =$$
$$\frac{5 \cdot 33}{5 \cdot 20} = \frac{5}{5} \cdot \frac{33}{20} = \frac{33}{20}$$

Thus, $w = \frac{33}{20}$.

Check. We repeat the calculations. The answer checks.

State. Jermaine and Oleta walked a total of $\frac{33}{20}$ mi.

11. $2\,7\,\boxed{5}\,3$

The digit 5 names the number of tens.

12. $6075 = 6$ thousands $+\ 0$ hundreds $+\ 7$ tens $+\ 5$ ones, or 6 thousands $+\ 7$ tens $+\ 5$ ones

13. A word name for 29,500 is twenty nine thousand, five hundred.

14. We can think of the figure as being divided into 16 equal parts, each the size of the smaller area shaded. Then the larger shaded area is the equivalent of 4 smaller shaded areas. Thus, $\frac{5}{16}$ of the figure is shaded.

15.
$$\begin{array}{r} \overset{1\ 1}{3\,7\,5} \\ +\ 2\,4\,8 \\ \hline 6\,2\,3 \end{array}$$

16. $29 + (-37)$

The absolute values are 29 and 37. The difference is 8. Since the negative number has the larger absolute value, the answer is negative.

$$29 + (-37) = -8$$

17. $\frac{3}{8} + \frac{1}{24} = \frac{3}{8} \cdot \frac{3}{3} + \frac{1}{24} = \frac{9}{24} + \frac{1}{24} = \frac{10}{24} = \frac{2 \cdot 5}{2 \cdot 12} = \frac{2}{2} \cdot \frac{5}{12} = \frac{5}{12}$

18.
$$\begin{array}{r} 2\ \ \ \frac{3}{4} = 2\frac{3}{4} \\ +5\ \boxed{\frac{1}{2} \cdot \frac{2}{2}} = +5\frac{2}{4} \\ \hline 7\frac{5}{4} = 7 + \frac{5}{4} \\ = 7 + 1\frac{1}{4} \\ = 8\frac{1}{4} \end{array}$$

19.
$$\begin{array}{r} 7\,4\,6\,9 \\ -\ 2\,3\,4\,5 \\ \hline 5\,1\,2\,4 \end{array}$$

20. $-9 - (-25) = -9 + 25 = 16$

21. $\frac{4}{t} - \frac{9}{t} = \frac{4-9}{t} = \frac{-5}{t}$, or $-\frac{5}{t}$

22.
$$\begin{array}{r} 2\ \boxed{\frac{1}{3} \cdot \frac{2}{2}} = 2\frac{2}{6} \\ -1\ \ \frac{1}{6} = -1\frac{1}{6} \\ \hline 1\frac{1}{6} \end{array}$$

23.
$$\begin{array}{r} \overset{6\ 6}{2\,7\,8} \\ \times\ 1\,8 \\ \hline 2\,2\,2\,4 \\ 2\,7\,8\,0 \\ \hline 5\,0\,0\,4 \end{array}$$

24. $29(-5) = -145$

25. $\frac{9}{10} \cdot \frac{5}{3} = \frac{9 \cdot 5}{10 \cdot 3} = \frac{3 \cdot 3 \cdot 5}{2 \cdot 5 \cdot 3} = \frac{3 \cdot 5}{3 \cdot 5} \cdot \frac{3}{2} = \frac{3}{2}$

26. $18 \cdot \left(-\frac{5}{6}\right) = -\frac{18 \cdot 5}{6} = -\frac{3 \cdot 6 \cdot 5}{6 \cdot 1} = \frac{6}{6} \cdot \left(-\frac{3 \cdot 5}{1}\right) = -15$

27. $2\frac{1}{3} \cdot 3\frac{1}{7} = \frac{7}{3} \cdot \frac{22}{7} = \frac{7 \cdot 22}{3 \cdot 7} = \frac{7}{7} \cdot \frac{22}{3} = \frac{22}{3} = 7\frac{1}{3}$

28.
```
         4 8
   1 5 | 7 3 1
         6 0 0
         1 3 1
         1 2 0
           1 1
```
The answer is 48 R 11.

29.
```
           5 6
   4 5 | 2 5 3 1
         2 2 5 0
           2 8 1
           2 7 0
             1 1
```
The answer is 56 R 11.

30. The remainder is 11 and the divisor is 45, so a mixed numeral for the answer is $56\frac{11}{45}$.

31. $\frac{2}{5} \div \left(-\frac{7}{10}\right) = \frac{2}{5} \cdot \left(-\frac{10}{7}\right) = -\frac{2 \cdot 10}{5 \cdot 7} = -\frac{2 \cdot 2 \cdot 5}{5 \cdot 7} = \frac{5}{5} \cdot \left(-\frac{2 \cdot 2}{7}\right) = -\frac{4}{7}$

32. $2\frac{1}{5} \div \frac{3}{10} = \frac{11}{5} \div \frac{3}{10} = \frac{11}{5} \cdot \frac{10}{3} = \frac{11 \cdot 10}{5 \cdot 3} = \frac{11 \cdot 2 \cdot 5}{5 \cdot 3} = \frac{5}{5} \cdot \frac{11 \cdot 2}{3} = \frac{22}{3} = 7\frac{1}{3}$

33. Round 38,478 to the nearest hundred.

$3\,8,4\,\boxed{7}\,8$
$\uparrow$

The digit 4 is in the hundreds place. Consider the next digit to the right. Since the digit, 7, is 5 or higher, round 4 hundreds up to 5 hundreds. Then change all digits to the right of the hundreds digit to zero.

The answer is 38,500.

34. $24 = 2 \cdot 2 \cdot 2 \cdot 3 = 2^3 \cdot 3$
$36 = 2 \cdot 2 \cdot 3 \cdot 3 = 2^2 \cdot 3^2$

We form the LCM using the greatest power of each factor. The LCM is $2^3 \cdot 3^2$, or 72.

35. 4296 is even because the ones digit is even. Because $4 + 2 + 9 + 6 = 21$ and 21 is divisible by 3, then 4296 is divisible by 3. Thus, 4296 is divisible by 6.

36. We make a list of factorizations.
$16 = 1 \cdot 16$
$16 = 2 \cdot 8$
$16 = 4 \cdot 4$

Factors: 1, 2, 4, 8, 16

37. The LCD is 30.
$\frac{4}{5} \cdot \frac{6}{6} = \frac{24}{30}$
$\frac{4}{6} \cdot \frac{5}{5} = \frac{20}{30}$

Since $24 > 20$, it follows that $\frac{24}{30} > \frac{20}{30}$, so $\frac{4}{5} > \frac{4}{6}$.

38. The LCD is 84.
$\frac{-3}{7} \cdot \frac{12}{12} = \frac{-36}{84}$
$\frac{-5}{12} \cdot \frac{7}{7} = \frac{-35}{84}$

Since $-36 < -35$, it follows that $\frac{-36}{84} < \frac{-35}{84}$, so $-\frac{3}{7} < -\frac{5}{12}$.

39. $\frac{36}{45} = \frac{4 \cdot 9}{5 \cdot 9} = \frac{4}{5} \cdot \frac{9}{9} = \frac{4}{5} \cdot 1 = \frac{4}{5}$

40. $-\frac{420}{30} = -\frac{14 \cdot 30}{30 \cdot 1} = -\frac{14}{1} \cdot \frac{30}{30} = -14$

41. $7\frac{3}{10} = \frac{73}{10}$ $(7 \cdot 10 = 70, 70 + 3 = 73)$

42. First consider $\frac{17}{3}$.
```
       5
   3 | 1 7
       1 5
         2
```
Since $\frac{17}{3} = 5\frac{2}{3}$, we have $-\frac{17}{3} = -5\frac{2}{3}$.

43. $x + 37 = 92$
$x + 37 - 37 = 92 - 37$
$x = 55$
The solution is 55.

44. $x + \frac{7}{9} = \frac{4}{3}$
$x + \frac{7}{9} - \frac{7}{9} = \frac{4}{3} - \frac{7}{9}$
$x = \frac{4}{3} \cdot \frac{3}{3} - \frac{7}{9}$
$x = \frac{12}{9} - \frac{7}{9}$
$x = \frac{5}{9}$
The solution is $\frac{5}{9}$.

45. $\frac{7}{9} \cdot t = -\frac{4}{3}$
$t = -\frac{4}{3} \div \frac{7}{9}$
$t = -\frac{4}{3} \cdot \frac{9}{7} = -\frac{4 \cdot 9}{3 \cdot 7} = -\frac{4 \cdot 3 \cdot 3}{3 \cdot 7}$
$= \frac{3}{3} \cdot \left(-\frac{4 \cdot 3}{7}\right) = -\frac{12}{7}$
The solution is $-\frac{12}{7}$.

46.
$$\frac{5}{7} = \frac{1}{3} + 4a$$
$$\frac{5}{7} - \frac{1}{3} = \frac{1}{3} + 4a - \frac{1}{3}$$
$$\frac{5}{7} \cdot \frac{3}{3} - \frac{1}{3} \cdot \frac{7}{7} = 4a$$
$$\frac{15}{21} - \frac{7}{21} = 4a$$
$$\frac{8}{21} = 4a$$
$$\frac{1}{4} \cdot \frac{8}{21} = \frac{1}{4} \cdot 4a$$
$$\frac{1 \cdot 8}{4 \cdot 21} = a$$
$$\frac{1 \cdot 2 \cdot \not{4}}{\not{4} \cdot 21} = a$$
$$\frac{2}{21} = a$$

The solution is $\frac{2}{21}$.

47. $\dfrac{t + p}{3} = \dfrac{-4 + 16}{3} = \dfrac{12}{3} = 4$

48. $7(b - 5) = 7 \cdot b - 7 \cdot 5 = 7b - 35$

49. $-3(x - 2 + z) = -3 \cdot x - (-3) \cdot 2 - 3 \cdot z =$
$-3x - (-6) - 3z = -3x + 6 - 3z$

50. $x - 5 - 7x - 4 = (1 - 7)x + (-5 - 4) = -6x + (-9) =$
$-6x - 9$

51. $P = 2(l + w)$
$$= 2\left(11 \text{ in.} + 8\frac{1}{2} \text{ in.}\right)$$
$$= 2\left(19\frac{1}{2} \text{ in.}\right)$$
$$= 2 \cdot \frac{39}{2} \text{ in.}$$
$$= \frac{\not{2} \cdot 39}{\not{2} \cdot 1} \text{ in.}$$
$$= 39 \text{ in.}$$

52. $P = 4s = 4 \cdot 12 \text{ ft} = 48 \text{ ft}$

53. $A = \frac{1}{2} \cdot b \cdot h$
$$= \frac{1}{2} \cdot \frac{5}{2} \text{ ft} \cdot 3 \text{ ft}$$
$$= \frac{1 \cdot 5 \cdot 3}{2 \cdot 2} \text{ ft}^2$$
$$= \frac{15}{4} \text{ ft}^2$$

54. First we find the area of the rectangular portion of the figure.
$$A = l \cdot w$$
$$= 20 \text{ ft} \cdot 10 \text{ ft}$$
$$= 200 \text{ ft}^2$$
Next we find the area of the triangular portion. The base is 20 ft and the height is 15 ft − 10 ft, or 5 ft.

$$A = \frac{1}{2} \cdot b \cdot h$$
$$= \frac{1}{2} \cdot 20 \text{ ft} \cdot 5 \text{ ft}$$
$$= \frac{1 \cdot 20 \cdot 5}{2} \text{ ft}^2$$
$$= 50 \text{ ft}^2$$

Finally we add to find the total area.
$$200 \text{ ft}^2 + 50 \text{ ft}^2 = 250 \text{ ft}^2$$

55.
$$7x - \frac{2}{3}(x - 6) = 6\frac{5}{7}$$
$$7x - \frac{2}{3}x + \frac{2}{3} \cdot 6 = \frac{47}{7}$$
$$\frac{21}{3}x - \frac{2}{3}x + 4 = \frac{47}{7}$$
$$\frac{19}{3}x + 4 = \frac{47}{7}$$
$$\frac{19}{3}x + 4 - 4 = \frac{47}{7} - 4$$
$$\frac{19}{3}x = \frac{47}{7} - \frac{28}{7}$$
$$\frac{19}{3}x = \frac{19}{7}$$
$$\frac{3}{19} \cdot \frac{19}{3}x = \frac{3}{19} \cdot \frac{19}{7}$$
$$x = \frac{3 \cdot \not{19}}{\not{19} \cdot 7}$$
$$x = \frac{3}{7}$$

The solution is $\frac{3}{7}$.

56. Area of each floor including elevator/stairwell:
$$A = l \cdot w = 25 \text{ m} \cdot 22\frac{1}{2} \text{ m}$$
$$= 25 \text{ m} \cdot \frac{45}{2} \text{ m}$$
$$= \frac{25 \cdot 45}{2} \text{ m}^2$$
$$= \frac{1125}{2} \text{ m}^2$$

Area of each elevator/stairwell:
$$A = l \cdot w = 5 \text{ m} \cdot 4\frac{1}{2} \text{ m}$$
$$= 5 \text{ m} \cdot \frac{9}{2} \text{ m}$$
$$= \frac{5 \cdot 9}{2} \text{ m}^2$$
$$= \frac{45}{2} \text{ m}^2$$

Area of each floor excluding elevator/stairwell:
$$\frac{1125}{2} \text{ m}^2 - \frac{45}{2} \text{ m}^2 = \frac{1080}{2} \text{ m}^2 = 540 \text{ m}^2$$

Total area:
$$7 \cdot 540 \text{ m}^2 = 3780 \text{ m}^2$$

Chapter 5

Decimal Notation

1. 63.05

   a) Write a word name for
      the whole number. ┌─────────────┐
                         │ Sixty-three │
                         └─────────────┘
                         Sixty-three
   b) Write "and" for the
      decimal point.   ┌─────┐
                         │ and │
                         └─────┘
   c) Write a word name for
      the number to the right   Sixty-three
      of the decimal point,   and
      followed by the place   ┌────────────────┐
                               │ five hundredths │
      value of the last digit. └────────────────┘

   A word name for 63.05 is sixty-three and five hundredths.

3. 26.59

   a) Write a word name for
      the whole number. ┌────────────┐
                         │ Twenty-six │
                         └────────────┘
                         Twenty-six
   b) Write "and" for the
      decimal point.   ┌─────┐
                         │ and │
                         └─────┘
   c) Write a word name for
      the number to the right   Twenty-six
      of the decimal point,   and
      followed by the place   ┌───────────┐
                               │ fifty-nine │
      value of the last digit. │ hundredths │
                               └───────────┘

   A word name for 26.59 is twenty-six and fifty-nine hundredths.

5. A word name for 8.35 is eight and thirty-five hundredths.

7. A word name for 24.6875 is twenty-four and six thousand eight hundred seventy-five ten-thousandths.

9. 5.63

   a) Write a word name for
      the whole number. ┌──────┐
                         │ Five │
                         └──────┘
                         Five
   b) Write "and" for the
      decimal point.   ┌─────┐
                         │ and │
                         └─────┘
   c) Write a word name for
      the number to the right   Five
      of the decimal point,   and
      followed by the place   ┌─────────────┐
                               │ sixty-three │
      value of the last digit. │ hundredths │
                               └─────────────┘

   A word name for 5.63 is five and sixty-three hundredths.

11. Write "and 95 cents" as "and $\frac{95}{100}$ dollars." A word name for $524.95 is five hundred twenty-four and $\frac{95}{100}$ dollars.

13. Write "and 72 cents" as "and $\frac{72}{100}$ dollars." A word name for $36.72 is thirty-six and $\frac{72}{100}$ dollars.

15. 7.3   7.3. $\frac{73}{10}$
             └↑
          1 place   1 zero

   $7.3 = \frac{73}{10}$

   To write 7.3 as a mixed numeral, we rewrite the whole number part and express the rest in fraction form.

   $$7.3 = 7\frac{3}{10}$$

17. 203.6   203.6. $\frac{2036}{10}$
              └↑
          1 place   1 zero

   $203.6 = \frac{2036}{10}$

   To write 203.6 as a mixed numeral, we rewrite the whole number part and express the rest in fraction form.

   $$203.6 = 203\frac{6}{10}$$

19. −2.703   −2.703. $\frac{-2073}{1000}$
               └──↑
          3 places   3 zeros

   $-2.703 = \frac{-2073}{1000}$, or $-\frac{2703}{1000}$

   To write −2.703 as a mixed numeral, we rewrite the whole number part and express the rest in fraction form.

   $$-2.703 = -2\frac{703}{1000}$$

21. 0.0109   0.0109. $\frac{109}{10,000}$
               └──·↑
          4 places   4 zeros

   $0.0109 = \frac{109}{10,000}$

   Since the whole number part of 0.0109 is zero, we cannot express this number as a mixed numeral.

23. -4.0003 $\underset{\text{4 places}}{-4.000\underleftarrow{3.}}$ $\underset{\text{4 zeros}}{\dfrac{-40,003}{10,000}}$

$$-4.0003 = -\dfrac{40,003}{10,000}$$

To write -4.0003 as a mixed numeral, we rewrite the whole number part and express the rest in fraction form.

$$-4.0003 = -4\dfrac{3}{10,000}$$

25. -0.0207 $\underset{\text{4 places}}{-0.020\underleftarrow{7.}}$ $\underset{\text{4 zeros}}{-\dfrac{207}{10,000}}$

$$-0.0207 = -\dfrac{207}{10,000}$$

Since the whole number part of -0.0207 is zero, we cannot express this number as a mixed numeral.

27. 70.00105 $\underset{\text{5 places}}{70.0010\underleftarrow{5.}}$ $\underset{\text{5 zeros}}{\dfrac{7,000,105}{100,000}}$

$$70.00105 = -\dfrac{7,000,105}{100,000}$$

To write $70,00105$ as a mixed numeral, we rewrite the whole number part and express the rest in fraction form.

$$70.00105 = 70\dfrac{105}{100,000}$$

29. $\dfrac{3}{10}$ $0.\underrightarrow{3}.$

$\underset{\text{1 zero}}{}$ $\underset{\text{Move 1 place.}}{}$

$$\dfrac{3}{10} = 0.3$$

31. $-\dfrac{59}{100}$ $-0.\underleftarrow{59}.$

$\underset{\text{2 zeros}}{}$ $\underset{\text{Move 2 places.}}{}$

$$-\dfrac{59}{100} = -0.59$$

33. $\dfrac{3798}{1000}$ $3.\underleftarrow{798}.$

$\underset{\text{3 zeros}}{}$ $\underset{\text{Move 3 places.}}{}$

$$\dfrac{3798}{1000} = 3.798$$

35. $\dfrac{78}{10,000}$ $0.\underleftarrow{0078}.$

$\underset{\text{4 zeros}}{}$ $\underset{\text{Move 4 places.}}{}$

$$\dfrac{78}{10,000} = 0.0078$$

37. $\dfrac{-18}{100,000}$ $-0.\underleftarrow{00018}.$

$\underset{\text{5 zeros}}{}$ $\underset{\text{Move 5 places.}}{}$

$$\dfrac{-18}{100,000} = -0.00018$$

39. $\dfrac{486,197}{1,000,000}$ $0.\underleftarrow{486197}.$

$\underset{\text{6 zeros}}{}$ $\underset{\text{Move 6 places.}}{}$

$$\dfrac{486,197}{1,000,000} = 0.486197$$

41. $7\dfrac{13}{1000} = 7 + \dfrac{13}{1000} = 7 \text{ and } \dfrac{13}{1000} = 7.013$

43. $-8\dfrac{431}{1000}$

First consider $8\dfrac{431}{1000}$:

$$8\dfrac{431}{1000} = 8 + \dfrac{431}{1000} = 8 \text{ and } \dfrac{431}{1000} = 8.431$$

Since $8\dfrac{431}{1000} = 8.431$, we have $-8\dfrac{431}{1000} = -8.431$.

45. $2\dfrac{1739}{10,000} = 2 + \dfrac{1739}{10,000} = 2 \text{ and } \dfrac{1739}{10,000} = 2.1739$

47. $8\dfrac{953,073}{1,000,000} = 8 + \dfrac{953,073}{1,000,000} =$

$8 \text{ and } \dfrac{953,073}{1,000,000} = 8.953073$

49. To compare two numbers in decimal notation, start at the left and compare corresponding digits moving from left to right. When two digits differ, the number with the larger digit is the larger of the two numbers.

0.06

$\downarrow$ Different; 5 is larger than 0.

0.58

Thus, 0.58 is larger.

51. 0.403

$\updownarrow$ Starting at the left, these digits are the first to differ; 1 is larger than 0.

0.410

Thus, 0.410 is larger.

53. -5.046

$\updownarrow$ Starting at the left, these digits are the first to differ, and 3 is smaller than 6.

-5.043

Thus, -5.043 is larger.

55. 234.07

$\updownarrow$ Starting at the left, these digits are the first to differ, and 5 is larger than 4.

235.07

Thus, 235.07 is larger.

57. $\frac{7}{100} = 0.07$ so we compare 0.007 and 0.07.

0.007

↑ Starting at the left, these digits are the first to differ, and 7 is larger than 0.

0.07

Thus, 0.07 or $\frac{7}{100}$ is larger.

59. −0.872

↑ Starting at the left, these digits are the first to differ, and 2 is smaller than 3.

−0.873

Thus, −0.872 is larger.

61. 0.2⌐3⌐ Hundredths digit is 4 or lower.
↓ Round down.
0.2

63. −0.3⌐7⌐2 Hundredths digit is 5 or higher.
↓ Round from −0.372 to −0.4.
−0.4

65. 2.9⌐5⌐1 Hundredths digit is 5 or higher.
↓ Round up.
3.0

(When we make the tenths digit a 10, we carry 1 to the ones place.)

67. −327.2⌐3⌐47 Hundredths digit is 4 or lower.
↓ Round from −327.2347 to −327.2.
−327.2

69. 0.89⌐3⌐ Thousandths digit is 4 or lower.
↓ Round down.
0.89

71. −0.66⌐6⌐6 Thousandths digit is 5 or higher.
↓ Round from −0.6666 to −0.67.
−0.67

73. 0.99⌐5⌐2 Thousandths digit is 5 or higher.
↓ Round up.
1.00

(When we make the hundredths digit a 10, we carry 1 to the tenths place. This then requires us to carry 1 to the ones place.)

75. −0.03⌐4⌐88 Thousandths digit is 4 or lower.
↓ Round from −0.03488 to −0.03.
−0.03

77. 0.572⌐4⌐ Ten-thousandths digit is 4 or lower.
↓ Round down.
0.572

79. 17.001⌐5⌐ Ten-thousandths digit is 5 or higher.
↓ Round up.
17.002

81. −20.202⌐0⌐2 Ten-thousandths digit is 4 or lower.
↓ Round −20.20202 to −20.202.
−20.202

83. 9.984⌐8⌐ Ten-thousandths digit is 5 or higher.
↓ Round up.
9.985

85. 809.4⌐7⌐321 Hundredths digit is 5 or higher.
↓ Round up.
809.5

87. 809.47⌐3⌐21 Thousandths digit is 4 or lower.
↓ Round down.
809.47

89. Discussion and Writing Exercise

91.
```
  1 1
  6 8 1
+ 1 4 9
-------
  8 3 0
```

93.
```
  1 16
  2̸ 0̸ 7
−   8 5
-------
  1 8 2
```

95. $\frac{37}{55} - \frac{49}{55} = \frac{37-49}{55} = \frac{-12}{55}$, or $-\frac{12}{55}$

97.
```
        18
    3 8̸ 9 13
  3 4,9̸ 0̸ 3̸
−    1 9 4 5
-----------
  3 2,9 5 8
```

99. Discussion and Writing Exercise

101. All of the numbers except −1.09 have 3 decimal places, so we add a zero to −1.09 so that each number has 3 decimal places. Then we start at the left and compare corresponding digits moving from left to right. Keep in mind that, when we compare two negative numbers and when digits differ, the number with the smaller digit is the larger of the two numbers. The numbers, listed from smallest to largest, are −1.09, −1.009, −0.989, −0.898, and −0.098.

103. 6.78346⌐123⌐ ←Drop all decimal places past the fifth place.

The answer is 6.78346.

105. 99.99999 |9999| ←Drop all decimal places
                              past the fifth place.

The answer is 99.99999.

107. From the graph we see that the years for which the vertical bars lie above +0.4° are 1983, 1988, 1990, 1991, 1992, 1995, 1997, 1998, 1999, 2000, 2001, 2002, 2003, and 2004.

109. From the graph we see that the last year for which the corresponding vertical bar line below 0.0° is 1985.

Exercise Set 5.2

1.
$$
\begin{array}{r}
\overset{1}{4\,2}6.2\,5 \\
+\quad 3\,8.1\,2 \\
\hline
4\,6\,4.3\,7
\end{array}
$$
Add hundredths.
Add tenths.
Write a decimal point in the answer.
Add ones.
Add tens.
Add hundreds.

3.
$$
\begin{array}{r}
\overset{1\ \ 1}{6\,5\,9}.4\,0\,3 \\
+\ 9\,1\,6.8\,1\,2 \\
\hline
1\,5\,7\,6.2\,1\,5
\end{array}
$$
Add thousandths.
Add hundredths.
Add tenths.
Write a decimal point in the answer.
Add ones.
Add tens.
Add hundreds.

5.
$$
\begin{array}{r}
\overset{1}{\ \ }\quad 9.1\,0\overset{1}{4} \\
+\ 1\,2\,3.4\,5\,6 \\
\hline
1\,3\,2.5\,6\,0
\end{array}
$$

7. Line up the decimal points.
$$
\begin{array}{r}
\overset{1}{2}.0\,0\,6 \\
+\ 5.8\,1\,7 \\
\hline
7.8\,2\,3
\end{array}
$$

9. Line up the decimal points.
$$
\begin{array}{r}
2\,0.0\,1\,2\,4 \\
+\ 3\,0.0\,1\,2\,4 \\
\hline
5\,0.0\,2\,4\,8
\end{array}
$$

11. Line up the decimal points.
$$
\begin{array}{r}
0.8\,3\,0 \\
+\ 0.0\,0\,5 \\
\hline
0.8\,3\,5
\end{array}
$$
Writing an extra zero

13. Line up the decimal points.
$$
\begin{array}{r}
\overset{1}{\ }\quad 0.3\,4\,0 \\
3.5\,0\,0 \\
0.1\,2\,7 \\
+\ 7\,6\,8.0\,0\,0 \\
\hline
7\,7\,1.9\,6\,7
\end{array}
$$
Writing an extra zero
Writing 2 extra zeros

Writing in the decimal point
and 3 extra zeros

Adding

15.
$$
\begin{array}{r}
\overset{1\ \ \ 1\ 1}{1\,7}.0\,0\,0\,0 \\
3.2\,4\,0\,0 \\
0.2\,5\,6\,0 \\
+\ \ \ 0.3\,6\,8\,9 \\
\hline
2\,0.8\,6\,4\,9
\end{array}
$$
Writing in the decimal point.
You may find it helpful to
write extra zeros.

17.
$$
\begin{array}{r}
\overset{1\ 2\,1\ \ \ \ \ \ 1}{\quad}\ 2.7\,0\,3\,0 \\
7\,8.3\,3\,0\,0 \\
2\,8.0\,0\,0\,9 \\
+\ 1\,1\,8.4\,3\,4\,1 \\
\hline
2\,2\,7.4\,6\,8\,0
\end{array}
$$

19.
$$
\begin{array}{r}
4\,7.5\,9\,6 \\
-\ \ 6.2\,1\,5 \\
\hline
4\,1.3\,8\,1
\end{array}
$$
Subtract thousandths.
Subtract hundredths.
Subtract tenths.
Write a decimal point in the answer.
Subtract ones.
Subtract tens.

21.
$$
\begin{array}{r}
\overset{4\ 11\ 2\ 11}{5\,\cancel{1}.\cancel{3}\,\cancel{1}} \\
-\quad 2.2\,9 \\
\hline
4\,9.0\,2
\end{array}
$$
Borrow tenths to subtract hundredths.
Subtract hundredths.
Subtract tenths.
Write a decimal point in the answer.
Borrow tens to subtract ones.
Subtract ones.
Subtract tens.

23.
$$
\begin{array}{r}
\overset{5\ 9\ 10}{3.6\,\cancel{0}\,\cancel{0}} \\
-\ 0.0\,3\,6 \\
\hline
3.5\,6\,4
\end{array}
$$
Writing 2 extra zeros

25.
$$
\begin{array}{r}
\overset{8\ \overset{11}{\cancel{1}}\ 13}{\cancel{9}\,\cancel{2}.\cancel{3}\,4\,1} \\
-\quad 6.4\,2 \\
\hline
8\,5.9\,2\,1
\end{array}
$$

27.
$$
\begin{array}{r}
\overset{2\ 9\ 10\ 6\ 14}{3.\cancel{0}\,\cancel{0}\,\cancel{7}\,\cancel{4}} \\
-\ 1.3\,4\,0\,8 \\
\hline
1.6\,6\,6\,6
\end{array}
$$

29.
$$
\begin{array}{r}
\overset{6\ 9\ 10}{6.0\,\cancel{7}\,\cancel{0}\,\cancel{0}} \\
-\ 2.0\,0\,7\,8 \\
\hline
4.0\,6\,2\,2
\end{array}
$$
Writing 2 extra zeros

31. Line up the decimal points. Write an extra zero if desired.
$$
\begin{array}{r}
\overset{\ \ \ \ \ 11\ 13}{\overset{2\ 9\ \cancel{1}\ \cancel{3}\ 10}{3.\cancel{0}\,\cancel{2}\,4\,\cancel{0}}} \\
-\quad 0.2\,4\,1 \\
\hline
2\,9.9\,9\,9
\end{array}
$$

33.
$$
\begin{array}{r}
\overset{3\ 10}{3\,\cancel{4}.\cancel{0}\,7} \\
-\ 3\,0.7 \\
\hline
3.3\,7
\end{array}
$$

35.
$$
\begin{array}{r}
\overset{4\ 10}{8.4\,\cancel{5}\,\cancel{0}} \\
-\ 7.4\,0\,5 \\
\hline
1.0\,4\,5
\end{array}
$$

37.
$$\begin{array}{r} \overset{5\ 10}{\cancel{6}.\cancel{0}\,0\,3} \\ -\ 2.3 \\ \hline 3.7\,0\,3 \end{array}$$

39.
$$\begin{array}{r} \overset{1\ 9\ 9\ 9\ 10}{2.\cancel{0}\,\cancel{0}\,\cancel{0}\,\cancel{0}} \\ -1.0\,9\,0\,8 \\ \hline 0.9\,0\,9\,2 \end{array}$$
Writing in the decimal point and 4 extra zeros

Subtracting

41.
$$\begin{array}{r} \overset{\quad\ \ 13}{\overset{1\ \ \cancel{3}\ 9\ 10}{6\,\cancel{2}\,4.\cancel{0}\,\cancel{0}}} \\ -\ \ \ 1\,8.7\,9 \\ \hline 6\,0\,5.2\,1 \end{array}$$

43.
$$\begin{array}{r} 5\,7.8\,0\,3 \\ -\ \ \ 4.6 \\ \hline 5\,3.2\,0\,3 \end{array}$$

45.
$$\begin{array}{r} \overset{\qquad\ 6\ 10}{2\,6\,3.7\,\cancel{0}} \\ -1\,0\,2.0\,8 \\ \hline 1\,6\,1.6\,2 \end{array}$$

47.
$$\begin{array}{r} \overset{\ \ 4\ 9\ 9\ 10}{4\,5.\cancel{0}\,\cancel{0}\,\cancel{0}} \\ -\ \ 0.9\,9\,9 \\ \hline 4\,4.0\,0\,1 \end{array}$$

49. $-5.02 + 1.73$ A positive and a negative number

   a) $|-5.02| = 5.02$, $|1.73| = 1.73$, and $|-5.02| > |-1.73|$, so the answer is negative.

   b)
$$\begin{array}{r} \overset{4\ 9\ 12}{\cancel{5}.\cancel{0}\,\cancel{2}} \\ -\ 1.7\,3 \\ \hline 3.2\,9 \end{array}$$
Find the difference in the absolute values.

   c) $-5.02 + 1.73 = -3.29$

51. $12.9 - 15.4 = 12.9 + (-15.4)$
We add the opposite of 15.4. We have a positive and a negative number.

   a) $|12.9| = 12.9$, $|-15.4| = 15.4$, and $|15.4| > |12.9|$, so the answer is negative.

   b)
$$\begin{array}{r} \overset{4\ 14}{1\,\cancel{5}.\cancel{4}} \\ -\ 1\,2.9 \\ \hline 2.5 \end{array}$$
Finding the difference in the absolute values

   c) $12.9 - 15.4 = -2.5$

53. $-2.9 + (-4.3)$ Two negative numbers

   a)
$$\begin{array}{r} \overset{1}{2}.9 \\ +\ 4.3 \\ \hline 7.2 \end{array}$$
Adding the absolute values

   b) $-2.9 + (-4.3) = -7.2$ The sum of two negative numbers is negative.

55. $-4.301 + 7.68$ A negative and a positive number

   a) $|-4.301| = 4.301$, $|7.68| = 7.68$, and $|7.68| > |-4.301|$, so the answer is positive.

   b)
$$\begin{array}{r} \overset{\quad\ 7\ 10}{7.6\,\cancel{8}\,\cancel{0}} \\ -\ 4.3\,0\,1 \\ \hline 3.3\,7\,9 \end{array}$$
Finding the difference in the absolute values

   c) $-4.301 + 7.68 = 3.379$

57.
$$\begin{aligned} &-12.9 - 3.7 \\ =\ &-12.9 + (-3.7) &&\text{Adding the opposite of } 3.7 \\ =\ &-16.6 &&\text{The sum of two negatives is negative.} \end{aligned}$$

59.
$$\begin{aligned} &-2.1 - (-4.6) \\ =\ &-2.1 + 4.6 &&\text{Adding the opposite of } -4.6 \\ =\ &2.5 &&\text{Subtracting absolute values. Since 4.6 has the larger absolute value, the answer is positive.} \end{aligned}$$

61.
$$\begin{aligned} &14.301 + (-17.82) \\ =\ &-3.519 &&\text{Subtracting absolute values. Since } -17.82 \text{ has the larger absolute value, the answer is negative.} \end{aligned}$$

63.
$$\begin{aligned} &7.201 - (-2.4) \\ =\ &7.201 + 2.4 &&\text{Adding the opposite of } -2.4 \\ =\ &9.601 &&\text{Adding} \end{aligned}$$

65.
$$\begin{aligned} &96.9 + (-21.4) \\ =\ &75.5 &&\text{Subtracting absolute values. Since 96.9 has the larger absolute value, the answer is positive.} \end{aligned}$$

67.
$$\begin{aligned} &-8.9 - (-12.7) \\ =\ &-8.9 + 12.7 &&\text{Adding the opposite of } -12.7 \\ =\ &3.8 &&\text{Subtracting absolute values. Since 12.7 has the larger absolute value, the answer is positive.} \end{aligned}$$

69.
$$\begin{aligned} &-4.9 - 5.392 \\ =\ &-4.9 + (-5.392) &&\text{Adding the opposite of } 5.392 \\ =\ &-10.292 &&\text{The sum of two negatives is negative.} \end{aligned}$$

71.
$$\begin{aligned} &14.7 - 23.5 \\ =\ &14.7 + (-23.5) &&\text{Adding the opposite of } 23.5 \\ =\ &-8.8 &&\text{Subtracting absolute values. Since } -23.5 \text{ has the larger absolute value, the answer is negative.} \end{aligned}$$

73.
$$\begin{aligned} &1.8x + 3.9x \\ =\ &(1.8 + 3.9)x &&\text{Using the distributive law} \\ =\ &5.7x &&\text{Adding} \end{aligned}$$

75.
$$\begin{aligned} &17.59a - 12.73a \\ =\ &(17.59 - 12.73)a \\ =\ &4.86a \end{aligned}$$

77.
$$\begin{aligned} &15.2t + 7.9 + 5.9t \\ =\ &15.2t + 5.9t + 7.9 &&\text{Using the commutative law} \\ =\ &(15.2 + 5.9)t + 7.9 &&\text{Using the distributive law} \\ =\ &21.1t + 7.9 \end{aligned}$$

79. $5.217x - 8.134x$

$= (5.217 - 8.134)x$ Using the distributive law

$= (5.217 + (-8.134))x$ Adding the opposite of 8.134

$= -2.917x$ Subtracting absolute values. The coefficient is negative since -8.134 has the larger absolute value.

81. $4.906y - 7.1 + 3.2y$

$= 4.906y + 3.2y - 7.1$

$= (4.906 + 3.2)y - 7.1$

$= 8.106y - 7.1$

83. $4.8x + 1.9y - 5.7x + 1.2y$

$= 4.8x + 1.9y + (-5.7x) + 1.2y$ Rewriting as addition

$= 4.8x + (-5.7x) + 1.9y + 1.2y$ Using the commutative law

$= (4.8 + (-5.7))x + (1.9 + 1.2)y$

$= -0.9x + 3.1y$

85. $4.9 - 3.9t + 2.3 - 4.5t$

$= 4.9 + (-3.9t) + 2.3 + (-4.5t)$

$= 4.9 + 2.3 + (-3.9t) + (-4.5t)$

$= (4.9 + 2.3) + (-3.9 + (-4.5))t$

$= 7.2 + (-8.4t)$

$= 7.2 - 8.4t$

87. Discussion and Writing Exercise

89. $\dfrac{3}{5} \cdot \dfrac{4}{7} = \dfrac{3 \cdot 4}{5 \cdot 7} = \dfrac{12}{35}$

91. $\dfrac{3}{10} \cdot \dfrac{21}{100} = \dfrac{3 \cdot 21}{10 \cdot 100} = \dfrac{63}{1000}$

93. $5 - 3x^2$

$= 5 - 3 \cdot 2^2$ Substituting 2 for x

$= 5 - 3 \cdot 4$

$= 5 - 12$

$= -7$

95. Discussion and Writing Exercise

97. $-3.928 - 4.39a + 7.4b - 8.073 + 2.0001a -$
$9.931b - 9.8799a + 12.897b$

$= -3.928 - 8.073 - 4.39a + 2.0001a - 9.8799a +$
$7.4b - 9.931b + 12.897b$

$= -12.001 - 12.2698a + 10.366b$

99. $39.123a - 42.458b - 72.457a + 31.462b -$
$59.491 + 37.927a$

$= 39.123a - 72.457a + 37.927a - 42.458b +$
$31.462b - 59.491$

$= 4.593a - 10.996b - 59.491$

101. First "undo" the incorrect subtraction by adding 349.2 to the incorrect answer:

$$-836.9 + 349.2 = -487.7$$

Now add 349.2 to -487.7:

$$-487.7 + 349.2 = -138.5$$

The correct answer is -138.5.

103.
$$
\begin{array}{r}
\overset{\scriptstyle 3\ 13}{9\,3.\,a\,\cancel{4}\,\cancel{3}} \\
-\,8\,7.\,9\,6\,9 \\
\hline
5.\,2\,7\,4
\end{array}
$$

We need to borrow 1 tenth in order to subtract hundredths. Then when we subtract 9 tenths from $(a-1)$ tenths we get 2. Since $11 - 9 = 2$, we know that we had to borrow a one in order to subtract tenths and this was added to 1 tenth, or $a - 1$. Then if $a - 1 = 1$, we know that $a = 2$.

Exercise Set 5.3

1.
$$
\begin{array}{r}
6.\,8 \\
\times\quad 7 \\
\hline
4\,7.\,6
\end{array}
$$
(1 decimal place)
(0 decimal places)
(1 decimal place)

3.
$$
\begin{array}{r}
0.\,8\,4 \\
\times\quad 8 \\
\hline
6.\,7\,2
\end{array}
$$
(2 decimal places)
(0 decimal places)
(2 decimal places)

5.
$$
\begin{array}{r}
6.\,3 \\
\times\,0.\,0\,4 \\
\hline
0.\,2\,5\,2
\end{array}
$$
(1 decimal place)
(2 decimal places)
(3 decimal places)

7.
$$
\begin{array}{r}
2\,8.\,6 \\
\times\,0.\,0\,9 \\
\hline
2.\,5\,7\,4
\end{array}
$$
(1 decimal place)
(2 decimal places)

9. 10×42.63 42.6.3

1 zero   Move 1 place to the right.

$10 \times 42.63 = 426.3$

11. $-1000 \times 783.686852 = -(1000 \times 783.686852)$

$-(1000 \times 783.686852)$ -783.686852

3 zeros   Move 3 places to the right.

$-1000 \times 783.686852 = -783,686.852$

13. -7.8×100 $-7.80.$

2 zeros   Move 2 places to the right.

$-7.8 \times 100 = -780$

15. 0.1×79.18 7.9.18

1 decimal place   Move 1 place to the left.

$0.1 \times 79.18 = 7.918$

17. 0.001×97.68 0.097.68

3 decimal places   Move 3 places to the left.

$0.001 \times 97.68 = 0.09768$

19. $28.7 \times (-0.01) = -(28.7 \times 0.01)$

$-(28.7 \times 0.01)$ $-0.28.7$

2 decimal places   Move 2 places to the left.

$28.7 \times (-0.01) = -0.287$

21.
```
    2. 7 3      (2 decimal places)
  ×    1 6      (0 decimal places)
  ─────────
    1 6 3 8
    2 7 3 0
  ─────────
    4 3. 6 8    (2 decimal places)
```

23.
```
    0. 9 8 4    (3 decimal places)
  ×      3. 3   (1 decimal place)
  ──────────
      2 9 5 2
      2 9 5 2 0
  ──────────
    3. 2 4 7 2  (4 decimal places)
```

25. We multiply the absolute values.
```
      3 7. 4    (1 decimal place)
  ×      2. 4   (1 decimal place)
  ──────────
      1 4 9 6
      7 4 8 0
  ──────────
      8 9. 7 6  (2 decimal places)
```
Since the product of two negative numbers is positive, the answer is 89.76.

27. We multiply the absolute values.
```
      7 4 9     (0 decimal places)
  ×    0. 4 3   (2 decimal places)
  ──────────
      2 2 4 7
      2 9 9 6 0
  ──────────
    3 2 2. 0 7  (2 decimal places)
```
Since the product of a positive number and a negative number is negative, the answer is −322.07.

29.
```
    0. 8 7      (2 decimal places)
  ×    6 4      (0 decimal places)
  ─────────
      3 4 8
    5 2 2 0
  ─────────
    5 5. 6 8    (2 decimal places)
```

31.
```
    4 6. 5 0    (2 decimal places)
  ×      7 5    (0 decimal places)
  ──────────
    2 3 2 5 0
    3 2 5 5 0 0
  ──────────
  3 4 8 7. 5 0  (2 decimal places)
```
Since the last decimal place is 0, we could also write this answer as 3487.5.

33. We multiply the absolute values.
```
    0. 2 3 1    (3 decimal places)
  ×      0. 5   (1 decimal place)
  ──────────
    0. 1 1 5 5  (4 decimal places)
```
Since the product of two negative numbers is positive, the answer is 0.1155.

35. $9.42 \times (-1000) = -(9.42 \times 1000)$

$-(9.42 \times 1\underline{000})$ $-9.420.$

3 zeros   Move 3 places to the right.

$9.42 \times (-1000) = -9420$

37. $-95.3 \times (-0.0001) = 95.3 \times 0.0001$

$95.3 \times 0.\underline{0001}$ $0.0095.3$

4 decimal places   Move 4 places to the left.

$-95.3 \times (-0.0001) = 0.00953$

39. Move 2 places to the right.

$57.06.¢$

Change from $ sign in front to ¢ sign at end.

$\$57.06 = 5706¢$

41. Move 2 places to the right.

$0.95.¢$

Change from $ sign in front to ¢ sign at end.

$\$0.95 = 95¢$

43. Move 2 places to the right.

$0.01.¢$

Change from $ sign in front to ¢ sign at end.

$\$0.01 = 1¢$

45. Move 2 places to the left.

$0.72.¢$

Change from ¢ sign at end to $ sign in front.

$72¢ = \$0.72$

47. Move 2 places to the left.

$0.02.¢$

Change from ¢ sign at end to $ sign in front.

$2¢ = \$0.02$

49. Move 2 places to the left.

$63.99.¢$

Change from ¢ sign at end to $ sign in front.

$6399¢ = \$63.99$

51. 3.156 billion = $3.156 \times 1,\underbrace{000,000,000}_{9 \text{ zeros}}$

3.156000000.

Move 9 places to the right.

3.156 billion = $3,156,000,000$

53. 63.1 trillion = $63.1 \times 1,\underbrace{000,000,000,000}_{12 \text{ zeros}}$

63.1000000000000.

Move 12 places to the right.

63.1 trillion = $63,100,000,000,000$

55. 11.98 million = $11.98 \times 1,\underbrace{000,000}_{6 \text{ zeros}}$

$11.980000.

Move 6 places to the right.

11.98 million = $11,980,000$

57. $P + Prt$
$= 10,000 + 10,000(0.04)(2.5)$ Substituting
$= 10,000 + 400(2.5)$ Multiplying and dividing
$= 10,000 + 1000$ in order from left to right
$= 11,000$ Adding

59. $vt + at^2$
$= 10(1.5) + 4.9(1.5)^2$
$= 10(1.5) + 4.9(1.5)(1.5)$
$= 10(1.5) + 4.9(2.25)$ Squaring first
$= 15 + 11.025$
$= 26.025$

61. a) $P = 2l + 2w$
      $= 2(12.5) + 2(9.5)$
      $= 25 + 19$
      $= 44$

    The perimeter is 44 ft.

   b) $A = l \cdot w$
      $= (12.5)(9.5)$
      $= 118.75$

    The area is 118.75 ft^2.

63. a) $P = 2l + 2w$
      $= 2(10.5) + 2(8.4)$
      $= 21 + 16.8$
      $= 37.8$

    The perimeter is 37.8 m.

   b) $A = l \cdot w$
      $= (10.5)(8.4)$
      $= 88.2$

    The area is 88.2 m^2.

65. In 2010, $t = 2010 - 2000 = 10$. Substitute 10 for t.
    $5.06t + 9.7 = 5.06(10) + 9.7$
                   $= 50.6 + 9.7$
                   $= 60.3$

In 2010 there will be approximately 60.3 billion e-mails each day in North America.

67. Discussion and Writing Exercise

69. $-162 \div 6 = -27$
(The signs are different, so the quotient is negative.)

71. $-1035 \div (-15) = 69$
(The signs are the same, so the quotient is positive.)

73. $-525 \div 25 = -21$
(The signs are different, so the quotient is negative.)

75. $-7050 \div 50 = -141$
(The signs are different, so the quotient is negative.)

77. Discussion and Writing Exercise

79. $85 \times 9.46 \times 10^{12}$
$= 85 \times 9.46 \times 10 \times 10 \times 10 \times 10 \times 10 \times 10 \times$
        $10 \times 10 \times 10 \times 10 \times 10 \times 10$
$= 8.04.1 \times 1,000,000,000,000$
$= (804.1 \times 1000) \times 1,000,000,000$
$= 804,100 \times 1,000,000,000$
$= 804,100 \times 1$ billion

Regulus is 804,100 billion km from Earth.

81. Use a calculator.
    $d + vt + at^2$
$= 79.2 + 3.029(7.355) + 4.9(7.355)^2$
$= 79.2 + 3.029(7.355) + 4.9(54.096025)$
$= 79.2 + 22.278295 + 265.0705225$
$= 101.478295 + 265.0705225$
$= 366.5488175$

83. Use a calculator.
    $0.5(b_1 + b_2)h = 0.5(9.7 \text{ cm} + 13.4 \text{ cm})(6.32 \text{ cm})$
                   $= 0.5(23.1 \text{ cm})(6.32 \text{ cm})$
                   $= 72.996 \text{ cm}^2$

85. (1 trillion) $\cdot$ (1 billion)
$= 1,\underbrace{000,000,000,000}_{12 \text{ zeros}} \times 1,\underbrace{000,000,000}_{9 \text{ zeros}}$
$= 1,\underbrace{000,000,000,000,000,000,000}_{21 \text{ zeros}}$
$= 10^{21}$

87. For the British number 6.6 billion, we have:
    6.6 billion
$= 6.6 \times 1,000,000 \times 1,000,000$
$= 6.6 \times 1,000,000,000,000$
$= 6,600,000,000,000$ Moving the decimal point
                               12 places to the right

89. The period from April 20 to May 20 consists of 30 days, so the "customer charge" is 30 × $0.374, or $11.22. The "energy charge" for the first 250 kilowatt-hours is 250 × $0.1174, or $29.35.

We subtract to find the number of kilowatt-hours in excess of 250: 480 − 250 = 230. Then the "energy charge" for the 230 kilowatt-hours in excess of 250 kilowatt-hours is 230 × $0.09079, or $20.88 (rounding to the nearest cent).

Finally, we add to find the total bill:

$11.22 + $29.35 + $20.88 = $61.45.

Exercise Set 5.4

1.
```
      12.6
  5 ) 63.0
      50
      13
      10
       30      Write an extra 0.
       30
        0
```

3.
```
     23.78
  4 ) 95.12
     8000
     1512
     1200
      312
      280
       32
       32
        0
```
Divide as though dividing whole numbers. Place the decimal point directly above the decimal point in the dividend.

5.
```
       7.48
  12 ) 89.76
       8400
        576
        480
         96
         96
          0
```

7.
```
        7.2
  33 ) 237.6
       2310
         66
         66
          0
```

9. We first consider 5.4 ÷ 6.
```
      0.9
  6 ) 5.4
      54
       0
```
Since a positive number divided by a negative number is negative, the answer is −0.9.

11. We first consider 9.144 ÷ 8.
```
      1.143
  8 ) 9.144
      8000
      1144
       800
       344
       320
        24
        24
         0
```
Since a negative number divided by a positive number is negative, the answer is −1.143.

13.
```
              140.
  0.06ʌ) 8.40ʌ
          600
          240
          240
            0
```
Multiply the divisor by 100 (move the decimal point 2 places). Multiply the same way in the dividend (move 2 places). Then divide.

15.
```
            40.
  2.6ʌ) 104.0ʌ
        1040
           0
```
Put a decimal point at the end of the whole number. Multiply the divisor by 10 (move the decimal point 1 place). Multiply the same way in the dividend (move 1 place), adding an extra 0. Then divide.

17. We first consider 1.8 ÷ 12.
```
       0.15
  12 ) 1.80
       12
       60
       60
        0
```
Divide as though dividing whole numbers. Place the decimal point directly above the decimal point in the dividend.

Since a positive number divided by a negative number is negative, the answer is −0.15.

19.
```
            48.
  2.7ʌ) 129.6ʌ
        1080
         216
         216
           0
```

21.
```
            3.2
  8.5ʌ) 27.2ʌ0
        255
        170    Write an extra zero.
        170
          0
```

23. We first consider 5 ÷ 8.
```
      0.625
  8 ) 5.000
      48
       20      Write an extra 0.
       16
        40     Write an extra 0.
        40
         0
```
Since a negative number divided by a negative number is positive, the answer is 0.625.

25.
```
        0.2 6
0.4 7∧⟌0. 1 2∧2 2
        9 4 0
        ‾‾‾‾‾
        2 8 2
        2 8 2
        ‾‾‾‾‾
          0
```

27.
```
          2.3 4
0.0 3 2∧⟌0.0 7 4∧8 8
         6 4 0 0
         ‾‾‾‾‾‾‾
         1 0 8 8
           9 6 0
         ‾‾‾‾‾‾‾
           1 2 8
           1 2 8
         ‾‾‾‾‾‾‾
             0
```

29. We first consider $24.969 \div 82$.
```
        0.3 0 4 5
8 2⟌2 4.9 6 9 0
      2 4 6 0 0
      ‾‾‾‾‾‾‾
          3 6 9
          3 2 8
          ‾‾‾‾‾    Write an extra 0.
            4 1 0
            4 1 0
          ‾‾‾‾‾‾‾
              0
```

Since a negative number divided by a positive number is negative, the answer is -0.3045.

31. $\dfrac{-213.4567}{100}$ $-2.13.4567$

2 zeros   Move 2 places to the left.

$\dfrac{-213.4567}{100} = -2.134567$

33. $\dfrac{1.0237}{0.001}$ $1.023.7$

3 decimal places   Move 3 places to the right.

$\dfrac{1.0237}{0.001} = 1023.7$

35. $\dfrac{92.36}{-0.01} = \dfrac{-92.36}{0.01}$

$\dfrac{-92.36}{0.01}$ -92.36

2 decimal places   Move 2 places to the right.

$\dfrac{92.36}{-0.01} = \dfrac{-92.36}{0.01} = -9236$

37. $\dfrac{0.8172}{10}$ $0.0.8172$

1 zero   Move 1 place to the left.

$\dfrac{0.8172}{10} = 0.08172$

39. $\dfrac{0.97}{0.1}$ $0.9.7$

1 decimal place   Move 1 place to the right.

$\dfrac{0.97}{0.1} = 9.7$

41. $\dfrac{52.7}{-1000} = \dfrac{-52.7}{1000}$

$\dfrac{-52.7}{1000}$ $-0.052.7$

3 zeros   Move 3 places to the left.

$\dfrac{52.7}{-1000} = \dfrac{-52.7}{1000} = -0.0527$

43. $\dfrac{75.3}{-0.001} = -\dfrac{75.3}{0.001}$

$-\dfrac{75.3}{0.001}$ $-75.300.$

3 decimal places   Move 3 places to the right.

$\dfrac{75.3}{-0.001} = -75,300$

45. $\dfrac{-75.3}{1000}$ $-0.075.3$

3 zeros   Move 3 places to the left.

$\dfrac{-75.3}{1000} = -0.0753$

47. $14 \times (82.6 + 67.9)$
$= 14 \times (150.5)$ Doing the calculation inside the parentheses
$= 2107$ Multiplying

49. $0.003 + 3.03 \div (-0.01) = 0.003 - 303$ Dividing first
$= -302.997$ Subtracting

51. $(4.9 - 18.6) \times 13$
$= -13.7 \times 13$ Doing the calculation inside the parentheses
$= -178.1$ Multiplying

53. $210.3 - 4.24 \times 1.01$
$= 210.3 - 4.2824$ Multiplying
$= 206.0176$ Subtracting

55. $12 \div (-0.03) - 12 \times 0.03^2$
$= 12 \div (-0.03) - 12 \times 0.0009$ Evaluating the exponential expression
$= -400 - 0.0108$ Dividing and multiplying in order from left to right
$= -400.0108$ Subtracting

57. $(4 - 2.5)^2 \div 100 + 0.1 \times 6.5$

$= (1.5)^2 \div 100 + 0.1 \times 6.5$ Doing the calculation inside the parentheses

$= 2.25 \div 100 + 0.1 \times 6.5$ Evaluating the exponential expression

$= 0.0225 + 0.65$ Dividing and multiplying in order from left to right

$= 0.6725$ Adding

59. $6 \times 0.9 - 0.1 \div 4 + 0.2^3$

$= 6 \times 0.9 - 0.1 \div 4 + 0.008$ Evaluating the exponential expression

$= 5.4 - 0.025 + 0.008$ Multiplying and dividing in order from left to right

$= 5.383$ Subtracting and adding in order from left to right

61. $12^2 \div (12 + 2.4) - [(2 - 2.4) \div 0.8]$

$= 12^2 \div (12 + 2.4) - [-0.4 \div 0.8]$ Doing the calculations in the innermost parentheses first

$= 12^2 \div 14.4 - [-0.5]$ Doing the calculations inside the parentheses

$= 12^2 \div 14.4 + 0.5$ Simplifying

$= 144 \div 14.4 + 0.5$ Evaluating the exponential expression

$= 10 + 0.5$ Dividing

$= 10.5$ Adding

63. We add the amounts and divide by the number of addends, 5.

$$\frac{131.8 + 168.7 + 230.2 + 250.0 + 251.0}{5}$$

$$= \frac{1031.7}{5} = 206.34$$

The average amount paid per year in individual income tax over the five-year period was $206.34 billion.

65. We add the amounts and divide by the number of addends, 5.

$$\frac{83.51 + 81.71 + 81.71 + 81.59 + 81.25}{5}$$

$$= \frac{409.77}{5} = 81.954$$

The average life expectancy for the given countries is 81.954 yr.

67. We add the amounts and divide by the number of addends, 5.

$$\frac{33.5 + 31.1 + 16.0 + 13.8 + 12.3}{5}$$

$$= \frac{106.7}{5} = 21.34$$

The average length of the tunnels is 21.34 mi.

69. Discussion and Writing Exercise

71. $\dfrac{33}{44} = \dfrac{3 \cdot 11}{4 \cdot 11} = \dfrac{3}{4} \cdot \dfrac{11}{11} = \dfrac{3}{4}$

73. $-\dfrac{27}{18} = -\dfrac{3 \cdot 9}{2 \cdot 9} = -\dfrac{3}{2} \cdot \dfrac{9}{9} = -\dfrac{3}{2}$

75. $\dfrac{9a}{27} = \dfrac{9 \cdot a}{9 \cdot 3} = \dfrac{9}{9} \cdot \dfrac{a}{3} = \dfrac{a}{3}$

77. $\dfrac{4r}{20r} = \dfrac{4 \cdot r \cdot 1}{5 \cdot 4 \cdot r} = \dfrac{4 \cdot r}{4 \cdot r} \cdot \dfrac{1}{5} = \dfrac{1}{5}$

79. Discussion and Writing Exercise

81. Use a calculator.

$7.434 \div (-1.2) \times 9.5 + 1.47^2$

$= 7.434 \div (-1.2) \times 9.5 + 2.1609$ Evaluating the exponential expression

$= -6.195 \times 9.5 + 2.1609$

$= -58.8525 + 2.1609$ Multiplying and dividing in order from left to right

$= -56.6916$ Adding

83. Use a calculator.

$9.0534 - 2.041^2 \times 0.731 \div 1.043^2$

$= 9.0534 - 4.165681 \times 0.731 \div 1.087849$ Evaluating the exponential expressions

$= 9.0534 - 3.045112811 \div 1.087849$

$= 9.0534 - 2.799205415$ Multiplying and dividing in order from left to right

$= 6.254194585$ Subtracting

85. $439.57 \times 0.01 \div 1000 \cdot x = 4.3957$

$4.3957 \div 1000 \cdot x = 4.3957$

$0.0043957 \cdot x = 4.3957$

$x = \dfrac{4.3957}{0.0043957}$

$x = 1000$

The solution is 1000.

87. $0.0329 \div 0.001 \times 10^4 \div x = 3290$

$0.0329 \div 0.001 \times 10,000 \div x = 3290$

$32.9 \times 10,000 \div x = 3290$

$329,000 \div x = 3290$

We need to divide 329,000 by a number that moves the decimal point 2 places to the left. Thus, we need to divide by 100. The solution is 100.

89. We divide. Note that 5.6 million $= 5.6 \times 1,000,000 = 5,600,000$.

```
                5.71
980,000)5,600,000.00
        4 900 000
        7 000 000
        6 860 000
          140 000 0
          980 000 0
          420 000 0
```

Rounding to the nearest tenth, we have 5.7 rating points.

91. The period from August 20 to September 20 consists of 31 days.

The "customer charge" is $31 \times \$0.374 = \$11.59$ (rounded to the nearest cent). The "energy charge" for the first 250 kilowatt-hours is $250 \times \$0.1174 = \$29.35$.

Subtract to find the "energy charge" for the kilowatt-hours in excess of 250:

$59.10 - $11.59 - $29.35 = $18.16

Divide to find the number of kilowatt-hours in excess of 250:

$18.16 \div $0.09079 = 200$ (rounded to the nearest hour)

The total number of kilowatt-hours of electricity used is $250 + 200 = 450$ kwh.

Exercise Set 5.5

1. Since $\dfrac{3}{8}$ means $3 \div 8$ we have:

$$
\begin{array}{r}
0.375 \\
8\overline{)3.000} \\
\underline{24} \\
60 \\
\underline{56} \\
40 \\
\underline{40} \\
0
\end{array}
$$

$\dfrac{3}{8} = 0.375$

3. Since $\dfrac{-1}{2}$ is negative, we divide 1 by 2 and make the results negative.

$$
\begin{array}{r}
0.5 \\
2\overline{)1.0} \\
\underline{10} \\
0
\end{array}
$$

Thus, $\dfrac{-1}{2} = -0.5$.

5. Since $\dfrac{3}{25}$ means $3 \div 25$, we have:

$$
\begin{array}{r}
0.12 \\
25\overline{)3.00} \\
\underline{25} \\
50 \\
\underline{50} \\
0
\end{array}
$$

$\dfrac{3}{25} = 0.12$

7. Since $\dfrac{9}{40}$ means $9 \div 40$, we have:

$$
\begin{array}{r}
0.225 \\
40\overline{)9.000} \\
\underline{80} \\
100 \\
\underline{80} \\
200 \\
\underline{200} \\
0
\end{array}
$$

$\dfrac{9}{40} = 0.225$

9. Since $\dfrac{13}{25}$ means $13 \div 25$, we have:

$$
\begin{array}{r}
0.52 \\
25\overline{)13.00} \\
\underline{125} \\
50 \\
\underline{50} \\
0
\end{array}
$$

$\dfrac{13}{25} = 0.52$

11. Since $\dfrac{-17}{20}$ is negative, we divide 17 by 20 and make the result negative.

$$
\begin{array}{r}
0.85 \\
20\overline{)17.00} \\
\underline{160} \\
100 \\
\underline{100} \\
0
\end{array}
$$

Thus, $\dfrac{-17}{20} = -0.85$.

13. Since $-\dfrac{9}{16}$ is negative, we divide 9 by 16 and make the result negative.

$$
\begin{array}{r}
0.5625 \\
16\overline{)9.0000} \\
\underline{80} \\
100 \\
\underline{96} \\
40 \\
\underline{32} \\
80 \\
\underline{80} \\
0
\end{array}
$$

Thus, $-\dfrac{9}{16} = -0.5625$.

15. Since $\dfrac{7}{5}$ means $7 \div 5$, we have:

$$
\begin{array}{r}
1.4 \\
5\overline{)7.0} \\
\underline{5} \\
20 \\
\underline{20} \\
0
\end{array}
$$

$\dfrac{7}{5} = 1.4$

17. Since $\dfrac{28}{25}$ means $28 \div 25$, we have:

$$
\begin{array}{r}
1.12 \\
25\overline{)28.00} \\
\underline{25} \\
30 \\
\underline{25} \\
50 \\
\underline{50} \\
0
\end{array}
$$

$\dfrac{28}{25} = 1.12$

19. Since $\dfrac{11}{-8}$ is negative, we divide 11 by 8 and make the result negative.

$$
\begin{array}{r}
1.375 \\
8\overline{\smash{\big)}\,11.000} \\
\underline{8} \\
30 \\
\underline{24} \\
60 \\
\underline{56} \\
40 \\
\underline{40} \\
0
\end{array}
$$

Thus, $\dfrac{11}{-8} = -1.375$.

21. Since $-\dfrac{39}{40}$ is negative, we divide 39 by 40 and make the result negative.

$$
\begin{array}{r}
0.975 \\
40\overline{\smash{\big)}\,39.000} \\
\underline{360} \\
300 \\
\underline{280} \\
200 \\
\underline{200} \\
0
\end{array}
$$

Thus, $-\dfrac{39}{40} = -0.975$.

23. Since $\dfrac{121}{200}$ means $121 \div 200$, we have:

$$
\begin{array}{r}
0.605 \\
200\overline{\smash{\big)}\,121.000} \\
\underline{1200} \\
1000 \\
\underline{1000} \\
0
\end{array}
$$

$$\dfrac{121}{200} = 0.605$$

25. Since $\dfrac{8}{15}$ means $8 \div 15$, we have:

$$
\begin{array}{r}
0.533 \\
15\overline{\smash{\big)}\,8.000} \\
\underline{75} \\
50 \\
\underline{45} \\
50 \\
\underline{45} \\
5
\end{array}
$$

Since 5 keeps reappearing as a remainder, the digits repeat and

$$\dfrac{8}{15} = 0.533\ldots \text{ or } 0.5\overline{3}.$$

27. Since $\dfrac{1}{3}$ means $1 \div 3$, we have:

$$
\begin{array}{r}
0.333 \\
3\overline{\smash{\big)}\,1.000} \\
\underline{9} \\
10 \\
\underline{9} \\
10 \\
\underline{9} \\
1
\end{array}
$$

Since 1 keeps reappearing as a remainder, the digits repeat and

$$\dfrac{1}{3} = 0.333\ldots \text{ or } 0.\overline{3}.$$

29. Since $\dfrac{-4}{3}$ is negative, we divide by 4 and 3 and make the result negative.

$$
\begin{array}{r}
1.33 \\
3\overline{\smash{\big)}\,4.00} \\
\underline{3} \\
10 \\
\underline{9} \\
10 \\
\underline{9} \\
1
\end{array}
$$

Since 1 keeps reappearing as a remainder, the digits repeat and

$$\dfrac{4}{3} = 1.333\ldots \text{ or } 1.\overline{3}.$$

Thus, $\dfrac{-4}{3} = -1.\overline{3}$.

31. Since $\dfrac{7}{6}$ means $7 \div 6$, we have:

$$
\begin{array}{r}
1.166 \\
6\overline{\smash{\big)}\,7.000} \\
\underline{6} \\
10 \\
\underline{6} \\
40 \\
\underline{36} \\
40 \\
\underline{36} \\
4
\end{array}
$$

Since 4 keeps reappearing as a remainder, the digits repeat and

$$\dfrac{7}{6} = 1.166\ldots \text{ or } 1.1\overline{6}.$$

33. Since $-\dfrac{14}{11}$ is negative, we divide 14 by 11 and make the result negative.

$$
\begin{array}{r}
1.2727 \\
11\,\overline{)\,14.0000} \\
\underline{11} \\
30 \\
\underline{22} \\
80 \\
\underline{77} \\
30 \\
\underline{22} \\
80 \\
\underline{77} \\
3
\end{array}
$$

Since 3 and 8 keep reappearing as remainders, the sequence of digits "27" repeats in the quotient and $\dfrac{14}{11} = 1.2727\ldots$, or $1.\overline{27}$.

Thus, $-\dfrac{14}{11} = -1.\overline{27}$.

35. Since $-\dfrac{5}{12}$ is negative, we divide 5 by 12 and make the result negative.

$$
\begin{array}{r}
0.4166 \\
12\,\overline{)\,5.0000} \\
\underline{48} \\
20 \\
\underline{12} \\
80 \\
\underline{72} \\
80 \\
\underline{72} \\
8
\end{array}
$$

Since 8 keeps reappearing as a remainder, the digits repeat and

$$\frac{5}{12} = 0.4166\ldots, \text{ or } 0.41\overline{6}.$$

Thus, $\dfrac{-5}{12} = -0.41\overline{6}.$

37. Since $\dfrac{127}{500}$ means $127 \div 500$, we have:

$$
\begin{array}{r}
0.254 \\
500\,\overline{)\,127.000} \\
\underline{1000} \\
2700 \\
\underline{2500} \\
2000 \\
\underline{2000} \\
0
\end{array}
$$

$$\frac{127}{500} = 0.254$$

39. Since $\dfrac{4}{33}$ means $4 \div 33$, we have:

$$
\begin{array}{r}
0.1212 \\
33\,\overline{)\,4.0000} \\
\underline{33} \\
70 \\
\underline{66} \\
40 \\
\underline{33} \\
70 \\
\underline{66} \\
4
\end{array}
$$

Since 7 and 4 keep reappearing as remainders, the sequence of digits "12" repeats in the quotient and

$$\frac{4}{33} = 0.1212\ldots, \text{ or } 0.\overline{12}.$$

41. Since $-\dfrac{12}{55}$ is negative, we divide 12 by 55 and make the result negative.

$$
\begin{array}{r}
0.21818 \\
55\,\overline{)\,12.00000} \\
\underline{110} \\
100 \\
\underline{55} \\
450 \\
\underline{440} \\
100 \\
\underline{55} \\
450 \\
\underline{440} \\
10
\end{array}
$$

Since 10 and 45 keep reappearing as remainders, the sequence of digits "18" repeats in the quotient and $\dfrac{12}{55} = 0.21818\ldots$, or $0.2\overline{18}$.

Thus, $\dfrac{-12}{55} = -0.2\overline{18}.$

43. Since $\dfrac{35}{111}$ means $35 \div 111$, we have:

$$
\begin{array}{r}
0.315315 \\
111\,\overline{)\,35.000000} \\
\underline{333} \\
170 \\
\underline{111} \\
590 \\
\underline{555} \\
350 \\
\underline{333} \\
170 \\
\underline{111} \\
590 \\
\underline{555} \\
35
\end{array}
$$

Since 17, 59, and 35 keep reappearing as remainders, the sequence of digits "315" repeats in the quotient and

$$\frac{35}{111} = 0.315315\ldots, \text{ or } 0.\overline{315}.$$

45. Since $\frac{4}{7}$ means $4 \div 7$, we have:

```
      0.5 7 1 4 2 8
7 ) 4.0 0 0 0 0 0
      3 5
      ___
        5 0
        4 9
        ___
          1 0
            7
          ___
          3 0
          2 8
          ___
            2 0
            1 4
            ___
              6 0
              5 6
              ___
                4
```

Since we have already divided 7 into 4, the sequence of digits "571428" repeats in the quotient and
$$\frac{4}{7} = 0.571428571428\ldots, \text{ or } 0.\overline{571428}.$$

47. Since $\frac{-37}{25}$ is negative, we divide 37 by 25 and make the result negative.

```
       1.4 8
2 5 ) 3 7.0 0
      2 5
      ___
      1 2 0
      1 0 0
      _____
        2 0 0
        2 0 0
        _____
            0
```

Thus, $\frac{-37}{25} = -1.48$.

49. In Example 4 we see that $\frac{4}{11} = 0.\overline{36}$.

Round $0.\,3\,\boxed{6}\,3\,6\ldots$ to the nearest tenth.

     Hundredths digit is 6 or more.

$0.\,4$ Round up.

Round $0.\,3\,\underline{6}\,\boxed{3}\,6\ldots$ to the nearest hundredth.

     Thousandths digit is 4 or less.

$0.\,3\,6$ Round down.

Round $0.\,3\,6\,\underline{3}\,\boxed{6}\ldots$ to the nearest thousandth.

     Ten-thousandths digit is 5 or more.

$0.\,3\,6\,4$ Round up.

51. First we find decimal notation for $-\frac{5}{3}$.

```
     1.6 6
3 ) 5.0 0
    3
    ___
    2 0
    1 8
    ___
      2 0
      1 8
      ___
        2
```

$\frac{5}{3} = 1.\overline{6}$, so $-\frac{5}{3} = -1.\overline{6}$

Round $-1.\,\underline{6}\,\boxed{6}\,6\,6\ldots$ to the nearest tenth.

     Hundredths digit is 5 or more.

$-1.\,7$ Round to -1.7.

Round $-1.\,6\,\underline{6}\,\boxed{6}\,6\ldots$ to the nearest hundredth.

     Thousandths digit is 5 or more.

$-1.\,6\,7$ Round to -1.67.

Round $-1.\,6\,6\,\underline{6}\,\boxed{6}\ldots$ to the nearest thousandth.

     Ten-thousandths digit is 5 or more.

$-1.\,6\,6\,7$ Round to -1.667.

53. First we find decimal notation for $\frac{-8}{17}$.

```
       0.4 7 0 5 8 8
1 7 ) 8.0 0 0 0 0
      6 8
      ___
      1 2 0
      1 1 9
      _____
          1 0 0
            8 5
          _____
          1 5 0
          1 3 6
          _____
            1 4 0
            1 3 6
            _____
                4
```

The digits repeat eventually but we have enough decimal places now to be able to round as instructed.

We have $\frac{8}{17} \approx 0.47059$, so $\frac{-8}{17} \approx -0.47059$.

Round $-0.\,4\,\boxed{7}\,0\,5\,9\ldots$ to the nearest tenth.

     Hundredths digit is 5 or more.

$-0.\,5$ Round to -0.5.

Round $-0.\,4\,\underline{7}\,\boxed{0}\,5\,9\ldots$ to the nearest hundredth.

     Thousandths digit is 4 or less.

$-0.\,4\,7$ Round to -0.47.

Round $-0.\,4\,7\,\underline{0}\,\boxed{5}\,9\ldots$ to the nearest thousandth.

     Ten-thousandths digit is 5 or more.

$-0.\,4\,7\,1$ Round to -0.471.

55. First find decimal notation for $\frac{7}{12}$.

```
       0.5 8 3 3
1 2 ) 7.0 0 0 0
      6 0
      ___
      1 0 0
        9 6
      _____
        4 0
        3 6
        ___
          4 0
          3 6
          ___
            4
```

$$\frac{7}{12} = 0.58\overline{3}$$

Round 0. 5 $\boxed{8}$ 3 3 ... to the nearest tenth.

                              Hundredths digit is 5 or more.

    0. 6   Round up.

Round 0. 5 8 $\boxed{3}$ 3 ... to the nearest hundredth.

                            Thousandths digit is 4 or less.

    0. 5 8   Round down.

Round 0. 5 8 3 $\boxed{3}$ ... to the nearest thousandth.

                            Ten-thousandths digit is 4 or less.

    0. 5 8 3   Round down.

57. First find decimal notation for $\dfrac{29}{-150}$.

```
        0. 1 9 3 3
150 | 2 9. 0 0 0 0
      1 5 0
      ─────
      1 4 0 0
      1 3 5 0
      ───────
          5 0 0
          4 5 0
          ─────
            5 0 0
            4 5 0
            ─────
              5 0
```

We have $\dfrac{29}{150} = 0.19\overline{3}$, so $\dfrac{29}{-150} = -0.19\overline{3}$.

Round $-0.$ 1 $\boxed{9}$ 3 3 ... to the nearest tenth.

                            Hundredths digit is 5 or more.

    $-0.$ 2   Round to -0.2.

Round $-0.$ 1 9 $\boxed{3}$ 3 ... to the nearest hundredth.

                            Thousandths digit is 3 or less.

    $-0.$ 1 9   Round to -0.19.

Round $-0.$ 1 9 3 $\boxed{3}$ ... to the nearest thousandth.

                            Ten-thousandths digit is 3 or less.

    $-0.$ 1 9 3   Round to -0.193.

59. First find decimal notation for $\dfrac{7}{-9}$.

```
      0. 7 7
9 | 7. 0 0
    6 3
    ───
      7 0
      6 3
      ───
        7
```

We have $\dfrac{7}{9} = 0.\overline{7}$, so $\dfrac{7}{-9} = -0.\overline{7}$.

Round $-0.$ 7 $\boxed{7}$ 7 7 ... to the nearest tenth.

                            Hundredths digit is 5 or more.

    $-0.$ 8   Round to -0.8.

Round $-0.$ 7 $\boxed{7}$ 7 ... to the nearest hundredth.

                            Thousandths digit is 5 or more.

    $-0.$ 7 8   Round to -0.78.

Round $-0.$ 7 7 7 $\boxed{7}$ ... to the nearest thousandth.

                            Ten-thousandths digit is 5 or more.

    $-0.$ 7 7 8   Round to -0.778.

61. We will use the first method discussed in the text.

$$\frac{7}{8}(10.84) = \frac{7}{8} \times \frac{10.84}{1} = \frac{7 \times 10.84}{8} = \frac{75.88}{8} = 9.485$$

63. We will use the third method discussed in the text.

$$\frac{47}{9}(-79.95) = \frac{47}{9} \cdot \left(-\frac{7995}{100}\right)$$

$$= -\frac{47 \cdot 7995}{9 \cdot 100}$$

$$= -\frac{47 \cdot \cancel{3} \cdot \cancel{5} \cdot 533}{\cancel{3} \cdot 3 \cdot \cancel{5} \cdot 20}$$

$$= -\frac{25,051}{60}$$

$$= -417.51\overline{6}$$

65. We will use the first method discussed in the text.

$$\left(\frac{1}{6}\right)0.0765 + \left(\frac{3}{4}\right)0.1124 = \frac{1}{6} \times \frac{0.0765}{1} + \frac{3}{4} \times \frac{0.1124}{1}$$

$$= \frac{0.0765}{6} + \frac{3 \times 0.1124}{4}$$

$$= \frac{0.0765}{6} + \frac{0.3372}{4}$$

$$= 0.01275 + 0.0843$$

$$= 0.09705$$

67. We will use the third method discussed in the text.

$$\frac{3}{4} \times 2.56 - \frac{7}{8} \times 3.94$$

$$= \frac{3}{4} \times \frac{256}{100} - \frac{7}{8} \times \frac{394}{100}$$

$$= \frac{768}{400} - \frac{2758}{800}$$

$$= \frac{768}{400} \cdot \frac{2}{2} - \frac{2758}{800}$$

$$= \frac{1536}{800} - \frac{2758}{800}$$

$$= \frac{-1222}{800} = -\frac{1222}{800}$$

$$= -\frac{2 \cdot 611}{2 \cdot 400} = \frac{2}{2} \cdot \left(-\frac{611}{400}\right)$$

$$= -\frac{611}{400}, \text{ or } -1.5275$$

69. We will use the second method discussed in the text.

$$5.2 \times 1\frac{7}{8} \div 0.4 = 5.2 \times 1.875 \div 0.4$$

$$= 9.75 \div 0.4$$

$$= 24.375$$

71. *Familiarize*. We draw a picture and recall that the formula for the area A of a triangle with base b and height h is $A = \frac{1}{2} \times b \times h$.

1.8 m

1.2 m

Translate. We substitute 1.2 for b and 1.8 for h.
$$A = \frac{1}{2} \times b \times h = \frac{1}{2} \times 1.2 \times 1.8$$
Solve. We carry out the computation.
$$A = \frac{1}{2} \times 1.2 \times 1.8$$
$$= \frac{1.2}{2} \times 1.8 \qquad \text{Multiplying } \frac{1}{2} \text{ and } 1.2$$
$$= 0.6 \times 1.8 \qquad \text{Dividing}$$
$$= 1.08 \qquad \text{Multiplying}$$

Check. We repeat the calculations using a different method.
$$\frac{1}{2} \times 1.2 \times 1.8 = 0.5 \times (1.2 \times 1.8) = 0.5 \times 2.16 = 1.08$$
Our answer checks.

State. The area of the shawl is 1.08 m^2.

73. *Familiarize*. We draw a picture and recall that the formula for the area A of a triangle with base b and height h is $A = \frac{1}{2} \times b \times h$.

3.4 cm

3.4 cm

Translate. We substitute 3.4 for b and 3.4 for h.
$$A = \frac{1}{2} \times b \times h = \frac{1}{2} \times 3.4 \times 3.4$$
Solve. We carry out the computation.
$$A = \frac{1}{2} \times 3.4 \times 3.4$$
$$= \frac{3.4}{2} \times 3.4 \qquad \text{Multiplying } \frac{1}{2} \text{ and } 3.4$$
$$= 1.7 \times 3.4 \qquad \text{Dividing}$$
$$= 5.78 \qquad \text{Multiplying}$$

Check. We repeat the calculations using a different method.
$$\frac{1}{2} \times 3.4 \times 3.4 = 0.5 \times (3.4 \times 3.4) = 0.5 \times 11.56 = 5.78$$
Our answer checks.

State. The area of the stamp is 5.78 cm^2.

75. *Familiarize*. First combine the lengths of the two 19.5-in. segments: 19.5 in. + 19.5 in. = 39 in. Now we can think of the area as the sum of the areas of two triangles, one with base 39 in. and height 11.25 in. and the other with base 39 in. and height 29.31 in.

Translate. We use the formula $A = \frac{1}{2}bh$ twice and add.
$$A = \frac{1}{2} \times 39 \times 11.25 + \frac{1}{2} \times 39 \times 29.31$$
Solve. We carry out the computation.
$$A = \frac{1}{2} \times 39 \times 11.25 + \frac{1}{2} \times 39 \times 29.31$$
$$= \frac{39}{2} \times 11.25 + \frac{39}{2} \times 29.31 \qquad \text{Multiplying } \frac{1}{2} \text{ and } 39 \text{ twice}$$
$$= 19.5 \times 11.25 + 19.5 \times 29.31 \quad \text{Dividing}$$
$$= 219.375 + 571.545$$
$$= 790.92$$

Check. We repeat the calculation using a different method.
$$\frac{1}{2} \times 39 \times 11.25 + \frac{1}{2} \times 39 \times 29.31$$
$$= 0.5 \times (39 \times 11.25) + 0.5 \times (39 \times 29.31)$$
$$= 0.5 \times 438.75 + 0.5 \times 1143.09$$
$$= 219.375 + 571.545$$
$$= 790.92$$
The answer checks.

State. The area of the kite is 790.92 in^2.

77. Discussion and Writing Exercise

79. 3 5 7 [2]

The ones digit is 4 or less so we round down to 3570.

81. 7 8, 9 [5] 1

The tens digit is 5 or more so we round up to 79,000.

83. $\frac{n}{1} = n$, for any integer n.

Thus, $\frac{95}{-1} = \frac{-95}{1} = -95$.

85.
$$9 - 4 + 2 \div (-1) \cdot 6$$
$$= 9 - 4 - 2 \cdot 6 \qquad \text{Multiplying and dividing in}$$
$$= 9 - 4 - 12 \qquad \text{order from left to right}$$
$$= 5 - 12 \qquad \text{Adding and subtracting in}$$
$$= -7 \qquad \text{order from left to right}$$

87. Discussion and Writing Exercise

89. Using a calculator we find that
$$\frac{1}{7} = 1 \div 7 = 0.\overline{142857}.$$

91. Using a calculator we find that
$$\frac{3}{7} = 3 \div 7 = 0.\overline{428571}.$$

93. Using a calculator we find that
$$\frac{5}{7} = 5 \div 7 = 0.\overline{714285}.$$

95. Using a calculator we find that
$$\frac{1}{9} = 1 \div 9 = 0.\overline{1}.$$

97. Using a calculator we find that
$$\frac{1}{999} = 0.\overline{001}.$$

99. We substitute $\frac{22}{7}$ for π and 2.1 for r.

$$A = \pi r^2$$
$$= \frac{22}{7}(2.1)^2$$
$$= \frac{22}{7}(4.41)$$
$$= \frac{22 \times 4.41}{7}$$
$$= \frac{97.02}{7}$$
$$= 13.86 \text{ cm}^2$$

101. We substitute 3.14 for π and $\frac{3}{4}$ for r.

$$A = \pi r^2$$
$$= 3.14\left(\frac{3}{4}\right)^2$$
$$= 3.14\left(\frac{9}{16}\right)$$
$$= \frac{3.14 \times 9}{16}$$
$$= \frac{28.26}{16}$$
$$= 1.76625 \text{ ft}^2$$

When the calculation is done using the π key on a calculator, the result is 1.767145868 ft^2.

103. Discussion and Writing Exercise

Exercise Set 5.6

1. We are estimating the sum

$$\$279 + \$149.99.$$

We round both numbers to the nearest ten. The estimate is

$$\$280 + \$150 = \$430.$$

3. We are estimating the difference

$$\$279 - \$149.99.$$

We round both numbers to the nearest ten. The estimate is

$$\$280 - \$150 = \$130.$$

5. We are estimating the product

$$6 \times \$79.95.$$

We round $79.95 to the nearest ten. The estimate is

$$6 \times \$80 = \$480.$$

7. We are estimating the quotient

$$\$830 \div \$79.95.$$

We round $830 to the nearest hundred and $79.95 to the nearest ten. The estimate is

$$\$800 \div \$80 = 10 \text{ sets.}$$

9. This is about $0.0 + 1.3 + 0.3$, so the answer is about 1.6.

11. This is about $6 + 0 + 0$, so the answer is about 6.

13. This is about $52 + 1 + 7$, so the answer is about 60.

15. This is about $2.7 - 0.4$, so the answer is about 2.3.

17. This is about $200 - 20$, so the answer is about 180.

19. This is about 50×8, rounding 49 to the nearest ten and 7.89 to the nearest one, so the answer is about 400. Answer (a) is correct.

21. This is about 100×0.08, rounding 98.4 to the nearest ten and 0.083 to the nearest hundredth, so the answer is about 8. Answer (c) is correct.

23. This is about $4 \div 4$, so the answer is about 1. Answer (b) is correct.

25. This is about $75 \div 25$, so the answer is about 3. Answer (b) is correct.

27. We estimate the quotient $1760 \div 8.625$.

$$1800 \div 9 = 200$$

We estimate that 200 posts will be needed. Answers may vary depending on how the rounding was done.

29. Discussion and Writing Exercise

31. The decimal $0.57\overline{3}$ is an example of a <u>repeating</u> decimal.

33. The sentence $5(3+8) = 5 \cdot 3 + 5 \cdot 8$ illustrates the <u>distributive</u> law.

35. The number 1 is the <u>multiplicative</u> identity.

37. The least common <u>denominator</u> of two or more fractions is the least common <u>multiple</u> of their denominators.

39. Discussion and Writing Exercise

41. We round each factor to the nearest ten. The estimate is $180 \times 60 = 10,800$. The estimate is close to the result given, so the decimal point was placed correctly.

43. We round each number on the left to the nearest one. The estimate is $19 - 1 \times 4 = 19 - 4 = 15$. The estimate is not close to the result given, so the decimal point was not placed correctly.

45. a) Observe that $2^{13} = 8192 \approx 8000$, $156,876.8 \approx 160,000$, and $8000 \times 20 = 160,000$. Thus, we want to find the product of 2^{13} and a number that is approximately 20. Since $0.37 + 18.78 = 19.15 \approx 20$, we add inside the parentheses and then multiply:

$$(0.37 + 18.78) \times 2^{13} = 156,876.8$$

We can use a calculator to confirm this result.

b) Observe that $312.84 \approx 6 \cdot 50$. We start by multiplying 6.4 and 51.2, getting 327.68. Then we can use a calculator to find that if we add 2.56 to this product and then subtract 17.4, we have the desired result. Thus, we have

$$2.56 + 6.4 \times 51.2 - 17.4 = 312.84.$$

Exercise Set 5.7

1. $5x = 27$

$\dfrac{5x}{5} = \dfrac{27}{5}$ Dividing both sides by 5

$x = 5.4$

Check: $5x = 27$

$\overline{5(5.4)\ ?\ 27}$

$27\ \Big|\ 27$ TRUE

The solution is 5.4.

3. $x + 15.7 = 3.1$

$x + 15.7 - 15.7 = 3.1 - 15.7$ Adding -15.7 to (or
 subtracting 15.7 from) both sides

$x = -12.6$

Check: $x + 15.7 = 3.1$

$\overline{-12.6 + 15.7\ ?\ 3.1}$

$3.1\ \Big|\ 3.1$ TRUE

The solution is -12.6.

5. $5x - 8 = 22$

$5x - 8 + 8 = 22 + 8$ Adding 8 to both sides

$5x = 30$

$\dfrac{5x}{5} = \dfrac{30}{5}$ Dividing both sides by 5

$x = 6$

The solution is 6.

7. $6.9x - 8.4 = 4.02$

$6.9x - 8.4 + 8.4 = 4.02 + 8.4$ Adding 8.4 to both sides

$6.9x = 12.42$

$\dfrac{6.9x}{6.9} = \dfrac{12.42}{6.9}$ Dividing both sides by 6.9

$x = 1.8$

The solution is 1.8.

9. $21.6 + 4.1t = 6.43$

$21.6 + 4.1t - 21.6 = 6.43 - 21.6$ Subtracting 21.6
 from both sides

$4.1t = -15.17$

$\dfrac{4.1t}{4.1} = \dfrac{-15.17}{4.1}$

 Dividing both sides
 by 4.1

$t = -3.7$

The solution is -3.7.

11. $-26.05 = 7.5x + 9.2$

$-26.05 - 9.2 = 7.5x + 9.2 - 9.2$

$-35.25 = 7.5x$

$-4.7 = x$

The solution is -4.7.

13. $-4.2x + 3.04 = -4.1$

$-4.2x + 3.04 - 3.04 = -4.1 - 3.04$

$-4.2x = -7.14$

$\dfrac{-4.2x}{-4.2} = \dfrac{-7.14}{-4.2}$

$x = 1.7$

The solution is 1.7.

15. $-3.05 = 7.24 - 3.5t$

$-3.05 - 7.24 = 7.24 - 3.5t - 7.24$

$-10.29 = -3.5t$

$\dfrac{-10.29}{-3.5} = \dfrac{-3.5t}{-3.5}$

$2.94 = t$

The solution is 2.94.

17. $9x - 2 = 5x + 34$

$9x - 2 + 2 = 5x + 34 + 2$ Adding 2 to both sides

$9x = 5x + 36$

$9x - 5x = 5x + 36 - 5x$ Subtracting $5x$ from
 both sides

$4x = 36$

$\dfrac{4x}{4} = \dfrac{36}{4}$ Dividing both sides by 4

$x = 9$

Check: $9x - 2 = 5x + 34$

$\overline{9 \cdot 9 - 2\ ?\ 5 \cdot 9 + 34}$

$81 - 2\ \Big|\ 45 + 34$

$79\ \Big|\ 79$ TRUE

The solution is 9.

19. $2x + 6 = 7x - 10$

$2x + 6 - 6 = 7x - 10 - 6$ Subtracting 6 from
 both sides

$2x = 7x - 16$

$2x - 7x = 7x - 16 - 7x$ Subtracting $7x$ from
 both sides

$-5x = -16$

$\dfrac{-5x}{-5} = \dfrac{-16}{-5}$ Dividing both sides by -5

$x = 3.2$

Check: $2x + 6 = 7x - 10$

$\overline{2(3.2) + 6\ ?\ 7(3.2) - 10}$

$6.4 + 6\ \Big|\ 22.4 - 10$

$12.4\ \Big|\ 12.4$ TRUE

The solution is 3.2.

21.
$$5y - 3 = 4 + 9y$$
$$5y - 3 + 3 = 4 + 9y + 3$$
$$5y = 9y + 7$$
$$5y - 9y = 9y + 7 - 9y$$
$$-4y = 7$$
$$\frac{-4y}{-4} = \frac{7}{-4}$$
$$y = -1.75$$
The solution is -1.75.

23.
$$5.9x + 67 = 7.6x + 16$$
$$5.9x + 67 - 16 = 7.6x + 16 - 16$$
$$5.9x + 51 = 7.6x$$
$$5.9x + 51 - 5.9x = 7.6x - 5.9x$$
$$51 = 1.7x$$
$$\frac{51}{1.7} = \frac{1.7x}{1.7}$$
$$30 = x$$
The solution is 30.

25.
$$7.8a + 2 = 2.4a + 19.28$$
$$7.8a + 2 - 2 = 2.4a + 19.28 - 2$$
$$7.8a = 2.4a + 17.28$$
$$7.8a - 2.4a = 2.4a + 17.28 - 2.4a$$
$$5.4a = 17.28$$
$$\frac{5.4a}{5.4} = \frac{17.28}{5.4}$$
$$a = 3.2$$
The solution is 3.2

27.
$$6(x + 2) = 4x + 30$$
$$6x + 12 = 4x + 30 \quad \text{Using the distributive law}$$
$$6x + 12 - 12 = 4x + 30 - 12$$
$$6x = 4x + 18$$
$$6x - 4x = 4x + 18 - 4x$$
$$2x = 18$$
$$\frac{2x}{2} = \frac{18}{2}$$
$$x = 9$$

Check:
$$\begin{array}{c} 6(x+2) = 4x + 30 \\ \hline 6(9+2) \ ? \ 4 \cdot 9 + 30 \\ 6(11) \ \big| \ 36 + 30 \\ 66 \ \big| \ 66 \qquad \text{TRUE} \end{array}$$

The solution is 9.

29.
$$5(x + 3) = 15x - 6$$
$$5x + 15 = 15x - 6 \quad \text{Using the distributive law}$$
$$5x + 15 - 15 = 15x - 6 - 15$$
$$5x = 15x - 21$$
$$5x - 15x = 15x - 21 - 15x$$
$$-10x = -21$$
$$\frac{-10x}{-10} = \frac{-21}{-10}$$
$$x = 2.1$$

Check:
$$\begin{array}{c} 5(x+3) = 15x - 6 \\ \hline 5(2.1+3) \ ? \ 15(2.1) - 6 \\ 5(5.1) \ \big| \ 31.5 - 6 \\ 25.5 \ \big| \ 25.5 \qquad \text{TRUE} \end{array}$$

The solution is 2.1.

31.
$$7a - 9 = 15(a - 3)$$
$$7a - 9 = 15a - 45 \quad \text{Using the distributive law}$$
$$7a - 9 + 9 = 15a - 45 + 9$$
$$7a = 15a - 36$$
$$7a - 15a = 15a - 36 - 15a$$
$$-8a = -36$$
$$\frac{-8a}{-8} = \frac{-36}{-8}$$
$$a = 4.5$$
The solution is 4.5.

33.
$$2.9(x + 8.1) = 7.8x - 3.95$$
$$2.9x + 23.49 = 7.8x - 3.95$$
$$2.9x + 23.49 - 23.49 = 7.8x - 3.95 - 23.49$$
$$2.9x = 7.8x - 27.44$$
$$2.9x - 7.8x = 7.8x - 27.44 - 7.8x$$
$$-4.9x = -27.44$$
$$\frac{-4.9x}{-4.9} = \frac{-27.44}{-4.9}$$
$$x = 5.6$$
The solution is 5.6.

35.
$$-6.21 - 4.3t = 9.8(t + 2.1)$$
$$-6.21 - 4.3t = 9.8t + 20.58$$
$$-6.21 - 4.3t + 6.21 = 9.8t + 20.58 + 6.21$$
$$-4.3t = 9.8t + 26.79$$
$$-4.3t - 9.8t = 26.79$$
$$-14.1t = 26.79$$
$$\frac{-14.1t}{-14.1} = \frac{26.79}{-14.1}$$
$$t = -1.9$$
The solution is -1.9.

37. $4(x-2) - 9 = 2x + 9$

$4x - 8 - 9 = 2x + 9$

$4x - 17 = 2x + 9$

$4x - 17 + 17 = 2x + 9 + 17$

$4x = 2x + 26$

$4x - 2x = 2x + 26 - 2x$

$2x = 26$

$\dfrac{2x}{2} = \dfrac{26}{2}$

$x = 13$

The solution is 13.

39. $43(7 - 2x) + 34 = 50(x - 4.1) + 744$

$301 - 86x + 34 = 50x - 205 + 744$

$-86x + 335 = 50x + 539$

$-86x + 335 - 335 = 50x + 539 - 335$

$-86x = 50x + 204$

$-86x - 50x = 50x + 204 - 50x$

$-136x = 204$

$\dfrac{-136x}{-136} = \dfrac{204}{-136}$

$x = -1.5$

The solution is -1.5.

41. Discussion and Writing Exercise

43. We use the formula $A = \dfrac{1}{2} \cdot b \cdot h$ and substitute 7 m for b and 4 m for h.

$A = \dfrac{1}{2} \cdot b \cdot h$

$= \dfrac{1}{2} \cdot 7 \text{ m} \cdot 4 \text{ m}$

$= \dfrac{7 \cdot 4}{2} \text{ m}^2$

$= 14 \text{ m}^2$

45. We use the formula $A = \dfrac{1}{2} \cdot b \cdot h$ and substitute 5 in. for b and 5 in. for h.

$A = \dfrac{1}{2} \cdot b \cdot h$

$= \dfrac{1}{2} \cdot 5 \text{ in.} \cdot 5 \text{ in.}$

$= \dfrac{5 \cdot 5}{2} \text{ in}^2$

$= \dfrac{25}{2} \text{ in}^2, \text{ or } 12.5 \text{ in}^2$

47. The area of the figure is the sum of the areas of two triangles, each with base 5 ft and height 1 ft. Then we have

$A = \dfrac{1}{2} \cdot b \cdot h + \dfrac{1}{2} \cdot b \cdot h$

$= \dfrac{1}{2} \cdot 5 \text{ ft} \cdot 1 \text{ ft} + \dfrac{1}{2} \cdot 5 \text{ ft} \cdot 1 \text{ ft}$

$= \dfrac{5 \cdot 1}{2} \text{ ft}^2 + \dfrac{5 \cdot 1}{2} \text{ ft}^2$

$= \dfrac{5}{2} \text{ ft}^2 + \dfrac{5}{2} \text{ ft}^2$

$= 5 \text{ ft}^2$

49. $\dfrac{3}{25} - \dfrac{7}{10} = \dfrac{3}{25} \cdot \dfrac{2}{2} - \dfrac{7}{10} \cdot \dfrac{5}{5}$ The LCM is 50.

$= \dfrac{6}{50} - \dfrac{35}{50}$

$= -\dfrac{29}{50}$

51. We add in order from left to right.

$-17 + 24 + (-9) = 7 + (-9) = -2$

53. Discussion and Writing Exercise

55. $7.035(4.91x - 8.21) + 17.401 =$
$23.902x - 7.372815$

$34.54185x - 57.75735 + 17.401 =$
$23.902x - 7.372815$

$34.54185x - 40.35635 =$
$23.902x - 7.372815$

$34.54185x - 40.35635 - 23.902x =$
$23.902x - 7.372815 - 23.902x$

$10.63985x - 40.35635 = -7.372815$

$10.63985x - 40.35635 + 40.35635 =$
$-7.372815 + 40.35635$

$10.63985x = 32.983535$

$\dfrac{10.63985x}{10.63985} = \dfrac{32.983535}{10.63985}$

$x = 3.1$

The solution is 3.1.

57. $5(x - 4.2) + 3[2x - 5(x + 7)] =$
$39 + 2(7.5 - 6x) + 3x$

$5(x - 4.2) + 3[2x - 5x - 35] =$
$39 + 2(7.5 - 6x) + 3x$

$5(x - 4.2) + 3[-3x - 35] = 39 + 2(7.5 - 6x) + 3x$

$5x - 21 - 9x - 105 = 39 + 15 - 12x + 3x$

$-4x - 126 = 54 - 9x$

$-4x - 126 + 9x = 54 - 9x + 9x$

$5x - 126 = 54$

$5x - 126 + 126 = 54 + 126$

$5x = 180$

$\dfrac{5x}{5} = \dfrac{180}{5}$

$x = 36$

The solution is 36.

59. $3.5(4.8x - 2.9) + 4.5 = 9.4x - 3.4(x - 1.9)$

$16.8x - 10.15 + 4.5 = 9.4x - 3.4x + 6.46$

$16.8x - 5.65 = 6x + 6.46$

$16.8x - 5.65 - 6x = 6x + 6.46 - 6x$

$10.8x - 5.65 = 6.46$

$10.8x - 5.65 + 5.65 = 6.46 + 5.65$

$10.8x = 12.11$

$\dfrac{10.8x}{10.8} = \dfrac{12.11}{10.8}$

$x \approx 1.1212963$

The solution is approximately 1.1212963.

Exercise Set 5.8

1. *Familiarize.* Repeated addition fits this situation. We let C = the cost of 7 jackets.

$$\underbrace{\boxed{\$32.98} + \boxed{\$32.98} + \cdots + \boxed{\$32.98}}_{7 \text{ addends}}$$

Translate.

$$\underbrace{\text{Price}}_{\downarrow} \quad \underbrace{\text{times}}_{\downarrow} \quad \underbrace{\text{Number}}_{\downarrow} \quad \underbrace{\text{is}}_{\downarrow} \quad \underbrace{\text{Total}}_{\downarrow}$$

Price per jacket	times	Number of jackets	is	Total cost
32.98	×	7	=	C

Solve. We carry out the multiplication.

$$\begin{array}{r} 3\,2.\,9\,8 \\ \times \qquad 7 \\ \hline 2\,3\,0.\,8\,6 \end{array}$$

Thus, $C = 230.86$.

Check. We obtain a partial check by rounding and estimating:

$$32.98 \times 7 \approx 30 \times 7 = 210 \approx 230.86.$$

State. Seven jackets cost $230.86.

3. *Familiarize.* Repeated addition fits this situation. We let c = the cost of 20.4 gal of gasoline, in dollars.

Translate.

Cost per gallon	times	Number of gallons	is	Total cost
2.249	·	20.4	=	c

Solve. We carry out the multiplication.

$$\begin{array}{r} 2.\,2\,4\,9 \\ \times \qquad 2\,0.\,4 \\ \hline 8\,9\,9\,6 \\ 4\,4\,9\,8\,0\,0 \\ \hline 4\,5.\,8\,7\,9\,6 \end{array}$$

Thus, $c = 45.8796$.

Check. We obtain a partial check by rounding and estimating:

$$2.249 \times 20.4 \approx 2.25 \times 20 = 45 \approx 45.8796.$$

State. We round $45.8796 to the nearest cent and find that the cost of the gasoline is $45.88.

5. *Familiarize.* We visualize the situation. We let n = the new temperature.

98.6°	4.2°
n	

Translate. We are combining amounts.

Normal body temperature	plus	Degrees temperature rises	is	New temperature
98.6	+	4.2	=	n

Solve. To solve the equation we carry out the addition.

$$\begin{array}{r} {}^{1} \\ 9\,8.6 \\ +\quad 4.2 \\ \hline 1\,0\,2.8 \end{array}$$

Thus, $n = 102.8$.

Check. We can check by repeating the addition. We can also check by rounding:

$$98.6 + 4.2 \approx 99 + 4 = 103 \approx 102.8$$

State. The new temperature was 102.8°F.

7. *Familiarize.* We visualize the situation. Let w = each winner's share.

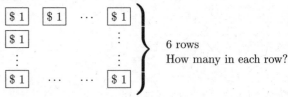

Translate.

Total prize	÷	Number of winners	=	Each winner's share
127,315	÷	6	=	w

Solve. We carry out the division.

```
        2 1, 2 1 9. 1 6 6
    6 ⟌ 1 2 7, 3 1 5. 0 0 0
        1 2 0 0 0 0
        ─────────
            7 3 1 5
            6 0 0 0
        ─────────
            1 3 1 5
            1 2 0 0
        ─────────
              1 1 5
                6 0
            ─────────
                5 5
                5 4
              ─────────
                  1 0
                   6
                ─────────
                  4 0
                  3 6
                ─────────
                  4 0
                  3 6
                ─────────
                    4
```

Rounding to the nearest cent, or hundredth, we get $w = 21,219.17$.

Check. We can repeat the calculation. The answer checks.

State. Each winner's share is $21,219.17.

9. *Familiarize*. Let $A =$ the area, in sq cm, and $P =$ the perimeter, in cm.

Translate. We use the formulas $A = l \cdot w$ and $P = l + w + l + w$ and substitute 3.25 for l and 2.5 for w.

$$A = l \cdot w = (3.25) \cdot (2.5)$$
$$P = l + w + l + w = 3.25 + 2.5 + 3.25 + 2.5$$

Solve. To find the area we carry out the multiplication.

```
      3. 2 5
    ×   2. 5
    ─────────
    1 6 2 5
    6 5 0 0
    ─────────
    8. 1 2 5
```

Thus, $A = 8.125$

To find the perimeter we carry out the addition.

```
      3. 2 5
      2. 5
      3. 2 5
    + 2. 5
    ─────────
    1 1. 5 0
```

Then $P = 11.5$.

Check. We can obtain partial checks by estimating.

$(3.25) \times (2.5) \approx 3 \times 3 \approx 9 \approx 8.125$

$3.25 + 2.5 + 3.25 + 2.5 \approx 3 + 3 + 3 + 3 = 12 \approx 11.5$

The answers check.

State. The area of the stamp is 8.125 sq cm, and the perimeter is 11.5 cm.

11. *Familiarize*. We visualize the situation. We let $m =$ the odometer reading at the end of the trip.

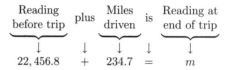

Translate. We are combining amounts.

Reading before trip	plus	Miles driven	is	Reading at end of trip
↓	↓	↓	↓	↓
22,456.8	+	234.7	=	m

Solve. To solve the equation we carry out the addition.

```
        1  1
      2 2,4 5 6.8
    +     2 3 4.7
    ───────────────
      2 2,6 9 1.5
```

Thus, $m = 22,691.5$.

Check. We can check by repeating the addition. We can also check by rounding:

$22,456.8 + 234.7 \approx 22,460 + 230 = 22,690 \approx 22,691.5$

State. The odometer reading at the end of the trip was 22,691.5.

13. *Familiarize*. Let $c =$ the amount of change.

Translate. We subtract the price of the DVD and the amount of the sales tax from $50.

$$c = 50 - 29.24 - 1.61$$

Solve. We carry out the subtractions in order from left to right.

```
      4 9 9 10              1 10
      5̶ 0̶. 0̶ 0̶             2̶ 0̶. 7 6
    − 2 9. 2 4           −    1. 6 1
    ───────────         ─────────────
      2 0. 7 6             1 9. 1 5
```

Check. If we add the cost of the DVD and the amount of the sales tax to the change we should get $50.

$19.15 + 29.24 + 1.61 = 48.39 + 1.61 = 50$

The answer checks.

State. Andrew received $19.15 in change.

15. *Familiarize*. We visualize the situation. We let $d =$ the number of degrees Wanda's temperature dropped.

103.2°F	
99.7°F	d

Translate. This is a "take-away" situation.

Original temperature	minus	New temperature	is	Drop in temperature
↓	↓	↓	↓	↓
103.2	−	99.7	=	d

Solve. We carry out the subtraction.

$$\begin{array}{r} {\scriptstyle 12} \\ {\scriptstyle 9\ \ \not{2}\ 12} \\ \not{1}\,\not{0}\,\not{3}.\,\not{2} \\ -\ \ 9\,9.7 \\ \hline 3.5 \end{array}$$

Thus, $d = 3.5$.

Check. We check by adding 3.5 to 99.7 to get 103.2. The answer checks.

State. Wanda's temperature dropped 3.5°F.

17. Familiarize. Let $c =$ the cost per serving.

Translate.

Price per serving	is	Price per pound	divided by	Number of servings
↓	↓	↓	↓	↓
c	=	16.95	÷	3

Solve. We carry out the division.

$$\begin{array}{r} 5.6\,5 \\ 3\overline{)1\,6.9\,5} \\ 1\,5\,0\,0 \\ \hline 1\,9\,5 \\ 1\,8\,0 \\ \hline 1\,5 \\ 1\,5 \\ \hline 0 \end{array}$$

Thus, $c = 5.65$.

Check. We can check by multiplying: $3 \times 5.65 = 16.95$. The answer checks.

State. The cost per serving is $5.65.

19. Familiarize. We are combining amounts. Let $d =$ the total number of gallons of liquids the average U.S. citizen drinks each year.

Translate.

Soft drinks	+ Water	+ Milk	+ Coffee	+ Fruit juice	= Total
↓	↓	↓	↓	↓	↓
49.0	+ 41.2	+ 25.3	+ 24.8	+ 7.8	= d

Solve. We carry out the addition.

$$\begin{array}{r} {\scriptstyle 2\ \ 2} \\ 49.0 \\ 41.2 \\ 25.3 \\ 24.8 \\ +\ 7.8 \\ \hline 148.1 \end{array}$$

Thus, $d = 148.1$.

Check. We repeat the calculation. The answer checks.

State. The average U.S. citizen drinks 148.1 gallons of liquids each year.

21. Familiarize. This is a two-step problem. First, we find the number of miles that have been driven between fillups. This is a "how-much-more" situation. We let $n =$ the number of miles driven.

Translate and Solve.

First odometer reading	plus	Number of miles driven	is	Second odometer reading
↓	↓	↓	↓	↓
26,342.8	+	n	=	26,736.7

To solve the equation we subtract 26,342.8 on both sides.

$n = 26,736.7 - 26,342.8$
$n = 393.9$

$$\begin{array}{r} 2\,6,7\,3\,6.7 \\ -\ 2\,6,3\,4\,2.8 \\ \hline 3\,9\,3.9 \end{array}$$

Second, we divide the total number of miles driven by the number of gallons. This gives us $m =$ the number of miles per gallon.

$$393.9 \div 19.5 = m$$

To find the number m, we divide.

$$\begin{array}{r} 2\,0.2 \\ 19.5_\wedge\overline{)3\,9\,3.9_\wedge 0} \\ 3\,9\,0\,0 \\ \hline 3\,9\,0 \\ 3\,9\,0 \\ \hline 0 \end{array}$$

Thus, $m = 20.2$.

Check. To check, we first multiply the number of miles per gallon times the number of gallons:

$$19.5 \times 20.2 = 393.9$$

Then we add 393.9 to 26,342.8:

$$26,342.8 + 393.9 = 26,736.7$$

The number 20.2 checks.

State. The van gets 20.2 miles per gallon.

23. Familiarize. Let $a =$ the amount of insulin consumed in a week, in cubic centimeters. Note that 38 units of insulin corresponds to 0.38 cc. (See Example 2.)

Translate.

Amount used each day	times	Number of days in a week	is	Total amount used
↓	↓	↓	↓	↓
0.38	×	7	=	a

Solve. We carry out the multiplication.

$$\begin{array}{r} 0.3\,8 \\ \times\ \ 7 \\ \hline 2.6\,6 \end{array}$$

Thus, $a = 2.66$.

Check. We can approximate the product.

$$0.38 \times 7 \approx 0.4 \times 7 = 2.8 \approx 2.66$$

The answer checks.

State. Phil consumes 2.66 cc of insulin in a week.

25. Familiarize. This is a multi-step problem. We find the total area of the poster and the area devoted to the painting and then we subtract to find the area not devoted to the painting.

Translate and Solve. First we use the formula $A = l \times w$ to find the total area of the poster.

$$A = l \times w$$
$$= 27.4 \text{ in.} \times 19.3 \text{ in.}$$
$$= 528.82 \text{ in}^2$$

$$
\begin{array}{r}
2\,7.\,4 \\
\times\;1\,9.\,3 \\
\hline
8\,2\,2 \\
2\,4\,6\,6\,0 \\
2\,7\,4\,0\,0 \\
\hline
5\,2\,8.\,8\,2
\end{array}
$$

Next we use the formula $A = l \times w$ again to find the area of the painting.

$$A = l \times w$$
$$= 18.8 \text{ in.} \times 15.7 \text{ in.}$$
$$= 295.16 \text{ in}^2$$

$$
\begin{array}{r}
1\,8.\,8 \\
\times\;1\,5.\,7 \\
\hline
1\,3\,1\,6 \\
9\,4\,0\,0 \\
1\,8\,8\,0\,0 \\
\hline
2\,9\,5.\,1\,6
\end{array}
$$

Let a = the area not devoted to the painting. We subtract to find this area.

$$a = 528.82 \text{ in}^2 - 295.16 \text{ in}^2$$
$$= 233.66 \text{ in}^2$$

$$
\begin{array}{r}
{}^{4\;\;12}\;\;{}^{7\;\;12}\\
\cancel{5}\,\cancel{2}\,8.\,\cancel{8}\,\cancel{2} \\
-\;2\,9\,5.\,1\,6 \\
\hline
2\,3\,3.\,6\,6
\end{array}
$$

Check. We repeat the calculations. The answer checks.

State. The area not devoted to the painting is 233.66 in².

27. Familiarize. Let n = the area available for notes. The total area available is twice the area of one side of the card. Recall that the formula for the area of a rectangle with length l and width w is $A = l \times w$.

Translate.

$$
\underbrace{\text{Total area}}_{n} \;\;\underset{=}{\text{is}}\;\; \underset{2\times}{\text{twice}}\;\; \underbrace{\text{Area of one side}}_{12.7 \times 7.6}
$$

Solve. We find the product.

$$n = 2 \times 12.7 \times 7.6$$
$$= 25.4 \times 7.6$$
$$= 193.04$$

$$
\begin{array}{r}
2\,5.\,4 \\
\times\;\;\;7.\,6 \\
\hline
1\,5\,2\,4 \\
1\,7\,7\,8\,0 \\
\hline
1\,9\,3.\,0\,4
\end{array}
$$

Check. We can approximate the product.

$$2 \times 12.7 \times 7.6 \approx 2 \times 13 \times 7.5 = 195 \approx 193.04$$

The answer checks.

State. The area available for notes is 193.04 cm².

29. Familiarize. This is a two-step problem. First, we find the number of games that can be played in one hour. Think of an array containing 60 minutes (1 hour = 60 minutes) with 1.5 minutes in each row. We want to find how many rows there are. We let g represent this number.

Translate and Solve. We think (Number of minutes) ÷ (Number of minutes per game) = (Number of games).

$$60 \div 1.5 = g$$

To solve the equation we carry out the division.

$$
\begin{array}{r}
4\,0. \\
1.5_{\wedge}\overline{\smash{)}\,6\,0.\,0_{\wedge}} \\
\underline{6\,0\,0} \\
0 \\
\underline{0} \\
0
\end{array}
$$

Thus, $g = 40$.

Second, we find the cost t of playing 40 video games. Repeated addition fits this situation. (We express 75¢ as \$0.75.)

$$
\underbrace{\begin{array}{c}\text{Cost of}\\\text{one game}\end{array}}_{0.75} \;\;\underset{\times}{\text{times}}\;\; \underbrace{\begin{array}{c}\text{Number of}\\\text{games played}\end{array}}_{40} \;\;\underset{=}{\text{is}}\;\; \underbrace{\begin{array}{c}\text{Total}\\\text{cost}\end{array}}_{t}
$$

To solve the equation we carry out the multiplication.

$$
\begin{array}{r}
0.\,7\,5 \\
\times\;\;\;\;4\,0 \\
\hline
3\,0.\,0\,0
\end{array}
$$

Thus, $t = 30$.

Check. To check, we first divide the total cost by the cost per game to find the number of games played:

$$30 \div 0.75 = 40$$

Then we multiply 40 by 1.5 to find the total time:

$$1.5 \times 40 = 60$$

The number 30 checks.

State. It costs \$30 to play video games for one hour.

31. Familiarize. This is a multistep problem. First we find the sum s of the two 0.8 cm segments. Then we use this length to find d.

Translate and Solve.

$$
\underbrace{\begin{array}{c}\text{Length of one}\\\text{small segment}\end{array}}_{0.8} \;\;\underset{+}{\text{plus}}\;\; \underbrace{\begin{array}{c}\text{Length of other}\\\text{small segment}\end{array}}_{0.8} \;\;\underset{=}{\text{is}}\;\; \underbrace{\begin{array}{c}\text{Total}\\\text{length}\end{array}}_{s}
$$

To solve we carry out the addition.

$$
\begin{array}{r}
{}^{1}\\
0.\,8 \\
+\;0.\,8 \\
\hline
1.\,6
\end{array}
$$

Thus, $s = 1.6$.

Now we find d.

$$
\underbrace{\begin{array}{c}\text{Total length of}\\\text{smaller segments}\end{array}}_{1.6} \;\;\underset{+}{\text{plus}}\;\; \underbrace{\begin{array}{c}\text{length}\\\text{of } d\end{array}}_{d} \;\;\underset{=}{\text{is 3.91 cm}}\;\; \underset{3.91}{}
$$

To solve we subtract 1.6 on both sides of the equation.

$$d = 3.91 - 1.6$$
$$d = 2.31$$

$$
\begin{array}{r}
3.\,9\,1 \\
-\;1.\,6\,0 \\
\hline
2.\,3\,1
\end{array}
$$

Check. We repeat the calculations.

State. The length d is 2.31 cm.

33. *Familiarize*. We make and label a drawing. The question deals with a rectangle and a circle, so we also list the relevant area formulas. We let d = the amount of decking needed.

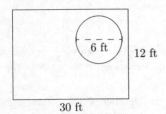

Area of a rectangle with length l and width w:
$A = l \times w$

Area of a circle with radius r: $A = \pi r^2$, where $\pi \approx 3.14$

Translate. We subtract the area of the circle from the area of the rectangle. Recall that a circle's radius is half of its diameter.

Area of rectangle	minus	Area of circle	is	Area covered by decking
↓	↓	↓ ↓	↓	↓
30×12	$-$	$3.14\left(\dfrac{6}{2}\right)^2$	$=$	d

Solve. We carry out the computations.

$$30 \times 12 - 3.14\left(\frac{6}{2}\right)^2 = d$$
$$30 \times 12 - 3.14(3)^2 = d$$
$$30 \times 12 - 3.14 \times 9 + d$$
$$360 - 28.26 = d$$
$$331.74 = d$$

Check. We can repeat the calculations. Also note that 331.74 is less than the area of the yard but more than the area of the flower garden. This agrees with the impression given by our drawing.

State. The amount of decking needed is 331.74 ft^2.

35. *Familiarize*. This is a multistep problem. First we find the number of minutes in excess of 450. Then we find the charge for the excess minutes. Finally we add this charge to the monthly charge for 450 minutes to find the total cost for the month. Let m = the number of minutes in excess of 450.

Translate and Solve. First we have a "how much more" situation.

First 450 minutes	plus	Excess minutes	is	Total minutes
↓	↓	↓	↓	↓
450	$+$	m	$=$	479

We subtract 450 on both sides of the equation.

$$450 + m = 479$$
$$450 + m - 450 = 479 - 450$$
$$m = 29$$

We see that 29 minutes are charged at the rate of $0.45 per minute. We multiply to find c, the cost of these minutes.

$$\begin{array}{r} 2\,9 \\ \times\,0.4\,5 \\ \hline 1\,4\,5 \\ 1\,1\,6\,0 \\ \hline 1\,3.0\,5 \end{array}$$

Thus, $c = 13.05$.

Finally we add the cost of the first 450 minutes and the cost of the additional 29 minutes to find t, the total cost for the month.

$$\begin{array}{r} 3\,9.9\,9 \\ +\,1\,3.0\,5 \\ \hline 5\,3.0\,4 \end{array}$$

Thus, $t = 53.04$.

Check. We can repeat the calculations. The answer checks.

State. The total cost for the month was $53.04.

37. *Familiarize*. Let m = the number of megabytes Nikki paid for. We will express 9 cents as $0.09.

Translate.

Processing fee	plus	$0.09	times	Number of megabytes	is	Total bill
↓	↓	↓	↓	↓	↓	↓
10	$+$	0.09	$\cdot$	m	$=$	88.75

Solve.

$$10 + 0.09 \cdot m = 88.75$$
$$10 + 0.09 \cdot m - 10 = 88.75 - 10$$
$$0.09 \cdot m = 78.75$$
$$\frac{0.09 \cdot m}{0.09} = \frac{78.75}{0.09}$$
$$m = 875$$

Check. $0.09 \cdot 875 = \$78.75$ and $\$78.75 + \$10 = \$88.75$, so the answer checks.

State. Nikki paid for 875 megabytes.

39. *Familiarize*. This is a multistep problem. First we find the amount charged for minutes in excess of 900. Then we find the number of minutes in excess of 900 and finally we find the total number of minutes used. Let c = the amount charged for the minutes in excess of 900.

Translate and Solve.

Access fee	plus	Additional fees	plus	Charge for excess minutes	is	Total bill
↓	↓	↓	↓	↓	↓	↓
59.99	$+$	5.79	$+$	c	$=$	89.78

We solve this equation.

$$59.99 + 5.79 + c = 89.78$$
$$65.78 + c = 89.78 \quad \text{Adding on the left side}$$
$$65.78 + c - 65.78 = 89.78 - 65.78$$
$$c = 24$$

We see that $24 was charged for the number of minutes used in excess of 900 minutes. Let m = the number of minutes in excess of 900. We will express 40 cents as $0.40.

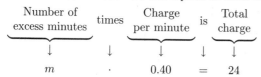

$$m \cdot 0.40 = 24$$

We solve this equation.

$$m \cdot 0.40 = 24$$
$$\frac{m \cdot 0.40}{0.40} = \frac{24}{0.40}$$
$$m = 60$$

Finally let t = the total number of minutes Jeff used. We add.

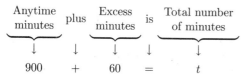

$$900 + 60 = t$$

We solve the equation.

$$900 + 60 = t$$
$$960 = t$$

Check. We can repeat the calculations. The answer checks.

State. Jeff used 960 minutes.

41. Familiarize. Let m = the number of minutes of video capture the sorority can buy. Then $m - 30$ = the number of minutes in excess of 30 minutes. We will express 50 cents as $0.50.

Translate.

Charge for first 30 minutes	plus	Number of minutes in excess of 30	times	Charge per minute	is	Total charge
↓	↓	↓	↓	↓	↓	↓
50	+	$(m-30)$	·	0.50	=	95

Solve.

$$50 + (m-30) \cdot 0.50 = 95$$
$$50 + 0.50m - 15 = 95$$
$$35 + 0.50m = 95$$
$$35 + 0.50m - 35 = 95 - 35$$
$$0.50m = 60$$
$$\frac{0.50m}{0.50} = \frac{60}{0.50}$$
$$m = 120$$

Check. If 120 min are captured, then the number of minutes in excess of 30 is $120 - 30$, or 90 minutes.

The charge for the excess minutes is $90 \cdot \$0.50$, or $45, and $45 + \$50 = \$95$. The answer checks.

State. The sorority can buy 120 min of video capture.

43. Familiarize. Let c = the number of credit card transactions processed. Then $c - 500$ = the number of transactions in excess of 500. We will express 10 cents as $0.10.

Translate.

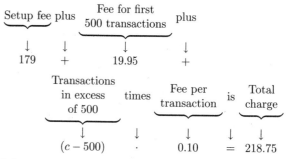

Transactions in excess of 500	times	Fee per transaction	is	Total charge
↓	↓	↓	↓	↓
$(c-500)$	·	0.10	=	218.75

Solve.

$$179 + 19.95 + (c - 500) \cdot 0.10 = 218.75$$
$$179 + 19.95 + 0.10c - 50 = 218.75$$
$$148.95 + 0.10c = 218.75$$
$$148.95 + 0.10c - 193.95 = 218.75 - 148.95$$
$$0.10c = 69.8$$
$$\frac{0.10c}{0.10} = \frac{69.8}{0.10}$$
$$c = 698$$

Check. The number of transactions in excess of 500 is $698 - 500$, or 198. At $0.10 each, the fee for processing 198 transactions is $\$0.10(198)$, or $19.80. Then the total charge is

$$\$179 + \$19.95 + \$19.80, \text{ or } \$218.75.$$

The answer checks.

State. VeriSign processed 698 credit card transactions.

45. Familiarize. This is a three-step problem. We will find the area S of a standard soccer field and the area F of a standard football field using the formula Area $= l \cdot w$. Then we will find E, the amount by which the area of a soccer field exceeds the area of a football field.

Translate and Solve.

$$S = l \cdot w = 114.9 \times 74.4 = 8548.56$$
$$F = l \cdot w = 120 \times 53.3 = 6396$$

Area of football field	plus	Excess area of soccer field	is	Area of soccer field
↓	↓	↓	↓	↓
6396	+	E	=	8548.56

To solve the equation we subtract 6396 on both sides.

$$E = 8548.56 - 6396$$
$$E = 2152.56$$

$$\begin{array}{r} {}^{4\ 14} \\ 8\ \not{5}\ \not{4}\ 8.5\ 6 \\ -\ 6\ 3\ 9\ 6.0\ 0 \\ \hline 2\ 1\ 5\ 2.5\ 6 \end{array}$$

Check. We can obtain a partial check by rounding and estimating:

$$114.9 \times 74.4 \approx 110 \times 75 = 8250 \approx 8548.56$$
$$120 \times 53.3 \approx 120 \times 50 = 6000 \approx 6396$$

$$8250 - 6000 = 2250 \approx 2152.56$$

State. The area of a soccer field is 2152.56 yd^2 greater than the area of a football field.

47. *Familiarize*. This is a multistep problem. First, we find the cost of the cheese. We let c = the cost of the cheese.

Translate and Solve.

$$\underbrace{\text{Number of pounds}}_{6} \;\; \underbrace{\text{times}}_{\cdot} \;\; \underbrace{\text{Price per pound}}_{\$4.79} \;\; \underbrace{\text{is}}_{=} \;\; \underbrace{\text{Cost of cheese}}_{c}$$

To solve the equation we carry out the multiplication.

$$\begin{array}{r} \$4.\,7\,9 \\ \times \qquad 6 \\ \hline \$2\,8.\,7\,4 \end{array}$$

Thus, $c = \$28.74$.

Next, we subtract to find how much money m is left to purchase seltzer.

$$\begin{array}{ll} m = \$40 - \$28.74 & \quad \overset{3\ 9\ 9\ 10}{\cancel{4\,0.\,0\,0}} \\ m = \$11.26 & \quad \underline{-\ 2\,8.\,7\,4} \\ & \quad 1\,1.\,2\,6 \end{array}$$

Finally, we divide the amount of money left over by the cost of a bottle of seltzer to find how many bottles can be purchased. We let b = the number of bottles of seltzer that can be purchased.

$$\$11.26 \div \$0.64 = b$$

To find b we carry out the division.

$$\begin{array}{r} 1\,7. \\ 0.6\,4_\wedge \overline{\big)\,1\,1.2\,6_\wedge} \\ \underline{6\,4\,0} \\ 4\,8\,6 \\ \underline{4\,4\,8} \\ 3\,8 \end{array}$$

We stop dividing at this point, because Frank cannot purchase a fraction of a bottle. Thus, $b = 17$ (rounded to the nearest 1).

Check. The cost of the seltzer is $17 \cdot \$0.64$ or $\$10.88$. The cost of the cheese is $6 \cdot \$4.79$, or $\$28.74$. Frank has spent a total of $\$10.88 + \$28.74$, or $\$39.62$. Frank has $\$40 - \$39.62$, or $\$0.38$ left over. This is not enough to purchase another bottle of seltzer, so our answer checks.

State. Frank should buy 17 bottles of seltzer.

49. Discussion and Writing Exercise

51. $\dfrac{0}{n} = 0$, for any integer n that is not 0.

Thus, $\dfrac{0}{-13} = 0$.

53. $\dfrac{8}{11} - \dfrac{4}{3} = \dfrac{8}{11} \cdot \dfrac{3}{3} - \dfrac{4}{3} \cdot \dfrac{11}{11}$ \qquad The LCM is 33.

$$= \dfrac{24}{33} - \dfrac{44}{33}$$

$$= \dfrac{-20}{33}, \text{ or } -\dfrac{20}{33}$$

55.

$$\begin{array}{rclcl} 4\dfrac{1}{3} & = & 4\dfrac{1}{3} \cdot \dfrac{2}{2} & = & 4\dfrac{2}{6} \\[2mm] +\,2\dfrac{1}{2} & = & +\,2\dfrac{1}{2} \cdot \dfrac{3}{3} & = & +\,2\dfrac{3}{6} \\ \hline & & & & 6\dfrac{5}{6} \end{array}$$

57. Discussion and Writing Exercise

59. Discussion and Writing Exercise

61. *Familiarize*. We will use the formula Distance = Speed × Time in the form Time = Distance ÷ Speed. Let t = the difference in the travel times for the two routes, in hours.

Translate.

$$\underbrace{\text{Time for faster route}}_{\frac{7.6}{65}} \;\; \underbrace{\text{minus}}_{-} \;\; \underbrace{\text{Time for slower route}}_{\frac{5.6}{50}} \;\; \underbrace{\text{is}}_{=} \;\; \underbrace{\text{Difference in times}}_{t}$$

Solve. We carry out the subtraction. First we multiply each fraction by 1 in the form $\dfrac{10}{10}$ to eliminate the decimals in the numerators.

$$\dfrac{7.6}{65} - \dfrac{5.6}{50} = t$$

$$\dfrac{7.6}{65} \cdot \dfrac{10}{10} - \dfrac{5.6}{50} \cdot \dfrac{10}{10} = t$$

$$\dfrac{76}{650} - \dfrac{56}{500} = t$$

$$\dfrac{76}{650} \cdot \dfrac{10}{10} - \dfrac{56}{500} \cdot \dfrac{13}{13} = t \qquad \text{The LCD is 6500.}$$

$$\dfrac{760}{6500} - \dfrac{728}{6500} = t$$

$$\dfrac{32}{6500} = t$$

$$\dfrac{\cancel{4} \cdot 8}{\cancel{4} \cdot 1625} = t$$

$$\dfrac{8}{1625} = t$$

The difference in times is $\dfrac{8}{1625}$ hr. We can convert this time to minutes as follows.

$$\dfrac{8}{1625} \text{ hr} = \dfrac{8}{1625} \text{ hr} \cdot \dfrac{60 \text{ min}}{1 \text{ hr}}$$

$$= \dfrac{8 \cdot 60}{1625} \cdot \dfrac{\text{hr}}{\text{hr}} \cdot \text{min}$$

$$= \dfrac{8 \cdot \cancel{5} \cdot 12}{\cancel{5} \cdot 325} \text{ min} \qquad \left(\dfrac{\text{hr}}{\text{hr}} = 1\right)$$

$$= \dfrac{96}{325} \text{ min}$$

We can convert this to seconds.

$$\dfrac{96}{325} \text{ min} = \dfrac{96}{365} \text{ min} \cdot \dfrac{60 \text{ sec}}{1 \text{ min}}$$

$$= \dfrac{96 \cdot 60}{325} \cdot \dfrac{\text{min}}{\text{min}} \cdot \text{sec}$$

$$= \dfrac{5760}{325} \text{ sec}$$

$$\approx 18 \text{ sec}$$

Check. We can repeat the calculations. The answer checks.

State. You can save $\dfrac{96}{325}$ min, or about 18 sec, by taking the faster route.

63. We must make some assumptions. First we assume that the figures are nested squares formed by connecting the midpoints of consecutive sides of the next larger square. Next assume that the shaded area is the same as the area of the innermost square. (It appears that if we folded the shaded area into the innermost square, it would exactly fill the square.) Finally assume that the length of a side of the innermost square is 5 cm. (If we project the vertices of the innermost square onto the corresponding sides of the largest square, it appears that the distance between each projection and the nearest vertex of the largest square is one-fourth the length of a side of the largest square. Thus, the distance between projections on each side of the largest square is $\dfrac{1}{2} \cdot 10$ cm, or 5 cm and, hence, the length of a side of the innermost square is 5 cm.) Then the area of the innermost square is 5 cm $\cdot$ 5 cm, or 25 cm^2, so the shaded area is 25 cm^2.

65. *Familiarize*. This is a multistep problem. First we will subtract the amounts of the contracts in successive years to find the yearly increases. Let $x =$ the increase from 2003 to 2004, $y =$ the increase from 2004 to 2005, and $z =$ the increase from 2005 to 2006, in millions.

Translate and Solve. We have three "take away" situations.
$$x = 12.1 - 10.3 = 1.8$$
$$y = 17.2 - 12.1 = 5.1$$
$$z = 22.6 - 17.2 = 5.4$$

Now we add the increases and divide by the number of addends, 3, to find the average increase. Let $a =$ the average yearly increase.
$$a = \frac{1.84 + 5.1 + 5.4}{3} = \frac{12.3}{3} = 4.1$$

Check. We check the calculations. The answer checks.

State. The average yearly increase in a contract was $4.1 million.

Chapter 5 Review Exercises

1. 6.59 million $= 6.59 \times 1,\underbrace{000,000}_{6 \text{ zeros}}$

6.590000.

Move 6 places to the right.

6.59 million $= 6,590,000$

2. 6.9 million $= 6.9 \times 1,\underbrace{000,000}_{6 \text{ zeros}}$

6.900000.

Move 6 places to the right.

6.9 million $= 6,900,000$

3. A word name for 3.47 is three and forty-seven hundredths.

4. A word name for 0.031 is thirty-one thousandths.

5. 0.09   0.09. $\dfrac{9}{100}$

2 places   Move 2 places.   2 zeros

$0.09 = \dfrac{9}{100}$

6. -4.561 $-4.561.$ $-\dfrac{4561}{1000}$

3 places   Move 3 places.   3 zeros

$-4.561 = -\dfrac{4561}{1000}$, and $-4.561 = -4\dfrac{561}{1000}$.

7. -0.089 $-0.089.$ $-\dfrac{89}{1000}$

3 places   Move 3 places.   3 zeros

$-0.089 = -\dfrac{89}{1000}$

8. 3.0227   3.0227. $\dfrac{30,227}{10,000}$

4 places   Move 4 places.   4 zeros

$3.0227 = \dfrac{30,227}{10,000}$, and $3.0227 = 3\dfrac{227}{10,000}$

9. $-\dfrac{34}{1000}$ $-0.034.$

3 zeros   Move 3 places.

$-\dfrac{34}{1000} = -0.034$

10. $\dfrac{42,603}{10,000}$ 4.2603.

4 zeros   Move 4 places.

$\dfrac{42,603}{10,000} = 4.2603$

11. $27\dfrac{91}{100} = 27 + \dfrac{91}{100} = 27$ and $\dfrac{91}{100} = 27.91$

12. $867\dfrac{6}{1000} = 867 + \dfrac{6}{1000} = 867$ and $\dfrac{6}{1000} = 867.006$

$867\dfrac{6}{1000} = 867.006$, so $-867\dfrac{6}{1000} = -867.006$.

13. 0.034

Starting at the left, these digits are the first to differ; 3 is larger than 1.

0.0185

Thus, 0.034 is larger.

14. -0.91

Starting at the left, these digits are the first to differ; 1 is smaller than 9.

-0.19

Thus, -0.19 is larger.

15.

17.4⎢2⎢87   Hundredths digit is 4 or lower.
                  Round down.
17.4

16.

17.428⎢7⎢   Ten-thousandths digit is 5 or higher.
                  Round up.
17.429

17.
```
      1
   2 3 6.2 3 1
   2 6 3.4
 +    0.1 9 8
 ───────────
   4 9 9.8 2 9
```

18.
```
         13
   2 17 5 ⌀ 15
   ⌀ 7.⌀ 4 ⌀
 −   8.4 9 7
 ───────────
   2 9.1 4 8
```

19.
```
     1 1
   2 1 9.3
       2.8
 +     7.0
 ─────────
   2 2 9.1
```

20.
```
        13 14   10
     6  ⌀  ⌀ 9 ⌀ 10
   7 4 5.0 1 ⌀ 9
 −   5 9.9 5 9 0
 ───────────────
   6 8 5.0 5 1 9
```

21. $-37.8 + (-19.5)$

Add the absolute values: $37.8 + 19.5 = 57.3$

Make the answer negative: $-37.8 + (-19.5) = -57.3$

22. $-7.52 - (-9.89) = -7.52 + 9.89 = 2.37$

23.
```
       4 8
   ×  0.2 7
   ───────
     3 3 6
     9 6 0
   ───────
   1 2.9 6
```

24. $-3.7(0.29) = -1.073$

25.
```
   2 4.6 8
 × 1 0 0 0
```
The number 1000 has 3 zeros so we move the decimal point in 24.68 three places to the right. The product is 24,680.

26.
```
        3.2
   2 5 ⟌ 8 0.0
        7 5
        ───
          5 0
          5 0
          ───
            0
```

27. First we consider $11.52 \div 7.2$.
```
          1.6
   7.2∧⟌ 1 1 5.∧2
         7 2 0
         ─────
           4 3 2
           4 3 2
           ─────
               0
```

Since we have a positive number divided by a negative number, the answer is -1.6.

28. $\dfrac{276.3}{1000}$

The number 1000 has 3 zeros, so we move the decimal point in the numerator 3 places to the left.

$$\frac{276.3}{1000} = 0.2763$$

29.
$$3.7x - 5.2y - 1.5x - 3.9y$$
$$= 3.7x + (-5.2y) + (-1.5x) + (-3.9y)$$
$$= 3.7x + (-1.5x) + (-5.2y) + (-3.9y)$$
$$= [3.7 + (-1.5)]x + [-5.2 + (-3.9)]y$$
$$= 2.2x - 9.1y$$

30.
$$7.94 - 3.89a + 4.63 + 1.05a$$
$$= 7.94 + (-3.89a) + 4.63 + 1.05a$$
$$= -3.89a + 1.05a + 7.94 + 4.63$$
$$= (-3.89 + 1.05)a + (7.94 + 4.63)$$
$$= -2.84a + 12.57$$

31.
$$P - Prt = 1000 - 1000(0.05)(1.5)$$
$$= 1000 - 50(1.5)$$
$$= 1000 - 75$$
$$= 925$$

32.
$$9 - 3.2(-1.5) + 5.2^2$$
$$= 9 - 3.2(-1.5) + 5.2(5.2)$$
$$= 9 - 3.2(-1.5) + 27.04$$
$$= 9 + 4.8 + 27.04$$
$$= 13.8 + 27.04$$
$$= 40.84$$

33. This is about $7.3 + 4.0$, so the sum is about 11.3.

34. This is about $50.0 \div 2.5$, so about 20 videotapes can be purchased.

35. Move 2 places to the left.

$15.49.¢

Change from ¢ sign at end to $ sign in front.

$1549¢ = \$15.49$

36. Round

2 4 8. 2 7 ⎢2⎢ 7 ... to the nearest hundredth.
                      Thousandths digit is 4 or less.
2 4 8. 2 7   Round down.

37. $\dfrac{13}{5} = \dfrac{13}{5} \cdot \dfrac{2}{2} = \dfrac{26}{10} = 2.6$

38. $\dfrac{32}{25} = \dfrac{32}{25} \cdot \dfrac{4}{4} = \dfrac{128}{100} = 1.28$

39.
$$4 \overline{\smash{\big)}\ 13.00} \quad \begin{array}{r} 3.25 \end{array}$$

$$\begin{array}{r} 3.\,2\,5 \\ 4\overline{\smash{\big)}\ 1\,3.\,0\,0} \\ \underline{1\ 2} \\ 1\ 0 \\ \underline{8} \\ 2\ 0 \\ \underline{2\ 0} \\ 0 \end{array}$$

$$\frac{13}{4} = 3.25$$

40. Since $-\dfrac{7}{6}$ is negative, we divide 7 by 6 and make the result negative.

$$\begin{array}{r} 1.\,1\,6\,6 \\ 6\overline{\smash{\big)}\ 7.\,0\,0\,0} \\ \underline{6} \\ 1\ 0 \\ \underline{6} \\ 4\ 0 \\ \underline{3\ 6} \\ 4\ 0 \\ \underline{3\ 6} \\ 4 \end{array}$$

Since 4 keeps reappearing as a remainder, the digits repeat and
$$\frac{7}{6} = 1.166\ldots, \text{ or } 1.1\overline{6}.$$
Thus, $-\dfrac{7}{6} = -1.1\overline{6}.$

41. $\dfrac{4}{15} \times 79.05 = \dfrac{4}{15} \times \dfrac{79.05}{1} = \dfrac{4 \times 79.05}{15 \times 1} = \dfrac{316.2}{15} = 21.08$

42.
$$t - 4.3 = -7.5$$
$$t - 4.3 + 4.3 = -7.5 + 4.3$$
$$t = -3.2$$
The solution is -3.2.

43.
$$4.1x + 5.6 = -6.7$$
$$4.1x + 5.6 - 5.6 = -6.7 - 5.6$$
$$4.1x = -12.3$$
$$\frac{4.1x}{4.1} = \frac{-12.3}{4.1}$$
$$x = -3$$
The solution is -3.

44.
$$6x - 11 = 8x + 4$$
$$6x - 11 + 11 = 8x + 4 + 11$$
$$6x = 8x + 15$$
$$6x - 8x = 8x + 15 - 8x$$
$$-2x = 15$$
$$\frac{-2x}{-2} = \frac{15}{-2}$$
$$x = -7.5$$
The solution is -7.5.

45.
$$3(x + 2) = 5x - 7$$
$$3x + 6 = 5x - 7$$
$$3x + 6 - 6 = 5x - 7 - 6$$
$$3x = 5x - 13$$
$$3x - 5x = 5x - 13 - 5x$$
$$-2x = -13$$
$$\frac{-2x}{-2} = \frac{-13}{-2}$$
$$x = 6.5$$
The solution is 6.5.

46. *Familiarize.* Let t = the number by which the number of telephone poles for every 100 people in the U.S. exceeds the number in Canada.

Translate. We have a "how much more" situation.

Number of poles in Canada	plus	How many more poles	is	Number of poles in U.S.
↓	↓	↓	↓	↓
40.65	+	t	=	51.81

Solve.
$$40.65 + t = 51.81$$
$$40.65 + t - 40.65 = 51.81 - 40.65$$
$$t = 11.16$$

Check. Since $40.65 + 11.16 = 51.81$, the answer checks.

State. There are 11.16 more telephone poles for every 100 people in the U.S. than in Canada.

47. *Familiarize.* Let h = Stacia's hourly wage.

Translate.

Hourly wage	times	Number of hours worked	is	Total earnings
↓	↓	↓	↓	↓
h	·	40	=	620.74

Solve.
$$h \cdot 40 = 620.74$$
$$\frac{h \cdot 40}{40} = \frac{620.74}{40}$$
$$h \approx 15.52$$

Check. $40 \cdot \$15.52 = \$620.80 \approx \$620.74$, so the answer checks. (Remember, we rounded the solution of the equation.)

State. Stacia earns $15.52 per hour.

48. *Familiarize.* We let a = the area of grass in the yard. Recall that the area of a rectangle with length l and width w is $A = l \times w$ and the area of a circle with radius r is $A = \pi r^2$, where $\pi \approx 3.14$.

Translate. We subtract the area of the base of the fountain from the area of the yard. Recall that a circle's radius is half of its diameter, or width.

$$\underbrace{\text{Area of}\atop\text{yard}}\quad\text{minus}\quad\underbrace{\text{Area of}\atop\text{fountain}}\quad\text{is}\quad\underbrace{\text{Area to}\atop\text{be seeded}}$$

$$\downarrow\qquad\downarrow\qquad\downarrow\qquad\downarrow\qquad\downarrow$$

$$20\times15\quad-\quad3.14\left(\frac{8}{2}\right)^2\quad=\quad a$$

Solve. We carry out the computations.

$$20\times15-3.14\left(\frac{8}{2}\right)^2=a$$
$$20\times15-3.14(4)^2=a$$
$$20\times15-3.14(16)=a$$
$$300-50.24=a$$
$$249.76=a$$

Check. We recheck the calculations. Our answer checks.

State. The area of grass in the yard is 249.76 ft^2.

49. Familiarize. Let a = the amount left in the account after the purchase was made.

Translate. We write a subtraction sentence.

$$a=6274.35-485.79$$

Solve. We carry out the subtraction.

$$\begin{array}{r}
{\scriptstyle 11\ 16\,13\ 12}\\
{\scriptstyle 5\ \ 1\ \ 6\ \ 3\ \ 2\ \ 15}\\
6\,2\,7\,4.\,3\,5\\
-\ \ 4\,8\,5.\,7\,9\\
\hline
5\,7\,8\,8.\,5\,6
\end{array}$$

Thus, $a = 5788.56$.

Check. $\$5788.56 + \$485.79 = \$6274.35$, so the answer checks.

State. There is $5788.56 left in the account.

50. Familiarize. This is a multistep problem. First we find the number of minutes in excess of 900. Then we find the charge for the excess minutes. Finally we add this charge to the monthly charge for 900 minutes to find the total cost for the month. Let m = the number of minutes in excess of 900.

Translate and Solve. First we have a "how much more" situation.

$$\underbrace{\text{First 900}\atop\text{minutes}}\quad\text{plus}\quad\underbrace{\text{Excess}\atop\text{minutes}}\quad\text{is}\quad\underbrace{\text{Total}\atop\text{minutes}}$$

$$\downarrow\qquad\downarrow\qquad\downarrow\qquad\downarrow\qquad\downarrow$$

$$900\quad+\quad m\quad=\quad946$$

We subtract 900 on both sides of the equation.

$$900+m=946$$
$$900+m-900=946-900$$
$$m=46$$

We see that 46 minutes are charged at the rate of $0.40 per minute. We multiply to find c, the cost of these minutes.

$$\begin{array}{r}
4\,6\\
\times\ 0.\,4\,0\\
\hline
1\,8.\,4\,0
\end{array}$$

Thus, $c = 18.40$.

Finally we add the cost of the first 900 minutes and the cost of the additional 46 minutes to find t, the total cost for the month.

$$\begin{array}{r}
5\,9.\,9\,9\\
+\ 1\,8.\,4\,0\\
\hline
7\,8.\,3\,9
\end{array}$$

Thus, $t = 78.39$.

Check. We can repeat the calculations. The answer checks.

State. The total cost for the month was $78.39.

51. Familiarize. Let m = the number of megabytes Cody paid for. We will express 2 cents as $0.02.

Translate.

$$\underbrace{\text{Processing}\atop\text{fee}}\ \text{plus}\ \$0.02\ \text{times}\ \underbrace{\text{Number of}\atop\text{megabytes}}\ \text{is}\ \underbrace{\text{Total}\atop\text{bill}}$$

$$\downarrow\quad\downarrow\quad\downarrow\quad\downarrow\quad\downarrow\quad\downarrow\quad\downarrow$$

$$10\ +\ 0.02\ \cdot\quad m\quad=\quad46.60$$

Solve.
$$10+0.02\cdot m=46.60$$
$$10+0.02\cdot m-10=46.60-10$$
$$0.02\cdot m=36.60$$
$$\frac{0.02\cdot m}{0.02}=\frac{36.60}{0.02}$$
$$m=1830$$

Check. $\$0.02\cdot1830=\$36.60$ and $\$36.60+\$10=\$46.60$, so the answer checks.

State. Cody paid for 1830 megabytes of storage.

52. Familiarize. This is a two-step problem. First, we find the number of miles that have been driven between fillups. This is a "how-much-more" situation. We let n = the number of miles driven.

Translate and Solve.

$$\underbrace{\text{First}\atop\text{odometer}\atop\text{reading}}\ \text{plus}\ \underbrace{\text{Number}\atop\text{of miles}\atop\text{driven}}\ \text{is}\ \underbrace{\text{Second}\atop\text{odometer}\atop\text{reading}}$$

$$\downarrow\qquad\downarrow\qquad\downarrow\qquad\downarrow\qquad\downarrow$$

$$36,057.1\ +\quad n\quad=\quad36,217.6$$

To solve the equation we subtract 36,057.1 on both sides.

$$n=36,217.6-36,057.1$$
$$n=160.5$$

$$\begin{array}{r}
3\,6,2\,1\,7.6\\
-\ 3\,6,0\,5\,7.1\\
\hline
1\,6\,0.5
\end{array}$$

Second, we divide the total number of miles driven by the number of gallons. This gives us m = the number of miles per gallon.

$$160.5\div11.1=m$$

To find the number m, we divide.

$$\begin{array}{r}
1\,4.\,4\,5\\
1\,1.1_\wedge\overline{\big)\,1\,6\,0.\,5_\wedge0\,0}\\
\underline{1\,1\,1\,0}\\
4\,9\,5\\
\underline{4\,4\,4}\\
5\,1\,0\\
\underline{4\,4\,4}\\
6\,6\,0\\
\underline{5\,5\,5}\\
1\,0\,5
\end{array}$$

Thus, $m \approx 14.5$.

Check. To check, we first multiply the number of miles per gallon times the number of gallons:

$$11.1 \times 14.5 = 160.95$$

Then we add 160.95 to 36,057.1:

$$36,057.1 + 160.95 = 36,218.05 \approx 36,217.6$$

The number 14.5 checks.

State. Inge gets 14.5 miles per gallon.

53. a) **Familiarize.** Let $s =$ the total consumption of seafood per person, in pounds, for the seven given years.

Translate. We add the seven amounts shown in the graph in the text.

$$s = 12.4 + 15.0 + 14.9 + 14.8 + 15.2 + 14.7 + 15.6$$

Solve. We carry out the addition.

```
    3 3
  1 2. 4
  1 5. 0
  1 4. 9
  1 4. 8
  1 5. 2
  1 4. 7
+ 1 5. 6
─────────
1 0 2. 6
```

Check. We repeat the calculation. The answer checks.

State. The total consumption of seafood per person for the seven given years was 102.6 lb.

b) We add the amounts and divide by the number of addends. From part (a) we know that the sum of the seven numbers is 102.6, so we have $102.6 \div 7$:

```
      1 4. 6 5
  7 │ 1 0 2. 6 0
      7 0 0
    ─────────
      3 2 6
      2 8 0
    ─────────
        4 6
        4 2
      ─────────
          4 0
          3 5
        ─────────
            5
```

Rounding to the nearest tenth, we find that the average seafood consumption per person was about 14.7 lb.

54. Familiarize. Let $d =$ the number of miles that an out-of-towner can travel for $15.23. We will express 95¢ as $0.95.

Translate.

Initial charge	plus	$0.95	times	Distance traveled	is	Fare
↓	↓	↓	↓	↓	↓	↓
7.25	+	0.95	·	d	=	15.23

Solve.

$$7.25 + 0.95 \cdot d = 15.23$$
$$7.25 + 0.95 \cdot d - 7.25 = 15.23 - 7.25$$
$$0.95 \cdot d = 7.98$$
$$\frac{0.95 \cdot d}{0.95} = \frac{7.98}{0.95}$$
$$d = 8.4$$

Check. $0.95 \cdot 8.4 = \$7.98$ and $\$7.98 + \$7.25 = \$15.23$, so the answer checks.

State. An out-of-towner can travel 8.4 mi for $15.23.

55. Familiarize. Let $c =$ the cost per serving.

Translate.

Cost	divided by	Number of servings	is	Cost per serving
↓	↓	↓	↓	↓
5.99	÷	4.5	=	c

Solve. We carry out the division.

```
          1. 3 3 1
  4.5∧│ 5.9∧9 0 0
        4 5 0
      ─────────
        1 4 9
        1 3 5
      ─────────
          1 4 0
          1 3 5
        ─────────
            5 0
            4 5
          ─────────
              5
```

Rounding to the nearest cent, we have $c \approx 1.33$.

Check. We find the cost of 4.5 servings at $1.33 per serving.

$$4.5 \cdot \$1.33 = \$5.985 \approx \$5.99$$

The answer checks

State. The ham costs about $1.33 per serving.

56. Familiarize. We will find the perimeter P of the room to determine how much crown molding is needed, and we will find the area A of the floor to determine how many square feet of bamboo tiles are needed. We will use the formulas $P = 2l + 2w$ and $A = l \cdot w$.

Translate.

$$P = 2 \cdot 14.5 \text{ ft} + 2 \cdot 16.25 \text{ ft}$$
$$A = 16.25 \text{ ft} \cdot 14.5 \text{ ft}$$

Solve. We carry out the calculations.

$$P = 2 \cdot 14.5 \text{ ft} + 2 \cdot 16.25 \text{ ft}$$
$$= 29 \text{ ft} + 32.5 \text{ ft}$$
$$= 61.5 \text{ ft}$$
$$A = 16.25 \text{ ft} \cdot 14.5 \text{ ft} = 235.625 \text{ ft}^2$$

Check. We can repeat the calculations. The answer checks.

State. 61.5 ft of crown molding and 235.625 ft^2 of bamboo tiles are needed.

57. *Discussion and Writing Exercise.* Since there are 20 nickels to a dollar, $\frac{3}{20}$ corresponds to 3 nickels, or 15¢, which is 0.15 dollars.

58. *Discussion and Writing Exercise.* In decimal notation, $\frac{1}{3}$ and $\frac{1}{6}$ both must be rounded before they can be multiplied. The best way to express $\frac{1}{3} \cdot \frac{1}{6}$ as a decimal is to multiply the fractions and then convert the product $\frac{1}{18}$ to decimal notation.

59. a) By trial we find the following true sentence.

$2.56 - 6.4 + 51.2 - 17.4 + 89.7 = 119.66$

b) By trial we find the following true sentence.

$(11.12 - 0.29)3^4 = 877.23$

60. First we find decimal notation for each fraction. Then we compare these numbers in decimal notation.

$-\frac{2}{3} = -0.\overline{6}, \quad -\frac{15}{19} \approx -0.789474, \quad -\frac{11}{13} = -0.\overline{846153},$

$\frac{-5}{7} = -0.\overline{714285}, \quad \frac{-13}{15} = -0.8\overline{6}, \quad \frac{-17}{20} = -0.85$

Arranging these numbers from smallest to largest and writing them in fraction notation, we have

$\frac{-13}{15}, \frac{-17}{20}, -\frac{11}{13}, \frac{15}{19}, \frac{-5}{7}, -\frac{2}{3}$

61. *Familiarize.* Let $m =$ the number of miles Quentin drove the car in 2006. Then $m - 10,000 =$ the number of miles in excess of 10,000. At \$396 per month, the leasing cost for 1 year, or 12 months, is $12 \cdot \$396$. We will express 20 cents as \$0.20.

Translate.

Leasing cost	plus	\$0.20	times	Miles over 10,000	is	Total bill
↓	↓	↓	↓	↓	↓	↓
$12 \cdot 396$	$+$	0.20	$\cdot$	$(m - 10,000)$	$=$	5952

Solve.

$12 \cdot 396 + 0.20 \cdot (m - 10,000) = 5952$

$4752 + 0.20 \cdot (m - 10,000) = 5952$

$4752 + 0.20m - 2000 = 5952$

$2752 + 0.20m = 5952$

$2752 + 0.20m - 2752 = 5952 - 2752$

$0.20m = 3200$

$\dfrac{0.20m}{0.20} = \dfrac{3200}{0.20}$

$m = 16,000$

Check. If Quentin drives 16,000 mi, then he drives $16,000 - 10,000$, or 6000 mi, in excess of 10,000. The charge for the excess miles is $\$0.20 \cdot 6000$, or \$1200. The leasing fee is $12 \cdot \$396$, or \$4752, and $\$4752 + \$1200 = \$5952$, so the answer checks.

State. Quentin drove the car 16,000 mi in 2006.

62. *Discussion and Writing Exercise.* The Sicilian pizza, at $\frac{4.4¢}{\text{in}^2}$, is a better buy than the round pizza which costs $\frac{5.5¢}{\text{in}^2}$.

Chapter 5 Test

---

1.   8.9 billion

$= 8.9 \times 1$ billion

$= 8.9 \times 1,000,000,000$ 9 zeros

$= 8,900,000,000$ Moving the decimal point 9 places to the right

2.   3.756 million

$= 3.756 \times 1$ million

$= 3.756 \times 1,000,000$ 6 zeros

$= 3,756,000$ Moving the decimal point 6 places to the right

3. 2.34

a) Write a word name for the whole number. | Two |

b) Write "and" for the decimal point.   Two | and |

c) Write a word name for the number to the right of the decimal point, followed by the place value of the last digit.   Two   and | thirty-four hundredths |

A word name for 2.34 is two and thirty-four hundredths.

4. 105.0005

a) Write a word name for the whole number. | One hundred five |

b) Write "and" for the decimal point.   One hundred five | and |

c) Write a word name for the number to the right of the decimal point, followed by the place value of the last digit.   One hundred five   and | five ten-thousandths |

A word name for 105.0005 is one hundred five and five ten-thousandths.

5. -0.3 $-0.3.$ $-\dfrac{3}{10}$

         ↳↑

     1 place   1 zero

$-0.3 = -\dfrac{3}{10}$

6. 2.<u>769</u>   2.769. $\dfrac{2769}{1000}$

3 places   Move 3 places.   3 zeros

$2.769 = \dfrac{2769}{1000}$

7. $\dfrac{74}{1000}$ 0.074.

3 zeros   Move 3 places.

$\dfrac{74}{1000} = 0.074$

8. $-\dfrac{37,047}{10,000}$ $-3.7047.$

4 zeros   Move 4 places.

$-\dfrac{37,047}{10,000} = -3.7047$

9. $756\dfrac{9}{100} = 756 + \dfrac{9}{100} = 756$ and $\dfrac{9}{100} = 756.09$

10. $91\dfrac{703}{1000} = 91 + \dfrac{703}{1000} = 91$ and $\dfrac{703}{1000} = 91.703$

11. To compare two positive numbers in decimal notation, start at the left and compare corresponding digits moving from left to right. When two digits differ, the number with the larger digit is the larger of the two numbers.

0.07

Different; 1 is larger than 0.

0.162

Thus, 0.162 is larger.

12. To compare two negative numbers in decimal notation, start at the left and compare corresponding digits moving from left to right. When two digits differ, the number with the smaller digit is the larger of the two numbers.

−0.173

Different; 1 is smaller than 2.

−0.25

Thus, −0.173 is larger.

13.

9.4 5 23   Hundredths digit is 5 or higher.
            Round up.
9.5

14.

9.452 3   Ten-thousandths digit is 4 or lower.
           Round down.
9.452

15.
```
      1
  4 0 2. 3
      2. 8 1
+     0. 1 0 9
  4 0 5. 2 1 9
```

16.
```
      0. 1 2 5    (3 decimal places)
×     0. 2 4      (2 decimal places)
      5 0 0
    2 5 0 0
  0.0 3 0 0 0     (5 decimal places)
```

17. $0.\underline{001} \times 213.45$ $0.213.45$

3 decimal places   Move 3 places to the left.

$0.001 \times 213.45 = 0.21345$

18.
```
        11
     4 1 10 8 11
     5 2. 0 9 1
  −    7. 3 4 5
     4 4. 7 4 6
```

19.
```
       1  1
   3 4 2. 9
       8. 1
 +     5. 3 7
   3 5 6. 3 7
```

20. $-9.5 + 7.3$

The absolute values are 9.5 and 7.3. The difference is 2.2. The negative number has the larger absolute value, so the answer is negative.

$-9.5 + 7.3 = -2.2$

21. We write extra zeros.
```
   1  9 9 9 10
   2. 0 0 0 0
 − 0. 0 0 5 4
   1. 9 9 4 6
```

22. $1\underline{000} \times 73.962$ $73.962.$

3 zeros   Move 3 places to the right.

$1000 \times 73.962 = 73,962$

23.
```
        4. 7 5
  4 ) 1 9. 0 0
      1 6
        3 0
        2 8
          2 0
          2 0
             0
```

24.
```
           3 0. 4
  3.3 ) 1 0 0.3 2
        9 9 0 0
          1 3 2
          1 3 2
              0
```

25. $\dfrac{-346.82}{1000}$ $-0.346.82$

3 zeros   Move 3 places to the left.

$\dfrac{-346.82}{1000} = -0.34682$

26. $\dfrac{346.82}{0.01}$ 346.82.

2 decimal places   Move 2 places to the right.

$\dfrac{346.82}{0.01} = 34{,}682$

27. Move 2 places to the right.

$179.82.¢

Change from $ sign in front to ¢ sign at end.

$179.82 = 17{,}982¢

28. $4.1x + 5.2 - 3.9y + 5.7x - 9.8$
$= 4.1x + 5.2 + (-3.9y) + 5.7x + (-9.8)$
$= 4.1x + 5.7x + (-3.9y) + 5.2 + (-9.8)$
$= (4.1 + 5.7)x + (-3.9y) + (5.2 + (-9.8))$
$= 9.8x - 3.9y - 4.6$

29. $2l + 4w + 2h = 2 \cdot 2.4 + 4 \cdot 1.3 + 2 \cdot 0.8$
$= 4.8 + 5.2 + 1.6$
$= 10.0 + 1.6$
$= 11.6$

30. $20 \div 5(-2)^2 - 8.4 = 20 \div 5 \cdot 4 - 8.4$
$= 4 \cdot 4 - 8.4$
$= 16 - 8.4$
$= 7.6$

31. *Familiarize*. Let g = the number of gallons of gasoline that can be bought with $20.

***Translate*.**

Price per gallon	times	Number of gallons	is	Total cost
↓	↓	↓	↓	↓
2.749	·	g	=	20

***Solve*.**

$2.749 \cdot g = 20$

$\dfrac{2.749 \cdot g}{2.749} = \dfrac{20}{2.749}$

$g \approx 7$ Rounding to the nearest gallon

***Check*.** $2.749 \cdot 7 \approx $19.24 \approx 20, so the answer checks. Remember that we rounded to the nearest gallon.

***State*.** About 7 gal of gasoline can be bought with $20.

32.   48.7474...

          Hundredths digit is 4 or lower.

48.7   Round down.

33. $\dfrac{8}{5} = \dfrac{8}{5} \cdot \dfrac{2}{2} = \dfrac{16}{10} = 1.6$

34. $\dfrac{21}{4} = \dfrac{21}{4} \cdot \dfrac{25}{25} = \dfrac{525}{100} = 5.25$

35. First consider $\dfrac{7}{16}$.

$$
\begin{array}{r}
0.4375 \\
16\overline{\smash{)}7.0000} \\
\underline{64} \\
60 \\
\underline{48} \\
120 \\
\underline{112} \\
80 \\
\underline{80} \\
0
\end{array}
$$

Since $\dfrac{7}{16} = 0.4375$, we have $-\dfrac{7}{16} = -0.4375$.

36.

$$
\begin{array}{r}
1.55 \\
9\overline{\smash{)}14.00} \\
\underline{9} \\
50 \\
\underline{45} \\
50 \\
\underline{45} \\
5
\end{array}
$$

Since 5 keeps reappearing as a remainder, the digit 5 repeats and

$$\dfrac{14}{9} = 1.55\ldots = 1.\overline{5}.$$

37.   1.5555...

          Thousandths digit is 5 or higher.

1.56   Round up.

38. $8.91 \times 22.457 \approx 9 \times 22 = 198$

39. $78.2209 \div 16.09 \approx 80 \div 20 = 4$

40. $\dfrac{3}{8} \times 45.6 - \dfrac{1}{5} \times 36.9$

$= \dfrac{3 \times 45.6}{8} - \dfrac{36.9}{5}$

$= \dfrac{136.8}{8} - \dfrac{36.9}{5}$

$= 17.1 - 7.38$

$= 9.72$

41. $17y - 3.12 = -58.2$
$17y - 3.12 + 3.12 = -58.2 + 3.12$
$17y = -55.08$
$\dfrac{17y}{17} = \dfrac{-55.08}{17}$
$y = -3.24$

The solution is -3.24.

42.
$$9t - 4 = 6t + 26$$
$$9t - 4 + 4 = 6t + 26 + 4$$
$$9t = 6t + 30$$
$$9t - 6t = 6t + 30 - 6t$$
$$3t = 30$$
$$\frac{3t}{3} = \frac{30}{3}$$
$$t = 10$$

The solution is 10.

43.
$$4 + 2(x - 3) = 7x - 9$$
$$4 + 2x - 6 = 7x - 9$$
$$2x - 2 = 7x - 9$$
$$2x - 2 + 2 = 7x - 9 + 2$$
$$2x = 7x - 7$$
$$2x - 7x = 7x - 7 - 7x$$
$$-5x = -7$$
$$\frac{-5x}{-5} = \frac{-7}{-5}$$
$$x = 1.4$$

The solution is 1.4.

44. We add the numbers and divide by the number of addends.
$$\frac{76.1 + 69.4 + 55.0 + 53.2 + 37.5}{5} = \frac{291.2}{5} = 58.24$$
The average number of passengers is 58.24 million.

45. **Familiarize.** This is a multistep problem. First we find the amount charged for minutes in excess of 2000. Then we find the number of minutes in excess of 2000 and finally we find the total number of minutes used. Let c = the amount charged for the minutes in excess of 2000.

Translate and Solve.

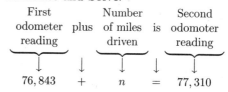

We solve this equation.
$$99.99 + c = 314.99$$
$$99.99 + c - 99.99 = 314.99 - 99.99$$
$$c = 215$$

We see that \$215 was charged for the number of minutes used in excess of 2000. Let m = the number of minutes in excess of 2000.

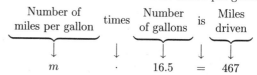

We solve this equation.
$$m \cdot 0.25 = 215$$
$$\frac{m \cdot 0.25}{0.25} = \frac{215}{0.25}$$
$$m = 860$$

Finally let t = the total number of minutes Trey used. We add.

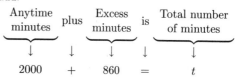

We solve the equation.
$$2000 + 860 = t$$
$$2860 = t$$

Check. We can repeat the calculations. The answer checks.

State. Trey used 2860 minutes.

46. **Familiarize.** This is a two-step problem. First we will find the number of miles that are driven between fillups. Then we find the gas mileage. Let n = the number of miles driven between fillups.

Translate and Solve.

First odometer reading	plus	Number of miles driven	is	Second odometer reading
↓	↓	↓	↓	↓
76,843	+	n	=	77,310

To solve the equation, we subtract 76,843 on both sides.
$$n = 77,310 - 76,843 = 467$$

Now let m = the number of miles driven per gallon.

Number of miles per gallon	times	Number of gallons	is	Miles driven
↓	↓	↓	↓	↓
m	·	16.5	=	467

We divide by 16.5 on both sides to find m.
$$m = 467 \div 16.5$$
$$m = 28.\overline{30}$$
$$m \approx 28.3 \quad \text{Rounding to the nearest tenth}$$

Check. First we multiply the number of miles per gallon by the number of gallons to find the number of miles driven:
$$16.5 \cdot 28.3 = 466.95 \approx 467$$

Then we add 467 mi to the first odometer reading:
$$76,843 + 467 = 77,310$$

This is the second odometer reading, so the answer checks.

State. The gas mileage is about 28.3 miles per gallon.

47. **Familiarize.** Let b = the balance after the purchases are made.

Translate. We subtract the amounts of the three purchases from the original balance:
$$b = 10,200 - 123.89 - 56.68 - 3446.98$$

Solve. We carry out the calculations.

$$b = 10,200 - 123.89 - 56.68 - 3446.98$$
$$= 10,076.11 - 56.68 - 3446.98$$
$$= 10,019.43 - 3446.98$$
$$= 6572.45$$

Check. We can find the total amount of the purchases and then subtract to find the new balance.

$$\$123.89 + \$56.68 + \$3446.98 = \$3627.55$$
$$\$10,200 - \$3627.55 = \$6572.45$$

The answer checks.

State. After the purchases were made, the balance was $6572.45.

48. Familiarize. Let c = the total cost of the copy paper.

Translate.

Cost per case	times	Number of cases	is	Total cost
↓	↓	↓	↓	↓
25.99	·	7	=	c

Solve. We carry out the multiplication.

$$\begin{array}{r} 2\,5.9\,9 \quad \text{(2 decimal places)} \\ \times \qquad 7 \\ \hline 1\,8\,1.9\,3 \quad \text{(2 decimal places)} \end{array}$$

Thus, $c = 181.93$.

Check. We can obtain a partial check by rounding and estimating:

$$25.99 \times 7 \approx 25 \times 7 = 175 \approx 181.93$$

State. The total cost of the copy paper is $181.93.

49. a) The product of two numbers greater than 0 and less than 1 is <u>always</u> less than 1.

b) The product of two numbers greater than 1 is <u>never</u> less than 1.

c) The product of a number greater than 1 and a number less than 1 is <u>sometimes</u> equal to 1.

d) The product of a number greater than 1 and a number less than 1 is <u>sometimes</u> equal to 0.

50. Familiarize. This is a multistep problem. First we will subtract the amounts of bottled water consumed in successive years to find the yearly increases. Let x = the increase from 2002 to 2003, y = the increase from 2003 to 2004, and z = the increase from 2004 to 2005, in gallons per person.

Translate and Solve. We have three "take away" situations.

$$x = 22.1 - 20.7 = 1.4$$
$$y = 23.8 - 22.1 = 1.7$$
$$z = 25.0 - 23.8 = 1.2$$

Now we add the increases and divide by the number of addends, 3, to find the average increase. Let a = the average yearly increase.

$$a = \frac{1.4 + 1.7 + 1.2}{3} = \frac{4.3}{3} = 1.4\overline{3}$$

Check. We check the calculations. The answer checks.

State. The average yearly increase in bottled water consumption was $1.4\overline{3}$ gal per person.

51. The cost to drive roundtrip is 2 · $0.32 · 320, or $204.80.

a) Since $189 < $204.80, it is more economical for an individual to fly.

b) The airfare for a couple is 2 · $189, or $378. Since $378 > $204.80, it is more economical for a couple to drive.

c) Since we found in part (b) that it is more economical for a couple to drive, it seems reasonable that it is also more economical for a family of 3 to drive. We can verify this by first finding the airfare for 3 people: 3 · $189 = $567. Since $567 > $204.80, we confirm that it is more economical for a family of 3 to drive.

Cumulative Review Chapters 1 - 5

1. A word name for 207,491 is two hundred seven thousand, four hundred ninety-one

2. 6.25 billion $= 6.25 \times 1,\underbrace{000,000,000}_{9 \text{ zeros}}$

6.250000000.
└────────↑

Move 9 places to the right.

6.25 billion $= 6,250,000,000$

3. 10.<u>09</u>   10.09. $\dfrac{1009}{100}$
                   └─↑
2 places   Move 2 places.   2 zeros

$$10.09 = \frac{1009}{100}$$

4. $4\dfrac{3}{8} = \dfrac{35}{8}$ $(4 \cdot 8 = 32, 32 + 3 = 35)$

5. $\dfrac{-35}{1000}$ $-0.035.$
                      ↑──┘
3 zeros   Move 3 places.

$$\frac{-35}{1000} = -0.035$$

6. Since 66 is even we know 2 is a factor. The sum of the digits is 12, so 3 is also a factor. Since 66 is even and divisible by 3, we know that 6 is also a factor. We make a list of factorizations to find all the factors.

$$66 = 1 \cdot 66$$
$$66 = 2 \cdot 33$$
$$66 = 3 \cdot 22$$
$$66 = 6 \cdot 11$$

The factors of 66 are 1, 2, 3, 6, 11, 22, 33, and 66.

7.
$$\begin{array}{r} 1\,1 \quad \leftarrow \text{11 is prime.} \\ 7\,\overline{)\,7\,7} \\ 2\,\overline{)\,1\,5\,4} \end{array}$$

$$154 = 2 \cdot 7 \cdot 11$$

8. $28 = 2 \cdot 2 \cdot 7 = 2^2 \cdot 7$
$35 = 5 \cdot 7$

We form the LCM using the greatest power of each factor. The LCM is $2^2 \cdot 5 \cdot 7$.

9. Round 6962.4721 to the nearest hundred.

6 9 $\boxed{6}$ 2.4 7 2 1

The digit 9 is in the hundreds place. Consider the next digit to the right. Since the digit, 6, is 5 or higher, round 9 hundreds up to 10 hundreds. Then change all digits to the right of the hundreds digit to zeros.

The answer is 7000.

10.

6962.47$\boxed{2}$1   Thousandths digit is 4 or lower.
                  Round down.
6962.47

11.

$3\,\boxed{\dfrac{2}{3} \cdot \dfrac{3}{3}} = \ 3\,\dfrac{6}{9}$

$+2\,\dfrac{5}{9} \qquad\qquad +2\,\dfrac{5}{9}$

$5\,\dfrac{11}{9} = 5 + \dfrac{11}{9}$

$= 5 + 1\dfrac{2}{9}$

$= 6\dfrac{2}{9}$

12.

$\overset{3}{}$
1 1 0. 8 6 3
0. 7 3
1 2 1. 9
+   1. 9 0 4
─────────
2 3 5. 3 9 7

13.

$\overset{1}{}$
5 2 4 9
2 1 5
+   3 1
─────────
5 4 9 5

14. $-\dfrac{4}{15} + \dfrac{7}{30} = -\dfrac{4}{15} \cdot \dfrac{2}{2} + \dfrac{7}{30} = -\dfrac{8}{30} + \dfrac{7}{30} = -\dfrac{1}{30}$

15. $-23 - 48 = -23 + (-48) = -71$

16.

$\overset{10}{}$
$\overset{8\ 9\ \cancel{0}\ 9\ 9\ 10}{\cancel{9\,0\,1\,0.\,0\,0}}$
$-\quad 5\ 6\ 3.\ 4\ 7$
─────────
8 4 4 6. 5 3

17. $\dfrac{8}{9} - \dfrac{7}{8} = \dfrac{8}{9} \cdot \dfrac{8}{8} - \dfrac{7}{8} \cdot \dfrac{9}{9} = \dfrac{64}{72} - \dfrac{63}{72} = \dfrac{1}{72}$

18.

$7\dfrac{1}{5} = \ 6\dfrac{6}{5}$

$-3\dfrac{4}{5} = -3\dfrac{4}{5}$

─────────

$3\dfrac{2}{5}$

19.

2 3. 9
$\times$ 0. 2
─────────
4. 7 8

20. $-\dfrac{3}{5} \times \dfrac{10}{21} = -\dfrac{3 \times 10}{5 \times 21}$

$= -\dfrac{3 \times 2 \times 5}{5 \times 3 \times 7}$

$= -\dfrac{2}{7} \times \dfrac{3 \times 5}{3 \times 5}$

$= -\dfrac{2}{7}$

21. $3\dfrac{2}{11} \cdot 4\dfrac{2}{7} = \dfrac{35}{11} \cdot \dfrac{30}{7}$

$= \dfrac{35 \cdot 30}{11 \cdot 7}$

$= \dfrac{5 \cdot 7 \cdot 30}{11 \cdot 7}$

$= \dfrac{7}{7} \cdot \dfrac{5 \cdot 30}{11}$

$= \dfrac{150}{11} = 13\dfrac{7}{11}$

22. $5 \cdot \dfrac{3}{10} = \dfrac{5 \cdot 3}{10} = \dfrac{5 \cdot 3}{2 \cdot 5} = \dfrac{5}{5} \cdot \dfrac{3}{2} = \dfrac{3}{2}$

23. $2\dfrac{4}{5} \div 1\dfrac{13}{15} = \dfrac{14}{5} \div \dfrac{28}{15}$

$= \dfrac{14}{5} \cdot \dfrac{15}{28}$

$= \dfrac{14 \cdot 15}{5 \cdot 28}$

$= \dfrac{14 \cdot 3 \cdot 5}{5 \cdot 2 \cdot 14}$

$= \dfrac{3}{2} \cdot \dfrac{14 \cdot 5}{14 \cdot 5}$

$= \dfrac{3}{2} = 1\dfrac{1}{2}$

24. $\dfrac{6}{5} \div \dfrac{7}{8} = \dfrac{6}{5} \cdot \dfrac{8}{7} = \dfrac{6 \cdot 8}{5 \cdot 7} = \dfrac{48}{35}$, or $1\dfrac{13}{35}$

25. $-43.795 \div 0.001 = \dfrac{-43.795}{0.001}$

$\dfrac{-43,795}{0.\underline{001}} \qquad\qquad -43.795.$

3 decimal places   Move 3 places to the right.

$-43.795 \div 0.001 = -43,795$

26.

$\phantom{2.1\sqrt{}}2\ 0.6$
$2.1_\wedge \overline{)4\,3\ .2_\wedge 6}$
$\underline{4\,2\ 0\,0}$
$1\ 2\ 6$
$\underline{1\ 2\ 6}$
0

27. $\dfrac{2}{3} = \dfrac{2}{3} \cdot \dfrac{7}{7} = \dfrac{14}{21}$

$\dfrac{5}{7} = \dfrac{5}{7} \cdot \dfrac{3}{3} = \dfrac{15}{21}$

Since $14 < 15$ it follows that $\dfrac{14}{21} < \dfrac{15}{21}$ and thus $\dfrac{2}{3} < \dfrac{5}{7}$.

28. -7 is to the left of -4 on the number line, so $-7 < -4$.

29. $a \div 3 \cdot b = 18 \div 3 \cdot 2 = 6 \cdot 2 = 12$

30. $4(x - y + 3) = 4 \cdot x - 4 \cdot y + 4 \cdot 3 = 4x - 4y + 12$

31. $\begin{aligned} -4p + 9 + 11p - 17 &= -4p + 11p + 9 - 17 \\ &= (-4 + 11)p + (9 - 17) \\ &= 7p - 8 \end{aligned}$

32. $\begin{aligned} x - 9 + 13x - 2 &= x + 13x - 9 - 2 \\ &= (1 + 13)x + (-9 - 2) \\ &= 14x - 11 \end{aligned}$

33. $\begin{aligned} 8.32 + x &= 9.1 \\ 8.32 + x - 8.32 &= 9.1 - 8.32 \\ x &= 0.78 \end{aligned}$

The solution is 0.78.

34. $\begin{aligned} -75 \cdot x &= 2100 \\ \frac{-75 \cdot x}{-75} &= \frac{2100}{-75} \\ x &= -28 \end{aligned}$

The solution is -28.

35. $\begin{aligned} y \cdot 9.47 &= 81.6314 \\ \frac{y \cdot 9.47}{9.47} &= \frac{81.6314}{9.47} \\ y &= 8.62 \end{aligned}$

The solution is 8.62.

36. $\begin{aligned} 1062 - y &= -368,313 \\ 1062 - y - 1062 &= -368,313 - 1062 \\ -y &= 369,375 \\ -1 \cdot (-y) &= -1 \cdot (-369,375) \\ y &= 369,375 \end{aligned}$

The solution is 369,375.

37. $\begin{aligned} t + \frac{5}{6} &= \frac{8}{9} \\ t + \frac{5}{6} - \frac{5}{6} &= \frac{8}{9} - \frac{5}{6} \\ t &= \frac{8}{9} \cdot \frac{2}{2} - \frac{5}{6} \cdot \frac{3}{3} \\ t &= \frac{16}{18} - \frac{15}{18} \\ t &= \frac{1}{18} \end{aligned}$

The solution is $\frac{1}{18}$.

38. $\begin{aligned} \frac{7}{8} \cdot t &= \frac{7}{16} \\ \frac{8}{7} \cdot \frac{7}{8} \cdot t &= \frac{8}{7} \cdot \frac{7}{16} \\ t &= \frac{8 \cdot 7}{7 \cdot 16} = \frac{8 \cdot 7 \cdot 1}{7 \cdot 2 \cdot 8} \\ t &= \frac{8 \cdot 7}{8 \cdot 7} \cdot \frac{1}{2} \\ t &= \frac{1}{2} \end{aligned}$

The solution is $\frac{1}{2}$.

39. $\begin{aligned} 2.4x - 7.1 &= 2.05 \\ 2.4x - 7.1 + 7.1 &= 2.05 + 7.1 \\ 2.4x &= 9.15 \\ \frac{2.4x}{2.4} &= \frac{9.15}{2.4} \\ x &= 3.8125 \end{aligned}$

The solution is 3.8125.

40. $\begin{aligned} 2(x - 3) &= 5x - 13 \\ 2x - 6 &= 5x - 13 \\ 2x - 6 - 5x &= 5x - 13 - 5x \\ -3x - 6 &= -13 \\ -3x - 6 + 6 &= -13 + 6 \\ -3x &= -7 \\ \frac{-3x}{-3} &= \frac{-7}{-3} \\ x &= \frac{7}{3} \end{aligned}$

The solution is $\frac{7}{3}$.

41. *Familiarize*. Let $m = $ the total number of minutes the four players spent in the penalty box. We are combining numbers, so we add.

Translate.

$$m = 3447 + 2755 + 2382 + 2361$$

Solve. We carry out the addition.

$$\begin{array}{r} {\scriptstyle 1\ 2\ 1} \\ 3\ 4\ 4\ 7 \\ 2\ 7\ 5\ 5 \\ 2\ 3\ 8\ 2 \\ +\quad 2\ 3\ 6\ 1 \\ \hline 1\ 0,9\ 4\ 5 \end{array}$$

We have $m = 10,945$.

Check. We repeat the addition. The answer checks.

State. The four players spent a total of 10,945 minutes in the penalty box.

42. *Familiarize*. Let $c = $ the cost of the motorcycle.

Translate.

$$\begin{array}{ccccc} \$450 & \text{is} & \frac{3}{10} & \text{of} & \underbrace{\text{the cost of the motorcycle}} \\ \downarrow & \downarrow & \downarrow & \downarrow & \downarrow \\ 450 & = & \frac{3}{10} & \cdot & c \end{array}$$

Solve.

$$450 = \frac{3}{10} \cdot c$$

$$\frac{10}{3} \cdot 450 = \frac{10}{3} \cdot \frac{3}{10} \cdot c$$

$$\frac{10 \cdot 450}{3} = c$$

$$\frac{10 \cdot \cancel{3} \cdot 150}{\cancel{3} \cdot 1} = c$$

$$1500 = c$$

Check. $\frac{3}{10}$ of \$1500 is $\frac{3}{10} \cdot \$1500 = \frac{3 \cdot \$1500}{10} = \frac{\$4500}{10} = $ \$450. The answer checks.

State. The motorcycle cost \$1500.

43. *Familiarize.* Let s = the number of seconds in a day.

Translate.

Seconds in a minute	times	Minutes in an hour	times	Hours in a day	is	Seconds in a day
↓	↓	↓	↓	↓	↓	↓
60	·	60	·	24	=	s

Solve. We carry out the multiplications.

$$60 \cdot 60 \cdot 24 = s$$

$$3600 \cdot 24 = s$$

$$86,400 = s$$

Check. We repeat the calculations. The answer checks.

State. There are 86,400 seconds in a day.

44. *Familiarize.* Let l = the amount of the loan.

Translate.

Amount of loan	is	$\frac{2}{3}$	of	\$4200
↓	↓	↓	↓	↓
l	=	$\frac{2}{3}$	·	4200

Solve. We carry out the multiplication.

$$l = \frac{2}{3} \cdot 4200 = \frac{2 \cdot 4200}{3}$$

$$l = \frac{2 \cdot \cancel{3} \cdot 1400}{\cancel{3} \cdot 1}$$

$$l = 2800$$

Check. $\frac{2800}{4200} = \frac{2 \cdot 1400}{3 \cdot 1400} = \frac{2}{3}$, so the answer checks.

State. The loan was for \$2800.

45. *Familiarize.* Let b = the balance in the account after the check is written.

Translate.

Original balance	minus	Amount of check	is	New balance
↓	↓	↓	↓	↓
314.79	−	56.02	=	b

Solve. We carry out the subtraction.

$$\begin{array}{r} \overset{\overset{10}{2\ \cancel{0}\ 14}}{\cancel{3}\ \cancel{1}\ \cancel{4}.\ 7\ 9} \\ -\quad 5\ 6.\ 0\ 2 \\ \hline 2\ 5\ 8.\ 7\ 7 \end{array}$$

We have $b = 258.77$.

Check. The amount of the check added to the new balance should equal the original balance. Since \$258.77+\$56.02 = \$314.79, the answer checks.

State. After the check is written the balance in the account is \$258.77.

46. *Familiarize.* Let p = the number of pounds of food sold. We are combining amounts, so we add.

Translate.

$$p = 1\frac{1}{2} + 2\frac{3}{4} + 2\frac{1}{4}$$

Solve. We carry out the addition.

$$\begin{array}{rcl} 1\boxed{\dfrac{1}{2} \cdot \dfrac{2}{2}} & = & 1\dfrac{2}{4} \\[2mm] 2\dfrac{3}{4} & = & 2\dfrac{3}{4} \\[2mm] +2\dfrac{1}{4} & = & +2\dfrac{1}{4} \\ \hline & & 5\dfrac{6}{4} = 5 + \dfrac{6}{4} \\[2mm] & & = 5 + 1\dfrac{2}{4} \\[2mm] & & = 6\dfrac{2}{4} \\[2mm] & & = 6\dfrac{1}{2} \end{array}$$

Check. We can repeat the calculation. The answer checks.

State. A total of $6\frac{1}{2}$ lb of food was sold.

47. *Familiarize.* Let A = the area of the sail. We will use the formula for the area of a triangle, $A = \frac{1}{2} \cdot b \cdot h$.

Translate. We substitute 11 ft for b and 16 ft for h.

$$A = \frac{1}{2} \cdot 11 \text{ ft} \cdot 16 \text{ ft}$$

Solve. We carry out the multiplication.

$$A = \frac{1}{2} \cdot 11 \text{ ft} \cdot 16 \text{ ft} = \frac{11 \cdot 16}{2} \cdot \text{ft} \cdot \text{ft}$$

$$A = 88 \text{ ft}^2$$

Check. We can repeat the calculation. The answer checks.

State. The area of the sail is 88 ft^2.

48. *Familiarize.* Let s = the number of square inches of steel left after the rectangle is punched out. This area is the area of a circle with diameter 9 in. less the area of a 4-in. by 5-in. rectangle. The formula for the area of a circle is $\pi \cdot r \cdot r$, where r is the radius and $\pi \approx 3.14$. The area of a rectangle is length × width. For a circle with $d = 9$ in., we have $r = \dfrac{9 \text{ in.}}{2}$, or 4.5 in.

Translate.

Area of circle	less	Area of rectangle	is	Area left over
↓	↓	↓	↓	↓
$3.14 \cdot 4.5 \cdot 4.5$	$-$	$4 \cdot 5$	$=$	s

Solve.

$$3.14 \cdot 4.5 \cdot 4.5 - 4 \cdot 5 = s$$
$$63.585 - 20 = s$$
$$43.585 = s$$

Check. We repeat the calculations. The answer checks.

State. 43.585 in² of steel will be left over.

49. Familiarize. Let $n =$ the number of 25.4-oz boxes in a 320-oz carton.

Translate.

Weight of a box	times	Number of boxes	is	320 oz
↓	↓	↓	↓	↓
25.4	·	n	$=$	320

Solve.

$$25.4 \cdot n = 320$$
$$\frac{25.4 \cdot n}{25.4} = \frac{320}{25.4}$$
$$n \approx 12.6$$

Since we are considering whole boxes, we see that the carton can contain at most 12 whole boxes.

Check. The weight of 12 boxes is $25.4 \cdot 12 = 304.8$ oz. Since $320 - 304.8 = 15.2$ oz, we see that the carton cannot hold another 25.4-oz box. Thus the answer checks.

State. The greatest number of boxes that can be inside the carton is 12 boxes.

50. Familiarize. Let $c =$ the cost per carton with the coupon.

Translate. We will divide the total purchase price by the number of cartons purchased.

$$c = \frac{3 \cdot 1.89}{4}$$

Solve. We carry out the calculation.

$$c = \frac{3 \cdot 1.89}{4} = \frac{5.67}{4} = 1.4175 \approx 1.42$$

Check. The total purchase price was $3 \cdot \$1.89$, or \$5.67. If 4 cartons had been purchased at \$1.42 each, the total price would have been $4 \cdot \$1.42$, or $\$5.68 \approx \$5.67$. The answer checks.

State. Rounded to the nearest cent, the cost per carton was about \$1.42.

51. First we convert the number of penalty minutes for each player to hours.

$$3447 \text{ min} \cdot \frac{1 \text{ hr}}{60 \text{ min}} = \frac{3447}{60} \cdot \frac{\text{min}}{\text{min}} \cdot \text{hr} = 57\frac{9}{20} \text{ hr}$$

$$2755 \text{ min} \cdot \frac{1 \text{ hr}}{60 \text{ min}} = \frac{2755}{60} \cdot \frac{\text{min}}{\text{min}} \cdot \text{hr} = 45\frac{11}{12} \text{ hr}$$

$$2382 \text{ min} \cdot \frac{1 \text{ hr}}{60 \text{ min}} = \frac{2382}{60} \cdot \frac{\text{min}}{\text{min}} \cdot \text{hr} = 39\frac{7}{10} \text{ hr}$$

$$2361 \text{ min} \cdot \frac{1 \text{ hr}}{60 \text{ min}} = \frac{2361}{60} \cdot \frac{\text{min}}{\text{min}} \cdot \text{hr} = 39\frac{7}{20} \text{ hr}$$

To find the average of these numbers we add them and then divide by the number of addends, 4.

$$\frac{57\dfrac{9}{20} + 45\dfrac{11}{12} + 39\dfrac{7}{10} + 39\dfrac{7}{20}}{4}$$

$$= \frac{57\dfrac{27}{60} + 45\dfrac{55}{60} + 39\dfrac{42}{60} + 39\dfrac{21}{60}}{4}$$

$$= \frac{180\dfrac{145}{60}}{4} = \frac{180 + 2\dfrac{25}{60}}{4}$$

$$= \frac{182\dfrac{25}{60}}{4} = \frac{10,945}{60} \div 4$$

$$= \frac{10,945}{60} \cdot \frac{1}{4} = \frac{10,945}{240} = 45\frac{145}{240}$$

$$= 45\frac{29}{48} \text{ hr}$$

To express the answer in terms of hours and minutes, we convert $\dfrac{29}{48}$ hr to minutes.

$$\frac{29}{48} \text{ hr} \cdot \frac{60 \text{ min}}{1 \text{ hr}} = \frac{29 \cdot 60}{48} \cdot \frac{\text{hr}}{\text{hr}} \cdot \text{min}$$

$$= \frac{1748}{48} \text{ min}$$

$$= 36.25 \text{ min}$$

Thus we can also express the average time as 45 hr, 36.25 min.

52. Familiarize. This is a multistep problem. First we let $c =$ the cost of membership for 6 months without the coupon.

Translate and Solve.

Membership fee	+	Monthly fee	·	Number of months	is	Total cost
↓	↓	↓	↓	↓	↓	↓
79	+	42.50	·	6	$=$	c

We solve the equation.

$$79 + 42.50 \cdot 6 = c$$
$$79 + 255 = c$$
$$334 = c$$

Now let $s =$ the amount Alayn saves if she uses the coupon. We have a "how much more" situation.

Cost with coupon	plus	How much more	is	Cost without coupon
↓	↓	↓	↓	↓
299	+	s	$=$	334

We solve the equation.

$$299 + s = 334$$
$$299 + s - 299 = 334 - 299$$
$$s = 35$$

Check. We repeat the calculations. The answer checks.

State. Alayn will save $35 if she uses the coupon.

Chapter 6

Introduction to Graphing and Statistics

Exercise Set 6.1

1. The smallest number in the Sodium column, 200, corresponds to Cinnamon Life, so Cinnamon Life cereal has the smallest amount of sodium per serving.

3. The smallest number in the Fat column, 0.7, corresponds to Kellogg's Complete, so Kellogg's Complete cereal has the smallest amount of fat per serving.

5. We add the numbers in the Fat column and divide by the number of entries, 5.
$$\frac{1.3 + 2.0 + 1.0 + 0.7 + 1.0}{5} = \frac{6.0}{5} = 1.2$$
The average amount of fat per serving in the cereals listed is 1.2 g.

7. Go down the Planet column to Jupiter. Then go across to the column headed Average Distance from Sun (in miles) and read the entry, 483,612,200. The average distance from the sun to Jupiter is 483,612,200 miles.

9. Go down the column headed Time of Revolution in Earth Time (in years) to 164.78. Then go across the Planet column. The entry there is Neptune, so Neptune has a time of revolution of 164.78 days.

11. The entries in the column headed Average Distance from Sun (in miles) that are greater than 500,000,000 correspond to the planets Saturn, Uranus, Neptune, and Pluto. These are the planets with an average distance from the sun that is greater than 500,000,000 mi.

13. Go down the column headed Actual Temperature (°F) to 80°. Then go across to the Relative Humidity column headed 60%. The entry is 92, so the apparent temperature is 92°F.

15. Go down the column headed Actual Temperature (°F) to 85°. Then go across the Relative Humidity column headed 90%. The entry is 108, so the apparent temperature is 108°F.

17. The number 100 appears in the columns headed Apparent Temperature (°F) 3 times, so there are 3 temperature-humidity combinations that given an apparent temperature of 100°.

19. Go down the Relative Humidity column headed 50% and find all the entries greater than 100. The last 4 entries are greater than 100. Then go across to the column headed Actual Temperature (°F) and read the temperatures that correspond to these entries. At 50% humidity, the actual temperatures 90° and higher give an apparent temperature above 100°.

21. Go down the column headed Actual Temperature (°F) to 95°. Then read across to locate the entries greater than 100. All of the entries except the first two are greater than 100. Go up from each entry to find the corresponding relative humidity. At an actual temperature of 95°, relative humidities of 30% and higher give an apparent temperature above 100°.

23. Go down the column headed Actual Temperature (°F) to 85°, then across to 94, and up to find that the corresponding relative humidity is 40%. Similarly, go down to 85°, across to 108, and up to 90%. At an actual temperature of 85°, the humidity would have to increase by
$$90\% - 40\%, \text{ or } 50\%$$
to raise the apparent temperature from 94° to 108°.

25. The number 1976 lies below the heading "1940," and the number 3849 lies below "1980." Thus, the cigarette consumption in 1940 was 1976 cigarettes per capita and in 1980 it was 3849 cigarettes per capita.

We subtract to find the increase from 1940 to 1980.
$$\begin{array}{r} \overset{17}{} \\ 2\,\,\overset{\not7}{}\,\,14 \\ \not3\,\not8\,\not4\,9 \\ -\,1\,9\,7\,6 \\ \hline 1\,8\,7\,3 \end{array}$$
Cigarette consumption increased 1873 cigarettes per capita from 1940 to 1980.

27. For 1920 to 1950 the average consumption is:
$$\frac{665 + 1485 + 1976 + 3552}{4} = \frac{7678}{4} \approx 1920$$
For 1970 to 2000 the average consumption is:
$$\frac{3985 + 3849 + 2817 + 2056}{4} = \frac{12,707}{4} \approx 3177$$
We subtract to find by how many cigarettes the second average exceeds the first:
$$\begin{array}{r} 2\,\,11 \\ \not3\,\not7\,7\,7 \\ -\,1\,9\,2\,0 \\ \hline 1\,2\,5\,7 \end{array}$$
The latter average exceeds the former by 1257 cigarettes per capita.

29. The world population in 1850 is represented by 1 symbol, so the population was 1 billion.

31. The 2070 (projected) population is represented by the most symbols, so the population will be largest in 2070.

33. The smallest increase in the number of symbols is represented by $\frac{1}{2}$ symbol from 1650 to 1850 (as opposed to 1 or more symbols for each of the other pairs). Then the growth was the least between these two years.

35. The world population in 1975 is represented by 4 symbols so it was 4×1 billion, or 4 billion people. The population in 2012 is represented by 7 symbols so it will be 7×1 billion, or 7 billion people. We subtract to find the difference:

$$7 \text{ billion} - 4 \text{ billion} = 3 \text{ billion}$$

37. The smallest portion of a symbol represents Africa, so the smallest amount of water, per person, is consumed in Africa.

39. North America is represented by $4\frac{3}{4}$ symbols so the water consumption, per person, in North America is
$$4\frac{3}{4} \times 10,000 = \frac{19}{4} \times 100,000 = \frac{1,900,000}{4} = 475,000 \text{ gal.}$$

41. From Exercise 39, we know that 475,000 gal of water are consumed, per person, in North America. Asia is represented by $1\frac{1}{2}$ symbols so the water consumption, per person, in Asia is $1\frac{1}{2} \times 100,000 = \frac{3}{2} \times 100,000 = \frac{300,000}{2} = 150,000$ gal.

We subtract to find how many more gallons are consumed, per person, in North America than in Asia:

$$475,000 - 150,000 = 325,000 \text{ gal}$$

43. We divide to determine the number of symbols required when each symbol represents 100 cups of coffee.

Germany: $1113 \div 100 = 11.13 \approx 11\frac{1}{10}$

United States: $615 \div 100 = 6.15 \approx 6\frac{1}{10}$

Switzerland: $1220 \div 100 = 12.2 = 12\frac{1}{5}$

France: $790 \div 100 = 7.9 = 7\frac{9}{10}$

Italy: $730 \div 100 = 7.3 = 7\frac{3}{10}$

Coffee Consumption

45. We divide to determine the number of symbols required when each symbol represents 10,000 elephants.

Cameroon: $20,050 \div 10,000 = 2.0050 \approx 2$

Zimbabwe: $49,800 \div 10,000 = 4.98 \approx 5$

Sudan: $19,800 \div 10,000 = 1.98 \approx 2$

Zaire: $110,200 \div 10,000 = 11.02 \approx 11$

Tanzania: $60,070 \div 10,000 = 6.007 \approx 6$

Botswana: $69,105 \div 10,000 = 6.9105 \approx 7$

Elephant Population

47. Discussion and Writing Exercise

49.
$$-\frac{3}{8} + \frac{5}{16} = -\frac{3}{8} \cdot \frac{2}{2} + \frac{5}{16}$$
$$= -\frac{6}{16} + \frac{5}{16}$$
$$= -\frac{1}{16}$$

51.
$$9x - 5 = -23$$
$$9x - 5 + 5 = -23 + 5 \quad \text{Adding 5 to both sides}$$
$$9x = -18$$
$$\frac{9x}{9} = \frac{-18}{9} \quad \text{Dividing both sides by 9}$$
$$x = -2$$
The solution is -2.

53.
$$-4x = 3x - 7$$
$$-4x - 3x = 3x - 7 - 3x \quad \text{Subtracting } 3x \text{ from both sides}$$
$$-7x = -7$$
$$\frac{-7x}{-7} = \frac{-7}{-7} \quad \text{Dividing both sides by } -7$$
$$x = 1$$
The solution is 1.

55. $\frac{29}{25}$ means $29 \div 25$.

```
      1.1 6
 2 5 │ 2 9.0 0
      2 5
      ─────
        4 0
        2 5
      ─────
        1 5 0
        1 5 0
        ─────
            0
```

$$\frac{29}{25} = 1.16$$

57. Discussion and Writing Exercise

59. *Familiarize.* Let $m =$ the number of minutes Alan spoke to each friend. Then the calls to Germany cost $\$0.06m$ and those to South Africa cost $\$0.17m$. The monthly fee is $\$0.59$, so the total bill is $\$0.06m + \$0.17m + \$0.59$, or $\$0.23m + \$0.59$.

Translate.

The total bill is $16.

$$0.23m + 0.59 = 16$$

Solve. We solve the equation.

$$0.23m + 0.59 = 16$$
$$0.23m + 0.59 - 0.59 = 16 - 0.59$$
$$0.23m = 15.41$$
$$\frac{0.23m}{0.23} = \frac{15.41}{0.23}$$
$$m = 67$$

Check. At \$0.06 per minute, the cost for 67 min is 67(\$0.06), or \$4.02. At \$0.17 per minute, the cost for 67 min is 67(\$0.17), or \$11.39. The total bill for the calls and the monthly fee is \$4.02 + \$11.39 + \$0.59, or \$16. The answer checks.

State. Alan spoke to each friend for 67 min.

61. Discussion and Writing Exercise

Exercise Set 6.2

1. Move to the right along the bar representing 1 cup of hot cocoa with skim milk. We read that there are about 190 calories in the cup of cocoa.

3. The longest bar is for 1 slice of chocolate cake with fudge frosting. Thus, it has the highest caloric content.

5. We locate 460 calories at the bottom of the graph and then go up until we reach a bar that ends at approximately 460 calories. Now go across to the left and read the dessert, 1 cup of premium chocolate ice cream.

7. From the graph we see that 1 cup of hot cocoa made with whole milk has about 310 calories and 1 cup of hot cocoa made with skim milk has about 190 calories. We subtract to find the difference:

$$310 - 190 = 120$$

The cocoa made with whole milk has about 120 more calories than the cocoa made with skim milk.

9. From Exercise 5 we know that 1 cup of premium ice cream has about 460 calories. We multiply to find the caloric content of 2 cups:

$$2 \times 460 = 920$$

Rae consumes about 920 calories.

11. From the graph we see that a 2-oz chocolate bar with peanuts contains about 270 calories. We multiply to find the number of extra calories Paul adds to his diet in 1 year:

$$365 \times 270 \text{ calories} = 98,550 \text{ calories}$$

Then we divide to determine the number of pounds he will gain:

$$\frac{98,550}{3500} \approx 28$$

Paul will gain about 28 pounds.

13. Find the bar representing men with bachelor's degrees in 1970 and read \$11,000 on the vertical scale. Do the same for the bar representing men with bachelor's degrees in 2003 and read \$58,000.

Subtract to find the amount of increase:

$$\$58,000 - \$11,000 = \$47,000$$

15. Find the bar representing women with a high school diploma in 1970 and read \$6000. Do the same for the bar representing women with a high school diploma in 2003 and read \$26,000.

Subtract to find the increase:

$$\$26,000 - \$6000 = \$20,000$$

17. From Exercise 13 we know that men with bachelor's degrees earned \$11,000 in 1970. Find the bar representing men with a high school diploma in 1970 and read \$7000.

Subtract to find the increase:

$$\$11,000 - \$7000 = \$4000$$

19. Find the bar representing women with bachelor's degrees in 2003 and read \$49,000. Do the same for the bar representing men with high school diplomas in 2003 and read \$34,000.

Subtract to find how much more the women earned:

$$\$49,000 - \$34,000 = \$15,000$$

21. On the horizontal scale in six equally spaced intervals indicate the names of the cities. Label this scale "City." Then label the vertical scale "Commuting Time (in minutes)." Note that the smallest time is 21.6 minutes and the largest is 39.0 minutes. We could start the vertical scale at 0 or we could start it at 20, using a jagged line to indicate the missing numbers. We choose the second option. Label the marks on the vertical scale by 5's. Finally, draw vertical bars above the cities to show the commuting times.

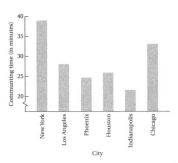

23. The shortest bar represents Indianapolis, so it has the least commuting time.

25. We add the commuting times and divide by the number of addends, 6:

$$\frac{39.0 + 28.1 + 24.7 + 25.9 + 21.6 + 33.1}{6} = \frac{172.4}{6} = 28.7\overline{3} \text{ min}$$

27. $650 - 70 = 580$ calories per hour

29. Find the calories burned doing office work for 8 hours:

$$180 \cdot 8 = 1440 \text{ calories}$$

Find the calories burned sleeping for 7 hours:

$$70 \cdot 7 = 490 \text{ calories}$$

Find the sum: $1440 + 490 = 1930$ calories

31. From the graph we read that the average driving distance was 256.9 yd in 1980 and 287.3 yd in 2004. We subtract to find the increase:

$$287.3 - 256.9 = 30.4$$

The driving distance in 2004 was 30.4 yd farther than in 1980.

33. Find 264 on the vertical scale and observe that the horizontal line representing 264 intersects the graph at the points corresponding to 1988 and 1995 on the horizontal scale. Thus, the average driving distance was about 264 yd in 1988 and in 1995.

35. First indicate the years on the horizontal scale and label it "Year." The smallest ozone level is 284.3 Dobson Units and the largest is 294.5 Dobson Units. We could start the vertical scale at 0, but the graph will be more compact and easier to read if we start at a higher number, say at 280. We do this, using a jagged line to indicate the missing numbers. Mark the vertical scale appropriately and label it "Dobson Units." Next, at the appropriate level above each year, mark the corresponding ozone level. Finally, draw line segments connecting the points. We title the graph "Ozone Level."

Ozone Level

37. The graph rises most sharply from 1997 to 1998. (The data in the table confirm that the greatest increase occurred between 1997 and 1998.) Thus, the increase in the ozone level was the greatest between 1997 and 1998.

39. We add the ozone levels from 2000 through 2004 and divide by the number of years, 4.

$$\frac{292.1 + 290.4 + 292.6 + 287.9 + 284.3}{5} = \frac{1447.3}{5} = 289.46$$

The average ozone level from 2000 through 2004 was about 289.46 Dobson Units.

41. The segment connecting 2000 and 2001 rises most steeply, so the increase was the greatest between 2000 and 2001. (The data in the table confirm this.)

43. The graph begins to rise between 2000 and 2001, so the number of Democratic governors began to rise between 2000 and 2001.

45. We add the numbers of Democratic governors from 1995 through 1999 and divide by the number of years, 5.

$$\frac{19 + 18 + 17 + 17 + 17}{5} = \frac{88}{5} = 17.6$$

The average number of Democratic governors from 1995 through 1999 was 17.6.

47. Discussion and Writing Exercise

49. *Familiarize*. We draw a picture. We let $n =$ the number of 12-oz bottles that can be filled.

12 oz in each row

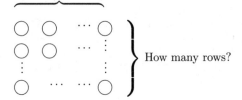

How many rows?

Translate and Solve. We translate to an equation and solve as follows:

$$408 \div 12 = n$$

$$
\begin{array}{r}
3\ 4 \\
1\ 2\ \overline{)\ 4\ 0\ 8} \\
3\ 6\ 0 \\
\hline
4\ 8 \\
4\ 8 \\
\hline
0
\end{array}
$$

Check. We check by multiplying the number of bottles by 12:

$$12 \times 34 = 408$$

State. 34 twelve-oz bottles can be filled.

51. *Familiarize*. We draw a picture. We let $n =$ the number of fluid ounces in a six-pack of Coca Cola.

12 oz in each row

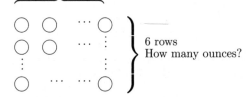

6 rows
How many ounces?

Translate and Solve. We translate to an equation and solve as follows.

$$12 \times 6 = n$$
$$72 = n$$

Check. We check by dividing the total number of ounces by 6: $72 \div 6 = 12$. The answer checks.

State. There are 72 fluid ounces in a six-pack of Coca Cola.

53. *Familiarize*. Let $n =$ the number.

Translate.

$$
\begin{array}{ccccc}
\frac{2}{3} & \text{of} & 75 & \text{is} & \underline{\text{what number?}} \\
\downarrow & \downarrow & \downarrow & \downarrow & \downarrow \\
\frac{2}{3} & \cdot & 75 & = & n
\end{array}
$$

Solve. We carry out the multiplication.

$$\frac{2}{3} \cdot 75 = n$$

$$\frac{2 \cdot 75}{3} = n$$

$$\frac{2 \cdot \cancel{3} \cdot 25}{\cancel{3} \cdot 1} = n$$

$$50 = n$$

Check. We repeat the calculation.

State. $\frac{2}{3}$ of 75 is 50.

55. $-9 = -2x + 3$

$-9 - 3 = -2x + 3 - 3$ Subtracting 3 from
                                        both sides

$-12 = -2x$

$\dfrac{-12}{-2} = \dfrac{-2x}{-2}$ Dividing both sides by -2

$6 = x$

The solution is 6.

57. Discussion and Writing Exercise

59. We draw the horizontal scale with equally spaced intervals, counting by hundreds. The vertical scale represents the years.

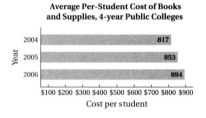

Average Per-Student Cost of Books and Supplies, 4-year Public Colleges

61. We draw the line graphs on a single graph and extend them. Then we observe that the line representing crossover SUV sales lies about 1 million units above the graph representing traditional SUV sales in 2009. Answers may vary depending on how the lines were extended.

Traditional and Crossover SUV Sales

63. Discussion and Writing Exercise

Exercise Set 6.3

1. To plot $(4, 4)$, we locate 4 on the first, or horizontal, axis. From there we go up 4 units and make a dot.

To plot $(-2, 4)$, we locate -2 on the first, or horizontal, axis. From there we go up 4 units and make a dot.

To plot $(5, -3)$, we locate 5 on the first, or horizontal, axis. From there we go down 3 units and make a dot.

To plot $(-5, -5)$, we locate -5 on the first, on the horizontal, axis. From there we go down 5 units and make a dot.

To plot $(0, 4)$, we locate 0 on the first, or horizontal, axis. From there we go up 4 units and make a dot.

To plot $(0, -4)$, we locate 0 on the first, or horizontal, axis. From there we go down 4 units and make a dot.

To plot $(3, 0)$, we locate 3 on the first, or horizontal, axis. Since the second coordinate is 0, we do not move up or down. We make a dot at the point we located on the first axis.

To plot $(-4, 0)$, we locate -4 on the first, or horizontal, axis. Since the second coordinate is 0, we do not move up or down. We make a dot at the point we located on the first axis.

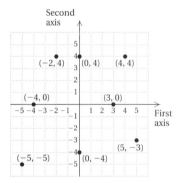

3. To plot $(-2, -4)$, we locate -2 on the first, or horizontal, axis. From there we go down 4 units and make a dot.

To plot $(5, -4)$, we locate 5 on the first, or horizontal, axis. From there we go down 4 units and make a dot.

To plot $\left(0, 3\frac{1}{2}\right)$, we locate 0 on the first, or horizontal, axis. From there we go up $3\frac{1}{2}$ units and make a dot.

To plot $\left(4, 3\frac{1}{2}\right)$, we locate 4 on the first, or horizontal, axis. From there we go up $3\frac{1}{2}$ units and make a dot.

To plot $(-1, -3)$, we locate -1 on the first, or horizontal, axis. From there we go down 3 units and make a dot.

To plot $(-1, 5)$, we locate 5 on the first, or horizontal, axis. From there we go up 5 units and make a dot.

To plot $(4, -1)$, we locate 4 on the first, or horizontal, axis. From there we go down 1 unit and make a dot.

To plot $(-2, 0)$, we locate -2 on the first, or horizontal axis. Since the second coordinate is 0, we do not move up or down. We make a dot at the point we located on the first axis.

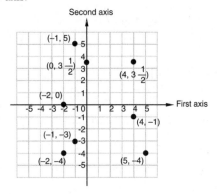

5.

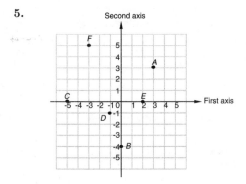

We look below point A to see that its first coordinate is 3. Looking to the left of point A, we find that its second coordinate is also 3. Thus, the coordinates of point A are $(3, 3)$.

We look above point B to see that its first coordinate is 0. Looking at the location of point B on the second, or vertical, axis, we find that its second coordinate is -4. Thus, the coordinates of point B are $(0, -4)$.

Looking at the location of point C on the first, or horizontal, axis, we see that the first coordinate of point C is -5. We look to the right of point C to see that its second coordinate is 0. Thus, the coordinates of point C are $(-5, 0)$.

We look above point D to see that its first coordinate is -1. Looking to the right of point D, we find that its second coordinate is also -1. Thus, the coordinates of point D are $(-1, -1)$.

Looking at the location of point E on the first, or horizontal, axis, we see that the first coordinate of point E is 2. We look to the left of point E to see that its second coordinate is 0. Thus, the coordinates of point E are $(2, 0)$.

We look below point F to see that its first coordinate is -3. Looking to the right of point F, we find that its second coordinate is 5. Thus, the coordinates of point F are $(-3, 5)$.

7.

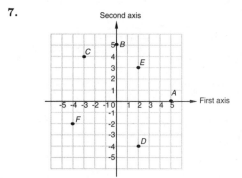

Looking at the location of point A on the first, or horizontal, axis, we see that the first coordinate of point A is 5. We look to the left of point A to see that its second coordinate is 0. Thus, the coordinates of point A are $(5, 0)$.

We look below point B to see that its first coordinate is 0. Looking at the location of point B on the second, or vertical, axis, we find that its second coordinate is 5. Thus, the coordinates of point B are $(0, 5)$.

We look below point C to see that its first coordinate is -3. Looking to the right of point C, we find that its second coordinate is 4. Thus, the coordinates of point C are $(-3, 4)$.

We look above point D to see that its first coordinate is 2. Looking to the left of point D, we find that its second coordinate is -4. Thus, the coordinates of point D are $(2, -4)$.

We look below point E to see that its first coordinate is 2. Looking to the left of point E, we find that its second coordinate is 3. Thus the coordinates of point E are $(2, 3)$.

We look above point F to see that its first coordinate is -4. Looking to the right of point F, we find that its second coordinate is -2. Thus, the coordinates of point F are $(-4, -2)$.

9. Since the first coordinate is negative and the second coordinate positive, the point $(-5, 3)$ is located in quadrant II.

11. Since the first coordinate is positive and the second coordinate negative, the point $(100, -1)$ is in quadrant IV.

13. Since both coordinates are negative, the point $(-6.5, -1.9)$ is in quadrant III.

15. Since both coordinates are positive, the point $\left(3\frac{7}{10}, 9\frac{1}{11}\right)$ is in quadrant I.

17. In quadrant IV, first coordinates are always <u>positive</u> and second coordinates are always <u>negative</u>.

19. In quadrant <u>III</u>, both coordinates are always negative.

21. In quadrant I, the first coordinate is always positive. The other quadrant where this occurs is IV. Thus, the statement should read as follows:

In quadrants I and <u>IV</u>, the first coordinate is always <u>positive</u>.

23. $y = 2x - 5$

$$\begin{array}{c|c} \hline 3 \ ? \ 2 \cdot 4 - 5 & \text{Substituting 4 for } x \text{ and 3 for } y \\ 8 - 5 & \text{(alphabetical order of variables)} \\ 3 \ \bigm| \ 3 & \text{TRUE} \end{array}$$

Since the equation becomes true, $(4, 3)$ is a solution.

25. $3x - y = 4$

$$\begin{array}{c|c} \hline 3 \cdot 2 - (-3) \ ? \ 4 & \text{Substituting 2 for } x \text{ and } -3 \text{ for } y \\ 6 + 3 & \\ 9 \ \bigm| \ 4 & \text{FALSE} \end{array}$$

Since the equation becomes false, $(2, -3)$ is not a solution.

27. $3c + 2d = -8$

$$\begin{array}{c|c} \hline 3(-2) + 2(-1) \ ? \ -8 & \text{Substituting } -2 \text{ for } c \text{ and } -1 \\ & \text{for } d \\ -6 - 2 & \\ -8 \ \bigm| \ -8 & \text{TRUE} \end{array}$$

Since the equation becomes true, $(-2, -1)$ is a solution.

29. $3x + y = 19$

$$\begin{array}{c|c} \hline 3 \cdot 5 + (-4) \ ? \ 19 & \text{Substituting 5 for } x \text{ and } -4 \text{ for } y \\ 15 - 4 & \\ 11 \ \bigm| \ 19 & \text{FALSE} \end{array}$$

Since the equation becomes false, $(5, -4)$ is not a solution.

31. $2q - 3p = 3$

$$\begin{array}{c|c} \hline 2 \cdot 6 - 3\left(2\frac{1}{3}\right) \ ? \ 3 & \text{Substituting } 2\frac{1}{3} \text{ for } p \text{ and 6 for } q \\ 12 - 3 \cdot \frac{7}{3} & \\ 12 - 7 & \\ 5 \ \bigm| \ 3 & \text{FALSE} \end{array}$$

The equation becomes false; $\left(2\frac{1}{3}, 6\right)$ is not a solution.

33. $y = 5x - 11.3$

$$\begin{array}{c|c} \hline 0.7 \ ? \ 5(2.4) - 11.3 & \text{Substituting 2.4 for } x \text{ and 0.7} \\ & \text{for } y \\ 12 - 11.3 & \\ 0.7 \ \bigm| \ 0.7 & \text{TRUE} \end{array}$$

Since the equation becomes true, $(2.4, 0.7)$ is a solution.

35. Discussion and Writing Exercise

37. $3x - 4 = 17$

$3x - 4 + 4 = 17 + 4$ Adding 4 to both sides

$3x = 21$

$\dfrac{3x}{3} = \dfrac{21}{3}$ Dividing both sides by 3

$x = 7$

The solution is 7.

39. $5(x - 2) = 3x - 4$

$5x - 10 = 3x - 4$

$5x - 10 + 10 = 3x - 4 + 10$ Adding 10 to both sides

$5x = 3x + 6$

$5x - 3x = 3x + 6 - 3x$ Subtracting $3x$ from both sides

$2x = 6$

$\dfrac{2x}{2} = \dfrac{6}{2}$ Dividing both sides by 2

$x = 3$

The solution is 3.

41. $-\dfrac{1}{9}t = \dfrac{2}{3}t$

$9\left(-\dfrac{1}{9}t\right) = 9 \cdot \dfrac{2}{3}t$ Clearing fractions

$-t = 6t$

$-t + t = 6t + t$ Adding t to both sides

$0 = 7t$

$\dfrac{0}{7} = \dfrac{7t}{7}$ Dividing both sides by 7

$0 = t$

The solution is 0.

43. $7\dfrac{2}{11}a - 5\dfrac{1}{3}a = \left(7\dfrac{2}{11} - 5\dfrac{1}{3}\right)a = \left(7\dfrac{6}{33} - 5\dfrac{11}{33}\right)a = \left(6\dfrac{39}{33} - 5\dfrac{11}{33}\right)a = 1\dfrac{28}{33}a$

45. Discussion and Writing Exercise

47. $5.2x + 6.1y = -4.821$

$$\begin{array}{c|c} \hline 5.2(-2.37) + 6.1(1.23) \ ? \ -4.821 & \\ -12.324 + 7.503 & \\ -4.821 \ \bigm| \ -4.821 & \text{TRUE} \end{array}$$

Since the equation becomes true, $(-2.37, 1.23)$ is a solution.

49.

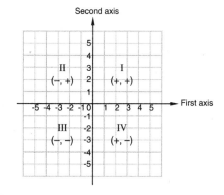

If the first coordinate is positive, then the point must be in quadrant I or quadrant IV.

51. If the first and second coordinates are equal, then the coordinates have the same sign. Thus, the point must be in quadrant I or quadrant III. (See the graph in Exercise 49.)

53.

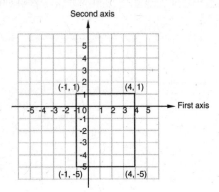

The coordinates of the fourth vertex are $(-1, -5)$.

55. Answers may vary. There are infinitely many points for which the sum of the coordinates is 7. Eight of them are shown below. All such points are shown on the following graph.

$(-1, 8)$ $-1 + 8 = 7$
$(0, 7)$ $0 + 7 = 7$
$(1, 6)$ $1 + 6 = 7$
$(2, 5)$ $2 + 5 = 7$
$(3, 4)$ $3 + 4 = 7$
$(4, 3)$ $4 + 3 = 7$
$(5, 2)$ $5 + 2 = 7$
$(6, 1)$ $6 + 1 = 7$

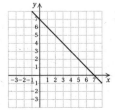

57.

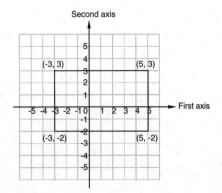

The length is 8, and the width is 5.

$P = 2l + 2w$

$P = 2 \cdot 8 + 2 \cdot 5 = 16 + 10 = 26$

Exercise Set 6.4

1. $x + y = 8$
$5 + y = 8$ Substituting 5 for x
$y = 3$ Subtracting 5 from both sides

The pair $(5, 3)$ is a solution of $x + y = 8$.

3. $2x + y = 7$
$2 \cdot 3 + y = 7$ Substituting 3 for x
$6 + y = 7$
$y = 1$ Subtracting 6 from both sides

The pair $(3, 1)$ is a solution of $2x + y = 7$.

5. $y = 3x - 1$
$y = 3 \cdot 5 - 1$ Substituting 5 for x
$y = 15 - 1$
$y = 14$

The pair $(5, 14)$ is a solution of $y = 3x - 1$.

7. $x + 3y = 1$
$10 + 3y = 1$ Substituting 10 for x
$3y = -9$ Subtracting 10 from both sides
$y = -3$ Dividing both sides by 3

The pair $(10, -3)$ is a solution of $x + 3y = 1$.

9. $2x + 5y = 17$
$2 \cdot 1 + 5y = 17$ Substituting 1 for x
$2 + 5y = 17$
$5y = 15$ Subtracting 2 from both sides
$y = 3$ Dividing both sides by 5

The pair $(1, 3)$ is a solution of $2x + 5y = 17$.

11. $3x - 2y = 8$
$3x - 2(-1) = 8$ Substituting -1 for y
$3x + 2 = 8$
$3x = 6$ Subtracting from both sides
$x = 2$ Dividing both sides by 3

The pair $(2, -1)$ is a solution of $3x - 2y = 8$.

13. To complete the pair $(\ ,3)$, we replace y with 3 and solve for x.
$x + y = 4$
$x + 3 = 4$ Substituting 3 for y
$x = 1$ Subtracting 3 from both sides

Thus, $(1, 3)$ is a solution of $x + y = 7$.

To complete the pair $(-1, \ )$, we replace x with -1 and solve for y.
$x + y = 4$
$-1 + y = 4$ Substituting 4 for x
$y = 5$ Adding 1 to both sides

Thus, $(-1, 5)$ is a solution of $x + y = 7$.

15. To complete the pair $(\ ,3)$, we replace y with 3 and solve for x.
$x - y = 4$
$x - 3 = 4$ Substituting 3 for y
$x = 7$ Adding 3 to both sides

Thus, $(7, 3)$ is a solution of $x - y = 4$.

To complete the pair $(10, \quad)$, we replace x with 10 and solve for y.

$$x - y = 4$$
$$10 - y = 4 \quad \text{Substituting 10 for } x$$
$$-y = -6 \quad \text{Subtracting 10 from both sides}$$
$$-1 \cdot y = -6 \quad \text{Recall that } -a = -1 \cdot a.$$
$$y = 6 \quad \text{Dividing both sides by } -1$$

Thus, $(10, 6)$ is a solution of $x - y = 4$.

17. To complete the pair $(3, \quad)$, we replace x with 3 and solve for y.

$$2x + 3y = 15$$
$$2 \cdot 3 + 3y = 15 \quad \text{Substituting 3 for } x$$
$$6 + 3y = 15$$
$$3y = 9 \quad \text{Subtracting 6 from both sides}$$
$$y = 3 \quad \text{Dividing both sides by 3}$$

Thus, $(3, 3)$ is a solution of $2x + 3y = 15$.

To complete the pair $(\quad, 1)$, we replace y with 1 and solve for x.

$$2x + 3y = 15$$
$$2x + 3 \cdot 1 = 15 \quad \text{Substituting 1 for } y$$
$$2x + 3 = 15$$
$$2x = 12 \quad \text{Subtracting 3 from both sides}$$
$$x = 6 \quad \text{Dividing both sides by 2}$$

Thus, $(6, 1)$ is a solution of $2x + 3y = 15$.

19. To complete the pair $(3, \quad)$, we replace x with 3 and solve for y.

$$3x + 5y = 14$$
$$3 \cdot 3 + 5y = 14 \quad \text{Substituting 3 for } x$$
$$9 + 5y = 14$$
$$5y = 5 \quad \text{Subtracting 9 from both sides}$$
$$y = 1 \quad \text{Dividing both sides by 5}$$

Thus, $(3, 1)$ is a solution of $3x + 5y = 14$.

To complete the pair $(\quad, 4)$, we replace y with 4 and solve for x.

$$3x + 5y = 14$$
$$3x + 5 \cdot 4 = 14 \quad \text{Substituting 4 for } y$$
$$3x + 20 = 14$$
$$3x = -6 \quad \text{Subtracting 20 from both sides}$$
$$x = -2 \quad \text{Dividing both sides by 3}$$

Thus, $(-2, 4)$ is a solution of $3x + 5y = 14$.

21. To complete the pair $(\quad, 4)$, we replace y with 4 and solve for x.

$$y = 4x$$
$$4 = 4x \quad \text{Substituting 4 for y}$$
$$1 = x \quad \text{Dividing both sides by 4}$$

Thus, $(1, 4)$ is a solution of $y = 4x$.

To complete the pair $(-2, \quad)$, we replace x with -2 and solve for y.

$$y = 4x$$
$$y = 4(-2) \quad \text{Substituting } -2 \text{ for } x$$
$$y = -8$$

Thus, $(-2, -8)$ is a solution of $y = 4x$.

23. To complete the pair $(0, \quad)$, we replace x with 0 and solve for y.

$$2x + 5y = 3$$
$$2 \cdot 0 + 5y = 3 \quad \text{Substituting 0 for } x$$
$$5y = 3$$
$$y = \frac{3}{5} \quad \text{Dividing both sides by 5}$$

Thus, $\left(0, \dfrac{3}{5}\right)$ is a solution of $2x + 5y = 3$.

To complete the pair $(\quad, 0)$, we replace y with 0 and solve for x.

$$2x + 5y = 3$$
$$2x + 5 \cdot 0 = 3 \quad \text{Substituting 0 for } y$$
$$2x = 3$$
$$x = \frac{3}{2} \quad \text{Dividing both sides by 2}$$

Thus, $\left(\dfrac{3}{2}, 0\right)$ is a solution of $2x + 5y = 3$.

25. We are free to choose any number as a replacement for x or y. To find one solution we choose to replace x with 0.

$$x + y = 9$$
$$0 + y = 9 \quad \text{Substituting 0 for } x$$
$$y = 9$$

Thus, $(0, 9)$ is one solution of $x + y = 9$.

To find a second solution we can replace y with 5.

$$x + y = 9$$
$$x + 5 = 9 \quad \text{Substituting 5 for } y$$
$$x = 4 \quad \text{Subtracting 5 from both sides}$$

Thus, $(4, 5)$ is a second solution of $x + y = 9$.

To find a third solution we can replace x with 10.

$$x + y = 9$$
$$10 + y = 9 \quad \text{Substituting 10 for } x$$
$$y = -1 \quad \text{Subtracting 10 from both sides}$$

Thus, $(10, -1)$ is a third solution of $x + y = 9$.

27. The solutions $(1, 4)$ and $(-2, -8)$ were found in Exercise 21. To find a third solution we can replace x with 0.

$$y = 4x$$
$$y = 4 \cdot 0 \quad \text{Substituting 0 for } x$$
$$y = 0$$

Thus, $(0, 0)$ is a third solution of $y = 4x$.

29. We are free to choose any number as a replacement for x or y. To find one solution we choose to replace x with 0.

$$3x + y = 13$$
$$3 \cdot 0 + y = 13 \quad \text{Substituting 0 for } x$$
$$0 + y = 13$$
$$y = 13$$

Thus, $(0, 13)$ is one solution of $3x + y = 13$.

To find a second solution we can replace y with 10.

$$3x + y = 13$$
$$3x + 10 = 13 \quad \text{Substituting 10 for } y$$
$$3x = 3 \quad \text{Subtracting 10 from both sides}$$
$$x = 1 \quad \text{Dividing both sides by 3}$$

Thus, $(1, 10)$ is a second solution of $3x + y = 13$.

To find a third solution we can replace x with 2.

$$3x + y = 13$$
$$3 \cdot 2 + y = 13 \quad \text{Substituting 2 for } x$$
$$6 + y = 13$$
$$y = 7 \quad \text{Subtracting 6 from both sides}$$

Thus, $(2,7)$ is a third solution of $3x + y = 13$.

31. We are free to choose any number as a replacement for x or y. Since y is isolated it is generally easiest to substitute for x and then calculate y. To find one solution we choose to replace x with -1.

$$y = 3x - 1$$
$$y = 3(-1) - 1 \quad \text{Substituting } -1 \text{ for } x$$
$$y = -3 - 1$$
$$y = -4$$

Thus, $(-1, -4)$ is one solution of $y = 3x - 1$.

To find a second solution we can replace x with 0.

$$y = 3x - 1$$
$$y = 3 \cdot 0 - 1 \quad \text{Substituting 0 for } x$$
$$y = -1$$

Thus, $(0, -1)$ is a second solution of $y = 3x - 1$.

To find a third solution we can replace x with 2.

$$y = 3x - 1$$
$$y = 3 \cdot 2 - 1 \quad \text{Substituting 2 for } x$$
$$y = 6 - 1$$
$$y = 5$$

Thus, $(2, 5)$ is a third solution of $y = 3x - 1$.

33. We are free to choose any number as a replacement for x or y. Since y is isolated it is generally easiest to substitute for x and then calculate y. To find one solution we choose to replace x with 0.

$$y = -7x$$
$$y = -7 \cdot 0 \quad \text{Substituting 0 for } x$$
$$y = 0$$

Thus, $(0, 0)$ is one solution of $y = -7x$.

To find a second solution we can replace x with 1.

$$y = -7x$$
$$y = -7 \cdot 1 \quad \text{Substituting 1 for } x$$
$$y = -7$$

Thus, $(1, -7)$ is a second solution of $y = -7x$.

To find a third solution we can replace x with -2.

$$y = -7x$$
$$y = -7(-2) \quad \text{Substituting } -2 \text{ for x}$$
$$y = 14$$

Thus, $(-2, 14)$ is a third solution of $y = -7x$.

35. We are free to choose any number as a replacement for x or y. Since x is isolated it is easiest to substitute for y and then calculate x. To find one solution we choose to replace y with -4.

$$4 + y = x$$
$$4 + (-4) = x \quad \text{Substituting } -4 \text{ for } y$$
$$0 = x$$

Thus, $(0, -4)$ is one solution of $4 + y = x$.

To find a second solution we can replace y with 0.

$$4 + y = x$$
$$4 + 0 = x \quad \text{Substituting 0 for } y$$
$$4 = x$$

Thus, $(4, 0)$ is a second solution of $4 + y = x$.

To find a third solution we can replace y with -3.

$$4 + y = x$$
$$4 + (-3) = x \quad \text{Substituting } -3 \text{ for } y$$
$$1 = x$$

Thus, $(1, -3)$ is a third solution of $4 + y = x$.

37. We are free to choose any number as a replacement for x or y. To find one solution we choose to replace x with 0.

$$3x + 2y = 12$$
$$3 \cdot 0 + 2y = 12 \quad \text{Substituting 0 for } x$$
$$2y = 12$$
$$y = 6 \quad \text{Dividing both sides by 2}$$

Thus, $(0, 6)$ is one solution of $3x + 2y = 12$.

To find a second solution we can replace y with 0.

$$3x + 2y = 12$$
$$3x + 2 \cdot 0 = 12 \quad \text{Substituting 0 for } y$$
$$3x = 12$$
$$x = 4 \quad \text{Dividing both sides by 3}$$

Thus, $(4, 0)$ is a second solution of $3x + 2y = 12$.

To find a third solution we can replace x with 1.

$$3x + 2y = 12$$
$$3 \cdot 1 + 2y = 12 \quad \text{Substituting 1 for } x$$
$$3 + 2y = 12$$
$$2y = 9 \quad \text{Subtracting 3 from both sides}$$
$$y = \frac{9}{2} \quad \text{Dividing both sides by 2}$$

Thus, $\left(1, \frac{9}{2}\right)$ is a third solution of $3x + 2y = 12$.

39. We are free to choose any number as a replacement for x or y. Since y is isolated it is generally easiest to substitute for x and then calculate y. Note that when x is a multiple of 3, fraction values for y are avoided. To find one solution we choose to replace x with -3.

$$y = \frac{1}{3}x + 2$$
$$y = \frac{1}{3}(-3) + 2 \quad \text{Substituting } -3 \text{ for } x$$
$$y = -1 + 2$$
$$y = 1$$

Thus, $(-3, 1)$ is one solution of $y = \frac{1}{3}x + 2$.

To find a second solution we can replace x with 0.

$$y = \frac{1}{3}x + 2$$
$$y = \frac{1}{3} \cdot 0 + 2 \quad \text{Substituting 0 for } x$$
$$y = 2$$

Thus, $(0, 2)$ is a second solution of $y = \frac{1}{3}x + 2$.

To find a third solution we can replace x with 3.

$$y = \frac{1}{3}x + 2$$

$$y = \frac{1}{3} \cdot 3 + 2 \quad \text{Substituting 3 for } x$$

$$y = 1 + 2$$

$$y = 3$$

Thus, $(3,3)$ is a third solution of $y = \frac{1}{3}x + 2$.

41. Graph: $x + y = 6$

We make a table of solutions. Then we plot the points, draw the line, and label it.

When $x = 0$: $\quad 0 + y = 6$
$\qquad\qquad\qquad\quad y = 6$

When $y = 2$: $\quad x + 2 = 6$
$\qquad\qquad\qquad\quad x = 4$

When $x = 4$: $\quad 4 + y = 6$
$\qquad\qquad\qquad\quad y = 2$

x	y $x + y = 6$	(x,y)
0	6	$(0,6)$
2	4	$(2,4)$
4	2	$(4,2)$

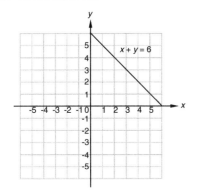

43. Graph: $x - 1 = y$

We make a table of solutions. Then we plot the points, draw the line, and label it.

If $x = 5$, then $y = 5 - 1 = 4$.
If $x = 0$, then $y = 0 - 1 = -1$.
If $x = -2$, then $y = -2 - 1 = -3$.

x	y $x - 1 = y$	(x,y)
5	4	$(5,4)$
0	-1	$(0,-1)$
-2	-3	$(-2,-3)$

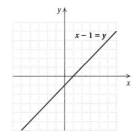

45. Graph: $y = x - 4$

We make a table of solutions. Then we plot the points, draw the line, and label it.

If $x = -1$, $y = -1 - 4 = -5$.
If $x = 0$, $y = 0 - 4 = -4$.
If $x = 3$, $y = 3 - 4 = -1$.

x	y $y = x - 4$	(x,y)
-1	-5	$(-1,-5)$
0	-4	$(0,-4)$
3	-1	$(3,-1)$

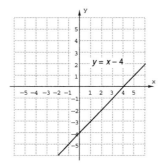

47. Graph: $y = \frac{1}{3}x$

We make a table of solutions. Note that when x is a multiple of 3, fraction values for y are avoided. We plot the points, draw the line, and label it.

If $x = -3$, $y = \frac{1}{3}(-3) = -1$.

If $x = 0$, $y = \frac{1}{3} \cdot 0 = 0$.

If $x = 3$, $y = \frac{1}{3} \cdot 3 = 1$.

x	y $y = \frac{1}{3}x$	(x,y)
-3	-1	$(-3,-1)$
0	0	$(0,0)$
3	1	$(3,1)$

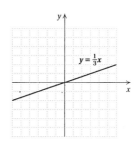

49. Graph: $y = x$

We make a table of solutions. Then we plot the points, draw the line, and label it.

If $x = -3$, $y = -3$.
If $x = 0$, $y = 0$.
If $x = 2$, $y = 2$.

x	y $y = x$	(x, y)
-3	-3	$(-3, -3)$
0	0	$(0, 0)$
2	2	$(2, 2)$

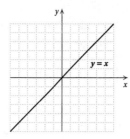

51. Graph: $y = 2x - 1$

We make a table of solutions. Then we plot the points, draw the line, and label it.

If $x = -1$, $y = 2(-1) - 1 = -2 - 1 = -3$.
If $x = 0$, $y = 2 \cdot 0 - 1 = -1$.
If $x = 2$, $y = 2 \cdot 2 - 1 = 4 - 1 = 3$.

x	y $y = 2x - 1$	(x, y)
-1	-3	$(-1, -3)$
0	-1	$(0, -1)$
2	3	$(2, 3)$

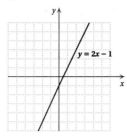

53. Graph: $y = 2x + 1$

We make a table of solutions. Then we plot the points, draw the line, and label it.

If $x = 0$, $y = 2 \cdot 0 + 1 = 0 + 1 = 1$.
If $x = 2$, $y = 2 \cdot 2 + 1 = 4 + 1 = 5$.
If $x = -2$, $y = 2(-2) + 1 = -4 + 1 = -3$.

x	y $y = 2x + 1$	(x, y)
0	1	$(0, 1)$
2	5	$(2, 5)$
-2	-3	$(-2, -3)$

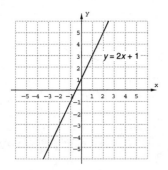

55. Graph: $y = \dfrac{2}{5}x$

We make a table of solutions. Note that when x is a multiple of 5, fraction values for y are avoided. We plot the points, draw the line, and label it.

If $x = -5$, $y = \dfrac{2}{5}(-5) = -2$.

If $x = 0$, $y = \dfrac{2}{5} \cdot 0 = 0$.

If $x = 5$, $y = \dfrac{2}{5} \cdot 5 = 2$.

x	y $y = \frac{2}{5}x$	(x, y)
-5	-2	$(-5, -2)$
0	0	$(0, 0)$
5	2	$(5, 2)$

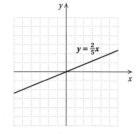

57. Graph: $y = -x + 4$

We make a table of solutions. Then we plot the points, draw the line, and label it.

If $x = -1$, $y = -(-1) + 4 = 1 + 4 = 5$.
If $x = 1$, $y = -1 + 4 = 3$.
If $x = 4$, $y = -4 + 4 = 0$.

x	y $y = -x + 4$	(x, y)
-1	5	$(-1, 5)$
1	3	$(1, 3)$
4	0	$(4, 0)$

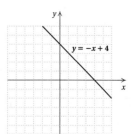

59. Graph: $y = \dfrac{2}{3}x + 1$

We make a table of solutions. Note that when x is a multiple of 3, fraction values for y are avoided. We plot the points, draw the line, and label it.

If $x = -3$, $y = \dfrac{2}{3}(-3) + 1 = -2 + 1 = -1$.

If $x = 0$, $y = \dfrac{2}{3} \cdot 0 + 1 = 0 + 1 = 1$.

If $x = 3$, $y = \dfrac{2}{3} \cdot 3 + 1 = 2 + 1 = 3$.

x	y $y = \frac{2}{3}x + 1$	(x, y)
-3	-1	$(-3, -1)$
0	1	$(0, 1)$
3	3	$(3, 3)$

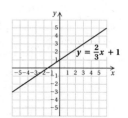

61. Graph: $y = 2 = 0 \cdot x + 2$

No matter what number we choose for x, y must be 2.

x	y $y = 2$	(x, y)
-3	2	$(-3, 2)$
0	2	$(0, 2)$
2	2	$(2, 2)$

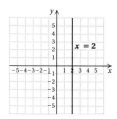

63. Graph: $x = 2$, or $x + 0 \cdot y = 2$

We make up a table with all 2's in the x-column.

x $x = 2$	y	(x, y)
2	-4	$(2, -4)$
2	-1	$(2, -1)$
2	3	$(2, 3)$

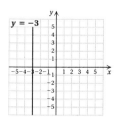

65. Graph: $x = -3$, or $x + 0 \cdot y = -3$

We make up a table with all -3's in the x-column.

x $x = -3$	y	(x, y)
-3	-2	$(-3, -2)$
-3	0	$(-3, 0)$
-3	4	$(-3, 4)$

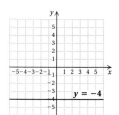

67. Graph: $y = -4 = 0 \cdot x - 4$

No matter what number we choose for x, y must be -4.

x	y $y = -4$	(x, y)
-3	-4	$(-3, -4)$
2	-4	$(2, -4)$
4	-4	$(4, -4)$

69. Discussion and Writing Exercise

71. We add the times and divide the number of addends, 5.

$$\frac{9 + 9\frac{1}{2} + 5\frac{1}{2} + 11\frac{1}{2} + 9\frac{1}{2}}{5} = \frac{45}{5} = 9$$

The average length of a tune is 9 min.

73. *Familiarize*. We let $y =$ the amount of vinegar needed to make $2\frac{1}{2}$ batches of chili.

Translate. We write a multiplication sentence that fits the situation.

$$y = 2\frac{1}{2} \cdot \left(\frac{3}{4}\right)$$

Solve. We do the computation.

$$y = 2\frac{1}{2} \cdot \left(\frac{3}{4}\right)$$

$$y = \frac{5}{2} \cdot \frac{3}{4} \qquad \text{Writing } 2\frac{1}{2} \text{ as } \frac{5}{2}$$

$$y = \frac{15}{8}, \text{ or } 1\frac{7}{8}$$

Check. We repeat the computation. The result checks.

State. $1\frac{7}{8}$ cups of vinegar are needed to make $2\frac{1}{2}$ batches of chili.

75.
$$\begin{aligned}
-8 - 5^2 \cdot 2(3 - 4) &= -8 - 5^2 \cdot 2(-1) \\
&= -8 - 5 \cdot 5 \cdot 2(-1) \\
&= -8 - 25 \cdot 2(-1) \\
&= -8 - 50(-1) \\
&= -8 + 50 \\
&= 42
\end{aligned}$$

77.
$$\begin{aligned}
4.8 - 1.5x &= 0.9 \\
4.8 - 1.5x - 4.8 &= 0.9 - 4.8 \qquad \text{Subtracting 4.8 from} \\
&\qquad\qquad\qquad\qquad \text{both sides} \\
-1.5x &= -3.9 \\
\frac{-1.5x}{-1.5} &= \frac{-3.9}{-1.5} \qquad \text{Dividing both} \\
&\qquad\qquad\qquad \text{sides by } -1.5 \\
x &= 2.6
\end{aligned}$$

The solution is 2.6.

79. Discussion and Writing Exercise

81. We are free to choose any number as a replacement for x or y. To find one solution we choose to replace x with 0.

$$\begin{aligned}
21x - 70y &= -14 \\
21 \cdot 0 - 70y &= -14 \\
0 - 70y &= -14 \\
-70y &= -14 \\
\frac{-70y}{-70} &= \frac{-14}{-70} \\
y &= 0.2
\end{aligned}$$

Thus, $(0, 0.2)$ is one solution of $21x - 70y = -14$.

To find a second solution we can replace x with -4.

$$\begin{aligned}
21x - 70y &= -14 \\
21(-4) - 70y &= -14 \\
-84 - 70y &= -14 \\
-70y &= -14 + 84 \\
-70y &= 70 \\
\frac{-70y}{-70} &= \frac{70}{-70} \\
y &= -1
\end{aligned}$$

Thus, $(-4, -1)$ is a solution of $21x - 70y = -14$.

To find a third solution we can replace x with 1.

$$21x - 70y = -14$$
$$21 \cdot 1 - 70y = -14$$
$$21 - 70y = -14$$
$$-70y = -14 - 21$$
$$-70y = -35$$
$$\frac{-70y}{-70} = \frac{-35}{-70}$$
$$y = 0.5$$

Thus, $(1, 0.5)$ is a solution of $21x - 70y = -14$. We graph the equation.

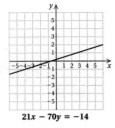

21x − 70y = −14

83. Replace x with -3:

$$50x + 75y = 180$$
$$50(-3) + 75y = 180$$
$$-150 + 75y = 180$$
$$75y = 330$$
$$y = 4.4$$

$(-3, 4.4)$ is one solution.

Replace y with 5:

$$50x + 75y = 180$$
$$50x + 75 \cdot 5 = 180$$
$$50x + 375 = 180$$
$$50x = -195$$
$$x = -3.9$$

$(-3.9, 5)$ is a second solution.

Replace x with 3:

$$50x + 75y = 180$$
$$50 \cdot 3 + 75y = 180$$
$$150 + 75y = 180$$
$$75y = 30$$
$$y = 0.4$$

$(3, 0.4)$ is a third solution.

We graph the equation.

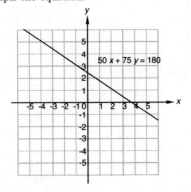

85. On the graph we locate three points, other than those already listed, whose coordinates are easily determined. We see that three solutions of $y = x + 2$ are $(2, 4)$, $(-3, -1)$, and $(-5, -3)$.

87. Answers may vary.

We substitute three negative values for x to find three solutions in the second quadrant.

If $x = -4$, $y = |-4| = 4$.
If $x = -2$, $y = |-2| = 2$.
If $x = -1$, $y = |-1| = 1$.

Thus, three solutions in the second quadrant are $(-4, 4)$, $(-2, 2)$, and $(-1, 1)$.

We substitute three positive values for x to find three solutions in the first quadrant.

If $x = 1$, $y = |1| = 1$.
If $x = 3$, $y = |3| = 3$.
If $x = 5$, $y = |5| = 5$.

Thus, three solutions in the first quadrant are $(1, 1)$, $(3, 3)$, and $(5, 5)$.

All of the solutions found above lie on the graph of $y = |x|$.

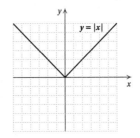

89.

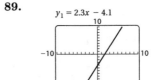

$y_1 = 2.3x - 4.1$

Exercise Set 6.5

1. To find the mean, add the numbers. Then divide by the number of addends.

$$\frac{17 + 19 + 29 + 18 + 14 + 29}{6} = \frac{126}{6} = 21$$

The mean is 21.

To find the median, first list the numbers in order from smallest to largest. Then locate the middle number.

$$14, 17, 18, 19, 29, 29$$
$$\uparrow$$
Middle number

The median is halfway between 18 and 19. It is the average of the two middle numbers:

$$\frac{18 + 19}{2} = \frac{37}{2} = 18.5$$

Find the mode:

The number that occurs most often is 29. The mode is 29.

3. To find the mean, add the numbers. Then divide by the number of addends.

$$\frac{5 + 37 + 20 + 20 + 35 + 5 + 25}{7} = \frac{147}{7} = 21$$

The mean is 21.

To find the median, first list the numbers in order from smallest to largest. Then locate the middle number.

$$5, 5, 20, 20, 25, 35, 37$$
$$\uparrow$$
Middle number

The median is 20.

Find the mode:

There are two numbers that occur most often, 5 and 20. Thus the modes are 5 and 20.

5. Find the mean:

$$\frac{4.3 + 7.4 + 1.2 + 5.7 + 7.4}{5} = \frac{26}{5} = 5.2$$

The mean is 5.2.

Find the median:

$$1.2, 4.3, 5.7, 7.4, 7.4$$
$$\uparrow$$
Middle number

The median is 5.7.

Find the mode:

The number that occurs most often is 7.4. The mode is 7.4.

7. Find the mean:

$$\frac{234 + 228 + 234 + 229 + 234 + 278}{6} = \frac{1437}{6} = 239.5$$

The mean is 239.5.

Find the median:

$$228, 229, 234, 234, 234, 278$$
$$\uparrow$$
Middle number

The median is halfway between 234 and 234. Although it seems clear that this is 234, we can compute it as follows:

$$\frac{234 + 234}{2} = \frac{468}{2} = 234$$

The median is 234.

Find the mode:

The number that occurs most often is 234. The mode is 234.

9. Find the mean:

$$\frac{1 + 1 + 3 + 3 + 16 + 1 + 0 + 2 + 3 + 1 + 11 + 1}{12} = \frac{43}{12} =$$
$$3.58\overline{3}$$

The mean is $3.58\overline{3}$.

Find the median:

$$0, 1, 1, 1, 1, 1, 2, 3, 3, 3, 11, 16$$
$$\uparrow$$
Middle number

The median is halfway between 1 and 2. It is the average of the two middle numbers.

$$\frac{1 + 2}{2} = \frac{3}{2}, \text{ or } 1.5$$

The median is 1.5.

Find the mode:

The number that occurs most often is 1. The mode is 1.

11. We divide the total number of miles, 279, by the number of gallons, 9.

$$\frac{279}{9} = 31$$

The mean was 31 miles per gallon.

13. To find the GPA we first add the grade point values for each hour taken. This is done by first multiplying the grade point value by the number of hours in the course and then adding as follows:

$$
\begin{array}{lll}
B & 3.00 \cdot 4 = & 12 \\
A & 4.00 \cdot 5 = & 20 \\
D & 1.00 \cdot 3 = & 3 \\
C & 2.00 \cdot 4 = & 8 \\
\hline
& & 43 \ \text{(Total)}
\end{array}
$$

The total number of hours taken is

$$4 + 5 + 3 + 4, \text{ or } 16.$$

We divide 43 by 16 and round to the nearest tenth.

$$\frac{43}{16} = 2.6875 \approx 2.7$$

The student's grade point average is 2.7.

15. Find the average price per pound:

$$\frac{\$6.99 + \$8.49 + \$8.99 + \$6.99 + \$9.49}{5} = \frac{\$40.95}{5} = \$8.19$$

The average price per pound of Atlantic salmon was $8.19.

Find the median price per pound:

List the prices in order:

$$\$6.99, \$6.99, \$8.49, \$8.99, \$9.49$$
$$\uparrow$$
Middle number

The median is $8.49.

Find the mode:

The number that occurs most often is $6.99. The mode is $6.99.

17. Adding the numbers in the Alcohol-Related Traffic Deaths column, we get 137,608. Then we divide by the number of items of data, 8:

$$\frac{137,608}{8} = 17,201$$

The mean number of alcohol-related traffic deaths for the years 1995-2002 was 17,201.

19. We can find the total of the five scores needed as follows:

$$80 + 80 + 80 + 80 + 80 = 400.$$

The total of the scores on the first four tests is

$$80 + 74 + 81 + 75 = 310.$$

Thus Rich needs to get at least

$$400 - 310, \text{ or } 90$$

to get a B. We can check this as follows:

$$\frac{80 + 74 + 81 + 75 + 90}{5} = \frac{400}{5} = 80.$$

21. We can find the total number of days needed as follows:

$$266 + 266 + 266 + 266 = 1064.$$

The total number of days for Marta's first three pregnancies is

$$270 + 259 + 272 = 801.$$

Thus, Marta's fourth pregnancy must last

$$1064 - 801 = 263 \text{ days}$$

in order to equal the worldwide average.

We can check this as follows:

$$\frac{270 + 259 + 272 + 263}{4} = \frac{1064}{4} = 266.$$

23. Compare the means of the two sets of data.

Bulb A: Mean = $(983 + 964 + 1214 + 1417 + 1211 + 1521 + 1084 + 1075 + 892 + 1423 + 949 + 1322)/12 = 1171.25$

Bulb B: Mean = $(979 + 1083 + 1344 + 984 + 1445 + 975 + 1492 + 1325 + 1283 + 1325 + 1352 + 1432)/12 \approx 1251.58$

Since the mean life of Bulb A is 1171.25 hr and of Bulb B is about 1251.58 hr, Bulb B is better.

25. Discussion and Writing Exercise

27.
```
      1 4
    × 1 4
    ─────
      5 6
    1 4 0
    ─────
    1 9 6
```

29.
```
      1. 4    (1 decimal place)
    × 1. 4    (1 decimal place)
    ─────
      5 6
    1 4 0
    ─────
    1. 9 6    (2 decimal places)
```

31.
```
      1 2. 8 6    (2 decimal places)
    × 1 7.5       (1 decimal place)
    ─────────
      6 4 3 0
    9 0 0 2 0
  1 2 8 6 0 0
  ───────────
  2 2 5. 0 5 0    (3 decimal places)
```

33.
$$\frac{4}{5} \cdot \frac{3}{28} = \frac{4 \cdot 3}{5 \cdot 28}$$
$$= \frac{4 \cdot 3}{5 \cdot 4 \cdot 7}$$
$$= \frac{4}{4} \cdot \frac{3}{5 \cdot 7}$$
$$= \frac{3}{35}$$

35. *Familiarize.* Let $h = $ the number of hours the disc jockey can work for $165. The amount charged for setting-up and working h hours is given by $40 + 50h$.

Translate.

$$\underbrace{\text{Amount charged}}_{\downarrow} \quad \underset{\downarrow}{\text{is}} \quad \underset{\downarrow}{\$165.}$$
$$40 + 50h \quad = \quad 165$$

Solve. We solve the equation.

$$40 + 50h = 165$$
$$40 + 50h - 40 = 165 - 40$$
$$50h = 125$$
$$h = \frac{125}{50} = 2.5$$

Check. At $50 per hour, the disc jockey charges 50(2.5), or $125, for working 2.5 hr. The total cost is $125 plus the set-up fee of $40, $125 + $40, or $165. The answer checks.

State. The disc jockey can work for 2.5 hr.

37. Discussion and Writing Exercise

39. Use a calculator to divide the total by the number of games.

$$\frac{4176}{23} \approx 181.57$$

Drop the amount to the right of the decimal point. The bowler's average was 181.

41. We can find the total number of home runs needed over Aaron's 22-yr career as follows:

$$22 \cdot 34\frac{7}{22} = 22 \cdot \frac{755}{22} = \frac{22 \cdot 755}{22} = \frac{22}{22} \cdot \frac{755}{1} = 755.$$

The total number of home runs during the first 21 years of Aaron's career was

$$21 \cdot 35\frac{10}{21} = 21 \cdot \frac{745}{21} = \frac{21 \cdot 745}{21} = \frac{21}{21} \cdot \frac{745}{1} = 745.$$

Then Aaron hit

$$755 - 745 = 10 \text{ home runs}$$

in his final year.

43. Amy's second offer: $\dfrac{\$3600 + \$3200}{2} = \$3400$

Jim's second offer: $\dfrac{\$3400 + \$3600}{2} = \$3500$

Amy's third offer: $\dfrac{\$3500 + \$3400}{2} = \$3450$

Jim's third offer: $\dfrac{\$3500 + \$3450}{2} = \$3475$

Amy will pay $3475 for the car.

45. A full gas tank holds 10.5 gallons, so $\dfrac{3}{4}$ of a tank holds $\dfrac{3}{4}(10.5)$, or 7.875 gal. Then, at an average of 61 miles per gallon, using 7.875 gal the car can be driven

$\qquad 61(7.875)$, or 480.375 mi.

We could also do this problem by first finding the number of miles the car can be driven using a full tank of gas:

$\qquad 61(10.5) = 640.5$ mi.

Then $\dfrac{3}{4}$ of this distance is

$\qquad \dfrac{3}{4}(640.5) = 480.375$ mi.

Exercise Set 6.6

1. We interpolate by finding the average of the data values for 17 hours of study and 21 hours of study.

$\qquad \dfrac{80 + 86}{2} = \dfrac{166}{2} = 83$

The missing data value is 83.

We could have also used a graph to find this value, as in Example 1.

3. Graph the data and use the graph to extrapolate. Draw a "representative" line through the data and beyond. To estimate the value for 2005, draw a vertical line up from 2005 to the representative line. Then go to the left and read a value of about $15.4 million.

5. Graph the data and use the graph to extrapolate. Draw a "representative" line through the data and beyond. To estimate the value for 2006, draw a vertical line up from 2006 to the representative line. Then go to the left and read a value of about $2.7 million.

7. We interpolate by finding the average of the data values for 2001 and 2003.

$\qquad \dfrac{50.6 + 59.2}{2} = \dfrac{109.8}{2} = 54.9$

The missing data value is $54.9 billion.

We could also have used a graph to find this value as in Example 1.

9. Since 1, 2, 3, 4, 5, or 6 are equally likely to occur, the probability that a 3 is rolled is $\dfrac{1}{6}$, or $0.1\overline{6}$.

11. Since 1, 2, 3, 4, 5, or 6 are equally likely to occur and 3 of these possibilities are odd numbers, the probability of rolling an odd number is:

$\qquad \dfrac{\text{Number of ways to roll an odd number}}{\text{Number of equally likely outcomes}}$

$\qquad = \dfrac{3}{6}$

$\qquad = \dfrac{1}{2}$, or 0.5

13. The probability that the card is the jack of spades is:

$\qquad \dfrac{\text{Number of ways to select the jack of spades}}{\text{Number of ways to select any card}}$

$\qquad = \dfrac{1}{52}$

15. The probability that an eight or six is selected is:

$\qquad \dfrac{\text{Number of ways to select an 8 or a 6}}{\text{Number of ways to select any card}}$

$\qquad = \dfrac{8}{52}$

$\qquad = \dfrac{2}{13}$

17. The probability of selecting a red picture card is:

$\qquad \dfrac{\text{Number of ways to select a red picture card}}{\text{Number of ways to select any card}}$

$\qquad = \dfrac{6}{52}$

$\qquad = \dfrac{3}{26}$

19. The probability that a cherry gumdrop is selected is:

$\qquad \dfrac{\text{Number of ways to select a cherry gumdrop}}{\text{Number of ways to select any gumdrop}}$

$\qquad = \dfrac{4}{39}$

21. The probability that a gumdrop that is not lime is selected is:

$\qquad \dfrac{\text{Number of ways to select a non-lime gumdrop}}{\text{Number of ways to select any gumdrop}}$

$\qquad = \dfrac{34}{39}$

23. Discussion and Writing Exercise

25. The set $\{1, 2, 3, 4, 5 \ldots\}$ is called the set of <u>natural</u> numbers.

27. To find the <u>mean</u> of a set of numbers, add the numbers and then divide by the number of items of data.

29. Values in between known values can be estimated using <u>interpolation</u>.

31. The statement $x + t = t + x$ illustrates the <u>commutative</u> law.

33. Discussion and Writing Exercise

35. The probability that each flip of the coin produces a head is $\dfrac{1}{2} \cdot \dfrac{1}{2} = \dfrac{1}{4}$, or 0.25.

37. The probability that each roll of the die produces a 3 is $\dfrac{1}{6} \cdot \dfrac{1}{6} = \dfrac{1}{36}$.

39. Discussion and Writing Exercise

Chapter 6 Review Exercises

1. Go down the FedEx Letter column to 3. Then go across to the column headed FedEx Priority Overnight and read the entry, $41.71. Thus the cost of a 3-lb FedEx Priority Overnight delivery is $41.71.

2. Go down the FedEx Letter Column to 10. Then go across to the column headed FedEx Standard Overnight and read the entry, $59.64. Thus the cost of a 10-lb FedEx Standard Overnight delivery is $59.64.

3. From the table we see that it costs $15.44 to send a 3-lb letter by FedEx 2Day delivery. Now we subtract to find the amount saved by using 2Day delivery:

$$\$41.71 - \$15.44 = \$26.27$$

4. From the table we see that it costs $27.58 to send a 10-lb letter by FedEx 2Day delivery. Now we subtract to find the amount saved by using 2Day delivery:

$$\$59.64 - \$27.58 = \$32.06$$

5. Within each category the price is the same for all packages up to 8 oz, so there is no difference in price between sending a 5-oz package FedEx Priority Overnight and sending an 8-oz package in the same way.

6. Cost for 4-lb package: $40.18

Cost for 5-lb package: $42.22

Total cost for a 4-lb package and a 5-lb package:

$$\$40.18 + \$42.22 = \$82.40$$

Weight of combined packages: 4 lb + 5 lb = 9 lb

Cost for 9-lb package: $56.30

Amount saved by sending both packages as one:

$$\$82.40 - \$56.30 = \$26.10$$

7. The Chicago police force is represented by 7 symbols, so there are 7×2000, or 14,000, officers.

8. $9000 \div 2000 = 4.5$, so we look for a city represented by about 4.5 symbols. It is Los Angeles.

9. Houston is represented by the smallest number of symbols, so it has the smallest police force.

10. First we find the number of symbols representing each police force. Answers may vary slightly depending on how partial symbols are counted.

New York: 17

Chicago: 7

Los Angeles: 4.8

Philadelphia: 3.6

Washington, D.C.: 2.6

Houston: 2.5

Now we find the average of these numbers.

$$\frac{17 + 7 + 4.8 + 3.6 + 2.6 + 2.5}{6} = \frac{37.5}{6} = 6.25$$

Finally we multiply to find the number of officers represented by 6.25 symbols.

$$6.25 \times 2000 = 12,500$$

The average size of the six police forces is 12,500 officers.

11. The longest bar is for MLB, so Major League Baseball has the greatest number of fans.

12. Move to the right along the bar for the NFL. We read that the National Football League has about 120 million fans.

13. We locate 100 at the bottom of the graph and then go up until we find a bar that ends at approximately 100. Now go across to the left and read that the NBA (National Basketball Association) has about 100 million fans.

14. We look for the bars that extend to 100 or beyond it. Those bars are for the NFL, the NBA, and MLB, so these are the sports with 100 million or more fans.

15. From the graph we see that MLB has about 175 million fans and NASCAR has about 75 million fans. We subtract to find how many more fans MLB has.

$$175 \text{ million} - 75 \text{ million} = 100 \text{ million fans}$$

16. From the graph we see that MLB, NASCAR, and the NHL have about 175 million, 75 million, and 80 million fans, respectively. We add to find the number of fans of NASCAR and the NHL combined.

$$75 \text{ million} + 80 \text{ million} = 155 \text{ million.}$$

Since 155 million is less than 175 million, it is true that there are more MLB fans than NASCAR and NHL fans combined.

We could also use the result of Exercise 15, that MLB has 100 million more fans than NASCAR. Since the NHL has fewer than 100 million fans, this indicates that the given statement is true.

17. The highest point on the graph lies above the Under 20 label on the horizontal scale, so the under 20 age group has the most accidents per 100 drivers.

18. Find the lowest point on the graph and then move across to the vertical scale to read that 12 accidents is the fewest number of accidents per 100 drivers in any age group.

19. From the graph we see that people 75 and over have 25 accidents per 100 drivers and those in the 65-74 age range have about 12 accidents per 100 drivers. We subtract to find the difference.

$$25 - 12 = 13 \text{ accidents per 100 drivers}$$

20. We see that the line is nearly horizontal (it rises and falls only slightly) from the 45-54 age group to the 65-74 age group. Thus the number of accidents stays basically the same from ages 45 to 74.

21. From the graph we see that people in the 25-34 age group have about 22 accidents per 100 drivers and those in the 20-24 age group have about 34. We subtract to find the difference.

$$34 - 22 = 12 \text{ accidents per 100 drivers}$$

22. From the graph we see that people in the 55-64 age group have about 12 accidents per 100 drivers. Then $3 \cdot 12 = 36$ and we see that people under 20 have about 36 accidents per 100 drivers, so people in this age group have about three times as many accidents as those in the 55-64 age group.

23. On the horizontal scale in seven equally spaced intervals indicate the years. Label this scale "Year." Then label the vertical scale "Cost of first-class postage." The smallest cost is 25¢ and the largest is 39¢, so we start the vertical scale at 0 and extend it to 40¢, labeling it by 5's. Finally, draw vertical bars above the years to show the cost of the postage.

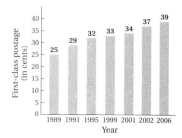

24. Prepare horizontal and vertical scales as described in Exercise 23. Then, at the appropriate level above each year, mark the corresponding postage. Finally, draw line segments connecting the points.

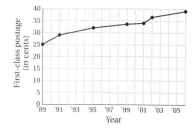

25. We look above point A to see that its first coordinate is -5. Looking to the right of point A, we find that its second coordinate is -1. Thus, the coordinates of point A are $(-5, -1)$.

26. We look below point B to see that its first coordinate is -2. Looking to the right of point B, we find that its second coordinate is 5. Thus, the coordinates of point B are $(-2, 5)$.

27. Looking at the location of point C on the x-axis, we see that its first coordinate is 3. Looking to the left of point C, we see that its second coordinate is 0. Thus, the coordinates of point C are $(3, 0)$.

28. We look above point D to see that its first coordinate is 4. Looking to the left of point D, we find that its second coordinate is -2. Thus, the coordinates of point D are $(4, -2)$.

29. To plot $(2, 5)$ we locate 2 on the x-axis. From there we go up 5 units and make a dot. See the graph in Exercise 31.

30. To plot $(0, -3)$ we locate 0 on the x-axis. From there we go down 3 units and make a dot. See the graph in Exercise 31.

31. To plot $(-4, -2)$ we locate -4 on the x-axis. From there we go down 2 units and make a dot.

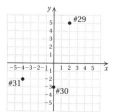

32. Since the first coordinate is positive and the second coordinate is negative, the point $(3, -8)$ is in quadrant IV.

33. Since both coordinates are negative, the point $(-20, -14)$ is in quadrant III.

34. Since both coordinates are positive, the point $\left(4\frac{9}{10}, 1\frac{3}{10}\right)$ is in quadrant I.

35.
$$2x + 4y = 10$$
$$2 \cdot 1 + 4y = 10$$
$$2 + 4y = 10$$
$$4y = 8$$
$$y = 2$$

The pair $(1, 2)$ is a solution of $2x + 4y = 10$.
$$2x + 4y = 10$$
$$2x + 4(-2) = 10$$
$$2x - 8 = 10$$
$$2x = 18$$
$$x = 9$$

The pair $(9, -2)$ is a solution of $2x + 4y = 10$.

36. Graph: $y = 2x - 5$

We make a table of values. Then we plot the points, draw the line, and label it.

When $x = 0$, $y = 2 \cdot 0 - 5 = 0 - 5 = -5$.

When $x = 2$, $y = 2 \cdot 2 - 5 = 4 - 5 = -1$.

When $x = 4$, $y = 2 \cdot 4 - 5 = 8 - 5 = 3$.

x	y = $2x - 5$	(x, y)
0	-5	$(0, -5)$
2	-1	$(2, -1)$
4	3	$(4, 3)$

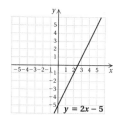

37. Graph: $y = -\dfrac{3}{4}x$

We make a table of values. Note that when x is a multiple of 4, fraction values for y are avoided. We plot the points, draw the line, and label it.

When $x = -4$, $y = -\frac{3}{4}(-4) = 3$.

When $x = 0$, $y = -\frac{3}{4} \cdot 0 = 0$.

When $x = 4$, $y = -\frac{3}{4} \cdot 4 = -3$.

x	$y = -\frac{3}{4}x$	(x, y)
-4	3	$(-4, 3)$
0	0	$(0, 0)$
4	-3	$(4, -3)$

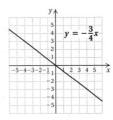

38. Graph: $x + y = 4$

We make a table of values. Then we plot the points, draw the line, and label it.

When $x = -1$: $-1 + y = 4$
                         $y = 5$

When $x = 2$: $2 + y = 4$
                         $y = 2$

When $x = 4$: $4 + y = 4$
                         $y = 0$

x	$x + y = 4$	(x, y)
-1	5	$(-1, 5)$
2	2	$(2, 2)$
4	0	$(4, 0)$

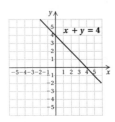

39. Graph: $x = -5$

Any ordered pair of the form $(-5, y)$ is a solution of the equation. The graph is a vertical line 5 units left of the y-axis.

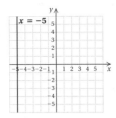

40. Graph: $y = 6$, or $0 \cdot x + y = 6$

No matter what number we choose for x, y must be 6. The graph is a horizontal line 6 units above the x-axis.

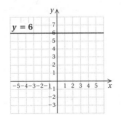

41. a) Mean: $\dfrac{26 + 51 + 34 + 26 + 43}{5} = \dfrac{180}{5} = 36$

   b) Median:
$$26, 26, 34, 43, 51$$
$$\uparrow$$
$$\text{Middle number}$$

The median is 34.

   c) Mode: The number 26 occurs most often. It is the mode.

42. a) Mean: $\dfrac{11 + 14 + 17 + 17 + 21 + 7 + 11}{7} = \dfrac{98}{7} = 14$

   b) Median:
$$7, 11, 11, 14, 17, 17, 21$$
$$\uparrow$$
$$\text{Middle number}$$

The median is 14.

   c) Mode: There are two numbers that occur most often, 11 and 17. They are the modes.

43. a) Mean:
$$\dfrac{500 + 25 + 470 + 190 + 470 + 280}{6} = \dfrac{1935}{6} = 322.5$$

   b) Median:
$$25, 190, 280, 470, 470, 500$$
$$\uparrow$$
$$\text{Middle number}$$

The median is halfway between 280 and 470. It is the average of the two middle numbers.
$$\dfrac{280 + 470}{2} = \dfrac{750}{2} = 375$$

   c) Mode: The number that occurs most often is 470. It is the mode.

44. a) Mean:
$$\dfrac{700 + 700 + 1900 + 2700 + 3000}{5} = \dfrac{9000}{5} = 1800$$

   b) Median:
$$700, 700, 1900, 2700, 3000$$
$$\uparrow$$
$$\text{Middle number}$$

The median is 1900.

   c) Mode: The number that occurs most often is 700. It is the mode.

45. a) Mean:
$$\dfrac{\$30,000 + \$75,000 + \$20,000 + \$25,000}{4} =$$
$$\dfrac{\$150,000}{4} = \$37,500$$

b) Median:

$$\$20,000, \$25,000, \$30,000, \$75,000$$
$$\uparrow$$
Middle number

The median is halfway between $25,000 and $30,000. It is the average of the two middle numbers.

$$\frac{\$25,000 + \$30,000}{2} = \frac{\$55,000}{2} = \$27,500$$

c) Mode: Each number occurs the same number of times, so there is no mode.

46. We can find the total of the four scores needed as follows:

$$90 + 90 + 90 + 90 = 360.$$

The total of the scores on the first three tests is

$$94 + 78 + 92 = 264.$$

Thus the student needs to get at least

$$360 - 264 = 96$$

to get an A. We can check this as follows:

$$\frac{94 + 78 + 92 + 96}{4} = \frac{360}{4} = 90.$$

47. To find the GPA we first add the grade point values for each hour taken. This is done by first multiplying the grade point value by the number of hours in the course and then adding as follows:

$$
\begin{array}{llll}
\text{A} & 4.0 \cdot 5 & = & 20 \\
\text{B} & 3.0 \cdot 3 & = & 9 \\
\text{C} & 2.0 \cdot 4 & = & 8 \\
\text{B} & 3.0 \cdot 3 & = & 9 \\
\text{B} & 3.0 \cdot 1 & = & 3 \\
\hline
& & & 49 \quad \text{(Total)}
\end{array}
$$

The total number of hours taken is

$$5 + 3 + 4 + 3 + 1, \text{ or } 16.$$

We divide 49 by 16 and round to the nearest tenth.

$$\frac{49}{16} = 3.0625 \approx 3.1$$

The student's grade point average is 3.1.

48. Battery A mean:

$$(38.9 + 39.3 + 40.4 + 53.1 + 41.7 + 38.0 + 36.8 + 47.7 +$$
$$48.1 + 38.2 + 46.9 + 47.4) \div 12 = \frac{516.5}{12} \approx 43.04$$

Battery B mean:

$$(39.3 + 38.6 + 38.8 + 37.4 + 47.6 + 37.9 + 46.9 + 37.8 +$$
$$38.1 + 47.9 + 50.1 + 38.2) \div 12 = \frac{498.6}{12} = 41.55$$

Because the mean time for Battery A is longer, it is the better battery.

49. Using the graph, we estimate the number on the vertical scale that corresponds to the point on the horizontal scale that is halfway between the 25 to 34 and the 35 to 44 age groups. We estimate that there are about 18 accidents per 100 drivers in the 30 to 44 age group.

50. $\dfrac{\text{Number of ways to select the five of clubs}}{\text{Number of ways to select any card}}$

$$= \frac{1}{52}$$

51. $\dfrac{\text{Number of ways to select a red card}}{\text{Number of ways to select any card}}$

$$= \frac{26}{52} = \frac{1}{2}$$

52. *Discussion and Writing Exercise.* It is possible for the mean of a set of numbers to be larger than all but one number in the set. To see this, note that the mean of the set $\{6, 8\}$ is 7, which is larger than all of the numbers in the set but one.

53. *Discussion and Writing Exercise.* The median of a set of four numbers can be in the set. For example, the median of the set $\{11, 15, 15, 17\}$ is 15, which is in the set.

54. $34x + 47y = 100$

Substitute 0 for x:

$$34 \cdot 0 + 47y = 100$$
$$47y = 100$$
$$y = \frac{100}{47}$$

One solution is $\left(0, \dfrac{100}{47}\right)$.

Now substitute 0 for y:

$$34x + 47 \cdot 0 = 100$$
$$34x = 100$$
$$x = \frac{100}{34} = \frac{50}{17}$$

Another solution is $\left(0, \dfrac{50}{17}\right)$.

Finally, substitute 2 for x.

$$34 \cdot 2 + 47y = 100$$
$$68 + 47y = 100$$
$$47y = 32$$
$$y = \frac{32}{47}$$

A third solution is $\left(2, \dfrac{32}{47}\right)$.

The ordered pairs found may vary. We plot these points and draw the graph.

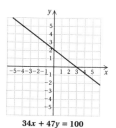

$34x + 47y = 100$

55. We find the total amount earned per hour and then divide by the number of items of data, 4 + 9, or 13.

$$\frac{4 \cdot \$12.35 + 9 \cdot \$11.15}{13} = \frac{\$149.75}{13} \approx \$11.52$$

The mean hourly wage is about \$11.52 per hour.

56. 298, 301, 305, a, 323, b, 390

Since the median is 316 and a is the middle number, we have $a = 316$.

Since the mean is 326, we know the sum of the seven numbers must be $7 \cdot 326$, or 2282. Then we have:

$$298 + 301 + 305 + 316 + 323 + b + 390 = 2282$$
$$1933 + b = 2282$$
$$1933 + b - 1933 = 2282 - 1933$$
$$b = 349$$

57. Graph: $1\frac{2}{3}x + \frac{3}{4}y = 2$

We make a table of values, plot points, and draw and label the graph.

x	y	(x, y)
0	$2\frac{2}{3}$	$\left(0, 2\frac{2}{3}\right)$
$1\frac{1}{5}$	0	$\left(1\frac{1}{5}, 0\right)$
3	-4	$(3, -4)$

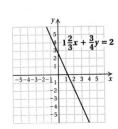

58. Graph: $\frac{3}{4}x - 2\frac{1}{2}y = 3$

We make a table of values, plot points, and draw and label the graph.

x	y	(x, y)
0	$-1\frac{1}{5}$	$\left(0, -1\frac{1}{5}\right)$
4	0	$(4, 0)$
1	$-\frac{9}{10}$	$\left(1, -\frac{9}{10}\right)$

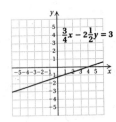

Chapter 6 Test

1. The largest number in the column headed "132 lb" is 249. It corresponds to hiking at 3 mph with a 20-lb load, so this is the activity that provides the greatest benefit in burned calories for a person who weighs 132 lb.

2. There are two numbers that are at least 250 in the column headed "154 lb." The least strenuous is hiking at 3 mph with a 10-lb load.

3. Since $600 \div 100 = 6$, we look for a country represented by 6 symbols. We find that it is Japan.

4. Since $1000 \div 100 = 10$, we look for a country represented by 10 symbols. We find that it is the United States.

5. The amount of waste generated per person per year in France is represented by 8 symbols, so each person generates $8 \cdot 100$, or 800 lb, of waste per year.

6. The amount of waste generated per person per year in Finland is represented by 4 symbols, so each person generates $4 \cdot 100$, or 400 lb, of waste per year.

7. First indicate the names of the animals in seven equally spaced intervals on the horizontal scale. Title this scale "Animals." Now note that the lowest speed is 28 mph and the highest is 225 mph. We start the vertical scaling at 0 and label the marks on the scale by 50's from 0 to 300. Title this scale "Maximum speed (in miles per hour)." Finally, draw vertical bars above the names of the animals to show the speeds.

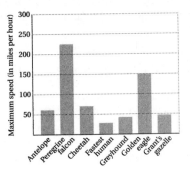

8. From the table or the bar graph, we see that the slowest speed is 28 mph and the fastest is 225 mph. Then the fastest speed exceeds the slowest by

225 − 28, or 197 mph.

9. The fastest human's maximum speed is 28 mph and a greyhound's maximum speed is 42 mph. We divide to find how many times faster the greyhound is.

$$\frac{42}{28} = \frac{2 \cdot 3 \cdot 7}{2 \cdot 2 \cdot 2} = \frac{2 \cdot 7}{2 \cdot 7} \cdot \frac{3}{2} = \frac{3}{2}, \text{ or } 1\frac{1}{2}$$

A greyhound can run $1\frac{1}{2}$ times faster than the fastest human.

10. The highest point on the graph corresponds to '01, so the average price was highest in 2001.

11. The line slants down most steeply between '01 and '02, so the price fell the most between 2001 and 2002.

12. We look from left to right along the line at 30. We see that the point on the line at 30 corresponds to '00, so the price was about \$30 in 2000.

13. Note that the line slants slightly downward from 2003 to 2004. We extend the line to 2005, continuing this trend. To estimate a value for 2005, go vertically up to the graph from 2005 and then go to the left and read a value of about \$26 on the vertical scale.

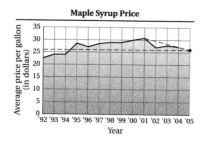

Maple Syrup Price

14. Since the first coordinate is negative and the second coordinate is positive, the point $\left(-\frac{1}{2}, 7\right)$ is in quadrant II.

15. Since both coordinates are negative, the point $(-5, -6)$ is in quadrant III.

16. We look below point A to see that its first coordinate is 3. Looking to the left of point A, we find that its second coordinate is 4. Thus, the coordinates of point A are $(3, 4)$.

17. We look above point B to see that its first coordinate is 0. Looking at the location of point B on the y-axis, we see that its second coordinate is -4. Thus, the coordinates of point B are $(0, -4)$.

18. We look below point C to see that its first coordinate is -4. Looking to the right of point C, we find that its second coordinate is 2. Thus, the coordinates of point C are $(-4, 2)$.

19.
$$y - 3x = -10$$
$$2 - 3x = -10$$
$$-3x = -12$$
$$x = 4$$

The pair $(4, 2)$ is a solution of $y - 3x = -10$.

20. Graph: $y = 2x - 2$

We make a table of values. Then we plot the points and draw and label the line.

When $x = -1$, $y = 2(-1) - 2 = -2 - 2 = -4$.

When $x = 0$, $y = 2 \cdot 0 - 2 = 0 - 2 = -2$.

When $x = 3$, $y = 2 \cdot 3 - 2 = 6 - 2 = 4$.

x	y $y = 2x - 2$	(x, y)
-1	-4	$(-1, -4)$
0	-2	$(0, -2)$
3	4	$(3, 4)$

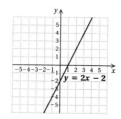

21. Graph: $y = -\frac{3}{2}x$

We make a table of values. Note that when x is a multiple of 2, fraction values for y are avoided. We plot the points and draw and label the line.

When $x = -2$, $y = -\frac{3}{2}(-2) = 3$.

When $x = 0$, $y = -\frac{3}{2} \cdot 0 = 0$.

When $x = 2$, $y = -\frac{3}{2} \cdot 2 = -3$.

x	y $y = -\frac{3}{2}x$	(x, y)
-2	3	$(-2, 3)$
0	0	$(0, 0)$
2	-3	$(2, -3)$

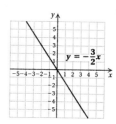

22. Graph: $x = -2$

Any ordered pair of the form $(-2, y)$ is a solution of the equation. The graph is a vertical line 2 units left of the y-axis.

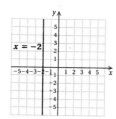

23. $\dfrac{45 + 49 + 52 + 54}{4} = \dfrac{200}{4} = 50$

24. $\dfrac{1 + 2 + 3 + 4 + 5}{5} = \dfrac{15}{5} = 3$

25. $\dfrac{3 + 17 + 17 + 18 + 18 + 20}{6} = \dfrac{93}{6} = 15.5$

26. Median:

$$45, 47, 54, 54$$
$$\uparrow$$
Middle number

The median is halfway between 47 and 54. It is the average of the two middle numbers.

$$\frac{47 + 54}{2} = \frac{101}{2} = 50.5$$

Mode: The number 54 occurs most often, so it is the mode.

27. Median:

$$1, 2, 3, 4, 5$$
$$\uparrow$$
Middle number

The median is 3.

Mode: Each number occurs the same number of times, so there is no mode.

28. Median:

$$3, 17, 17, 18, 18, 20$$
$$\uparrow$$

Middle number

The median is halfway between 17 and 18. It is the average of the two middle numbers.

$$\frac{17 + 18}{2} = \frac{35}{2} = 17.5$$

Mode: There are two numbers that occur most often, 17 and 18. They are the modes.

29. We divide the number of miles by the number of gallons.

$$\frac{432}{16} = 27 \text{ mpg}$$

30. The total of the four scores needed is

$$70 + 70 + 70 + 70 = 4 \cdot 70, \text{ or } 280.$$

The total of the scores on the first three tests is

$$68 + 71 + 65 = 204.$$

Thus the student needs to get at least

$$280 - 204, \text{ or } 76$$

on the fourth test.

31. We find the mean of each set of ratings.

Bar A:
$$\frac{9 + 10 + 8 + 10 + 9 + 7 + 6 + 9 + 10 + 7 + 8 + 8}{12} =$$

$$\frac{101}{12} \approx 8.417$$

Bar B:
$$\frac{10 + 6 + 8 + 9 + 10 + 10 + 8 + 7 + 6 + 9 + 10 + 8}{12} =$$

$$\frac{101}{12} \approx 8.417$$

Since the means are equal, the chocolate bars are of equal quality.

32. To find the GPA we first add the grade point values for each class taken. This is done by first multiplying the grade point value by the number of hours in the course and then adding as follows:

$$
\begin{array}{lll}
\text{B} & 3.0 \cdot 3 = & 9 \\
\text{A} & 4.0 \cdot 3 = & 12 \\
\text{C} & 2.0 \cdot 4 = & 8 \\
\text{B} & 3.0 \cdot 3 = & 9 \\
\text{B} & 3.0 \cdot 2 = & \underline{6} \\
& & 44 \text{ (Total)}
\end{array}
$$

The total number of hours taken is

$$3 + 3 + 4 + 3 + 2, \text{ or } 15.$$

We divide 44 by 15 and round to the nearest tenth.

$$\frac{44}{15} = 2.9\overline{3} \approx 2.9$$

The grade point average is 2.9.

33.

$$\frac{\text{Number of ways to select a month that begins with J}}{\text{Number of ways to select a month}}$$

$$= \frac{3}{12} = \frac{1}{4}$$

34. Graph: $\frac{1}{4}x + 3\frac{1}{2}y = 1$

We make a table of values, plot the points, and draw and label the graph.

x	y	(x, y)
-4	$\frac{4}{7}$	$\left(-4, \frac{4}{7}\right)$
0	$\frac{2}{7}$	$\left(0, \frac{2}{7}\right)$
4	0	$(4, 0)$

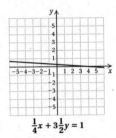

$\frac{1}{4}x + 3\frac{1}{2}y = 1$

35. Graph: $\frac{5}{6}x - 2\frac{1}{3}y = 1$

We make a table of values, plot the points, and draw and label the graph.

x	y	(x, y)
-3	$-\frac{3}{2}$	$\left(-3, -\frac{3}{2}\right)$
0	$-\frac{3}{7}$	$\left(0, -\frac{3}{7}\right)$
$\frac{6}{5}$	0	$\left(\frac{6}{5}, 0\right)$

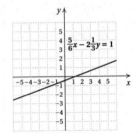

36.

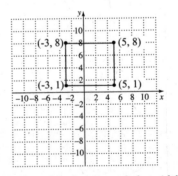

The length is 8 units and the width is 7 units.

$$A = l \cdot w$$

$$A = 8 \text{ units} \cdot 7 \text{ units} = 56 \text{ square units}$$

Cumulative Review Chapters 1 - 6

1. $\underbrace{7 \cdot 7 \cdot 7 \cdot 7}_{4 \text{ factors}} = 7^4$

2. Standard notation for 8 billion is 8,000,000,000.

3. *Familiarize.* Let $p =$ the number of pounds of peanuts and products containing peanuts the average American eats in one year.

Translate. We add the individual amounts to find p.

$$p = 2.7 + 1.5 + 1.2 + 0.7 + 0.1$$

Solve. We carry out the addition.

$$\begin{array}{r} \overset{2}{} \\ 2.\,7 \\ 1.\,5 \\ 1.\,2 \\ 0.\,7 \\ +\,0.\,1 \\ \hline 6.\,2 \end{array}$$

Thus, $p = 6.2$.

Check. We repeat the calculation. The answer checks.

State. The average American eats 6.2 lb of peanuts and products containing peanuts in one year.

4. $P = 2l + 2w$

$P = 2 \cdot 7 \text{ cm} + 2 \cdot 4 \text{ cm}$

$ = 14 \text{ cm} + 8 \text{ cm} = 22 \text{ cm}$

$A = l \cdot w$

$A = 7 \text{ cm} \cdot 4 \text{ cm}$

$ = 7 \cdot 4 \cdot \text{cm} \cdot \text{cm} = 28 \text{ cm}^2$

5. ***Familiarize***. Let $k =$ the number of kilowatt-hours, in billions, generated by American utility companies in the given year.

Translate. We add the individual amounts to find k.

$k = 1464 + 455 + 273 + 250 + 118 + 12$

Solve. We carry out the addition.

$$\begin{array}{r} \overset{1\;2\;2}{} \\ 1\,4\,6\,4 \\ 4\,5\,5 \\ 2\,7\,3 \\ 2\,5\,0 \\ 1\,1\,8 \\ +\quad 1\,2 \\ \hline 2\,5\,7\,2 \end{array}$$

Thus, $k = 2572$.

Check. We repeat the calculation. The answer checks.

State. American utility companies generated 2572 billion kilowatt-hours of electricity.

6. ***Familiarize***. Let $s =$ the number of cups of sugar that should be used for $\frac{1}{2}$ of the recipe.

Translate. We translate to a multiplication sentence.

$s = \frac{1}{2} \cdot \frac{3}{4}$

Solve. We carry out the multiplication.

$s = \frac{1}{2} \cdot \frac{3}{4} = \frac{1 \cdot 3}{2 \cdot 4} = \frac{3}{8}$

Check. We repeat the calculation. The answer checks.

State. $\frac{3}{8}$ cup of sugar should be used for $\frac{1}{2}$ of the recipe.

7. The integer 8 corresponds to winning 8 cases. The integer -7 corresponds to losing 7 cases.

8. Since 1 is to the right of -7 on the number line, we have $1 > -7$.

9. The LCD is 63.

$\frac{4}{9} \cdot \frac{7}{7} = \frac{28}{63}$, $\frac{3}{7} \cdot \frac{9}{9} = \frac{27}{63}$

Since $28 > 27$, it follows that $\frac{28}{63} > \frac{27}{63}$. Thus, $\frac{4}{9} > \frac{3}{7}$.

10. Since -4.8 is to the left of -4.09 on the number line, we have $-4.8 < -4.09$.

11. $-x = -(-9) = 9$

12. $-(-x) = -(-17) = 17$

13. $2x - y = 2 \cdot 3 - 8 = 6 - 8 = -2$

14. $6x + 4y - 8x - 3y = 6x - 8x + 4y - 3y$

$ = (6 - 8)x + (4 - 3)y$

$ = -2x + y$

15. Since 36 is even, we know that 2 is a factor. The sum of the digits is $3+6$, or 9, and since 9 is divisible by 3, then 36 is divisible by 3. Also, since 36 is both even and divisible by 3, it is divisible by 6. We make a list of factorizations.

$36 = 1 \cdot 36 \qquad 36 = 4 \cdot 9$

$36 = 2 \cdot 18 \qquad 36 = 6 \cdot 6$

$36 = 3 \cdot 12$

The factors of 36 are 1, 2, 3, 4, 6, 9, 12, 18, and 36.

16. 732 is even so it is divisible by 2. The sum of the digits is $7 + 3 + 2$, or 12, is divisible by 3, so 732 is divisible by 3. Then 732 is divisible by 6.

17. $\dfrac{-7}{x} = -\dfrac{7}{x} = \dfrac{7}{-x}$

18. $5(2a - 3b + 1) = 5 \cdot 2a - 5 \cdot 3b + 5 \cdot 1 = 10a - 15b + 5$

19. The object is divided into 7 equal parts, so the unit is $\frac{1}{7}$. The denominator is 7. There are 3 units shaded, so the numerator is 3. Thus, $\frac{3}{7}$ is shaded.

20. Since $35 \div 7 = 5$, we multiply by 1 using $\frac{5}{5}$.

$\frac{2}{7} = \frac{2}{7} \cdot \frac{5}{5} = \frac{10}{35}$

21.

$$\begin{array}{r} \overset{\quad 12}{4\;\overset{}{\not{5}}\;\overset{7}{\not{3}}\;\overset{16}{\not{6}}} \\ -\;3\;9\;8 \\ \hline 1\;3\;8 \end{array}$$

22.

$$\begin{array}{r} 2\,8 \\ \times\,1\,7 \\ \hline 1\,9\,6 \\ 2\,8\,0 \\ \hline 4\,7\,6 \end{array}$$

23. $63 \div (-7) = -9$ Check: $-9(-7) = 63$

24. $-18 + (-21)$

Add the absolute values: $18 + 21 = 39$

Make the answer negative: $-18 + (-21) = -39$

25. $\dfrac{3}{7} + \dfrac{2}{7} = \dfrac{3+2}{7} = \dfrac{5}{7}$

26. $\dfrac{3}{7} \div \dfrac{9}{5} = \dfrac{3}{7} \cdot \dfrac{5}{9} = \dfrac{3 \cdot 5}{7 \cdot 9} = \dfrac{3 \cdot 5}{7 \cdot 3 \cdot 3} = \dfrac{3}{3} \cdot \dfrac{5}{7 \cdot 3} = \dfrac{5}{21}$

27. The LCD is 18.

$$\dfrac{5}{6} - \dfrac{1}{9} = \dfrac{5}{6} \cdot \dfrac{3}{3} - \dfrac{1}{9} \cdot \dfrac{2}{2}$$

$$= \dfrac{15}{18} - \dfrac{2}{18} = \dfrac{15-2}{18}$$

$$= \dfrac{13}{18}$$

28. The LCD is 30.

$$\dfrac{-2}{15} + \dfrac{3}{10} = \dfrac{-2}{15} \cdot \dfrac{2}{2} + \dfrac{3}{10} \cdot \dfrac{3}{3}$$

$$= \dfrac{-4}{30} + \dfrac{9}{30}$$

$$= \dfrac{-4+9}{30} = \dfrac{5}{30}$$

$$= \dfrac{\cancel{5} \cdot 1}{\cancel{5} \cdot 6} = \dfrac{1}{6}$$

29. $\dfrac{8}{11} \cdot \dfrac{11}{8} = \dfrac{8 \cdot 11}{11 \cdot 8} = 1$

30.

$$3\boxed{\dfrac{1}{4} \cdot \dfrac{2}{2}} = 3\dfrac{2}{8}$$
$$+ 5\dfrac{7}{8} \qquad = + 5\dfrac{7}{8}$$
$$\overline{\qquad\qquad\qquad\qquad}$$
$$8\dfrac{9}{8} = 8 + \dfrac{9}{8}$$
$$= 8 + 1\dfrac{1}{8}$$
$$= 9\dfrac{1}{8}$$

31. $7\dfrac{2}{3}x - 5\dfrac{1}{4}x = \left(7\dfrac{2}{3} - 5\dfrac{1}{4}\right)x$

$$= \left(7\dfrac{8}{12} - 5\dfrac{3}{12}\right)x$$

$$= 2\dfrac{5}{12}x$$

32. $4\dfrac{1}{5} \cdot 3\dfrac{1}{7} = \dfrac{21}{5} \cdot \dfrac{22}{7}$

$$= \dfrac{21 \cdot 22}{5 \cdot 7}$$

$$= \dfrac{3 \cdot 7 \cdot 22}{5 \cdot 7} = \dfrac{66}{5}$$

$$= 13\dfrac{1}{5}$$

33.
$$\begin{array}{r} \scriptstyle 1\ \ 1 \\ 3\,9.\,7\,2 \\ +\ 4\,3.\,5\,6 \\ \hline 8\,3.\,2\,8 \end{array}$$

34.
$$\begin{array}{r} 6\,2.3\,4\,5 \\ 2\,1.4 _{\wedge}\!\overline{\smash{)}1\,3\,3\,4.1_{\wedge}8\,3\,0} \\ 1\,2\,8\,4\,0\,0\,0 \\ \hline 5\,0\,1\,8\,3 \\ 4\,2\,8\,0\,0 \\ \hline 7\,3\,8\,3 \\ 6\,4\,2\,0 \\ \hline 9\,6\,3 \\ 8\,5\,6 \\ \hline 1\,0\,7\,0 \\ 1\,0\,7\,0 \\ \hline 0 \end{array}$$

35. $17.4(-2.43)$

First we multiply the absolute values.

$$\begin{array}{r} 2.\,4\,3 \\ \times\ 1\,7.\,4 \\ \hline 9\,7\,2 \\ 1\,7\,0\,1\,0 \\ 2\,4\,3\,0\,0 \\ \hline 4\,2.\,2\,8\,2 \end{array}$$

Since the product of a positive number and a negative number is negative, the answer is -42.282.

36. $\dfrac{8t}{8t} = 1$

(Remember: $\dfrac{n}{n} = 1$, for any integer n that is not 0.)

37. $\dfrac{4x}{1} = 4x$

(Remember: $\dfrac{n}{1} = n$, for any integer n.)

38. $\dfrac{0}{7x} = 0$

(Remember: $\dfrac{0}{n} = 0$, for any integer n that is not 0.)

39.

$$x + \dfrac{2}{3} = -\dfrac{1}{5}$$

$$x + \dfrac{2}{3} - \dfrac{2}{3} = -\dfrac{1}{5} - \dfrac{2}{3}$$

$$x + 0 = -\dfrac{1}{5} \cdot \dfrac{3}{3} - \dfrac{2}{3} \cdot \dfrac{5}{5}$$

$$x = -\dfrac{3}{15} - \dfrac{10}{15}$$

$$x = -\dfrac{13}{15}$$

The solution is $-\dfrac{13}{15}$.

40.

$$\dfrac{3}{8}x + 2 = 11$$

$$\dfrac{3}{8}x + 2 - 2 = 11 - 2$$

$$\dfrac{3}{8}x = 9$$

$$\dfrac{8}{3} \cdot \dfrac{3}{8}x = \dfrac{8}{3} \cdot 9$$

$$x = \dfrac{8 \cdot 9}{3} = \dfrac{8 \cdot 3 \cdot \cancel{3}}{\cancel{3} \cdot 1}$$

$$x = 24$$

The solution is 24.

41.
$$3(x - 5) = 7x + 2$$
$$3x - 15 = 7x + 2$$
$$3x - 15 - 7x = 7x + 2 - 7x$$
$$-4x - 15 = 2$$
$$-4x - 15 + 15 = 2 + 15$$
$$-4x = 17$$
$$\frac{-4x}{-4} = \frac{17}{-4}$$
$$x = \frac{17}{-4}, \text{ or } -\frac{17}{4}$$

The solution is $-\frac{17}{4}$.

42. Since the first coordinate is negative and the second coordinate is positive, the point $(-4, 9)$ is in quadrant II.

43. Graph: $y = \frac{1}{2}x - 4$

We make a table of values. Note that when x is a multiple of 2, fraction values for y are avoided. We plot the points and draw and label the line.

When $x = -2$, $y = \frac{1}{2}(-2) - 4 = -1 - 4 = -5$.

When $x = 0$, $y = \frac{1}{2} \cdot 0 - 4 = 0 - 4 = -4$.

When $x = 4$, $y = \frac{1}{2} \cdot 4 - 4 = 2 - 4 = -2$.

x	$y = \frac{1}{2}x - 4$	(x, y)
-2	-5	$(-2, -5)$
0	-4	$(0, -4)$
4	-2	$(4, -2)$

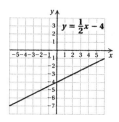

44. $\dfrac{19 + 39 + 34 + 52}{4} = \dfrac{144}{4} = 36$

45. $7, 9, 12, 35$
$\quad\quad\uparrow$
Middle number

The median is halfway between 9 and 12. It is the average of the two middle numbers.
$$\frac{9 + 12}{2} = \frac{21}{2} = 10.5$$

46. The number that occurs most often is 49. It is the mode.

47. $\dfrac{324 \text{ mi}}{12 \text{ gal}} = 27$ miles per gallon

48.
$$\left(\frac{3}{4}\right)^2 - \frac{1}{8} \cdot \left(3 - 1\frac{1}{2}\right)^2 = \left(\frac{3}{4}\right)^2 - \frac{1}{8} \cdot \left(\frac{6}{2} - \frac{3}{2}\right)^2$$
$$= \left(\frac{3}{4}\right)^2 - \frac{1}{8} \cdot \left(\frac{3}{2}\right)^2$$
$$= \frac{9}{16} - \frac{1}{8} \cdot \frac{9}{4}$$
$$= \frac{9}{16} - \frac{9}{32}$$
$$= \frac{9}{16} \cdot \frac{2}{2} - \frac{9}{32}$$
$$= \frac{18}{32} - \frac{9}{32}$$
$$= \frac{9}{32}$$

49. We could approach this problem in several ways. For instance, we could express all the numbers as mixed numerals, as fractions, or in decimal notation. We will use decimal notation and then convert the result to a mixed numeral.

$$-5\frac{42}{100} + \frac{355}{100} + \frac{89}{10} + \frac{17}{1000}$$
$$= -5.42 + 3.55 + 8.9 + 0.017$$
$$= -1.87 + 8.9 + 0.017$$
$$= 7.03 + 0.017$$
$$= 7.047$$
$$= 7\frac{47}{1000}$$

50. Since the sides of the square are 8 units long, a horizontal line through the center of the square extends 8/2, or 4 units to the left of $(2, 3)$, or to $(2 - 4, 3)$, or $(-2, 3)$. It also extends 4 units to the right of the center, or to $(2 + 4, 3)$, or $(6, 3)$.

Similarly, a vertical line through the center extends 4 units above and below the center to $(2, 3 + 4)$ and $(2, 3 - 4)$, or $(2, 7)$ and $(2, -1)$.

We make a drawing.

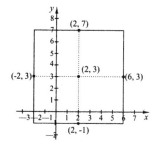

From the drawing we see that the vertices are $(-2, 7)$, $(6, 7)$, $(6, -1)$, and $(-2, -1)$.

Chapter 7

Ratio and Proportion

Exercise Set 7.1

---

1. The ratio of 4 to 5 is $\dfrac{4}{5}$.

3. The ratio of 178 to 572 is $\dfrac{178}{572}$.

5. The ratio of 0.4 to 12 is $\dfrac{0.4}{12}$.

7. The ratio of 3.8 to 7.4 is $\dfrac{3.8}{7.4}$.

9. The ratio of 56.78 to 98.35 is $\dfrac{56.78}{98.35}$.

11. The ratio of $8\frac{3}{4}$ to $9\frac{5}{6}$ is $\dfrac{8\frac{3}{4}}{9\frac{5}{6}}$.

13. If four of every five fatal accidents involving a Corvette do not involve another vehicle, then $5 - 4$, or 1, involves a Corvette and at least one other vehicle. Thus, the ratio of fatal accidents involving just a Corvette to those involving a Corvette and at least one other vehicle is $\dfrac{4}{1}$.

15. The ratio of physicians to residents in Connecticut was $\dfrac{362}{100,000}$.

The ratio of physicians to residents in Wyoming was $\dfrac{192}{100,000}$.

17. The ratio is $\dfrac{93.2}{1000}$.

19. The ratio of hits to at-bats is $\dfrac{163}{509}$.

The ratio of at-bats to hits is $\dfrac{509}{163}$.

21. The ratio of width to length is $\dfrac{60}{100}$.

The ratio of length to width is $\dfrac{100}{60}$.

23. The ratio of 4 to 6 is $\dfrac{4}{6} = \dfrac{2 \cdot 2}{2 \cdot 3} = \dfrac{2}{2} \cdot \dfrac{2}{3} = \dfrac{2}{3}$.

25. The ratio of 18 to 24 is $\dfrac{18}{24} = \dfrac{3 \cdot 6}{4 \cdot 6} = \dfrac{3}{4} \cdot \dfrac{6}{6} = \dfrac{3}{4}$.

27. The ratio of 4.8 to 10 is $\dfrac{4.8}{10} = \dfrac{4.8}{10} \cdot \dfrac{10}{10} = \dfrac{48}{100} = \dfrac{4 \cdot 12}{4 \cdot 25} = \dfrac{4}{4} \cdot \dfrac{12}{25} = \dfrac{12}{25}$.

29. The ratio of 2.8 to 3.6 is $\dfrac{2.8}{3.6} = \dfrac{2.8}{3.6} \cdot \dfrac{10}{10} = \dfrac{28}{36} = \dfrac{4 \cdot 7}{4 \cdot 9} = \dfrac{4}{4} \cdot \dfrac{7}{9} = \dfrac{7}{9}$.

31. The ratio is $\dfrac{20}{30} = \dfrac{2 \cdot 10}{3 \cdot 10} = \dfrac{2}{3} \cdot \dfrac{10}{10} = \dfrac{2}{3}$.

33. The ratio is $\dfrac{56}{100} = \dfrac{4 \cdot 14}{4 \cdot 25} = \dfrac{4}{4} \cdot \dfrac{14}{25} = \dfrac{14}{25}$.

35. The ratio is $\dfrac{128}{256} = \dfrac{1 \cdot 128}{2 \cdot 128} = \dfrac{1}{2} \cdot \dfrac{128}{128} = \dfrac{1}{2}$.

37. The ratio is $\dfrac{0.48}{0.64} = \dfrac{0.48}{0.64} \cdot \dfrac{100}{100} = \dfrac{48}{64} = \dfrac{3 \cdot 16}{4 \cdot 16} = \dfrac{3}{4} \cdot \dfrac{16}{16} = \dfrac{3}{4}$.

39. The ratio of length to width is $\dfrac{478}{213}$.

The ratio of width to length is $\dfrac{213}{478}$.

41. If 51 of every 100 people are females, then $100 - 51$, or 49, are males. Thus, the ratio of females to males is $\dfrac{51}{49}$. This ratio cannot be simplified further.

43.
$$\dfrac{3\frac{1}{5}}{4\frac{1}{10}} = \dfrac{\frac{16}{5}}{\frac{41}{10}} = \dfrac{16}{5} \div \dfrac{41}{10}$$
$$= \dfrac{16}{5} \cdot \dfrac{10}{41}$$
$$= \dfrac{16 \cdot 2 \cdot 5}{5 \cdot 41} = \dfrac{5}{5} \cdot \dfrac{16 \cdot 2}{41}$$
$$= \dfrac{32}{41}$$

45.
$$\dfrac{7\frac{1}{8}}{2\frac{3}{4}} = \dfrac{\frac{57}{8}}{\frac{11}{4}} = \dfrac{57}{8} \div \dfrac{11}{4}$$
$$= \dfrac{57}{8} \cdot \dfrac{4}{11}$$
$$= \dfrac{57 \cdot 4}{2 \cdot 4 \cdot 11} = \dfrac{4}{4} \cdot \dfrac{57}{2 \cdot 11}$$
$$= \dfrac{57}{22}$$

47.
$$\dfrac{8\frac{2}{9}}{7\frac{1}{6}} = \dfrac{\frac{74}{9}}{\frac{43}{6}} = \dfrac{74}{9} \div \dfrac{43}{6}$$
$$= \dfrac{74}{9} \cdot \dfrac{6}{43}$$
$$= \dfrac{74 \cdot 2 \cdot 3}{3 \cdot 3 \cdot 43} = \dfrac{3}{3} \cdot \dfrac{74 \cdot 2}{3 \cdot 43}$$
$$= \dfrac{148}{129}$$

49. Discussion and Writing Exercise

51. $-\dfrac{5}{6} \;\square\; -\dfrac{3}{4}$, or $\dfrac{-5}{6} \;\square\; \dfrac{-3}{4}$

The LCD is 12.

$\dfrac{-5}{6} \cdot \dfrac{2}{2} = \dfrac{-10}{12}, \; \dfrac{-3}{4} \cdot \dfrac{3}{3} = \dfrac{-9}{12}$

Since $-10 < -9$, it follows that $\dfrac{-10}{12} < \dfrac{-9}{12}$, or $-\dfrac{5}{6} < -\dfrac{3}{4}$.

53. $\dfrac{5}{9} \;\square\; \dfrac{6}{11}$

The LCD is 99.

$\dfrac{5}{9} \cdot \dfrac{11}{11} = \dfrac{55}{99}, \; \dfrac{6}{11} \cdot \dfrac{9}{9} = \dfrac{54}{99}$

Since $55 > 54$, it follows that $\dfrac{55}{99} > \dfrac{54}{99}$, or $\dfrac{5}{9} > \dfrac{6}{11}$.

55.
$$
\begin{array}{r}
5\,0 \\
4\,\overline{)2\,0\,0} \\
2\,0\,0 \\ \hline
0 \\
0 \\ \hline
0
\end{array}
$$

The answer is 50.

57.
$$
\begin{array}{r}
1\,4.\,5 \\
1\,6\,\overline{)2\,3\,2.\,0} \\
1\,6\,0 \\ \hline
7\,2 \\
6\,4 \\ \hline
8\,0 \\
8\,0 \\ \hline
0
\end{array}
$$

The answer is 14.5.

59. *Familiarize*. We let h = Rocky's excess height.

Translate. We have a "how much more" situation.

Height of daughter	plus	How much more height	is	Rocky's height
$\downarrow$	$\downarrow$	$\downarrow$	$\downarrow$	$\downarrow$
$180\dfrac{3}{4}$	$+$	h	$=$	$187\dfrac{1}{10}$

Solve. We solve the equation as follows:

$$h = 187\dfrac{1}{10} - 180\dfrac{3}{4}$$

$$187\boxed{\dfrac{1}{10} \cdot \dfrac{2}{2}} = 187\dfrac{2}{20}$$

$$180\boxed{\dfrac{3}{4} \cdot \dfrac{5}{5}} = 180\dfrac{15}{20}$$

$$
\begin{array}{r}
187\dfrac{1}{10} = \;\; 187\dfrac{2}{20} = \;\; 186\dfrac{22}{20} \\
-180\dfrac{3}{4} = -180\dfrac{15}{20} = -180\dfrac{15}{20} \\ \hline
6\dfrac{7}{20}
\end{array}
$$

Thus, $h = 6\dfrac{7}{20}$.

Check. We add Rocky's excess height to his daughter's height:

$$180\dfrac{3}{4} + 6\dfrac{7}{20} = 180\dfrac{15}{20} + 6\dfrac{7}{20} = 186\dfrac{22}{20} = 187\dfrac{2}{20} = 187\dfrac{1}{10}$$

The answer checks.

State. Rocky is $6\dfrac{7}{20}$ cm taller.

61. Discussion and Writing Exercise

63. $\dfrac{\$208,306,817}{\$2,191,886,898} \approx 0.0950353858$

The ratio is 0.0950353858 to 1.

65. We use a calculator.

The parents' total height is $187\dfrac{1}{10}$ cm $+ 168\dfrac{1}{4}$ cm $= 355\dfrac{7}{20}$ cm. The children's total height is $180\dfrac{3}{4}$ cm $+ 150\dfrac{7}{10}$ cm $= 331\dfrac{9}{20}$ cm. Then the ratio of the parents' total height to the children's total height is $\dfrac{355\dfrac{7}{20}}{331\dfrac{9}{20}} = \dfrac{7107}{6629}$.

67. We divide each number in the ratio by 5. Since $5 \div 5 = 1$, $10 \div 5 = 2$, and $15 \div 5 = 3$, we have $1 : 2 : 3$.

Exercise Set 7.2

1. $\dfrac{120 \text{ mi}}{3 \text{ hr}} = 40 \;\dfrac{\text{mi}}{\text{hr}}$, or 40 mph

3. $\dfrac{217 \text{ mi}}{29 \text{ sec}} \approx 7.48 \;\dfrac{\text{mi}}{\text{sec}}$

5. $\dfrac{300 \text{ mi}}{12.5 \text{ gal}} = 24$ mpg

7. $\dfrac{448.5 \text{ mi}}{19.5 \text{ gal}} = 23$ mpg

9. $\dfrac{32,270 \text{ people}}{0.75 \text{ sq mi}} \approx 43,027$ people/sq mi

11. $\dfrac{623 \text{ gal}}{1000 \text{ sq ft}} = 0.623$ gal/ft^2

13. $\dfrac{186,000 \text{ mi}}{1 \text{ sec}} = 186,000 \;\dfrac{\text{mi}}{\text{sec}}$

15. $\dfrac{310 \text{ km}}{2.5 \text{ hr}} = 124 \;\dfrac{\text{km}}{\text{hr}}$

17. $\dfrac{500 \text{ mi}}{20 \text{ hr}} = 25 \;\dfrac{\text{mi}}{\text{hr}}$

$\dfrac{20 \text{ hr}}{500 \text{ mi}} = 0.04 \;\dfrac{\text{hr}}{\text{mi}}$

19. $\dfrac{1465 \text{ points}}{80 \text{ games}} \approx 18.3 \;\dfrac{\text{points}}{\text{game}}$

21. $\dfrac{1500 \text{ beats}}{60 \text{ min}} = 25 \;\dfrac{\text{beats}}{\text{min}}$

23. $\dfrac{\$2.59}{13.5\text{ oz}} = \dfrac{259\cancel{c}}{13.5\text{ oz}} \approx 19.185\cancel{c}/\text{oz}$

$\dfrac{\$3.99}{25.4\text{ oz}} = \dfrac{399\cancel{c}}{25.4\text{ oz}} \approx 15.709\cancel{c}/\text{oz}$

The 25.4-oz size has the lower unit price.

25. $\dfrac{\$1.84}{16\text{ oz}} = \dfrac{184\cancel{c}}{16\text{ oz}} = 11.5\cancel{c}/\text{oz}$

$\dfrac{\$2.49}{18\text{ oz}} = \dfrac{249\cancel{c}}{18\text{ oz}} \approx 13.833\cancel{c}/\text{oz}$

The 16-oz size has the lower unit price.

27. $\dfrac{\$2.09}{11.5\text{ oz}} = \dfrac{209\cancel{c}}{11.5\text{ oz}} \approx 18.174\cancel{c}/\text{oz}$

$\dfrac{\$5.27}{34.5\text{ oz}} = \dfrac{527\cancel{c}}{34.5\text{ oz}} \approx 15.275\cancel{c}/\text{oz}$

The 34.5-oz size has the lower unit price.

29. $\dfrac{\$1.89}{18\text{ oz}} = \dfrac{189\cancel{c}}{18\text{ oz}} = 10.5\cancel{c}/\text{oz}$

$\dfrac{\$3.25}{28\text{ oz}} = \dfrac{325\cancel{c}}{28\text{ oz}} \approx 11.607\cancel{c}/\text{oz}$

$\dfrac{\$4.99}{40\text{ oz}} = \dfrac{499\cancel{c}}{40\text{ oz}} = 12.475\cancel{c}/\text{oz}$

$\dfrac{\$7.99}{64\text{ oz}} = \dfrac{799\cancel{c}}{64\text{ oz}} \approx 12.484\cancel{c}/\text{oz}$

The 18-oz size has the lowest unit price.

31. $\dfrac{\$4.29}{50\text{ oz}} = \dfrac{429\cancel{c}}{50\text{ oz}} = 8.58\cancel{c}/\text{oz}$

$\dfrac{\$5.29}{100\text{ oz}} = \dfrac{529\cancel{c}}{100\text{ oz}} = 5.29\cancel{c}/\text{oz}$

$\dfrac{\$10.49}{200\text{ oz}} = \dfrac{1049\cancel{c}}{200\text{ oz}} = 5.245\cancel{c}/\text{oz}$

$\dfrac{\$15.79}{300\text{ oz}} = \dfrac{1579\cancel{c}}{300\text{ oz}} \approx 5.263\cancel{c}/\text{oz}$

The 200 fl oz size has the lowest unit price.

33. Discussion and Writing Exercise

35. $\dfrac{3}{11} \ \square \ \dfrac{5}{13}$

The LCD is $11 \cdot 13$, or 143.

$\dfrac{3}{11} \cdot \dfrac{13}{13} = \dfrac{39}{143}, \ \dfrac{5}{13} \cdot \dfrac{11}{11} = \dfrac{55}{143}$

Since $39 < 55$, it follows that $\dfrac{39}{143} < \dfrac{55}{143}$, or $\dfrac{3}{11} < \dfrac{5}{13}$.

37. $\dfrac{4}{9} \ \square \ \dfrac{3}{7}$

The LCD is $9 \cdot 7$, or 63.

$\dfrac{4}{9} \cdot \dfrac{7}{7} = \dfrac{28}{63}, \ \dfrac{3}{7} \cdot \dfrac{9}{9} = \dfrac{27}{63}$

Since $28 > 27$, it follows that $\dfrac{28}{63} > \dfrac{27}{63}$, or $\dfrac{4}{9} > \dfrac{3}{7}$.

39. $-\dfrac{3}{10} \ \square \ -\dfrac{2}{17}$, or $\dfrac{-3}{10} \ \square \ \dfrac{-2}{7}$

The LCD is $10 \cdot 7$, or 70.

$\dfrac{-3}{10} \cdot \dfrac{7}{7} = \dfrac{-21}{70}, \ \dfrac{-2}{7} \cdot \dfrac{10}{10} = \dfrac{-20}{70}$

Since $-21 < -20$, it follows that $\dfrac{-21}{70} < \dfrac{-20}{70}$, or $-\dfrac{3}{10} < -\dfrac{2}{7}$.

41. *Familiarize*. We visualize the situation. We let p = the number by which the number of piano players exceeds the number of guitar players, in millions.

18.9 million	p
20.6 million	

Translate. This is a "how-much-more" situation.

Number of guitar players + Additional number of piano players = Number of piano players

$18.9 + p = 20.6$

Solve. To solve the equation we subtract 18.9 on both sides.

$p = 20.6 - 18.9$
$p = 1.7$

$\begin{array}{r} {\scriptstyle 1\ 9\ 16} \\ \cancel{2}\,0.\cancel{6} \\ -\ 1\ 8.\ 9 \\ \hline 1.\ 7 \end{array}$

Check. We repeat the calculation.

State. There are 1.7 million more piano players than guitar players.

43. Discussion and Writing Exercise

45. a) For the 6-oz container: $\dfrac{65\cancel{c}}{6\text{ oz}} \approx 10.83\cancel{c}/\text{oz}$

For the 5.5-oz container: $\dfrac{60\cancel{c}}{5.5\text{ oz}} \approx 10.91\cancel{c}/\text{oz}$

b) For the \$0.89 roll: $\dfrac{\$0.89}{78\text{ ft}^2} = \dfrac{89\cancel{c}}{78\text{ ft}^2} \approx 1.14\cancel{c}/\text{ft}^2$

For the \$0.79 roll: $\dfrac{\$0.79}{65\text{ ft}^2} = \dfrac{79\cancel{c}}{65\text{ ft}^2} \approx 1.22\cancel{c}/\text{ft}^2$

47. The unit price of a 64-oz carton that costs \$5 is

$\dfrac{\$5}{64\text{ oz}} = \dfrac{500\cancel{c}}{64\text{ oz}} \approx 7.8\cancel{c}/\text{oz}.$

The unit price of a 56-oz carton that costs \$4.35 is

$\dfrac{\$4.35}{56\text{ oz}} = \dfrac{435\cancel{c}}{56\text{ oz}} \approx 7.8\cancel{c}/\text{oz}.$

Thus, at first the unit price for the smaller carton was about the same as the unit price for the larger carton.

The unit price of a 56-oz carton that costs \$5 is

$\dfrac{\$5}{56\text{ oz}} = \dfrac{500\cancel{c}}{56\text{ oz}} \approx 8.9\cancel{c}/\text{oz}.$

We see that the unit price rose to about 8.9¢/oz after about a year.

49. From Exercise 14 we know that sound travels 1100 ft in 1 second, so in 1 minute, or 60 seconds, it travels $1100 \cdot 60$, or 66,000 ft.

We convert 25 mi to feet:

$25\text{ mi} = 25\text{ mi} \cdot \dfrac{5280\text{ ft}}{1\text{ mi}} = 25 \cdot 5280 \cdot \text{ ft} \cdot \dfrac{\text{mi}}{\text{mi}} =$

$132,000\text{ ft}$

Then $132{,}000 \text{ ft} \div 66{,}000 \text{ ft/min} =$

$132{,}000 \text{ ft} \cdot \dfrac{1}{66{,}000} \cdot \dfrac{\min}{\text{ft}} = \dfrac{132{,}000}{66{,}000} \cdot \dfrac{\text{ft}}{\text{ft}} \cdot \min =$

2 min.

Thus, it will take 2 min to hear the thunder.

From Exercise 13 we know that light travels at a speed of 186,000 mi/sec. Then

$25 \text{ mi} \div 186{,}000 \text{ mi/sec} = 25 \text{ mi} \cdot \dfrac{1}{186{,}000} \cdot \dfrac{\text{sec}}{\text{mi}} =$

$\dfrac{25}{186{,}000} \cdot \dfrac{\text{mi}}{\text{mi}} \cdot \text{sec} \approx 0.0001344 \text{ sec}$, or about

0.0000022 min.

Thus, it will take about 0.0000022 min to see the flash of light.

51. Number of cans in the case: $4 \cdot 6 = 24$ cans

Number of ounces of ginger ale in 24 12-oz cans:

$\qquad 24 \cdot 12 \text{ oz} = 288 \text{ oz}$

Unit price in ounces per dollar: $\dfrac{288 \text{ oz}}{\$11} \approx 26.2 \text{ oz/dollar}$.

Exercise Set 7.3

1. We can use cross products:

$5 \cdot 9 = 45 \quad$ 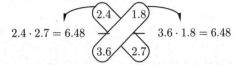 $\quad 6 \cdot 7 = 42$

Since the cross products are not the same, $45 \neq 42$, we know that the numbers are not proportional.

3. We can use cross products:

$1 \cdot 20 = 20 \quad$ ① ⑩ ② ⑳ $\quad 2 \cdot 10 = 20$

Since the cross products are the same, $20 = 20$, we know that $\dfrac{1}{2} = \dfrac{10}{20}$, so the numbers are proportional.

5. We can use cross products:

$2.4 \cdot 2.7 = 6.48 \quad$ 2.4 1.8 3.6 2.7 $\quad 3.6 \cdot 1.8 = 6.48$

Since the cross products are the same, $6.48 = 6.48$, we know that $\dfrac{2.4}{3.6} = \dfrac{1.8}{2.7}$, so the numbers are proportional.

7. We can use cross products:

$5\frac{1}{3} \cdot 9\frac{1}{2} = 50\frac{2}{3} \quad$ $5\frac{1}{3}$ $2\frac{1}{5}$ $8\frac{1}{4}$ $9\frac{1}{2}$ $\quad 8\frac{1}{4} \cdot 2\frac{1}{5} = 18\frac{3}{20}$

Since the cross products are not the same, $50\frac{2}{3} \neq 18\frac{3}{20}$, we know that the numbers are not proportional.

9. Tom Brady:

Completion rate $= \dfrac{334 \text{ completions}}{530 \text{ attempts}}$

$\qquad\qquad \approx 0.63 \dfrac{\text{completion}}{\text{attempt}}$

Trent Green:

Completion rate $= \dfrac{317 \text{ completions}}{507 \text{ attempts}}$

$\qquad\qquad \approx 0.63 \dfrac{\text{completion}}{\text{attempt}}$

Brett Favre:

Completion rate $= \dfrac{372 \text{ completions}}{605 \text{ attempts}}$

$\qquad\qquad \approx 0.61 \dfrac{\text{completion}}{\text{attempt}}$

Carson Palmer:

Completion rate $= \dfrac{345 \text{ completions}}{509 \text{ attempts}}$

$\qquad\qquad \approx 0.68 \dfrac{\text{completion}}{\text{attempt}}$

The completion rates for Brady and Green, rounded to the nearest hundredth, are the same.

11. $\qquad \dfrac{18}{4} = \dfrac{x}{10}$

$18 \cdot 10 = 4 \cdot x \qquad$ Equating cross products

$\dfrac{18 \cdot 10}{4} = \dfrac{4 \cdot x}{4} \qquad$ Dividing by 4

$\dfrac{18 \cdot 10}{4} = x$

$\dfrac{180}{4} = x \qquad$ Multiplying

$45 = x \qquad$ Dividing

13. $\qquad \dfrac{x}{8} = \dfrac{9}{6}$

$6 \cdot x = 8 \cdot 9 \qquad$ Equating cross products

$\dfrac{6 \cdot x}{6} = \dfrac{8 \cdot 9}{6} \qquad$ Dividing by 6

$x = \dfrac{8 \cdot 9}{6}$

$x = \dfrac{72}{6} \qquad$ Multiplying

$x = 12 \qquad$ Dividing

15. $\qquad \dfrac{t}{12} = \dfrac{5}{6}$

$6 \cdot t = 12 \cdot 5$

$\dfrac{6 \cdot t}{6} = \dfrac{12 \cdot 5}{6}$

$t = \dfrac{12 \cdot 5}{6}$

$t = \dfrac{60}{6}$

$t = 10$

17. $\dfrac{2}{5} = \dfrac{8}{n}$

$2 \cdot n = 5 \cdot 8$

$\dfrac{2 \cdot n}{2} = \dfrac{5 \cdot 8}{2}$

$n = \dfrac{5 \cdot 8}{2}$

$n = \dfrac{40}{2}$

$n = 20$

19. $\dfrac{n}{15} = \dfrac{10}{30}$

$30 \cdot n = 15 \cdot 10$

$\dfrac{30 \cdot n}{30} = \dfrac{15 \cdot 10}{30}$

$n = \dfrac{15 \cdot 10}{30}$

$n = \dfrac{150}{30}$

$n = 5$

21. $\dfrac{16}{12} = \dfrac{24}{x}$

$16 \cdot x = 12 \cdot 24$

$\dfrac{16 \cdot x}{16} = \dfrac{12 \cdot 24}{6}$

$x = \dfrac{12 \cdot 24}{16}$

$x = \dfrac{288}{16}$

$x = 18$

23. $\dfrac{6}{11} = \dfrac{12}{x}$

$6 \cdot x = 11 \cdot 12$

$\dfrac{6 \cdot x}{6} = \dfrac{11 \cdot 12}{6}$

$x = \dfrac{11 \cdot 12}{6}$

$x = \dfrac{132}{6}$

$x = 22$

25. $\dfrac{20}{7} = \dfrac{80}{x}$

$20 \cdot x = 7 \cdot 80$

$\dfrac{20 \cdot x}{20} = \dfrac{7 \cdot 80}{20}$

$x = \dfrac{7 \cdot 80}{20}$

$x = \dfrac{560}{20}$

$x = 28$

27. $\dfrac{12}{9} = \dfrac{x}{7}$

$12 \cdot 7 = 9 \cdot x$

$\dfrac{12 \cdot 7}{9} = \dfrac{9 \cdot x}{9}$

$\dfrac{12 \cdot 7}{9} = x$

$\dfrac{84}{9} = x$

$\dfrac{28}{3} = x$ Simplifying

$9\dfrac{1}{3} = x$ Writing a mixed numeral

29. $\dfrac{x}{13} = \dfrac{2}{9}$

$9 \cdot x = 13 \cdot 2$

$\dfrac{9 \cdot x}{9} = \dfrac{13 \cdot 2}{9}$

$x = \dfrac{13 \cdot 2}{9}$

$x = \dfrac{26}{9}$, or $2\dfrac{8}{9}$

31. $\dfrac{t}{0.16} = \dfrac{0.15}{0.40}$

$0.40 \times t = 0.16 \times 0.15$

$\dfrac{0.40 \times t}{0.40} = \dfrac{0.16 \times 0.15}{0.40}$

$t = \dfrac{0.16 \times 0.15}{0.40}$

$t = \dfrac{0.024}{0.40}$

$t = 0.06$

33. $\dfrac{100}{25} = \dfrac{20}{n}$

$100 \cdot n = 25 \cdot 20$

$\dfrac{100 \cdot n}{100} = \dfrac{25 \cdot 20}{100}$

$n = \dfrac{25 \cdot 20}{100}$

$n = \dfrac{500}{100}$

$n = 5$

35. $\dfrac{7}{\frac{1}{4}} = \dfrac{28}{x}$

$7 \cdot x = \dfrac{1}{4} \cdot 28$

$\dfrac{7 \cdot x}{7} = \dfrac{\frac{1}{4} \cdot 28}{7}$

$x = \dfrac{\frac{1}{4} \cdot 28}{7}$

$x = \dfrac{7}{7}$

$x = 1$

37. $\dfrac{\frac{1}{4}}{\frac{1}{2}} = \dfrac{\frac{1}{2}}{x}$

$\dfrac{1}{4} \cdot x = \dfrac{1}{2} \cdot \dfrac{1}{2}$

$\dfrac{\frac{1}{4} \cdot x}{\frac{1}{4}} = \dfrac{\frac{1}{2} \cdot \frac{1}{2}}{\frac{1}{4}}$

$x = \dfrac{\frac{1}{2} \cdot \frac{1}{2}}{\frac{1}{4}}$

$x = \dfrac{\frac{1}{4}}{\frac{1}{4}}$

$x = 1$

39. $\dfrac{1}{2} = \dfrac{7}{x}$

$1 \cdot x = 2 \cdot 7$

$x = \dfrac{2 \cdot 7}{1}$

$x = 14$

41. $\dfrac{\frac{2}{7}}{\frac{3}{4}} = \dfrac{\frac{5}{6}}{y}$

$\dfrac{2}{7} \cdot y = \dfrac{3}{4} \cdot \dfrac{5}{6}$

$y = \dfrac{3}{4} \cdot \dfrac{5}{6} \cdot \dfrac{7}{2}$ Dividing by $\dfrac{2}{7}$

$y = \dfrac{3}{4} \cdot \dfrac{5}{2 \cdot 3} \cdot \dfrac{7}{2}$

$y = \dfrac{5 \cdot 7}{4 \cdot 2 \cdot 2}$

$y = \dfrac{35}{16}$, or $2\dfrac{3}{16}$

43. $\dfrac{2\frac{1}{2}}{3\frac{1}{3}} = \dfrac{x}{4\frac{1}{4}}$

$2\dfrac{1}{2} \cdot 4\dfrac{1}{4} = 3\dfrac{1}{3} \cdot x$

$\dfrac{5}{2} \cdot \dfrac{17}{4} = \dfrac{10}{3} \cdot x$

$\dfrac{3}{10} \cdot \dfrac{5}{2} \cdot \dfrac{17}{4} = x$ Dividing by $\dfrac{10}{3}$

$\dfrac{3}{5 \cdot 2} \cdot \dfrac{5}{2} \cdot \dfrac{17}{4} = x$

$\dfrac{3 \cdot 17}{2 \cdot 2 \cdot 4} = x$

$\dfrac{51}{16} = x$, or

$3\dfrac{3}{16} = x$

45. $\dfrac{1.28}{3.76} = \dfrac{4.28}{y}$

$1.28 \times y = 3.76 \times 4.28$

$\dfrac{1.28 \times y}{1.28} = \dfrac{3.76 \times 4.28}{1.28}$

$y = \dfrac{3.76 \times 4.28}{1.28}$

$y = \dfrac{16.0928}{1.28}$

$y = 12.5725$

47. $\dfrac{10\frac{3}{8}}{12\frac{2}{3}} = \dfrac{5\frac{3}{4}}{y}$

$10\dfrac{3}{8} \cdot y = 12\dfrac{2}{3} \cdot 5\dfrac{3}{4}$

$\dfrac{83}{8} \cdot y = \dfrac{38}{3} \cdot \dfrac{23}{4}$

$y = \dfrac{38}{3} \cdot \dfrac{23}{4} \cdot \dfrac{8}{83}$ Dividing by $\dfrac{83}{3}$

$y = \dfrac{38}{3} \cdot \dfrac{23}{4} \cdot \dfrac{2 \cdot 4}{83}$

$y = \dfrac{38 \cdot 23 \cdot 2}{3 \cdot 83}$

$y = \dfrac{1748}{249}$, or $7\dfrac{5}{249}$

49. Discussion and Writing Exercise

51. A ratio is the quotient of two quantities.

53. To compute a mean of a set of numbers, we add the numbers and then divide by the number of addends.

55. The numbers -3 and 3 are called opposites.

57. The sentence $\dfrac{2}{5} \cdot \dfrac{4}{9} = \dfrac{4}{9} \cdot \dfrac{2}{5}$ illustrates the commutative law of multiplication.

59. Discussion and Writing Exercise

61. $\dfrac{1728}{5643} = \dfrac{836.4}{x}$

$1728 \cdot x = 5643 \cdot 836.4$

$\dfrac{1728x}{1728} = \dfrac{5643 \cdot 836.4}{1728}$

$x \approx 2731.4$ Using a calculator

The solution is approximately 2731.4.

63. $\dfrac{x}{4} = \dfrac{x-1}{6}$

$6x = 4(x-1)$

$6x = 4x - 4$

$6x - 4x = 4x - 4 - 4x$

$2x = -4$

$\dfrac{2x}{2} = \dfrac{-4}{2}$

$x = -2$

The solution is -2.

65. $\dfrac{a}{b} = \dfrac{c}{d} \Rightarrow bd \cdot \dfrac{a}{b} = bd \cdot \dfrac{c}{d} \Rightarrow da = bc \Rightarrow \dfrac{da}{ba} = \dfrac{bc}{ba} \Rightarrow \dfrac{d}{b} = \dfrac{c}{a}$

Exercise Set 7.4

1. _Familiarize_. Let h = the number of hours Lisa would have to study to receive a score of 92.

Translate. We translate to a proportion, keeping the number of hours in the numerators.

$$\text{Hours} \rightarrow \frac{9}{75} = \frac{h}{92} \leftarrow \text{Hours}$$
$$\text{Score} \rightarrow \qquad \qquad \leftarrow \text{Score}$$

Solve. We solve the proportion.

$$9 \cdot 92 = 75 \cdot h \quad \text{Equating cross products}$$
$$\frac{9 \cdot 92}{75} = \frac{75 \cdot h}{75}$$
$$\frac{9 \cdot 92}{75} = h$$
$$11.04 = h$$

Check. We substitute into the proportion and check cross products.

$$\frac{9}{75} = \frac{11.04}{92}$$
$$9 \cdot 92 = 828; \ 75 \cdot 11.04 = 828$$

State. Lisa would have to study 11.04 hr to get a score of 92.

3. _Familiarize_. Let c = the number of calories in 6 cups of cereal.

Translate. We translate to a proportion, keeping the number of calories in the numerators.

$$\text{Calories} \rightarrow \frac{110}{3/4} = \frac{c}{6} \leftarrow \text{Calories}$$
$$\text{Cups} \ \rightarrow \qquad \qquad \leftarrow \text{Cups}$$

Solve. We solve the proportion.

$$110 \cdot 6 = \frac{3}{4} \cdot c \quad \text{Equating cross products}$$
$$\frac{110 \cdot 6}{3/4} = \frac{\frac{3}{4} \cdot c}{3/4}$$
$$\frac{110 \cdot 6}{3/4} = c$$
$$110 \cdot 6 \cdot \frac{4}{3} = c$$
$$880 = c$$

Check. We substitute into the proportion and check cross products.

$$\frac{110}{3/4} = \frac{880}{6}$$
$$110 \cdot 6 = 660; \ \frac{3}{4} \cdot 880 = 660$$

The cross products are the same.

State. There are 880 calories in 6 cups of cereal.

5. _Familiarize_. Let n = the number of Americans who would be considered overweight.

Translate. We translate to a proportion.

$$\text{Overweight} \rightarrow \frac{60}{100} = \frac{n}{295,000,000} \leftarrow \text{Overweight}$$
$$\text{Total} \ \rightarrow \qquad \qquad \leftarrow \text{Total}$$

Solve. We solve the proportion.

$$60 \cdot 295,000,000 = 100 \cdot n \quad \text{Equating cross products}$$
$$\frac{60 \cdot 295,000,000}{100} = \frac{100 \cdot n}{100}$$
$$\frac{60 \cdot 295,000,000}{100} = n$$
$$177,000,000 = n$$

Check. We substitute in the proportion and check cross products.

$$\frac{60}{100} = \frac{177,000,000}{295,000,000}$$
$$60 \cdot 295,000,000 = 17,700,000,000$$
$$100 \cdot 177,700,000 = 17,700,000,000$$

The cross products are the same.

State. 177,000,000, or 177 million, Americans would be considered overweight.

7. _Familiarize_. Let g = the number of gallons of gasoline needed to travel 126 mi.

Translate. We translate to a proportion.

$$\text{Miles} \ \rightarrow \frac{84}{6.5} = \frac{126}{g} \leftarrow \text{Miles}$$
$$\text{Gallons} \rightarrow \qquad \qquad \leftarrow \text{Gallons}$$

Solve.

$$84 \cdot g = 6.5 \cdot 126 \quad \text{Equating cross products}$$
$$g = \frac{6.5 \cdot 126}{84} \quad \text{Dividing by 84}$$
$$g = \frac{819}{84}$$
$$g = 9.75$$

Check. We substitute in the proportion and check cross products.

$$\frac{84}{6.5} = \frac{126}{9.75}$$
$$84 \cdot 9.75 = 819; \ 6.5 \cdot 126 = 819$$

The cross products are the same.

State. 9.75 gallons of gasoline are needed to travel 126 mi.

9. _Familiarize_. Let d = the number of defective bulbs in a lot of 2500.

Translate. We translate to a proportion.

$$\text{Defective bulbs} \rightarrow \frac{7}{100} = \frac{d}{2500} \leftarrow \text{Defective bulbs}$$
$$\text{Bulbs in lot} \ \rightarrow \qquad \qquad \leftarrow \text{Bulbs in lot}$$

Solve.

$$7 \cdot 2500 = 100 \cdot d$$
$$\frac{7 \cdot 2500}{100} = d$$
$$\frac{7 \cdot 25 \cdot 100}{100} = d$$
$$7 \cdot 25 = d$$
$$175 = d$$

Check. We substitute in the proportion and check cross products.

$$\frac{7}{100} = \frac{175}{2500}$$

$7 \cdot 2500 = 17,500; \ 100 \cdot 175 = 17,500$

State. There will be 175 defective bulbs in a lot of 2500.

11. *Familiarize.* Let s = the amount of sap needed.

Translate. We translate to a proportion.

Sap $\rightarrow$ $\dfrac{38}{2} = \dfrac{s}{9}$ $\leftarrow$ Sap
Syrup $\rightarrow$ $\phantom{\dfrac{38}{2}}$ $\leftarrow$ Syrup

Solve. $\quad 38 \cdot 9 = 2 \cdot s$

$$\frac{38 \cdot 9}{2} = s$$

$$\frac{2 \cdot 19 \cdot 9}{2} = s$$

$$19 \cdot 9 = s$$

$$171 = s$$

Check. We substitute in the proportion and check cross products.

$$\frac{38}{2} = \frac{171}{9}$$

$38 \cdot 9 = 342; \ 2 \cdot 171 = 342$

State. 171 gal of sap is needed to produce 9 gal of syrup.

13. *Familiarize.* Let s = the number of square feet of siding that Fred can paint with 7 gal of paint.

Translate. We translate to a proportion.

Gallons $\rightarrow$ $\dfrac{3}{1275} = \dfrac{7}{s}$ $\leftarrow$ Gallons
Siding $\rightarrow$ $\phantom{\dfrac{3}{1275}}$ $\leftarrow$ Siding

Solve.

$$3 \cdot s = 1275 \cdot 7$$

$$s = \frac{1275 \cdot 7}{3}$$

$$s = \frac{3 \cdot 425 \cdot 7}{3}$$

$$s = 425 \cdot 7$$

$$s = 2975$$

Check. We find the number of square feet covered by 1 gallon of paint and then multiply that number by 7.

$1275 \div 3 = 425$ and $425 \cdot 7 = 2975$

The answer checks.

State. Fred can paint 2975 ft^2 of siding with 7 gal of paint.

15. *Familiarize.* Let p = the number of published pages in a 540-page manuscript.

Translate. We translate to a proportion.

Published pages $\rightarrow$ $\dfrac{5}{6} = \dfrac{p}{540}$ $\leftarrow$ Published pages
Manuscript $\rightarrow$ $\phantom{\dfrac{5}{6}}$ $\leftarrow$ Manuscript

Solve.

$$5 \cdot 540 = 6 \cdot p$$

$$\frac{5 \cdot 540}{6} = p$$

$$\frac{5 \cdot 6 \cdot 90}{6} = p$$

$$5 \cdot 90 = p$$

$$450 = p$$

Check. We substitute in the proportion and check cross products.

$$\frac{5}{6} = \frac{450}{540}$$

$5 \cdot 540 = 2700; \ 6 \cdot 450 = 2700$

The cross products are the same.

State. A 540-page manuscript will become 450 published pages.

17. a) *Familiarize.* Let g = the number of gallons of gasoline needed to drive 2690 mi.

Translate. We translate to a proportion.

Gallons $\rightarrow$ $\dfrac{15.5}{372} = \dfrac{g}{2690}$ $\leftarrow$ Gallons
Miles $\rightarrow$ $\phantom{\dfrac{15.5}{372}}$ $\leftarrow$ Miles

Solve.

$15.5 \cdot 2690 = 372 \cdot g \quad$ Equating cross products

$$\frac{15.5 \cdot 2690}{372} = g$$

$$112 \approx g$$

Check. We find how far the car can be driven on 1 gallon of gasoline and then divide to find the number of gallons required for a 2690-mi trip.

$372 \div 15.5 = 24$ and $2690 \div 24 \approx 112$

The answer checks.

State. It will take about 112 gal of gasoline to drive 2690 mi.

b) *Familiarize.* Let d = the number of miles the car can be driven on 140 gal of gasoline.

Translate. We translate to a proportion.

Gallons $\rightarrow$ $\dfrac{15.5}{372} = \dfrac{140}{d}$ $\leftarrow$ Gallons
Miles $\rightarrow$ $\phantom{\dfrac{15.5}{372}}$ $\leftarrow$ Miles

Solve.

$15.5 \cdot d = 372 \cdot 140 \quad$ Equating cross products

$$d = \frac{372 \cdot 140}{15.5}$$

$$d = 3360$$

Check. From the check in part (a) we know that the car can be driven 24 mi on 1 gal of gasoline. We multiply to find how far it can be driven on 140 gal.

$140 \cdot 24 = 3360$

The answer checks.

State. The car can be driven 3360 mi on 140 gal of gasoline.

19. *Familiarize.* Let m = the number of miles the car will be driven in 1 year. Note that 1 year = 12 months.

Translate.

Months $\rightarrow$ $\dfrac{8}{9000} = \dfrac{12}{m}$ $\leftarrow$ Months
Miles $\rightarrow$ $\phantom{\dfrac{8}{9000}}$ $\leftarrow$ Miles

Solve.

$$8 \cdot m = 9000 \cdot 12$$

$$m = \frac{9000 \cdot 12}{8}$$

$$m = \frac{2 \cdot 4500 \cdot 3 \cdot 4}{2 \cdot 4}$$

$$m = 4500 \cdot 3$$

$$m = 13,500$$

Check. We find the average number of miles driven in 1 month and then multiply to find the number of miles the car will be driven in 1 yr, or 12 months.

$$9000 \div 8 = 1125 \text{ and } 12 \cdot 1125 = 13,500$$

The answer checks.

State. At the given rate, the car will be driven 13,500 mi in one year.

21. Familiarize. Let z = the number of pounds of zinc in the alloy.

Translate. We translate to a proportion.

$$\text{Zinc} \rightarrow \frac{3}{13} = \frac{z}{520} \leftarrow \text{Zinc}$$
$$\text{Copper} \rightarrow \qquad\qquad \leftarrow \text{Copper}$$

Solve.

$$3 \cdot 520 = 13 \cdot z$$

$$\frac{3 \cdot 520}{13} = z$$

$$\frac{3 \cdot 13 \cdot 40}{13} = z$$

$$3 \cdot 40 = z$$

$$120 = z$$

Check. We substitute in the proportion and check cross products.

$$\frac{3}{13} = \frac{120}{520}$$

$$3 \cdot 520 = 1560; \ 13 \cdot 120 = 1560$$

The cross products are the same.

State. There are 120 lb of zinc in the alloy.

23. Familiarize. Let p = the number of gallons of paint Helen should buy.

Translate. We translate to a proportion.

$$\text{Area} \rightarrow \frac{950}{2} = \frac{30,000}{p} \leftarrow \text{Area}$$
$$\text{Paint} \rightarrow \qquad\qquad \leftarrow \text{Paint}$$

Solve.

$$950 \cdot p = 2 \cdot 30,000$$

$$p = \frac{2 \cdot 30,000}{950}$$

$$p = \frac{2 \cdot 50 \cdot 600}{19 \cdot 50}$$

$$p = \frac{2 \cdot 600}{19}$$

$$p = \frac{1200}{19}, \text{ or } 63\frac{3}{19}$$

Check. We find the area covered by 1 gal of paint and then divide to find the number of gallons needed to paint a 30,000-ft^2 wall.

$$950 \div 2 = 475 \text{ and } 30,000 \div 475 = 63\frac{3}{39}$$

The answer checks.

State. Since Helen is buying paint in one gallon cans, she will have to buy 64 cans of paint.

25. Familiarize. Let s = the number of ounces of grass seed needed for 5000 ft^2 of lawn.

Translate. We translate to a proportion.

$$\text{Seed} \rightarrow \frac{60}{3000} = \frac{s}{5000} \leftarrow \text{Seed}$$
$$\text{Area} \rightarrow \qquad\qquad \leftarrow \text{Area}$$

Solve.

$$60 \cdot 5000 = 3000 \cdot s$$

$$\frac{60 \cdot 5000}{3000} = s$$

$$100 = s$$

Check. We find the number of ounces of seed needed for 1 ft^2 of lawn and then multiply this number by 5000:

$$60 \div 3000 = 0.02 \text{ and } 5000(0.02) = 100$$

The answer checks.

State. 100 oz of grass seed would be needed to seed 5000 ft^2 of lawn.

27. Familiarize. Let D = the number of deer in the game preserve.

Translate. We translate to a proportion.

$$\begin{array}{l}\text{Deer tagged} \\ \text{originally}\end{array} \rightarrow \frac{318}{D} = \frac{56}{168} \begin{array}{l}\leftarrow \text{Tagged deer} \\ \text{caught later}\end{array}$$
$$\begin{array}{l}\text{Deer in game} \\ \text{preserve}\end{array} \rightarrow \qquad\qquad \begin{array}{l}\leftarrow \text{Deer caught} \\ \text{later}\end{array}$$

Solve.

$$318 \cdot 168 = 56 \cdot D$$

$$\frac{318 \cdot 168}{56} = D$$

$$954 = D$$

Check. We substitute in the proportion and check cross products.

$$\frac{318}{954} = \frac{56}{168}; \ 318 \cdot 168 = 53,424; \ 954 \cdot 56 = 53,424$$

Since the cross products are the same, the answer checks.

State. We estimate that there are 954 deer in the game preserve.

29. Familiarize. Let d = the actual distance between the cities.

Translate. We translate to a proportion.

$$\text{Map distance} \rightarrow \frac{1}{16.6} = \frac{3.5}{d} \leftarrow \text{Map distance}$$
$$\text{Actual distance} \rightarrow \qquad\qquad \leftarrow \text{Actual distance}$$

Solve.

$$1 \cdot d = 16.6 \cdot 3.5$$

$$d = 58.1$$

Check. We use a different approach. Since 1 in. represents 16.6 mi, we multiply 16.6 by 3.5:

$$3.5(16.6) = 58.1$$

The answer checks.

State. The cities are 58.1 mi apart.

31. Discussion and Writing Exercise

33.
$$\begin{array}{r} 1\ 0\ 1 \\ 2\,\overline{)\ 2\ 0\ 2} \\ 2\,\overline{)\ 4\ 0\ 4} \\ 2\,\overline{)\ 8\ 0\ 8} \end{array}$$

$808 = 2 \cdot 2 \cdot 2 \cdot 101$, or $2^3 \cdot 101$

35.
$$\begin{array}{r} 4\ 3\ 3 \\ 2\,\overline{)\ 8\ 6\ 6} \end{array}$$

$866 = 2 \cdot 433$

37.
$$\begin{array}{r} 1\ 0\ 1 \\ 5\,\overline{)\ 5\ 0\ 5} \\ 2\,\overline{)\ 1\ 0\ 1\ 0} \\ 2\,\overline{)\ 2\ 0\ 2\ 0} \end{array}$$

$2020 = 2 \cdot 2 \cdot 5 \cdot 101$, or $2^2 \cdot 5 \cdot 101$

39. First consider $13.11 \div 5.7$:

$$\begin{array}{r} 2\,.3 \\ 5.\,7_\wedge\,\overline{)\ 1\ 3\ .1_\wedge 1} \\ \underline{1\ 1\ 4\ 0} \\ 1\ 7\ 1 \\ \underline{1\ 7\ 1} \\ 0 \end{array}$$

Since a negative number divided by a positive number is negative, the answer is -2.3.

41. $-19.7 + 12.5$

The difference in absolute values is $19.7 - 12.5$, or 7.2. Since the negative number has the larger absolute value, the answer is negative.

$-19.7 + 12.5 = -7.2$

43. Discussion and Writing Exercise

45. a) **Familiarize**. Let a = the number of British pounds equivalent to 45 U.S. dollars.

Translate. We translate to a proportion.
$$\begin{array}{cc} \text{U.S.} \rightarrow \\ \text{British} \rightarrow \end{array} \frac{1}{0.544174} = \frac{45}{a} \begin{array}{l} \leftarrow \text{U.S.} \\ \leftarrow \text{British} \end{array}$$

Solve.
$1 \cdot a = 0.544174 \cdot 45$ Equating cross products
$a = 24.48783$

Check. We substitute in the proportion and check cross products.
$$\frac{1}{0.544174} = \frac{45}{24.48783}$$
$1 \cdot 24.48783 = 24.48783$; $0.544174 \cdot 45 = 24.48783$

The cross products are the same.

State. 45 U.S. dollars would be worth 24.48783 British pounds.

b) **Familiarize**. Let c = the cost of the car in U.S. dollars.

Translate. We translate to a proportion.
$$\begin{array}{cc} \text{U.S.} \rightarrow \\ \text{British} \rightarrow \end{array} \frac{1}{0.544174} = \frac{c}{8640} \begin{array}{l} \leftarrow \text{U.S.} \\ \leftarrow \text{British} \end{array}$$

Solve.
$1 \cdot 8640 = 0.544174 \cdot c$ Equating cross products
$\dfrac{1 \cdot 8640}{0.544174} = c$
$\$15,877.27 \approx c$

Check. We substitute in the proportion and check cross products.
$$\frac{1}{0.544174} = \frac{15,877.27}{8640}$$
$1 \cdot 8640 = 8640$; $0.544174 \cdot 15,877.27 \approx 8640$

The cross products are about the same. Remember that we rounded the value of c.

State. The car would cost $\$15,877.27$ in U.S. dollars.

47. **Familiarize**. Let f = the number of faculty positions required to maintain the current student-to-faculty ratio after the university expands.

Translate. We translate to a proportion.
$$\begin{array}{cc} \text{Students} \rightarrow \\ \text{Faculty} \rightarrow \end{array} \frac{850}{69} = \frac{1050}{f} \begin{array}{l} \leftarrow \text{Students} \\ \leftarrow \text{Faculty} \end{array}$$

Solve.
$850 \cdot f = 69 \cdot 1050$
$f = \dfrac{69 \cdot 1050}{850}$
$f = \dfrac{72,450}{850}$, or $85\dfrac{200}{850}$, or $85\dfrac{4}{17}$

Since it is impossible to create a fractional part of a position, we round up to the nearest whole position. Thus, 86 positions will be required after the university expands. We subtract to find how many new positions should be created:

$86 - 69 = 17$

Check. We substitute in the proportion and check cross products.
$$\frac{850}{69} = \frac{1050}{72,450/850}$$
$850 \cdot \dfrac{72,450}{850} = 72,450$; $69 \cdot 1050 = 72,450$

State. 17 new faculty positions should be created.

49. **Familiarize**. Let r = the number of earned runs Cy Young gave up in his career.

Translate. We translate to a proportion.
$$\begin{array}{cc} \text{Runs} \rightarrow \\ \text{Innings} \rightarrow \end{array} \frac{2.63}{9} = \frac{r}{7356} \begin{array}{l} \leftarrow \text{Runs} \\ \leftarrow \text{Innings} \end{array}$$

Solve.
$2.63 \cdot 7356 = 9 \cdot r$
$\dfrac{2.63 \cdot 7356}{9} = r$
$2150 \approx r$

Check. We substitute in the proportion and check cross products.
$$\frac{2.63}{9} = \frac{2150}{7356}$$
$2.63 \cdot 7356 = 19,346.28$ and $9 \cdot 2150 = 19,350 \approx 19,346.28$

State. Cy Young gave up 2150 earned runs in his career.

51. $1 + 3 + 2 = 6$, and $\$800/6 = \$133\frac{1}{3}$.

Then $1 \cdot \$133\frac{1}{3}$, or \$133.33, would be spent on a CD player; $3 \cdot \$133\frac{1}{3}$, or \$400, would be spent on a receiver; and $2 \cdot \$133\frac{1}{3}$, or \$266.67, would be spent on speakers.

Exercise Set 7.5

1. The ratio of h to 5 is the same as the ratio of 45 to 9. We have the proportion
$$\frac{h}{5} = \frac{45}{9}.$$
Solve: $9 \cdot h = 5 \cdot 45$ Equating cross-products
$\qquad h = \frac{5 \cdot 45}{9}$ Dividing by 9 on both sides
$\qquad h = 25$ Simplifying

The missing length h is 25.

3. The ratio of x to 2 is the same as the ratio of 2 to 3. We have the proportion
$$\frac{x}{2} = \frac{2}{3}.$$
Solve: $3 \cdot x = 2 \cdot 2$ Equating cross-products
$\qquad x = \frac{2 \cdot 2}{3}$ Dividing by 3 on both sides
$\qquad x = \frac{4}{3}$, or $1\frac{1}{3}$

The missing length x is $\frac{4}{3}$, or $1\frac{1}{3}$. We could also have used $\frac{x}{2} = \frac{1}{1\frac{1}{2}}$ to find x.

5. First we find x. The ratio of x to 9 is the same as the ratio of 6 to 8. We have the proportion
$$\frac{x}{9} = \frac{6}{8}.$$
Solve: $8 \cdot x = 9 \cdot 6$
$\qquad x = \frac{9 \cdot 6}{8}$
$\qquad x = \frac{27}{4}$, or $6\frac{3}{4}$

The missing length x is $\frac{27}{4}$, or $6\frac{3}{4}$.

Next we find y. The ratio of y to 12 is the same as the ratio of 6 to 8. We have the proportion
$$\frac{y}{12} = \frac{6}{8}.$$
Solve: $8 \cdot y = 12 \cdot 6$
$\qquad y = \frac{12 \cdot 6}{8}$
$\qquad y = 9$

The missing length y is 9.

7. First we find x. The ratio of x to 2.5 is the same as the ratio of 2.1 to 0.7. We have the proportion
$$\frac{x}{2.5} = \frac{2.1}{0.7}.$$
Solve: $0.7 \cdot x = 2.5 \cdot 2.1$
$\qquad x = \frac{2.5 \cdot 2.1}{0.7}$
$\qquad x = 7.5$

The missing length x is 7.5.

Next we find y. The ratio of y to 2.4 is the same as the ratio of 2.1 to 0.7. We have the proportion
$$\frac{y}{2.4} = \frac{2.1}{0.7}.$$
Solve: $0.7 \cdot y = 2.4 \cdot 2.1$
$\qquad y = \frac{2.4 \cdot 2.1}{0.7}$
$\qquad y = 7.2$

The missing length y is 7.2.

9. Referring to the drawing in the text, we see that the ratio of 6 to p is the same as the ratio of 5 to 56. We have the proportion
$$\frac{6}{p} = \frac{5}{56}.$$
Solve: $6 \cdot 56 = 5 \cdot p$
$\qquad \frac{6 \cdot 56}{5} = p$
$\qquad 67.2 = p$

The flagpole is 67.2 ft tall.

11. If we use the sun's rays to represent the third side of a triangle in a drawing of the situation, we see that we have similar triangles. We let $h =$ the height of a flag pole, in feet, that casts a 42-ft shadow.

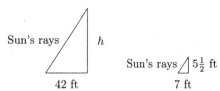

The ratio of h to $5\frac{1}{2}$ is the same as the ratio of 42 to 7.

We have the proportion
$$\frac{h}{5\frac{1}{2}} = \frac{42}{7}.$$

Solve: $7 \cdot h = 5\frac{1}{2} \cdot 42$

$7 \cdot h = \frac{11}{2} \cdot 42$

$7 \cdot h = 231$

$h = \frac{231}{7}$

$h = 33$

The flagpole is 33 ft tall.

13. If we use the sun's rays to represent the third side of a triangle in a drawing of the situation, we see that we have similar triangles. We let h = the height of the tree.

Sun's rays h

Sun's rays 8 ft

32 ft   9 ft

The ratio of h to 8 is the same as the ratio of 32 to 9. We have the proportion

$$\frac{h}{8} = \frac{32}{9}.$$

Solve: $9 \cdot h = 8 \cdot 32$

$h = \frac{8 \cdot 32}{9}$

$h = \frac{256}{9} = 28\frac{4}{9}$

The tree is $28\frac{4}{9}$ ft tall.

15. The ratio of h to 7 ft is the same as the ratio of 6 ft to 6 ft. We have the proportion

$$\frac{h}{7} = \frac{6}{6}.$$

Solve: $6 \cdot h = 7 \cdot 6$

$h = \frac{7 \cdot 6}{6}$

$h = 7$

The wall is 7 ft high.

17. Since the ratio of d to 25 ft is the same as the ratio of 40 ft to 10 ft, we have the proportion

$$\frac{d}{25} = \frac{40}{10}.$$

Solve: $10 \cdot d = 25 \cdot 40$

$d = \frac{25 \cdot 40}{10}$

$d = 100$

The distance across the river is 100 ft.

19. Width $\rightarrow$ $\dfrac{6}{9} = \dfrac{x}{6}$ $\leftarrow$ Width
Length $\rightarrow$ $\phantom{\dfrac{6}{9}}$ $\leftarrow$ Length

Solve: $\frac{2}{3} = \frac{x}{6}$ Rewriting $\frac{6}{9}$ as $\frac{2}{3}$

$2 \cdot 6 = 3 \cdot x$ Equating cross products

$\frac{2 \cdot 6}{3} = x$

$\frac{2 \cdot 2 \cdot 3}{3} = x$

$2 \cdot 2 = x$

$4 = x$

The missing length x is 4.

21. Width $\rightarrow$ $\dfrac{4}{7} = \dfrac{6}{x}$ $\leftarrow$ Width
Length $\rightarrow$ $\phantom{\dfrac{4}{7}}$ $\leftarrow$ Length

Solve: $4 \cdot x = 7 \cdot 6$ Equating cross products

$x = \frac{7 \cdot 6}{4}$

$x = \frac{7 \cdot 2 \cdot 3}{2 \cdot 2}$

$x = \frac{7 \cdot 3}{2}$

$x = \frac{21}{2}$, or $10\frac{1}{2}$

The missing length x is $10\frac{1}{2}$.

23. First we find x. The ratio of x to 8 is the same as the ratio of 3 to 4. We have the proportion

$$\frac{x}{8} = \frac{3}{4}.$$

Solve: $4 \cdot x = 8 \cdot 3$

$x = \frac{8 \cdot 3}{4}$

$x = 6$

The missing length x is 6.

Next we find y. The ratio of y to 7 is the same as the ratio of 3 to 4. We have the proportion

$$\frac{y}{7} = \frac{3}{4}.$$

Solve: $4 \cdot y = 7 \cdot 3$

$y = \frac{7 \cdot 3}{4}$

$y = \frac{21}{4}$, or $5\frac{1}{4}$, or 5.25

The missing length y is $\frac{21}{4}$, or $5\frac{1}{4}$, or 5.25.

Finally we find z. The ratio of z to 4 is the same as the ratio of 3 to 4. This statement tells us that z must be 3. We could also calculate this using the proportion

$$\frac{z}{4} = \frac{3}{4}.$$

The missing length z is 3.

25. First we find x. The ratio of x to 8 is the same as the ratio of 2 to 3. We have the proportion

$$\frac{x}{8} = \frac{2}{3}.$$

Solve: $3 \cdot x = 8 \cdot 2$

$$x = \frac{8 \cdot 2}{3}$$

$$x = \frac{16}{3}, \text{ or } 5\frac{1}{3}$$

The missing length x is $\frac{16}{3}$, or $5\frac{1}{3}$, or $5.\overline{3}$.

Next we find y. The ratio of y to 7 is the same as the ratio of 2 to 3. We have the proportion

$$\frac{y}{7} = \frac{2}{3}.$$

Solve: $3 \cdot y = 7 \cdot 2$

$$y = \frac{7 \cdot 2}{3}$$

$$y = \frac{14}{3} = 4\frac{2}{3}$$

The missing length y is $\frac{14}{3}$, or $4\frac{2}{3}$, or $4.\overline{6}$.

Finally we find z. The ratio of z to 8 is the same as the ratio of 2 to 3. We have the proportion

$$\frac{z}{9} = \frac{2}{3}.$$

This is the same proportion we solved above when we found x. Then the missing length z is $5\frac{1}{3}$, or $5.\overline{3}$.

27. Height $\rightarrow$ $\dfrac{h}{32} = \dfrac{5}{8}$ $\leftarrow$ Height
Width $\rightarrow$ $\phantom{\dfrac{h}{32}}$ $\phantom{\dfrac{5}{8}}$ $\leftarrow$ Width

Solve: $8 \cdot h = 32 \cdot 5$

$$h = \frac{32 \cdot 5}{8}$$

$$h = \frac{4 \cdot 8 \cdot 5}{8}$$

$$h = 4 \cdot 5$$

$$h = 20$$

The missing length is 20 ft.

29. The ratio of h to 15 is the same as the ratio of 116 to 12. We have the proportion

$$\frac{h}{19} = \frac{120}{15}.$$

Solve: $15 \cdot h = 19 \cdot 120$

$$h = \frac{19 \cdot 120}{15}$$

$$h = 152$$

The addition will be 152 ft high.

31. Discussion and Writing Exercise

33. *Familiarize.* This is a multistep problem.

First we find the total cost of the purchases. We let $c =$ this amount.

Translate and Solve.

Price of book	plus	Price of CD	plus	Price of sweatshirt	is	Total cost
↓	↓	↓	↓	↓	↓	↓
$49.95	+	$14.88	+	$29.95	=	c

To solve the equation we carry out the addition.

$$
\begin{array}{r}
^{2}\,4\,^{2}9.\,^{1}9\,5 \\
1\,4.\,8\,8 \\
+\,2\,9.\,9\,5 \\
\hline
9\,4.\,7\,8
\end{array}
$$

Thus, $c = \$94.78$.

Now we find how much more money the student needs to make these purchases. We let $m =$ this amount.

Money student has	plus	How much more money	is	Total cost of purchases
↓	↓	↓	↓	↓
$34.97	+	m	=	$94.78

To solve the equation we subtract 34.97 on both sides.

$$m = 94.78 - 34.97$$
$$m = 59.81$$

$$
\begin{array}{r}
^{8}\,\overset{13}{\cancel{9}}\,\overset{17}{\cancel{4}}.\,\overset{}{\cancel{7}}\,8 \\
-\,3\,4.\,9\,7 \\
\hline
5\,9.\,8\,1
\end{array}
$$

Check. We repeat the calculations.

State. The student needs $59.81 more to make the purchases.

35. First we multiply the absolute values.

$$
\begin{array}{r}
^{7}\,^{7}\,^{1} \\
^{3}\,^{3} \\
8\,0.\,8\,9\,2 \\
\times8.\,4 \\
\hline
3\,2\,3\,5\,6\,8 \\
6\,4\,7\,1\,3\,6\,0 \\
\hline
6\,7\,9.\,4\,9\,2\,8
\end{array}
$$

Since the product of a negative and a positive number is negative, the product is -679.4928.

37. $1\underline{00} \times 274.568 \qquad 274.56.8$

2 zeros   Move 2 places to the right.

$100 \times 274.568 = 27,456.8$

39. $\dfrac{17}{20} = \dfrac{17}{20} \cdot \dfrac{5}{5} = \dfrac{85}{100} = 0.85$

41. First we find decimal notation for $\dfrac{10}{11}$.

$$
\begin{array}{r}
0.\,9\,0\,9\,0 \\
11\overline{)1\,0.\,0\,0\,0} \\
\underline{9\,9} \\
1\,0\,0 \\
\underline{9\,9} \\
1\,0
\end{array}
$$

Because we are rounding to the nearest thousandth, we stop here.

$$\frac{10}{11} \approx 0.909, \text{ so } -\frac{10}{11} \approx -0.909.$$

43. Discussion and Writing Exercise

45.

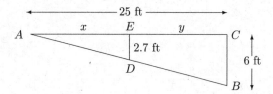

We note that triangle ADE is similar to triangle ABC and use this information to find the length x.

$$\frac{x}{25} = \frac{2.7}{6}$$
$$6 \cdot x = 25 \cdot 2.7$$
$$x = \frac{25 \cdot 2.7}{6}$$
$$x = 11.25$$

Thus the goalie should be 11.25 ft from point A. We subtract to find how far from the goal the goalie should be located.

$$25 - 11.25 = 13.75$$

The goalie should stand 13.75 ft from the goal.

47. From Exercise 29 we know that a height of 19 cm on the model corresponds to a height of 152 ft on the building. We let $h =$ the height of the model hoop. Then we translate to a proportion.

Model height → $\dfrac{19}{152} = \dfrac{h}{10}$ ← Model height

Actual height →   ← Actual height

Solve: $19 \cdot 10 = 152 \cdot h$ Equating cross products

$$\frac{19 \cdot 10}{152} = h$$
$$1.25 = h$$

The model hoop should be 1.25 cm high.

49. First we find x. We see from the drawing that the ratio of x to $19.35 + x$ is the same as the ratio of 0.3 to 16.8. We have the proportion

$$\frac{x}{19.35 + x} = \frac{0.3}{16.8}$$

Solve: $16.8 \cdot x = 0.3(19.35 + x)$

$$16.8x = 5.805 + 0.3x$$
$$16.5x = 5.805$$
$$x \approx 0.35$$

The missing length x is about 0.35.

Now we find y. The ratio of y to 22.4 is the same as the ratio of 0.3 to 16.8. We have the proportion

$$\frac{y}{22.4} = \frac{0.3}{16.8}.$$

Solve: $16.8 \cdot y = 22.4(0.3)$

$$y = \frac{22.4(0.3)}{16.8}$$
$$y = 0.4$$

The missing length y is 0.4.

Chapter 7 Review Exercises

1. The ratio of 47 to 84 is $\dfrac{47}{84}$.

2. The ratio of 46 to 1.27 is $\dfrac{46}{1.27}$.

3. The ratio of 83 to 100 is $\dfrac{83}{100}$.

4. The ratio of 0.72 to 197 is $\dfrac{0.72}{197}$.

5. a) The ratio of 12,480 to 16,640 is $\dfrac{12,480}{16,640}$.

We can simplify this ratio as follows:

$$\frac{12,480}{16,640} = \frac{3 \cdot 5 \cdot 8 \cdot 8 \cdot 13}{4 \cdot 5 \cdot 8 \cdot 8 \cdot 13} = \frac{3}{4} \cdot \frac{5 \cdot 8 \cdot 8 \cdot 13}{5 \cdot 8 \cdot 8 \cdot 13} = \frac{3}{4}$$

b) The total of both kinds of fish sold is 12,480 lb + 16,640 lb, or 29,120 lb. Then the ratio of salmon sold to the total amount of both kinds of fish sold is $\dfrac{16,640}{29,120}$.

We can simplify this ratio as follows:

$$\frac{16,640}{29,120} = \frac{4 \cdot 5 \cdot 8 \cdot 8 \cdot 13}{5 \cdot 7 \cdot 8 \cdot 8 \cdot 13} = \frac{4}{7} \cdot \frac{5 \cdot 8 \cdot 8 \cdot 13}{5 \cdot 8 \cdot 8 \cdot 13} = \frac{4}{7}$$

6. $\dfrac{9}{12} = \dfrac{3 \cdot 3}{4 \cdot 4} = \dfrac{3}{3} \cdot \dfrac{3}{4} = \dfrac{3}{4}$

7. $\dfrac{3.6}{6.4} = \dfrac{3.6}{6.4} \cdot \dfrac{10}{10} = \dfrac{36}{64} = \dfrac{4 \cdot 9}{4 \cdot 16} = \dfrac{4}{4} \cdot \dfrac{9}{16} = \dfrac{9}{16}$

8. $\dfrac{377 \text{ mi}}{14.5 \text{ gal}} = 26 \dfrac{\text{mi}}{\text{gal}}$, or 26 mpg

9. $\dfrac{472,500 \text{ revolutions}}{75 \text{ min}} = 6300 \dfrac{\text{revolutions}}{\text{min}}$, or 6300 rpm

10. $\dfrac{319 \text{ gal}}{500 \text{ ft}^2} = 0.638 \text{ gal/ft}^2$

11. $\dfrac{18 \text{ servings}}{25 \text{ lb}} = 0.72 \text{ serving/lb}$

12. $\dfrac{\$12.99}{300 \text{ tablets}} = \dfrac{1299\cancel{c}}{300 \text{ tablets}} = 4.33\cancel{c}/\text{tablet}$

13. $\dfrac{\$1.97}{13.9 \text{ oz}} = \dfrac{197\cancel{c}}{13.9 \text{ oz}} \approx 14.173\cancel{c}/\text{oz}$

14. In 8 rolls of towels with 60 sheets per roll there are $8 \cdot 60$, or 480 sheets.

$$\frac{\$6.38}{480 \text{ sheets}} = \frac{638\cancel{c}}{480 \text{ sheets}} \approx 1.329\cancel{c}/\text{sheet}$$

In 15 rolls of towels with 60 sheets per roll there are $15 \cdot 60$, or 900 sheets.

$$\frac{\$13.99}{900 \text{ sheets}} = \frac{1399\cancel{c}}{900 \text{ sheets}} \approx 1.554\cancel{c}/\text{sheet}$$

In 6 rolls of towels with 165 sheets per roll there are $6 \cdot 165$, or 990 sheets.

$$\frac{\$10.99}{990 \text{ sheets}} = \frac{1099\cancel{c}}{990 \text{ sheets}} \approx 1.110\cancel{c}/\text{sheet}$$

The package containing 6 big rolls with 165 sheets per roll has the lowest unit price.

15. $\dfrac{\$2.19}{32 \text{ oz}} = \dfrac{219¢}{32 \text{ oz}} \approx 6.844¢/\text{oz}$

$\dfrac{\$2.49}{48 \text{ oz}} = \dfrac{249¢}{48 \text{ oz}} \approx 5.188¢/\text{oz}$

$\dfrac{\$3.59}{64 \text{ oz}} = \dfrac{359¢}{64 \text{ oz}} \approx 5.609¢/\text{oz}$

$\dfrac{\$7.09}{128 \text{ oz}} = \dfrac{709¢}{128 \text{ oz}} \approx 5.539¢/\text{oz}$

The 48 oz package has the lowest unit price.

16. We can use cross products:

$9 \cdot 59 = 531 \quad \overset{9 \quad 36}{\underset{15 \quad 59}{\times}} \quad 15 \cdot 36 = 540$

Since the cross products are not the same, $531 \neq 540$, we know that the numbers are not proportional.

17. We can use cross products:

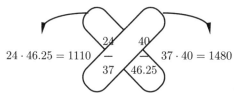

$24 \cdot 46.25 = 1110 \quad \overset{24 \quad 40}{\underset{37 \quad 46.25}{\times}} \quad 37 \cdot 40 = 1480$

Since the cross products are not the same, $1110 \neq 1480$, we know that the numbers are not proportional.

18. $\dfrac{8}{9} = \dfrac{x}{36}$

$8 \cdot 36 = 9 \cdot x \qquad$ Equating cross products

$\dfrac{8 \cdot 36}{9} = \dfrac{9 \cdot x}{9}$

$\dfrac{288}{9} = x$

$32 = x$

19. $\dfrac{6}{x} = \dfrac{48}{56}$

$6 \cdot 56 = x \cdot 48$

$\dfrac{6 \cdot 56}{48} = \dfrac{x \cdot 48}{48}$

$\dfrac{336}{48} = x$

$7 = x$

20. $\dfrac{120}{\frac{3}{7}} = \dfrac{7}{x}$

$120 \cdot x = \dfrac{3}{7} \cdot 7$

$120 \cdot x = 3$

$\dfrac{120 \cdot x}{120} = \dfrac{3}{120}$

$x = \dfrac{1}{40}$

21. $\dfrac{4.5}{120} = \dfrac{0.9}{x}$

$4.5 \cdot x = 120 \cdot 0.9$

$\dfrac{4.5 \cdot x}{4.5} = \dfrac{120 \cdot 0.9}{4.5}$

$x = \dfrac{108}{4.5}$

$x = 24$

22. *Familiarize.* Let $p =$ the price of 5 dozen eggs.

Translate. We translate to a proportion.

$\begin{array}{ll} \text{Eggs} \rightarrow & \dfrac{3}{2.67} = \dfrac{5}{p} \quad \leftarrow \text{Eggs} \\ \text{Price} \rightarrow & \qquad \qquad \leftarrow \text{Price} \end{array}$

Solve. We solve the proportion.

$\dfrac{3}{2.67} = \dfrac{5}{p}$

$3 \cdot p = 2.67 \cdot 5$

$\dfrac{3 \cdot p}{3} = \dfrac{2.67 \cdot 5}{3}$

$p = 4.45$

Check. We substitute in the proportion and check cross products.

$\dfrac{3}{2.67} = \dfrac{5}{4.45}$

$3 \cdot 4.45 = 13.35; \; 2.67 \cdot 5 = 13.35$

The cross products are the same, so the answer checks.

State. 5 dozen eggs would cost \$4.45.

23. *Familiarize.* Let $d =$ the number of defective circuits in a lot of 585.

Translate. We translate to a proportion.

$\begin{array}{ll} \text{Defective} \rightarrow & \dfrac{39}{65} = \dfrac{d}{585} \quad \leftarrow \text{Defective} \\ \text{Total circuits} \rightarrow & \qquad \qquad \leftarrow \text{Total circuits} \end{array}$

Solve. We solve the proportion.

$\dfrac{39}{65} = \dfrac{d}{585}$

$39 \cdot 585 = 65 \cdot d$

$\dfrac{39 \cdot 585}{65} = d$

$351 = d$

Check. We substitute in the proportion and check cross products.

$\dfrac{39}{65} = \dfrac{351}{585}$

$39 \cdot 585 = 22,815; \; 65 \cdot 351 = 22,815$

The cross products are the same, so the answer checks.

State. It would be expected that 351 defective circuits would occur in a lot of 585 circuits.

24. a) *Familiarize.* Let $a =$ the number of Euros equivalent to 250 U.S. dollars.

Translate. We translate to a proportion.

$\begin{array}{ll} \text{U.S. dollars} \rightarrow & \dfrac{1}{0.782459} = \dfrac{250}{a} \quad \leftarrow \text{U.S. dollars} \\ \text{Euros} \rightarrow & \qquad \qquad \leftarrow \text{Euros} \end{array}$

Solve.

$1 \cdot a = 0.782459 \cdot 250$ Equating cross products

$a = 195.61475$

Check. We substitute in the proportion and check cross products.

$$\frac{1}{0.782459} = \frac{250}{195.61475}$$

$1 \cdot 195.61475 = 195.61475; \ 0.782459 \cdot 250 = 195.61475$

The cross products are the same, so the answer checks.

State. 250 U.S. dollars would be worth 195.61475 Euros.

b) *Familiarize*. Let c = the cost of the sweatshirt in U.S. dollars.

Translate. We translate to a proportion.

$$\begin{array}{l}\text{U.S. dollars} \rightarrow \\ \text{Euros} \quad \rightarrow \end{array} \frac{1}{0.782459} = \frac{c}{50} \begin{array}{l} \leftarrow \text{U.S. dollars} \\ \leftarrow \quad \text{Euros}\end{array}$$

Solve.

$1 \cdot 50 = 0.782459 \cdot c$ Equating cross products

$$\frac{1 \cdot 50}{0.782459} = c$$

$63.90 \approx c$

Check. We substitute in the proportion and check cross products.

$$\frac{1}{0.782459} = \frac{63.90}{50}$$

$1 \cdot 50 = 50; \ 0.782459 \cdot 63.90 \approx 50$

The cross products are about the same. Remember that we rounded the value of c.

State. The sweatshirt cost $63.90 in U.S. dollars.

25. Familiarize. Let d = the number of miles the train will travel in 13 hr.

Translate. We translate to a proportion.

$$\begin{array}{l}\text{Miles} \rightarrow \\ \text{Hours} \rightarrow\end{array} \frac{448}{7} = \frac{d}{13} \begin{array}{l}\leftarrow \text{Miles} \\ \leftarrow \text{Hours}\end{array}$$

Solve.

$448 \cdot 13 = 7 \cdot d$ Equating cross products

$$\frac{448 \cdot 13}{7} = \frac{7 \cdot d}{7}$$

$832 = d$

Check. We find how far the train travels in 1 hr and then multiply by 13:

$448 \div 7 = 64$ and $64 \cdot 13 = 832$

The answer checks.

State. The train will travel 832 mi in 13 hr.

26. Familiarize. Let a = the number of acres required to produce 97.2 bushels of tomatoes.

Translate. We translate to a proportion.

$$\begin{array}{l}\text{Acres} \ \rightarrow \\ \text{Bushels} \rightarrow\end{array} \frac{15}{54} = \frac{a}{97.2}$$

Solve.

$15 \cdot 97.2 = 54 \cdot a$ Equating cross products

$$\frac{15 \cdot 97.2}{54} = \frac{54 \cdot a}{54}$$

$27 = a$

Check. We substitute in the proportion and check cross products.

$$\frac{15}{54} = \frac{27}{97.2}$$

$15 \cdot 97.2 = 1458; \ 54 \cdot 27 = 1458$

The answer checks.

State. 27 acres are required to produce 97.2 bushels of tomatoes.

27. Familiarize. Let g = the number of kilograms of garbage produced in San Diego in one day.

Translate. We translate to a proportion.

$$\begin{array}{l}\text{Garbage} \rightarrow \\ \text{People} \ \rightarrow\end{array} \frac{13}{5} = \frac{g}{1,266,753} \begin{array}{l}\leftarrow \text{Garbage} \\ \leftarrow \ \text{People}\end{array}$$

Solve.

$13 \cdot 1,266,753 = 5 \cdot g$ Equating cross products

$$\frac{13 \cdot 1,266,753}{5} = \frac{5 \cdot g}{5}$$

$3,293,558 \approx g$

Check. We can divide to find the amount of garbage produced by one person and then multiply to find the amount produced by 1,266,753 people.

$13 \div 5 = 2.6$ and $2.6 \cdot 1,266,753 = 3,293,557.8 \approx 3,293,558$. The answer checks.

State. About 3,293,558 kg of garbage is produced in San Diego in one day.

28. Familiarize. Let w = the number of inches of water to which $4\frac{1}{2}$ ft of snow melts.

Translate. We translate to a proportion.

$$\begin{array}{l}\text{Snow} \rightarrow \\ \text{Water} \rightarrow\end{array} \frac{1\frac{1}{2}}{2} = \frac{4\frac{1}{2}}{w} \begin{array}{l}\leftarrow \text{Snow} \\ \leftarrow \text{Water}\end{array}$$

Solve.

$1\frac{1}{2} \cdot w = 2 \cdot 4\frac{1}{2}$ Equating cross products

$\frac{3}{2} \cdot w = 2 \cdot \frac{9}{2}$

$\frac{3}{2} \cdot w = 9$

$w = 9 \div \frac{3}{2}$ Dividing by $\frac{3}{2}$ on both sides

$w = 9 \cdot \frac{2}{3}$

$w = \frac{9 \cdot 2}{3}$

$w = 6$

Check. We substitute in the proportion and check cross products.

$$\frac{1\frac{1}{2}}{2} = \frac{4\frac{1}{2}}{6}$$

$$1\frac{1}{2} \cdot 6 = \frac{3}{2} \cdot 6 = \frac{3 \cdot 6}{2} = 9; \quad 2 \cdot 4\frac{1}{2} = 2 \cdot \frac{9}{2} = \frac{2 \cdot 9}{2} = 9$$

The cross products are the same, so the answer checks.

State. $4\frac{1}{2}$ ft of snow will melt to 6 in. of water.

29. ***Familiarize.*** Let $l =$ the number of lawyers we would expect to find in Detroit.

Translate. We translate to a proportion.

$$\text{Lawyers} \rightarrow \frac{2.3}{1000} = \frac{l}{911,402} \leftarrow \text{Lawyers}$$
$$\text{Population} \rightarrow \qquad\qquad \leftarrow \text{Population}$$

Solve.

$$2.3 \cdot 911,402 = 1000 \cdot l \quad \text{Equating cross products}$$

$$\frac{2.3 \cdot 911,402}{1000} = \frac{1000 \cdot l}{1000}$$

$$2096 \approx l$$

Check. We substitute in the proportion and check cross products.

$$\frac{2.3}{1000} = \frac{2096}{911,402}$$

$2.3 \cdot 911,402 = 2,096,224.6; \quad 1000 \cdot 2096 = 2,096,000 \approx 2,096,224.6$

The answer checks.

State. We would expect that there would be about 2096 lawyers in Detroit.

30. The ratio of x to 7 is the same as the ratio of 6 to 9.

$$\frac{x}{7} = \frac{6}{9}$$

$$x \cdot 9 = 7 \cdot 6$$

$$x = \frac{7 \cdot 6}{9} = \frac{7 \cdot 2 \cdot 3}{3 \cdot 3}$$

$$x = \frac{7 \cdot 2}{3} \cdot \frac{3}{3} = \frac{7 \cdot 2}{3}$$

$$x = \frac{14}{3}, \text{ or } 4\frac{2}{3}$$

We could also have used the proportion $\frac{x}{7} = \frac{2}{3}$ to find x.

31. The ratio of x to 8 is the same as the ratio of 7 to 5.

$$\frac{x}{8} = \frac{7}{5}$$

$$x \cdot 5 = 8 \cdot 7$$

$$x = \frac{8 \cdot 7}{5}$$

$$x = \frac{56}{5}, \text{ or } 11\frac{1}{5}$$

The ratio of y to 9 is the same as the ratio of 7 to 5.

$$\frac{y}{9} = \frac{7}{5}$$

$$y \cdot 5 = 9 \cdot 7$$

$$y = \frac{9 \cdot 7}{5}$$

$$y = \frac{63}{5}, \text{ or } 12\frac{3}{5}$$

32. If we use the sun's rays to represent the third side of a triangle in a drawing of the situation, we see that we have similar triangles. We let $h =$ the height of the billboard, in feet.

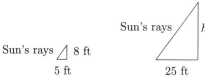

The ratio of h to 8 is the same as the ratio of 25 to 5.

$$\frac{h}{8} = \frac{25}{5}$$

$$h \cdot 5 = 8 \cdot 25$$

$$h = \frac{8 \cdot 25}{5}$$

$$h = 40$$

The billboard is 40 ft high.

33. The ratio of x to 2 is the same as the ratio of 9 to 6.

$$\frac{x}{2} = \frac{9}{6}$$

$$x \cdot 6 = 2 \cdot 9$$

$$x = \frac{2 \cdot 9}{6}$$

$$x = 3$$

The ratio of y to 6 is the same as the ratio of 9 to 6, so we see that $y = 9$.

The ratio of z to 5 is the same as the ratio of 9 to 6.

$$\frac{z}{5} = \frac{9}{6}$$

$$z \cdot 6 = 5 \cdot 9$$

$$z = \frac{5 \cdot 9}{6} = \frac{5 \cdot 3 \cdot 3}{2 \cdot 3}$$

$$z = \frac{5 \cdot 3}{2} \cdot \frac{3}{3} = \frac{5 \cdot 3}{2}$$

$$z = \frac{15}{2}, \text{ or } 7\frac{1}{2}$$

34. *Discussion and Writing Exercise.* In terms of cost, a low faculty-to-student ratio is less expensive than a high faculty-to-student ratio. In terms of quality of education and student satisfaction, a high faculty-to-student ratio is more desirable. A college president must balance the cost and quality issues.

35. *Discussion and Writing Exercise.* Leslie used 4 gal of gasoline to drive 92 mile. At the same rate, how many gallons would be needed to travel 368 mi?

36. ***Familiarize.*** Let $t =$ the number of minutes it will take Yancy to type a 7-page term paper.

Translate. We translate to a proportion.

$$\text{Time} \rightarrow \frac{10}{\frac{2}{3}} = \frac{t}{7} \leftarrow \text{Time}$$
$$\text{Pages} \rightarrow \qquad\qquad \leftarrow \text{Pages}$$

Solve.

$$10 \cdot 7 = \frac{2}{3} \cdot t$$

$$70 = \frac{2}{3} \cdot t$$

$$70 \div \frac{2}{3} = t$$

$$70 \cdot \frac{3}{2} = t$$

$$\frac{70 \cdot 3}{2} = t$$

$$105 = t$$

Check. We substitute in the proportion and check cross products.

$$\frac{10}{\frac{2}{3}} = \frac{105}{7}$$

$$10 \cdot 7 = 70; \quad \frac{2}{3} \cdot 105 = \frac{2 \cdot 105}{3} = 70$$

The cross products are the same, so the answer checks.

State. It would take 105 min, or 1 hr, 45 min, to type the term paper.

37. Let x = the number of purple beads in each bracelet. Then $40 - x$ = the number of lavender beads in each one. We use a proportion to find the number of each color of bead in a bracelet.

$$\frac{3}{5} = \frac{x}{40 - x}$$

$$3(40 - x) = 5x$$

$$120 - 3x = 5x$$

$$120 = 8x$$

$$15 = x$$

If there are 15 purple beads in each bracelet, then with 60 purple beads Sari can make 60/15, or 4 bracelets.

If there are 15 purple beads in each bracelet, then the number of lavender beads in each one is $40 - 15$, or 25. Thus, for 4 bracelets the number of lavender beads required is $4 \cdot 25$, or 100 lavender beads.

38. *Familiarize*. Let m = the number of minutes required to type 34,000 keystrokes. Recall that 1 hr = 60 min.

Translate. We translate to a proportion.

$$\begin{array}{l} \text{Keystrokes} \to 8500 \\ \text{Minutes} \to 60 \end{array} = \begin{array}{l} 34,000 \leftarrow \text{Keystrokes} \\ m \leftarrow \text{Minutes} \end{array}$$

Solve. We solve the proportion.

$$8500 \times m = 60 \times 34,000$$

$$\frac{8500 \times m}{8500} = \frac{60 \times 34,000}{8500}$$

$$m = 240$$

Check. We substitute in the proportion and check cross product.s

$$\frac{8500}{60} = \frac{34,000}{240}$$

$$8500 \times 240 = 2,040,000; \quad 60 \times 34,000 = 2,040,000$$

The cross products are the same, so the answer checks.

State. It would take 240 min to type 34,000 keystrokes.

39. First we divide to find how many gallons of finishing paint are needed.

$$4950 \div 450 = 11 \text{ gal}$$

Next we write and solve a proportion to find how many gallons of primer are needed. Let p = the amount of primer needed.

$$\begin{array}{l} \text{Finishing paint} \to \\ \text{Primer} \to \end{array} \frac{2}{3} = \frac{11}{p} \begin{array}{l} \leftarrow \text{Finishing paint} \\ \leftarrow \text{Primer} \end{array}$$

$$2 \cdot p = 3 \cdot 11$$

$$p = \frac{3 \cdot 11}{2}$$

$$p = \frac{33}{2}, \text{ or } 16.5$$

Thus, 11 gal of finishing paint and 16.5 gal of primer should be purchased.

Chapter 7 Test

1. The ratio of 85 to 97 is $\frac{85}{97}$.

2. The ratio of 0.34 to 124 is $\frac{0.34}{124}$.

3. $\dfrac{10 \text{ ft}}{16 \text{ sec}} = \dfrac{10}{16} \dfrac{\text{ft}}{\text{sec}} = 0.625 \text{ ft/sec}$

4. $\dfrac{319 \text{ mi}}{14.5 \text{ gal}} = \dfrac{319}{14.5} \dfrac{\text{mi}}{\text{gal}} = 22 \text{ mpg}$

5. $\dfrac{\$3.69}{33 \text{ oz}} = \dfrac{369\cancel{c}}{33 \text{ oz}} \approx 11.182\cancel{c}/\text{oz}$

$\dfrac{\$6.22}{87 \text{ oz}} = \dfrac{622\cancel{c}}{87 \text{ oz}} \approx 7.149\cancel{c}/\text{oz}$

$\dfrac{\$10.99}{131 \text{ oz}} = \dfrac{1099\cancel{c}}{131 \text{ oz}} \approx 8.389\cancel{c}/\text{oz}$

$\dfrac{\$17.99}{263 \text{ oz}} = \dfrac{1799\cancel{c}}{263 \text{ oz}} \approx 6.840\cancel{c}/\text{oz}$

The 263 oz package has the lowest unit price.

6. $\dfrac{0.32}{0.15} = \dfrac{0.32}{0.15} \cdot \dfrac{100}{100} = \dfrac{32}{15}$

7. We can use cross products:

$$7 \cdot 72 = 504 \quad \begin{array}{c} 7 \quad 63 \\ 8 \quad 72 \end{array} \quad 8 \cdot 63 = 504$$

Since the cross products are the same, $504 = 504$, we know that $\dfrac{7}{8} = \dfrac{63}{72}$, so the numbers are proportional.

8. We can use cross products:

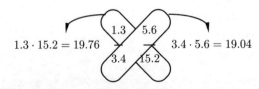

$$1.3 \cdot 15.2 = 19.76 \quad \begin{array}{c} 1.3 \quad 5.6 \\ 3.4 \quad 15.2 \end{array} \quad 3.4 \cdot 5.6 = 19.04$$

Since the cross products are not the same, $19.76 \neq 19.04$, we know that $\frac{1.3}{3.4} \neq \frac{5.6}{15.2}$, so the numbers are not proportional.

9. $\quad \frac{9}{4} = \frac{27}{x}$

$9 \cdot x = 4 \cdot 27$ Equating cross products

$\frac{9 \cdot x}{9} = \frac{4 \cdot 27}{9}$

$x = \frac{4 \cdot 9 \cdot 3}{9 \cdot 1}$

$x = 12$

10. $\quad \frac{150}{2.5} = \frac{x}{6}$

$150 \cdot 6 = 2.5 \cdot x$ Equating cross products

$\frac{150 \cdot 6}{2.5} = \frac{2.5 \cdot x}{2.5}$

$\frac{900}{2.5} = x$

$360 = x$

11. $\quad \frac{x}{100} = \frac{27}{64}$

$x \cdot 64 = 100 \cdot 27$ Equating cross products

$\frac{x \cdot 64}{64} = \frac{100 \cdot 27}{64}$

$x = \frac{2700}{64}$

$x = 42.1875$

12. $\quad \frac{68}{y} = \frac{17}{25}$

$68 \cdot 25 = y \cdot 17$ Equating cross products

$\frac{68 \cdot 25}{17} = \frac{y \cdot 17}{17}$

$\frac{4 \cdot \cancel{17} \cdot 25}{\cancel{17} \cdot 1} = y$

$100 = y$

13. *Familiarize.* Let $d =$ the actual distance between the cities.

Translate. We translate to a proportion.

Map distance $\rightarrow \dfrac{3}{225} = \dfrac{7}{d} \leftarrow$ Map distance
Actual distance $\rightarrow$ $\quad\quad\quad \leftarrow$ Actual distance

Solve.

$3 \cdot d = 225 \cdot 7$ Equating cross products

$\frac{3 \cdot d}{3} = \frac{225 \cdot 7}{3}$

$d = 525$ Multiplying and dividing

Check. We substitute in the proportion and check cross products.

$\frac{3}{225} = \frac{7}{525}$

$3 \cdot 525 = 1575; \ 225 \cdot 7 = 1575$

The cross products are the same, so the answer checks.

State. The cities are 525 mi apart.

14. If we use the sun's rays to represent the third side of the triangles in the drawing of the situation in the text, we see that we have similar triangles. The ratio of 3 to h is the same as the ratio of 5 to 110. We have the proportion

$\frac{3}{h} = \frac{5}{110}.$

Solve: $3 \cdot 110 = h \cdot 5$

$\frac{3 \cdot 110}{5} = \frac{h \cdot 5}{5}$

$66 = h$

The tower is 66 m high.

15. *Familiarize.* Let $d =$ the distance the boat would travel in 42 hr.

Translate. We translate to a proportion.

Distance $\rightarrow \dfrac{432}{12} = \dfrac{d}{42} \leftarrow$ Distance
Time $\rightarrow$ $\quad\quad\quad \leftarrow$ Time

Solve.

$432 \cdot 42 = 12 \cdot d$ Equating cross products

$\frac{432 \cdot 42}{12} = \frac{12 \cdot d}{12}$

$1512 = d$ Multiplying and dividing

Check. We substitute in the proportion and check cross products.

$\frac{432}{12} = \frac{1512}{42}$

$432 \cdot 42 = 18,144; \ 12 \cdot 1512 = 18,144$

The cross products are the same, so the answer checks.

State. The boat would travel 1512 km in 42 hr.

16. *Familiarize.* Let $m =$ the number of minutes the watch will lose in 24 hr.

Translate. We translate to a proportion.

Minutes lost $\rightarrow \dfrac{2}{10} = \dfrac{m}{24} \leftarrow$ Minutes lost
Hours $\rightarrow$ $\quad\quad\quad \leftarrow$ Hours

Solve.

$2 \cdot 24 = 10 \cdot m$ Equating cross products

$\frac{2 \cdot 24}{10} = \frac{10 \cdot m}{10}$

$4.8 = m$ Multiplying and dividing

Check. We substitute in the proportion and check cross products.

$\frac{2}{10} = \frac{4.8}{24}$

$2 \cdot 24 = 48; \ 10 \cdot 4.8 = 48$

The cross products are the same, so the answer checks.

State. The watch will lose 4.8 min in 24 hr.

17. *Familiarize.* Let $s =$ the number of students in the class.

Translate. We translate to a proportion.

Books $\rightarrow \dfrac{3}{5} = \dfrac{9}{s} \leftarrow$ Books
Students $\rightarrow$ $\quad\quad\quad \leftarrow$ Students

Solve.

$$3 \cdot s = 5 \cdot 9$$
$$\frac{3 \cdot s}{3} = \frac{5 \cdot 9}{3}$$
$$s = 15$$

Check. We substitute in the proportion and check cross products.

$$\frac{3}{5} = \frac{9}{15}$$
$$3 \cdot 15 = 45; \ 5 \cdot 9 = 45$$

The cross products are the same, so the answer checks.

State. There are 15 students in the class.

18. *Familiarize*. Let $t =$ the amount of tax paid on a \$135,000 home. We express 39 cents as \$0.39.

Translate. We translate to a proportion.

$$\text{Tax} \rightarrow \quad \frac{0.39}{1000} = \frac{t}{135,000} \quad \leftarrow \text{Tax}$$
$$\text{Property value} \rightarrow \qquad\qquad\qquad \leftarrow \text{Property value}$$

Solve.

$$0.39 \times 135,000 = 1000 \times t$$
$$\frac{0.39 \times 135,000}{1000} = t$$
$$52.65 = t$$

Check. We substitute in the proportion and check cross products.

$$\frac{0.39}{1000} = \frac{52.65}{135,000}$$
$$0.39 \times 135,000 = 52,650; \ 1000 \times 52.65 = 52,650$$

The cross products are the same, so the answer checks.

State. The owner of a \$135,000 home would pay \$52.65 in tax.

19. The ratio of 10 to x is the same as the ratio of 5 to 4. We have the proportion

$$\frac{10}{x} = \frac{5}{4}.$$

Solve: $10 \cdot 4 = x \cdot 5$
$$\frac{10 \cdot 4}{5} = \frac{x \cdot 5}{5}$$
$$8 = x$$

The ratio of 11 to y is the same as the ratio of 5 to 4. We have the proportion

$$\frac{11}{y} = \frac{5}{4}.$$

Solve: $11 \cdot 4 = y \cdot 5$
$$\frac{11 \cdot 4}{5} = \frac{y \cdot 5}{5}$$
$$8.8 = y$$

20. The ratio of 5 to x is the same as the ratio of 5 to 8. This tells us that $x = 8$.

The ratio of 5 to y is the same as the ratio of 5 to 8. This tells us that $y = 8$.

The ratio of 7.5 to z is the same as the ratio of 5 to 8. We have the proportion

$$\frac{7.5}{z} = \frac{5}{8}.$$

Solve: $7.5 \cdot 8 = z \cdot 5$
$$\frac{7.5 \cdot 8}{5} = \frac{z \cdot 5}{5}$$
$$12 = z$$

21.
$$\frac{x + 3}{4} = \frac{5x + 2}{8}$$
$$8(x + 3) = 4(5x + 2) \qquad \text{Equating cross products}$$
$$8x + 24 = 20x + 8$$
$$8x + 24 - 8x = 20x + 8 - 8x$$
$$24 = 12x + 8$$
$$24 - 8 = 12x + 8 - 8$$
$$16 = 12x$$
$$\frac{16}{12} = \frac{12x}{12}$$
$$\frac{4}{3} = x, \text{ or}$$
$$1.\overline{3} = x$$

The solution is $\frac{4}{3}$, or $1.\overline{3}$.

22.
$$\frac{4 - 6x}{5} = \frac{x + 3}{2}$$
$$2(4 - 6x) = 5(x + 3) \qquad \text{Equating cross products}$$
$$8 - 12x = 5x + 15$$
$$8 - 12x + 12x = 5x + 15 + 12x$$
$$8 = 17x + 15$$
$$8 - 15 = 17x + 15 - 15$$
$$-7 = 17x$$
$$\frac{-7}{17} = \frac{17x}{17}$$
$$\frac{-7}{17} = x$$

The solution is $\frac{-7}{17}$, or $-\frac{7}{17}$.

23. *Familiarize*. Let $y =$ the number of balls of burgundy yarn Lara needs.

Translate and Solve. First we translate to a proportion.

$$\text{Burgundy} \rightarrow \quad \frac{3}{2\frac{1}{2}} = \frac{y}{10} \quad \leftarrow \text{Burgundy}$$
$$\text{Black} \rightarrow \qquad\qquad\qquad \leftarrow \text{Black}$$

We solve the proportion.

$$3 \cdot 10 = 2\frac{1}{2} \cdot y$$
$$3 \cdot 10 = \frac{5}{2} \cdot y$$
$$\frac{2}{5} \cdot 3 \cdot 10 = \frac{2}{5} \cdot \frac{5}{2} \cdot y$$
$$\frac{2 \cdot 3 \cdot 10}{5} = y$$
$$12 = y$$

If each sweater requires $2\frac{1}{2}$ balls of black yarn and Lara has 10 balls of black yarn (and the proportionally correct number of balls of burgundy yarn), then the number of sweaters she can make is

$$10 \div 2\frac{1}{2} = 10 \div \frac{5}{2} = 10 \cdot \frac{2}{5} = \frac{10 \cdot 2}{5} = 4.$$

Check. We repeat the calculations. The answer checks.

State. Lara needs 12 balls of burgundy yarn. She can knit 4 sweaters.

24. **Familiarize**. Let $x =$ the amount the Johnsons will pay. Then $\$79.85 - x =$ the amount the Solominis will pay. The ratio of Johnsons to Solominis is $3/2$.

Translate. We translate to a proportion.

Johnsons $\rightarrow$ $\dfrac{3}{2} = \dfrac{x}{79.85 - x}$ $\leftarrow$ Johnsons' amount
Solominis $\rightarrow$ $\phantom{\dfrac{3}{2} = \dfrac{x}{79.85 - x}}$ $\leftarrow$ Solominis' amount

Solve.

$$3 \cdot (79.85 - x) = 2 \cdot x$$
$$239.55 - 3x = 2x$$
$$239.55 = 5x \qquad \text{Adding } 3x$$
$$\frac{239.55}{5} = \frac{5x}{5}$$
$$47.91 = x$$

Check. We substitute in the proportion and check cross products. If $x = 47.91$, then $79.85 - x = 79.85 - 47.91$, or 31.94.

$$\frac{3}{2} = \frac{47.91}{31.94}$$
$$3 \times 31.94 = 95.82; \ 2 \times 47.91 = 95.82$$

The cross products are the same, so the answer checks.

State. The Johnsons will pay $\$47.91$.

25. The ratio of 35 to w is the same as the ratio of 65 to $190 - 65$, or 125.

$$\frac{35}{w} = \frac{65}{125}$$
$$35 \cdot 125 = w \cdot 65$$
$$\frac{35 \cdot 125}{65} = \frac{w \cdot 65}{65}$$
$$\frac{\cancel{5} \cdot 7 \cdot 125}{\cancel{5} \cdot 13} = w$$
$$\frac{875}{13} = w, \text{ or}$$
$$67.308 \approx w$$

The width of the river is $\dfrac{875}{13}$ ft, or about 67.308 ft.

26. Let $d =$ the distance labeled with a "?" in the drawing in the text. Note that 8 yd $= 8 \cdot 1$ yd $= 8 \cdot 3$ ft $= 24$ ft. Then the ratio of 10 to 24 is the same as the ratio of $18 - d$ to 18.

$$\frac{10}{24} = \frac{18 - d}{18}$$
$$10 \cdot 18 = 24 \cdot (18 - d)$$
$$180 = 432 - 24d$$
$$-252 = -24d \qquad \text{Subtracting } 432$$
$$10.5 = d \qquad \text{Dividing by } -24$$

The goalie should stand 10.5 ft in front of the goal.

1.
$$\begin{array}{r} \overset{1\ 1}{}\overset{2\ 2}{} \\ 1\ 3\ 7.\ 1\ 8\ 6 \\ 2\ 3.\ 0\ 1\ 9 \\ +\ 4\ 8\ 3.\ 2\ 9\ 7 \\ \hline 6\ 4\ 3.\ 5\ 0\ 2 \end{array}$$

2.
$$3 \boxed{\frac{4}{5} \cdot \frac{2}{2}} = 3\frac{8}{10}$$
$$+8\frac{7}{10} = +8\frac{7}{10}$$
$$\phantom{+8\frac{7}{10}} \ 11\frac{15}{10} = 11 + \frac{15}{10}$$
$$= 11 + 1\frac{5}{10}$$
$$= 12\frac{5}{10}$$
$$= 12\frac{1}{2}$$

3. The LCD is 140.
$$\frac{6}{35} + \frac{5}{28} = \frac{6}{35} \cdot \frac{4}{4} + \frac{5}{28} \cdot \frac{5}{5}$$
$$= \frac{24}{140} + \frac{25}{140}$$
$$= \frac{49}{140} = \frac{7 \cdot 7}{7 \cdot 20}$$
$$= \frac{7}{7} \cdot \frac{7}{20} = \frac{7}{20}$$

4.
$$\begin{array}{r} \overset{12}{}\overset{14}{} \\ \overset{8}{}\overset{2}{\cancel{9}}\ \overset{4}{\cancel{3}}\ \overset{10}{\cancel{5}}.\overset{4}{\cancel{0}}\ \overset{10}{\cancel{5}}\ \cancel{0} \\ -\ \ \ \ 6\ 6.\ 8\ 3\ 4 \\ \hline 1\ 8\ 6\ 8.\ 2\ 1\ 6 \end{array}$$ Writing an extra zero

5. $-32 - (-15) = -32 + 15 = -17$

6. The LCD is 60.
$$\frac{4}{15} - \frac{3}{20} = \frac{4}{15} \cdot \frac{4}{4} - \frac{3}{20} \cdot \frac{3}{3}$$
$$= \frac{16}{60} - \frac{9}{60}$$
$$= \frac{7}{60}$$

7.
$$\begin{array}{r} 3\ 7.\ 6\ 4 \\ \times \ \ \ \ 5.\ 9 \\ \hline 3\ 3\ 8\ 7\ 6 \\ 1\ 8\ 8\ 2\ 0\ 0 \\ \hline 2\ 2\ 2.\ 0\ 7\ 6 \end{array}$$

8. $-43(15) = -645$

9. $7\frac{4}{5} \cdot 3\frac{1}{2} = \frac{39}{5} \cdot \frac{7}{2} = \frac{39 \cdot 7}{5 \cdot 2} = \frac{273}{10} = 27\frac{3}{10}$

10.
$$\begin{array}{r} 4\ 3. \\ 2.3_\wedge \overline{\smash{)}\ 9\ 8.\ 9_\wedge} \\ 9\ 2\ 0 \\ \hline 6\ 9 \\ 6\ 9 \\ \hline 0 \end{array}$$

11. First we find $306 \div 6$.

$$\begin{array}{r} 5\,1 \\ 6\overline{)3\,0\,6} \\ \underline{3\,0\,0} \\ 6 \\ \underline{6} \\ 0 \end{array}$$

We have $306 \div 6 = 51$, so $-306 \div 6 = -51$.

12. $\dfrac{7}{11} \div \dfrac{14}{33} = \dfrac{7}{11} \cdot \dfrac{33}{14} = \dfrac{7 \cdot 33}{11 \cdot 14} = \dfrac{7 \cdot 3 \cdot 11}{11 \cdot 2 \cdot 7} = \dfrac{7 \cdot 11}{7 \cdot 11} \cdot \dfrac{3}{2} = \dfrac{3}{2}$

13. $30{,}074 = 3$ ten thousands $+ 0$ thousands $+ 0$ hundreds $+ 7$ tens $+ 4$ ones, or 3 ten thousands $+ 7$ tens $+ 4$ ones

14. A word name for 120.07 is one hundred twenty and seven hundredths.

15. To compare two positive numbers in decimal notation, start at the left and compare corresponding digits moving from left to right. When two digits differ, the number with the larger digit is the larger of the two numbers.

0.7

$\updownarrow$　Different; 7 is larger than 0.

0.698

Thus, 0.7 is larger.

16. To compare two negative numbers in decimal notation, start at the left and compare corresponding digits moving from left to right. When two digits differ, the number with the smaller digit is the larger of the two numbers.

−0.799

$\updownarrow$　Different; 7 is smaller than 8.

−0.8

Thus, −0.799 is larger.

17. $\begin{array}{r} 3 \\ 3\overline{)9} \\ 2\overline{)1\,8} \\ 2\overline{)3\,6} \\ 2\overline{)7\,2} \\ 2\overline{)1\,4\,4} \end{array}$ ← 3 is prime.

Thus, $144 = 2 \cdot 2 \cdot 2 \cdot 2 \cdot 3 \cdot 3$, or $2^4 \cdot 3^2$.

18. $42 = 2 \cdot 3 \cdot 7$

$78 = 2 \cdot 3 \cdot 13$

The LCM is $2 \cdot 3 \cdot 7 \cdot 13$, or 546.

19. The rectangle is divided into 8 equal parts. The unit is $\dfrac{1}{8}$. The denominator is 8. We have 5 parts shaded. This tells us that the numerator is 5. Thus, $\dfrac{5}{8}$ is shaded.

20. $\dfrac{108}{128} = \dfrac{2 \cdot 2 \cdot 3 \cdot 3 \cdot 3}{2 \cdot 2 \cdot 2 \cdot 2 \cdot 2 \cdot 2 \cdot 2}$

$= \dfrac{2 \cdot 2}{2 \cdot 2} \cdot \dfrac{3 \cdot 3 \cdot 3}{2 \cdot 2 \cdot 2 \cdot 2 \cdot 2}$

$= \dfrac{27}{32}$

21. $\dfrac{3}{5} \times 9.53 = 0.6 \times 9.53 = 5.718$

22. $7.2 \div 0.4(-1.5) + (1.2)^2 = 7.2 \div 0.4(-1.5) + 1.44$

$= 18(-1.5) + 1.44$

$= -27 + 1.44$

$= -25.56$

23. $\dfrac{23 + 49 + 52 + 71}{4} = \dfrac{195}{4} = 48.75$

24. We can use cross products:

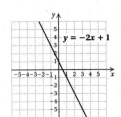

$6 \cdot 196 = 1176$　　　　$8 \cdot 14 = 112$

Since the cross products are not the same, $1176 \neq 112$, then $\dfrac{6}{8} \neq \dfrac{14}{196}$ and the numbers are not proportional.

25. Graph $y = -2x + 1$.

We make a table of values. Then we plot the points and draw and label the line.

If x is -2, then $y = -2(-2) + 1 = 4 + 1 = 5$.

If x is 0, then $y = -2 \cdot 0 + 1 = 0 + 1 = 1$.

If x is 3, then $y = -2 \cdot 3 + 1 = -6 + 1 = -5$.

x	$y = -2x + 1$	(x, y)
−2	5	(−2, 5)
0	1	(0, 1)
3	−5	(3, −5)

26. $\dfrac{t - 7}{w} = \dfrac{-3 - 7}{-2} = \dfrac{-10}{-2} = 5$

27. $\dfrac{14}{25} = \dfrac{x}{54}$

$14 \cdot 54 = 25 \cdot x$　　Equating cross products

$\dfrac{14 \cdot 54}{25} = \dfrac{25 \cdot x}{25}$

$30.24 = x$

The solution is 30.24, or $30\dfrac{6}{25}$.

28. $-423 = 16 \cdot t$

$\dfrac{-423}{16} = \dfrac{16 \cdot t}{16}$

$\dfrac{-423}{16} = t$, or

$-26.4375 = t$

The solution is $\dfrac{-423}{16}$, or -26.4375.

29.
$$9x - 7 = -43$$
$$9x - 7 + 7 = -43 + 7$$
$$9x = -36$$
$$\frac{9x}{9} = \frac{-36}{9}$$
$$x = -4$$

The solution is -4.

30. $2(x - 3) + 9 = 5x - 6$
$$2x - 6 + 9 = 5x - 6$$
$$2x + 3 = 5x - 6$$
$$2x + 3 - 2x = 5x - 6 - 2x$$
$$3 = 3x - 6$$
$$3 + 6 = 3x - 6 + 6$$
$$9 = 3x$$
$$\frac{9}{3} = \frac{3x}{3}$$
$$3 = x$$

The solution is 3.

31.
$$34.56 + n = 67.9$$
$$34.56 + n - 34.56 = 67.9 - 34.56$$
$$n = 33.34$$

The solution is 33.34.

32. $\frac{2}{3} \cdot x = \frac{16}{27}$
$$x = \frac{16}{27} \div \frac{2}{3}$$
$$x = \frac{16}{27} \cdot \frac{3}{2} = \frac{16 \cdot 3}{27 \cdot 2} = \frac{2 \cdot 8 \cdot 3}{3 \cdot 9 \cdot 2}$$
$$x = \frac{2 \cdot 3}{2 \cdot 3} \cdot \frac{8}{9} = \frac{8}{9}$$

The solution is $\frac{8}{9}$.

33. Familiarize. Let d = the number of miles the truck could travel in 3.8 hr.

Translate. We translate to a proportion.
$$\text{Distance} \rightarrow \frac{143}{2.6} = \frac{d}{3.8} \leftarrow \text{Distance}$$
$$\text{Time} \rightarrow \qquad\qquad \leftarrow \text{Time}$$

Solve.
$$143 \times 3.8 = 2.6 \times d$$
$$\frac{143 \times 3.8}{2.6} = d$$
$$209 = d$$

Check. We substitute in the proportion and check cross products.
$$\frac{143}{2.6} = \frac{209}{3.8}$$
$$143 \times 3.8 = 543.4; \ 2.6 \times 209 = 543.4$$

The cross products are the same, so the answer checks.

State. The truck could travel 209 mi in 3.8 hr.

34. Familiarize. Let t = the number of minutes required to stamp out 1295 washers.

Translate. We translate to a proportion.
$$\text{Washers} \rightarrow \frac{925}{5} = \frac{1295}{t} \leftarrow \text{Washers}$$
$$\text{Time} \rightarrow \qquad\qquad \leftarrow \text{Time}$$

Solve.
$$925 \cdot t = 5 \cdot 1295 \quad \text{Equating cross products}$$
$$\frac{925 \cdot t}{925} = \frac{5 \cdot 1295}{925}$$
$$t = 7$$

Check. The number of washers that can be stamped out in 1 min is $925 \div 5$, or 185, so in 7 min $7 \cdot 185$, or 1295, washers can be stamped out. The answer checks.

State. it will take 7 min to stamp out 1295 washers.

35. Familiarize. Let t = the total mileage.

Translate.

Mileage of first trip	plus	Mileage of second trip	plus	Mileage of third trip	is	Total mileage
347.6	+	249.8	+	379.5	=	t

$$
\begin{array}{r}
^{1\ 2\ 1} \\
3\ 4\ 7.6 \\
2\ 4\ 9.8 \\
+\ 3\ 7\ 9.5 \\
\hline
9\ 7\ 6.9
\end{array}
$$

Check. We repeat the calculation. The answer checks.

State. The total mileage was 976.9 mi.

36. $\frac{319 \text{ mi}}{14.5 \text{ gal}} = 22$ mi/gal, or 22 mpg

37. If we consider 6 ft to be the base of each triangle, then the height is $\frac{4 \text{ ft}}{2}$, or 2 ft. First we will find the area of one triangle.
$$A = \frac{1}{2} \cdot b \cdot h$$
$$= \frac{1}{2} \cdot 6 \text{ ft} \cdot 2 \text{ ft}$$
$$= \frac{6 \cdot 2}{2} \cdot \text{ft} \cdot \text{ft}$$
$$= 6 \text{ ft}^2$$

Then the area of two triangles is $2 \cdot 6$ ft^2, or 12 ft^2.

38 a), b), c) **Familiarize.** The plane has $2 + 16$, or 18 tires.

Translate and Solve. Let c = the cost of a set of new tires for one plane. We multiply the number of tires by the cost of a tire to find c.
$$c = 18 \cdot \$20,000 = \$360,000$$

Next we multiply the cost of new tires for one plane by the number of planes in the fleet to find the cost of new tires, t, for all the planes.
$$t = 400 \cdot \$360,000 = \$144,000,000$$

Finally, we multiply the cost of tires for all the planes by the number of months in a year to find the cost y of tires for an entire year.

$$y = 12 \cdot \$144,000,000 = \$1,728,000,000$$

Check. We repeat all the calculations. The answers check.

State. (a) The cost of new tires for one plane is $360,000; (b) The cost of new tires for all of the planes is $144,000,000; (c) The total cost of tires for an entire year would be $1,728,000,000.

39. Familiarize. Let j = the number of cups of juice left over.

Translate. This is a "how much more" situation.

Juice used	plus	Juice left over	is	Amount of juice in can
$\downarrow$	$\downarrow$	$\downarrow$	$\downarrow$	$\downarrow$
$3\frac{1}{2}$	$+$	j	$=$	$5\frac{3}{4}$

Solve.

$$3\frac{1}{2} + j = 5\frac{3}{4}$$

$$3\frac{1}{2} + j - 3\frac{1}{2} = 5\frac{3}{4} - 3\frac{1}{2}$$

$$j = 5\frac{3}{4} - 3\frac{2}{4} \qquad \left(\frac{1}{2} = \frac{2}{4}\right)$$

$$j = 2\frac{1}{4}$$

Check. $3\frac{1}{2} + 2\frac{1}{4} = 3\frac{2}{4} + 2\frac{1}{4} = 5\frac{3}{4}$, so the answer checks.

State. There are $2\frac{1}{4}$ cups of juice left over.

40. $88 \text{ ft} = 88 \text{ ft} \cdot \dfrac{1 \text{ mi}}{5280 \text{ ft}} = \dfrac{88}{5280} \cdot \dfrac{\text{ft}}{\text{ft}} \cdot 1 \text{ mi} = \dfrac{1}{60} \text{ mi}$

$1 \text{ sec} = 1 \text{ sec} \cdot \dfrac{1 \text{ min}}{60 \text{ sec}} \cdot \dfrac{1 \text{ hr}}{60 \text{ min}} = \dfrac{1 \cdot 1}{60 \cdot 60} \cdot \dfrac{\text{sec}}{\text{sec}} \cdot \dfrac{\text{min}}{\text{min}} \cdot 1 \text{ hr} = \dfrac{1}{3600} \text{ hr}$

Then $\dfrac{88 \text{ ft}}{1 \text{ sec}} = \dfrac{\frac{1}{60} \text{ mi}}{\frac{1}{3600} \text{ hr}} = \dfrac{1}{60} \cdot \dfrac{3600}{1} \dfrac{\text{mi}}{\text{hr}} = \dfrac{3600}{60} \text{ mph} = 60 \text{ mph.}$

41. We find the unit prices. Recall that 1 lb = 16 oz.

$$\dfrac{\$2.07}{12 \text{ oz}} = \dfrac{207\cancel{c}}{12 \text{ oz}} = 17.25\cancel{c}/\text{oz}$$

$$\dfrac{\$2.79}{16 \text{ oz}} = \dfrac{279\cancel{c}}{16 \text{ oz}} = 17.4375\cancel{c}/\text{oz}$$

The 12-oz bag is the better buy.

42. Let x = the amount Hans will spend in $2 \cdot 16$, or 32 weeks. We use a proportion.

$$\begin{array}{c} \text{Expenses} \rightarrow \\ \text{Weeks} \rightarrow \end{array} \dfrac{150}{3} = \dfrac{x}{32} \begin{array}{c} \leftarrow \text{Expenses} \\ \leftarrow \text{Weeks} \end{array}$$

We solve the proportion.

$$150 \cdot 32 = 3 \cdot x$$

$$\dfrac{150 \cdot 32}{3} = \dfrac{3 \cdot x}{3}$$

$$1600 = x$$

Since $1600 exceeds the $1200 budget, the budget for incidental expenses will not be adequate if Hans continues to spend at the current rate.

Now let w = the number of weeks it will take for $1200 to be exhausted when it is spent at the rate of $150 per 3 weeks. We use another proportion.

$$\begin{array}{c} \text{Expenses} \rightarrow \\ \text{Weeks} \rightarrow \end{array} \dfrac{150}{3} = \dfrac{1200}{w} \begin{array}{c} \leftarrow \text{Expenses} \\ \leftarrow \text{Weeks} \end{array}$$

We solve the proportion.

$$150 \cdot w = 3 \cdot 1200$$

$$\dfrac{150 \cdot w}{150} = \dfrac{3 \cdot 1200}{150}$$

$$w = 24$$

The money will be exhausted in 24 weeks.

We subtract to find how much more will be needed:

$$\$1600 - \$1200 = \$400.$$

43. Let c = the amount to be spent on the CD player. Then we have

$$\dfrac{1}{2} = \dfrac{c}{250}$$

$$1 \cdot 250 = 2 \cdot c$$

$$\dfrac{1 \cdot 250}{2} = \dfrac{2 \cdot c}{2}$$

$$125 = c$$

Thus, $125 should be spent on the CD player.

The receiver should cost 3 times as much as the CD player, so the amount that should be spent on the receiver is

$$3 \cdot \$125, \text{ or } \$375.$$

Then the total cost of the sound system is

$$\$125 + \$375 + \$250, \text{ or } \$750.$$

44. $1 + 3 + 2 = 6$, and $\$4800/6 = \$800$.

Then $1 \cdot \$800$, or $800 per month is allocated to Leisure & Incidental Expenses; $3 \cdot \$800$, or $2400 per month is allocated to Food & Other Necessary Expenses; and $2 \cdot \$800$, or $1600 per month is allocated to Debt Payments.

We multiply to find the amounts allotted to Debts and to Leisure & Incidental Expenses in a year, or 12 months.

Debts: $12 \cdot \$1600 = \$19,200$

Leisure & Incidental Expenses: $12 \cdot \$800 = \$9600$

Chapter 8

Percent Notation

Exercise Set 8.1

1. $90\% = \dfrac{90}{100}$ A ratio of 90 to 100

$90\% = 90 \times \dfrac{1}{100}$ Replacing % with $\times \dfrac{1}{100}$

$90\% = 90 \times 0.01$ Replacing % with $\times 0.01$

3. $12.5\% = \dfrac{12.5}{100}$ A ratio of 12.5 to 100

$12.5\% = 12.5 \times \dfrac{1}{100}$ Replacing % with $\times \dfrac{1}{100}$

$12.5\% = 12.5 \times 0.01$ Replacing % with $\times 0.01$

5. 67%

  a) Replace the percent symbol with $\times 0.01$.

    67×0.01

  b) Move the decimal point two places to the left.

    $0.67.$

Thus, $67\% = 0.67$.

7. 45.6%

  a) Replace the percent symbol with $\times 0.01$.

    45.6×0.01

  b) Move the decimal point two places to the left.

    $0.45.6$

Thus, $45.6\% = 0.456$.

9. 59.01%

  a) Replace the percent symbol with $\times 0.01$.

    59.01×0.01

  b) Move the decimal point two places to the left.

    $0.59.01$

Thus, $59.01\% = 0.5901$.

11. 10%

  a) Replace the percent symbol with $\times 0.01$.

    10×0.01

  b) Move the decimal point two places to the left.

    $0.10.$

Thus, $10\% = 0.1$.

13. 1%

  a) Replace the percent symbol with $\times 0.01$.

    1×0.01

  b) Move the decimal point two places to the left.

    $0.01.$

Thus, $1\% = 0.01$.

15. 200%

  a) Replace the percent symbol with $\times 0.01$.

    200×0.01

  b) Move the decimal point two places to the left.

    $2.00.$

Thus, $200\% = 2$.

17. 0.1%

  a) Replace the percent symbol with $\times 0.01$.

    0.1×0.01

  b) Move the decimal point two places to the left.

    $0.00.1$

Thus, $0.1\% = 0.001$.

19. 0.09%

  a) Replace the percent symbol with $\times 0.01$.

    0.09×0.01

  b) Move the decimal point two places to the left.

    $0.00.09$

Thus, $0.09\% = 0.0009$.

21. 0.18%

  a) Replace the percent symbol with $\times 0.01$.

    0.18×0.01

  b) Move the decimal point two places to the left.

    $0.00.18$

Thus, $0.18\% = 0.0018$.

23. 23.19%

  a) Replace the percent symbol with $\times 0.01$.

    23.19×0.01

  b) Move the decimal point two places to the left.

    $0.23.19$

Thus, $23.19\% = 0.2319$.

25. $14\frac{7}{8}\%$

   a) Convert $14\frac{7}{8}$ to decimal notation and replace the percent symbol with $\times 0.01$.

   $$14.875 \times 0.01$$

   b) Move the decimal point two places to the left.

   $$0.14.875$$

   Thus, $14\frac{7}{8}\% = 0.14875$.

27. $56\frac{1}{2}\%$

   a) Convert $56\frac{1}{2}$ to decimal notation and replace the percent symbol with $\times 0.01$.

   $$56.5 \times 0.01$$

   b) Move the decimal point two places to the left.

   $$0.56.5$$

   Thus, $56\frac{1}{2}\% = 0.565$.

29. 9%

   a) Replace the percent symbol with $\times 0.01$.

   $$9 \times 0.01$$

   b) Move the decimal point two places to the left.

   $$0.09.$$

   Thus, $9\% = 0.09$.

   58%

   a) Replace the percent symbol with $\times 0.01$.

   $$58 \times 0.01$$

   b) Move the decimal point two places to the left.

   $$0.58.$$

   Thus, $58\% = 0.58$.

31. 44%

   a) Replace the percent symbol with $\times 0.01$.

   $$44 \times 0.01$$

   b) Move the decimal point two places to the left.

   $$0.44.$$

   Thus, $44\% = 0.44$.

33. 36%

   a) Replace the percent symbol with $\times 0.01$.

   $$36 \times 0.01$$

   b) Move the decimal point two places to the left.

   $$0.36.$$

   Thus, $36\% = 0.36$.

35. 0.47

   a) Move the decimal point two places to the right.

   $$0.47.$$

   b) Write a percent symbol: 47%

   Thus, $0.47 = 47\%$.

37. 0.03

   a) Move the decimal point two places to the right.

   $$0.03.$$

   b) Write a percent symbol: 3%

   Thus, $0.03 = 3\%$.

39. 8.7

   a) Move the decimal point two places to the right.

   $$8.70.$$

   b) Write a percent symbol: 870%

   Thus, $8.7 = 870\%$.

41. 0.334

   a) Move the decimal point two places to the right.

   $$0.33.4$$

   b) Write a percent symbol: 33.4%

   Thus, $0.334 = 33.4\%$.

43. 0.75

   a) Move the decimal point two places to the right.

   $$0.75.$$

   b) Write a percent symbol: 75%

   Thus, $0.75 = 75\%$.

45. 0.4

   a) Move the decimal point two places to the right.

   $$0.40.$$

   b) Write a percent symbol: 40%

   Thus, $0.4 = 40\%$.

47. 0.006

   a) Move the decimal point two places to the right.

   $$0.00.6$$

   b) Write a percent symbol: 0.6%

   Thus, $0.006 = 0.6\%$.

49. 0.017

a) Move the decimal point two places to the right.

0.01.7

b) Write a percent symbol: 1.7%

Thus, 0.017 = 1.7%.

51. 0.2718

a) Move the decimal point two places to the right.

0.27.18

b) Write a percent symbol: 27.18%

Thus, 0.2718 = 27.18%.

53. 0.0239

a) Move the decimal point two places to the right.

0.02.39

b) Write a percent symbol: 2.39%

Thus, 0.0239 = 2.39%.

55. 0.69

a) Move the decimal point two places to the right.

0.69.

b) Write a percent symbol: 69%

Thus, 0.69 = 69%.

57. 0.177

a) Move the decimal point two places to the right.

0.17.7

b) Write a percent symbol: 17.7%

Thus, 0.117 = 17.7%.

59. 0.26

a) Move the decimal point two places to the right.

0.26.

b) Write a percent symbol: 26%

Thus, 0.26 = 26%.

0.38

a) Move the decimal point two places to the right.

0.38.

b) Write a percent symbol: 38%

Thus, 0.38 = 38%.

61. We use the definition of percent as a ratio.

$$\frac{41}{100} = 41\%$$

63. We use the definition of percent as a ratio.

$$\frac{5}{100} = 5\%$$

65. We multiply by 1 to get 100 in the denominator.

$$\frac{2}{10} \cdot \frac{10}{10} = \frac{20}{100} = 20\%$$

67. We multiply by 1 to get 100 in the denominator.

$$\frac{7}{25} \cdot \frac{4}{4} = \frac{28}{100} = 28\%$$

69. We multiply by 1 to get 100 in the denominator.

$$\frac{1}{2} \cdot \frac{50}{50} = \frac{50}{100} = 50\%$$

71. Find decimal notation by division.

$$\frac{0.875}{8 \overline{)7.000}}$$

$$\frac{7}{8} = 0.875$$

Convert to percent notation.

0.87.5

$$\frac{7}{8} = 87.5\%, \text{ or } 87\frac{1}{2}\%$$

73. $\frac{4}{5} = \frac{4}{5} \cdot \frac{20}{20} = \frac{80}{100} = 80\%$

75. Find decimal notation by division.

$$\frac{0.666}{3 \overline{)2.000}}$$

We get a repeating decimal: $\frac{2}{3} = 0.66\overline{6}$

Convert to percent notation.

0.66.$\overline{6}$

$$\frac{2}{3} = 66.\overline{6}\%, \text{ or } 66\frac{2}{3}\%$$

77.
$$6\overline{\smash{\big)}1.0\,0\,0}$$
$$\begin{array}{r} 0.1\,6\,6 \\ \underline{6} \\ 4\,0 \\ \underline{3\,6} \\ 4\,0 \\ \underline{3\,6} \\ 4 \end{array}$$

We get a repeating decimal: $\dfrac{1}{6} = 0.16\overline{6}$

Convert to percent notation.

$$0.16.\overline{6}$$
$$\underset{\underrightarrow{}}{}$$

$\dfrac{1}{6} = 16.\overline{6}\%$, or $16\dfrac{2}{3}\%$

79.
$$16\overline{\smash{\big)}3.0\,0\,0\,0}$$
$$\begin{array}{r} 0.1\,8\,7\,5 \\ \underline{1\,6} \\ 1\,4\,0 \\ \underline{1\,2\,8} \\ 1\,2\,0 \\ \underline{1\,1\,2} \\ 8\,0 \\ \underline{8\,0} \\ 0 \end{array}$$

$\dfrac{3}{16} = 0.1875$

Convert to percent notation.

$$0.18.75$$
$$\underset{\underrightarrow{}}{}$$

$\dfrac{3}{16} = 18.75\%$, or $18\dfrac{3}{4}\%$

81. $\dfrac{3}{20} = \dfrac{3}{20} \cdot \dfrac{5}{5} = \dfrac{15}{100} = 15\%$

83. $\dfrac{29}{50} = \dfrac{29}{50} \cdot \dfrac{2}{2} = \dfrac{58}{100} = 58\%$

85. $\dfrac{11}{50} = \dfrac{11}{50} \cdot \dfrac{2}{2} = \dfrac{22}{100} = 22\%$

87. $\dfrac{1}{20} = \dfrac{1}{20} \cdot \dfrac{5}{5} = \dfrac{5}{100} = 5\%$

89. $\dfrac{9}{100} = 9\%$

91. $\dfrac{2}{5} = \dfrac{2}{5} \cdot \dfrac{20}{20} = \dfrac{40}{100} = 40\%$;

$\dfrac{9}{50} = \dfrac{9}{50} \cdot \dfrac{2}{2} = \dfrac{18}{100} = 18\%$

93. $85\% = \dfrac{85}{100}$ Definition of percent

$$= \dfrac{5 \cdot 17}{5 \cdot 20} \left.\begin{array}{c} \\ \\ \\ \end{array}\right\}$$
$$= \dfrac{5}{5} \cdot \dfrac{17}{20} \quad \text{Simplifying}$$
$$= \dfrac{17}{20}$$

95. $62.5\% = \dfrac{62.5}{100}$ Definition of percent

$$= \dfrac{62.5}{100} \cdot \dfrac{10}{10} \quad \begin{array}{l}\text{Multiplying by 1 to elim-}\\ \text{inate the decimal point}\\ \text{in the numerator}\end{array}$$

$$= \dfrac{625}{1000}$$

$$= \dfrac{5 \cdot 125}{8 \cdot 125} \left.\begin{array}{c} \\ \\ \\ \\ \end{array}\right.$$
$$= \dfrac{5}{8} \cdot \dfrac{125}{125} \left.\begin{array}{c} \\ \end{array}\right\} \text{Simplifying}$$
$$= \dfrac{5}{8}$$

97. $33\dfrac{1}{3}\% = \dfrac{100}{3}\% \quad \begin{array}{l}\text{Converting from mixed nu-}\\ \text{meral to fractional notation}\end{array}$

$$= \dfrac{100}{3} \times \dfrac{1}{100} \quad \text{Definition of percent}$$

$$= \dfrac{100 \cdot 1}{3 \cdot 100} \quad \text{Multiplying}$$

$$= \dfrac{1}{3} \cdot \dfrac{100}{100} \left.\begin{array}{c} \\ \\ \end{array}\right\} \text{Simplifying}$$
$$= \dfrac{1}{3}$$

99. $16.\overline{6}\% = 16\dfrac{2}{3}\% \qquad \left(16.\overline{6} = 16\dfrac{2}{3}\right)$

$$= \dfrac{50}{3}\% \quad \begin{array}{l}\text{Converting from mixed nu-}\\ \text{meral to fractional notation}\end{array}$$

$$= \dfrac{50}{3} \times \dfrac{1}{100} \quad \text{Definition of percent}$$

$$= \dfrac{50 \cdot 1}{3 \cdot 50 \cdot 2} \quad \text{Multiplying}$$

$$= \dfrac{1}{2 \cdot 3} \cdot \dfrac{50}{50} \left.\begin{array}{c} \\ \\ \end{array}\right\} \text{Simplifying}$$
$$= \dfrac{1}{6}$$

101. $7.25\% = \dfrac{7.25}{100} = \dfrac{7.25}{100} \cdot \dfrac{100}{100}$

$$= \dfrac{725}{10,000} = \dfrac{29 \cdot 25}{400 \cdot 25} = \dfrac{29}{400} \cdot \dfrac{25}{25}$$

$$= \dfrac{29}{400}$$

103. $0.8\% = \dfrac{0.8}{100} = \dfrac{0.8}{100} \cdot \dfrac{10}{10}$

$$= \dfrac{8}{1000} = \dfrac{1 \cdot 8}{125 \cdot 8} = \dfrac{1}{125} \cdot \dfrac{8}{8}$$

$$= \dfrac{1}{125}$$

105. $150\% = \dfrac{150}{100} = \dfrac{3 \cdot 50}{2 \cdot 50} = \dfrac{3}{2} \cdot \dfrac{50}{50} = \dfrac{3}{2}$

107. Note that $33.\overline{3}\% = 33\frac{1}{3}\%$ and proceed as in Exercise 97;
$33.\overline{3}\% = \frac{1}{3}$.

109. $8\% = \frac{8}{100}$

$= \frac{4 \cdot 2}{4 \cdot 25} = \frac{4}{4} \cdot \frac{2}{25}$

$= \frac{2}{25}$

111. $60\% = \frac{60}{100}$

$= \frac{20 \cdot 3}{20 \cdot 5} = \frac{20}{20} \cdot \frac{3}{5}$

$= \frac{3}{5}$

113. $2\% = \frac{2}{100}$

$= \frac{1 \cdot 2}{50 \cdot 2} = \frac{1}{50} \cdot \frac{2}{2}$

$= \frac{1}{50}$

115. $35\% = \frac{35}{100}$

$= \frac{7 \cdot 5}{20 \cdot 5} = \frac{7}{20} \cdot \frac{5}{5}$

$= \frac{7}{20}$

117. $47\% = \frac{47}{100}$

119. $\frac{1}{8} = 1 \div 8$

$$\begin{array}{r} 0.1\,2\,5 \\ 8\,\overline{)1.0\,0\,0} \\ \underline{8} \\ 2\,0 \\ \underline{1\,6} \\ 4\,0 \\ \underline{4\,0} \\ 0 \end{array}$$

$\frac{1}{8} = 0.125 = 12\frac{1}{2}\%$, **or 12.5%**

$\frac{1}{6} = 1 \div 6$

$$\begin{array}{r} 0.1\,6\,6 \\ 6\,\overline{)1.0\,0\,0} \\ \underline{6} \\ 4\,0 \\ \underline{3\,6} \\ 4\,0 \\ \underline{3\,6} \\ 4 \end{array}$$

We get a repeating decimal: $0.1\overline{6}$

$0.16.\overline{6}$ $0.1\overline{6} = 16.\overline{6}\%$

$\frac{1}{6} = 0.1\overline{6} = 16.\overline{6}\%$, **or** $16\frac{2}{3}\%$

$20\% = \frac{20}{100} = \frac{1}{5} \cdot \frac{20}{20} = \frac{1}{5}$

$0.20.$ $20\% = 0.2$

$\frac{1}{5} = 0.2 = 20\%$

$0.25.$ $0.25 = 25\%$

$25\% = \frac{25}{100} = \frac{1}{4} \cdot \frac{25}{25} = \frac{1}{4}$

$\frac{1}{4} = 0.25 = 25\%$

$33\frac{1}{3}\% = \frac{100}{3}\% = \frac{100}{3} \times \frac{1}{100} = \frac{100}{300} = \frac{1}{3} \cdot \frac{100}{100} = \frac{1}{3}$

$0.33.\overline{3}$ $33.\overline{3}\% = 0.33\overline{3}$, or $0.\overline{3}$

$\frac{1}{3} = 0.\overline{3} = 33\frac{1}{3}\%$, **or** $33.\overline{3}\%$

$37.5\% = \frac{37.5}{100} = \frac{37.5}{100} \cdot \frac{10}{10} = \frac{375}{1000} = \frac{3}{8} \cdot \frac{125}{125} = \frac{3}{8}$

$0.37.5$ $37.5\% = 0.375$

$\frac{3}{8} = 0.375 = 37\frac{1}{2}\%$, **or** 37.5%

$40\% = \frac{40}{100} = \frac{2}{5} \cdot \frac{20}{20} = \frac{2}{5}$

$0.40.$ $40\% = 0.4$

$\frac{2}{5} = 0.4 = 40\%$

$\frac{1}{2} = \frac{1}{2} \cdot \frac{5}{5} = \frac{5}{10} = 0.5$

$\frac{1}{2} = \frac{1}{2} \cdot \frac{50}{50} = \frac{50}{100} = 5\%$

$\frac{1}{2} = 0.5 = 50\%$

121.   0.50. $0.5 = 50\%$

$$50\% = \frac{50}{100} = \frac{1}{2} \cdot \frac{50}{50} = \frac{1}{2}$$

$$\frac{1}{2} = 0.5 = 50\%$$

$$\frac{1}{3} = 1 \div 3$$

$$\begin{array}{r} 0.3 \\ 3\overline{)1.0} \\ \underline{9} \\ 1 \end{array}$$

We get a repeating decimal: $0.\overline{3}$

   $0.33.\overline{3}$ $0.\overline{3} = 33.\overline{3}\%$

$$\frac{1}{3} = 0.\overline{3} = 33.\overline{3}\%, \text{ or } 33\frac{1}{3}\%$$

$$25\% = \frac{25}{100} = \frac{25}{25} \cdot \frac{1}{4} = \frac{1}{4}$$

0.25. $25\% = 0.25$

$$\frac{1}{4} = 0.25 = 25\%$$

$$16\frac{2}{3}\% = \frac{50}{3}\% = \frac{50}{3} \times \frac{1}{100} = \frac{50 \cdot 1}{3 \cdot 2 \cdot 50} = \frac{50}{50} \cdot \frac{1}{6} = \frac{1}{6}$$

$$\frac{1}{6} = 1 \div 6$$

$$\begin{array}{r} 0.1\,6 \\ 6\overline{)1.0\,0} \\ \underline{6} \\ 4\,0 \\ \underline{3\,6} \\ 4 \end{array}$$

We get a repeating decimal: $0.1\overline{6}$

$$\frac{1}{6} = 0.1\overline{6} = 16\frac{2}{3}\%, \text{ or } 16.\overline{6}\%$$

0.12.5 $0.125 = 12.5\%$

$$12.5\% = \frac{12.5}{100} = \frac{12.5}{100} \cdot \frac{10}{10} = \frac{125}{1000} = \frac{125}{125} \cdot \frac{1}{8} = \frac{1}{8}$$

$$\frac{1}{8} = 0.125 = 12.5\%, \text{ or } 12\frac{1}{2}\%$$

$$\frac{3}{4} = \frac{3}{4} \cdot \frac{25}{25} = \frac{75}{100} = 75\%$$

0.75. $75\% = 0.75$

$$\frac{3}{4} = 0.75 = 75\%$$

$0.8\overline{3} = 0.83.\overline{3}$ $0.8\overline{3} = 83.\overline{3}\%$

$$83.\overline{3}\% = 83\frac{1}{3}\% = \frac{250}{3}\% = \frac{250}{3} \times \frac{1}{100} = \frac{5 \cdot 50}{3 \cdot 2 \cdot 50} =$$
$$\frac{5}{6} \cdot \frac{50}{50} = \frac{5}{6}$$

$$\frac{5}{6} = 0.8\overline{3} = 83.\overline{3}\%, \text{ or } 83\frac{1}{3}\%$$

$$\frac{3}{8} = 3 \div 8$$

$$\begin{array}{r} 0.3\,7\,5 \\ 8\overline{)3.0\,0\,0} \\ \underline{2\,4} \\ 6\,0 \\ \underline{5\,6} \\ 4\,0 \\ \underline{4\,0} \\ 0 \end{array}$$

$$\frac{3}{8} = 0.375$$

0.37.5 $0.375 = 37.5\%$

$$\frac{3}{8} = 0.375 = 37.5\%, \text{ or } 37\frac{1}{2}\%$$

123. Discussion and Writing Exercise

125. $13 \cdot x = 910$

$$\frac{13 \cdot x}{13} = \frac{910}{13}$$
$$x = 70$$

127. $0.05 \times b = 20$

$$\frac{0.05 \times b}{0.05} = \frac{20}{0.05}$$
$$b = 400$$

129. $\dfrac{24}{37} = \dfrac{15}{x}$

$24 \cdot x = 37 \cdot 15$ Equating cross products

$$x = \frac{37 \cdot 15}{24}$$
$$x = 23.125$$

131. $\dfrac{9}{10} = \dfrac{x}{5}$

$$9 \cdot 5 = 10 \cdot x$$
$$\frac{9 \cdot 5}{10} = x$$
$$\frac{45}{10} = x$$
$$\frac{9}{2} = x, \text{ or}$$
$$4.5 = x$$

133. Discussion and Writing Exercise

135. We use a calculator to find decimal notation for $\frac{41}{369}$. Then we perform the conversion to percent notation, moving the decimal point two places to the right and writing a % symbol.

$$\frac{41}{369} = 0.\overline{1} = 0.11\overline{1} = 11.\overline{1}\%$$

137. $\frac{14}{9}\% = \frac{14}{9} \times \frac{1}{100} = \frac{14}{900} = \frac{2 \cdot 7}{2 \cdot 450} = \frac{7}{450}$

To find decimal notation for $\frac{7}{450}$ we divide.

```
        .0 1 5 5
4 5 0 [7.0 0 0 0
        4 5 0
        2 5 0 0
        2 2 5 0
          2 5 0 0
          2 2 5 0
            2 5 0
```

We get a repeating decimal: $\frac{14}{9}\% = 0.01\overline{5}$.

139. Since the circle is divided into 100 sections, we can think of it as a pie cut into 100 equally sized pieces. We shade a wedge equal in size to 34 of these pieces to represent 34%. Then we shade wedges equal in size to 22 of these pieces to represent each of the three categories represented by 22%.

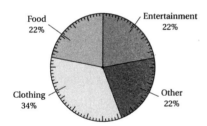

Food 22% · Entertainment 22% · Clothing 34% · Other 22%

Exercise Set 8.2

1. What is 32% of 78?

$a = 0.32 \times 78$

3. 89 is what percent of 99?

$89 = p \times 99$

5. 13 is 25% of what?

$13 = 0.25 \times b$

7. What is 85% of 276?

Translate: $a = 0.85 \cdot 276$

Solve: The letter is by itself. To solve the equation we multiply.

```
        2 7 6
      × 0. 8 5
      1 3 8 0
    2 2 0 8 0
a = 2 3 4. 6 0
```

234.6 is 85% of 276. The answer is 234.6.

9. 150% of 30 is what?

Translate: $1.5 \times 30 = a$

Solve: We multiply.

```
          3 0
      × 1. 5
      1 5 0
      3 0 0
a = 4 5. 0
```

150% of 30 is 45. The answer is 45.

11. What is 6% of \$300?

Translate: $a = 0.06 \cdot \$300$

Solve: We multiply.

```
        $ 3 0 0
      × 0. 0 6
a = $ 1 8. 0 0
```

\$18 is 6% of \$300. The answer is \$18.

13. 3.8% of 50 is what?

Translate: $0.038 \cdot 50 = a$

Solve: We multiply.

```
          5 0
    × 0. 0 3 8
        4 0 0
      1 5 0 0
a = 1. 9 0 0
```

3.8% of 50 is 1.9. The answer is 1.9.

15. \$39 is what percent of \$50?

Translate: $39 = n \times 50$

Solve: To solve the equation we divide on both sides by 50 and convert the answer to percent notation.

$$n \cdot 50 = 39$$
$$\frac{n \cdot 50}{50} = \frac{39}{50}$$
$$n = 0.78 = 78\%$$

\$39 is 78% of \$50. The answer is 78%.

17. 20 is what percent of 10?

Translate: $20 = n \times 10$

Solve: To solve the equation we divide on both sides by 10 and convert the answer to percent notation.

$$n \cdot 10 = 20$$
$$\frac{n \cdot 10}{10} = \frac{20}{10}$$
$$n = 2 = 200\%$$

20 is 200% of 10. The answer is 200%.

19. What percent of \$300 is \$150?

Translate: $n \times 300 = 150$

Solve: $n \cdot 300 = 150$

$$\frac{n \cdot 300}{300} = \frac{150}{300}$$
$$n = 0.5 = 50\%$$

50% of \$300 is \$150. The answer is 50%.

21. What percent of 80 is 100?

Translate: $n \times 80 = 100$

Solve: $n \cdot 80 = 100$

$$\frac{n \cdot 80}{80} = \frac{100}{80}$$

$$n = 1.25 = 125\%$$

125% of 80 is 100. The answer is 125%.

23. 20 is 50% of what?

Translate: $20 = 0.5 \times b$

Solve: To solve the equation we divide on both sides by 0.5:

$$\frac{20}{0.5} = \frac{0.5 \times b}{0.5}$$

$$\frac{20}{0.5} = b$$

$$40 = b$$

$$
\begin{array}{r}
4\,0. \\
0.5_{\wedge}\overline{)\,2\,0.\,0_{\wedge}} \\
2\,0\,0 \\
\hline
0 \\
0 \\
\hline
0
\end{array}
$$

20 is 50% of 40. The answer is 40.

25. 40% of what is $32?

Translate: $0.4 \times b = 32$

Solve: To solve the equation we divide both sides by 0.4:

$$\frac{0.40 \cdot b}{0.4} = \frac{32}{0.4}$$

$$b = \frac{32}{0.4}$$

$$b = 80$$

$$
\begin{array}{r}
8\,0. \\
0.4_{\wedge}\overline{)\,3\,2.\,0_{\wedge}} \\
3\,2\,0 \\
\hline
0 \\
0 \\
\hline
0
\end{array}
$$

40% of $80 is $32. The answer is $80.

27. 56.32 is 64% of what?

Translate: $56.32 = 0.64 \cdot b$

Solve: $\dfrac{56.32}{0.64} = \dfrac{0.64 \cdot b}{0.64}$

$$\frac{56.32}{0.64} = b$$

$$88 = b$$

$$
\begin{array}{r}
8\,8. \\
0.64_{\wedge}\overline{)\,5\,6.\,3\,2_{\wedge}} \\
5\,1\,2\,0 \\
\hline
5\,1\,2 \\
5\,1\,2 \\
\hline
0
\end{array}
$$

56.32 is 64% of 88. The answer is 88.

29. 70% of what is 14?

Translate: $0.7 \cdot b = 14$

Solve: $\dfrac{0.7 \cdot b}{0.7} = \dfrac{14}{0.7}$

$$b = \frac{14}{0.7}$$

$$b = 20$$

$$
\begin{array}{r}
2\,0. \\
0.7_{\wedge}\overline{)\,1\,4.\,0_{\wedge}} \\
1\,4\,0 \\
\hline
0 \\
0 \\
\hline
0
\end{array}
$$

70% of 20 is 14. The answer is 20.

31. What is $62\frac{1}{2}\%$ of 10?

Translate: $a = 0.625 \cdot 10$

Solve: $a = 0.625 \cdot 10$

$$a = 6.25 \qquad \text{Multiplying}$$

6.25 is $62\frac{1}{2}\%$ of 10. The answer is 6.25.

33. What is 8.3% of $10,200?

Translate: $a = 0.083 \cdot 10,200$

Solve: $a = 8.3\% \cdot 10,200$

$a = 0.083 \cdot 10,200$

$a = 846.6 \qquad \text{Multiplying}$

$846.60 is 8.3% of $10,200. The answer is $846.60.

35. $66\frac{2}{3}\%$ of what is 27.4?

Translate: $0.\overline{6} \cdot b = 27.4$

Solve: We convert $0.\overline{6}$ to $\frac{2}{3}$. Then we divide by $\frac{2}{3}$, or multiply by $\frac{3}{2}$ on both sides.

$$\frac{2}{3} \cdot b = 27.4$$

$$\frac{3}{2} \cdot \frac{2}{3} \cdot b = \frac{3}{2} \cdot 27.4$$

$$b = 41.1$$

$66\frac{2}{3}\%$ of 41.1 is 27.4. The answer is 41.1.

37. Discussion and Writing Exercise

39. $\quad 0.\underline{623} \quad = \quad \dfrac{623}{1000}$

3 decimal places $\qquad$ 3 zeros

41. $\quad 2.\underline{37} \quad = \quad \dfrac{237}{100}$

2 decimal places $\qquad$ 2 zeros

43. $\dfrac{39}{100} \qquad 0.39.$

2 zeros $\quad$ Move 2 places

$$\frac{39}{100} = 0.39$$

45. *Familiarize*. Let m = the number of mochas that can be made with 12 oz of chocolate. Repeated subtraction, or division, will work here.

***Translate*.** The problem can be translated to the following equation:

$$m = 12 \div \frac{2}{3}.$$

***Solve*.** We carry out the division.

$$m = 12 \div \frac{2}{3}$$
$$= 12 \cdot \frac{3}{2}$$
$$= \frac{12 \cdot 3}{1 \cdot 2}$$
$$= \frac{2 \cdot 6 \cdot 3}{1 \cdot 2}$$
$$= \frac{2}{2} \cdot \frac{6 \cdot 3}{1}$$
$$= 18$$

***Check*.** If each of 18 mochas contains $\frac{2}{3}$ oz of chocolate, then the total amount of chocolate used is $18 \cdot \frac{2}{3} =$ $\frac{18 \cdot 2}{3} = \frac{\cancel{3} \cdot 6 \cdot 2}{\cancel{3} \cdot 1} = 12$.

The answer checks.

***State*.** 18 mochas can be made with 12 oz of chocolate.

47. Discussion and Writing Exercise

49. Estimate: Round 7.75% to 8% and \$10,880 to \$11,000. Then translate:

What is 8% of \$11,000?
$\downarrow \quad \downarrow \quad \downarrow \quad \downarrow \qquad \downarrow$
$a \quad = \quad 0.08 \quad \cdot \quad 11,000$

Carrying out the multiplication, we have

$a = 880.$

We estimate that 7.75% of \$10,880 is about \$880.

Calculate: First we translate:

What is 7.75% of \$10,880?
$\downarrow \quad \downarrow \quad \downarrow \quad \downarrow \qquad \downarrow$
$a \quad = \quad 0.0775 \quad \cdot \quad 10,880$

Carrying out the multiplication, we have

$a = 843.2.$

Thus, we calculate that 7.75% of \$10,880 is \$843.20.

51. We reword the problem as two questions, translate each to an equation, and solve the equations.

What is 40% of 270?   What is 50% of 270?
$\downarrow \ \downarrow \ \downarrow \ \downarrow \ \downarrow$ $\downarrow \ \downarrow \ \downarrow \ \downarrow \ \downarrow$
$a \ = \ 0.4 \ \times \ 270$ $a \ = \ 0.5 \ \times \ 270$
$a \ = \ 108$ $a \ = \ 135$

108 tons to 135 tons of the trash is recyclable.

Exercise Set 8.3

1. What is 37% of 74?

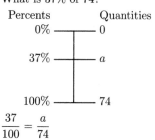

$$\frac{37}{100} = \frac{a}{74}$$

3. 4.3 is what percent of 5.9?

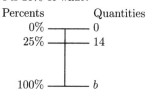

$$\frac{N}{100} = \frac{4.3}{5.9}$$

5. 14 is 25% of what?

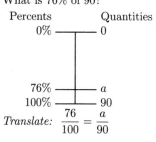

$$\frac{25}{100} = \frac{14}{b}$$

7. What is 76% of 90?

Percents   Quantities
0% —— 0

76% —— a
100% —— 90

Translate: $\dfrac{76}{100} = \dfrac{a}{90}$

Solve: $76 \cdot 90 = 100 \cdot a$ Equating cross-products

$$\frac{76 \cdot 90}{100} = \frac{100 \cdot a}{100} \quad \text{Dividing by 100}$$

$$\frac{6840}{100} = a$$

$$68.4 = a \qquad \text{Simplifying}$$

68.4 is 76% of 90. The answer is 68.4.

9. 70% of 660 is what?

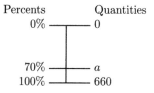

Translate: $\dfrac{70}{100} = \dfrac{a}{660}$

Solve: $70 \cdot 660 = 100 \cdot a$ Equating cross-products

$\dfrac{70 \cdot 660}{100} = \dfrac{100 \cdot a}{100}$ Dividing by 100

$\dfrac{46,200}{100} = a$

$462 = a$ Simplifying

70% of 660 is 462. The answer is 462.

11. What is 4% of 1000?

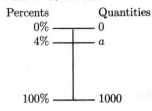

Translate: $\dfrac{4}{100} = \dfrac{a}{1000}$

Solve: $4 \cdot 1000 = 100 \cdot a$

$\dfrac{4 \cdot 1000}{100} = \dfrac{100 \cdot a}{100}$

$\dfrac{4000}{100} = a$

$40 = a$

40 is 4% of 1000. The answer is 40.

13. 4.8% of 60 is what?

Percents   Quantities
0% ——— 0
4.8% ——— a

100% ——— 60

Translate: $\dfrac{4.8}{100} = \dfrac{a}{60}$

Solve: $4.8 \cdot 60 = 100 \cdot a$

$\dfrac{4.8 \cdot 60}{100} = \dfrac{100 \cdot a}{100}$

$\dfrac{288}{100} = a$

$2.88 = a$

4.8% of 60 is 2.88. The answer is 2.88.

15. $24 is what percent of $96?

Percents   Quantities
0% ——— 0
N% ——— $24

100% ——— $96

Translate: $\dfrac{N}{100} = \dfrac{24}{96}$

Solve: $96 \cdot N = 100 \cdot 24$

$\dfrac{96N}{96} = \dfrac{100 \cdot 24}{96}$

$N = \dfrac{100 \cdot 24}{96}$

$N = 25$

$24 is 25% of $96. The answer is 25%.

17. 102 is what percent of 100?

Percents   Quantities
0% ——— 0

100% ——— 100
N% ——— 102

Translate: $\dfrac{N}{100} = \dfrac{102}{100}$

Solve: $100 \cdot N = 100 \cdot 102$

$\dfrac{100 \cdot N}{100} = \dfrac{100 \cdot 102}{100}$

$N = \dfrac{100 \cdot 102}{100}$

$N = 102$

102 is 102% of 100. The answer is 102%.

19. What percent of $480 is $120?

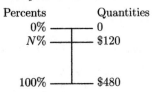

Translate: $\dfrac{N}{100} = \dfrac{120}{480}$

Solve: $480 \cdot N = 100 \cdot 120$

$\dfrac{480 \cdot N}{480} = \dfrac{100 \cdot 120}{480}$

$N = \dfrac{100 \cdot 120}{480}$

$N = 25$

25% of $480 is $120. The answer is 25%.

21. What percent of 160 is 150?

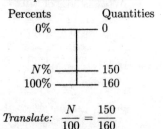

Translate: $\dfrac{N}{100} = \dfrac{150}{160}$

Solve: $160 \cdot N = 100 \cdot 150$

$$\frac{160 \cdot N}{160} = \frac{100 \cdot 150}{160}$$

$$N = \frac{100 \cdot 150}{160}$$

$$N = 93.75$$

93.75% of 160 is 150. The answer is 93.75%.

23. $18 is 25% of what?

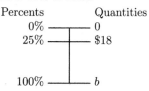

Percents   Quantities
0% ——— 0
25% ——— $18
100% ——— b

Translate: $\dfrac{25}{100} = \dfrac{18}{b}$

Solve: $25 \cdot b = 100 \cdot 18$

$$\frac{25 \cdot b}{b} = \frac{100 \cdot 18}{25}$$

$$b = \frac{100 \cdot 18}{25}$$

$$b = 72$$

$18 is 25% of $72. The answer is $72.

25. 60% of what is $54.

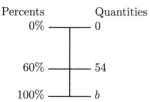

Percents   Quantities
0% ——— 0
60% ——— 54
100% ——— b

Translate: $\dfrac{60}{100} = \dfrac{54}{b}$

Solve: $60 \cdot b = 100 \cdot 54$

$$\frac{60 \cdot b}{b} = \frac{100 \cdot 54}{60}$$

$$b = \frac{100 \cdot 54}{60}$$

$$b = 90$$

60% of 90 is 54. The answer is 90.

27. 65.12 is 74% of what?

Percents   Quantities
0% ——— 0
74% ——— 65.12
100% ——— b

Translate: $\dfrac{74}{100} = \dfrac{65.12}{b}$

Solve: $74 \cdot b = 100 \cdot 65.12$

$$\frac{74 \cdot b}{74} = \frac{100 \cdot 65.12}{74}$$

$$b = \frac{100 \cdot 65.12}{74}$$

$$b = 88$$

65.12 is 74% of 88. The answer is 88.

29. 80% of what is 16?

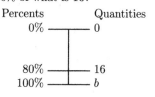

Percents   Quantities
0% ——— 0
80% ——— 16
100% ——— b

Translate: $\dfrac{80}{100} = \dfrac{16}{b}$

Solve: $80 \cdot b = 100 \cdot 16$

$$\frac{80 \cdot b}{80} = \frac{100 \cdot 16}{80}$$

$$b = \frac{100 \cdot 16}{80}$$

$$b = 20$$

80% of 20 is 16. The answer is 20.

31. What is $62\frac{1}{2}$% of 40?

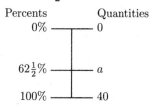

Percents   Quantities
0% ——— 0
$62\frac{1}{2}$% ——— a
100% ——— 40

Translate: $\dfrac{62\frac{1}{2}}{100} = \dfrac{a}{40}$

Solve: $62\frac{1}{2} \cdot 40 = 100 \cdot a$

$$\frac{125}{2} \cdot \frac{40}{1} = 100 \cdot a$$

$$2500 = 100 \cdot a$$

$$\frac{2500}{100} = \frac{100 \cdot a}{100}$$

$$25 = a$$

25 is $62\frac{1}{2}$% of 40. The answer is 25.

33. What is 9.4% of $8300?

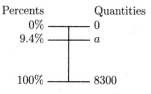

Percents   Quantities
0% ——— 0
9.4% ——— a
100% ——— 8300

Translate: $\dfrac{9.4}{100} = \dfrac{a}{8300}$

Solve: $9.4 \cdot 8300 = 100 \cdot a$

$$\frac{9.4 \cdot 8300}{100} = \frac{100 \cdot a}{100}$$

$$\frac{78,020}{100} = a$$

$$780.2 = a$$

$780.20 is 9.4% of $8300. The answer is $780.20.

35.

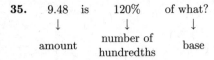

9.48   is   120%   of what?

amount   number of   base
         hundredths

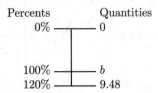

Percents   Quantities
0% ─────────── 0

100% ────────── b
120% ────────── 9.48

Translate: $\dfrac{120}{100} = \dfrac{9.48}{b}$

Solve: $120 \cdot b = 100 \cdot 9.48$

$$\frac{120b}{120} = \frac{948}{120}$$

$$b = 7.9$$

9.48 is 120% of 7.9. The answer is 7.9.

37. Discussion and Writing Exercise

39. Graph: $y = -\dfrac{1}{2}x$

We make a table of solutions. Note that when x is a multiple of 2, fractional values for y are avoided. Next we plot the points, draw the line, and label it.

When $x = -4$, $y = -\dfrac{1}{2}(-4) = \dfrac{4}{2} = 2$.

When $x = 0$, $y = -\dfrac{1}{2} \cdot 0 = 0$.

When $x = 2$, $y = -\dfrac{1}{2} \cdot 2 = -\dfrac{2}{2} = -1$.

x	$y = -\frac{1}{2}x$	(x,y)
-4	2	$(-4, 2)$
0	0	$(0, 0)$
2	-1	$(2, -1)$

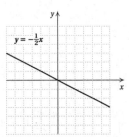

41. Graph: $y = 2x - 4$

We make a table of solutions. Then we plot the points, draw the line, and label it.

When $x = -1$, $y = 2(-1) - 4 = -2 - 4 = -6$.
When $x = 1$, $y = 2 \cdot 1 - 4 = 2 - 4 = -2$.
When $x = 4$, $y = 2 \cdot 4 - 4 = 8 - 4 = 4$.

x	$y = 2x - 4$	(x,y)
-1	-6	$(-1, -6)$
1	-2	$(1, -2)$
4	4	$(4, 4)$

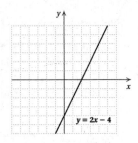

43. *Familiarize*. Let q = the number of quarts of liquid ingredients the recipe calls for.

***Translate*.**

Butter- plus Skim plus Oil is Total liquid
milk   milk   ingredients

$$\frac{1}{2} \quad + \quad \frac{1}{3} \quad + \quad \frac{1}{16} \quad = \quad q$$

***Solve*.** We carry out the addition. The LCM of the denominators is 48, so the LCD is 48.

$$\frac{1}{2} \cdot \frac{24}{24} + \frac{1}{3} \cdot \frac{16}{16} + \frac{1}{16} \cdot \frac{3}{3} = q$$

$$\frac{24}{48} + \frac{16}{48} + \frac{3}{48} = q$$

$$\frac{43}{48} = q$$

***Check*.** We repeat the calculation. The answer checks.

***State*.** The recipe calls for $\dfrac{43}{48}$ qt of liquid ingredients.

45. $0.05x = 40$

$$\frac{0.05x}{0.05} = \frac{40}{0.05} \quad \text{Dividing both sides by 0.05}$$

$$x = 800$$

The solution is 800.

47. $1.3n = 10.4$

$$\frac{1.3n}{1.3} = \frac{10.4}{1.3} \quad \text{Dividing both sides by 1.3}$$

$$n = 8$$

The solution is 8.

49. Discussion and Writing Exercise

51. Estimate: Round 8.85% to 10% and $12,640 to $12,000.

What is   10%   of 12,000?

amount   number of   base
         hundredths

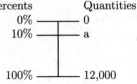

Percents   Quantities
0% ─────────── 0
10% ────────── a

100% ────────── 12,000

Translate: $\dfrac{10}{100} = \dfrac{a}{12,000}$

Solve: $10 \cdot 12,000 = 100 \cdot a$

$$\frac{120,000}{100} = \frac{100a}{100}$$

$$1200 = a$$

We estimate that 8.85% of $12,640 is $1200. (Answers may vary.)

Calculate:

What is 8.85% of 12,640?
↓ ↓ ↓
amount number of hundredths base

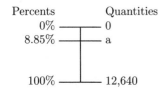

Translate: $\dfrac{8.85}{100} = \dfrac{a}{12,640}$

Solve: $8.85 \cdot 12,640 = 100 \cdot a$

$$\frac{111,864}{100} = \frac{100a}{100}$$

$$1118.64 = a$$

We calculate that 8.85% of $12,640 is $1118.64.

53. First we find 30% of 80:

$$\frac{30}{100} = \frac{a}{80}$$

$$30 \cdot 80 = 100 \cdot a$$

$$\frac{2400}{100} = \frac{100a}{100}$$

$$24 = a$$

Then we answer the question "24 is what percent of 120?"

$$\frac{N}{100} = \frac{24}{120}$$

$$120 \cdot N = 100 \cdot 24$$

$$\frac{120N}{120} = \frac{2400}{120}$$

$$N = 20$$

Thus, 30% of 80 is 20% of 120.

55. First we find 26% of 135:

$$\frac{26}{100} = \frac{a}{135}$$

$$26 \cdot 135 = 100 \cdot a$$

$$\frac{3510}{100} = \frac{100a}{100}$$

$$35.1 = a$$

Then we answer the question "What percent of 90 is 35.1?"

$$\frac{N}{100} = \frac{35.1}{90}$$

$$90 \cdot N = 100 \cdot 35.1$$

$$\frac{90N}{90} = \frac{3510}{90}$$

$$N = 39$$

Thus, 39% of 90 is the same as 26% of 135.

<hr>

Exercise Set 8.4

1. *Familiarize*. Let $w =$ the number of wild horses in Nevada.

***Translate*.** We translate to a percent equation.

What number is 48.4% of 27, 369?
↓ ↓ ↓ ↓ ↓
$w \qquad = 0.484 \cdot 27,369$

***Solve*.** We multiply.

$$w = 0.484 \cdot 27,369 = 13,246.596 \approx 13,247$$

***Check*.** We can repeat the calculations. We can also do a partial check by estimating: $48.4\% \cdot 27,369 \approx 50\% \cdot 27,00 = 13,500$. Since 13,500 is close to 13,247, our answer is reasonable.

***State*.** There are about 13,247 wild horses in Nevada.

3. *Familiarize*. Let $t =$ the value of a Nissan 350Z after three years and $f =$ the value after five years.

***Translate*.** We translate to two proportions.

$$\frac{62}{100} = \frac{t}{34,000} \text{ and } \frac{52}{100} = \frac{f}{34,000}$$

***Solve*.** We solve each proportion.

$$\frac{62}{100} = \frac{t}{34,000}$$

$$62 \cdot 34,000 = 100 \cdot t$$

$$\frac{62 \cdot 34,000}{100} = \frac{100 \cdot t}{100}$$

$$21,080 = t$$

$$\frac{52}{100} = \frac{f}{34,000}$$

$$52 \cdot 34,000 = 100 \cdot f$$

$$\frac{52 \cdot 34,000}{100} = \frac{100 \cdot f}{100}$$

$$17,680 = f$$

***Check*.** We can repeat the calculations. We can also do a partial check by estimating.

$$62\% \cdot 34,000 \approx 60\% \cdot 35,000 = 21,000 \approx 21,080$$

$$52\% \cdot 34,000 \approx 50\% \cdot 34,000 = 17,000 \approx 17,680$$

The answers check.

***State*.** The value of a Nissan 350Z will be $21,080 after three years; the value after five years will be $17,680.

5. *Familiarize*. Let $x =$ the number of people in the U.S. who are overweight and $y =$ the number who are obese, in millions.

***Translate*.** We translate to percent equations.

What number is 60% of 294?
↓ ↓ ↓ ↓ ↓
$x \qquad = 0.6 \cdot 294$

What number is 25% of 294?
↓ ↓ ↓ ↓ ↓
$y \qquad = 0.25 \cdot 294$

Solve.

$$x = 0.6 \cdot 294 = 176.4$$
$$y = 0.25 \cdot 294 = 73.5$$

Check. We can repeat the calculations. Also note that $60\% \cdot 294 \approx 60\% \cdot 300 = 180$ and $25\% \cdot 294 \approx 25\% \cdot 300 = 75$. Since 176.4 is close to 180 and 73.5 is close to 75, the answers check.

State. 176.4 million, or 176,400,000 people in the U.S. are overweight and 73.5 million, or 73,500,000 are obese.

7. **Familiarize.** First we find the amount of the solution that is acid. We let a = this amount.

   Translate. We translate to a percent equation.

   What is 3% of 680?

   $a = 0.03$ of 680

   Solve. We multiply.

   $$a = 3\% \times 680 = 0.03 \times 680 = 20.4$$

   Now we find the amount that is water. We let w = this amount.

Total amount	minus	Amount of acid	is	Amount of water
680	−	20.4	=	w

   To solve the equation we carry out the subtraction.

   $$w = 680 - 20.4 = 659.6$$

   Check. We can repeat the calculations. Also, observe that, since 3% of the solution is acid, 97% is water. Because 97% of $680 = 0.97 \times 680 = 659.6$, our answer checks.

   State. The solution contains 20.4 mL of acid and 659.6 mL of water.

9. **Familiarize.** Let n = the number of miles of the Mississippi River that are navigable.

   Translate. We translate to a proportion.

   $$\frac{77}{100} = \frac{n}{2348}$$

   Solve.

   $$\frac{77}{100} = \frac{n}{2348}$$
   $$77 \cdot 2348 = 100 \cdot n$$
   $$\frac{77 \cdot 2348}{100} = \frac{100 \cdot n}{100}$$
   $$1808 \approx n$$

   Check. We can repeat the calculations. Also note that $\frac{1808}{2348} \approx \frac{1800}{2400} = 0.75 = 75\% \approx 77\%$. The answer checks.

   State. About 1808 miles of the Mississippi River are navigable.

11. **Familiarize.** Let h = the number of Hispanic people in the U.S. in 2003, in millions.

    Translate. We translate to a proportion.

    $$\frac{13.7}{100} = \frac{h}{291}$$

Solve.

$$\frac{13.7}{100} = \frac{h}{291}$$
$$13.7 \cdot 291 = 100 \cdot h$$
$$\frac{13.7 \cdot 291}{100} = \frac{100 \cdot h}{100}$$
$$39.867 = h$$

Check. We can repeat the calculation. Also note that $\frac{39.867}{291} \approx \frac{40}{300} = 13.\overline{3}\% \approx 13.7\%$. The answer checks.

State. In 2003 about 39.867 million, or 39,867,000, Hispanic people lived in the United States.

13. **Familiarize.** First we find the number of items Christina got correct. Let b represent this number.

    Translate. We translate to a percent equation.

    What number is 91% of 40?

    $b = 0.91 \cdot 40$

    Solve. We multiply.

    $$b = 0.91 \cdot 40 = 36.4$$

    We subtract to find the number of items Christina got incorrect:

    $$40 - 36.4 = 3.6$$

    Check. We can repeat the calculation. Also note that $91\% \cdot 40 \approx 90\% \cdot 40 = 36 \approx 36.4$. The answer checks.

    State. Christina got 36.4 items correct and 3.6 items incorrect.

15. **Familiarize.** Let a = the number of items on the test.

    Translate. We translate to a proportion.

    $$\frac{86}{100} = \frac{81.7}{a}$$

    Solve.

    $$\frac{86}{100} = \frac{81.7}{a}$$
    $$86 \cdot a = 100 \cdot 81.7$$
    $$\frac{86 \cdot a}{86} = \frac{100 \cdot 81.7}{86}$$
    $$a = 95$$

    Check. We can repeat the calculation. Also note that $\frac{81.7}{95} \approx \frac{82}{100} = 82\% \approx 86\%$. The answer checks.

    State. There were 95 items on the test.

17. **Familiarize.** We let n = the percent of time that television sets are on.

    Translate. We translate to a percent equation.

    2190 is what percent of 8760?

    $2190 = n \times 8760$

    Solve. We divide on both sides by 8760 and convert the result to percent notation.

$$2190 = n \times 8760$$

$$\frac{2190}{8760} = \frac{n \times 8760}{8760}$$

$$0.25 = n$$

$$25\% = n$$

Check. To check we find 25% of 8760:

$25\% \times 8760 = 0.25 \times 8760 = 2190$. The answer checks.

State. Television sets are on for 25% of the year.

19. First we find the maximum heart rate for a 25 year old person.

Familiarize. Note that $220 - 25 = 195$. We let $x =$ the maximum heart rate for a 25 year old person.

Translate. We translate to a percent equation.

What is 85% of 195?

$\downarrow \quad \downarrow \quad \downarrow \quad \downarrow \quad \downarrow$

$x \quad = 0.85 \times \quad 195$

Solve. We multiply.

$$x = 0.85 \times 195 = 165.75 \approx 166$$

Check. We can repeat the calculations. Also, 85% of $195 \approx 0.85 \times 200 = 170 \approx 166$. The answer checks.

State. The maximum heart rate for a 25 year old person is 166 beats per minute.

Next we find the maximum heart rate for a 36 year old person.

Familiarize. Note that $220 - 36 = 184$. We let $x =$ the maximum heart rate for a 36 year old person.

Translate. We translate to a percent equation.

What is 85% of 184?

$\downarrow \quad \downarrow \quad \downarrow \quad \downarrow \quad \downarrow$

$x \quad = 0.85 \times \quad 184$

Solve. We multiply.

$$x = 0.85 \times 184 = 156.4 \approx 156$$

Check. We can repeat the calculations. Also, 85% of $184 \approx 0.9 \times 180 = 162 \approx 156$. The answer checks.

State. The maximum heart rate for a 36 year old person is 156 beats per minute.

Next we find the maximum heart rate for a 48 year old person.

Familiarize. Note that $220 - 48 = 172$. We let $x =$ the maximum heart rate for a 48 year old person.

Translate. We translate to a percent equation.

What is 85% of 172?

$\downarrow \quad \downarrow \quad \downarrow \quad \downarrow \quad \downarrow$

$x \quad = 0.85 \times \quad 172$

Solve. We multiply.

$$x = 0.85 \times 172 = 146.2 \approx 146$$

Check. We can repeat the calculations. Also, 85% of $172 \approx 0.9 \times 170 = 153 \approx 146$. The answer checks.

State. The maximum heart rate for a 48 year old person is 146 beats per minute.

We find the maximum heart rate for a 55 year old person.

Familiarize. Note that $220 - 55 = 165$. We let $x =$ the maximum heart rate for a 55 year old person.

Translate. We translate to a percent equation.

What is 85% of 165?

$\downarrow \quad \downarrow \quad \downarrow \quad \downarrow \quad \downarrow$

$x \quad = 0.85 \times \quad 165$

Solve. We multiply.

$$x = 0.85 \times 165 = 140.25 \approx 140$$

Check. We can repeat the calculations. Also, 85% of $165 \approx 0.9 \times 160 = 144 \approx 140$. The answer checks.

State. The maximum heart rate for a 55 year old person is 140 beats per minute.

Finally we find the maximum heart rate for a 76 year old person.

Familiarize. Note that $220 - 76 = 144$. We let $x =$ the maximum heart rate for a 76 year old person.

Translate. We translate to a percent equation.

What is 85% of 144?

$\downarrow \quad \downarrow \quad \downarrow \quad \downarrow \quad \downarrow$

$x \quad = 0.85 \times \quad 144$

Solve. We multiply.

$$x = 0.85 \times 144 = 122.4 \approx 122$$

Check. We can repeat the calculations. Also, 85% of $144 \approx 0.9 \times 140 = 126 \approx 122$. The answer checks.

State. The maximum heart rate for a 76 year old person is 122 beats per minute.

21. Familiarize. Use the drawing in the text to visualize the situation. Note that the increase in the amount was $16.

Let $n =$ the percent of increase.

Translate. We translate to a percent equation.

$16 is what percent of $200?

$\downarrow \quad \downarrow \qquad \downarrow \qquad \downarrow \quad \downarrow$

$16 = \qquad n \qquad \times \quad 200$

Solve. We divide by 200 on both sides and convert the result to percent notation.

$$16 = n \times 200$$

$$\frac{16}{200} = \frac{n \times 200}{200}$$

$$0.08 = n$$

$$8\% = n$$

Check. Find 8% of 200: $8\% \times 200 = 0.08 \times 200 = 16$. Since this is the amount of the increase, the answer checks.

State. The percent of increase was 8%.

23. Familiarize. We use the drawing in the text to visualize the situation. Note that the reduction is $18.

We let $n =$ the percent of decrease.

Translate. We translate to a percent equation.

$18 is what percent of $90?

$\downarrow \quad \downarrow \qquad \downarrow \qquad \downarrow \quad \downarrow$

$18 = \qquad n \qquad \times \quad 90$

Solve. To solve the equation, we divide on both sides by 90 and convert the result to percent notation.

$$\frac{18}{90} = \frac{n \times 90}{90}$$
$$0.2 = n$$
$$20\% = n$$

Check. We find 20% of 90: $20\% \times 90 = 0.2 \times 90 = 18$. Since this is the price decrease, the answer checks.

State. The percent of decrease was 20%.

25. **Familiarize**. We note that the amount of the reduction can be found and then subtracted from the old bill. Let b = the amount of the new bill.

Translate. We translate to a percent equation.

What is the old bill minus 50% of the old bill?

$$b = 106 - 0.5 \times 106$$

Solve.
$$b = 106 - 0.5 \times 106$$
$$b = 106 - 53$$
$$b = 53$$

Check. To check we note that the new bill is $100\% - 50\%$, or 50%, of the old bill. Since $0.5 \times 106 = 53$, the answer checks.

State. The new cooling bill would be $53.

27. **Familiarize**. First we find the amount of the decrease.

$$\begin{array}{r} 8\,9.9\,5 \\ -6\,5.4\,9 \\ \hline 2\,4.4\,6 \end{array}$$

Let p = the percent of decrease.

Translate. We translate to a percent equation.

$24.46 is what percent of $89.95

$$24.46 = p \cdot 89.95$$

Solve. We divide by 89.95 on both sides and convert to percent notation.

$$\frac{24.46}{89.95} = \frac{p \cdot 89.95}{89.95}$$
$$0.27 \approx p$$
$$27\% \approx p$$

Check. We find 27% of 89.95: $27\% \cdot 89.95 = 0.27 \cdot 89.95 \approx 24.29$. This is approximately the amount of the decrease, so the answer checks. (Remember that we rounded the percent.)

State. The percent of decrease is about 27%.

29. **Familiarize**. We note that the amount of the raise can be found and then added to the old salary. A drawing helps us visualize the situation.

$28,600	$?
100%	5%

We let x = the new salary.

Translate. We translate to a percent equation.

What is the old salary plus 5% of the old salary?

$$x = 28{,}600 + 0.05 \times 28{,}600$$

Solve. We simplify.

$$x = 28{,}600 + 0.05 \times 28{,}600$$
$$= 28{,}600 + 1430 \qquad \text{The raise is \$1430.}$$
$$= 30{,}030$$

Check. To check, we note that the new salary is 100% of the old salary plus 5% of the old salary, or 105% of the old salary. Since $1.05 \times 28{,}600 = 30{,}030$, our answer checks.

State. The new salary is $30,030.

31. **Familiarize**. Let d = the amount of depreciation the first year.

Translate. We translate to a proportion.

$$\frac{25}{100} = \frac{d}{21{,}566}$$

Solve.

$$\frac{25}{100} = \frac{d}{21{,}566}$$
$$25 \cdot 21{,}566 = 100 \cdot d$$
$$\frac{25 \cdot 21{,}566}{100} = d$$
$$5391.50 = d$$

Now we subtract to find the depreciated value after 1 year.

$$\begin{array}{r} 2\,1,5\,6\,6.0\,0 \\ -\quad 5\,3\,9\,1.5\,0 \\ \hline 1\,6,1\,7\,4.5\,0 \end{array}$$

The second year the car depreciates 25% of the value after 1 year. We use a proportion to find this amount, a.

$$\frac{25}{100} = \frac{a}{16{,}174.50}$$
$$25 \cdot 16{,}174.50 = 100 \cdot a$$
$$\frac{25 \cdot 16{,}174.50}{100} = a$$
$$4043.63 \approx a$$

Now we subtract to find the value of the car after 2 years.

$$\begin{array}{r} 1\,6,1\,7\,4.5\,0 \\ -\quad 4\,0\,4\,3.6\,3 \\ \hline 1\,2,1\,3\,0.8\,7 \end{array}$$

Check. We can repeat the calculations. Also note that after 1 year the value of the car will be $100\% - 25\%$, or 75%, of the original value:

$$75\% \times \$21{,}566 = \$16{,}174.50$$

After 2 years the value of the car will be $100\% - 25\%$, or 75%, of the value after 1 year:

$$75\% \times \$16{,}174.50 \approx \$12{,}130.88$$

The slight discrepancy in this amount is due to rounding. The answers check.

State. After 1 year the value of the car will be $16,174.50. After 2 years, its value will be $12,130.87.

33. *Familiarize*. This is a multistep problem. First we find the area of a cross-section of a finished board and of a rough board using the formula $A = l \cdot w$. Then we find the amount of wood removed in planing and drying and finally we find the percent of wood removed. Let $f =$ the area of a cross-section of a finished board and let $r =$ the area of a cross-section of a rough board.

Translate. We find the areas.
$$f = 3\frac{1}{2} \cdot 1\frac{1}{2}$$
$$r = 4 \cdot 2$$

Solve. We carry out the multiplications.
$$f = 3\frac{1}{2} \cdot 1\frac{1}{2} = \frac{7}{2} \cdot \frac{3}{2} = \frac{21}{4}$$
$$r = 4 \cdot 2 = 8$$

Now we subtract to find the amount of wood removed in planing and drying.
$$8 - \frac{21}{4} = \frac{32}{4} - \frac{21}{4} = \frac{11}{4}$$

Finally we find p, the percent of wood removed in planing and drying.

$$\underbrace{\frac{11}{4}}_{\downarrow} \quad \underbrace{\text{is}}_{\downarrow} \quad \underbrace{\text{what percent}}_{\downarrow} \quad \underbrace{\text{of}}_{\downarrow} \quad \underbrace{8?}_{\downarrow}$$
$$\frac{11}{4} \quad = \quad\quad p \quad\quad \cdot \quad 8$$

We solve the equation.
$$\frac{11}{4} = p \cdot 8$$
$$\frac{1}{8} \cdot \frac{11}{4} = p$$
$$\frac{11}{32} = p$$
$$0.34375 = p$$
$$34.375\% = p, \text{ or}$$
$$34\frac{3}{8}\% = p$$

Check. We repeat the calculations. The answer checks.

State. 34.375%, or $34\frac{3}{8}\%$, of the wood is removed in planing and drying.

35. *Familiarize*. First we subtract to find the amount of the increase.
$$\begin{array}{r} 7\ 3\ 5 \\ -\ 4\ 3\ 0 \\ \hline 3\ 0\ 5 \end{array}$$

Now let $p =$ the percent of increase.

Translate. We translate to an equation.
$$\underbrace{305}_{\downarrow} \quad \underbrace{\text{is}}_{\downarrow} \quad \underbrace{\text{what percent}}_{\downarrow} \quad \underbrace{\text{of}}_{\downarrow} \quad \underbrace{430?}_{\downarrow}$$
$$305 \quad = \quad\quad p \quad\quad \cdot \quad 430$$

Solve.
$$305 = p \cdot 430$$
$$\frac{305}{430} = \frac{p \cdot 430}{430}$$
$$0.71 \approx p$$
$$71\% \approx p$$

Check. We can repeat the calculations. Also note that $171\% \cdot 430 = 735.3 \approx 735$. The answer checks.

State. The percent of increase is about 71%.

37. *Familiarize*. Let $a =$ the amount of the increase.

Translate. We translate to a proportion.
$$\frac{100}{100} = \frac{a}{780}$$

Solve.
$$\frac{100}{100} = \frac{a}{780}$$
$$100 \cdot 780 = 100 \cdot a$$
$$\frac{100 \cdot 780}{100} = a$$
$$780 = a$$

Now we add to find the higher rate:
$$\begin{array}{r} 7\ 8\ 0 \\ +\ 7\ 8\ 0 \\ \hline 1\ 5\ 6\ 0 \end{array}$$

Check. We can repeat the calculations. Also note that $200\% \cdot \$780 = \$1560$. The answer checks.

State. The rate for smokers is $1560.

39. *Familiarize*. First we subtract to find the amount of the increase.
$$\begin{array}{r} 2\ 9\ 5\ 5 \\ -1\ 6\ 4\ 5 \\ \hline 1\ 3\ 1\ 0 \end{array}$$

Now let $p =$ the percent of increase.

Translate. We translate to an equation.
$$\underbrace{1310}_{\downarrow} \quad \underbrace{\text{is}}_{\downarrow} \quad \underbrace{\text{what percent}}_{\downarrow} \quad \underbrace{\text{of}}_{\downarrow} \quad \underbrace{1645?}_{\downarrow}$$
$$1310 \quad = \quad\quad p \quad\quad \cdot \quad 1645$$

Solve.
$$1310 = p \cdot 1645$$
$$\frac{1310}{1645} = p$$
$$0.80 \approx p$$
$$80\% \approx p$$

Check. We can repeat the calculations. Also note that $180\% \cdot 1645 = 2961 \approx 2955$. The answer checks.

State. The percent of increase is about 80%.

41. *Familiarize*. First we subtract to find the amount of change.
$$\begin{array}{r} 6\ 4\ 8,8\ 1\ 8 \\ -\ 5\ 5\ 0,0\ 4\ 3 \\ \hline 9\ 8,7\ 7\ 5 \end{array}$$

Now let $p =$ the percent of change.

Translate. We translate to a proportion.
$$\frac{p}{100} = \frac{98,775}{550,043}$$

Solve.

$$\frac{p}{100} = \frac{98,775}{550,043}$$

$$p \cdot 550,043 = 100 \cdot 98,775$$

$$p = \frac{100 \cdot 98,775}{550,043}$$

$$p \approx 18.0$$

Check. We can repeat the calculations. Also note that $118\% \cdot 550,043 \approx 649,051 \approx 648,818$. The answer checks.

State. The population of Alaska increased by 98,775. This was an 18% increase.

43. **Familiarize.** First we subtract to find the population in 1990.

$$\begin{array}{r} 9\,1\,7,6\,2\,1 \\ -1\,1\,8,5\,5\,6 \\ \hline 7\,9\,9,0\,6\,5 \end{array}$$

Now let p = the percent of change.

Translate. We translate to an equation.

118,556 is what percent of 799,065?

$$118,556 = p \cdot 799,065$$

Solve.

$$118,556 = p \cdot 799,065$$

$$\frac{118,556}{799,065} = p$$

$$0.148 \approx p$$

$$14.8\% \approx p$$

Check. We can repeat the calculations. Also note that $114.8\% \cdot 799,065 \approx 917,327 \approx 917,621$. The answer checks.

State. The population of Montana was 799,065 in 1990. The population had increased by about 14.8% in 2003.

45. **Familiarize.** First we add to find the population in 2003.

$$\begin{array}{r} 3,2\,9\,4,3\,9\,4 \\ +1,2\,5\,6,2\,9\,4 \\ \hline 4,5\,5\,0,6\,8\,8 \end{array}$$

Now let p = the percent of change.

Translate. We translate to a proportion.

$$\frac{p}{100} = \frac{1,256,294}{3,294,394}$$

Solve.

$$\frac{p}{100} = \frac{1,256,294}{3,294,394}$$

$$p \cdot 3,294,394 = 100 \cdot 1,256,294$$

$$p = \frac{100 \cdot 1,256,294}{3,294,394}$$

$$p \approx 38.1$$

Check. We can repeat the calculations. Also note that $138.1\% \cdot 3,294,394 \approx 4,549,558 \approx 4,550,688$. The answer checks.

State. The population of Colorado in 2003 was 4,550,688. The population had increased by about 38.1% in 2003.

47. a) **Familiarize.** First we find the amount of the decrease.

$$\begin{array}{r} 1,0\,2\,8,0\,0\,0 \\ -9\,5\,1,0\,0\,0 \\ \hline 7\,7,0\,0\,0 \end{array}$$

Let N = the percent of decrease.

Translate. We translate to a proportion.

$$\frac{N}{100} = \frac{77,000}{1,028,000}$$

Solve.

$$\frac{N}{100} = \frac{77,000}{1,028,000}$$

$$N \cdot 1,028,000 = 100 \cdot 77,000$$

$$\frac{N \cdot 1,028,000}{1,028,000} = \frac{100 \cdot 77,000}{1,028,000}$$

$$N \approx 7.5$$

Check. We can repeat the calculations. Also note that $7.5\% \cdot 1,028,000 \approx 7.5\% \cdot 1,000,000 = 75,000 \approx 77,000$. The answer checks.

State. The percent of decrease was about 7.5%

b) **Familiarize.** First we find the amount of the decrease in the next decade. Let b represent this number.

Translate. We translate to a proportion.

$$\frac{7.5}{100} = \frac{b}{951,000}$$

Solve.

$$\frac{7.5}{100} = \frac{b}{951,000}$$

$$7.5 \cdot 951,000 = 100 \cdot b$$

$$\frac{7.5 \cdot 951,000}{100} = \frac{100 \cdot b}{100}$$

$$71,325 \approx b$$

We subtract to find the population in 2010:

$$951,000 - 71,325 = 879,675$$

Check. We can repeat the calculations. Also note that the population in 2010 will be $100\% - 7.5\%$, or 92.5%, of the 2000 population and $92.5\% \cdot 951,000 \approx 879,675$. The answer checks.

State. In 2010 the population will be 879,675.

49. **Familiarize.** Since the car depreciates 25% in the first year, its value after the first year is $100\% - 25\%$, or 75%, of the original value. To find the decrease in value, we ask:

$27,300 is 75% of what?

Let b = the original cost.

Translate. We translate to an equation.

$27,300 is 75% of what?

$$\$27,300 = 0.75 \times b$$

Solve.

$$27,300 = 0.75 \times b$$

$$\frac{27,300}{0.75} = \frac{0.75 \times b}{0.75}$$

$$36,400 = b$$

Check. We find 25% of 36,400 and then subtract this amount from 36,400:

$$0.25 \times 36,400 = 9100 \text{ and}$$

$$36,400 - 9100 = 27,300$$

The answer checks.

State. The original cost was $36,400.

51. a) **Familiarize**. Let $p =$ the percent of U.S. births that are breech.

Translate. We translate to an equation.

120,000 is what percent of 4,000,000?

$$\downarrow \quad \downarrow \qquad \downarrow \qquad \downarrow \qquad \downarrow$$
$$120,000 = \qquad p \qquad \times \; 4,000,000$$

Solve.

$$120,000 = p \times 4,000,000$$

$$\frac{120,000}{4,000,000} = \frac{p \times 4,000,000}{4,000,000}$$

$$0.03 = p$$

$$3\% = p$$

Check. We repeat the calculation.

State. 3% of U.S. births are breech.

b) **Familiarize**. Let $s =$ the percent of cases that showed success with FAS.

Translate. We translate to an equation.

34 is what percent of 38?

$$\downarrow \downarrow \qquad \downarrow \qquad \downarrow \downarrow$$
$$34 = \qquad s \qquad \times \; 38$$

Solve.

$$34 = s \times 38$$

$$\frac{34}{38} = \frac{s \times 38}{38}$$

$$0.895 \approx s$$

$$89.5\% \approx s$$

Check. We repeat the calculation.

State. About 89.5% of the cases showed success with FAS.

c) **Familiarize**. We will use the success rate of 89.5% from part (b). Let $b =$ the number of breech babies that could be turned yearly.

Translate. We translate to an equation.

What is 89.5% of 120,000?

$$\downarrow \quad \downarrow \quad \downarrow \quad \downarrow \qquad \downarrow$$
$$b \; = 0.895 \times \; 120,000$$

Solve. We multiply.

$$b = 0.895 \times 120,000 = 107,400$$

Check. We repeat the calculation.

State. About 107,400 breech babies could be turned yearly.

d) **Familiarize**. We will use the success rate of 89.5% from part (b). Let $c =$ the number of C-sections per year that are due to breech position.

Translate. We translate to an equation.

2000 is 89.5% of what?

$$\downarrow \quad \downarrow \qquad \downarrow \quad \downarrow \qquad \downarrow$$
$$2000 = 0.895 \times \qquad c$$

Solve.

$$2000 = 0.895 \times c$$

$$\frac{2000}{0.895} = \frac{0.895 \times c}{0.895}$$

$$2235 \approx c$$

Check. We repeat the calculation.

State. About 2235 C-sections per year are due to breech position.

53. **Familiarize**. First we use the formula $A = l \times w$ to find the area of the strike zone:

$$A = 30 \times 17 = 510 \text{ in}^2$$

When a 2-in. border is added to the outside of the strike zone, the dimensions of the larger zone are 19 in. by 34 in. The area of this zone is

$$A = 34 \times 21 = 714 \text{ in}^2$$

We subtract to find the increase in area:

$$714 \text{ in}^2 - 510 \text{ in}^2 = 204 \text{ in}^2$$

We let $p =$ the percent of increase in the area.

Translate. We translate to a proportion.

204 is what percent of 510?

$$\downarrow \quad \downarrow \qquad \downarrow \qquad \downarrow \quad \downarrow$$
$$204 = \qquad P \qquad \times \; 510$$

Solve. We divide by 510 on both sides and convert to percent notation.

$$\frac{204}{510} = \frac{p \times 510}{510}$$

$$0.4 = p$$

$$40\% = p$$

Check. We repeat the calculations.

State. The area of the strike zone is increased by 40%.

55. Discussion and Writing Exercise

57. $\dfrac{25}{11} = 25 \div 11$

$$
\begin{array}{r}
2.\,2\,7 \\
11\overline{)2\,5.\,0\,0} \\
\underline{2\,2} \\
3\,0 \\
\underline{2\,2} \\
8\,0 \\
\underline{7\,7} \\
3
\end{array}
$$

Since the remainders begin to repeat, we have a repeating decimal.

$$\frac{25}{11} = 2.\overline{27}$$

59. $\dfrac{27}{8} = 27 \div 8$

$$
\begin{array}{r}
3.375 \\
8\overline{)27.000} \\
\underline{24} \\
30 \\
\underline{24} \\
60 \\
\underline{56} \\
40 \\
\underline{40} \\
0
\end{array}
$$

$\dfrac{27}{8} = 3.375$

We could also do this conversion as follows:

$$\frac{27}{8} = \frac{27}{8} \cdot \frac{125}{125} = \frac{3375}{1000} = 3.375$$

61. $\dfrac{23}{25} = \dfrac{23}{25} \cdot \dfrac{4}{4} = \dfrac{92}{100} = 0.92$

63. $\dfrac{14}{32} = 14 \div 32$

$$
\begin{array}{r}
0.4375 \\
32\overline{)14.0000} \\
\underline{128} \\
120 \\
\underline{96} \\
240 \\
\underline{224} \\
160 \\
\underline{160} \\
0
\end{array}
$$

$\dfrac{14}{32} = 0.4375$

(Note that we could have simplified the fraction first, getting $\dfrac{7}{16}$ and then found the quotient $7 \div 16$.)

65. Since 10,000 has 4 zeros, we move the decimal point in the number in the numerator 4 places to the left.

$$\frac{34,809}{10,000} = 3.4809$$

67. We add the lengths of the sides.

Perimeter
$= 8 \text{ cm} + 4 \text{ cm} + 4 \text{ cm} + 8 \text{ cm} + 4 \text{ cm} + 4 \text{ cm}$
$= (8 + 4 + 4 + 8 + 4 + 4) \text{ cm}$
$= 32 \text{ cm}$

69. We add the lengths of the sides.

Perimeter $= 6 \text{ in.} + 12 \text{ in.} + 6 \text{ in.} + 12 \text{ in.}$
$ = (6 + 12 + 6 + 12) \text{ in.}$
$ = 36 \text{ in.}$

71. Discussion and Writing Exercise

73. *Familiarize*. We will express 4 ft, 8 in. as 56 in. (4 ft + 8 in. $= 4 \cdot 12$ in. $+ 8$ in. $= 48$ in. $+ 8$ in. $= 56$ in.) We let $h =$ Cynthia's final adult height.

Translate. We translate to an equation.

$$\underset{\displaystyle 56}{\underbrace{56 \text{ in.}}} \; \underset{\displaystyle =}{\text{is}} \; \underset{\displaystyle 0.844}{84.4\%} \; \underset{\displaystyle \times}{\text{of}} \; \underset{\displaystyle h}{\text{what?}}$$

Solve.

$$56 = 0.844 \times h$$
$$\frac{56}{0.844} = \frac{0.844 \times h}{0.844}$$
$$66 \approx h$$

Check. We find 84.4% of 66: $0.844 \times 66 \approx 56$. The answer checks.

State. Cynthia's final adult height will be about 66 in., or 5 ft 6 in.

75. *Familiarize*. If p is 120% of q, then $p = 1.2q$. Let $n =$ the percent of p that q represents.

Translate. We translate to an equation. We use $1.2q$ for p.

$$\underset{\displaystyle q}{q} \; \underset{\displaystyle =}{\text{is}} \; \underset{\displaystyle n}{\underbrace{\text{what percent}}} \; \underset{\displaystyle \times}{\text{of}} \; \underset{\displaystyle 1.2q}{p?}$$

Solve.

$$q = n \times 1.2q$$
$$\frac{q}{1.2q} = \frac{n \times 1.2q}{1.2q}$$
$$\frac{1}{1.2} = n$$
$$0.8\overline{3} = n$$
$$83.\overline{3}\%, \text{ or } 83\frac{1}{3}\% = n$$

Check. We find $83\frac{1}{3}\%$ of $1.2q$:

$$0.8\overline{3} \times 1.2q = q$$

The answer checks.

State. q is $83.\overline{3}\%$, or $83\frac{1}{3}\%$, of p.

77. *Familiarize*. Let h and m represent the household incomes in Hawaii and Montana, respectively, in 2002-2003. From the table we see that the household income in Hawaii in 2003-2004 was the 2002-2003 income plus 6.6% of that income, or $h + 0.066h$, or $1.066h$. Similarly, the income in Montana in 2003-2004 is the 2002-2003 income less 3.6% of that income, or $m - 0.036m$, or $0.964m$.

Translate and Solve. First we write two equations.

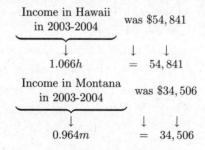

Now we solve the equations.

$$1.066h = 54,841 \qquad 0.964m = 34,506$$

$$\frac{1.066h}{1.066} = \frac{54,841}{1.066} \qquad \frac{0.964m}{0.964} = \frac{34,506}{0.964}$$

$$h \approx 51,445.59 \qquad m \approx 35,794.61$$

Finally we subtract to find how much higher the income was in Hawaii.

$$
\begin{array}{r}
1013 \\
4\cancel{0}\cancel{3}14415 \\
\cancel{5}\,\cancel{1},\,\cancel{4}\,\cancel{4}\,\cancel{5}.\,\cancel{5}\,9 \\
-\;3\,5,\;7\;9\;4.\;6\;1 \\
\hline
1\;5,\;6\;5\;0.\;9\;8
\end{array}
$$

Check. We repeat the calculations. The answer checks.

State. In 2002-2003 the household income in Hawaii was $15,650.98 higher than in Montana.

79. Let S = the original salary. After a 3% raise, the salary becomes $103\% \cdot S$, or $1.03S$. After a 6% raise, the new salary is $1.06\% \cdot 1.03S$, or $1.06(1.03S)$. Finally, after a 9% raise, the salary is $109\% \cdot 1.06(1.03S)$, or $1.09(1.06)(1.03S)$. Multiplying, we get $1.09(1.06)(1.03S) = 1.190062S$. This is equivalent to $119.0062\% \cdot S$, so the original salary has increased by 19.0062%, or about 19%.

81. Discussion and Writing Exercise

Exercise Set 8.5

1. The sales tax on an item costing $279 is

<u>Sales tax rate</u> $\times$ <u>Purchase price</u>

$$\downarrow \qquad \downarrow \qquad \downarrow$$
$$7\% \qquad \times \qquad \$279,$$

or 0.07×279, or 19.53. Thus the tax is $19.53.

3. The sales tax on an item costing $49.99 is

<u>Sales tax rate</u> $\times$ <u>Purchase price</u>

$$\downarrow \qquad \downarrow \qquad \downarrow$$
$$5.3\% \qquad \times \qquad \$49.99,$$

or 0.053×49.99, or about 2.65. Thus the tax is $2.65.

5. a) We first find the cost of the telephones. It is

$$5 \times \$69 = \$345.$$

b) The sales tax on items costing $345 is

<u>Sales tax rate</u> $\times$ <u>Purchase price</u>

$$\downarrow \qquad \downarrow \qquad \downarrow$$
$$4.75\% \qquad \times \qquad \$345,$$

or 0.0475×345, or about 16.39. Thus the tax is $16.39.

c) The total price is given by the purchase price plus the sales tax:

$$\$345 + \$16.39 = \$361.39.$$

To check, note that the total price is the purchase price plus 4.75% of the purchase price. Thus the total price is 104.75% of the purchase price. Since $1.0475 \times \$345 = \$361.3875 \approx \$361.39$, we have a check. The total price is $361.39.

7. *Rephrase:* $\underbrace{\text{Sales}}_{\text{tax}}$ is $\underbrace{\text{what}}_{\text{percent}}$ of $\underbrace{\text{purchase}}_{\text{price?}}$

$$\downarrow \qquad \downarrow \qquad \downarrow \qquad \downarrow \qquad \downarrow$$

Translate: $48 \quad = \quad r \quad \times \quad 960$

To solve the equation, we divide on both sides by 960.

$$\frac{48}{960} = \frac{r \times 960}{960}$$

$$0.05 = r$$

$$5\% = r$$

The sales tax rate is 5%.

9. *Rephrase:* $\underbrace{\text{Sales}}_{\text{tax}}$ is $\underbrace{\text{what}}_{\text{percent}}$ of $\underbrace{\text{purchase}}_{\text{price?}}$

$$\downarrow \qquad \downarrow \qquad \downarrow \qquad \downarrow \qquad \downarrow$$

Translate: $35.80 \quad = \quad r \quad \times \quad 895$

To solve the equation, we divide on both sides by 895.

$$\frac{35.80}{895} = \frac{r \times 895}{895}$$

$$0.04 = r$$

$$4\% = r$$

The sales tax rate is 4%.

11. *Rephrase:* $\underbrace{\text{Sales tax}}$ is 5% of what?

$$\downarrow \qquad \downarrow \quad \downarrow \quad \downarrow \quad \downarrow$$

Translate: $100 \quad = 5\% \times \quad b, \quad$ or

$\quad 100 \quad = 0.05 \times \quad b$

To solve the equation, we divide on both sides by 0.05.

$$\frac{100}{0.05} = \frac{0.05 \times b}{0.05}$$

$$2000 = b$$

$$
\begin{array}{r}
2\;0\;0\;0\;. \\
0.0\,5_{\wedge}\overline{\smash{)}\,1\;0\;0.0\;0_{\wedge}} \\
\underline{1\;0\;0\;0\;0} \\
0
\end{array}
$$

The purchase price is $2000.

13. *Rephrase:* $\underbrace{\text{Sales tax}}$ is 3.5% of what?

$$\downarrow \qquad \downarrow \quad \downarrow \quad \downarrow \quad \downarrow$$

Translate: $28 \quad = 3.5\% \times \quad b, \quad$ or

$\quad 28 \quad = 0.035 \times \quad b$

To solve the equation, we divide on both sides by 0.035.

$$\frac{28}{0.035} = \frac{0.035 \times b}{0.035}$$

$$800 = b$$

$$
\begin{array}{r}
8\;0\;0\;. \\
0.0\,3\,5_{\wedge}\overline{\smash{)}\,2\;8.0\;0\;0_{\wedge}} \\
\underline{2\;8\;0\;0\;0} \\
0
\end{array}
$$

The purchase price is $800.

15. a) We first find the cost of the shower units. It is

$$2 \times \$332.50 = \$665.$$

b) The total tax rate is the city tax rate plus the state tax rate, or $2\% + 6.25\% = 8.25\%$. The sales tax paid on items costing $665 is

$$\underbrace{\text{Sales tax rate}}_{} \times \underbrace{\text{Purchase price}}_{}$$
$$\downarrow \qquad \downarrow \qquad \downarrow$$
$$8.25\% \quad \times \quad \$665,$$

or 0.0825×665, or about 54.86. Thus the tax is $54.86.

c) The total price is given by the purchase price plus the sales tax:

$$\$665 + \$54.86 = \$719.86.$$

To check, note that the total price is the purchase price plus 8.25% of the purchase price. Thus the total price is 108.25% of the purchase price. Since $1.0825 \times 665 \approx 719.86$, we have a check. The total amount paid for the 2 shower units is $719.86.

17. *Rephrase:* $\underbrace{\text{Sales}}_{} $ is $\underbrace{\text{what}}_{}$ of $\underbrace{\text{purchase}}_{}$
             tax   percent   price?

$$\downarrow \quad \downarrow \quad \downarrow \quad \downarrow \quad \downarrow$$
Translate: $1030.40 = \quad r \quad \times \quad 18,400$

To solve the equation, we divide on both sides by 18,400.

$$\frac{1030.40}{18,400} = \frac{r \times 18,400}{18,400}$$
$$0.056 = r$$
$$5.6\% = r$$

The sales tax rate is 5.6%.

19. Commission = Commission rate $\times$ Sales
$$C \quad = \quad 6\% \quad \times 45,000$$
This tells us what to do. We multiply.
$$\begin{array}{r} 45,000 \\ \times \quad 0.06 \\ \hline 2\,7\,0\,0.00 \end{array} \quad (6\% = 0.06)$$
The commission is $2700.

21. Commission = Commission rate $\times$ Sales
$$120 \quad = \quad r \quad \times 2400$$
To solve this equation we divide on both sides by 2400:
$$\frac{120}{2400} = \frac{r \times 2400}{2400}$$
We can divide, but this time we simplify by removing a factor of 1:
$$r = \frac{120}{2400} = \frac{1}{20} \cdot \frac{120}{120} = \frac{1}{20} = 0.05 = 5\%$$
The commission rate is 5%.

23. Commission = Commission rate $\times$ Sales
$$392 \quad = \quad 40\% \quad \times \quad S$$
To solve this equation we divide on both sides by 0.4:
$$\frac{392}{0.4} = \frac{0.4 \times S}{0.4}$$
$$980 = S$$

$$\begin{array}{r} 980. \\ 0.4\,\overline{)392.0_\wedge} \\ 3600 \\ \hline 320 \\ 320 \\ \hline 0 \\ 0 \\ \hline 0 \end{array}$$

$980 worth of artwork was sold.

25. Commission = Commission rate $\times$ Sales
$$C \quad = \quad 6\% \quad \times 98,000$$
This tells us what to do. We multiply.
$$\begin{array}{r} 98,000 \\ \times \quad 0.06 \\ \hline 5\,8\,8\,0.00 \end{array} \quad (6\% = 0.06)$$
The commission is $5880.

27. Commission = Commission rate $\times$ Sales
$$280.80 \quad = \quad r \quad \times 2340$$
To solve this equation we divide on both sides by 2340.
$$\frac{280.80}{2340} = \frac{r \times 2340}{2340}$$
$$0.12 = r$$
$$12\% = r$$

$$\begin{array}{r} 0.12 \\ 2340\,\overline{)280.80} \\ 2340 \\ \hline 4680 \\ 4680 \\ \hline 0 \end{array}$$

The commission rate is 12%.

29. First we find the commission on the first $2000 of sales.
Commission = Commission rate $\times$ Sales
$$C \quad = \quad 5\% \quad \times 2000$$
This tells us what to do. We multiply.
$$\begin{array}{r} 2000 \\ \times \quad 0.05 \\ \hline 1\,0\,0.00 \end{array}$$
The commission on the first $2000 of sales is $100.

Next we subtract to find the amount of sales over $2000.
$$\$6000 - \$2000 = \$4000$$
Miguel had $4000 in sales over $2000.

Then we find the commission on the sales over $2000.
Commission = Commission rate $\times$ Sales
$$C \quad = \quad 8\% \quad \times 4000$$
This tells us what to do. We multiply.
$$\begin{array}{r} 4000 \\ \times \quad 0.08 \\ \hline 3\,2\,0.00 \end{array}$$
The commission on the sales over $2000 is $320.

Finally we add to find the total commission.
$$\$100 + \$320 = \$420$$
The total commission is $420.

31. Discount = Rate of discount $\times$ Marked price
$$D \quad = \quad 10\% \quad \times \quad \$300$$
Convert 10% to decimal notation and multiply.
$$\begin{array}{r} 300 \\ \times \quad 0.1 \\ \hline 3\,0.0 \end{array} \quad (10\% = 0.10 = 0.1)$$
The discount is $30.

Sale price = Marked price $-$ Discount
$$S \quad = \quad 300 \quad - \quad 30$$

We subtract:
$$\begin{array}{r} 3\,0\,0 \\ -\ \ 3\,0 \\ \hline 2\,7\,0 \end{array}$$

To check, note that the sale price is 90% of the marked price: $0.9 \times 300 = 270$.

The sale price is $270.

33. Discount = Rate of discount × Marked price
$$D\ \ =\ \ \ \ \ \ 15\%\ \ \ \ \times\ \ \ \ \ \$17$$

Convert 15% to decimal notation and multiply.
$$\begin{array}{r} 1\,7 \\ \times\,0.1\,5 \\ \hline 8\,5 \\ 1\,7\,0 \\ \hline 2.5\,5 \end{array} \qquad (15\% = 0.15)$$

The discount is $2.55.

Sale price = Marked price − Discount
$$S\ \ =\ \ \ \ \ 17\ \ \ -\ \ \ 2.55$$

We subtract:
$$\begin{array}{r} 1\,7.0\,0 \\ -\ \ 2.5\,5 \\ \hline 1\,4.4\,5 \end{array}$$

To check, note that the sale price is 85% of the marked price: $0.85 \times 17 = 14.45$.

The sale price is $14.45.

35. Discount = Rate of discount × Marked price
$$12.50\ \ =\ \ \ \ \ 10\%\ \ \ \ \times\ \ \ \ M$$

To solve the equation we divide on both sides by 0.1.
$$\frac{12.50}{0.1} = \frac{0.1 \times M}{0.1}$$
$$125 = M$$

The marked price is $125.

Sale price = Marked price − Discount
$$S\ \ =\ \ \ \ 125.00\ \ \ -\ \ \ 12.50$$

We subtract:
$$\begin{array}{r} 1\,2\,5.0\,0 \\ -\ \ 1\,2.5\,0 \\ \hline 1\,1\,2.5\,0 \end{array}$$

To check, note that the sale price is 90% of the marked price: $0.9 \times 125 = 112.50$.

The sale price is $112.50.

37. Discount = Rate of discount × Marked price
$$240\ \ =\ \ \ \ \ r\ \ \ \ \times\ \ \ \ \ 600$$

To solve the equation we divide on both sides by 600.
$$\frac{240}{600} = \frac{r \times 600}{600}$$
We can simplify by removing a factor of 1:
$$r = \frac{240}{600} = \frac{2}{5} \cdot \frac{120}{120} = \frac{2}{5} = 0.4 = 40\%$$
The rate of discount is 40%.

Sale price = Marked price − Discount
$$S\ \ =\ \ \ \ \ 600\ \ \ -\ \ \ 240$$

We subtract:
$$\begin{array}{r} 6\,0\,0 \\ -\,2\,4\,0 \\ \hline 3\,6\,0 \end{array}$$

To check, note that a 40% discount rate means that 60% of the marked price is paid. Since $\frac{360}{600} = 0.6$, or 60%, we have a check.

The sale price is $360.

39. Discount = Marked price − Sale price
$$D\ \ =\ \ \ \ 179.99\ \ \ -\ \ 149.99$$

We subtract:
$$\begin{array}{r} 1\,7\,9.9\,9 \\ -\,1\,4\,9.9\,9 \\ \hline 3\,0.0\,0 \end{array}$$

The discount is $30.

Discount = Rate of discount × Marked price
$$30\ \ =\ \ \ \ \ R\ \ \ \ \times\ \ \ \ 179.99$$

To solve the equation we divide on both sides by 179.99.
$$\frac{30}{179.99} = \frac{R \times 179.99}{179.99}$$
$$0.167 \approx R$$
$$16.7\% \approx R$$

To check, note that a discount rate of 16.7% means that 83.3% of the marked price is paid: $0.833 \times 179.99 = 149.93167 \approx 149.99$. Since that is the sale price, the answer checks.

The rate of discount is 16.7%.

41. Discount = Marked price − Sale price
$$200\ \ =\ \ \ \ \ M\ \ \ \ -\ \ \ \ 349$$

We add 349 on both sides of the equation:
$$200 + 349 = M$$
$$549 = M$$

The marked price is $549.

Discount = Rate of discount × Marked price
$$200\ \ =\ \ \ \ \ R\ \ \ \ \times\ \ \ \ \ 549$$

To solve the equation we divide on both sides by 549.
$$\frac{200}{549} = \frac{R \times 549}{549}$$
$$0.364 \approx R$$
$$36.4\% \approx R$$

To check note that a discount rate of 36.4% means that 63.6% of the marked price is paid: $0.636 \times 549 = 349.164 \approx 349$. Since this is the sale price, the answer checks.

The rate of discount is 36.4%.

43. Discussion and Writing Exercise

45.
$$\frac{x}{12} = \frac{24}{16}$$
$$16 \cdot x = 12 \cdot 24 \qquad \text{Equating cross-products}$$
$$x = \frac{12 \cdot 24}{16} \qquad \text{Dividing by 16 on both sides}$$
$$x = \frac{288}{16}$$
$$x = 18$$

The solution is 18.

47. Graph: $y = \frac{4}{3}x$

We make a table of solutions. Note that when x is a multiple of 3, fractional values for y are avoided. Next we plot the points, draw the line and label it.

When $x = -3$, $y = \frac{4}{3}(-3) = -\frac{12}{3} = -4$.

When $x = 0$, $y = \frac{4}{3} \cdot 0 = 0$.

When $x = 3$, $y = \frac{4}{3} \cdot 3 = \frac{12}{3} = 4$.

x	y $y = \frac{4}{3}x$	(x,y)
-3	-4	$(-3,-4)$
0	0	$(0,0)$
3	4	$(3,4)$

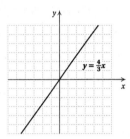

49. $\frac{5}{9} = 5 \div 9$

$$
\begin{array}{r}
0.\,5\,5 \\
9\,\overline{)5.\,0\,0} \\
\underline{4\,5} \\
5\,0 \\
\underline{4\,5} \\
5
\end{array}
$$

We get a repeating decimal.

$\frac{5}{9} = 0.\overline{5}$

51.
$$80\left(1 + \frac{0.06}{2}\right)^2$$

$= 80(1 + 0.03)^2$ Dividing

$= 80(1.03)^2$ Adding

$= 80(1.0609)$ Evaluating the exponential expression

$= 84.872$

53. Discussion and Writing Exercise

55. For a 39¢, or $0.39, tax we have:

$$
\underbrace{\text{Cigarette tax}}_{\downarrow \atop 0.39} \underset{\downarrow}{\text{is}} \underbrace{\text{what percent}}_{\downarrow \atop r} \underset{\downarrow}{\text{of}} \underbrace{\text{price per pack?}}_{\downarrow \atop 2.70}
$$

To solve the equation, we divide both sides by 2.70.

$$\frac{0.39}{2.70} = \frac{r \times 2.70}{2.70}$$

$$0.1\overline{4} = r$$

$$14.4\% \approx r$$

For a 69¢, or $0.69, tax we have:

$$
\underbrace{\text{Cigarette tax}}_{\downarrow \atop 0.69} \underset{\downarrow}{\text{is}} \underbrace{\text{what percent}}_{\downarrow \atop r} \underset{\downarrow}{\text{of}} \underbrace{\text{price per pack?}}_{\downarrow \atop 2.70}
$$

To solve the equation, we divide both sides by 2.70.

$$\frac{0.69}{2.70} = \frac{r \times 2.70}{2.70}$$

$$0.2\overline{5} = r$$

$$25.6\% \approx r$$

As a percentage of the price per pack, cigarette taxes range from 14.4% to 25.6%.

57. *Familiarize*. The subscription price is $100\% - 29.7\%$, or 70.3% of the newsstand price. Let $p =$ the newsstand price.

Translate.

$$
\underbrace{\text{Subscription price}}_{\downarrow \atop 1.89} \underset{\downarrow \atop =}{\text{is}} \underset{\downarrow \atop 70.3\%}{70.3\%} \underset{\downarrow \atop \times}{\text{of}} \underbrace{\text{newsstand price}}_{\downarrow \atop p}
$$

Solve.

$$1.89 = 70.3\% \times p$$

$$1.89 = 0.703 \times p$$

$$\frac{1.89}{0.703} = \frac{0.703 \times p}{0.703}$$

$$2.69 \approx p$$

Check. We find 29.7% of 2.69 and then subtract this amount from 2.69:

$0.297 \times 2.69 \approx 0.80$, $2.69 - 0.80 = 1.89$.

Since we get the subscription price, the answer checks.

State. The newsstand price was $2.69.

59. Commission = Commission rate × Sales

$\quad\quad C \quad\quad = \quad\quad 7.5\% \quad\quad \times\ 98,500$

We multiply.

$$
\begin{array}{r}
9\,8,5\,0\,0 \\
\times\quad 0.0\,7\,5 \\
\hline
4\,9\,2\,5\,0\,0 \\
6\,8\,9\,5\,0\,0\,0 \\
\hline
7\,3\,8\,7.5\,0\,0
\end{array}
$$

$(7.5\% = 0.075)$

The commission is $7387.50.

We subtract to find how much the seller gets for the house after paying the commission.

$\$98,500 - \$7387.50 = \$91,112.50$

Exercise Set 8.6

1. $I = P \cdot r \cdot t$

$= \$200 \times 4\% \times 1$

$= \$200 \times 0.04$

$= \$8$

3. $I = P \cdot r \cdot t$

$= \$2000 \times 8.4\% \times \frac{1}{2}$

$= \dfrac{\$2000 \times 0.084}{2}$

$= \$84$

5. $I = P \cdot r \cdot t$

$\qquad = \$4300 \times 10.56\% \times \dfrac{1}{4}$

$\qquad = \dfrac{\$4300 \times 0.1056}{4}$

$\qquad = \$113.52$

7. $I = P \cdot r \cdot t$

$\qquad = \$20,000 \times 4\dfrac{5}{8}\% \times 1$

$\qquad = \$20,000 \times 0.04625$

$\qquad = \$925$

9. $I = P \cdot r \cdot t$

$\qquad = \$50,000 \times 5\dfrac{3}{8}\% \times \dfrac{1}{4}$

$\qquad = \dfrac{\$50,000 \times 0.05375}{4}$

$\qquad \approx \$671.88$

11. a) We express 60 days as a fractional part of a year and find the interest.

$\qquad I = P \cdot r \cdot t$

$\qquad = \$10,000 \times 9\% \times \dfrac{60}{365}$

$\qquad = \$10,000 \times 0.09 \times \dfrac{60}{365}$

$\qquad \approx \$147.95 \quad$ Using a calculator

The interest due for 60 days is $147.95.

b) The total amount that must be paid after 60 days is the principal plus the interest.

$\qquad 10,000 + 147.95 = 10,147.95$

The total amount due is $10,147.95.

13. a) We express 90 days as a fractional part of a year and find the interest.

$\qquad I = P \cdot r \cdot t$

$\qquad = \$6500 \times 5\% \times \dfrac{90}{365}$

$\qquad = \$6500 \times 0.05 \times \dfrac{90}{365}$

$\qquad \approx \$80.14 \quad$ Using a calculator

The interest due for 90 days is $80.14.

b) The total amount that must be paid after 90 days is the principal plus the interest.

$\qquad 6500 + 80.14 = 6580.14$

The total amount due is $6580.14.

15. a) We express 30 days as a fractional part of a year and find the interest.

$\qquad I = P \cdot r \cdot t$

$\qquad = \$5600 \times 10\% \times \dfrac{30}{365}$

$\qquad = \$5600 \times 0.1 \times \dfrac{30}{365}$

$\qquad \approx \$46.03 \quad$ Using a calculator

The interest due for 30 days is $46.03.

b) The total amount that must be paid after 30 days is the principal plus the interest.

$\qquad 5600 + 46.03 = 5646.03$

The total amount due is $5646.03.

17. a) After 1 year, the account will contain 105% of $400.

$\qquad 1.05 \times \$400 = \$420$

$$
\begin{array}{r}
4\ 0\ 0 \\
\times\quad 1.0\ 5 \\
\hline
2\ 0\ 0\ 0 \\
4\ 0\ 0\ 0\ 0 \\
\hline
4\ 2\ 0.0\ 0
\end{array}
$$

b) At the end of the second year, the account will contain 1.05% of $420.

$\qquad 1.05 \times \$420 = \$441$

$$
\begin{array}{r}
4\ 2\ 0 \\
\times\quad 1.0\ 5 \\
\hline
2\ 1\ 0\ 0 \\
4\ 2\ 0\ 0\ 0 \\
\hline
4\ 4\ 1.0\ 0
\end{array}
$$

The amount in the account after 2 years is $441.

(Note that we could have used the formula

$A = P \cdot \left(1 + \dfrac{r}{n}\right)^{n \cdot t}$, substituting $400 for P, 5% for r,

1 for n, and 2 for t.)

19. We use the compound interest formula, substituting $2000 for P, 8.8% for r, 1 for n, and 4 for t.

$\qquad A = P \cdot \left(1 + \dfrac{r}{n}\right)^{n \cdot t}$

$\qquad = \$2000 \cdot \left(1 + \dfrac{8.8\%}{1}\right)^{1 \cdot 4}$

$\qquad = \$2000 \cdot (1 + 0.088)^4$

$\qquad = \$2000 \cdot (1.088)^4$

$\qquad \approx \$2802.50$

The amount in the account after 4 years is $2802.50.

21. We use the compound interest formula, substituting $4300 for P, 10.56% for r, 1 for n, and 6 for t.

$\qquad A = P \cdot \left(1 + \dfrac{r}{n}\right)^{n \cdot t}$

$\qquad = \$4300 \cdot \left(1 + \dfrac{10.56\%}{1}\right)^{1 \cdot 6}$

$\qquad = \$4300 \cdot (1 + 0.1056)^6$

$\qquad = \$4300 \cdot (1.1056)^6$

$\qquad \approx \$7853.38$

The amount in the account after 4 years is $7853.38.

23. We use the compound interest formula, substituting $20,000 for P, $6\frac{5}{8}\%$ for r, 1 for n, and 25 for t.

$$A = P \cdot \left(1 + \frac{r}{n}\right)^{n \cdot t}$$

$$= \$20,000 \cdot \left(1 + \frac{6\frac{5}{8}\%}{1}\right)^{1 \cdot 25}$$

$$= \$20,000 \cdot (1 + 0.06625)^{25}$$

$$= \$20,000 \cdot (1.06625)^{25}$$

$$\approx \$99,427.40$$

The amount in the account after 25 years is $99,427.40.

25. We use the compound interest formula, substituting $4000 for P, 6% for r, 2 for n, and 1 for t.

$$A = P \cdot \left(1 + \frac{r}{n}\right)^{n \cdot t}$$

$$= \$4000 \cdot \left(1 + \frac{6\%}{2}\right)^{2 \cdot 1}$$

$$= \$4000 \cdot \left(1 + \frac{0.06}{2}\right)^{2}$$

$$= \$4000 \cdot (1.03)^{2}$$

$$= \$4243.60$$

The amount in the account after 1 year is $4243.60.

27. We use the compound interest formula, substituting $20,000 for P, 8.8% for r, 2 for n, and 4 for t.

$$A = P \cdot \left(1 + \frac{r}{n}\right)^{n \cdot t}$$

$$= \$20,000 \cdot \left(1 + \frac{8.8\%}{2}\right)^{2 \cdot 4}$$

$$= \$20,000 \cdot \left(1 + \frac{0.088}{2}\right)^{8}$$

$$= \$20,000 \cdot (1.044)^{8}$$

$$\approx \$28,225.00$$

The amount in the account after 4 years is $28,225.00.

29. We use the compound interest formula, substituting $5000 for P, 10.56% for r, 2 for n, and 6 for t.

$$A = P \cdot \left(1 + \frac{r}{n}\right)^{n \cdot t}$$

$$= \$5000 \cdot \left(1 + \frac{10.56\%}{2}\right)^{2 \cdot 6}$$

$$= \$5000 \cdot \left(1 + \frac{0.1056}{2}\right)^{12}$$

$$= \$5000 \cdot (1.0528)^{12}$$

$$\approx \$9270.87$$

The amount in the account after 6 years is $9270.87.

31. We use the compound interest formula, substituting $20,000 for P, $7\frac{5}{8}\%$ for r, 2 for n, and 25 for t.

$$A = P \cdot \left(1 + \frac{r}{n}\right)^{n \cdot t}$$

$$= \$20,000 \cdot \left(1 + \frac{7\frac{5}{8}\%}{2}\right)^{2 \cdot 25}$$

$$= \$20,000 \cdot \left(1 + \frac{0.07625}{2}\right)^{50}$$

$$= \$20,000 \cdot (1.038125)^{50}$$

$$\approx \$129,871.09$$

The amount in the account after 25 years is $129,871.09.

33. We use the compound interest formula, substituting $4000 for P, 6% for r, 12 for n, and $\frac{5}{12}$ for t.

$$A = P \cdot \left(1 + \frac{r}{n}\right)^{n \cdot t}$$

$$= \$4000 \cdot \left(1 + \frac{6\%}{2}\right)^{12 \cdot \frac{5}{12}}$$

$$= \$4000 \cdot \left(1 + \frac{0.06}{12}\right)^{5}$$

$$= \$4000 \cdot (1.005)^{5}$$

$$\approx \$4101.01$$

The amount in the account after 5 months is $4101.01.

35. We use the compound interest formula, substituting $1200 for P, 10% for r, 4 for n, and 1 for t.

$$A = P \cdot \left(1 + \frac{r}{n}\right)^{n \cdot t}$$

$$= \$1200 \cdot \left(1 + \frac{10\%}{4}\right)^{4 \cdot 1}$$

$$= \$1200 \cdot \left(1 + \frac{0.1}{4}\right)^{4}$$

$$= \$1200 \cdot (1.025)^{4}$$

$$\approx \$1324.58$$

The amount in the account after 1 year is $1324.58.

37. We use the compound interest formula, substituting $20,000 for P, 6% for r, 365 for n, and $\frac{50}{365}$ for t.

$$A = P \cdot \left(1 + \frac{r}{n}\right)^{n \cdot t}$$

$$= \$20,000 \left(1 + \frac{6\%}{365}\right)^{365 \cdot \frac{50}{365}}$$

$$= \$20,000 \left(1 + \frac{0.06}{365}\right)^{50}$$

$$\approx \$20,165.05$$

After 50 days Emilio is owed $20,165.05.

39. Discussion and Writing Exercise

41. If the product of two numbers is 1, they are <u>reciprocals</u> of each other.

43. The number 0 is the <u>additive</u> identity.

45. The distance around an object is its <u>perimeter</u>.

47. A natural number that has exactly two different factors, only itself and 1, is called a <u>prime</u> number.

49. Discussion and Writing Exercise

51. For a principle P invested at 9% compounded monthly, to find the amount in the account at the end of 1 year we would multiply P by $(1 + 0.09/12)^{12}$. Since $(1 + 0.09/12)^{12} = 1.0075^{12} \approx 1.0938$, the effective yield is approximately 9.38%.

53. At the end of 1 year, the $20,000 spent on the car has been reduced in value by 30% of $20,000, or $0.3 \times \$20,000$, or $6000.

If the $20,000 is invested at 9%, compounded daily, the amount in the account at the end of 1 year is

$$A = \$20,000\left(1 + \frac{0.09}{365}\right)^{365}$$
$$\approx \$20,000(1.000246575)^{365}$$
$$\approx \$20,000(1.094162144)$$
$$\approx \$21,883.24.$$

Then the $20,000 has increased in value by $21,883.24 − $20,000, or $1883.24. All together, the Coniglios have saved the $6000 they would have lost on the value of the car plus the $1883.24 increase in the value of the $20,000 invested at 9%, compounded daily. That is, they have saved $6000 + $1883.24, or $7883.24.

Exercise Set 8.7

1. a) We multiply the balance by 2%:

$$0.02 \times \$4876.54 = \$97.5308.$$

Antonio's minimum payment, rounded to the nearest dollar, is $98.

b) We find the amount of interest on $4876.54 at 21.3% for one month.

$$I = P \cdot r \cdot t$$
$$= \$4876.54 \times 0.213 \times \frac{1}{12}$$
$$\approx \$86.56$$

We subtract to find the amount applied to decrease the principal in the first payment.

$$\$98 - \$86.56 = \$11.44$$

The principal is decreased by $11.44 with the first payment.

c) We find the amount of interest on $4876.54 at 12.6% for one month.

$$I = P \cdot r \cdot t$$
$$= \$4876.54 \times 0.126 \times \frac{1}{12}$$
$$\approx \$51.20$$

We subtract to find the amount applied to decrease the principal in the first payment.

$$\$98 - \$51.20 = \$46.80.$$

The principal is decreased by $46.80 with the first payment.

d) With the 12.6% rate the principal was decreased by $46.80 − $11.44, or $35.36 more than at the 21.3% rate. This also means that the interest at 12.6% is $35.36 less than at 21.3%.

3. a) We find the interest on $44,560 at 3.37% for one month.

$$I = P \cdot r \cdot t$$
$$= \$44,560 \times 0.0337 \times \frac{1}{12}$$
$$\approx \$125.14$$

The amount of interest in the first payment is $125.14.
We subtract to find amount applied to the principal.

$$\$437.93 - \$125.14 = \$312.79$$

With the first payment the principal will decrease by $312.79.

b) We find the interest on $44,560 at 4.75% for one month.

$$I = P \cdot r \cdot t$$
$$= \$44,560 \times 0.0475 \times \frac{1}{12}$$
$$\approx \$176.38$$

At 4.75% the additional interest in the first payment is $176.38 − $125.14 = $51.24.

c) For the 3.37% loan there will be 120 payments of $437.93:

$$120 \times \$437.93 = \$52,551.60$$

The total interest at this rate is

$$\$52,551.60 - \$44,560 = \$7991.60.$$

For the 4.75% loan there will be 120 payments of $467.20:

$$120 \times \$467.20 = \$56,064$$

The total interest at this rate is

$$\$56,064 - \$44,560 = \$11,504$$

At 4.75% Grace would pay

$$\$11,504 - \$7991.60 = \$3521.40$$

more in interest than at 3.37%.

5. a) We find the interest on $164,000 at $6\frac{1}{4}$%, or 6.25% for one month.

$$I = P \cdot r \cdot t$$
$$= \$164,000 \times 0.0625 \times \frac{1}{12}$$
$$\approx \$854.17$$

The amount applied to the principal is

$$\$1009.78 - \$854.17 = \$155.61.$$

b) The total paid will be

$$360 \times \$1009.78 = \$363,520.80.$$

Then the total amount of interest paid is

$$\$363,520.80 - \$164,000 = \$199,520.80.$$

c) We subtract to find the new principal after the first payment.

$$\$164,000 - \$155.61 = \$163,844.39$$

Now we find the interest on $163,844.39 at $6\frac{1}{4}\%$ for one month.

$$I = P \cdot r \cdot t$$

$$= \$163,844.39 \times 0.0625 \times \frac{1}{12}$$

$$\approx \$853.36$$

We subtract to find the amount applied to the principal.

$$\$1009.78 - \$853.36 = \$156.42$$

7. a) From Exercise 5(a) we know that the amount of interest in the first payment is $854.17. The amount applied to the principal is

$$\$1406.17 - \$854.17 = \$552$$

b) The total paid will be

$$180 \times \$1406.17 = \$253,110.60.$$

Then the total amount of interest paid is

$$\$253,110.60 - \$164,000 = \$89,110.60.$$

c) On the 15-yr loan the Martinez family will pay

$$\$199,520.80 - \$89,110.60 = \$110,410.20$$

less in interest than on the 30-yr loan.

9. Interest in first payment:

$$I = P \cdot r \cdot t$$

$$= \$100,000 \times 0.0698 \times \frac{1}{12}$$

$$\approx \$581.67$$

Amount of principal in first payment:

$$\$663.96 - \$581.67 = \$82.29$$

Principal after first payment:

$$\$100,000 - \$82.29 = \$99,917.71$$

Interest in second payment:

$$I = P \cdot r \cdot t$$

$$= \$99,917.71 \times 0.0698 \times \frac{1}{12}$$

$$\approx \$581.19$$

Amount of principal in second payment:

$$\$663.96 - \$581.19 = \$82.77$$

Principal after second payment:

$$\$99,917.71 - \$82.77 = \$99,834.94$$

11. Interest in first payment:

$$I = P \cdot r \cdot t$$

$$= \$100,000 \times 0.0804 \times \frac{1}{12}$$

$$\approx \$670.00$$

Amount of principal in first payment:

$$\$957.96 - \$670.00 = \$287.96$$

Principal after first payment:

$$\$100,000 - \$287.96 = \$99,712.04$$

Interest in second payment:

$$I = P \cdot r \cdot t$$

$$= \$99,712.04 \times 0.0804 \times \frac{1}{12}$$

$$\approx \$668.07$$

Amount of principal in second payment:

$$\$957.96 - \$668.07 = \$289.89$$

Principal after second payment:

$$\$99,712.04 - \$289.89 = \$99,422.15$$

13. Interest in first payment:

$$I = P \cdot r \cdot t$$

$$= \$150,000 \times 0.0724 \times \frac{1}{12}$$

$$= \$905.00$$

Amount of principal in first payment:

$$\$1022.25 - \$905.00 = \$117.25$$

Principal after first payment:

$$\$150,000 - \$117.25 = \$149,882.75$$

Interest in second payment:

$$I = P \cdot r \cdot t$$

$$= \$149,882.75 \times 0.0724 \times \frac{1}{12}$$

$$\approx \$904.29$$

Amount of principal in second payment:

$$\$1022.25 - \$904.29 = \$117.96$$

Principal after second payment:

$$\$149,882.75 - \$117.96 = \$149,764.79$$

15. Interest in first payment:

$$I = P \cdot r \cdot t$$

$$= \$200,000 \times 0.0724 \times \frac{1}{12}$$

$$\approx \$1206.67$$

Amount of principal in first payment:

$$\$1824.60 - \$1206.67 = \$617.93$$

Principal after first payment:

$$\$200,000 - \$617.93 = \$199,382.07$$

Interest in second payment:

$$I = P \cdot r \cdot t$$

$$= \$199,382.07 \times 0.0724 \times \frac{1}{12}$$

$$\approx \$1202.94$$

Amount of principal in second payment:

$$\$1824.60 - \$1202.94 = \$621.66$$

Principal after second payment:

$$\$199,382.07 - \$621.66 = \$198,760.41$$

17. a) The down payment is 10% of $23,950, or
$$0.1 \times \$23,950 = \$2395.$$
The amount borrowed is
$$\$23,950 - \$2395 = \$21,555$$
b) Interest in first payment:
$$I = P \cdot r \cdot t$$
$$= \$21,555 \times 0.029 \times \frac{1}{12}$$
$$\approx \$52.09$$
The amount of the first payment that is applied to reduce the principal is
$$\$454.06 - \$52.09 = \$401.97.$$
c) The total amount paid is
$$48 \cdot \$454.06 = \$21,794.88$$
Then the total interest paid is
$$\$21,794.88 - \$21,555 = \$239.88.$$

19. a) The down payment is 5% of $11,900:
$$0.05 \times \$11,900 = \$595$$
We subtract to find the amount borrowed:
$$\$11,900 - \$595 = \$11,305$$
b) We find the interest on $11,305 at 9.3% for one month.
$$I = P \cdot r \cdot t$$
$$= \$11,305 \times 0.093 \times \frac{1}{12}$$
$$\approx \$87.61$$
We subtract to find the amount applied to reduce the principal:
$$\$361.08 - \$87.61 = \$273.47$$
c) There will be 36 payments of $361.08:
$$36 \times \$361.08 = \$12,998.88$$
The total interest paid will be
$$\$12,998.88 - \$11,305 = \$1693.88$$

21. Discussion and Writing Exercise

23. $A = \frac{1}{2} \cdot b \cdot h$
$$= \frac{1}{2} \cdot 10 \text{ cm} \cdot 8 \text{ cm}$$
$$= \frac{10 \cdot 8}{2} \text{ cm}^2$$
$$= 40 \text{ cm}^2$$

25. $A = l \cdot w$
$$= 7 \text{ m} \cdot 3 \text{ m}$$
$$= (7 \cdot 3) \text{ m}^2$$
$$= 21 \text{ m}^2$$

27. $A = s \cdot s$
$$= 6 \text{ in.} \times 6 \text{ in.}$$
$$= (6 \times 6) \text{ in}^2$$
$$= 36 \text{ in}^2$$

29. Discussion and Writing Exercise

31. The monthly payment at $5\frac{1}{2}\%$ is $979.68 − $868.72, or $110.96, less than the payment at $6\frac{5}{8}\%$. The total savings each month is approximately $111. We can divide the cost of the refinancing by this monthly savings to determine the number of months it will take to recoup the $1200 refinancing charge: $1200 ÷ $111 ≈ 11. It will take the Sawyers approximately 11 months to recoup the refinance charge.

Chapter 8 Review Exercises

1. Move the decimal point two places to the right and write a percent symbol.
$$0.56 = 56\%$$

2. Move the decimal point two places to the right and write a percent symbol.
$$0.017 = 1.7\%$$

3. First we divide to find decimal notation.

```
      0.3 7 5
  8 | 3.0 0 0
      2 4
      ---
        6 0
        5 6
        ---
          4 0
          4 0
          ---
            0
```

$$\frac{3}{8} = 0.375$$
Now convert 0.375 to percent notation by moving the decimal point two places to the right and writing a percent symbol.
$$\frac{3}{8} = 37.5\%$$

4. First we divide to find decimal notation.

```
      0.3 3 3
  3 | 1.0 0 0
      9
      ---
      1 0
        9
      ---
      1 0
        9
      ---
        1
```

We get a repeating decimal: $\frac{1}{3} = 0.33\overline{3}$. We convert $0.33\overline{3}$ to percent notation by moving the decimal point two places to the right and writing a percent symbol.
$$\frac{1}{3} = 33.\overline{3}\%, \text{ or } 33\frac{1}{3}\%$$

5. 73.5%

a) Replace the percent symbol with ×0.01.

73.5 × 0.01

b) Move the decimal point two places to the left.

0.73.5

Thus, 73.5% = 0.735.

6. $6\frac{1}{2}\% = 6.5\%$

a) Replace the percent symbol with ×0.01.

6.5 × 0.01

b) Move the decimal point two places to the left.

0.06.5

Thus, 6.5% = 0.065.

7. $24\% = \frac{24}{100} = \frac{4 \cdot 6}{4 \cdot 25} = \frac{4}{4} \cdot \frac{6}{25} = \frac{6}{25}$

8. $6.3\% = \frac{6.3}{100} = \frac{6.3}{100} \cdot \frac{10}{10} = \frac{63}{1000}$

9. *Translate.* $30.6 = p \times 90$

Solve. We divide by 90 on both sides and convert to percent notation.

$$30.6 = p \times 90$$
$$\frac{30.6}{90} = \frac{p \times 90}{90}$$
$$0.34 = p$$
$$34\% = p$$

30.6 is 34% of 90.

10. *Translate.* $63 = 0.84 \times n$

Solve. We divide by 0.84 on both sides.

$$63 = 0.84 \times n$$
$$\frac{63}{0.84} = \frac{0.84 \times n}{0.84}$$
$$\frac{63}{0.84} = n$$
$$75 = n$$

63 is 84% of 75.

11. *Translate.* $y = 0.385 \times 168$

Solve. We multiply.

```
      1 6 8
    × 0. 3 8 5
        8 4 0
    1 3 4 4 0
    5 0 4 0 0
    6 4. 6 8 0
```

64.68 is $38\frac{1}{2}\%$ of 168.

12. 24 percent of what is 16.8?

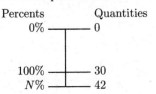

Translate: $\dfrac{24}{100} = \dfrac{16.8}{b}$

Solve: $24 \cdot b = 100 \cdot 16.8$

$$\frac{24 \cdot b}{24} = \frac{100 \cdot 16.8}{24}$$
$$b = \frac{100 \cdot 16.8}{24}$$
$$b = 70$$

24% of 70 is 16.8. The answer is 16.8.

13. 42 is what percent of 30?

Percents	Quantities
0%	0
100%	30
N%	42

Translate: $\dfrac{N}{100} = \dfrac{42}{30}$

Solve: $30 \cdot N = 100 \cdot 42$

$$\frac{30 \cdot N}{30} = \frac{100 \cdot 42}{30}$$
$$N = \frac{4200}{30}$$
$$N = 140$$

42 is 140% of 30. The answer is 140%.

14. What is 10.5% of 84?

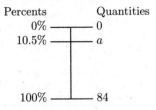

Translate: $\dfrac{10.5}{100} = \dfrac{a}{84}$

Solve: $10.5 \cdot 84 = 100 \cdot a$

$$\frac{10.5 \cdot 84}{100} = \frac{100 \cdot a}{100}$$
$$\frac{882}{100} = a$$
$$8.82 = a$$

8.82 is 10.5% of 84. The answer is 8.82.

15. Familiarize. Let c = the number of students who would choose chocolate as their favorite ice cream and b = the number who would choose butter pecan.

Translate. We translate to two equations.

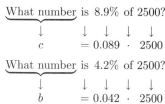

$$c = 0.089 \cdot 2500$$

$$b = 0.042 \cdot 2500$$

Solve. We multiply.

$$c = 0.089 \cdot 2500 = 222.5 \approx 223$$

$$b = 0.042 \cdot 2500 = 105$$

Check. We can repeat the calculation. We can also do partial checks by estimating.

$$0.089 \cdot 2500 \approx 0.1 \cdot 2500 = 250;$$

$$0.042 \cdot 2500 \approx 0.04 \cdot 2500 = 100$$

Since 250 is close to 223 and 100 is closer to 105, our answers seem reasonable.

State. 223 students would choose chocolate as their favorite ice cream and 105 would choose butter pecan.

16. Familiarize. Let p = the percent of people in the U.S. who take at least one kind of prescription drug per day.

Translate. We translate to a proportion.

$$\frac{p}{100} = \frac{123.64}{295}$$

Solve. We equate cross products.

$$p \cdot 295 = 100 \cdot 123.64$$

$$\frac{p \cdot 295}{295} = \frac{100 \cdot 123.64}{295}$$

$$p \approx 0.42$$

$$p \approx 42\%$$

Check. $42\% \cdot 295$ million $= 0.42 \times 295$ million $= 123.9$ million ≈ 123.64 million. The answer seems reasonable.

State. In the U.S. about 42% of the people take at least one kind of prescription drug per day.

17. Familiarize. Let w = the total output of water from the body per day.

Translate.

200 mL is 8% of what number?

$$200 = 0.08 \cdot w$$

Solve.

$$200 = 0.08 \cdot w$$

$$\frac{200}{0.08} = \frac{0.08 \cdot w}{0.08}$$

$$2500 = w$$

Check. $0.08 \cdot 2500 = 200$, so the answer checks.

State. The total output of water from the body is 2500 mL per day.

18. Familiarize. First we subtract to find the amount of the increase.

$$\begin{array}{r} {}^{7}\;{}^{14} \\ \not{8}\;\not{4} \\ -\;7\;5 \\ \hline 9 \end{array}$$

Now let p = the percent of increase.

Translate. We translate to a proportion.

$$\frac{p}{100} = \frac{9}{75}$$

Solve. We equate cross products.

$$p \cdot 75 = 100 \cdot 9$$

$$\frac{p \cdot 75}{75} = \frac{100 \cdot 9}{75}$$

$$p = 12$$

Check. $12\% \cdot 75 = 0.12 \cdot 75 = 9$, the amount of the increase, so the answer checks.

State. Jason's score increased 12%.

19. Familiarize. Let s = the new score. Note that the new score is the original score plus 15% of the original score.

New score is Original score plus 15% of Original score

$$s = 81 + 0.15 \cdot 81$$

Solve. We carry out the computation.

$$s = 81 + 0.15 \cdot 81 = 81 + 12.15 = 93.15$$

Check. We repeat the calculation. The answer checks.

State. Jenny's new score was 93.15.

20. The meals tax is

Meal tax rate $\times$ Cost of meal

$$4\frac{1}{2}\% \times \$320,$$

or $0.045 \times \$320$, or $\$14.40$.

21. Sales tax is what percent of purchase price?

$$378 = r \times 7560$$

To solve the equation, we divide on both sides by 7560.

$$\frac{378}{7560} = \frac{r \times 7560}{7560}$$

$$0.05 = r$$

$$5\% = r$$

The sales tax rate is 5%.

22. Commission $=$ Commission rate $\times$ Sales

$$753.50 = r \times 6850$$

To solve this equation, we divide on both sides by 6850.

$$\frac{753.50}{6850} = \frac{r \times 6850}{6850}$$

$$0.11 = r$$

$$11\% = r$$

The commission rate is 11%.

23. Discount = Rate of discount × Marked price
$$D \quad = \quad 12\% \quad \times \quad \$350$$

Convert 12% to decimal notation and multiply.

$$
\begin{array}{r}
3\,5\,0 \\
\times\,0.\,1\,2 \\
\hline
7\,0\,0 \\
3\,5\,0\,0 \\
\hline
4\,2.\,0\,0
\end{array}
$$

The discount is $42.

Sale price = Marked price − Discount
$$S \quad = \quad \$350 \quad - \quad \$42$$

We subtract:

$$
\begin{array}{r}
{}^{4}\;{}^{10} \\
3\,\cancel{5}\,\cancel{0} \\
-\quad 4\,2 \\
\hline
3\,0\,8
\end{array}
$$

The sale price is $308.

24. Discount = Rate of discount × Marked price
$$D \quad = \quad 14\% \quad \times \quad \$305$$

Convert 14% to decimal notation and multiply.

$$
\begin{array}{r}
3\,0\,5 \\
\times\,0.\,1\,4 \\
\hline
1\,2\,2\,0 \\
3\,0\,5\,0 \\
\hline
4\,2.\,7\,0
\end{array}
$$

The discount is $42.70.

Sale price = Marked price − Discount
$$S \quad = \quad \$305 \quad - \quad \$42.70$$

We subtract:

$$
\begin{array}{r}
{}^{2}\;{}^{10}\;{}^{4}\;{}^{10} \\
\cancel{3}\,\cancel{0}\,\cancel{5}.\,\cancel{0}\,0 \\
-\quad 4\,2.\,7\,0 \\
\hline
2\,6\,2.\,3\,0
\end{array}
$$

The sale price is $262.30.

25. Commission = Commission rate × Sales
$$C \quad = \quad 7\% \quad \times 42,000$$

We convert 7% to decimal notation and multiply.

$$
\begin{array}{r}
4\,2,\,0\,0\,0 \\
\times\quad 0.\,0\,7 \\
\hline
2\,9\,4\,0.\,0\,0
\end{array}
$$

The commission is $2940.

26. First we subtract to find the discount.

$$
\begin{array}{r}
{}^{3}\;{}^{18} \\
\cancel{4}\,\cancel{8}\,9.\,9\,9 \\
-\,3\,9\,9.\,6\,9 \\
\hline
9\,0.\,3\,0
\end{array}
$$

Discount = Rate of discount × Marked price
$$90.30 \quad = \quad r \quad \times \quad 489.99$$

We divide on both sides by 489.99.

$$\frac{90.30}{489.99} = \frac{r \times 489.99}{489.99}$$

$$0.184 \approx r$$

$$18.4\% \approx r$$

The rate of discount is about 18.4%.

27. $I = P \cdot r \cdot t$

$$= \$1800 \times 6\% \times \frac{1}{3}$$

$$= \$1800 \times 0.06 \times \frac{1}{3}$$

$$= \$36$$

28. a) $I = P \cdot r \cdot t$

$$= \$24,000 \times 10\% \times \frac{60}{365}$$

$$= \$24,000 \times 0.1 \times \frac{60}{365}$$

$$\approx \$394.52$$

b) $\$24,000 + \$394.52 = \$24,394.52$

29. $I = P \cdot r \cdot t$

$$= \$2200 \times 5.5\% \times 1$$

$$= \$2200 \times 0.055 \times 1$$

$$= \$121$$

30. $A = P \cdot \left(1 + \dfrac{r}{n}\right)^{n \cdot t}$

$$= \$7500 \cdot \left(1 + \frac{12\%}{12}\right)^{12 \cdot \frac{1}{4}}$$

$$= \$7500 \cdot \left(1 + \frac{0.12}{12}\right)^{3}$$

$$= \$7500 \cdot (1 + 0.01)^{3}$$

$$= \$7500 \cdot (1.01)^{3}$$

$$\approx \$7727.26$$

31. $A = P \cdot \left(1 + \dfrac{r}{n}\right)^{n \cdot t}$

$$= \$8000 \cdot \left(1 + \frac{9\%}{1}\right)^{1 \cdot 2}$$

$$= \$8000 \cdot (1 + 0.09)^{2}$$

$$= \$8000 \cdot (1.09)^{2}$$

$$= \$9504.80$$

32. a) 2% of $6428.74 = 0.02 × $6428.74 ≈ $129

b) $I = P \cdot r \cdot t$

$$= \$6428.74 \times 0.187 \times \frac{1}{12}$$

$$\approx \$100.18$$

The amount of interest is $100.18.

$129 − $100.18 = $28.82, so the principal is reduced by $28.82.

c) $I = P \cdot r \cdot t$

$$= \$6428.74 \times 0.132 \times \frac{1}{12}$$

$$\approx \$70.72$$

The amount of interest is $70.72.

$129 − $70.72 = $58.28, so the principal is reduced by $58.28 with the lower interest rate.

d) With the 13.2% rate the principal was decreased by $58.28 - $28.82, or $29.46, more than at the 18.7% rate. This also means that the interest at 13.2% is $29.46 less than at 18.7%.

33. *Discussion and Writing Exercise.* No; the 10% discount was based on the original price rather than on the sale price.

34. *Discussion and Writing Exercise.* A 40% discount is better. When successive discounts are taken, each is based on the previous discounted price rather than on the original price. A 20% discount followed by a 22% discount is the same as a 37.6% discount off the original price.

35. *Familiarize.* First we subtract to find the amount of the increase. We could also do this subtraction on a calculator.

$$
\begin{array}{r}
{\overset{14}{}}{\overset{13}{}} \\
2\;\overset{\;}{4}\;\overset{\;}{3}\;10\;8\;\overset{\;}{3} \\
3,\;3\;4\;0,\;9\;3\;9 \\
-\;2,\;9\;6\;3,\;6\;8\;1 \\
\hline
5\;7\;7,\;2\;5\;8
\end{array}
$$

Now let p = the percent of increase.

Translate.

577,258 is what percent of 2,963,681?

577,258 = p · 2,963,681

Solve.

$$577{,}258 = p \cdot 2{,}963{,}681$$
$$\frac{577{,}258}{2{,}963{,}681} = p$$
$$0.195 \approx p$$
$$19.5\% \approx p$$

Check. $19.5\% \cdot 2{,}963{,}681 = 0.195 \cdot 2{,}963{,}681 = 577{,}917.795 \approx 577{,}258$, so the answer seems reasonable.

State. The total land area of the United States increased about 19.5%.

36. *Familiarize.* Let d = the original price of the dress. After the 40% discount, the sale price is 60% of d, or $0.6d$. Let p = the percent by which the sale price must be increased to return to the original price.

Translate.

Sale price plus what percent of sale price is original price?

$0.6d$ + p · $0.6d$ = d

Solve.

$$0.6d + p \cdot 0.6d = d$$
$$(1+p)(0.6d) = d \quad \text{Factoring on the left}$$
$$\frac{(1+p)(0.6d)}{0.6d} = \frac{d}{0.6d}$$
$$1+p = \frac{1}{0.6} \cdot \frac{d}{d}$$
$$1+p = 1.66\overline{6}$$
$$p = 1.66\overline{6} - 1$$
$$p = 0.66\overline{6}$$
$$p = 66.\overline{6}\%, \text{ or } 66\frac{2}{3}\%$$

Check. Suppose the dress cost $100. Then the sale price is 60% of $100, or $60. Now $66\frac{2}{3}\% \cdot \$60 = \$40$ and $60 + $40 = $100, the original price. Since the answer checks for this specific price, it seems to be reasonable.

State. The sale price must be increased $66\frac{2}{3}\%$ after the sale to return to the original price.

37. The markup is $20\% \cdot \$200 = 0.2 \cdot \$200 = \$40$, and the marked up price is $200 + $40, or $240.

After 30 days:

Discount = Rate of discount × Marked price
D = 30% × $240

We convert 30% to decimal notation and multiply.

$$
\begin{array}{r}
2\;4\;0 \\
\times\;0.\;3 \\
\hline
7\;2.\;0
\end{array}
$$

Final price = Marked price − Discount
S = $240 − $72

We subtract.

$$
\begin{array}{r}
\overset{13}{} \\
1\;\overset{\;}{8}\;10 \\
2\;4\;0 \\
-\;\;\;7\;2 \\
\hline
1\;6\;8
\end{array}
$$

The final selling price was $168.

Chapter 8 Test

1. 6.4%

a) Replace the percent symbol with ×0.01.

6.4 × 0.01

b) Move the decimal point two places to the left.

0.06.4

Thus, 6.4% = 0.064.

2. 0.38

a) Move the decimal point two places to the right.

0.38.

b) Write a percent symbol: 38%

Thus, $0.38 = 38\%$.

3.

$$
\begin{array}{r}
1.3\,7\,5 \\
8\overline{\smash{)}11.0\,0\,0} \\
\underline{8} \\
3\,0 \\
\underline{2\,4} \\
6\,0 \\
\underline{5\,6} \\
4\,0 \\
\underline{4\,0} \\
0
\end{array}
$$

$\dfrac{11}{8} = 1.375$

Convert to percent notation.

1.37.5

$\dfrac{11}{8} = 137.5\%$, or $137\dfrac{1}{2}\%$

4. $65\% = \dfrac{65}{100}$ Definition of percent

$\left. \begin{aligned} &= \dfrac{5 \cdot 13}{5 \cdot 20} \\ &= \dfrac{5}{5} \cdot \dfrac{13}{20} \\ &= \dfrac{13}{20} \end{aligned} \right\}$ Simplifying

5. Translate: What is 40% of 55?

$$a = 0.4 \cdot 55$$

Solve: We multiply.

$a = 0.4 \cdot 55 = 22$

The answer is 22.

6. What percent of 80 is 65?

Percents	Quantities
0%	0
N%	65
100%	80

Translate: $\dfrac{N}{100} = \dfrac{65}{80}$

Solve: $80 \cdot N = 100 \cdot 65$

$\dfrac{80 \cdot N}{80} = \dfrac{100 \cdot 65}{80}$

$N = \dfrac{6500}{80}$

$N = 81.25$

The answer is 81.25%.

7. *Familiarize.* Let $x =$ the number of passengers in the 25-34 age group. Let $y =$ the number of passengers in the 35-44 age group.

Translate. We will translate to two equations.

What number is 16% of 2500?

$$x = 16\% \cdot 2500$$

What number is 23% of 2500?

$$y = 23\% \cdot 2500$$

Solve. To solve each equation we convert percent notation to decimal notation and multiply.

$$x = 16\% \cdot 2500 = 0.16 \cdot 2500 = 400$$

$$y = 23\% \cdot 2500 = 0.23 \cdot 2500 = 575$$

Check. We repeat the calculations. The answers check.

State. There are 400 passengers in the 25-34 age group and 575 passengers in the 35-44 age group.

8. *Familiarize.* First we find the amount of the increase.

$$
\begin{array}{r}
2,5\,4\,7,\,3\,8\,9 \\
-\ 1,7\,2\,2,\,8\,5\,0 \\
\hline
8\,2\,4,\,5\,3\,9
\end{array}
$$

Translate. We translate to a proportion.

$\dfrac{N}{100} = \dfrac{824,539}{1,722,850}$

Solve. We begin by equating cross products.

$$N \times 1,722,850 = 100 \times 824,539$$

$$\dfrac{N \times 1,722,850}{1,722,850} = \dfrac{100 \times 824,539}{1,722,850}$$

$$N = \dfrac{82,453,900}{1,722,850}$$

$$N \approx 47.9$$

Check. We find 47.9% of 1,722,850.

$$0.479 \times 1,722,850 = 825,245.15 \approx 824,539$$

Since we rounded the percent, we have a good check.

State. The population increased about 47.9%.

9. *Familiarize.* We first find the amount of decrease, in billions of dollars.

$$
\begin{array}{r}
5.5 \\
-\ 2.7 \\
\hline
2.8
\end{array}
$$

Let $p =$ the percent of decrease.

Translate. We translate to an equation.

2.8 is what percent of 5.5?

$$2.8 = p \cdot 5.5$$

Solve.

$$2.8 = p \cdot 5.5$$

$$\frac{2.8}{5.5} = \frac{p \cdot 5.5}{5.5}$$

$$0.50\overline{90} = p$$

$$50.\overline{90}\% = p$$

Check. Note that $50.\overline{90}\% \approx 51\%$. With a decrease of approximately 51%, the profit in 2000 should be about $100\% - 51\%$, or 49%, of the profit in 1999. Since $49\% \cdot 5.5 = 0.49 \cdot 5.5 = 2.695 \approx 2.7$, the answer checks.

State. The percent of decrease was $50.\overline{90}\%$.

10. **Familiarize.** Let p = the percent of people who have ever lived who are alive today. Note that the population numbers are given in billions.

Translate. We translate to an equation.

6.6 is what percent of 120?

$$6.6 = p \cdot 120$$

Solve.

$$6.6 = p \cdot 120$$

$$\frac{6.6}{120} = \frac{p \cdot 120}{120}$$

$$0.055 = p$$

$$5.5\% = p$$

Check. We find 5.5% of 120:

$$5.5\% \cdot 120 = 0.055 \cdot 120 = 6.6$$

The answer checks.

State. 5.5% of the people who have ever lived are alive today.

11. The sales tax on an item costing $324 is

Sales tax rate × Purchase price

$$5\% \times \$324,$$

or 0.05×324, or 16.2. Thus the tax is $16.20.

The total price is given by the purchase price plus the sales tax:

$$\$324 + \$16.20 = \$340.20$$

12. Commission = Commission rate × Sales

$$C = 15\% \times 4200$$
$$C = 0.15 \times 4200$$
$$C = 630$$

The commission is $630.

13. Discount = Rate of discount × Marked price

$$D = 20\% \times \$200$$

Convert 20% to decimal notation and multiply.

$$\begin{array}{r} 2\,0\,0 \\ \times \ \ 0.\,2 \\ \hline 4\,0.\,0 \end{array} \qquad (20\% = 0.20 = 0.2)$$

The discount is $40.

Sale price = Marked price − Discount

$$S = 200 - 40$$

We subtract:

$$\begin{array}{r} 2\,0\,0 \\ - \ \ 4\,0 \\ \hline 1\,6\,0 \end{array}$$

To check, note that the sale price is 80% of the marked price: $0.8 \times 200 = 160$.

The sale price is $160.

14. $I = P \cdot r \cdot t = \$120 \times 7.1\% \times 1$

$$= \$120 \times 0.071 \times 1$$

$$= \$8.52$$

15. $I = P \cdot r \cdot t = \$5200 \times 6\% \times \dfrac{1}{2}$

$$= \$5200 \times 0.06 \times \frac{1}{2}$$

$$= \$312 \times \frac{1}{2}$$

$$= \$156$$

The interest earned is $156. The amount in the account is the principal plus the interest: $\$5200 + \$156 = \$5356$.

16. $A = P \cdot \left(1 + \dfrac{r}{n}\right)^{n \cdot t}$

$$= \$1000 \cdot \left(1 + \frac{5\frac{3}{8}\%}{1}\right)^{1 \cdot 2}$$

$$= \$1000 \cdot \left(1 + \frac{0.05375}{1}\right)^{2}$$

$$= \$1000(1.05375)^2$$

$$\approx \$1110.39$$

17. $A = P \cdot \left(1 + \dfrac{r}{n}\right)^{n \cdot t}$

$$= \$10,000 \cdot \left(1 + \frac{4.9\%}{12}\right)^{12 \cdot 3}$$

$$= \$10,000 \cdot \left(1 + \frac{0.049}{12}\right)^{36}$$

$$\approx \$11,580.07$$

18. **Registered nurses:** We add to find the number of jobs in 2012; $2.3 + 0.6 = 2.9$, so we project that there will be 2.9 million jobs for registered nurses in 2012. We solve an equation to find the percent of increase, p.

0.6 is what percent of 2.3?

$$0.6 = p \cdot 2.3$$

Solve.

$$\frac{0.6}{2.3} = \frac{p \cdot 2.3}{2.3}$$

$$0.261 \approx p$$

$$26.1\% \approx p$$

Post-secondary teachers: We subtract to find the change; $2.2 - 1.6 = 0.6$, so we project that the change will be 0.6 million. We solve an equation to find the percent of increase, p.

$$0.6 \text{ is } \underbrace{\text{what percent}} \text{ of } 1.6?$$

$$\downarrow \ \downarrow \qquad \downarrow \qquad \downarrow \ \downarrow$$

$$0.6 \ = \qquad p \qquad \cdot \ 1.6$$

Solve.

$$\frac{0.6}{1.6} = \frac{p \cdot 1.6}{1.6}$$

$$0.375 = p$$

$$37.5\% = p$$

Food preparation and service workers: We subtract to find the number of jobs in 2002; $2.4 - 0.4 = 2.0$, so there were 2.0 million jobs for food preparation and service workers in 2002. We solve an equation to find the percent of increase, p.

$$0.4 \text{ is } \underbrace{\text{what percent}} \text{ of } 2.0?$$

$$\downarrow \ \downarrow \qquad \downarrow \qquad \downarrow \ \downarrow$$

$$0.4 \ = \qquad p \qquad \cdot \ 2.0$$

Solve.

$$\frac{0.4}{2.0} = \frac{p \cdot 2.0}{2.0}$$

$$0.2 = p$$

$$20\% = p$$

Restaurant servers: Let $n =$ the number of jobs in 2002, in millions. If the number of jobs increases by 19.0%, then the new number of jobs is 100% of $n + 19\%$ of n, or 119% of n, or $1.19 \cdot n$. We solve an equation to find n.

$$\underbrace{\text{The number of jobs in 2002 increased by 19\%}} \text{ is } \underbrace{2.5 \text{ million}}$$

$$\downarrow \qquad\qquad \downarrow \qquad \downarrow$$

$$1.19 \cdot n \qquad = \qquad 2.5$$

Solve.

$$\frac{1.19 \cdot n}{1.19} = \frac{2.5}{1.19}$$

$$n \approx 2.1$$

There were 2.1 million jobs for restaurant servers in 2002.

We subtract to find the change; $2.5 - 2.1 = 0.4$, so we project that the change will be 0.4 million.

19. Discount = Marked price − Sale price

$$D \qquad = \qquad 1950 \qquad - \quad 1675$$

We subtract: $1\,9\,5\,0$
                  $-\,1\,6\,7\,5$
                  $\overline{\ \ \ 2\,7\,5}$

The discount is $275.

Discount = Rate of discount × Marked price

$$275 \quad = \qquad R \qquad \times \quad 1950$$

To solve the equation we divide on both sides by 1950.

$$\frac{275}{1950} = \frac{R \times 1950}{1950}$$

$$0.141 \approx R$$

$$14.1\% \approx R$$

To check, note that a discount rate of 14.1% means that 85.9% of the marked price is paid: $0.859 \times 1950 = 1675.05 \approx 1675$. Since that is the sale price, the answer checks.

The rate of discount is about 14.1%.

20. To find the principal after the first payment, we first use the formula $I = P \cdot r \cdot t$ to find the amount of interest paid in the first payment.

$$I = P \cdot r \cdot t = \$120{,}000 \cdot 0.074 \cdot \frac{1}{12} \approx \$740.$$

Then the amount of the principal applied to the first payment is

$$\$830.86 - \$740 = \$90.86.$$

Finally, we find that the principal after the first payment is

$$\$120{,}000 - \$90.86 = \$119{,}909.14.$$

To find the principal after the second payment, we first use the formula $I = P \cdot r \cdot t$ to find the amount of interest paid in the second payment.

$$I = \$119{,}909.14 \cdot 0.074 \cdot \frac{1}{12} \approx \$739.44$$

Then the amount of the principal applied to the second payment is

$$\$830.86 - \$739.44 = \$91.42.$$

Finally, we find that the principal after the second payment is

$$\$119{,}909.14 - \$91.42 = \$119{,}817.72.$$

21. *Familiarize.* Let $p =$ the price for which a realtor would have to sell the house in order for Juan and Marie to receive $180,000 from the sale. The realtor's commission would be $7.5\% \cdot p$, or $0.075 \cdot p$, and Juan and Marie would receive 100% of $p - 7.5\%$ of p, or 92.5% of p, or $0.925 \cdot p$.

Translate.

$$\underbrace{\text{Amount Juan and Marie receive}} \text{ is } \$180{,}000$$

$$\downarrow \qquad\qquad \downarrow \quad \downarrow$$

$$0.925 \cdot p \qquad = \quad 180{,}000$$

Solve.

$$\frac{0.925 \cdot p}{0.925} = \frac{180{,}000}{0.925}$$

$$p \approx 194{,}600 \quad \text{Rounding to the nearest hundred}$$

Check. 7.5% of $194,600 = 0.075 \cdot \$194{,}600 = \$14{,}595$ and $\$194{,}600 - \$14{,}595 = \$180{,}005 \approx \$180{,}000$. The answer checks.

State. A realtor would need to sell the house for about $194,600.

22. First we find the commission.

Commission = Commission rate ×   Sales

$$C \quad = \qquad 16\% \qquad\quad \times \ \$15{,}000$$

$$C \quad = \qquad 0.16 \qquad\quad \times \ \$15{,}000$$

$$C \quad = \qquad \$2400$$

Now we find the amount in the account after 6 months.

$$A = P \cdot \left(1 + \frac{r}{n}\right)^{n \cdot t}$$

$$= \$2400 \cdot \left(1 + \frac{12\%}{4}\right)^{4 \cdot \frac{1}{2}}$$

$$= \$2400 \cdot \left(1 + \frac{0.12}{4}\right)^{2}$$

$$= \$2400 \cdot (1 + 0.03)^{2}$$

$$= \$2400 \cdot (1.03)^{2}$$

$$= \$2400(1.0609)$$

$$= \$2546.16$$

Cumulative Review Chapters 1 - 8

1. 0.0<u>91</u>   0.091. $\frac{91}{1000}$

            $\llcorner\!\!\uparrow$

3 places   Move 3 places.   3 zeros

$$0.091 = \frac{91}{1000}$$

2. $\frac{13}{6} = 13 \div 6$

```
        2.1 6 6
  6 | 1 3.0 0 0
      1 2
      ──
        1 0
          6
        ──
          4 0
          3 6
          ──
            4 0
            3 6
            ──
              4
```

Since 4 keeps reappearing as a remainder, the digits repeat and

$$\frac{13}{6} = 2.166\ldots, \text{ or } 2.1\overline{6}.$$

3. 3%

a) Replace the percent symbol with ×0.01.

     3×0.01

b) Move the decimal point two places to the left.

     $3\% = 0.03$

4. First divide to find decimal notation.

```
        1.1 2 5
  8 | 9.0 0 0
      8
      ──
      1 0
        8
      ──
        2 0
        1 6
        ──
          4 0
          4 0
          ──
            0
```

$$\frac{9}{8} = 1.125$$

Convert to percent notation by moving the decimal point two places to the right and writing a percent symbol.

$$\frac{9}{8} = 112.5\%$$

5. $\frac{0.6}{8} = \frac{0.6}{8} \cdot \frac{10}{10} = \frac{6}{80} = \frac{2 \cdot 3}{2 \cdot 40} = \frac{2}{2} \cdot \frac{3}{40} = \frac{3}{40}$

6. $\frac{350 \text{ km}}{15 \text{ hr}} = \frac{350}{15} \cdot \frac{\text{km}}{\text{hr}} = \frac{5 \cdot 70}{3 \cdot 5} \cdot \frac{\text{km}}{\text{hr}} = \frac{5}{5} \cdot \frac{70}{3} \cdot \frac{\text{km}}{\text{hr}} = $

$\frac{70}{3} \text{ km/h, or } 23\frac{1}{3} \text{ km/h}$

7. The LCD is 56.

$$\frac{5}{7} \cdot \frac{8}{8} = \frac{40}{56}$$

$$\frac{6}{8} \cdot \frac{7}{7} = \frac{42}{56}$$

Since $40 < 42$, it follows that $\frac{40}{56} < \frac{42}{56}$, so $\frac{5}{7} < \frac{6}{8}$.

8. -3.78 lies to the right of -37.8 on the number line, so we have $-3.78 > -37.8$.

9. $263,961 + 32,090 + 127.89 \approx 264,000 + 32,100 + 100 = 296,200$

10. $73,510 - 23,450 \approx 73,500 - 23,500 = 50,000$

11. $46 - [4(6 + 4 \div 2) + 2 \times 3 - 5]$

$$= 46 - [4(6 + 2) + 2 \times 3 - 5]$$

$$= 46 - [4(8) + 2 \times 3 - 5]$$

$$= 46 - [32 + 6 - 5]$$

$$= 46 - [38 - 5]$$

$$= 46 - 33$$

$$= 13$$

12. $5x - 9 - 7x - 5 = 5x - 7x - 9 - 5$

$$= (5 - 7)x + (-9 - 5)$$

$$= -2x + (-14)$$

$$= -2x - 14$$

13. $\frac{6}{5} + 1\frac{5}{6} = \frac{6}{5} + \frac{11}{6}$

$$= \frac{6}{5} \cdot \frac{6}{6} + \frac{11}{6} \cdot \frac{5}{5}$$

$$= \frac{36}{30} + \frac{55}{30}$$

$$= \frac{91}{30}, \text{ or } 3\frac{1}{30}$$

14. $-46.9 + 32.7$

The absolute values are 46.9 and 32.7. We find the difference.

```
    4 6. 9
  - 3 2. 7
  ───────
    1 4. 2
```

The negative number has the larger absolute value so the answer is negative.

$$-46.9 + 32.7 = -14.2$$

15.
$$\begin{array}{r} \overset{1\ 1\ \ 1\ 1\ 1}{4\ 8\ 7,0\ 9\ 4} \\ 6,9\ 3\ 6 \\ +\ \ \ 2\ 1,1\ 2\ 0 \\ \hline 5\ 1\ 5,1\ 5\ 0 \end{array}$$

16.
$$\begin{array}{r} \overset{4\ \ 9\ 10}{3\ 5.\cancel{0}\ \cancel{0}} \\ -\ 3\ 4.\ 9\ 8 \\ \hline 0.\ 0\ 2 \end{array}$$

17. $3\frac{1}{3} - 2\frac{2}{3} = 2\frac{4}{3} - 2\frac{2}{3} = \frac{2}{3}$

18. $-\dfrac{8}{9} - \dfrac{6}{7} = -\dfrac{8}{9} \cdot \dfrac{7}{7} - \dfrac{6}{7} \cdot \dfrac{9}{9}$

$\qquad = -\dfrac{56}{63} - \dfrac{54}{63}$

$\qquad = -\dfrac{110}{63},\ \text{or}\ -1\dfrac{47}{63}$

19. $\dfrac{7}{9} \cdot \dfrac{3}{14} = \dfrac{7\cdot 3}{9\cdot 14} = \dfrac{7\cdot 3\cdot 1}{3\cdot 3\cdot 2\cdot 7} = \dfrac{3\cdot 7}{3\cdot 7}\cdot \dfrac{1}{3\cdot 2} = \dfrac{1}{3\cdot 2} = \dfrac{1}{6}$

20. $(-32)(-4)(-3) = 128(-3) = -384$

21.
$$\begin{array}{r} 4\ 6.\ 0\ 1\ 2 \\ \times\ \ \ \ 0.\ 0\ 3 \\ \hline 1.\ 3\ 8\ 0\ 3\ 6 \end{array}$$

22. $6\dfrac{3}{5} \div 4\dfrac{2}{5} = \dfrac{33}{5} \div \dfrac{22}{5}$

$\qquad = \dfrac{33}{5} \div \dfrac{5}{22}$

$\qquad = \dfrac{33\cdot 5}{5\cdot 22}$

$\qquad = \dfrac{3\cdot 11\cdot 5}{5\cdot 2\cdot 11}$

$\qquad = \dfrac{5\cdot 11}{5\cdot 11}\cdot \dfrac{3}{2}$

$\qquad = \dfrac{3}{2} = 1\dfrac{1}{2}$

23. First we find $431.2 \div 35.2$.
$$\begin{array}{r} 1\ 2.2\ 5 \\ 3\ 5.2_{\wedge}\overline{\smash{)}\ 4\ 3\ 1.\ 2_{\wedge}0\ 0} \\ \underline{3\ 5\ 2\ 0} \\ 7\ 9\ 2 \\ \underline{7\ 0\ 4} \\ 8\ 8\ 0 \\ \underline{7\ 0\ 4} \\ 1\ 7\ 6\ 0 \\ \underline{1\ 7\ 6\ 0} \\ 0 \end{array}$$

Then $431.2 \div (-35.2) = -12.25$.

24.
$$\begin{array}{r} 1\ 2\ 3 \\ 1\ 5\overline{\smash{)}\ 1\ 8\ 5\ 0} \\ \underline{1\ 5\ 0\ 0} \\ 3\ 5\ 0 \\ \underline{3\ 0\ 0} \\ 5\ 0 \\ \underline{4\ 5} \\ 5 \end{array}$$

The answer is 123 R 5.

25. $18 \cdot x = 1710$

$\qquad \dfrac{18 \cdot x}{18} = \dfrac{1710}{18}$

$\qquad\qquad x = 95$

The solution is 95.

26. $\qquad y + 142.87 = 151$

$\qquad y + 142.87 - 142.87 = 151 - 142.87$

$\qquad\qquad\qquad\qquad y = 8.13$

The solution is 8.13.

27. $\qquad \dfrac{3}{7}x - 5 = 16$

$\qquad \dfrac{3}{7}x - 5 + 5 = 16 + 5$

$\qquad\qquad \dfrac{3}{7}x = 21$

$\qquad \dfrac{7}{3} \cdot \dfrac{3}{7}x = \dfrac{7}{3} \cdot 21$

$\qquad\qquad x = \dfrac{7 \cdot 21}{3} = \dfrac{7 \cdot \cancel{3} \cdot 7}{\cancel{3} \cdot 1}$

$\qquad\qquad x = 49$

The solution is 49.

28. $\qquad \dfrac{3}{4} + x = \dfrac{5}{6}$

$\qquad \dfrac{3}{4} + x - \dfrac{3}{4} = \dfrac{5}{6} - \dfrac{3}{4}$

$\qquad\qquad x = \dfrac{5}{6} \cdot \dfrac{2}{2} - \dfrac{3}{4} \cdot \dfrac{3}{3}$

$\qquad\qquad x = \dfrac{10}{12} - \dfrac{9}{12}$

$\qquad\qquad x = \dfrac{1}{12}$

The solution is $\dfrac{1}{12}$.

29. $\quad 3(x - 7) + 2 = 12x - 3$

$\qquad 3x - 21 + 2 = 12x - 3$

$\qquad\quad 3x - 19 = 12x - 3$

$\quad 3x - 19 - 3x = 12x - 3 - 3x$

$\qquad\qquad -19 = 9x - 3$

$\qquad -19 + 3 = 9x - 3 + 3$

$\qquad\qquad -16 = 9x$

$\qquad\quad -\dfrac{16}{9} = \dfrac{9x}{9}$

$\qquad\quad -\dfrac{16}{9} = x,\ \text{or}$

$\qquad\quad -1\dfrac{7}{9} = x$

The solution is $-1\dfrac{7}{9}$.

30. $\dfrac{16}{n} = \dfrac{21}{11}$

$16 \cdot 11 = n \cdot 21$ Equating cross products

$\dfrac{16 \cdot 11}{21} = \dfrac{n \cdot 21}{21}$

$\dfrac{176}{21} = n$, or

$8\dfrac{8}{21} = n$

The solution is $\dfrac{176}{21}$, or $8\dfrac{8}{21}$.

31. Both coordinates are negative, so $(-3, -1)$ lies in quadrant III.

32. Graph $y = -\dfrac{3}{5}x$.

We make a table of solutions, plot the points, and draw and label the line. Note that, by selecting multiples of 5 for x-values, we avoid fraction values for y.

When $x = -5$, $y = -\dfrac{3}{5}(-5) = 3$.

When $x = 0$, $y = -\dfrac{3}{5} \cdot 0 = 0$.

When $x = 5$, $y = -\dfrac{3}{5} \cdot 5 = -3$.

x	$y = -\dfrac{3}{5}x$	(x, y)
-5	3	$(-5, 3)$
0	0	$(0, 0)$
5	-3	$(5, -3)$

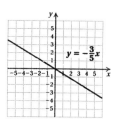

33. $\dfrac{19 + 29 + 34 + 39 + 45}{5} = \dfrac{166}{5} = 33.2$.

34. $7, 7, 12, 15, 21$
$\uparrow$
Middle number

The median is 12.

35. $P = 4s$

$ = 4 \cdot 15$ in.

$ = 60$ in.

36. $A = l \cdot w$

$ = 90 \text{ yd} \cdot 40 \text{ yd}$

$ = 90 \cdot 40 \cdot \text{yd} \cdot \text{yd}$

$ = 3600 \text{ yd}^2$

37. $\dfrac{\$3.60}{12 \text{ oz}} = \dfrac{360\cent}{12 \text{ oz}} = 30\cent/\text{oz}$

38. Familiarize. Let $d =$ the distance the bus would travel in 8 hr, in km.

Translate. We translate to a proportion.

$\begin{array}{r} \text{Distance} \to \\ \text{Time} \to \end{array} \dfrac{456}{6} = \dfrac{d}{8} \begin{array}{l} \leftarrow \text{Distance} \\ \leftarrow \text{Time} \end{array}$

Solve. We begin by equating cross products.

$456 \cdot 8 = 6 \cdot d$

$\dfrac{456 \cdot 8}{6} = d$

$608 = d$

Check. We substitute into the proportion and check cross products.

$\dfrac{456}{6} = \dfrac{608}{8}$

$456 \cdot 8 = 3648$; $6 \cdot 608 = 3648$

The cross products are the same, so the answer checks.

State. The bus could travel 608 km in 8 hr.

39. Familiarize. Note that the increase was $117 - 103$, or 14 yen. Let $p =$ the percent of increase.

Translate.

14 is what percent of 103?

$14 = p \cdot 103$

Solve. We divide both sides by 103.

$14 = p \cdot 103$

$\dfrac{14}{103} = \dfrac{p \cdot 103}{103}$

$0.136 \approx p$

$13.6\% \approx p$

Check. We find 13.6% of 103.

$0.136 \times 103 = 14.008 \approx 14$

Since we rounded the percent, this approximation is close enough to be a good check.

State. The percent of increase was about 13.6%.

40. Familiarize. Let $t =$ the number of telephone calls and $v =$ the number of voice mails received.

Translate. We translate to two equations.

What number is 27.4% of 150?

$t = 0.274 \times 150$

What number is 11.6% of 150?

$v = 0.116 \times 150$

Solve. We carry out the multiplications.

$t = 0.274 \times 150 = 41.1 \approx 41$

$v = 0.116 \times 150 = 17.4 \approx 17$

Check. We can repeat the calculations. The answers check.

State. Also 41 communications will be telephone calls, and about 17 will be voice mail.

41. *Familiarize.* Let $p =$ the number of pieces of ribbon that can be cut.

Translate.

$$\underbrace{\text{Length of a piece}}_{1\frac{4}{5}} \; \underbrace{\text{times}}_{\cdot} \; \underbrace{\text{Number of pieces}}_{p} \; \underbrace{\text{is}}_{=} \; \underbrace{\text{Total length}}_{9}$$

Solve.

$$1\frac{4}{5} \cdot p = 9$$
$$\frac{9}{5} \cdot p = 9$$
$$p = 9 \div \frac{9}{5}$$
$$p = 9 \cdot \frac{5}{9} = \frac{9 \cdot 5}{9 \cdot 1}$$
$$p = \frac{9}{9} \cdot \frac{5}{1}$$
$$p = 5$$

Check. $1\frac{4}{5} \cdot 5 = \frac{9}{5} \cdot 5 = 9$, so the answer checks.

State. 5 pieces can be cut.

42. *Familiarize.* Let $d =$ the total distance Bobbie walked, in km.

Translate.

$$\underbrace{\text{Distance to school}}_{\frac{7}{10}} \; \underbrace{\text{plus}}_{+} \; \underbrace{\text{Distance to library}}_{\frac{8}{10}} \; \underbrace{\text{is}}_{=} \; \underbrace{\text{Total distance}}_{d}$$

Solve. We carry out the addition.
$$\frac{7}{10} + \frac{8}{10} = \frac{15}{10} = \frac{3 \cdot 5}{2 \cdot 5} = \frac{3}{2} \cdot \frac{5}{5} = \frac{3}{2} = 1\frac{1}{2}$$
Check. We repeat the calculation. The answer checks.

State. Bobbie walked $1\frac{1}{2}$ km.

43. *Familiarize.* First we find the gas mileage in the mountains.
$$\frac{240 \text{ mi}}{7\frac{1}{2} \text{ gal}} = \frac{240 \text{ mi}}{\frac{15}{2} \text{ gal}} = 240 \cdot \frac{2}{15} \text{ mpg}$$
$$= \frac{240 \cdot 2}{15} \text{ mpg}$$
$$= \frac{\cancel{15} \cdot 16 \cdot 2}{\cancel{15} \cdot 1} \text{ mpg}$$
$$= 32 \text{ mi/gal}$$

When the car left the mountains for the plains, the mileage increased by $36 - 32$, or 4 mpg. Let $p =$ the percent of increase.

Translate.

$$\underbrace{4}_{4} \; \text{is} \underbrace{\text{what percent}}_{} \text{of } 32?$$
$$4 = \quad p \quad \cdot 32$$

Solve.
$$4 = p \cdot 32$$
$$\frac{4}{32} = \frac{p \cdot 32}{32}$$
$$0.125 = p$$
$$12.5\% = p$$
Check. 12.5% of 32 is $0.125 \cdot 32 = 4$. The answer checks.

State. The gas mileage increased 12.5%.

44. The car traveled $36 \cdot 5$, or 180 mi, across the plains. Then the total distance traveled was $240 + 180$, or 420 mi, and the amount of gas used was $7\frac{1}{2} + 5$, or $12\frac{1}{2}$ gal.

$$\frac{420 \text{ mi}}{12\frac{1}{2} \text{ gal}} = \frac{420 \text{ mi}}{\frac{25}{2} \text{ gal}} = 420 \cdot \frac{2}{25} \text{ mpg}$$
$$= \frac{420 \cdot 2}{25} \text{ mpg}$$
$$= 33.6 \text{ mpg}$$

45. a) Lot 1 pays $\frac{1}{5} \cdot \frac{1}{5}$, or $\frac{1}{25}$ of the cost.

Lot 2 pays $\frac{1}{25} + \frac{1}{4} \cdot \frac{1}{5}$, or $\frac{1}{25} + \frac{1}{20}$, or $\frac{4}{100} + \frac{5}{100}$, or $\frac{9}{100}$ of the cost.

Lot 3 pays $\frac{9}{100} + \frac{1}{3} \cdot \frac{1}{5}$, or $\frac{9}{100} + \frac{1}{15}$, or $\frac{27}{300} + \frac{20}{300}$, or $\frac{47}{300}$ of the cost.

Lot 4 pays $\frac{47}{300} + \frac{1}{2} \cdot \frac{1}{5}$, or $\frac{47}{300} + \frac{1}{10}$, or $\frac{47}{300} + \frac{30}{300}$, or $\frac{77}{300}$ of the cost.

Lot 5 pays $\frac{77}{300} + \frac{1}{5}$, or $\frac{77}{300} + \frac{60}{300}$, or $\frac{137}{300}$ of the cost.

b) Lot 1: $\frac{1}{25} = 4\%$

Lot 2: $\frac{9}{100} = 9\%$

Lot 3: $\frac{47}{300} = 0.15\overline{6} = 15.\overline{6}\%$, or $15\frac{2}{3}\%$

Lot 4: $\frac{77}{300} = 0.25\overline{6} = 25.\overline{6}\%$, or $25\frac{2}{3}\%$

Lot 5: $\frac{137}{300} = 0.45\overline{6} = 45.\overline{6}\%$, or $45\frac{2}{3}\%$

c) Total cost paid on lots 3, 4, and 5:
$$\frac{47}{300} + \frac{77}{300} + \frac{137}{300} = \frac{261}{300} = 0.87 = 87\%$$

46. After each 10% discount, the new price is 90% of the previous price. Let $p =$ the original price.

Price after first discount: $0.9p$

Price after second discount: $0.9(0.9p) = 0.81p$

Price after third discount: $0.9(0.81p) = 0.729p$

Price after fourth discount: $0.9(0.729p) = 0.6561p$

Price after fifth discount: $0.9(0.6561p) = 0.59049p$

Price after sixth discount: $0.9(0.59049p) = 0.531441p$

Price after seventh discount: $0.9(0.531441p) = 0.4782969$

It takes 7 successive 10% discounts to lower the price of an item to below 50% of its original price.

Chapter 9

Geometry and Measurement

1. 1 yd = 3 ft

This is the relation stated on page 586 of the text.

3. $1 \text{ in.} = 1 \text{ in.} \times \dfrac{1 \text{ ft}}{12 \text{ in.}}$ Multiplying by 1 using

$\dfrac{1 \text{ ft}}{12 \text{ in.}}$ to eliminate in.

$= \dfrac{1 \text{ in.}}{12 \text{ in.}} \times 1 \text{ ft}$

$= \dfrac{1}{12} \times \dfrac{\text{in.}}{\text{in.}} \times 1 \text{ ft}$

$= \dfrac{1}{12} \times 1 \text{ ft}$ The $\dfrac{\text{in.}}{\text{in.}}$ acts like 1,

so we can omit it.

$= \dfrac{1}{12} \text{ ft}$

5. 1 mi = 5280 ft   See page 586 of the text.

$= 5280 \text{ ft} \times \dfrac{1 \text{ yd}}{3 \text{ ft}}$ Multiplying by 1 using $\dfrac{1 \text{ yd}}{3 \text{ ft}}$

$= \dfrac{5280}{3} \times 1 \text{ yd}$

$= 1760 \times 1 \text{ yd}$

$= 1760 \text{ yd}$

7. 3 yd = 3 × 1 yd

$= 3 \times 3 \text{ ft}$ Substituting 3 ft for 1 yd

$= 9 \text{ ft}$ Multiplying

9. $84 \text{ in.} = \dfrac{84 \text{ in.}}{1} \times \dfrac{1 \text{ ft}}{12 \text{ in.}}$ Multiplying by 1 using

$\dfrac{1 \text{ ft}}{12 \text{ in.}}$

$= \dfrac{84}{12} \times 1 \text{ ft}$

$= 7 \times 1 \text{ ft}$

$= 7 \text{ ft}$

11. $48 \text{ ft} = 48 \text{ in.} \times \dfrac{1 \text{ yd}}{3 \text{ ft}}$ Multiplying by 1 using

$\dfrac{1 \text{ yd}}{3 \text{ ft}}$

$= \dfrac{48}{3} \times 1 \text{ yd}$

$= 16 \times 1 \text{ yd}$

$= 16 \text{ yd}$

13. 5 mi = 5 × 1 mi

$= 5 \times 5280 \text{ ft}$ Substituting 5280 ft for 1 mi

$= 5 \times 5280 \text{ ft} \times \dfrac{1 \text{ yd}}{3 \text{ ft}}$ Multiplying by 1 using

$\dfrac{1 \text{ yd}}{3 \text{ ft}}$

$= \dfrac{26,400}{3} \times 1 \text{ yd}$

$= 8800 \times 1 \text{ yd}$

$= 8800 \text{ yd}$

15. $48 \text{ in.} = \dfrac{48 \text{ in.}}{1} \times \dfrac{1 \text{ ft}}{12 \text{ in.}}$ Multiplying by 1 using

$\dfrac{1 \text{ ft}}{12 \text{ in.}}$

$= \dfrac{48}{12} \times 1 \text{ ft}$

$= 4 \times 1 \text{ ft}$

$= 4 \text{ ft}$

17. $11,616 \text{ ft} = 11,616 \text{ ft} \times \dfrac{1 \text{ mi}}{5280 \text{ ft}}$ Multiplying by 1

using $\dfrac{1 \text{ mi}}{5280 \text{ ft}}$

$= \dfrac{11,616}{5280} \times 1 \text{ mi}$

$= 2.2 \times 1 \text{ mi}$

$= 2.2 \text{ mi}$

19. $15,840 \text{ ft} = 15,840 \text{ ft} \times \dfrac{1 \text{ mi}}{5280 \text{ ft}}$ Multiplying by 1

using $\dfrac{1 \text{ mi}}{5280 \text{ ft}}$

$= \dfrac{15,840}{5280} \times 1 \text{ mi}$

$= 3 \times 1 \text{ mi}$

$= 3 \text{ mi}$

21. $7\dfrac{1}{2} \text{ ft} = 7\dfrac{1}{2} \text{ ft} \times \dfrac{1 \text{ yd}}{3 \text{ ft}}$

$= \dfrac{15}{2} \text{ ft} \times \dfrac{1 \text{ yd}}{3 \text{ ft}}$

$= \dfrac{15}{6} \times 1 \text{ yd}$

$= \dfrac{5}{2} \times 1 \text{ yd}$

$= \dfrac{5}{2} \text{ yd, or } 2\dfrac{1}{2} \text{ yd}$

23. $36 \text{ in.} = 36 \text{ in.} \times \dfrac{1 \text{ ft}}{12 \text{ in.}}$

$= \dfrac{36}{12} \times 1 \text{ ft}$

$= 3 \times 1 \text{ ft}$

$= 3 \text{ ft}$

25. 1760 yd = 1760 × 1 yd

   = 1760 × 3 ft

   $= 1760 \times 3 \text{ ft} \times \dfrac{1 \text{ mi}}{5280 \text{ ft}}$

   $= \dfrac{5280}{5280} \times 1 \text{ mi}$

   = 1 × 1 mi

   = 1 mi

27. $3520 \text{ yd} = 3520 \text{ yd} \times \dfrac{3 \text{ ft}}{1 \text{ yd}} \times \dfrac{1 \text{ mi}}{5280 \text{ ft}}$

   $= \dfrac{10,560}{5280} \times 1 \text{ mi}$

   = 2 × 1 mi

   = 2 mi

29. 25 mi = 25 × 1 mi

   = 25 × 5280 ft

   = 132,000 ft

31. 2 mi = 2 × 1 mi

   = 2 × 5280 ft

   $= 2 \times 5280 \text{ ft} \times \dfrac{12 \text{ in.}}{1 \text{ ft}}$

   = 126,720 in.

33. a) 1 km = _____ m

Think: To go from km to m in the table is a move of 3 places to the right. Thus, we move the decimal point 3 places to the right. This requires writing three additional zeros.

   1   1.000.

1 km = 1000 m

b) 1 m = _____ km

Think: To go from m to km in the table is a move of 3 places to the left. Thus, we move the decimal point 3 places to the left. This requires writing two additional zeros.

   1   0.001.

1 m = 0.001 km

35. a) 1 dam = _____ m

Think: To go from dam to m in the table is a move of 1 place to the right. Thus, we move the decimal point 1 place to the right. This requires writing an additional zero.

   1   1.0.

1 dam = 10 m

b) 1 m = _____ dam

Think: To go from m to dam in the table is a move of 1 place to the left. Thus, we move the decimal point 1 place to the left.

   1   0.1.

1 m = 0.1 dam

37. a) 1 cm = _____ m

Think: To go from cm to m in the table is a move of 2 places to the left. Thus, we move the decimal point 2 places to the left. This requires writing an additional zero.

   1   0.01.

1 cm = 0.01 m

b) 1 m = _____ cm

Think: To go from m to cm in the table is a move of 2 places to the right. Thus, we move the decimal point 2 places to the right. This requires writing two additional zeros.

   1   1.00.

1 m = 100 cm

39. 8.3 km = _____ m

Think: To go from km to m in the table is a move of 3 places to the right. Thus, we move the decimal point 3 places to the right. This requires writing two additional zeros.

   8.3   8.300.

8.3 km = 8300 m

41. 98 cm = _____ m

Think: To go from cm to m in the table is a move of 2 places to the left. Thus, we move the decimal point 2 places to the left.

   98   0.98.

98 cm = 0.98 m

43. 8921 m = _____ km

Think: To go from m to km in the table is a move of 3 places to the left. Thus, we move the decimal point 3 places to the left.

   8921   8.921.

8921 m = 8.921 km

45. 32.17 m = _____ km

Think: To go from m to km in the table is a move of 3 places to the left. Thus, we move the decimal point 3 places to the left. This requires writing an additional zero.

   32.17   0.032.17

32.17 m = 0.03217 km

47. 289 m = _____ cm

Think: To go from m to cm in the table is a move of 2 places to the right. Thus, we move the decimal point 2 places to the right. This requires writing two additional zeros.

   289   289.00.

289 m = 28,900 cm

49. 477 cm = _____ m

Think: To go from cm to m in the table is a move of 2 places to the left. Thus, we move the decimal point 2 places to the left.

477   4.77.
     ↑_|

477 cm = 4.77 m

51. 6.88 m = _____ cm

Think: To go from m to cm in the table is a move of 2 places to the right. Thus, we move the decimal point 2 places to the right.

6.88   6.88.
      |_↑

6.88 m = 688 cm

53. 1 mm = _____ cm

Think: To go from mm to cm in the table is a move of 1 place to the left. Thus, we move the decimal point 1 place to the left.

1   0.1.
   ↑_|

1 mm = 0.1 cm

55. 1 km = _____ cm

Think: To go from km to cm in the table is a move of 5 places to the right. Thus, we move the decimal point 5 places to the right. This requires writing five additional zeros.

1   1.00000.
   |____↑

1 km = 100,000 cm

57. 14.2 cm = _____ mm

Think: To go from cm to mm in the table is a move of 1 place to the right. Thus, we move the decimal point 1 place to the right.

14.2   14.2.
      |_↑

14.2 cm = 142 mm

59. 8.2 mm = _____ cm

Think: To go from mm to cm in the table is a move of 1 place to the left. Thus, we move the decimal point 1 place to the left.

8.2   0.8.2
     ↑_|

8.2 mm = 0.82 cm

61. 4500 mm = _____ cm

Think: To go from mm to cm in the table is a move of 1 place to the left. Thus, we move the decimal point 1 place to the left.

4500   450.0.
      ↑_|

4500 mm = 450 cm

63. 0.024 mm = _____ m

Think: To go from mm to m in the table is a move of 3 places to the left. Thus, we move the decimal point 3 places to the left. This requires writing three additional zeros.

0.024   0.000.024
       ↑___|

0.024 mm = 0.000024 m

65. 6.88 m = _____ dam

Think: To go from m to dam in the table is a move of 1 place to the left. Thus, we move the decimal point 1 place to the left.

6.88   0.6.88
      ↑_|

6.88 m = 0.688 dam

67. 2.3 dam = _____ dm

Think: To go from dam to dm in the table is a move of 2 places to the right. Thus, we move the decimal point 2 places to the right. This requires writing an additional zero.

2.3   2.30.
     |_↑

2.3 dam = 230 dm

69. $10 \text{ km} \approx 10 \text{ km} \times \dfrac{0.621 \text{ mi}}{1 \text{ km}} \approx 6.21 \text{ mi}$

71. $14 \text{ in.} \approx 14 \text{ in.} \times \dfrac{2.54 \text{ cm}}{1 \text{ in.}} \approx 35.56 \text{ cm}$

73. $65 \text{ mph} = 65\dfrac{\text{mi}}{\text{hr}} = 65 \times \dfrac{1 \text{ mi}}{\text{hr}} \approx 65 \times \dfrac{1.609 \text{ km}}{\text{hr}} =$ 104.585 km/h

75. $94 \text{ ft} = 94 \times 1 \text{ ft} \approx 94 \times 0.305 \text{ m} = 28.67 \text{ m}$

77. $180 \text{ cm} \approx 180 \text{ cm} \times \dfrac{1 \text{ in.}}{2.54 \text{ cm}} \approx \dfrac{180}{2.54} \text{ in.} \approx 70.866 \text{ in.}$

79. $36 \text{ yd} = 36 \times 1 \text{ yd} \approx 36 \times 0.914 \text{ m} = 32.904 \text{ m}$

81. Discussion and Writing Exercise

83.
$$-7x - 9x = 24$$
$$-16x = 24 \qquad \text{Collecting like terms}$$
$$\frac{-16x}{-16} = \frac{24}{-16} \qquad \text{Dividing both sides by } -16$$
$$x = \frac{3 \cdot 8}{-2 \cdot 8} = \frac{3}{-2} \cdot \frac{8}{8}$$
$$x = \frac{3}{-2}, \text{ or } -\frac{3}{2}$$

The solution is $-\dfrac{3}{2}$.

85. Let c represent the cost of 7 calculators. We translate to a proportion.

$$\text{Number} \rightarrow \frac{3}{43.50} = \frac{7}{c} \leftarrow \text{Number}$$
$$\text{Cost} \rightarrow \qquad\qquad\qquad \leftarrow \text{Cost}$$

Solve: $3 \cdot c = 43.50 \cdot 7$ Equating cross-products

$$c = \frac{43.50 \cdot 7}{3} \qquad \text{Dividing by 3 on both sides}$$

$$c = \frac{304.50}{3} \qquad \text{Multiplying}$$

$$c = 101.50 \qquad \text{Dividing}$$

Seven calculators would cost $101.50.

87. a) Multiply by 100 to move the decimal point two places to the right.

0.47.

b) Write a percent symbol: 47%

Thus, 0.47 = 47%.

89. Perimeter = 12 ft + 16 ft + 12 ft + 16 ft = 56 ft

Area = $l \cdot w = 12 \cdot 16 = 192$ sq ft

91. Discussion and Writing Exercise

93. Since a meter is just over a yard we place the decimal point as follows: 1.0.

95. Since a centimeter is about 0.3937 inch, we place the decimal point as follows: 1.40.

97.   300 cubits = 300×1 cubit

$\approx 300 \times 18$ in.

≈ 5400 in.

50 cubits = 50×1 cubit

$\approx 50 \times 18$ in.

≈ 900 in.

30 cubits = 30×1 cubit

$\approx 30 \times 18$ in.

≈ 540 in.

In inches, the length of Noah's ark was about 5400 in., the breadth was about 900 in., and the height was about 540 in.

Now we convert these dimensions to feet.

$$540 0 \text{ in.} = 5400 \text{ in.} \times \frac{1 \text{ ft}}{12 \text{ in.}}$$

$$= \frac{5400}{12} \times \frac{\text{in.}}{\text{in.}} \times 1 \text{ ft}$$

$$= 450 \times 1 \text{ ft}$$

$$= 450 \text{ ft}$$

$$900 \text{ in.} = 900 \text{ in.} \times \frac{1 \text{ ft}}{12 \text{ in.}}$$

$$= \frac{900}{12} \times \frac{\text{in.}}{\text{in.}} \times 1 \text{ ft}$$

$$= 75 \times 1 \text{ ft}$$

$$= 75 \text{ ft}$$

$$540 \text{ in.} = 540 \text{ in.} \times \frac{1 \text{ ft}}{12 \text{ in.}}$$

$$= \frac{540}{12} \times \frac{\text{in.}}{\text{in.}} \times 1 \text{ ft}$$

$$= 45 \times 1 \text{ ft}$$

$$= 45 \text{ ft}$$

In feet, the length of Noah's ark was 450 ft, the breadth was 75 ft, and the height was 45 ft.

99. $2 \text{ mi} \approx 2 \text{ mi} \times \dfrac{1.609 \text{ km}}{1 \text{ mi}} \times \dfrac{100,000 \text{ cm}}{1 \text{ km}} \approx$

$321,800 \times 1 \text{ cm} \approx 321,800 \text{ cm}$

101. We find the number of meters of tape that are used in 30 min of playing time.

$$1\frac{7}{8} \frac{\text{in.}}{\text{sec}}$$

$$\approx 1.875 \frac{\text{in.}}{\text{sec}} \times \frac{1 \text{ m}}{39.37 \text{ in.}} \times \frac{60 \text{ sec}}{1 \text{ min}} \times 30 \text{ min}$$

$$\approx \frac{3375}{39.37} \text{ m}$$

$$\approx 85.725 \text{ m}$$

About 85.725 m of tape is used for a 60-min cassette.

103. First find the height, in inches, of the stack of $1 bills.

1.382 $\times$ the distance to the moon

$= 1.382 \times 238,866$ mi

$= 330,112.812 \times 1$ mi

$= 330,112.812 \times 5280$ ft

$= 1,742,995,647 \times 1$ ft

$= 1,742,995,647 \times 12$ in.

$= 20,915,947,770$ in.

We divide this measure by 5.103 trillion to find the thickness, in inches, of a $1 bill.

$20,915,947,770 \div 5,103,000,000,000 \approx 0.0041$

A $1 bill is approximately 0.0041 in. thick.

105. Since 1 m is larger than 1 yd (1 m $\approx$ 3.3 ft $\approx$ 1.1 yd), we have 35 yd < 35 m.

107. Since 1 km $\approx$ 0.621 mi, then 18 km $\approx$ 18 $\times$ 0.6 mi $\approx$ 10.8 m, so 9 mi < 18 km.

109. Since 1 in. $\approx$ 2.54 cm, then 30 in. $\approx$ 30 $\times$ 2.5 cm $\approx$ 70 cm, so 30 in. < 90 cm.

Exercise Set 9.2

1. $5 \text{ yd}^2 = 5 \cdot 9 \text{ ft}^2$ Substituting 9 ft^2 for 1 yd^2

$\qquad = 45 \text{ ft}^2$

(Had we preferred to use canceling, we could have multiplied 5 yd^2 by $\dfrac{9 \text{ ft}^2}{1 \text{ yd}^2}$.)

3. $7 \text{ ft}^2 = 7 \cdot 144 \text{ in}^2$ Substituting 144 in^2 for 1 ft^2

$\phantom{7 \text{ ft}^2} = 1008 \text{ in}^2$

(Had we preferred to use canceling, we could have multiplied 7 ft^2 by $\dfrac{144 \text{ in}^2}{1 \text{ ft}^2}$.)

5. $432 \text{ in}^2 = 432 \text{ in}^2 \times \dfrac{1 \text{ ft}^2}{144 \text{ in}^2}$

$\phantom{432 \text{ in}^2} = \dfrac{432}{144} \times \text{ ft}^2$

$\phantom{432 \text{ in}^2} = 3 \text{ ft}^2$

7. $22 \text{ yd}^2 = 22 \times 9 \text{ ft}^2$ Substituting 9 ft^2 for 1 yd^2

$\phantom{22 \text{ yd}^2} = 198 \text{ ft}^2$

9. $15 \text{ ft}^2 = 15 \times 144 \text{ in}^2$ Substituting 144 in^2 for 1 ft^2

$\phantom{15 \text{ ft}^2} = 2160 \text{ in}^2$

11. $20 \text{ mi}^2 = 20 \cdot 640 \text{ acres}$ Substituting 640 acres for 1 mi^2

$\phantom{20 \text{ mi}^2} = 12{,}800 \text{ acres}$

13. $69 \text{ ft}^2 = 69 \text{ ft}^2 \times \dfrac{1 \text{ yd}^2}{9 \text{ ft}^2}$

$\phantom{69 \text{ ft}^2} = \dfrac{69}{9} \times \text{ yd}^2$

$\phantom{69 \text{ ft}^2} = \dfrac{23}{3} \text{ yd}^2, \text{ or } 7\dfrac{2}{3} \text{ yd}^2$

15. $720 \text{ in}^2 = 720 \text{ in}^2 \times \dfrac{1 \text{ ft}^2}{144 \text{ in}^2}$

$\phantom{720 \text{ in}^2} = \dfrac{720}{144} \times \text{ ft}^2$

$\phantom{720 \text{ in}^2} = 5 \text{ ft}^2$

17. $1 \text{ in}^2 = 1 \text{ in}^2 \times \dfrac{1 \text{ ft}^2}{144 \text{ in}^2}$

$\phantom{1 \text{ in}^2} = \dfrac{1}{144} \times \text{ ft}^2$

$\phantom{1 \text{ in}^2} = \dfrac{1}{144} \text{ ft}^2$

19. $1 \text{ acre} = 1 \text{ acre} \cdot \dfrac{1 \text{ mi}^2}{640 \text{ acres}}$

$\phantom{1 \text{ acre}} = \dfrac{1}{640} \cdot \text{ mi}^2$

$\phantom{1 \text{ acre}} = \dfrac{1}{640} \text{ mi}^2, \text{ or } 0.0015625 \text{ mi}^2$

21. $19 \text{ km}^2 = \underline{\hspace{1cm}} \text{ m}^2$

Think: A kilometer is 1000 times as big as a meter, so 1 km^2 is $1{,}000{,}000$ times as big as 1 m^2. We shift the decimal point six places to the right.

19   19.000000.

$\underline{}\uparrow$

$19 \text{ km}^2 = 19{,}000{,}000 \text{ m}^2$

23. $6.31 \text{ m}^2 = \underline{\hspace{1cm}} \text{ cm}^2$

Think: A meter is 100 times as big as a centimeter, so 1 m^2 is $10{,}000$ times as big as 1 cm^2. We shift the decimal point 4 places to the right.

6.31   6.3100.

$\underline{}\uparrow$

$6.31 \text{ m}^2 = 63{,}100 \text{ cm}^2$

25. $6.5432 \text{ mm}^2 = \underline{\hspace{1cm}} \text{ cm}^2$

Think: To convert from mm to cm, we shift the decimal point one place to the left. To convert from mm^2 to cm^2, we shift the decimal point two places to the left.

6.5432   0.06.5432

$\uparrow\underline{}$

$6.5432 \text{ mm}^2 = 0.065432 \text{ cm}^2$

27. $349 \text{ cm}^2 = \underline{\hspace{1cm}} \text{ m}^2$

Think: To convert from cm to m, we shift the decimal point two places to the left. To convert from cm^2 to m^2, we shift the decimal point four places to the left.

349   0.0349.

$\uparrow\underline{}$

$349 \text{ cm}^2 = 0.0349 \text{ m}^2$

29. $250{,}000 \text{ mm}^2 = \underline{\hspace{1cm}} \text{ cm}^2$

Think: To convert from mm to cm, we shift the decimal point one place to the left. To convert from mm^2 to cm^2, we shift the decimal point two places to the left.

250,000   2500.00.

$\uparrow\underline{}$

$250{,}000 \text{ mm}^2 = 2500 \text{ cm}^2$

31. $472{,}800 \text{ m}^2 = \underline{\hspace{1cm}} \text{ km}^2$

Think: To convert from m to km, we shift the decimal point three places to the left. To convert from m^2 to km^2, we shift the decimal point six places to the left.

472,800   0.472800.

$\uparrow\underline{}$

$472{,}800 \text{ m}^2 = 0.4728 \text{ km}^2$

33. First we convert 3 in. to feet.

$3 \text{ in.} = 3 \text{ in.} \times \dfrac{1 \text{ ft}}{12 \text{ in.}}$

$\phantom{3 \text{ in.}} = \dfrac{3}{12} \times \text{ ft}$

$\phantom{3 \text{ in.}} = \dfrac{1}{4} \text{ ft}$

Then we find the area using the formula for the area of a rectangle.

$A = l \cdot w$

$ = 4 \text{ ft} \cdot \dfrac{1}{4} \text{ ft}$

$ = \dfrac{4}{4} \text{ ft}^2$

$ = 1 \text{ ft}^2$

35. We convert 4 in. and 7 yd to feet.

$$4 \text{ in.} = 4 \text{ in.} \times \frac{1 \text{ ft}}{12 \text{ in.}}$$

$$= \frac{4}{12} \times \text{ ft}$$

$$= \frac{1}{3} \text{ ft}$$

$$7 \text{ yd} = 7 \cdot 3 \text{ ft} = 21 \text{ ft}$$

Then we find the area using the formula for the area of a trapezoid.

$$A = \frac{1}{2} \cdot h \cdot (a + b)$$

$$= \frac{1}{2} \cdot \frac{1}{3} \text{ ft} \cdot (5 + 21) \text{ ft}$$

$$= \frac{26}{2} \text{ ft}^2$$

$$= \frac{13}{3} \text{ ft}^2, \text{ or } 4\frac{1}{3} \text{ ft}^2$$

37. The area of the small triangle is

$$A = \frac{1}{2} \cdot b \cdot h$$

$$= \frac{1}{2} \cdot 2 \text{ cm} \cdot 2 \text{ cm}$$

$$= 2 \text{ cm}^2.$$

Then the total area of the 6 small triangles is

$$6 \cdot 2 \text{ cm}^2 = 12 \text{ cm}^2.$$

The area of the large triangle is

$$A = \frac{1}{2} \cdot b \cdot h$$

$$= \frac{1}{2} \cdot 12 \text{ cm} \cdot 12 \text{ cm}$$

$$= 72 \text{ cm}^2.$$

The total area is $12 \text{ cm}^2 + 72 \text{ cm}^2 = 84 \text{ cm}^2.$

39. Discussion and Writing Exercise

41. We first find the interest for 1 year:

$$5\% \times 700 = 0.05 \times 700 = 35$$

Then we multiply that amount by $\frac{1}{2}$:

$$\frac{1}{2} \times 35 = \frac{35}{2} = 17.5$$

The interest for $\frac{1}{2}$ yr is $17.50.

43. We first find the interest for 1 year:

$$6\% \times 450 = 27$$

Then we multiply that amount by $\frac{1}{4}$:

$$\frac{1}{4} \times 27 = 6.75$$

The interest for $\frac{1}{4}$ year is $6.75.

45. We first find the interest for 1 year:

$$12\% \times 1800 = 216$$

Then we multiply that amount by $\frac{60}{365}$:

$$\frac{60}{365} \times 216 \approx 35.51$$

The interest for 60 days is $35.51.

47. Discussion and Writing Exercise

49. First we convert 8 in. to ft.

$$8 \text{ in.} = 8 \text{ in.} \times \frac{1 \text{ ft}}{12 \text{ in.}}$$

$$= \frac{8}{12} \times \text{ ft}$$

$$= \frac{2}{3} \text{ ft}$$

Then we find the area of each tile.

$$A = s \cdot s$$

$$= \frac{2}{3} \text{ ft} \cdot \frac{2}{3} \text{ ft}$$

$$= \frac{4}{9} \text{ ft}^2$$

Next we find the area of the dance floor.

$$A = l \cdot w$$

$$= 18 \text{ ft} \cdot 42 \text{ ft}$$

$$= 756 \text{ ft}^2$$

We divide to find the number of tiles needed.

$$756 \div \frac{4}{9} = 756 \cdot \frac{9}{4} = \frac{756 \cdot 9}{4} = \frac{4 \cdot 189 \cdot 9}{4 \cdot 1} = 1701$$

Thus, 1701 tiles are needed.

Now we find the area of the floor.

$$A = l \cdot w$$

$$= 30 \text{ ft} \cdot 60 \text{ ft}$$

$$= 1800 \text{ ft}^2$$

We use an equation to find the percent of the floor area that is covered by the dance area.

$$\underbrace{756 \text{ is what percent of } 1800?}$$

$$756 = \qquad n \quad \times \quad 1800$$

To solve the equation we divide on both sides by 1800 and convert to percent notation.

$$756 = n \times 1800$$

$$\frac{756}{1800} = n$$

$$0.42 = n$$

$$42\% = n$$

The dance area is 42% of the area of the floor.

51. $1 \text{ in}^2 = 1 \text{ in.} \times 1 \text{ in.}$

$$\approx 2.54 \text{ cm} \times 2.54 \text{ cm}$$

$$\approx 6.4516 \text{ cm}^2$$

53. $1 \text{ acre} = 43,560 \text{ ft}^2$

$$\approx 43,560 \text{ ft}^2 \times \frac{1 \text{ m}}{3.3 \text{ ft}} \times \frac{1 \text{ m}}{3.3 \text{ ft}}$$

$$\approx \frac{43,560}{3.3(3.3)} \times \text{ ft} \times \text{ ft} \times \frac{\text{m}}{\text{ft}} \times \frac{\text{m}}{\text{ft}}$$

$$\approx 4000 \text{ m}^2$$

55. First we convert 2 m to centimeters. Think: Meters are 100 times as large as centimeters (100 cm = 1 m). We move the decimal point two places to the right.

$$2 \text{ m} = 200 \text{ cm}$$

Next we convert 10 in. to centimeters.

$$10 \text{ in.} \approx 10 \times 2.54 \text{ cm} \approx 25.4 \text{ cm}$$

Now we find the area of the scarf.

$$A = l \cdot w$$
$$= 200 \text{ cm} \cdot 25.4 \text{ cm}$$
$$= 5080 \text{ cm}^2$$

The area of the scarf is 5080 cm².

57. Each cube has 6 square sides. If the area of Janie's cube is 54 cm², then the area of each side is 54 cm²/6, or 9 cm². Since each side is a square, the length of a side is 3 cm (3 cm · 3 cm = 9 cm²). Then each side of Norm's cube is 2 · 3 cm, or 6 cm, and the area of each side is 6 cm · 6 cm, or 36 cm². Then the area of Norm's cube is 6 · 36 cm², or 216 cm².

Exercise Set 9.3

1. $A = b \cdot h$ Area of a parallelogram
$= 10 \text{ cm} \cdot 5 \text{ cm}$ Substituting 10 cm for b and 5 cm for h
$= 50 \text{ cm}^2$

3. $A = \frac{1}{2} \cdot h \cdot (a + b)$ Area of a trapezoid

$= \frac{1}{2} \cdot 8 \text{ ft} \cdot (6 + 20) \text{ ft}$ Substituting 8 ft for h, 6 ft for a, and 20 ft for b

$= \frac{8 \cdot 26}{2} \text{ ft}^2$

$= \frac{\cancel{2} \cdot 4 \cdot 26}{1 \cdot \cancel{2}} \text{ ft}^2$

$= 104 \text{ ft}^2$

5. $A = b \cdot h$ Area of a parallelogram
$= 8 \text{ m} \cdot 8 \text{ m}$ Substituting 8 m for b and 8 m for h
$= 64 \text{ m}^2$

7. $A = b \cdot h$ Area of a parallelogram
$= 6.9 \text{ cm} \cdot 10.5 \text{ cm}$ Substituting 6.9 cm for b and 10.5 cm for h
$= 72.45 \text{ cm}^2$

9. $A = \frac{1}{2} \cdot h \cdot (a + b)$ Area of a trapezoid

$= \frac{1}{2} \cdot 9 \text{ mi} \cdot (13 + 19) \text{ mi}$ Substituting 9 mi for h, 13 mi for a, and 19 mi for b

$= \frac{9 \cdot 32}{2} \text{ mi}^2$

$= \frac{9 \cdot \cancel{2} \cdot 16}{1 \cdot \cancel{2}} \text{ mi}^2$

$= 144 \text{ mi}^2$

11. $A = b \cdot h$ Area of a parallelogram

$= 12\frac{1}{4} \text{ ft} \cdot 4\frac{1}{2} \text{ ft}$ Substituting $12\frac{1}{4}$ ft for b and $4\frac{1}{2}$ ft for h

$= \frac{49}{4} \cdot \frac{9}{2} \cdot \text{ ft}^2$

$= \frac{441}{8} \text{ ft}^2$

$= 55\frac{1}{8} \text{ ft}^2$

13. $A = \frac{1}{2} \cdot h \cdot (a + b)$ Area of a trapezoid

$= \frac{1}{2} \cdot 7 \text{ m} \cdot (9 + 5) \text{ m}$ Substituting 7 m for h, 9 m for a, and 5 m for b

$= \frac{7 \cdot 14}{2} \text{ m}^2$

$= \frac{7 \cdot \cancel{2} \cdot 7}{1 \cdot \cancel{2}} \text{ m}^2$

$= 49 \text{ m}^2$

15. $A = b \cdot h$ Area of a parallelogram
$= 1.6 \text{ cm} \cdot 1.2 \text{ cm}$ Substituting 1.6 cm for b and 1.2 cm for h
$= 1.92 \text{ cm}^2$

17. $A = \frac{1}{2} \cdot h \cdot (a + b)$ Area of a trapezoid

$= \frac{1}{2} \cdot 8 \text{ yd} \cdot (9.1 + 7.9) \text{ yd}$ Substituting 8 yd for h, 9.1 yd for a, and 7.9 yd for b

$= \frac{8 \cdot 17}{2} \text{ yd}^2$

$= \frac{\cancel{2} \cdot 4 \cdot 17}{1 \cdot \cancel{2}} \text{ yd}^2$

$= 68 \text{ yd}^2$

19. $d = 2 \cdot r$
$= 2 \cdot 7 \text{ cm} = 14 \text{ cm}$

21. $d = 2 \cdot r$

$= 2 \cdot \frac{7}{8} \text{ in.} = \frac{14}{8} \text{ in.} = \frac{7}{4} \text{ in., or } 1\frac{3}{4} \text{ in.}$

23. $r = \frac{d}{2}$

$= \frac{20 \text{ ft}}{2} = 10 \text{ ft}$

25. $r = \frac{d}{2}$

$= \frac{1.4 \text{ cm}}{2} = 0.7 \text{ cm}$

27. $C = 2 \cdot \pi \cdot r$

$\approx 2 \cdot \frac{22}{7} \cdot 7 \text{ cm} \approx \frac{2 \cdot 22 \cdot 7}{7} \text{ cm} \approx 44 \text{ cm}$

29. $C = 2 \cdot \pi \cdot r$

$\approx 2 \cdot \frac{22}{7} \cdot \frac{7}{8} \text{ in.} \approx \frac{2 \cdot 22 \cdot 7}{7 \cdot 8} \text{ in.} \approx \frac{2 \cdot 2 \cdot 11 \cdot 7}{7 \cdot 2 \cdot 2 \cdot 2} \text{ in.} \approx$

$\frac{11}{2} \text{ in., or } 5\frac{1}{2} \text{ in.}$

31. $C = \pi \cdot d$
$\approx 3.14 \cdot 20$ ft ≈ 62.8 ft

33. $C = \pi \cdot d$
$\approx 3.14 \cdot 1.4$ cm ≈ 4.396 cm

35. $A = \pi \cdot r \cdot r$
$\approx \dfrac{22}{7} \cdot 7$ cm $\cdot 7$ cm $\approx \dfrac{22}{7} \cdot 49$ cm$^2 \approx 154$ cm^2

37. $A = \pi \cdot r \cdot r$
$\approx \dfrac{22}{7} \cdot \dfrac{7}{8}$ in. $\cdot \dfrac{7}{8}$ in. $\approx \dfrac{2 \cdot 11 \cdot 7 \cdot 7}{7 \cdot 2 \cdot 4 \cdot 8}$ in$^2 \approx$
$\dfrac{77}{32}$ in^2, or $2\dfrac{13}{32}$ in^2

39. $A = \pi \cdot r \cdot r$
$\approx 3.14 \cdot 10$ ft $\cdot 10$ ft $\left(r = \dfrac{d}{2}; r = \dfrac{20 \text{ ft}}{2} = 10 \text{ ft}\right)$
$\approx 3.14 \cdot 100$ ft^2
≈ 314 ft^2

41. $A = \pi \cdot r \cdot r$
$\approx 3.14 \cdot 0.7$ cm $\cdot 0.7$ cm
$\left(r = \dfrac{d}{2}; r = \dfrac{1.4 \text{ cm}}{2} = 0.7 \text{ cm}\right)$
$\approx 3.14 \cdot 0.49$ cm$^2 \approx 1.5386$ cm^2

43. Find the area of the small round pizza. Note that $d = 10$ in., so $r = \dfrac{10 \text{ in.}}{2} = 5$ in.
$A = \pi \cdot r \cdot r$
$\approx 3.14 \cdot 5$ in. $\cdot 5$ in.
≈ 78.5 in^2

Now find the area of the small square pizza.
$A = s \cdot s$
$= 8.5$ in. $\cdot 8.5$ in.
$= 72.25$ in^2

We see that the small round pizza has more area. We subtract to find how much more:
$$78.5 \text{ in}^2 - 72.25 \text{ in}^2 = 6.25 \text{ in}^2.$$

The small round pizza has 6.25 in^2 more area than the small square pizza.

45. Find the area of the large round pizza. Note that $d = 14$ in., so $r = \dfrac{14 \text{ in.}}{2} = 7$ in.
$A = \pi \cdot r \cdot r$
$= 3.14 \cdot 7$ in. $\cdot 7$ in.
≈ 153.86 in^2

Now find the area of the large square pizza.
$A = s \cdot s$
$= 12$ in. $\cdot 12$ in.
$= 144$ in^2

The large round pizza has more area. We subtract to find how much more:
$$153.86 \text{ in}^2 - 144 \text{ in}^2 = 9.86 \text{ in}^2.$$

The large round pizza has 9.86 in^2 more area than the large square pizza.

47. In Exercise 43 we found that the area of one small round pizza is about 78.5 in^2, so the area of two small round pizzas is $2(78.5 \text{ in}^2)$, or 157 in^2. In Exercise 45 we found that the area of one large square pizza is 144 in^2. Thus, two small round pizzas have more area. We subtract to find how much more:
$$157 \text{ in}^2 - 144 \text{ in}^2 = 13 \text{ in}^2.$$

Two small round pizzas have 13 in^2 more area than one large square pizza.

49. Find the area of a medium square pizza.
$A = s \cdot s$
$= 10$ in. $\cdot 10$ in.
$= 100$ in^2

Then the area of two medium square pizzas is $2 \cdot 100$ in^2, or 200 in^2.

Now find the area of one extra large round pizza. Note that $d = 16$ in., so $r = \dfrac{16 \text{ in.}}{2} = 8$ in.
$A = \pi \cdot r \cdot r$
$\approx 3.14 \cdot 8$ in. $\cdot 8$ in.
≈ 200.96 in^2

We see that one extra large round pizza has more area. We subtract to find how much more:
$$200.96 \text{ in}^2 - 200 \text{ in}^2 = 0.96 \text{ in}^2.$$

One extra large round pizza has 0.96 in^2 more area than two medium square pizzas.

51. $r = \dfrac{d}{2}$
$r = \dfrac{14 \text{ ft}}{2} = 7$ ft
$A = \pi \cdot r \cdot r$
$A \approx 3.14 \cdot 7$ ft $\cdot 7$ ft $= 153.86$ ft^2

The area of the trampoline is about 153.86 ft^2.

53. $A = \pi \cdot r \cdot r$
$\approx 3.14 \cdot 7$ mi $\cdot 7$ mi ≈ 153.86 mi^2

The broadcast area is about 153.86 mi^2.

55. The tree's circumference is 47.1 in.
$C = \pi \cdot d$
$47.1 \approx 3.14 \cdot d$
$\dfrac{47.1}{3.14} \approx d$
$15 \approx d$

The tree's diameter is about 15 in.

57. Find the area of the larger circle (pool plus walk). Its diameter is 1 yd + 20 yd + 1 yd, or 22 yd. Thus its radius is $\dfrac{22}{2}$ yd, or 11 yd.

$A = \pi \cdot r \cdot r$
$A \approx 3.14 \cdot 11$ yd $\cdot 11$ yd $= 379.94$ yd^2

Find the area of the pool. Its diameter is 20 yd. Thus its radius is $\frac{20}{2}$ yd, or 10 yd.

$$A = \pi \cdot r \cdot r$$
$$A \approx 3.14 \cdot 10 \text{ yd} \cdot 10 \text{ yd} = 314 \text{ yd}^2$$

We subtract to find the area of the walk:

$$A = 379.94 \text{ yd}^2 - 314 \text{ yd}^2$$
$$A = 65.94 \text{ yd}^2$$

The area of the walk is 65.94 yd^2.

59. The perimeter consists of the circumferences of three semicircles, each with diameter 8 ft, and one side of a square of length 8 ft. We first find the circumference of one semicircle. This is one-half the circumference of a circle with diameter 8 ft:

$$\frac{1}{2} \cdot \pi \cdot d \approx \frac{1}{2} \cdot 3.14 \cdot 8 \text{ ft} = 12.56 \text{ ft}$$

Then we multiply by 3:

$$3 \cdot (12.56 \text{ ft}) = 37.68 \text{ ft}$$

Finally we add the circumferences of the semicircles and the length of the side of the square:

$$37.68 \text{ ft} + 8 \text{ ft} = 45.68 \text{ ft}$$

The perimeter is 45.68 ft.

61. The perimeter consists of three-fourths of the perimeter of a square with side of length 10 yd and the circumference of a semicircle with diameter 10 yd. First we find three-fourths of the perimeter of the square:

$$\frac{3}{4} \cdot 4 \cdot s = \frac{3}{4} \cdot 4 \cdot 10 \text{ yd} = 30 \text{ yd}$$

Then we find one-half of the circumference of a circle with diameter 10 yd:

$$\frac{1}{2} \cdot \pi \cdot d \approx \frac{1}{2} \cdot 3.14 \cdot 10 \text{ yd} = 15.7 \text{ yd}$$

Then we add:

$$30 \text{ yd} + 15.7 \text{ yd} = 45.7 \text{ yd}$$

The perimeter is 45.7 yd.

63. The shaded region consists of a circle of radius 8 m, with two circles each of diameter 8 m, removed. First we find the area of the large circle:

$$A = \pi \cdot r \cdot r \approx 3.14 \cdot 8 \text{ m} \cdot 8 \text{ m} = 200.96 \text{ m}^2$$

Then we find the area of one of the small circles:
The radius is $\frac{8 \text{ m}}{2} = 4$ m.

$$A = \pi \cdot r \cdot r \approx 3.14 \cdot 4 \text{ m} \cdot 4 \text{ m} = 50.24 \text{ m}^2$$

We multiply this area by 2 to find the area of the two small circles:

$$2 \cdot 50.24 \text{ m}^2 = 100.48 \text{ m}^2$$

Finally we subtract to find the area of the shaded region:

$$200.96 \text{ m}^2 - 100.48 \text{ m}^2 = 100.48 \text{ m}^2$$

The area of the shaded region is 100.48 m^2.

65. The shaded region consists of one-half of a circle with diameter 2.8 cm and a triangle with base 2.8 cm and height 2.8 cm. First we find the area of the semicircle. The radius is $\frac{2.8 \text{ cm}}{2} = 1.4$ cm.

$$A = \frac{1}{2} \cdot \pi \cdot r \cdot r \approx \frac{1}{2} \cdot 3.14 \cdot 1.4 \text{ cm} \cdot 1.4 \text{ cm} = 3.0772 \text{ cm}^2$$

Then we find the area of the triangle:

$$A = \frac{1}{2} \cdot b \cdot h = \frac{1}{2} \cdot 2.8 \text{ cm} \cdot 2.8 \text{ cm} = 3.92 \text{ cm}^2$$

Finally we add to find the area of the shaded region:

$$3.0772 \text{ cm}^2 + 3.92 \text{ cm}^2 = 6.9972 \text{ cm}^2$$

The area of the shaded region is 6.9972 cm^2.

67. Discussion and Writing Exercise

69. $9.25\% = \frac{9.25}{100}$

$$= \frac{9.25}{100} \cdot \frac{100}{100}$$
$$= \frac{925}{10,000}$$
$$= \frac{25 \cdot 37}{25 \cdot 400}$$
$$= \frac{25}{25} \cdot \frac{37}{400}$$
$$= \frac{37}{400}$$

71. a) First find decimal notation by division.

```
      1.3 7 5
 8 ⟌ 1 1.0 0 0
     8
     ---
     3 0
     2 4
     ---
       6 0
       5 6
       ---
         4 0
         4 0
         ---
           0
```

$\frac{11}{8} = 1.375$

b) Convert the decimal notation to percent notation. Move the decimal point two places to the right and write a % symbol.

1.37.5

$\frac{11}{8} = 137.5\%$

73. a) First find decimal notation by division.

$$
\begin{array}{r}
1.2\,5 \\
4\,\overline{)\,5.0\,0} \\
\underline{4} \\
1\,0 \\
\underline{8} \\
2\,0 \\
\underline{2\,0} \\
0
\end{array}
$$

$$\frac{5}{4} = 1.25$$

b) Convert the decimal notation to percent notation. Move the decimal point two places to the right and write a % symbol.

1.25.

$$\frac{5}{4} = 125\%$$

75. Discussion and Writing Exercise

77. If we remove the top and bottom and "unroll" the can, we have two circles, each with a diameter of 2.5 in., and a rectangle whose length is the circumference of the can and whose width is the height of the can. First we find the area of the circles. Their radius is $\dfrac{d}{2} = \dfrac{2.5\text{ in.}}{2} = 1.25$ in. Then the area of each circle is

$$
\begin{aligned}
A &= \pi \cdot r \cdot r \\
&\approx 3.14 \cdot 1.25 \text{ in.} \cdot 1.25 \text{ in.} \approx 4.90625 \text{ in}^2.
\end{aligned}
$$

The area of the two circles is $2 \cdot 4.90625$ in^2 = 9.8125 in^2.

Now we find the area of the rectangle. The circumference of the can is

$$
\begin{aligned}
C &= \pi \cdot d \\
&\approx 3.14 \cdot 2.5 \text{ in.} \approx 7.85 \text{ in.}
\end{aligned}
$$

The area of a rectangle with length 7.85 in. and width 3.5 in. is

$$
\begin{aligned}
A &= l \cdot w \\
&= 7.85 \text{ in.} \cdot 3.5 \text{ in.} = 27.475 \text{ in}^2.
\end{aligned}
$$

Then the surface area of the can is

9.8125 in^2 + 27.475 in^2 = 37.2875 in^2.

79. Discussion and Writing Exercise

81. The height of the stack of tennis balls is three times the diameter of one ball, or $3 \cdot d$.

The circumference of the can is about the circumference of one ball, or $\pi \cdot d$.

The circumference of the can is greater than the height of the stack of balls, because $\pi > 3$.

83. Discussion and Writing Exercise

Exercise Set 9.4

1. $V = l \cdot w \cdot h$

$V = 10 \text{ cm} \cdot 5 \text{ cm} \cdot 5 \text{ cm}$

$V = 50 \cdot 5 \text{ cm}^3$

$V = 250 \text{ cm}^3$

3. $V = l \cdot w \cdot h$

$V = 9 \text{ in.} \cdot 3 \text{ in.} \cdot 5 \text{ in.}$

$V = 27 \cdot 5 \text{ in}^3$

$V = 135 \text{ in}^3$

5. $V = l \cdot w \cdot h$

$V = 10 \text{ m} \cdot 5 \text{ m} \cdot 1.5 \text{ m}$

$V = 50 \cdot 1.5 \text{ m}^3$

$V = 75 \text{ m}^3$

7. $V = l \cdot w \cdot h$

$V = 6\dfrac{1}{2} \text{ yd} \cdot 5\dfrac{1}{2} \text{ yd} \cdot 10 \text{ yd}$

$V = \dfrac{13}{2} \cdot \dfrac{11}{2} \cdot 10 \text{ yd}^3$

$V = \dfrac{1430}{4} \text{ yd}^3 = \dfrac{2 \cdot 715}{2 \cdot 2} \text{ yd}^3$

$V = \dfrac{715}{2} \text{ yd}^3$

$V = 357\dfrac{1}{2} \text{ yd}^3$

9. $V = Bh = \pi \cdot r^2 \cdot h$

$\approx 3.14 \times 10 \text{ ft} \times 10 \text{ ft} \times 13 \text{ ft}$

$\approx 4082 \text{ ft}^3$

11. $V = Bh = \pi \cdot r^2 \cdot h$

$\approx 3.14 \times 4 \text{ cm} \times 4 \text{ cm} \times 7.5 \text{ cm}$

$\approx 376.8 \text{ cm}^3$

13. $V = Bh = \pi \cdot r^2 \cdot h$

$\approx \dfrac{22}{7} \times 210 \text{ yd} \times 210 \text{ yd} \times 300 \text{ yd}$

$\approx 41{,}580{,}000 \text{ yd}^3$

15. $V = \dfrac{4}{3} \cdot \pi \cdot r^3$

$\approx \dfrac{4}{3} \times 3.14 \times (100 \text{ in.})^3$

$\approx \dfrac{4 \times 3.14 \times 1{,}000{,}000 \text{ in}^3}{3}$

$\approx 4{,}186{,}666.\overline{6} \text{ in}^3$, or $4{,}186{,}666\dfrac{2}{3} \text{ in}^3$

17. $V = \dfrac{4}{3} \cdot \pi \cdot r^3$

$\approx \dfrac{4}{3} \times 3.14 \times (3.1 \text{ m})^3$

$\approx \dfrac{4 \times 3.14 \times 29.791 \text{ m}^3}{3}$

$\approx 124.725 \text{ m}^3$

19. $V = \dfrac{4}{3} \cdot \pi \cdot r^3$

$\approx \dfrac{4}{3} \times \dfrac{22}{7} \times (7 \text{ km})^3$

$\approx \dfrac{4 \times 22 \times 343 \text{ km}^3}{3 \times 7}$

$\approx 1437\dfrac{1}{3} \text{ km}^3$

21. $1 \text{ L} = 1000 \text{ mL} = 1000 \text{ cm}^3$

These conversion relations appear in the text on page 624.

23. $59 \text{ L} = 59 \times (1 \text{ L})$
$= 59 \times (1000 \text{ mL})$
$= 59,000 \text{ mL}$

25. $49 \text{ mL} = 49 \times (1 \text{ mL})$
$= 49 \times (0.001 \text{ L})$
$= 0.049 \text{ L}$

27. $27.3 \text{ L} = 27.3 \times (1 \text{ L})$
$= 27.3 \times (1000 \text{ cm}^3)$
$= 27,300 \text{ cm}^3$

29. $5 \text{ gal} = 5 \times 1 \text{ gal}$
$= 5 \times 4 \text{ qt}$
$= 5 \times 4 \times 1 \text{ qt}$
$= 5 \times 4 \times 2 \text{ pt}$
$= 40 \text{ pt}$

31. $10 \text{ qt} = 10 \times 1 \text{ qt}$
$= 10 \times 2 \text{ pt}$
$= 10 \times 2 \times 1 \text{ pt}$
$= 10 \times 2 \times 16 \text{ oz}$
$= 320 \text{ oz}$

33. $24 \text{ oz} = 24 \text{ oz} \times \dfrac{1 \text{ cup}}{8 \text{ oz}}$

$= \dfrac{24}{8} \cdot 1 \text{ cup}$

$= 3 \text{ cups}$

35. $10 \text{ gal} = 10 \text{ gal} \times \dfrac{4 \text{ qt}}{1 \text{ gal}}$

$= 10 \cdot 4 \text{ qt}$
$= 40 \text{ qt}$

37. First we convert 3 gal to quarts:

$3 \text{ gal} = 3 \text{ gal} \times \dfrac{4 \text{ qt}}{1 \text{ gal}}$

$= 3 \cdot 4 \text{ qt}$
$= 12 \text{ qt}$

Next we convert 12 qt to pints:

$12 \text{ qt} = 12 \text{ qt} \times \dfrac{2 \text{ pt}}{1 \text{ qt}}$

$= 12 \cdot 2 \text{ pt}$
$= 24 \text{ pt}$

Finally we convert 24 pt to cups:

$24 \text{ pt} = 24 \text{ pt} \times \dfrac{2 \text{ cups}}{1 \text{ pt}}$

$= 24 \cdot 2 \text{ cups}$
$= 48 \text{ cups}$

39. First we convert 15 pt to quarts:

$15 \text{ pt} = 15 \text{ pt} \times \dfrac{1 \text{ qt}}{2 \text{ pt}}$

$= \dfrac{15}{2} \cdot 1 \text{ qt}$

$= 7.5 \text{ qt}$

Then we convert 7.5 qt to gallons:

$7.5 \text{ qt} = 7.5 \text{ qt} \times \dfrac{1 \text{ gal}}{4 \text{ qt}}$

$= \dfrac{7.5}{4} \cdot 1 \text{ gal}$

$= 1.875 \text{ gal, or } 1\dfrac{7}{8} \text{ gal}$

41. We must find the radius of the base in order to use the formula for the volume of a circular cylinder.

$r = \dfrac{d}{2} = \dfrac{0.7 \text{ yd}}{2} = 0.35 \text{ yd}$

$V = Bh = \pi \cdot r^2 \cdot h$

$\approx 3.14 \times 0.35 \text{ yd} \times 0.35 \text{ yd} \times 1.1 \text{yd}$

$\approx 0.423115 \text{ yd}^3$

43. We must find the radius of the silo in order to use the formula for the volume of a circular cylinder.

$r = \dfrac{d}{2} = \dfrac{6 \text{ m}}{2} = 3 \text{ m}$

$V = Bh = \pi \cdot r^2 \cdot h$

$\approx 3.14 \times 3 \text{ m} \times 3 \text{ m} \times 13 \text{ m}$

$\approx 367.38 \text{ m}^3$

45. First we find the radius of the ball:

$r = \dfrac{d}{2} = \dfrac{6.5 \text{ cm}}{2} = 3.25 \text{ cm}$

Then we find the volume, using the formula for the volume of a sphere.

$V = \dfrac{4}{3} \cdot \pi \cdot r^3$

$\approx \dfrac{4}{3} \cdot 3.14 \cdot (3.25 \text{ cm})^3$

$\approx 143.72 \text{ cm}^3$

47. $V = \dfrac{4}{3} \cdot \pi \cdot r^3$

$\approx \dfrac{4}{3} \cdot 3.14 \cdot (3980 \text{ mi})^3$

$\approx 263,947,530,000 \text{ mi}^3$

49. First we find the radius of the pressure sphere:

$r = \dfrac{d}{2} = \dfrac{2 \text{ m}}{2} = 1 \text{ m}$

Then we find the volume of the pressure sphere.

$V = \dfrac{4}{3} \cdot \pi \cdot r^3$

$\approx \dfrac{4}{3} \cdot 3.14 \cdot (1 \text{ m})^3$

$\approx 4.187 \text{ m}^3$

51. First we find the radius of the can.
$$r = \frac{d}{2} = \frac{6.5 \text{ cm}}{2} = 3.25 \text{ cm}$$

The height of the can is the length of the diameters of 3 tennis balls.
$$h = 3(6.5 \text{ cm}) = 19.5 \text{ cm}$$

Now we find the volume.
$$\begin{aligned} V = Bh &= \pi \cdot r^2 \cdot h \\ &\approx 3.14 \times 3.25 \text{ cm} \times 3.25 \text{ cm} \times 19.5 \text{ cm} \\ &\approx 646.74 \text{ cm}^3 \end{aligned}$$

53. First convert 32 oz to gallons:

$$\begin{aligned} 32 \text{ oz} &= 32 \text{ oz} \cdot \frac{1 \text{ pt}}{16 \text{ oz}} \cdot \frac{1 \text{ qt}}{2 \text{ pt}} \cdot \frac{1 \text{ gal}}{4 \text{ qt}} \\ &= \frac{32}{16 \cdot 2 \cdot 4} \times 1 \text{ gal} = 0.25 \text{ gal} \end{aligned}$$

Thus 0.25 gal per day are wasted by one person. We multiply to find how many gallons are wasted in a week (7 days) by one person:

$$7 \times 0.25 \text{ gal} = 1.75 \text{ gal}$$

Next we multiply to find how many gallons are wasted in a month (30 days) by one person:

$$30 \times 0.25 \text{ gal} = 7.5 \text{ gal}$$

We multiply again to find how many gallons are wasted in a year (365 days) by one person:

$$365 \times 0.25 \text{ gal} = 91.25 \text{ gal}$$

To find how much water is wasted in this country in a year, we multiply 91.25 gal by 261 million:

$$\begin{aligned} 261,000,000 \times 91.25 \text{ gal} &= 23,816,250,000 \text{ gal} \\ &\approx 24,000,000,000 \text{ gal} \end{aligned}$$

55. First find a volume of the storage compartment.
$$V = l \cdot w \cdot h = 9.83 \text{ ft} \cdot 5.67 \text{ ft} \cdot 5.83 \text{ ft} = 324.941463 \text{ ft}^3$$
Next find the volume of the attic:
$$V = l \cdot w \cdot h = 1.5 \text{ ft} \cdot 5.67 \text{ ft} \cdot 2.5 \text{ ft} = 21.2625 \text{ ft}^3$$
Finally, add to find the total volume of the compartment.
$$324.941463 \text{ ft}^3 + 21.2625 \text{ ft}^3 = 346.203963 \text{ ft}^3 \approx 346.2 \text{ ft}^3$$

57. Discussion and Writing Exercise

59. Interest = (Interest for 1 year) $\times \dfrac{1}{2}$

$$\begin{aligned} &= (8\% \times \$600) \times \frac{1}{2} \\ &= 0.08 \times \$600 \times \frac{1}{2} \\ &= \$48 \times \frac{1}{2} \\ &= \$24 \end{aligned}$$

The interest is $24.

61. Let c represent the cost of 12 pens. We translate to a proportion and solve.

$$\begin{array}{c} \text{Pens} \rightarrow \\ \text{Cost} \rightarrow \end{array} \frac{9}{\$8.01} = \frac{12}{c} \begin{array}{c} \leftarrow \text{Pens} \\ \leftarrow \text{Cost} \end{array}$$

$9 \cdot c = \$8.01 \cdot 12$ Equating cross-products
$$c = \frac{\$8.01 \cdot 12}{9}$$
$$c = \frac{\$96.12}{9}$$
$$c = \$10.68$$

12 pens would cost $10.68.

63.
$$\begin{aligned} -5y + 3 &= -12y - 4 \\ -5y + 3 + 12y &= -12y - 4 + 12y \quad \text{Adding } 12y \text{ on both} \\ &\qquad\qquad\qquad\qquad\qquad \text{sides} \\ 7y + 3 &= -4 \\ 7y + 3 - 3 &= -4 - 3 \quad \text{Subtracting 3 on both sides} \\ 7y &= -7 \\ \frac{7y}{7} &= \frac{-7}{7} \quad \text{Dividing by 7 on both sides} \\ y &= -1 \end{aligned}$$

The solution is -1.

65.
$$\begin{aligned} \frac{9}{5}C + 32 &= \frac{9}{5} \cdot 15 + 32 \quad \text{Substituting 15 for } C \\ &= \frac{9 \cdot 3 \cdot 5}{5} + 32 \\ &= 9 \cdot 3 + 32 \\ &= 27 + 32 \\ &= 59 \end{aligned}$$

67. Discussion and Writing Exercise

69. First find the volume of one balloon. Since the diameter is 7 ft, the radius is $\dfrac{7 \text{ ft}}{2}$, is $\dfrac{7}{2}$ ft. We use the formula for the volume of a sphere.

$$\begin{aligned} V &= \frac{4}{3} \cdot \pi \cdot r^3 \\ &\approx \frac{4}{3} \cdot \frac{22}{7} \cdot \left(\frac{7}{2} \text{ ft}\right)^3 \\ &\approx \frac{4 \cdot 22 \cdot 7 \cdot 7 \cdot 7 \cdot \text{ ft}^3}{3 \cdot 7 \cdot 2 \cdot 2 \cdot 2} \\ &\approx \frac{2 \cdot 2 \cdot 2 \cdot 11 \cdot 7 \cdot 7 \cdot 7 \cdot \text{ ft}^3}{3 \cdot 7 \cdot 2 \cdot 2 \cdot 2} \\ &\approx \frac{2 \cdot 2 \cdot 2 \cdot 7}{2 \cdot 2 \cdot 2 \cdot 7} \cdot \frac{11 \cdot 7 \cdot 7 \cdot \text{ ft}^3}{3} \\ &\approx 179.67 \text{ ft}^3 \end{aligned}$$

Now find the volume of 42 balloons.
$$42(179.67 \text{ ft})^3 \approx 7546 \text{ ft}^3$$

71. Both cases are rectangular solids. First we find the volume of an audiocassette case:
$$7 \text{ cm} \cdot 10.75 \text{ cm} \cdot 1.5 \text{ cm} = 112.875 \text{ cm}^3$$

It holds $\dfrac{90 \text{ min}}{112.875 \text{ cm}^3}$, or about $0.797 \dfrac{\text{min}}{\text{cm}^3}$.

Next we find the volume of a compact-disc case:

$$12.4 \text{ cm} \cdot 14.1 \text{ cm} \cdot 1 \text{ cm} = 174.84 \text{ cm}^3$$

It holds $\dfrac{50 \text{ min}}{174.84 \text{ cm}^3}$, or about $0.286 \dfrac{\text{min}}{\text{cm}^3}$.

An audiocassette case holds more music per cubic centimeter.

73. Let r = the radius of the basketball. Then a side of the cube-shaped box that is just large enough to hold the basketball is the length of the diameter of the ball, or $2r$. The volume of the box is given by

$$V = l \cdot w \cdot h$$
$$= 2r \cdot 2r \cdot 2r = 8r^3.$$

We find r^3 using the formula for the volume of a sphere. Since the volume of the ball is 2304π cm^3 and its radius is r, we have

$$V = \frac{4}{3} \cdot \pi \cdot r^3$$
$$2304\pi \text{ cm}^3 = \frac{4}{3} \cdot \pi \cdot r^3$$
$$\frac{3}{4\pi} \cdot 2304\pi \text{ cm}^3 = \frac{3}{4\pi} \cdot \frac{4}{3} \cdot \pi \cdot r^3$$
$$1728 \text{ cm}^3 = r^3$$

Then the volume of the box is
$8r^3 = 8 \cdot 1728 \text{ cm}^3 = 13,824 \text{ cm}^3.$

75. First find the volume of one one-dollar bill in cubic inches:
$$V = l \cdot w \cdot h$$
$$= 6.0625 \text{ in.} \cdot 2.3125 \text{ in.} \cdot 0.0041 \text{ in.}$$
$$\approx 0.05748 \text{ in}^3 \quad \text{Rounding}$$

Then multiply to find the volume of one million one-dollar bills in cubic inches:

$1,000,000 \cdot 0.05748 \text{ in}^3 = 57,480 \text{ in}^3$

The volume of one million one-dollar bills is about $57,480 \text{ in}^3$.

Exercise Set 9.5

1. The angle can be named in six different ways:

angle GHI, angle IHG, angle H, $\angle GHI$, $\angle IHG$, or $\angle H$.

3. Another name for $\angle 1$ is $\angle BDA$ or $\angle ADB$.

5. Place the $\triangle$ of the protractor at the vertex of the angle, and line up one of the sides at $0°$. We choose the horizontal side. Since $0°$ is on the inside scale, we check where the other side of the angle crosses the inside scale. It crosses at $10°$. Thus, the measure of the angle is $10°$.

7. Place the $\triangle$ of the protractor at the vertex of the angle, point B. Line up one of the sides at $0°$. We choose the side that contains point A. Since $0°$ is on the outside scale, we check where the other side crosses the outside scale. It crosses at $180°$. Thus, the measure of the angle is $180°$.

9. Place the $\triangle$ of the protractor at the vertex of the angle, and line up one of the sides at $0°$. We choose the horizontal side. Since $0°$ is on the inside scale, we check where the other side crosses the inside scale. It crosses at $90°$. Thus, the measure of the angle is $90°$.

11. Every circle graph contains a total of $360°$.

10% of the circle is a $0.1(360°)$, or $36°$ angle;

45% of the circle is a $0.45(360°)$, or $162°$ angle;

35% of the circle is a $0.35(360°)$, or $126°$ angle;

To draw a $36°$ angle, draw a radius and use a protractor to mark off the angle. From that mark, draw a segment to the center of the circle to complete the angle. From that segment draw a $162°$ angle and, from the segment that completes that angle, draw a $126°$ angle. To confirm that the remainder of the circle is $36°$, we measure it using a protractor.

Snacking Habits

Occasionally 45% 162°
Never 10% 36°
Heavily 10% 36°
Moderately 35% 126°

Arrangement of sections may vary.

13. Every circle graph contains a total of $360°$.

3% of the circle is a $0.03(360°)$, or $10.8°$ angle;

25% of the circle is a $0.25(360°)$, or $90°$ angle;

27% of the circle is a $0.27(360°)$, or $97.2°$ angle;

18% of the circle is a $0.18(360°)$, or $64.8°$ angle.

To draw a $10.8°$ angle, draw a radius and use a protractor to mark off the angle. From that segment draw a $90°$ angle. From the segment that completes that angle draw a $97.2°$ angle. From the segment that completes that angle draw another $97.2°$ angle. To confirm that the remainder of the circle is $64.8°$, we measure it using a protractor.

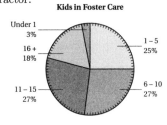

Kids in Foster Care

Under 1 3%
16+ 18%
1 – 5 25%
11 – 15 27%
6 – 10 27%

15. The measure of the angle in Exercise 5 is $10°$. Since its measure is greater than $0°$ and less than $90°$, it is an acute angle.

17. The measure of the angle in Exercise 7 is $180°$. It is a straight angle.

19. The measure of the angle in Exercise 9 is $90°$. It is a right angle.

21. The measure of the angle in Margin Exercise 1 is 30°. Since its measure is greater than 0° and less than 90°, it is an acute angle.

23. The measure of the angle in Margin Exercise 4 is 126°. Since its measure is greater than 90° and less than 180°, it is an obtuse angle.

25. $\angle 1$ and $\angle 3$ are vertical angles.

$\angle 2$ and $\angle 4$ are vertical angles.

27. $\angle GME$ and $\angle AMC$ are vertical angles. (We could also name them $\angle EMG$ and $\angle CMA$.)

$\angle AMG$ and $\angle EMC$ are vertical angles. (We could also name them $\angle GMA$ and $\angle CME$.)

29. $m\angle 2 = \underline{m\angle 4}$

$m\angle 3 = \underline{m\angle 1}$

31. $m\angle AMC = \underline{m\angle GME}$ (or $\underline{m\angle EMG}$)

$m\angle AMG = \underline{m\angle CME}$ (or $\underline{m\angle EMC}$)

33. Two angles are complementary if the sum of their measures is 90°.

$$90° - 11° = 79°.$$

The measure of a complement is 79°.

35. Two angles are complementary if the sum of their measures is 90°.

$$90° - 67° = 23°.$$

The measure of a complement is 23°.

37. Two angles are complementary if the sum of their measures is 90°.

$$90° - 58° = 32°.$$

The measure of a complement is 32°.

39. Two angles are complementary if the sum of their measures is 90°.

$$90° - 29° = 61°.$$

The measure of a complement is 61°.

41. Two angles are supplementary if the sum of their measures is 180°.

$$180° - 3° = 177°.$$

The measure of a supplement is 177°.

43. Two angles are supplementary if the sum of their measures is 180°.

$$180° - 139° = 41°.$$

The measure of a supplement is 41°.

45. Two angles are supplementary if the sum of their measures is 180°.

$$180° - 75° = 105°.$$

The measure of a supplement is 105°.

47. Two angles are supplementary if the sum of their measures is 180°.

$$180° - 104° = 76°.$$

The measure of a supplement is 76°.

49. All the sides are of different lengths. The triangle is a scalene triangle.

One angle is an obtuse angle. The triangle is an obtuse triangle.

51. All the sides are of different lengths. The triangle is a scalene triangle.

One angle is a right angle. The triangle is a right triangle.

53. All the sides are the same length. The triangle is an equilateral triangle.

All three angles are acute. The triangle is an acute triangle.

55. All the sides are of different lengths. The triangle is a scalene triangle.

One angle is an obtuse angle. The triangle is an obtuse triangle.

57. $m(\angle A) + m(\angle B) + m(\angle C) = 180°$

$$42° + 92° + x = 180°$$
$$134° + x = 180°$$
$$x = 180° - 134°$$
$$x = 46°$$

59. $31° + 29° + x = 180°$

$$60° + x = 180°$$
$$x = 180° - 60°$$
$$x = 120°$$

61. Discussion and Writing Exercise

63. $I = P \cdot r \cdot t$

$= \$2000 \cdot 8\% \cdot 1$

$= \$2000 \cdot 0.08 \cdot 1$

$= \$160$

65. $I = P \cdot r \cdot t$

$= \$4000 \cdot 7.4\% \cdot \dfrac{1}{2}$

$= \$4000 \cdot 0.074 \cdot \dfrac{1}{2}$

$= \$148$

67. $A = P \cdot \left(1 + \dfrac{r}{n}\right)^{n \cdot t}$

$= \$25,000 \cdot \left(1 + \dfrac{6\%}{2}\right)^{2 \cdot 5}$

$= \$25,000 \cdot \left(1 + \dfrac{0.06}{2}\right)^{10}$

$= \$25,000(1.03)^{10}$

$\approx \$33,597.91$

69. $A = P \cdot \left(1 + \dfrac{r}{n}\right)^{n \cdot t}$

$= \$150,000 \cdot \left(1 + \dfrac{7.4\%}{2}\right)^{2 \cdot 20}$

$= \$150,000 \cdot \left(1 + \dfrac{0.074}{2}\right)^{40}$

$= \$150,000(1.037)^{40}$

$\approx \$641,566.26$

71. $2^2 + 3^2 + 4^2$

$= 4 + 9 + 16$ Evaluating the exponential expressions

$= 13 + 16$ Adding in order

$= 29$ from left to right

73. Discussion and Writing Exercise

75. We find $m \angle 2$:

$\angle 1 + m \angle 2 + m \angle 3 = 180°$

$79.8° + m \angle 2 + 33.07° = 180°$

$112.87° + m \angle 2 = 180°$

$m \angle 2 = 180° - 112.87°$

$m \angle 2 = 67.13°$

The measure of angle 2 is $67.13°$.

$\angle 1$ and $\angle 4$ are vertical angles, so $m \angle 4 = m \angle 1 = 79.8°$.

$\angle 2$ and $\angle 5$ are vertical angles, so $m \angle 5 = m \angle 2 = 67.13°$.

$\angle 3$ and $\angle 6$ are vertical angles, so $m \angle 6 = m \angle 1 = 33.07°$.

77. $\angle ACB$ and $\angle ACD$ are complementary angles. Since $m \angle ACD = 40°$ and $90° - 40° = 50°$, we have $m \angle ACB = 50°$.

Now consider triangle ABC. We know that the sum of the measures of the angles is $180°$. Then

$m \angle ABC + m \angle BCA + m \angle CAB = 180°$

$50° + 90° + m \angle CAB = 180°$

$140° + m \angle CAB = 180°$

$m \angle CAB = 180° - 140°$

$m \angle CAB = 40°,$

so $m \angle CAB = 40°$.

To find $m \angle EBC$ we first find $m \angle CEB$. We note that $\angle DEC$ and $\angle CEB$ are supplementary angles. Since $m \angle DEC = 100°$ and $180° - 100° = 80°$, we have $m \angle CEB = 80°$. Now consider triangle BCE. We know that the sum of the measures of the angles is $180°$. Note that $\angle ACB$ can also be named $\angle BCE$. Then

$m \angle BCE + m \angle CEB + m \angle EBC = 180°$

$50° + 80° + m \angle EBC = 180°$

$130° + m \angle EBC = 180°$

$m \angle EBC = 180° - 130°$

$m \angle EBC = 50°,$

so $m \angle EBC = 50°$.

$\angle EBA$ and $\angle EBC$ are complementary angles. Since $m \angle EBC = 50°$ and $90° - 50° = 40°$, we have $m \angle EBA = 40°$.

$\angle AEB$ and $\angle DEC$ are vertical angles, so $m \angle AEB = m \angle DEC = 100°$.

To find $m \angle ADB$ we first find $m \angle EDC$. Consider triangle CDE. We know that the sum of the measures of the angles is $180°$. Then

$m \angle DEC + m \angle ECD + m \angle EDC = 180°$

$100° + 40° + m \angle EDC = 180°$

$140° + m \angle EDC = 180°$

$m \angle EDC = 180° - 140°$

$m \angle EDC = 40°,$

so $m \angle EDC = 40°$. We now note that $\angle ADB$ and $\angle EDC$ are complementary angles. Since $m \angle EDC = 40°$ and $90° - 40° = 50°$, we have $m \angle ADB = 50°$.

Exercise Set 9.6

1. The square roots of 16 are 4 and -4, because $4^2 = 16$ and $(-4)^2 = 16$.

3. The square roots of 121 are 11 and -11, because $11^2 = 121$ and $(-11)^2 = 121$.

5. The square roots of 169 are 13 and -13, because $13^2 = 169$ and $(-13)^2 = 169$.

7. The square roots of 2500 are 50 and -50, because $50^2 = 2500$ and $(-50)^2 = 2500$.

9. $\sqrt{64} = 8$

The square root of 64 is 8, because $8^2 = 64$ and 8 is positive.

11. $\sqrt{81} = 9$

The square root of 81 is 9, because $9^2 = 81$ and 9 is positive.

13. $\sqrt{225} = 15$

The square root of 225 is 15, because $15^2 = 225$ and 15 is positive.

15. $\sqrt{625} = 25$

The square root of 625 is 25 because $25^2 = 625$ and 25 is positive.

17. $\sqrt{400} = 20$

The square root of 400 is 20, because $20^2 = 400$ and 20 is positive.

19. $\sqrt{10,000} = 100$

The square root of 10,000 is 100 because $100^2 = 10,000$ and 100 is positive.

21. $\sqrt{48} \approx 6.928$

23. $\sqrt{8} \approx 2.828$

25. $\sqrt{3} \approx 1.732$

27. $\sqrt{12} \approx 3.464$

29. $\sqrt{19} \approx 4.359$

31. $\sqrt{110} \approx 10.488$

33. $a^2 + b^2 = c^2$ Pythagorean equation
$9^2 + 12^2 = c^2$ Substituting
$81 + 144 = c^2$
$225 = c^2$
$15 = c$

35. $a^2 + b^2 = c^2$ Pythagorean equation
$7^2 + 7^2 = c^2$ Substituting
$49 + 49 = c^2$
$98 = c^2$
$\sqrt{98} = c$ Exact answer
$9.899 \approx c$ Approximation

37. $a^2 + b^2 = c^2$
$a^2 + 12^2 = 13^2$
$a^2 + 144 = 169$
$a^2 + 144 - 144 = 169 - 144$
$a^2 = 169 - 144$
$a^2 = 25$
$a = 5$

39. $a^2 + b^2 = c^2$
$6^2 + b^2 = 9^2$
$36 + b^2 = 81$
$36 + b^2 - 36 = 81 - 36$
$b^2 = 81 - 36$
$b^2 = 45$
$b = \sqrt{45}$ Exact answer
$b \approx 6.708$ Approximation

41. $a^2 + b^2 = c^2$
$10^2 + 24^2 = c^2$
$100 + 576 = c^2$
$676 = c^2$
$26 = c$

43. $a^2 + b^2 = c^2$
$9^2 + b^2 = 15^2$
$81 + b^2 = 225$
$81 + b^2 - 81 = 225 - 81$
$b^2 = 225 - 81$
$b^2 = 144$
$b = 12$

45. $a^2 + b^2 = c^2$
$4^2 + 5^2 = c^2$
$16 + 25 = c^2$
$41 = c^2$
$\sqrt{41} = c$ Exact answer
$6.403 \approx c$ Approximation

47. $a^2 + b^2 = c^2$
$1^2 + b^2 = 32^2$
$1 + b^2 = 1024$
$1 + b^2 - 1 = 1024 - 1$
$b^2 = 1024 - 1$
$b^2 = 1023$
$b = \sqrt{1023}$ Exact answer
$b \approx 31.984$ Approximation

49. *Familiarize.* We first make a drawing. In it we see a right triangle. We let $s =$ the length of the string of lights.

Translate. We substitute 8 for a, 12 for b, and s for c in the Pythagorean equation.

$$a^2 + b^2 = c^2$$
$$8^2 + 12^2 = s^2$$

Solve. We solve the equation for w.

$$64 + 144 = s^2$$
$$208 = s^2$$
$$\sqrt{208} = s \qquad \text{Exact answer}$$
$$14.4 \approx s \qquad \text{Approximation}$$

Check. $8^2 + 12^2 = 64 + 144 = 208 = (\sqrt{208})^2$

State. The length of the string of lights is $\sqrt{208}$ ft, or about 14.4 ft.

51. *Familiarize.* We refer to the drawing in the text. We let $d =$ the distance from home to second base, in feet.

Translate. We substitute 65 for a, 65 for b, and d for c in the Pythagorean equation.

$$a^2 + b^2 = c^2$$
$$65^2 + 65^2 = d^2$$

Solve. We solve the equation for d.

$$4225 + 4225 = d^2$$
$$8450 = d^2$$
$$\sqrt{8450} = d$$
$$91.9 \approx d$$

Check. $65^2 + 65^2 = 4225 + 4225 = 8450 = (\sqrt{8450})^2$

State. The distance from home to second base is $\sqrt{8450}$ ft, or about 91.9 ft.

53. *Familiarize.* We refer to the drawing in the text.

Translate. We substitute in the Pythagorean equation.

$$a^2 + b^2 = c^2$$
$$20^2 + h^2 = 30^2$$

Solve. We solve the equation for h.

$$400 + h^2 = 900$$
$$400 + h^2 - 400 = 900 - 400$$
$$h^2 = 900 - 400$$
$$h^2 = 500$$
$$h = \sqrt{500}$$
$$h \approx 22.4$$

Check. $20^2 + (\sqrt{500})^2 = 400 + 500 = 900 = 30^2$

State. The height of the tree is $\sqrt{500}$ ft, or about 22.4 ft.

55. ***Familiarize.*** We refer to the drawing in the text. We let h = the plane's horizontal distance from the airport.

Translate. We substitute 4100 for a, h for b, and 15,100 for c in the Pythagorean equation.

$$a^2 + b^2 = c^2$$
$$4100^2 + h^2 = 15,100^2$$

Solve. We solve the equation for h.

$$16,810,000 + h^2 = 228,010,000$$
$$h^2 = 228,010,000 - 16,810,000$$
$$h^2 = 211,200,000$$
$$h = \sqrt{211,200,000}$$
$$h \approx 14,532.7$$

Check. $4100^2 + (\sqrt{211,200,000})^2 = 16,810,000 +$

$211,200,000 = 228,010,000 = 15,100^2$

State. The plane's horizontal distance from the airport is $\sqrt{211,200,000}$ ft, or about 14,532.7 ft.

57. Discussion and Writing Exercise

59. ***Familiarize.*** Let f = the amount the family spends for food.

Translate. We rephrase the question and translate.

What is 26% of \$1800?
$\downarrow$ $\quad$ $\downarrow$ $\downarrow$ $\downarrow$ $\quad$ $\downarrow$
f $\quad$ = 26% × 1800

Solve. Convert 26% to decimal notation and multiply.

$$f = 26\% \times 1800 = 0.26 \times 1800 = 468$$

Check. The answer seems reasonable since we are finding about one-fourth of 2000, which is 500. We can also repeat the calculation. The answer checks.

State. The family spends \$468 for food.

61. ***Familiarize.*** Let s = the number of students who are seniors.

Translate. We rephrase the question and translate.

What is 17.5% of 1850?
$\downarrow$ $\quad$ $\downarrow$ $\downarrow$ $\downarrow$ $\quad$ $\downarrow$
s $\quad$ = 17.5% × 1850

Solve. Convert 17.5% to decimal notation and multiply.

$$s = 17.5\% \times 1850 = 0.175 \times 1850 = 323.75 \approx 324$$

Check. We repeat the calculation. The answer checks.

State. About 324 students are seniors.

63. $2^3 = 2 \cdot 2 \cdot 2 = 8$

65. $4^3 = 4 \cdot 4 \cdot 4 = 64$

67. Discussion and Writing Exercise

69. We add some labels to the drawing of the polygon.

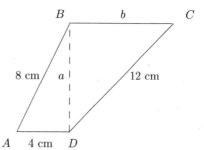

First we consider triangle ABD. We use the Pythagorean equation to find the length a.

$$a^2 + b^2 = c^2$$
$$a^2 + 4^2 = 8^2$$
$$a^2 + 16 = 64$$
$$a^2 + 16 - 16 = 64 - 16$$
$$a^2 = 48$$
$$a = \sqrt{48}$$
$$a \approx 6.93$$

Now we consider triangle BCD. We use the Pythagorean equation again to find the length b. We will use the exact value of a^2, 48, in this calculation.

$$a^2 + b^2 = c^2$$
$$48 + b^2 = 12^2$$
$$48 + b^2 = 144$$
$$48 + b^2 - 48 = 144 - 48$$
$$b^2 = 96$$
$$b = \sqrt{96}$$
$$b \approx 9.80$$

Next we find the area of each triangle.

For triangle ABD: $A = \dfrac{1}{2} \cdot b \cdot h$

$$= \frac{1}{2} \cdot 4 \text{ cm} \cdot 6.93 \text{ cm}$$
$$= \frac{27.72}{2} \text{ cm}^2$$
$$= 13.86 \text{ cm}^2$$

For triangle BCD: $A = \dfrac{1}{2} \cdot b \cdot h$

$$= \frac{1}{2} \cdot 9.80 \text{ cm} \cdot 6.93 \text{ cm}$$
$$= \frac{67.914}{2} \text{ cm}^2$$
$$= 33.957 \text{ cm}^2$$
$$\approx 33.96 \text{ cm}^2$$

We add these two areas to find the area of the polygon:

$$13.86 \text{ cm}^2 + 33.96 \text{ cm}^2 = 47.82 \text{ cm}^2$$

The area is about 47.82 cm². (Answers may vary slightly depending on when the rounding was done.)

71. ***Familiarize.*** Let d = the diagonal measurement of the screen in inches.

Translate. We substitute in the Pythagorean equation. We express the dimensions in decimal notation.

$$a^2 + b^2 = c^2$$
$$(31.75)^2 + (56.5)^2 = c^2$$

Solve. We solve the equation.
$$1008.0625 + 3192.25 = c^2$$
$$4200.3125 = c^2$$
$$64.8 \approx c$$

Check. We repeat the calculation. The answer checks.

State. The screen measures about 64.8 in. diagonally.

73. We let h = the height of a 19-in. screen and w = its width. We consider the following similar triangles.

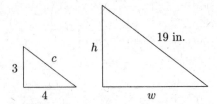

First we use the Pythagorean equation to find the length c.
$$a^2 + b^2 = c^2$$
$$3^2 + 4^2 = c^2$$
$$9 + 16 = c^2$$
$$25 = c^2$$
$$5 = c$$

Now we use proportions to find w and h.

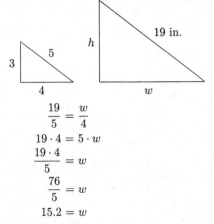

$$\frac{19}{5} = \frac{w}{4}$$
$$19 \cdot 4 = 5 \cdot w$$
$$\frac{19 \cdot 4}{5} = w$$
$$\frac{76}{5} = w$$
$$15.2 = w$$

The width of the screen is 15.2 in.
$$\frac{19}{5} = \frac{h}{3}$$
$$19 \cdot 3 = 5 \cdot h$$
$$\frac{19 \cdot 3}{5} = h$$
$$\frac{57}{5} = h$$
$$11.4 = h$$

The height of the screen is 11.4 in.

Exercise Set 9.7

---

1. 1 lb = 16 oz

This conversion relation is given in the text on page 655.

3. 8000 lb $= 8000 \text{ lb} \times \dfrac{1 \text{ T}}{2000 \text{ lb}}$ Writing 1 with tons on the top and pounds on the bottom

$\qquad = \dfrac{8000}{2000} \text{ T}$

$\qquad = 4 \text{ T}$

5. 3 lb $= 3 \times 1$ lb

$\qquad = 3 \times 16$ oz   Substituting 16 oz for 1 lb

$\qquad = 48$ oz

7. 4.5 T $= 4.5 \times 1$ T

$\qquad = 4.5 \times 2000$ lb   Substituting 2000 lb for 1 T

$\qquad = 9000$ lb

9. 4800 lb $= 4800 \text{ lb} \times \dfrac{1 \text{ T}}{2000 \text{ lb}}$ Writing 1 with tons on the top and pounds on the bottom

$\qquad = \dfrac{4800}{2000} \text{ T}$

$\qquad = 2.4 \text{ T}$

11. 72 oz $= 72 \text{ oz} \times \dfrac{1 \text{ lb}}{16 \text{ oz}}$

$\qquad = \dfrac{72}{16} \text{ lb}$

$\qquad = 4.5 \text{ lb}$

13. 4 kg = _____ g

Think: A kilogram is 1000 times the mass of a gram. Thus, we move the decimal point 3 places to the right.

4   4.000.

4 kg = 4000 g

15. 1 g = _____ kg

Think: It takes 1000 grams to have 1 kilogram. Thus, we move the decimal point 3 places to the left.

1   0.001.

1 g = 0.001 kg

17. 1 cg = _____ g

Think: It takes 100 centigrams to have 1 gram. Thus, we move the decimal point 2 places to the left.

1   0.01.

1 cg = 0.01 g

19. 1 g = _____ mg

Think: A gram is 1000 times the mass of a milligram. Thus, we move the decimal point 3 places to the right.

1   1.000.

1 g = 1000 mg

21. 1 g = _____ dg

Think: A gram is 10 times the mass of a decigram. Thus, we move the decimal point 1 place to the right.

1   1.0.

1 g = 10 dg

23. Complete: 934 kg = _____ g

Think: A kilogram is 1000 times the mass of a gram. Thus, we move the decimal point 3 places to the right.

934   934.000.

934 kg = 934,000 g

25. Complete: 6345 g = _____ kg

Think: It takes 1000 grams to have 1 kilogram. Thus, we move the decimal point 3 places to the left.

6345   6.345.

6345 g = 6.345 kg

27. 897 mg = _____ kg

Think: It takes 1,000,000 milligrams to have 1 kilogram. Thus, we move the decimal point 6 places to the left.

897   0.000897.

897 mg = 0.000897 kg

29. 7.32 kg = _____ g

Think: A kilogram is 1000 times the mass of a gram. Thus, we move the decimal point 3 places to the right.

7.32   7.320.

7.32 kg = 7320 g

31. Complete: 9350 g = _____ kg

Think: It takes 1000 grams to have 1 kilogram. Thus, we move the decimal point 3 places to the left.

9350   9.350.

9350 g = 9.35 kg

33. Complete: 69 mg = _____ cg

Think: It takes 10 milligrams to have 1 centigram. Thus, we move the decimal point 1 place to the left.

69   6.9.

69 mg = 6.9 cg

35. Complete: 8 kg = _____ cg

Think: A kilogram is 100,000 times the mass of a centigram. Thus, we move the decimal point 5 places to the right.

8   8.00000.

8 kg = 800,000 cg

37. 1 t = 1000 kg

This conversion relation is given in the text on page 656.

39. Complete: 3.4 cg = _____ dag

Think: It takes 1000 centigrams to have 1 dekagram. Thus, we move the decimal point 3 places to the left.

3.4   0.003.4

3.4 cg = 0.0034 dag

41. By laying a straightedge horizontally between the scales on page 658, we see that 178°F ≈ 80°C.

43. By laying a straightedge horizontally between the scales on page 658, we see that 140°F ≈ 60°C.

45. By laying a straightedge horizontally between the scales on page 658, we see that 68°F ≈ 20°C.

47. By laying a straightedge horizontally between the scales on page 658, we see that 10°F ≈ −10°C.

49. By laying a straightedge horizontally between the scales on page 658, we see that 86°C ≈ 190°F.

51. By laying a straightedge horizontally between the scales on page 658, we see that 58°C ≈ 140°F.

53. By laying a straightedge horizontally between the scales on page 658, we see that −10°C ≈ 10°F.

55. By laying a straightedge horizontally between the scales on page 658, we see that 5°C ≈ 40°F.

57. $F = \dfrac{9}{5} \cdot C + 32$

$F = \dfrac{9}{5} \cdot 30 + 32$

$= 54 + 32$

$= 86$

Thus, 30°C = 86°F.

59. $F = \dfrac{9}{5} \cdot C + 32$

$F = \dfrac{9}{5} \cdot 40 + 32$

$= 72 + 32$

$= 104$

Thus, 40°C = 104°F.

61. $F = \dfrac{9}{5} \cdot C + 32$

$F = \dfrac{9}{5} \cdot 3000 + 32$

$= 5400 + 32$

$= 5432$

Thus, 3000°C = 5432°F.

63. $C = \dfrac{5}{9} \cdot (F - 32)$

$\quad C = \dfrac{5}{9} \cdot (77 - 32)$

$\quad\quad = \dfrac{5}{9} \cdot 45$

$\quad\quad = 25$

Thus, $77°F = 25°C$.

65. $C = \dfrac{5}{9} \cdot (F - 32)$

$\quad C = \dfrac{5}{9} \cdot (131 - 32)$

$\quad\quad = \dfrac{5}{9} \cdot 99$

$\quad\quad = 55$

Thus, $131°F = 55°C$.

67. $C = \dfrac{5}{9} \cdot (F - 32)$

$\quad C = \dfrac{5}{9} \cdot (98.6 - 32)$

$\quad\quad = \dfrac{5}{9} \cdot 66.6$

$\quad\quad = 37$

Thus, $98.6°F = 37°C$.

69. a) $C = \dfrac{F - 32}{1.8}$

$\quad\quad C = \dfrac{136 - 32}{1.8}$

$\quad\quad\quad = \dfrac{104}{1.8}$

$\quad\quad\quad = 57.\overline{7}$

Thus, $136°F = 57.\overline{7}°C$.

$\quad F = \dfrac{9}{5} \cdot C + 32$

$\quad F = \dfrac{9}{5} \cdot 56\dfrac{2}{3} + 32$

$\quad\quad = \dfrac{9}{5} \cdot \dfrac{170}{3} + 32$

$\quad\quad = 102 + 32$

$\quad\quad = 134$

Thus, $56\dfrac{2}{3}°C = 134°F$.

b) $136°F - 134°F = 2°F$

The world record is $2°F$ higher than the U. S. record.

71. Discussion and Writing Exercise

73. When interest is paid on interest, it is called <u>compound</u> interest.

75. The <u>median</u> of a set of data is the middle number if there is an odd number of data items.

77. In <u>similar</u> triangles, the lengths of their corresponding sides have the same ratio.

79. A natural number, other than 1, that is not prime is <u>composite</u>.

81. Discussion and Writing Exercise

83. $1 \text{ lb} = 1\,\cancel{\text{lb}} \times \dfrac{453.6\,\cancel{\text{g}}}{1\,\cancel{\text{lb}}} \cdot \dfrac{1 \text{ kg}}{1000\,\cancel{\text{g}}} = \dfrac{453.6}{1000} \times 1 \text{ kg} = 0.4536 \text{ kg}$

85. a) First we find how many milligrams the Golden Jubilee Diamond weighs.

$\quad 545.67 \text{ carats} = 545.67 \times 1 \text{ carat}$

$\quad\quad\quad\quad\quad\quad\quad = 545.67 \times 200 \text{ mg}$

$\quad\quad\quad\quad\quad\quad\quad = 109,134 \text{ mg}$

To go from mg to g in the table is a move of 3 places to the left. Thus, we move the decimal point 3 places to the left:

$\quad 545.67 \text{ carats} = 109{,}134 \text{ mg} = 109.134 \text{ g}$

b) First we find how many milligrams the Hope Diamond weighs.

$\quad 45.52 \text{ carats} = 45.52 \times 1 \text{ carat}$

$\quad\quad\quad\quad\quad\quad\quad = 45.52 \times 200 \text{ mg}$

$\quad\quad\quad\quad\quad\quad\quad = 9104 \text{ mg}$

To go from mg to g in the table is a move of 3 places to the left. Thus, we move the decimal point 3 places to the left:

$\quad 45.52 \text{ carats} = 9104 \text{ mg} = 9.104 \text{ g}$

c) Golden Jubilee Diamond:

$\quad 109.134 \text{ g} = 109.134 \text{ g} \times \dfrac{1 \text{ lb}}{453.6 \text{ g}} \times \dfrac{16 \text{ oz}}{1 \text{ lb}}$

$\quad\quad\quad\quad = \dfrac{109.134 \times 16}{453.6} \times \dfrac{\text{g}}{\text{g}} \times \dfrac{\text{lb}}{\text{lb}} \times 1 \text{ oz}$

$\quad\quad\quad\quad \approx 3.85 \text{ oz}$

Hope Diamond:

$\quad 9.104 \text{ g} = 9.104 \text{ g} \times \dfrac{1 \text{ lb}}{453.6 \text{ g}} \times \dfrac{16 \text{ oz}}{1 \text{ lb}}$

$\quad\quad\quad = \dfrac{9.104 \times 16}{453.6} \times \dfrac{\text{g}}{\text{g}} \times \dfrac{\text{lb}}{\text{lb}} \times 1 \text{ oz}$

$\quad\quad\quad \approx 0.321 \text{ oz}$

87. First we find the volume of a sphere with diameter $5\dfrac{1}{2}$ cm.

$\quad r = \dfrac{d}{2}$

$\quad\quad = \left(5\dfrac{1}{2}\right) \div 2 = \dfrac{11}{2} \div 2$

$\quad\quad = \dfrac{11}{2} \cdot \dfrac{1}{2} = \dfrac{11}{4} \text{ cm}$

$\quad V = \dfrac{4}{3} \cdot \pi \cdot r^3$

$\quad\quad \approx \dfrac{4}{3} \cdot 3.14 \cdot \left(\dfrac{11}{4} \text{ cm}\right)^3$

$\quad\quad \approx \dfrac{4 \cdot 3.14 \cdot 1331 \text{ cm}^3}{3 \cdot 64}$

$\quad\quad \approx 87.07 \text{ cm}^3$

Now find the volume of a sphere with diameter 4 cm.

$\quad r = \dfrac{d}{2} = \dfrac{4 \text{ cm}}{2} = 2 \text{ cm}$

$$V = \frac{4}{3} \cdot \pi \cdot r^3$$

$$\approx \frac{4}{3} \cdot 3.14 \cdot (2 \text{ cm})^3$$

$$\approx \frac{4 \cdot 3.14 \cdot 8 \text{ cm}^3}{3}$$

$$\approx 33.49 \text{ cm}^3$$

Next we average the volumes of the two spheres.

$$\frac{87.07 \text{ cm}^3 + 33.49 \text{ cm}^3}{2} = \frac{120.56 \text{ cm}^3}{2} = 60.28 \text{ cm}^3$$

Since 1 cm^3 of water has a mass of 1 g, the mass of the egg is about 60 g.

89. First we find the volume of the shot put in cubic centimeters.

$$r = \frac{d}{2} = \frac{4.5 \text{ in.}}{2} = 2.25 \text{ in.}$$

2.25 in. = 2.25 × 1 in. ≈ 2.25 × 2.54 cm ≈ 5.715 cm

$$V = \frac{4}{3} \cdot \pi \cdot r^3$$

$$\approx \frac{4}{3} \cdot 3.14 \cdot (5.715 \text{ cm})^3$$

$$\approx 781.476 \text{ cm}^3$$

On page 672 of the text we are told that 1 kg ≈ 2.2 lb. Since 1 kg = 1000 g, we know that 1000 g ≈ 2.2 lb. We use this fact to convert the weight of the shot put to grams.

$$8.8 \text{ lb} \approx 8.8 \text{ lb} \times \frac{1000 \text{ g}}{2.2 \text{ lb}} \approx \frac{8800}{2.2} \text{ g} = 4000 \text{ g}$$

Finally, we divide to find the mass per cubic centimeter.

$$\frac{4000 \text{ g}}{781.476 \text{ cm}^3} \approx 5.1 \text{ g/cm}^3$$

Thus, the mass per cubic centimeter of the shot put is about 5.1 g/cm^3.

91. First convert $15\frac{17}{10}$ lb to ounces.

$$15\frac{17}{20} \text{ lb} = 15\frac{17}{20} \times 1 \text{ lb}$$

$$= \frac{317}{20} \times 16 \text{ oz}$$

$$= \frac{317 \times 16}{20} \text{ oz}$$

$$= \frac{1268}{5} \text{ oz}$$

Now we divide to find the number of packages in the shrink-wrapped brick of boxes.

$$\frac{1268}{5} \div 1\frac{3}{4} = \frac{1268}{5} \div \frac{7}{4}$$

$$= \frac{1268}{5} \cdot \frac{4}{7}$$

$$= \frac{5072}{35}$$

$$= 144\frac{32}{35}$$

We see that the weight of the brick consists of the weight of 144 boxes of pudding plus the weight of $\frac{32}{35}$ of another box. Since the brick does not contain partial boxes of pudding, we know that there are 144 boxes in the brick.

The additional weight is the weight of the plastic shrink wrap. It is $\frac{32}{35}$ of $1\frac{3}{4}$ oz:

$$\frac{32}{35} \cdot 1\frac{3}{4} = \frac{32}{35} \cdot \frac{7}{4}$$

$$= \frac{4 \cdot 8 \cdot 7}{5 \cdot 7 \cdot 4}$$

$$= \frac{8}{5}, \text{ or } 1.6$$

The weight of the shrink wrap is 1.6 oz. We could also convert this to pounds.

$$1.6 \text{ oz} = 1.6 \text{ oz} \cdot \frac{1 \text{ lb}}{16 \text{ oz}}$$

$$= \frac{1.6}{16} \times 1 \text{ lb}$$

$$= 0.1 \text{ lb}$$

Exercise Set 9.8

1. To convert L to mL, move the decimal point 3 places to the right: 2.0 L = 2000 mL.

3. If Rick takes 4 mg of doxazosin each day for 30 days, he takes a total of 30 · 4, or 120 mg. Now we convert 120 mg to grams. There are 1000 mg in 1 gram, so we move the decimal point 3 places to the left.

120   0.120.

120 mg = 0.12 g

Rick takes 0.12 g of doxazosin.

5. First convert 500 mg to grams by moving the decimal point three places to the left: 500 mg = 0.5 g.

Then divide to determine the number of 500 mg tablets that would have to be taken.

$$\begin{array}{r} 4. \\ 0.5\overline{)2.0} \\ \underline{2\,0} \\ 0 \end{array}$$

The patient would have to take 4 tablets per day.

7. a) 90 mcg × 2 × 4 = 720 mcg

b) 90 mcg = 90 × 0.000001 g = 0.00009 g = 0.09 mg

$$\frac{17 \text{ mg}}{0.09 \text{ mg}} \approx 189 \text{ actuations}$$

c) Find the number of actuations needed for a 4-month period. (Let 1 month = 30 days.)

$$2 \text{ actuations} \cdot \frac{4}{1 \text{ day}} \cdot \frac{30 \text{ days}}{1 \text{ month}} \cdot 4 \text{ month} =$$

960 actuations

From part (b) we know that one inhaler contains approximately 189 actuations. We divide to find the number of inhalers needed.

$$\frac{960}{189} \approx 5.08$$

Danielle will need 6 inhalers.

9. $0.5 \text{ L} \approx 0.5 \text{ qt} = 0.5 \times 32 \text{ oz} = 16 \text{ oz}$

The 16 oz bottle comes closest to filling the prescription.

11. a) First we multiply to find the amount of the drug originally prescribed in mg.

$$14 \text{ tablets} \cdot 30 \text{ mg/tablet} = 420 \text{ mg}$$

Then we convert 420 mg to grams.

$$420 \text{ mg} = 420 \times 0.001 \text{ g} = 0.42 \text{ g}$$

b) We can divide to find the number of 15 mg doses in the original prescription.

$$420 \text{ mg} \div 15 \text{ mg/dose} = 28 \text{ doses}$$

We could also observe that 15 mg is half of 30 mg, so each of Joanne's original 14 doses can be split in half, yielding $2 \cdot 14$, or 28, doses of 15 mg per dose.

13. We can use a proportion. Let $a =$ the amount of amoxicillin suspension that would have to be administered, in mL.

$$\begin{array}{l} \text{Amoxicillin} \rightarrow \dfrac{125}{5} = \dfrac{150}{a} \leftarrow \text{Amoxicillin} \\ \text{Suspension} \rightarrow \phantom{\dfrac{125}{}} \phantom{\dfrac{150}{}} \leftarrow \text{Suspension} \end{array}$$

We equate cross products and solve for a.

$$\frac{125}{5} = \frac{150}{a}$$
$$125 \cdot a = 5 \cdot 150$$
$$a = \frac{5 \cdot 150}{125}$$
$$a = 6$$

The parent should administer 6 mL of the amoxicillin suspension.

15. $1 \text{ mg} = 0.001 \text{ g}$
$$= 0.001 \times 1 \text{ g}$$
$$= 0.001 \times 1,000,000 \text{ mcg}$$
$$= 1000 \text{ mcg}$$

17. $325 \text{ mcg} = 325 \times 1 \text{ mcg}$
$$= 325 \times \frac{1}{1,000,000} \text{ g}$$
$$= 0.000325 \text{ g}$$
$$= 0.325 \text{ mg}$$

19. $0.25 \text{ mg} = 0.25 \times 1 \text{ mg}$
$$= 0.25 \times 0.001 \text{ g}$$
$$= 0.00025 \text{ g}$$
$$= 0.00025 \times 1 \text{ g}$$
$$= 0.00025 \times 1,000,000 \text{ mcg}$$
$$= 250 \text{ mcg}$$

There are 250 mcg in the dose of medication.

21. $0.125 \text{ mg} = 0.000125 \text{ g}$
$$= 0.000125 \times 1 \text{ g}$$
$$= 0.000125 \times 1,000,000 \text{ mcg}$$
$$= 125 \text{ mcg}$$

23. We multiply to find the number of milligrams that will be ingested.

$$\begin{array}{r} 0.125 \\ \times 7 \\ \hline 0.875 \end{array}$$

The patient will ingest 0.875 mg of Triazolam. Now convert 0.875 mg to micrograms.

$$0.875 \text{ mg} = 0.000875 \text{ g}$$
$$= 0.000875 \times 1 \text{ g}$$
$$= 0.000875 \times 1,000,000 \text{ mcg}$$
$$= 875 \text{ mcg}$$

25. Discussion and Writing Exercise

27.
$$\begin{array}{r} 5789 \\ -2431 \\ \hline 3358 \end{array}$$

29.
$$\begin{array}{r} {\scriptstyle 3\ 10} \\ 4\cancel{0}97 \\ -3243 \\ \hline 854 \end{array}$$

31. $7x + 9 - 2x - 1 = 7x - 2x + 9 - 1$
$$= (7 - 2)x + 9 - 1$$
$$= 5x + 8$$

33. $8t - 5 - t - 4 = 8t - t - 5 - 4$
$$= (8 - 1)t - 5 - 4$$
$$= 7t - 9$$

35. Discussion and Writing Exercise

37. a) First we find the total amount of medication the patient took in mg. Three doses a day for one week is $3 \cdot 7$, or 21 doses. At 200 mg per dose, we have

$$21 \text{ doses} \cdot 200 \text{ mg/dose} = 4200 \text{ mg}.$$

Two doses a day for one week is $2 \cdot 7$, or 14 doses.

At 200 mg per dose, we have

$$14 \text{ doses} \cdot 200 \text{ mg/dose} = 2800 \text{ mg}.$$

Returning to three doses per day for one week we have 21 doses again. At 100 mg per dose, we have

$$21 \text{ doses} \cdot 100 \text{ mg/dose} = 2100 \text{ mg}.$$

We add to find the total number of milligrams:

$$4200 \text{ mg} + 2800 \text{ mg} + 2100 \text{ mg} = 9100 \text{ mg}.$$

Now we convert to grams.

$$9100 \text{ mg} = 9100 \times 1 \text{ mg}$$
$$= 9100 \times 0.001 \text{ g}$$
$$= 9.1 \text{ g}$$

Altogether, 9.1 g of medication are used.

b) First we add to find the number of doses. Then we divide the total dosage by the number of doses to find the average dosage size.

$$21 + 14 + 21 = 56 \text{ doses}$$
$$\frac{9100 \text{ mg}}{56 \text{ doses}} = 162.5 \text{ mg per dose}$$

(We could also express this as 0.1625 g per dose.)

39. We will compare the daily cost of each drug.

Since naproxen sodium lasts 12 hr, 2 doses per day are required (24 hr ÷ 12 hr = 2). At 220 mg per dose, a total of 2·220 mg, or 440 mg, are required daily. Now we convert 440 mg to grams.

$$440 \text{ mg} = 440 \times 0.001 \text{ g} = 0.44 \text{ g}$$

In a bottle containing 44 g naproxen sodium, the number of daily doses is

$$44 \text{ g} \div 0.44 \text{ g/dose} = 100 \text{ doses}.$$

At $11 per bottle, the cost of a daily dose of naproxen sodium is

$$\$11 \div 100 = \$0.11.$$

Since each dose of ibuprofen lasts 6 hours, 4 doses per day are required (24 hr ÷ 6 hr = 4). At 2 · 200 mg, or 400 mg, per dose, then 4 · 400 mg, or 1600 mg, are required daily. Now we convert 1600 mg to grams.

$$1600 \text{ mg} = 1600 \times 0.001 \text{ g} = 1.6 \text{ g}$$

In a bottle containing 72 g of ibuprofen, the number of daily doses is

$$72 \text{ g} \div 1.6 \text{ g/dose} = 45 \text{ doses}.$$

At $11.24 per bottle, the cost of a daily dose of ibuprofen is

$$\$11.24 \div 45 \approx \$0.25.$$

Since the daily cost of naproxen sodium is less than the daily cost of ibuprofen, the naproxen sodium is a better purchase.

Chapter 9 Review Exercises

1. $10 \text{ ft} = 10 \cancel{\text{ft}} \times \dfrac{1 \text{ yd}}{3 \cancel{\text{ft}}}$

$= \dfrac{10}{3} \times 1 \text{ yd}$

$= 3\dfrac{1}{3} \text{ yd, or } 3.\overline{3} \text{ yd}$

2. $\dfrac{5}{6} \text{ yd} = \dfrac{5}{6} \times 1 \text{ yd}$

$= \dfrac{5}{6} \times 36 \text{ in.}$

$= \dfrac{5 \times 36}{6} \times 1 \text{ in.}$

$= 30 \text{ in.}$

3. 1.7 mm = _____ cm

Think: To go from mm to cm in the table is a move of 1 place to the left. Thus, we move the decimal point 1 place to the left.

1.7   0.1.7

1.7 mm = 0.17 cm

4. 6 m = _____ km

Think: To go from m to km in the table is a move of 3 places to the left. Thus, we move the decimal point 3 places to the left.

6   0.006.

6 m = 0.006 km

5. 4 km = _____ cm

Think: To go from km to cm in the table is a move of 5 places to the right. Thus, we move the decimal point 5 places to the right.

4   4.00000.

4 km = 400,000 cm

6. $14 \text{ in.} = 14 \cancel{\text{in.}} \times \dfrac{1 \text{ ft}}{12 \cancel{\text{in.}}}$

$= \dfrac{14}{12} \times 1 \text{ ft}$

$= \dfrac{7}{6} \text{ ft, or } 1\dfrac{1}{6} \text{ ft}$

7. $5 \text{ lb} = 5 \times 1 \text{ lb}$

$= 5 \times 16 \text{ oz}$

$= 80 \text{ oz}$

8. Complete: 3 g = _____ kg

Think: To go from g to kg in the table is a move of 3 places to the left. Thus, we move the decimal point 3 places to the left.

3   0.003.

3 g = 0.003 kg

9. $50 \text{ qt} = 50 \cancel{\text{qt}} \times \dfrac{1 \text{ gal}}{4 \cancel{\text{qt}}}$

$= \dfrac{50}{4} \times 1 \text{ gal}$

$= 12.5 \text{ gal}$

10. $28 \text{ gal} = 28 \times 1 \text{ gal}$

$= 28 \times 4 \text{ qt}$

$= 28 \times 4 \times 1 \text{ qt}$

$= 28 \times 4 \times 2 \text{ pt}$

$= 224 \text{ pt}$

11. $60 \text{ mL} = 60 \times 1 \text{ mL}$

$= 60 \times 0.001 \text{ L}$

$= 0.06 \text{ L}$

12. $0.4 \text{ L} = 0.4 \times 1 \text{ L}$

$= 0.4 \times 1000 \text{ mL}$

$= 400 \text{ mL}$

13. $0.7 \text{ T} = 0.7 \times 1 \text{ T}$

$= 0.7 \times 2000 \text{ lb}$

$= 1400 \text{ lb}$

14. 0.2 g = _____ mg

Think: A gram is 1000 times the mass of a milligram. Thus, we move the decimal point 3 places to the right.

0.2   0.200.
     └___↑

0.2 g = 200 mg

15. Complete: 4.7 kg = _____ g

Think: A kilogram is 1000 times the mass of a gram. Thus, we move the decimal point 3 places to the right.

4.7   4.700.
     └___↑

4 kg = 4700 g

16. 4 cg = _____ g

Think: It takes 100 centigrams to have 1 gram. Thus, we move the decimal point 2 places to the left.

4   0.04.
   ↑__┘

4 cg = 0.04 g

17. $4 \text{ yd}^2 = 4 \times 1 \text{ yd}^2$

$= 4 \times 9 \text{ ft}^2$

$= 36 \text{ ft}^2$

18. $0.7 \text{ km}^2 = $ _____ m^2

Think: To go from km to m in the diagram is a move of 3 places to the right. So we move the decimal point $2 \cdot 3$, or 6 places to the right.

0.7   0.700000.
     └_____↑

$0.7 \text{ km}^2 = 700{,}000 \text{ m}^2$

19. $1008 \text{ in}^2 = 1008 \text{ in}^2 \times \dfrac{1 \text{ ft}^2}{144 \text{ in}^2}$

$= \dfrac{1008}{144} \times 1 \text{ ft}^2$

$= 7 \text{ ft}^2$

20. $570 \text{ cm}^2 = $ _____ m^2

Think: To go from cm to m in the diagram is a move of 2 places to the left. So we move the decimal point $2 \cdot 2$, or 4 places to the left.

570   0.0570.
    ↑__┘

$570 \text{ cm}^2 = 0.057 \text{ m}^2$

21. $C = 2 \cdot \pi \cdot r$

$\approx 2 \cdot 3.14 \cdot 5 \text{ m} \approx 31.4 \text{ m}$

22. $r = \dfrac{d}{2} = \dfrac{\frac{28}{11} \text{ in.}}{2} = \dfrac{28}{11} \text{ in.} \cdot \dfrac{1}{2}$

$= \dfrac{28}{11 \cdot 2} \text{ in.} = \dfrac{2 \cdot 14}{11 \cdot 2} \text{ in.}$

$= \dfrac{14}{11} \text{ in., or } 1\dfrac{3}{11} \text{ in.}$

23. $d = 2 \cdot r = 2 \cdot 12 \text{ m} = 24 \text{ m}$

24. The shortest distance is composed of two semicircular lengths and two straight 85.56 yd segments. The two semicircles are equivalent to a single circle with an 85.56 yd diameter. First we find the circumference of the circle.

$C = \pi \cdot d$

$\approx 3.14 \cdot 85.56 \text{ yd} \approx 268.6584 \text{ yd}$

Then the total distance is 268.6584 yd + 85.56 yd+ 85.5 yd = 439.7784 yd.

25. $A = \dfrac{1}{2} \cdot h \cdot (a + b)$ Area of a trapezoid

$= \dfrac{1}{2} \cdot 3 \text{ in.} \cdot (7 + 5) \text{ in.}$

$= \dfrac{3 \cdot 12}{2} \text{ in}^2$

$= \dfrac{3 \cdot 2 \cdot 6}{2 \cdot 1} \text{ in}^2$

$= 18 \text{ in}^2$

26. $A = b \cdot h$ Area of a parallelogram

$= 7.1 \text{ ft} \cdot 4.2 \text{ ft}$

$= 29.82 \text{ ft}^2$

27. $A = b \cdot h$

$A = 12 \text{ cm} \cdot 5 \text{ cm}$

$A = 60 \text{ cm}^2$

28. $A = \dfrac{1}{2} \cdot h \cdot (a + b)$

$A = \dfrac{1}{2} \cdot 5 \text{ mm} \cdot (4 + 10) \text{ mm}$

$A = \dfrac{5 \cdot 14}{2} \text{ mm}^2$

$A = 35 \text{ mm}^2$

29. $A = \pi \cdot r \cdot r$

$\approx \dfrac{22}{7} \cdot 7 \text{ ft} \cdot 7 \text{ ft} \approx \dfrac{22}{7} \cdot 49 \text{ ft}^2 = 154 \text{ ft}^2$

30. $A = \pi \cdot r \cdot r$

$\approx 3.14 \cdot 10 \text{ cm} \cdot 10 \text{ cm} = 3.14 \cdot 100 \text{ cm}^2 = 314 \text{ cm}^2$

31. The shaded area is the area of a circle with radius of 21 ft less the area of a circle with a diameter of 21 ft. The radius of the smaller circle is $\dfrac{21 \text{ ft}}{2}$, or 10.5 ft.

$A = \pi \cdot 21 \text{ ft} \cdot 21 \text{ ft} - \pi \cdot 10.5 \text{ ft} \cdot 10.5 \text{ ft}$

$A \approx 3.14 \cdot 21 \text{ ft} \cdot 21 \text{ ft} - 3.14 \cdot 10.5 \text{ ft} \cdot 10.5 \text{ ft}$

$A = 1384.74 \text{ ft}^2 - 346.185 \text{ ft}^2$

$A = 1038.555 \text{ ft}^2$

32. The window is composed of half of a circle with radius 2 ft and of a rectangle with length 5 ft and width twice the radius of the half circle, or $2 \cdot 2$ ft, or 4 ft. To find the area of the window we add one-half the area of a circle with radius 2 ft and the area of a rectangle with length 5 ft and width 4 ft.

$$A = \frac{1}{2} \cdot \pi \cdot 2 \text{ ft} \cdot 2 \text{ ft} + 5 \text{ ft} \cdot 4 \text{ ft}$$

$$\approx \frac{1}{2} \cdot 3.14 \cdot 2 \text{ ft} \cdot 2 \text{ ft} + 5 \text{ ft} \cdot 4 \text{ ft}$$

$$= \frac{3 \cdot 14 \cdot 2 \cdot \cancel{2}}{\cancel{2}} \text{ ft}^2 + 20 \text{ ft}^2$$

$$= 6.28 \text{ ft}^2 + 20 \text{ ft}^2$$

$$= 26.28 \text{ ft}^2$$

33. Two angles are complementary if the sum of their measures is 90°.

$$90° - 41° = 49°.$$

The measure of a complement of $\angle BAC$ is 49°.

Two angles are supplementary if the sum of their measures is 180°.

$$180° - 41° = 139°.$$

The measure of a supplement is 139°.

34. $\angle PNM$ and $\angle SNR$ are vertical angles.

$\angle MNR$ and $\angle PNS$ are vertical angles.

35.
$$30° + 90° + x = 180°$$
$$120° + x = 180°$$
$$x = 180° - 120°$$
$$x = 60°$$

36. All the sides are of different lengths. The triangle is a scalene triangle.

37. One angle is a right angle. The triangle is a right triangle.

38.
$$V = l \cdot w \cdot h$$
$$V = 12 \text{ m} \cdot 3 \text{ m} \cdot 2.6 \text{ m}$$
$$V = 36 \cdot 2.6 \text{ m}^3$$
$$V = 93.6 \text{ m}^3$$

39.
$$V = l \cdot w \cdot h$$
$$V = 4.6 \text{ cm} \cdot 3 \text{ cm} \cdot 14 \text{ cm}$$
$$V = 13.8 \cdot 14 \text{ cm}^3$$
$$V = 193.2 \text{ cm}^3$$

40.
$$V = B \cdot h = \pi \cdot r^2 \cdot h$$
$$\approx 3.14 \times 10 \text{ cm} \times 10 \text{ cm} \times 90 \text{ cm}$$
$$= 28,260 \text{ cm}^3$$

41.
$$V = \frac{4}{3} \cdot \pi \cdot r^3$$
$$\approx \frac{4}{3} \times 3.14 \times (2 \text{ yd})^3$$
$$= \frac{4 \times 3.14 \times 8 \text{ yd}^3}{3}$$
$$= 33.49\overline{3} \text{ yd}^3$$

This result can also be expressed as $33\frac{37}{75}$ yd^3.

42.
$$V = B \cdot h = \pi \cdot r^2 \cdot h$$
$$\approx 3.14 \times 5 \text{ cm} \times 5 \text{ cm} \times 12 \text{ cm}$$
$$= 942 \text{ cm}^3$$

43. $\sqrt{64} = 8$ because $8 \cdot 8 = 64$.

44.
$$a^2 + b^2 = c^2$$
$$15^2 + 25^2 = c^2$$
$$225 + 625 = c^2$$
$$850 = c^2$$
$$\sqrt{850} = c \qquad \text{Exact answer}$$
$$29.155 \approx c \qquad \text{Approximation}$$

45.
$$a^2 + b^2 = c^2$$
$$4^2 + b^2 = 10^2$$
$$16 + b^2 = 100$$
$$b^2 = 100 - 16$$
$$b^2 = 84$$
$$b = \sqrt{84} \qquad \text{Exact answer}$$
$$b \approx 9.165 \qquad \text{Approximation}$$

46.
$$a^2 + b^2 = c^2$$
$$5^2 + 8^2 = c^2$$
$$25 + 64 = c^2$$
$$89 = c^2$$
$$\sqrt{89} = c$$
$$9.434 \approx c$$
$$c = \sqrt{89} \text{ ft, or approximately } 9.434 \text{ ft.}$$

47.
$$a^2 + b^2 = c^2$$
$$a^2 + 18^2 = 20^2$$
$$a^2 + 324 = 400$$
$$a^2 = 400 - 324$$
$$a^2 = 76$$
$$a = \sqrt{76} \approx 8.718$$
$$a = \sqrt{76} \text{ cm, or approximately } 8.718 \text{ cm}$$

48. *Familiarize*. Referring to the drawing in the text, we see that we have a right triangle and that $h = $ the height of the tree.

***Translate*.** Substitute 40 for a, h for b, and 60 for c in the Pythagorean equation.

$$a^2 + b^2 = c^2$$
$$40^2 + h^2 = 60^2$$

***Solve*.** We solve the equation for h.

$$1600 + h^2 = 3600$$
$$h^2 = 2000$$
$$h = \sqrt{2000} \approx 44.7$$

***Check*.** $40^2 + (\sqrt{2000})^2 = 1600 + 2000 = 3600 = 60^2$

***State*.** The tree is $\sqrt{2000}$ ft, or approximately 44.7 ft tall.

49.
$$F = 1.8 \cdot C + 32$$
$$F = 1.8 \cdot 35 + 32$$
$$= 63 + 32$$
$$= 95$$

Thus, $35°$C $= 95°$F.

50. $C = \dfrac{5}{9} \cdot (F - 32)$

$C = \dfrac{5}{9} \cdot (68 - 32)$

$= \dfrac{5}{9} \cdot 36$

$= 20$

Thus, $68°F = 20°C$.

51. We use a proportion. Let $a =$ the number of mL of amox-icillin to be administered.

$\dfrac{125}{200} = \dfrac{5}{a}$

$125 \cdot a = 200 \cdot 5$

$a = \dfrac{200 \cdot 5}{125}$

$a = \dfrac{25 \cdot 8 \cdot 5}{25 \cdot 5 \cdot 1} = \dfrac{25 \cdot 5}{25 \cdot 5} \cdot \dfrac{8}{1}$

$a = 8$

The parent should administer 8 mL of amoxicillin.

52. $3 \text{ L} = 3 \times 1 \text{ L}$

$= 3 \times 1000 \text{ mL}$

$= 3000 \text{ mL}$

53. $0.5 \text{ mg} = 0.0005 \text{ g}$

$= 0.0005 \times 1 \text{ g}$

$= 0.0005 \times 1,000,000 \text{ mcg}$

$= 500 \text{ mcg}$

54. *Discussion and Writing Exercise.* A square is a parallel-ogram because it is a four-sided figure with two pairs of parallel sides.

55. *Discussion and Writing Exercise.* No, the sum of the three angles of a triangle is $180°$. If there are two $90°$ angles, totaling $180°$, the third angle would be $0°$. A triangle cannot have an angle of $0°$.

56. *Discussion and Writing Exercise.* Since 1 m is slightly more than 1 yd, it follows that 1 m^3 is larger than 1 yd^3. Since $1 \text{ yd}^3 = 27 \text{ ft}^3$, we see that 1 m^3 is larger than 27 ft^3.

57. *Discussion and Writing Exercise.* Since 1 kg is about 2.2 lb and 32 oz is $\dfrac{32}{16}$, or 2 lb, 1 kg weighs more than 32 oz.

58. The largest possible diameter of the pizza is 35 cm. Then the radius is given by

$r = \dfrac{d}{2} = \dfrac{35 \text{ cm}}{2} = 17.5 \text{ cm}.$

Now we find the area of the pizza.

$A = \pi \cdot r \cdot r$

$\approx 3.14 \cdot 17.5 \text{ cm} \cdot 17.5 \text{ cm}$

$\approx 961.625 \text{ cm}^2$

(If we use the π key on a calculator, the result is about 962.113 cm^2.)

59. First we convert 26 miles to yards.

$26 \text{ mi} = 26 \times 1 \text{ mi}$

$= 26 \times 1760 \text{ yd}$

$= 45,760 \text{ yd}$

Then 26 mi, 385 yd is equivalent to $45,760 \text{ yd} + 385 \text{ yd}$, or $46,145 \text{ yd}$. We divide to find the number of laps required.

$\dfrac{46,145}{440} = 104.875 \text{ laps}$

60. *Familiarize.* This is a multistep problem. First we find the area of a cross-section of a finished board and of a rough board using the formula $A = l \cdot w$. Then we find the amount of wood removed in planing and drying and finally we find the percent of wood removed. Let $f =$ the area of a cross-section of a finished board and let $r =$ the area of a cross-section of a rough board.

Translate. We find the areas.

$f = 3\dfrac{1}{2} \cdot 1\dfrac{1}{2}$

$r = 4 \cdot 2$

Solve. We carry out the multiplications.

$f = 3\dfrac{1}{2} \cdot 1\dfrac{1}{2} = \dfrac{7}{2} \cdot \dfrac{3}{2} = \dfrac{21}{4}$

$r = 4 \cdot 2 = 8$

Now we subtract to find the amount of wood removed in planing and drying.

$8 - \dfrac{21}{4} = \dfrac{32}{4} - \dfrac{21}{4} = \dfrac{11}{4}$

Finally we find p, the percent of wood removed in planing and drying.

$\dfrac{11}{4}$ is what percent of 8?

$\dfrac{11}{4} = p \cdot 8$

We solve the equation.

$\dfrac{11}{4} = p \cdot 8$

$\dfrac{1}{8} \cdot \dfrac{11}{4} = p$

$\dfrac{11}{32} = p$

$0.34375 = p$

$34.375\% = p$, or

$34\dfrac{3}{8}\% = p$

Check. We repeat the calculations. The answer checks.

State. 34.375%, or $34\dfrac{3}{8}\%$, of the wood is removed in planing and drying.

61. The depth of the water will be $10 \text{ ft} - 1 \text{ ft}$, or 9 ft. Then the volume of the water is given by

$V = l \cdot w \cdot h$

$= 100 \text{ ft} \cdot 50 \text{ ft} \cdot 9 \text{ ft} = 45,000 \text{ ft}^3.$

Now we divide to find how many 1000 ft^3 are in $45,000 \text{ ft}^3$:

$$\frac{45,000}{1000} = 45$$

Finally we multiply to find the cost of the water.

$$\$2.25 \cdot 45 = \$101.25$$

Chapter 9 Test

1. $8 \text{ ft} = 8 \times 1 \text{ ft}$

$= 8 \times 12 \text{ in.}$

$= 96 \text{ in.}$

2. $280 \text{ cm} = \underline{\hspace{1cm}} \text{ m}$

Think: To go from cm to m in the table is a move of 2 places to the left. Thus, we move the decimal point 2 places to the left.

$280 \text{ cm} = 2.8 \text{ m}$

3. $2 \text{ yd}^2 = 2 \cdot 9 \text{ ft}^2 = 18 \text{ ft}^2$

4. $5 \text{ km} = \underline{\hspace{1cm}} \text{ m}$

Think: To go from km to m in the table is a move of 3 places to the right. Thus, we move the decimal point 3 places to the right.

5   5.000.

$5 \text{ km} = 5000 \text{ m}$

5. $9.1 \text{ mm} = \underline{\hspace{1cm}} \text{ cm}$

Think: To go from mm to cm in the table is a move of 1 place to the left. Thus, we move the decimal point 1 place to the left.

9.1   0.9.1

$9.1 \text{ mm} = 0.91 \text{ cm}$

6. $4520 \text{ m}^2 = \underline{\hspace{1cm}} \text{ km}^2$

Think: To convert from m to km, we shift the decimal point three places to the left. To convert from m^2 to km^2, we shift the decimal point six places to the left.

$4520 \text{ m}^2 = 0.00452 \text{ km}^2$

7. $2983 \text{ mL} = 2983 \times 1 \text{ mL}$

$= 2983 \times 0.001 \text{ L}$

$= 2.983 \text{ L}$

8. $3.8 \text{ kg} = \underline{\hspace{1cm}} \text{ g}$

Think: To go from kg to g in the table is a move of 3 places to the right. Thus, we move the decimal point 3 places to the right.

$3.8 \text{ kg} = 3800 \text{ g}$

9. $10 \text{ gal} = 10 \times 1 \text{ gal} = 10 \times 4 \text{ qt}$

$= 10 \times 4 \times 1 \text{ qt} = 10 \times 4 \times 2 \text{ pt}$

$= 10 \times 4 \times 2 \times 1 \text{ pt} = 10 \times 4 \times 2 \times 16 \text{ oz}$

$= 1280 \text{ oz}$

10. $0.69 \text{ L} = 0.69 \times 1 \text{ L}$

$= 0.69 \times 1000 \text{ mL}$

$= 690 \text{ mL}$

11. $9 \text{ lb} = 9 \times 1 \text{ lb}$

$= 9 \times 16 \text{ oz}$

$= 144 \text{ oz}$

12. $4.11 \text{ T} = 4.11 \times 1 \text{ T}$

$= 4.11 \times 2000 \text{ lb}$

$= 8220 \text{ lb}$

13. $r = \dfrac{d}{2} = \dfrac{16 \text{ cm}}{2} = 8 \text{ cm}$

14. $A = \pi \cdot r \cdot r$

$\approx 3.14 \cdot 4 \text{ m} \cdot 4 \text{ m}$

$= 3.14 \cdot 16 \text{ m}^2$

$= 50.24 \text{ m}^2$

15. $C = 2 \cdot \pi \cdot r$

$\approx 2 \cdot \dfrac{22}{7} \cdot 14 \text{ ft}$

$= \dfrac{2 \cdot 22 \cdot 14}{7} \text{ ft}$

$= \dfrac{2 \cdot 22 \cdot 2 \cdot 7}{7 \cdot 1} \text{ ft}$

$= 88 \text{ ft}$

16. $A = b \cdot h$

$= 10 \text{ cm} \cdot 2.5 \text{ cm}$

$= 25 \text{ cm}^2$

17. The region is the area of a rectangle that is 8 m by 6 m less the area of a semicircle with diameter 6 m. Note that the area of the semicircle is one-half the area of a circle with diameter 6 m.

First we find the area of the rectangle.

$A = l \cdot w$

$= 8 \text{ m} \cdot 6 \text{ m}$

$= 48 \text{ m}^2$

Now find the area of the semicircle. The radius is $\dfrac{6 \text{ m}}{2}$, or 3 m.

$A = \dfrac{1}{2} \cdot \pi \cdot r \cdot r$

$\approx \dfrac{1}{2} \cdot 3.14 \cdot 3 \text{ m} \cdot 3 \text{ m}$

$= \dfrac{3.14 \cdot 3 \cdot 3}{2} \text{ m}^2$

$= 14.13 \text{ m}^2$

Finally, we subtract to find the area of the shaded region.

$48 \text{ m}^2 - 14.13 \text{ m}^2 = 33.87 \text{ m}^2$

18. $A = \frac{1}{2} \cdot h \cdot (a + b)$

$ = \frac{1}{2} \cdot 3 \text{ ft} \cdot (8 \text{ ft} + 4 \text{ ft})$

$ = \frac{1}{2} \cdot 3 \text{ ft} \cdot 12 \text{ ft}$

$ = \frac{3 \cdot 12}{2} \text{ ft}^2$

$ = 18 \text{ ft}^2$

19. First we find the area of the red card stock. It is the area of a circle with radius $\frac{4 \text{ in.}}{2}$, or 2 in.

$A = \pi \cdot r \cdot r$

$ \approx 3.14 \cdot 2 \text{ in.} \cdot 2 \text{ in.} \approx 12.56 \text{ in}^2$

Next we find the area of the photo. It is the area of a circle with radius $\frac{3 \text{ in.}}{2}$, or 1.5 in.

$A = \pi \cdot r \cdot r$

$ \approx 3.14 \cdot 1.5 \text{ in.} \cdot 1.5 \text{ in.} \approx 7.065 \text{ in}^2$

Finally we subtract to find the area of the card stock that borders the photo.

$12.56 \text{ in}^2 - 7.065 \text{ in}^2 = 5.495 \text{ in}^2$

20. $m \angle CAD = 65°$

$180 - 65° = 115°$, so the measure of a supplement is 115°.

21. $\angle GAF$ and $\angle CAD$ are vertical angles, so $m \angle GAF = m \angle CAD = 65°$.

22. $V = l \cdot w \cdot h$

$ = 10\frac{1}{2} \text{ in.} \cdot 8 \text{ in.} \cdot 5 \text{ in.}$

$ = \frac{21}{2} \text{ in.} \cdot 8 \text{ in.} \cdot 5 \text{ in.}$

$ = \frac{21 \cdot 8 \cdot 5}{2} \text{ in}^3$

$ = \frac{21 \cdot 2 \cdot 4 \cdot 5}{2 \cdot 1} \text{ in}^3$

$ = 420 \text{ in}^3$

23. $V = \pi \cdot r^2 \cdot h$

$ \approx 3.14 \times 5 \text{ ft} \times 5 \text{ ft} \times 8 \text{ ft}$

$ = 628 \text{ ft}^3$

24. $V = \frac{4}{3} \cdot \pi \cdot r^3$

$ \approx \frac{4}{3} \times 3.14 \times (10 \text{ yd})^3$

$ = 4186.\overline{6} \text{ yd}^3$

25. $V = l \cdot w \cdot h$

$ = 3 \text{ m} \cdot 2 \text{ m} \cdot 5 \text{ m}$

$ = 6 \cdot 5 \text{ m}^3$

$ = 30 \text{ m}^3$

26. First we divide to find the number of doses the patient receives.

$\frac{48}{8} = 6 \text{ doses}$

Then the patient receives 6(0.5 L), or 3 L of dextrose solution. We convert 3 L to milliliters.

$3 \text{ L} = 3 \times 1 \text{ L} = 3 \times 1000 \text{ mL} = 3000 \text{ mL}$

27. $a^2 + b^2 = c^2$

$1^2 + 1^2 = c^2$

$1 + 1 = c^2$

$2 = c^2$

$\sqrt{2} = c \qquad \text{Exact answer}$

$1.414 \approx c \qquad \text{Approximation}$

28. $a^2 + b^2 = c^2$

$7^2 + b^2 = 10^2$

$49 + b^2 = 100$

$49 + b^2 - 49 = 100 - 49$

$b^2 = 51$

$b = \sqrt{51} \qquad \text{Exact answer}$

$b \approx 7.141 \qquad \text{Approximation}$

29. $\sqrt{121} = 11$, because $11^2 = 121$ and 11 is positive.

30. $C = \frac{5}{9} \cdot (F - 32)$

$C = \frac{5}{9} (32 - 32)$

$ = \frac{5}{9} \cdot 0$

$ = 0$

Thus, $32°\text{ F} = 0°\text{ C}$.

31. Let $s = $ the number of mL's of the solution Pat should take. We use a proportion.

$$\begin{array}{c} \text{Solution} \rightarrow \\ \text{Medication} \rightarrow \end{array} \frac{5}{25} = \frac{s}{30} \begin{array}{c} \leftarrow \text{Solution} \\ \leftarrow \text{Medication} \end{array}$$

We solve the proportion.

$\frac{5}{25} = \frac{s}{30}$

$5 \cdot 30 = 25 \cdot s$

$\frac{5 \cdot 30}{25} = s$

$6 = s$

Pat should take 6 mL of the solution.

32. Let $x = m \angle SMC$. Then $90 - x = $ the measure of the complement of $\angle SMC$. Then we have

$x = 3(90 - x)$

$x = 270 - 3x$

$4x = 270$

$x = 62.5$

Thus, $m \angle SMC = 62.5°$. Now we subtract to find the measure of its supplement.

$180° - 62.5° = 112.5°$.

33. First we find the total volume of wood ordered.

For 25 pieces, 2 in. by 4 in. by 8 ft:

$$8 \text{ ft} = 8 \times 1 \text{ ft} = 8 \times 12 \text{ in.} = 96 \text{ in.}$$

$$V = 25 \cdot 2 \text{ in.} \cdot 4 \text{ in.} \cdot 96 \text{ in.} = 19,200 \text{ in}^3$$

For 32 pieces, 2 in. by 6 in. by 10 ft:

$$10 \text{ ft} = 10 \times 1 \text{ ft} = 10 \times 12 \text{ in.} = 120 \text{ in.}$$

$$V = 32 \cdot 2 \text{ in.} \cdot 6 \text{ in.} \cdot 120 \text{ in.} = 46,080 \text{ in}^3$$

For 24 pieces, 2 in. by 8 in. by 12 ft:

$$12 \text{ ft} = 12 \times 1 \text{ ft} = 12 \times 12 \text{ in.} = 144 \text{ in.}$$

$$V = 24 \cdot 2 \text{ in.} \cdot 8 \text{ in.} \cdot 144 \text{ in.} = 55,296 \text{ in}^3$$

Total volume:

$$19,200 \text{ in}^3 + 46,080 \text{ in}^3 + 55,296 \text{ in}^3 = 120,576 \text{ in}^3$$

Now we find the volume of one board foot:

$$V = 12 \text{ in.} \cdot 12 \text{ in.} \cdot 1 \text{ in} = 144 \text{ in}^3$$

Next we divide to find the number of board feet ordered:

$$\frac{120,576}{144} = 837.\overline{3} \text{ board feet}$$

We divide again to find how many thousands are in $837.\overline{3}$:

$$\frac{837.\overline{3}}{1000} = 0.837\overline{3}$$

Finally we multiply to find the cost of the order:

$$0.837\overline{3} \cdot \$225 = \$188.40$$

34. First we convert 2.6 in. and 3 in. to feet.

$$2.6 \text{ in.} = 2.6 \text{ in.} \cdot \frac{1 \text{ ft}}{12 \text{ in.}}$$

$$= \frac{2.6}{12} \cdot \frac{\text{in.}}{\text{in.}} \cdot 1 \text{ ft}$$

$$= \frac{2.6}{12} \text{ ft}$$

Next we convert 3 in. to feet.

$$3 \text{ in.} = 3 \text{ in.} \cdot \frac{1 \text{ ft}}{12 \text{ in.}}$$

$$= \frac{3}{12} \cdot \frac{\text{in.}}{\text{in.}} \cdot 1 \text{ ft}$$

$$= \frac{1}{4} \text{ ft}$$

Now we find the volume.

$$V = l \cdot w \cdot h$$

$$= 12 \text{ ft} \cdot \frac{1}{4} \text{ ft} \cdot \frac{2.6}{12} \text{ ft}$$

$$= \frac{12 \cdot 2.6}{4 \cdot 12} \text{ ft}^3$$

$$= 0.65 \text{ ft}^3$$

35. First we find the radius of the cylinder.

$$r = \frac{d}{2} = \frac{\frac{3}{4} \text{ in.}}{2} = \frac{3}{4} \text{ in.} \cdot \frac{1}{2} = \frac{3}{8} \text{ in.}$$

Now we convert $\frac{3}{8}$ in. to feet.

$$\frac{3}{8} \text{ in.} = \frac{3}{8} \text{ in.} \cdot \frac{1 \text{ ft}}{12 \text{ in.}}$$

$$= \frac{3}{8 \cdot 12} \cdot \frac{\text{in.}}{\text{in.}} \cdot 1 \text{ ft}$$

$$= \frac{1}{32} \text{ ft}$$

Finally, we find the volume.

$$V = B \cdot h = \pi \cdot r^2 \cdot h$$

$$\approx 3.14 \cdot \frac{1}{32} \text{ ft} \cdot \frac{1}{32} \text{ ft} \cdot 18 \text{ ft}$$

$$= \frac{3.14 \cdot 18}{32 \cdot 32} \text{ ft}^3$$

$$\approx 0.055 \text{ ft}^3$$

Cumulative Review Chapters 1 - 9

1. The LCD is 6.

$$\begin{array}{r} 4\boxed{\frac{2}{3} \cdot \frac{2}{2}} = 4\frac{4}{6} \\ +5\boxed{\frac{1}{2} \cdot \frac{3}{3}} = +5\frac{3}{6} \\ \hline 9\frac{7}{6} = 9 + \frac{7}{6} \\ = 9 + 1\frac{1}{6} \\ = 10\frac{1}{6} \end{array}$$

2.

$$\left(\frac{1}{4}\right)^2 \div \left(\frac{1}{2}\right)^3 \times 2^4 + (10.3)(4)$$

$$= \frac{1}{16} \div \frac{1}{8} \times 16 + (10.3)(4)$$

$$= \frac{1}{16} \cdot \frac{8}{1} \times 16 + (10.3)(4)$$

$$= \frac{8}{16} \times 16 + (10.3)(4)$$

$$= \frac{8 \times 16}{16} + (10.3)(4)$$

$$= \frac{8 \times \cancel{16}}{\cancel{16} \times 1} + (10.3)(4)$$

$$= 8 + 41.2$$

$$= 49.2$$

3.

$$\begin{array}{r} \overset{11}{}\overset{9}{\cancel{1}}\overset{14}{} \\ \overset{\cancel{1}}{1}\,\overset{\cancel{9}}{2}\,\overset{\cancel{4}}{0}.\overset{10}{\cancel{5}}\,\cancel{0} \\ -\;3\,2.\,9\;8 \\ \hline 8\,7.\,5\;2 \end{array}$$

4. First we find $27,148 \div 22$.

$$
\begin{array}{r}
1\,2\,3\,4 \\
2\,2\,\overline{)2\,7{,}1\,4\,8} \\
2\,2\,0\,0\,0 \\
\hline
5\,1\,4\,8 \\
4\,4\,0\,0 \\
\hline
7\,4\,8 \\
6\,6\,0 \\
\hline
8\,8 \\
8\,8 \\
\hline
0
\end{array}
$$

$27,148 \div 22 = 1234$, so $-27,148 \div 22 = -1234$

5.
$$
\begin{aligned}
&14 \div [33 \div 11 + 8 \times 2 - (15 - 3)] \\
&= 14 \div [33 \div 11 + 8 \times 2 - 12] \\
&= 14 \div [3 + 8 \times 2 - 12] \\
&= 14 \div [3 + 16 - 12] \\
&= 14 \div [19 - 12] \\
&= 14 \div 7 \\
&= 2
\end{aligned}
$$

6.
$$
\begin{aligned}
&8^3 + 45 \cdot 24 - 9^2 \div 3 \\
&= 512 + 45 \cdot 24 - 81 \div 3 \\
&= 512 + 1080 - 81 \div 3 \\
&= 512 + 1080 - 27 \\
&= 1592 - 27 \\
&= 1565
\end{aligned}
$$

7. $1.2 = \dfrac{12}{10} = \dfrac{2 \cdot 6}{2 \cdot 5} = \dfrac{2}{2} \cdot \dfrac{6}{5} = \dfrac{6}{5}$

8. $\dfrac{9}{20} = \dfrac{9}{20} \cdot \dfrac{5}{5} = \dfrac{45}{100} = 45\%$

9. $\dfrac{5}{6} = \dfrac{5}{6} \cdot \dfrac{4}{4} = \dfrac{20}{24}$

$\dfrac{7}{8} = \dfrac{7}{8} \cdot \dfrac{3}{3} = \dfrac{21}{24}$

Since $20 < 21$, $\dfrac{20}{24} < \dfrac{21}{24}$ and thus $\dfrac{5}{6} < \dfrac{7}{8}$.

10. $\dfrac{5}{12} = \dfrac{5}{12} \cdot \dfrac{5}{5} = \dfrac{25}{60}$

$\dfrac{3}{10} = \dfrac{3}{10} \cdot \dfrac{6}{6} = \dfrac{18}{60}$

Since $25 > 18$, $\dfrac{25}{60} > \dfrac{18}{60}$ and thus $\dfrac{5}{12} > \dfrac{3}{10}$.

11. $6 \text{ oz} = 6 \; \cancel{oz} \times \dfrac{1 \text{ lb}}{16 \; \cancel{oz}}$

$= \dfrac{6}{16} \times 1 \text{ lb}$

$= \dfrac{3}{8} \text{ lb}$

12. $F = 1.8 \cdot C + 32 = 1.8 \cdot 100 + 32 = 180 + 32 = 212$

Thus, $100°C = 212°F$.

13.
$$
\begin{aligned}
0.087 \text{ L} &= 0.087 \times 1 \text{ L} \\
&= 0.087 \times 1000 \text{ mL} \\
&= 87 \text{ mL}
\end{aligned}
$$

14.
$$
\begin{aligned}
2.5 \text{ yd} &= 2.5 \times 1 \text{ yd} \\
&= 2.5 \times 36 \text{ in.} \\
&= 90 \text{ in.}
\end{aligned}
$$

15.
$$
\begin{aligned}
3 \text{ yd}^2 &= 3 \times 1 \text{ yd}^2 \\
&= 3 \times 9 \text{ ft}^2 \\
&= 27 \text{ ft}^2
\end{aligned}
$$

16. We move the decimal point 2 places to the left.

$$37 \text{ cm} = 0.37 \text{ m}$$

17. Perimeter $= 17 \text{ m} + 15 \text{ m} + 20 \text{ m} = 52 \text{ m}$

$$
\begin{aligned}
A &= \frac{1}{2} \cdot b \cdot h \\
&= \frac{1}{2} \cdot 20 \text{ m} \cdot 12 \text{ m} \\
&= \frac{20 \cdot 12}{2} \text{ m}^2 \\
&= 120 \text{ m}^2
\end{aligned}
$$

18.
$$
\begin{aligned}
12a - 7 - 3a - 9 &= (12 - 3)a + (-7 - 9) \\
&= 9a + (-16) \\
&= 9a - 16
\end{aligned}
$$

19. Graph $y = -\dfrac{1}{3}x + 2$.

We make a table of solutions, plot the points, and draw and label the line.

If x is -3, then $y = -\dfrac{1}{3}(-3) + 2 = 1 + 2 = 3$.

If x is 0, then $y = -\dfrac{1}{3} \cdot 0 + 2 = 0 + 2 = 2$.

If x is 3, then $y = -\dfrac{1}{3} \cdot 3 + 2 = -1 + 2 = 1$.

x	y	(x, y)
-3	3	$(-3, 3)$
0	2	$(0, 2)$
3	1	$(3, 1)$

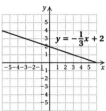

20. We go to the top of the bar representing 1998 and then go across to the left to read that Medicare spent about $110 million on PATH clients in 1998.

21. Locate 200 on the left and go across to the right to the first bar whose top is above this level. We see that Medicare spending on PATH clients was first projected to exceed $200 million in 2005.

22.
$$\frac{12}{15} = \frac{x}{18}$$
$$12 \cdot 18 = 15 \cdot x \qquad \text{Equating cross products}$$
$$\frac{12 \cdot 18}{15} = \frac{15 \cdot x}{15}$$
$$\frac{\cancel{3} \cdot 4 \cdot 18}{\cancel{3} \cdot 5} = x$$
$$\frac{72}{5} = x, \text{ or}$$
$$14\frac{2}{5} = x$$

The solution is $\frac{72}{5}$, or $14\frac{2}{5}$, or 14.4.

23.
$$1 - 7x = 4 - (x + 9)$$
$$1 - 7x = 4 - x - 9$$
$$1 - 7x = -x - 5$$
$$1 - 7x + x = -x - 5 + x$$
$$1 - 6x = -5$$
$$1 - 6x - 1 = -5 - 1$$
$$-6x = -6$$
$$\frac{-6x}{-6} = \frac{-6}{-6}$$
$$x = 1$$

The solution is 1.

24.
$$-15x = 280$$
$$\frac{-15x}{-15} = \frac{280}{-15}$$
$$x = -\frac{\cancel{5} \cdot 56}{3 \cdot \cancel{5}}$$
$$x = -\frac{56}{3}, \text{ or } -18\frac{2}{3}$$

25.
$$x + \frac{3}{4} = \frac{7}{8}$$
$$x + \frac{3}{4} - \frac{3}{4} = \frac{7}{8} - \frac{3}{4}$$
$$x = \frac{7}{8} - \frac{3}{4} \cdot \frac{2}{2}$$
$$x = \frac{7}{8} - \frac{6}{8}$$
$$x = \frac{1}{8}$$

The solution is $\frac{1}{8}$.

26. The ratio of the amount spent in Florida to the total amount spent is $\frac{18.2}{79.3}$.

The ratio of the total amount spent to the amount spent in Florida is $\frac{79.3}{18.2}$.

27. First we find the discount.
$$\text{Discount} = \text{Marked price} - \text{Sale price}$$
$$D = 100 - 58.99$$

We subtract:

$$\begin{array}{r} 1\,0\,0.\,0\,0 \\ -\ \ 5\,8.\,9\,9 \\ \hline 4\,1.\,0\,1 \end{array}$$

The discount is $41.01.

Now we find the rate of discount.

$$\text{Discount} = \text{Rate of discount} \times \text{Marked price}$$
$$41.01 = R \times 100$$

To solve the equation we divide on both sides by 100.
$$\frac{41.01}{100} = \frac{R \times 100}{100}$$
$$0.4101 \approx R$$
$$41.01\% \approx R$$

To check, note that a discount rate of 41.01% means that 58.99% of the marked price is paid: $0.5899 \times 100 = 58.99$. Since this is the sale price, the answer checks. The rate of discount is about 41.01%.

28. $\dfrac{49 + 53 + 60 + 62 + 69}{5} = \dfrac{293}{5} = 58.6$

29. $I = P \cdot r \cdot t$
$$= \$800 \cdot 0.12 \cdot \frac{1}{4}$$
$$= \$24$$

30. *Familiarize.* We first make a drawing. In it we see a right triangle. We let $r =$ the length of the rope, in meters.

Translate. We substitute 15 for a, 8 for b, and r for c in the Pythagorean equation.
$$a^2 + b^2 = c^2$$
$$15^2 + 8^2 = r^2$$

Solve. We solve the equation for w.
$$225 + 64 = r^2$$
$$289 = r^2$$
$$\sqrt{289} = r$$
$$17 = r$$

Check. $15^2 + 8^2 = 225 + 64 = 289 = (17)^2$

State. The length of the rope is 17 m.

31. $\underbrace{\text{Sales tax}}_{\downarrow}$ is $\underbrace{\text{what percent}}_{\downarrow \quad \downarrow}$ of $\underbrace{\text{purchase price?}}_{\downarrow}$
$$\$0.33 = r \times \$5.50$$

To solve the equation we divide by 5.50 on both sides.
$$\frac{0.33}{5.50} = \frac{r \times 5.50}{5.50}$$
$$0.06 = r$$
$$6\% = r$$

The sales tax rate is 6%.

32. Familiarize. Let f = the number of yards of fabric remaining on the bolt.

Translate.

$$\underbrace{\text{Original amount}}_{10\frac{3}{4}} \;\; \underset{-}{\text{minus}} \;\; \underbrace{\text{Amount purchased}}_{8\frac{5}{8}} \;\; \underset{=}{\text{is}} \;\; \underbrace{\text{Amount remaining}}_{f}$$

Solve. We carry out the subtraction.

$$10\;\boxed{\frac{3}{4}\cdot\frac{2}{2}} = 10\frac{6}{8}$$
$$-8\;\frac{5}{8} \qquad = -8\frac{5}{8}$$
$$\rule{3cm}{0.4pt}$$
$$2\frac{1}{8}$$

Thus, $f = 2\frac{1}{8}$.

Check. $8\frac{5}{8} + 2\frac{1}{8} = 10\frac{6}{8} = 10\frac{3}{4}$, so the answer checks.

State. $2\frac{1}{8}$ yd of fabric remains on the bolt.

33. Familiarize. Let c = the cost of the gasoline. We express 178.9¢ as \$1.789.

Translate.

$$\underbrace{\text{Cost per gallon}}_{\$1.789} \;\; \underset{\times}{\text{times}} \;\; \underbrace{\text{Number of gallons}}_{15.6} \;\; \underset{=}{\text{is}} \;\; \underbrace{\text{Total cost}}_{c}$$

Solve. We carry out the multiplication.

$$\begin{array}{r} \$1.7\,8\,9 \\ \times\;\;1\,5.6 \\ \hline 1\,0\,7\,3\,4 \\ 8\,9\,4\,5\,0 \\ 1\,7\,8\,9\,0\,0 \\ \hline \$2\,7.9\,0\,8\,4 \end{array}$$

Thus, $c \approx \$27.91$.

Check. We repeat the calculation. The answer checks.

State. The gasoline cost \$27.91.

34. $\dfrac{\$4.99}{20\text{ qt}} = \dfrac{499¢}{20\text{ qt}} = 24.95¢/\text{qt}$

$\dfrac{\$1.99}{8\text{ qt}} = \dfrac{199¢}{8\text{ qt}} = 24.875¢/\text{qt}$

The 8-qt box has the lower unit price.

35. Familiarize. Let d = the distance Maria walked, in km. She walks $\frac{1}{4}$ of the distance from the dormitory to the library and then turns and walks the same distance back to the dormitory, so she walks a total of $\frac{1}{4} + \frac{1}{4}$, or $\frac{1}{2}$, of the distance from the dormitory to the library.

Translate.

$$\underbrace{\text{Distance walked}}_{d} \;\; \underset{=}{\text{is}} \;\; \underset{\frac{1}{2}}{\tfrac{1}{2}} \;\; \underset{\cdot}{\text{of}} \;\; \underset{\frac{7}{10}}{\tfrac{7}{10}} \text{ km}$$

Solve. We carry out the multiplication.

$$d = \frac{1}{2}\cdot\frac{7}{10} = \frac{7}{20}$$

Check. We repeat the calculation. The answer checks.

State. Maria walked $\frac{7}{20}$ km.

36. $\angle DBA$ and $\angle ABC$ are vertical angles, so $m\angle DBA = m\angle ABC = 148°$.

37. $\angle ABD$ and $\angle DBE$ are supplementary angles.
$$m\angle ABD + m\angle DBE = 180°$$
$$43° + m\angle DBE = 180°$$
$$m\angle DBE = 180° - 43°$$
$$m\angle DBE = 137°$$

38. The area of the entire lot is 75 ft·200 ft, or 15,000 ft². The portion of the lot that does not contain the sidewalks is a rectangle with dimensions $75 - 3$ by $200 - 3$, or 72 ft by 197 ft. Its area is 197 ft · 72 ft, or 14,184 ft².

Then the area of the sidewalks is $15,000 - 14,184$, or 816 ft².

To find the volume of the snow, we first convert 4 in. to feet.

$$4\text{ in.} = 4\text{ in.} \times \frac{1\text{ ft}}{12\text{ in.}} = \frac{4}{12} \times 1\text{ ft} = \frac{1}{3}\text{ ft}$$

Then the volume of the snow is given by $l\cdot w\cdot h$ where $l\cdot w$ = the area of the sidewalks, 816 ft², and h = the depth of the snow, $\frac{1}{3}$ ft:

$$816\text{ ft}^2 \cdot \frac{1}{3}\text{ ft} = 272\text{ ft}^3$$

39. First we convert 108 in. to feet:
$$108\text{ in.} = 180\text{ in.} \times \frac{1\text{ ft}}{12\text{ in.}} = \frac{108}{12} \times \frac{\text{in.}}{\text{in.}} \times \text{ft} = 9\text{ ft}$$

Next we find the lengthwise perimeter of the box:
$$P = 2\cdot l + 2\cdot w$$
$$= 2\cdot 3\text{ ft} + 2\cdot 1\text{ ft}$$
$$= 6\text{ ft} + 2\text{ ft} = 8\text{ ft}$$

Now we find the widthwise perimeter of the box:
$$P = 2\cdot l + 2\cdot w$$
$$= 2\cdot 2\text{ ft} + 2\cdot 1\text{ ft}$$
$$= 4\text{ ft} + 2\text{ ft} = 6\text{ ft}$$

The sum of the two perimeters is
$$8\text{ ft} + 6\text{ ft} = 14\text{ ft}.$$

Since 14 ft exceeds 9 ft, or 108 in., the box will not be accepted for shipping.

(There are two other ways in which we could designate lengthwise and widthwise perimeter, but the sum is greater than 108 in. in each case.)

Chapter 10

Polynomials

---

1. $(3x + 7) + (-7x + 3) = (3 - 7)x + (7 + 3) = -4x + 10$

3. $(-9x + 7) + (x^2 + x - 2) =$
$x^2 + (-9 + 1)x + (7 - 2) = x^2 - 8x + 5$

5. $(x^2 - 7) + (x^2 + 7) = (1 + 1)x^2 + (-7 + 7) = 2x^2$

7. $(6t^4 + 4t^3 - 1) + (5t^2 - t + 1) =$
$6t^4 + 4t^3 + 5t^2 - t + (-1 + 1) =$
$6t^4 + 4t^3 + 5t^2 - t$

9. $(2 + 4x + 6x^2 + 7x^3) + (5 - 4x + 6x^2 - 7x^3) =$
$(2 + 5) + (4 - 4)x + (6 + 6)x^2 + (7 - 7)x^3 =$
$7 + 0x + 12x^2 + 0x^3 = 7 + 12x^2$, or $12x^2 + 7$

11. $(9x^8 - 7x^4 + 2x^2 + 5) + (8x^7 + 4x^4 - 2x) =$
$9x^8 + 8x^7 + (-7 + 4)x^4 + 2x^2 - 2x + 5 =$
$9x^8 + 8x^7 - 3x^4 + 2x^2 - 2x + 5$

13. $(8t^4 + 6t^3 - t^2 + 3t) + (5t^4 - 2t^3 + t - 3) =$
$(8 + 5)t^4 + (6 - 2)t^3 - t^2 + (3 + 1)t - 3 =$
$13t^4 + 4t^3 - t^2 + 4t - 3$

15. $(-5x^4y^3 + 7x^3y^2 - 4xy^2) + (2x^3y^3 - 3x^3y^2 - 5xy) =$
$-5x^4y^3 + 2x^3y^3 + (7 - 3)x^3y^2 - 4xy^2 - 5xy =$
$-5x^4y^3 + 2x^3y^3 + 4x^3y^2 - 4xy^2 - 5xy$

17. $(8a^3b^2 + 5a^2b^2 + 6ab^2) + (5a^3b^2 - a^2b^2 - 4a^2b) =$
$(8 + 5)a^3b^2 + (5 - 1)a^2b^2 + 6ab^2 - 4a^2b =$
$13a^3b^2 + 4a^2b^2 + 6ab^2 - 4a^2b$

19. $(17.5abc^3 + 4.3a^2bc) + (-4.9a^2bc - 5.2abc) =$
$17.5abc^3 + (4.3 - 4.9)a^2bc - 5.2abc =$
$17.5abc^3 - 0.6a^2bc - 5.2abc$

21. Two equivalent expressions for the additive inverse of $-5x$ are

a) $-(-5x)$ and

b) $5x$. (Changing the sign)

23. Two equivalent expressions for the additive inverse of $-x^2 + 13x - 7$ are

a) $-(-x^2 + 13x - 7)$ and

b) $x^2 - 13x + 7$. (Changing the sign of every term)

25. Two equivalent expressions for the additive inverse of $12x^4 - 3x^3 + 3$ are

a) $-(12x^4 - 3x^3 + 3)$ and

b) $-12x^4 + 3x^3 - 3$. (Changing the sign of every term)

27. We change the sign of every term inside parentheses.
$-(3x - 5) = -3x + 5$

29. We change the sign of every term inside parentheses.
$-(4x^2 - 3x + 2) = -4x^2 + 3x - 2$

31. We change the sign of every term inside parentheses.
$-\left(-4x^4 + 6x^2 + \dfrac{3}{4}x - 8\right) = 4x^4 - 6x^2 - \dfrac{3}{4}x + 8$

33. $(3x + 2) - (-4x + 3) = 3x + 2 + 4x - 3$

$\qquad$ Changing the sign of every
$\qquad$ term inside parentheses

$\qquad = 7x - 1$

35. $(9t^2 + 7t + 5) - (5t^2 + t - 1)$
$= 9t^2 + 7t + 5 - 5t^2 - t + 1$
$= 4t^2 + 6t + 6$

37. $(-8x + 2) - (x^2 + x - 3) = -8x + 2 - x^2 - x + 3$
$\qquad\qquad = -x^2 - 9x + 5$

39. $(7a^2 + 5a - 9) - (2a^2 + 7)$
$= 7a^2 + 5a - 9 - 2a^2 - 7$
$= 5a^2 + 5a - 16$

41. $(8x^4 + 3x^3 - 1) - (4x^2 - 3x + 5)$
$= 8x^4 + 3x^3 - 1 - 4x^2 + 3x - 5$
$= 8x^4 + 3x^3 - 4x^2 + 3x - 6$

43. $(1.2x^3 + 4.5x^2 - 3.8x) - (-3.4x^3 - 4.7x^2 + 23)$
$= 1.2x^3 + 4.5x^2 - 3.8x + 3.4x^3 + 4.7x^2 - 23$
$= 4.6x^3 + 9.2x^2 - 3.8x - 23$

45. $\left(\dfrac{5}{8}x^3 - \dfrac{1}{4}x - \dfrac{1}{3}\right) - \left(-\dfrac{1}{8}x^3 + \dfrac{1}{4}x - \dfrac{1}{3}\right)$

$= \dfrac{5}{8}x^3 - \dfrac{1}{4}x - \dfrac{1}{3} + \dfrac{1}{8}x^3 - \dfrac{1}{4}x + \dfrac{1}{3}$

$= \dfrac{6}{8}x^3 - \dfrac{2}{4}x$

$= \dfrac{3}{4}x^3 - \dfrac{1}{2}x$

47. $(9x^3y^3 + 8x^2y^2 + 7xy) - (3x^3y^3 - 2x^2y + 3xy)$
$= 9x^3y^3 + 8x^2y^2 + 7xy - 3x^3y^3 + 2x^2y - 3xy$
$= 6x^3y^3 + 8x^2y^2 + 2x^2y + 4xy$

49. $-7x + 5 = -7 \cdot 4 + 5 = -28 + 5 = -23$

51. $2x^2 - 5x + 7 = 2 \cdot 4^2 - 5 \cdot 4 + 7 = 2 \cdot 16 - 20 + 7 =$
$32 - 20 + 7 = 19$

53. $x^3 - 5x^2 + x = 4^3 - 5 \cdot 4^2 + 4 = 64 - 5 \cdot 16 + 4 =$
$64 - 80 + 4 = -12$

55. $2x + 9 = 2(-1) + 9 = -2 + 9 = 7$

57. $x^2 - 2x + 1 = (-1)^2 - 2(-1) + 1 = 1 + 2 + 1 = 4$

59. $-3x^3 + 7x^2 - 3x - 2 =$
$-3(-1)^3 + 7(-1)^2 - 3(-1) - 2 =$
$-3(-1) + 7 \cdot 1 + 3 - 2 = 3 + 7 + 3 - 2 = 11$

61. We evaluate the polynomial for $t = 8$:
$$16t^2 = 16(8)^2 = 16 \cdot 64 = 1024$$
The cliff is 1024 ft high.

63. We evaluate the polynomial for $n = 92$.
$$-0.01096n^2 + 4n + 548$$
$$= -0.01096(92)^2 + 4(92) + 548$$
$$\approx 823 \qquad \text{Using a calculator}$$
In Chicago there are about 823 min of daylight 92 days after December 21.

65. $-x^3 + 80x^2 - 1723x + 22,669$
$$= -35^3 + 80 \cdot 35^2 - 1723 \cdot 35 + 22,669$$
$$= -42,875 + 80 \cdot 1225 - 1723 \cdot 35 + 22,669$$
$$= -42,875 + 98,000 - 60,305 + 22,669$$
$$= 17,489 \text{ degrees}$$

67. We evaluate the polynomial for $a = 18$:
$$0.4a^2 - 40a + 1039 = 0.4(18)^2 - 40(18) + 1039$$
$$= 0.4(324) - 40(18) + 1039$$
$$= 129.6 - 720 + 1039$$
$$= 448.6$$
The daily number of accidents involving 18-year-old drivers is 448.6, or about 449.

69. We evaluate the polynomial for $x = 75$:
$$280x - 0.4x^2 = 280(75) - 0.4(75)^2$$
$$= 280(75) - 0.4(5625)$$
$$= 21,000 - 2250$$
$$= 18,750$$
The total revenue from the sale of 75 stereos is \$18,750.

71. We evaluate the polynomial for $x = 500$:
$$5000 + 0.6x^2 = 5000 + 0.6(500)^2$$
$$= 5000 + 0.6(250,000)$$
$$= 5000 + 150,000$$
$$= 155,000$$
The total cost of producing 500 stereos is \$155,000.

73. Discussion and Writing Exercise

75. $\dfrac{7 \text{ servings}}{10 \text{ lb}} = \dfrac{7}{10}\dfrac{\text{servings}}{\text{lb}}$, or $0.7\dfrac{\text{servings}}{\text{lb}}$

77. The sales tax is
$$\underbrace{\text{Sales tax rate}}_{\downarrow} \cdot \underbrace{\text{Purchase price}}_{\downarrow} \cdot$$
$$\quad\; 6\% \qquad\qquad \$1350,$$
or $0.06 \cdot 1350$, or 81. Thus, the tax is \$81.

79. $A = \pi \cdot r \cdot r$
$$\approx 3.14 \cdot 20 \text{ cm} \cdot 20 \text{ cm}$$
$$\approx 1256 \text{ cm}^2$$

81.
```
          7
      3 ⌐ 2 1
    2 ⌐ 4 2
  2 ⌐ 8 4
2 ⌐ 1 6 8
```
$168 = 2 \cdot 2 \cdot 2 \cdot 3 \cdot 7$

83.
```
          7
      7 ⌐ 4 9
    5 ⌐ 2 4 5
  3 ⌐ 7 3 5
```
$735 = 3 \cdot 5 \cdot 7 \cdot 7$

85. Discussion and Writing Exercise

87. First we need to determine n. Ground Hog Day is February 2 so we need to find the number of days from December 21 to February 2. We will do this on a month-by-month basis. The number of days from December 21 to December 31 is $31 - 21$, or 10. There are 31 days in January and we must also include the first 2 days of February. Thus, $n = 10 + 31 + 2 = 43$. Now we evaluate the polynomial for $n = 43$.
$$-0.0085n^2 + 3.1014n + 593$$
$$= -0.0085(43)^2 + 3.1014(43) + 593$$
$$\approx 711 \qquad \text{Using a calculator}$$
On Ground Hog Day there are about 711 min of daylight in Los Angeles.

89. First we evaluate the polynomial for values of x representing 1970, 1975, 1980, and so on through 2010. The results are shown in the table below.

Year, x	Number of degrees
1970, 0	22,669
1975, 5	15,929
1980, 10	12,439
1985, 15	11,449
1990, 20	12,209
1995, 25	13,969
2000, 30	15,979
2005, 35	17,489
2010, 40	17,749

We use the data in the table to draw a vertical bar graph.

91. a) We evaluate the polynomial for $t = 1$:
$$0.5t^4 + 3.45t^3 - 96.65t^2 + 347.7t$$
$$= 0.5(1)^4 + 3.45(1)^3 - 96.65(1)^2 + 347.7(1)$$
$$= 0.5 + 3.45 - 96.65 + 347.7$$
$$= 255$$

There will be 255 mg of ibuprofen in the bloodstream 1 hr after 400 mg is swallowed.

b) We evaluate the polynomial for $t = 2$:
$$0.5t^4 + 3.45t^3 - 96.65t^2 + 347.7t$$
$$= 0.5(2)^4 + 3.45(2)^3 - 96.65(2)^2 + 347.7(2)$$
$$= 0.5(16) + 3.45(8) - 96.65(4) + 347.7(2)$$
$$= 8 + 27.6 - 386.6 + 695.4$$
$$= 344.4$$

There will be 344.4 mg of ibuprofen in the bloodstream 2 hr after 400 mg is swallowed.

c) We evaluate the polynomial for $t = 6$:
$$0.5t^4 + 3.45t^3 - 96.65t^2 + 347.7t$$
$$= 0.5(6)^4 + 3.45(6)^3 - 96.65(6)^2 + 347.7(6)$$
$$= 0.5(1296) + 3.45(216) - 96.65(36) + 347.7(6)$$
$$= 648 + 745.2 - 3479.4 + 2086.2$$
$$= 0$$

There will be 0 mg of ibuprofen in the bloodstream 6 hr after 400 mg is swallowed.

93. $(7y^2 - 5y + 6) - (3y^2 + 8y - 12) + (8y^2 - 10y + 3)$
$= 7y^2 - 5y + 6 - 3y^2 - 8y + 12 + 8y^2 - 10y + 3$
$= 12y^2 - 23y + 21$

95. $(-y^4 - 7y^3 + y^2) + (-2y^4 + 5y - 2) - (-6y^3 + y^2)$
$= -y^4 - 7y^3 + y^2 - 2y^4 + 5y - 2 + 6y^3 - y^2$
$= -3y^4 - y^3 + 5y - 2$

97. $9x^4 + 3x^4 = 12x^4$, $5x^2 + 0 = 5x^2$, $-7x^3 + 2x^3 = -5x^3$, and $-9 + (-7) = -16$ so we have $9x^4 + \underline{3x^4} + 5x^2 - 7x^3 + \underline{2x^3} - 9 + (\underline{-7}) = 12x^4 - 5x^3 + 5x^2 - 16$

The order of the answers may vary.

Exercise Set 10.2

1. $(4a)(7a) = (4 \cdot 7)(a \cdot a)$
$= 28a^2$

3. $(-4x)(15x) = (-4 \cdot 15)(x \cdot x)$
$= -60x^2$

5. $(7x^5)(4x^3) = (7 \cdot 4)(x^5 \cdot x^3)$
$= 28x^8$

7. $(-0.1x^6)(0.7x^3) = (-0.1 \cdot 0.7)(x^6 \cdot x^3)$
$= -0.07x^9$

9. $(5x^2y^3)(7x^4y^9) = (5 \cdot 7)(x^2 \cdot x^4)(y^3 \cdot y^9)$
$= 35x^6y^{12}$

11. $(4a^3b^4c^2)(3a^5b^4) = (4 \cdot 3)(a^3 \cdot a^5)(b^4 \cdot b^4)(c^2)$
$= 12a^8b^8c^2$

13. $(3x^2)(-4x^3)(2x^6) = (3)(-4)(2)(x^2 \cdot x^3 \cdot x^6) = -24x^{11}$

15. $3x(-x + 7) = 3x(-x) + 3x \cdot 7$
$= -3x^2 + 21x$

17. $-3x(x - 2) = -3x(x) - 3x(-2)$
$= -3x^2 + 6x$

19. $x^2(x^3 + 1) = x^2 \cdot x^3 + x^2 \cdot 1$
$= x^5 + x^2$

21. $5x(2x^2 - 6x + 1) = 5x \cdot 2x^2 + 5x(-6x) + 5x \cdot 1$
$= 10x^3 - 30x^2 + 5x$

23. $4xy(3x^2 + 2y) = 4xy \cdot 3x^2 + 4xy \cdot 2y$
$= 12x^3y + 8xy^2$

25. $3a^2b(4a^5b^2 - 3a^2b^2) = 3a^2b \cdot 4a^5b^2 - 3a^2b \cdot 3a^2b^2$
$= 12a^7b^3 - 9a^4b^3$

27. Factor $2x + 8$.
The prime factorization of $2x$ is $2 \cdot x$.
The prime factorization of 8 is $2 \cdot 2 \cdot 2$.
The largest common factor is 2.
$2x + 8 = 2 \cdot x + 2 \cdot 4$
$= 2(x + 4)$

29. Factor $7a - 35$.
The prime factorization of $7a$ is $7 \cdot a$.
The prime factorization of 35 is $5 \cdot 7$.
The largest common factor is 7.
$7a - 35 = 7 \cdot a - 7 \cdot 5$
$= 7(a - 5)$

31. Factor $28x + 21y$.
The prime factorization of $28x$ is $2 \cdot 2 \cdot 7 \cdot x$.
The prime factorization of $21y$ is $3 \cdot 7 \cdot y$.
The largest common factor is 7.
$28x + 21y = 7 \cdot 4x + 7 \cdot 3y$
$= 7(4x + 3y)$

33. Factor $9a - 27b + 81$.
The prime factorization of $9a$ is $3 \cdot 3 \cdot a$.
The prime factorization of $27b$ is $3 \cdot 3 \cdot 3 \cdot b$.
The prime factorization of 81 is $3 \cdot 3 \cdot 3 \cdot 3$.
The largest common factor is $3 \cdot 3$, or 9.
$9a - 27b + 81 = 9 \cdot a - 9 \cdot 3b + 9 \cdot 9$
$= 9(a - 3b + 9)$

35. Factor $18 - 6m$.

The prime factorization of 18 is $2 \cdot 3 \cdot 3$.

The prime factorization of $6m$ is $2 \cdot 3 \cdot m$.

The largest common factor is $2 \cdot 3$, or 6.

$$18 - 6m = 6 \cdot 3 - 6 \cdot m$$
$$= 6(3 - m)$$

37. Factor $-16 - 8x + 40y$.

The prime factorization of 16 is $2 \cdot 2 \cdot 2 \cdot 2$.

The prime factorization of $8x$ is $2 \cdot 2 \cdot 2 \cdot x$.

The prime factorization of $40y$ is $2 \cdot 2 \cdot 2 \cdot 5 \cdot y$.

The largest common factor is $2 \cdot 2 \cdot 2$, or 8. Since the first term is negative, we will factor out -8.

$$-16 - 8x + 40y = -8 \cdot 2 - 8 \cdot x - 8(-5y)$$
$$= -8(2 + x - 5y)$$

39. Factor $9x^5 + 9x$.

The prime factorization of $9x^5$ is $3 \cdot 3 \cdot x \cdot x \cdot x \cdot x \cdot x$.

The prime factorization of $9x$ is $3 \cdot 3 \cdot x$.

The largest common factor is $3 \cdot 3 \cdot x$, or $9x$.

$$9x^5 + 9x = 9x \cdot x^4 + 9x \cdot 1$$
$$= 9x(x^4 + 1)$$

41. Factor $a^3 - 8a^2$.

The prime factorization of a^3 is $a \cdot a \cdot a$.

The prime factorization of $8a^2$ is $2 \cdot 2 \cdot 2 \cdot a \cdot a$.

The largest common factor is $a \cdot a$, or a^2.

$$a^3 - 8a^2 = a^2 \cdot a - a^2 \cdot 8$$
$$= a^2(a - 8)$$

43. Factor $8x^3 - 6x^2 + 2x$.

The prime factorization of $8x^3$ is $2 \cdot 2 \cdot 2 \cdot x \cdot x \cdot x$.

The prime factorization of $6x^2$ is $2 \cdot 3 \cdot x \cdot x$.

The prime factorization of $2x$ is $2 \cdot x$.

The largest common factor is $2 \cdot x$, or $2x$.

$$8x^3 - 6x^2 + 2x = 2x \cdot 4x^2 - 2x \cdot 3x + 2x \cdot 1$$
$$= 2x(4x^2 - 3x + 1)$$

45. Factor $12a^4b^3 + 18a^5b^2$.

The prime factorization of $12a^4b^3$ is $2 \cdot 2 \cdot 3 \cdot a \cdot a \cdot a \cdot a \cdot b \cdot b \cdot b$.

The prime factorization of $18a^5b^2$ is $2 \cdot 3 \cdot 3 \cdot a \cdot a \cdot a \cdot a \cdot a \cdot b \cdot b$.

The largest common factor is $2 \cdot 3 \cdot a \cdot a \cdot a \cdot a \cdot b \cdot b$, or $6a^4b^2$.

$$12a^4b^3 + 18a^5b^2 = 6a^4b^2 \cdot 2b + 6a^4b^2 \cdot 3a$$
$$= 6a^4b^2(2b + 3a)$$

47. Discussion and Writing Exercise

49. *Familiarize*. The tennis court is a rectangle that measures 27 ft by 78 ft. We make a drawing.

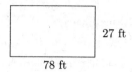

Translate. We substitute in the formula for the area of a rectangle.

$$P = 2 \cdot (l + w) = 2 \cdot (78 \text{ ft} + 27 \text{ ft})$$

Solve. We carry out the calculation.

$$P = 2 \cdot (78 \text{ ft} + 27 \text{ ft})$$
$$= 2 \cdot 105 \text{ ft}$$
$$= 210 \text{ ft}$$

Check. We repeat the calculation.

State. The perimeter of the tennis court is 210 ft.

51. *Familiarize*. Let $m =$ the old truck's mileage.

Translate.

$$\underbrace{\text{21 miles per gallon}}_{21} \ \underset{=}{\text{is}} \ \underbrace{\text{old truck's mileage}}_{m} \ \underset{+}{\text{plus}} \ \underbrace{20\% \text{ of}}_{20\% \cdot} \ \underbrace{\text{old truck's mileage.}}_{m}$$

Solve. We convert 20% to decimal notation and solve the equation.

$$21 = m + 0.2m$$
$$21 = 1.2m$$
$$\frac{21}{1.2} = \frac{1.2m}{1.2}$$
$$17.5 = m$$

Check. Note that the new truck's mileage is 100% of the old truck's mileage plus 20% of the old truck's mileage, or 120% of the old truck's mileage. Since $1.2(17.5) = 21$, the answer checks.

State. The old truck got 17.5 miles per gallon.

53. *Familiarize*. The number of fish that were not trout is $8 - 3$, or 5. Let $n =$ the percent that were not trout.

Translate.

$$\underset{5 =}{5} \ \text{is} \ \underbrace{\text{what percent}}_{n} \ \text{of} \ \underset{\cdot \ 8}{8}?$$

Solve.

$$5 = n \cdot 8$$
$$\frac{5}{8} = n$$
$$0.625 = n$$
$$62.5\% = n$$

Check. We could solve the same problem as a proportion. Also note that 62.5% of 8 is 5, so the answer checks.

State. 62.5% of the fish were not trout.

55. $-57 \cdot 48 = -2736$

(The numbers have different signs, so the product is negative.)

57. Discussion and Writing Exercise

59. $391x^{391} + 299x^{299} = 23x^{299} \cdot 17x^{92} + 23x^{299} \cdot 13$
$$= 23x^{299}(17x^{92} + 13)$$

61. $84a^7b^9c^{11} - 42a^8b^6c^{10} + 49a^9b^7c^8$
$$= 7a^7b^6c^8 \cdot 12b^3c^3 - 7a^7b^6c^8 \cdot 6ac^2 + 7a^7b^6c^8 \cdot 7a^2b$$
$$= 7a^7b^6c^8(12b^3c^3 - 6ac^2 + 7a^2b)$$

Exercise Set 10.3

1. $(x+6)(x+2) = (x+6)x + (x+6)2$
$$= x \cdot x + 6 \cdot x + x \cdot 2 + 6 \cdot 2$$
$$= x^2 + 6x + 2x + 12$$
$$= x^2 + 8x + 12$$

3. $(x+5)(x-2) = (x+5)x + (x+5)(-2)$
$$= x \cdot x + 5 \cdot x + x(-2) + 5(-2)$$
$$= x^2 + 5x - 2x - 10$$
$$= x^2 + 3x - 10$$

5. $(x+6)(x-2) = x \cdot x - 2 \cdot x + 6 \cdot x + 6(-2)$
$$= x^2 - 2x + 6x - 12$$
$$= x^2 + 4x - 12$$

7. $(x-7)(x-3) = x \cdot x - 3 \cdot x - 7 \cdot x + (-7)(-3)$
$$= x^2 - 3x - 7x + 21$$
$$= x^2 - 10x + 21$$

9. $(x+5)(x-5) = x \cdot x - 5 \cdot x + 5 \cdot x + 5(-5)$
$$= x^2 - 5x + 5x - 25$$
$$= x^2 - 25$$

11. $(3+x)(6+2x) = 3 \cdot 6 + 3 \cdot 2x + 6 \cdot x + x \cdot 2x$
$$= 18 + 6x + 6x + 2x^2$$
$$= 18 + 12x + 2x^2$$

13. $(3x-4)(3x-4) = 3x \cdot 3x - 4 \cdot 3x - 4 \cdot 3x + (-4)(-4)$
$$= 9x^2 - 12x - 12x + 16$$
$$= 9x^2 - 24x + 16$$

15. $\left(x - \frac{5}{2}\right)\left(x + \frac{2}{5}\right) = x \cdot x + \frac{2}{5} \cdot x - \frac{5}{2} \cdot x - \frac{5}{2} \cdot \frac{2}{5}$
$$= x^2 + \frac{2}{5}x - \frac{5}{2}x - \frac{5 \cdot 2}{2 \cdot 5}$$
$$= x^2 + \frac{4}{10}x - \frac{25}{10}x - 1$$
$$= x^2 - \frac{21}{10}x - 1$$

17. $(x^2 + x - 3)(x + 1)$
$$= (x^2 + x - 3)x + (x^2 + x - 3)(1)$$
$$= x^2 \cdot x + x \cdot x - 3 \cdot x + x^2 \cdot 1 + x \cdot 1 - 3 \cdot 1$$
$$= x^3 + x^2 - 3x + x^2 + x - 3$$
$$= x^3 + 2x^2 - 2x - 3$$

19. $(2x+1)(2x^2 + 6x + 1)$
$$= 2x(2x^2 + 6x + 1) + 1(2x^2 + 6x + 1)$$
$$= 2x \cdot 2x^2 + 2x \cdot 6x + 2x \cdot 1 + 1 \cdot 2x^2 + 1 \cdot 6x + 1 \cdot 1$$
$$= 4x^3 + 12x^2 + 2x + 2x^2 + 6x + 1$$
$$= 4x^3 + 14x^2 + 8x + 1$$

21. $(y^2 - 3)(3y^2 - 6y + 2)$
$$= y^2(3y^2 - 6y + 2) - 3(3y^2 - 6y + 2)$$
$$= y^2 \cdot 3y^2 + y^2(-6y) + y^2 \cdot 2 - 3 \cdot 3y^2 - 3(-6y) - 3 \cdot 2$$
$$= 3y^4 - 6y^3 + 2y^2 - 9y^2 + 18y - 6$$
$$= 3y^4 - 6y^3 - 7y^2 + 18y - 6$$

23. $(x^3 + x^2)(x^3 + x^2 - x)$
$$= x^3(x^3 + x^2 - x) + x^2(x^3 + x^2 - x)$$
$$= x^3 \cdot x^3 + x^3 \cdot x^2 + x^3(-x) + x^2 \cdot x^3 + x^2 \cdot x^2 + x^2(-x)$$
$$= x^6 + x^5 - x^4 + x^5 + x^4 - x^3$$
$$= x^6 + 2x^5 - x^3$$

25.
$$
\begin{array}{rl}
2t^2 - t - 4 & \\
3t^2 + 2t - 1 & \\
\hline
- 2t^2 + t + 4 & \text{Multiplying by } -1 \\
4t^3 - 2t^2 - 8t & \text{Multiplying by } 2t \\
6t^4 - 3t^3 - 12t^2 & \text{Multiplying by } 3t^2 \\
\hline
6t^4 + t^3 - 16t^2 - 7t + 4 &
\end{array}
$$

27.
$$
\begin{array}{rl}
x -x^3 +x^5 & \\
-1 +x^2 +x^4 & \text{Rewriting in ascending order} \\
\hline
x^5 - x^7 + x^9 & \text{Multiplying by } x^4 \\
x^3 - x^5 + x^7 & \text{Multiplying by } x^2 \\
-x + x^3 - x^5 & \text{Multiplying by } -1 \\
\hline
-x + 2x^3 - x^5 + x^9 &
\end{array}
$$

29. Discussion and Writing Exercise

31. *Familiarize.* We label the width of the sidewalk s.

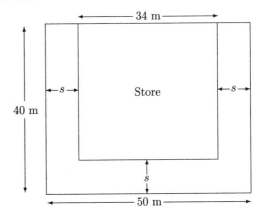

Translate. This is a two-step problem. We first find the width of the sidewalk. Then we find the area of the sidewalk.

$$
\underbrace{\text{Sidewalk width}}_{\downarrow \atop s} \quad \underbrace{\text{plus}}_{\downarrow \atop +} \quad \underbrace{\text{Store width}}_{\downarrow \atop 34} \quad \underbrace{\text{plus}}_{\downarrow \atop +} \quad \underbrace{\text{Sidewalk width}}_{\downarrow \atop s} \quad \underbrace{\text{is 50 m.}}_{\downarrow \quad \downarrow \atop = \quad 50}
$$

Area of sidewalk	is	Total area	minus	Area of store
↓	↓	↓	↓	↓
A	$=(40 \text{ m}) \times (50 \text{ m})$		$-$	$(34 \text{ m}) \times (40 \text{ m} - s)$

Solve. We solve the first equation.

$$s + 34 + s = 50$$
$$2s + 34 = 50 \quad \text{Collecting like terms}$$
$$2s = 16 \quad \text{Subtracting 34 on both sides}$$
$$s = 8 \quad \text{Dividing by 2 on both sides}$$

Thus the width of the sidewalk is 8 m.

Then we solve the second equation.

$$A = (40 \text{ m}) \times (50 \text{ m}) - (34 \text{ m}) \times (40 \text{ m} - s)$$
$$A = (40 \text{ m}) \times (50 \text{ m}) - (34 \text{ m}) \times (40 \text{ m} - 8 \text{ m})$$
$$\qquad\qquad\qquad\qquad \text{Substituting 8 m for } s$$
$$A = (40 \text{ m}) \times (50 \text{ m}) - (34 \text{ m}) \times (32 \text{ m})$$
$$A = (40 \times 50 \times \text{ m} \times \text{ m}) - (34 \times 32 \times \text{ m} \times \text{ m})$$
$$A = 2000 \text{ m}^2 - 1088 \text{ m}^2$$
$$A = 912 \text{ m}^2$$

Check. We repeat the calculations.

State. The area of the sidewalk is 912 m^2.

33. *Translate*:

What percent of 24 is 32?

↓   ↓ ↓ ↓ ↓

n $\times$ 24 $=$ 32

Solve: We divide by 24 on both sides and convert the result to percent notation.

$$n \times 24 = 32$$
$$\frac{n \times 24}{24} = \frac{32}{24}$$
$$n = 1.\overline{3} = 133.\overline{3}\%, \text{ or } 133\frac{1}{3}\%$$

Thus, 24 is 133.$\overline{3}$% of 32. The answer is 133.$\overline{3}$%, or $133\frac{1}{3}$%.

35. We rephrase the question and translate.

What percent of 162 is 103?

↓   ↓ ↓ ↓ ↓

n $\times$ 162 $=$ 103

To solve, we divide on both sides by 162 and convert to percent notation.

$$n \times 162 = 103$$
$$n = \frac{103}{162}$$
$$n \approx 0.636$$
$$n \approx 63.6\%$$

37. $-5 + (-12) = -17$

(Add the absolute values and make the answer negative.)

39. $17 + (-24) = -7$

(Find the difference in absolute values. The negative integer has the larger absolute value, so the answer is negative.)

41. Discussion and Writing Exercise

43. For $x = 5$;

$$(x^2 + 2x - 3)(x^2 + 4)$$
$$= (5^2 + 2 \cdot 5 - 3)(5^2 + 4)$$
$$= (25 + 10 - 3)(25 + 4)$$
$$= 32 \cdot 29$$
$$= 928$$

$$x^4 + 2x^3 + x^2 + 8x - 12$$
$$= 5^4 + 2 \cdot 5^3 + 5^2 + 8 \cdot 5 - 12$$
$$= 625 + 250 + 25 + 40 - 12$$
$$= 928$$

The expressions have the same value for $x = 5$.

For $x = 3.5$;

$$(x^2 + 2x - 3)(x^2 + 4)$$
$$= [(3.5)^2 + 2 \cdot 3.5 - 3][(3.5)^2 + 4]$$
$$= (12.25 + 7 - 3)(12.25 + 4)$$
$$= 16.25 \cdot 16.25$$
$$= 264.0625$$

$$x^4 + 2x^3 + x^2 + 8x - 12$$
$$= (3.5)^4 + 2(3.5)^3 + (3.5)^2 + 8(3.5) - 12$$
$$= 150.0625 + 85.75 + 12.25 + 28 - 12$$
$$= 264.0625$$

The expressions have the same value for $x = 3.5$.

For $x = -1.2$;

$$(x^2 + 2x - 3)(x^2 + 4)$$
$$= [(-1.2)^2 + 2(-1.2) - 3][(-1.2)^2 + 4]$$
$$= (1.44 - 2.4 - 3)(1.44 + 4)$$
$$= -3.96(5.44)$$
$$= -21.5424$$

$$x^4 + 2x^3 + x^2 + 8x - 12$$
$$= (-1.2)^4 + 2(-1.2)^3 + (-1.2)^2 + 8(-1.2) - 12$$
$$= 2.0736 - 3.456 + 1.44 - 9.6 - 12$$
$$= -21.5424$$

The expressions have the same value for $x = -1.2$.

45.

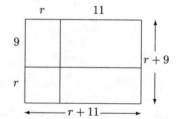

a) The length and width of the figure can be expressed as $r + 11$ and $r + 9$, respectively. The perimeter of this figure (a rectangle) is given by $P = 2 \cdot (l + w) = 2 \cdot (r + 11 + r + 9) = 2 \cdot (2r + 20) = 4r + 40$.

b) The area of this figure is given by $A = l \cdot w = (r+11)(r+9) = (r+11)r+(r+11)9 = r^2+11r+9r+99 = r^2+20r+99$.

47. The shaded area is the area of the large rectangle less the area of the small rectangle:

$$4t(21t+8) - 2t(3t-4) = 84t^2 + 32t - 6t^2 + 8t$$
$$= 78t^2 + 40t$$

49. $(3x-5)^2 = (3x-5)(3x-5) = 3x(3x-5) - 5(3x-5) = 9x^2 - 15x - 15x + 25 = 9x^2 - 30x + 25$

Exercise Set 10.4

---

1. $4^0 = 1$ $(b^0 = 1,$ for any nonzero number b.)

3. $3.14^0 = 1$ $(b^0 = 1,$ for any nonzero number b.)

5. $(-19.57)^1 = -19.57$

7. $(-98.6)^0 = 1$

9. $x^0 = 1, \ x \neq 0$

11. $(3x-17)^0 = (3 \cdot 10 - 17)^0$ Substituting
$\qquad\qquad = (30-17)^0$ Multiplying
$\qquad\qquad = 13^0$
$\qquad\qquad = 1$

13. $(5x-3)^1 = (5 \cdot 4 - 3)^1$ Substituting
$\qquad\qquad = (20-3)^1$ Multiplying
$\qquad\qquad = 17^1$
$\qquad\qquad = 17$

15. $(4m-19)^0 = (4 \cdot 3 - 19)^0$
$\qquad\qquad = (12-19)^0$
$\qquad\qquad = (-7)^0$
$\qquad\qquad = 1$

17. $3x^0 + 4 = 3(-2)^0 + 4$
$\qquad\qquad = 3 \cdot 1 + 4$
$\qquad\qquad = 3 + 4$
$\qquad\qquad = 7$

19. $(3x)^0 + 4 = [3(-2)]^0 + 4$
$\qquad\qquad = (-6)^0 + 4$
$\qquad\qquad = 1 + 4$
$\qquad\qquad = 5$

21. $(5-3x^0)^1 = (5 - 3 \cdot 19^0)^1$
$\qquad\qquad = (5 - 3 \cdot 1)^1$
$\qquad\qquad = (5-3)^1$
$\qquad\qquad = 2^1$
$\qquad\qquad = 2$

23. $3^{-2} = \dfrac{1}{3^2} = \dfrac{1}{9}$

25. $10^{-4} = \dfrac{1}{10^4} = \dfrac{1}{10,000}$

27. $t^{-4} = \dfrac{1}{t^4}$

29. $(-5)^{-2} = \dfrac{1}{(-5)^2} = \dfrac{1}{25}$

31. $3x^{-7} = 3\left(\dfrac{1}{x^7}\right) = \dfrac{3}{x^7}$

33. $\dfrac{x}{y^{-4}} = xy^4$ Instead of dividing by y^{-4}, multiply by y^4.

35. $\dfrac{r^5}{t^{-3}} = r^5 t^3$ Instead of dividing by t^{-3}, multiply by t^3.

37. $-7a^{-9} = -7\left(\dfrac{1}{a^9}\right) = \dfrac{-7}{a^9},$ or $-\dfrac{7}{a^9}$

39. $\left(\dfrac{2}{5}\right)^{-2} = \left(\dfrac{5}{2}\right)^2 = \dfrac{5}{2} \cdot \dfrac{5}{2} = \dfrac{25}{4}$

41. $\left(\dfrac{5}{a}\right)^{-3} = \left(\dfrac{a}{5}\right)^3 = \dfrac{a}{5} \cdot \dfrac{a}{5} \cdot \dfrac{a}{5} = \dfrac{a^3}{125}$

43. $\dfrac{1}{7^3} = 7^{-3}$

45. $\dfrac{9}{x^3} = 9x^{-3}$

47. $x^{-2} \cdot x = x^{-2+1} = x^{-1} = \dfrac{1}{x}$

49. $x^4 \cdot x^{-4} = x^{4+(-4)} = x^0 = 1,$ assuming $x \neq 0$

51. $t^{-4} \cdot t^{-11} = t^{-4+(-11)} = t^{-15} = \dfrac{1}{t^{15}}$

53. $(3a^2 b^{-7})(2ab^9)$
$\qquad = 3 \cdot 2 \cdot a^2 \cdot a \cdot b^{-7} \cdot b^9$ Using the commutative and associative laws
$\qquad = 6a^{2+1}b^{-7+9}$ Using the product rule
$\qquad = 6a^3 b^2$

55. $(-2x^{-3}y^8)(3xy^{-2})$
$\qquad = -2 \cdot 3 \cdot x^{-3} \cdot x \cdot y^8 \cdot y^{-2}$ Using the commutative and associative laws
$\qquad = -6x^{-3+1}y^{8+(-2)}$ Using the product rule
$\qquad = -6x^{-2}y^6$
$\qquad = -6\left(\dfrac{1}{x^2}\right)y^6$
$\qquad = -\dfrac{6y^6}{x^2}$

57. $(3a^{-4}bc^2)(2a^{-2}b^{-5}c)$
$\qquad = 3 \cdot 2 \cdot a^{-4} \cdot a^{-2} \cdot b \cdot b^{-5} \cdot c^2 \cdot c$
$\qquad = 6a^{-4+(-2)}b^{1+(-5)}c^{2+1}$
$\qquad = 6a^{-6}b^{-4}c^3$
$\qquad = \dfrac{6c^3}{a^6 b^4}$

59. Discussion and Writing Exercise

61. A <u>binomial</u> is a polynomial with two terms.

63. In the metric system, the <u>gram</u> is the basic unit of mass.

65. An <u>angle</u> is a set of points consisting of two rays.

67. The <u>perimeter</u> of a polygon is the sum of the lengths of its sides.

69. Discussion and Writing Exercise

71. Discussion and Writing Exercise

73. Use a calculator.

For $x = -3$:
$$\frac{5^x}{5^{x+1}} = \frac{5^{-3}}{5^{-3+1}} = \frac{5^{-3}}{5^{-2}} = \frac{1}{5}, \text{ or } 0.2$$

For $x = -30$:
$$\frac{5^x}{5^{x+1}} = \frac{5^{-30}}{5^{-30+1}} = \frac{5^{-30}}{5^{-29}} = \frac{1}{5}, \text{ or } 0.2$$

75. $(y^{2x})(y^{3x}) = y^{2x+3x} = y^{5x}$

77. $\dfrac{a^{6t}(a^{7t})}{a^{9t}} = \dfrac{a^{6t+7t}}{a^{9t}} = \dfrac{a^{13t}}{a^{9t}} = a^{13t} \cdot a^{-9t} = a^{13t+(-9t)} = a^{4t}$

Exercise Set 10.5

1. 2.8,000,000,000.

↑⌐_____⌐ 10 places

Large number, so the exponent is positive.
$28,000,000,000 = 2.8 \times 10^{10}$

3. 9.07,000,000,000,000,000.

↑⌐_____⌐ 17 places

Large number, so the exponent is positive.
$907,000,000,000,000,000 = 9.07 \times 10^{17}$

5. 0.000003.04

⌐____⌐↑ 6 places

Small number, so the exponent is negative.
$0.00000304 = 3.04 \times 10^{-6}$

7. 0.00000001. 8

⌐____⌐↑ 8 places

Small number, so the exponent is negative.
$0.000000018 = 1.8 \times 10^{-8}$

9. 1.00,000,000,000.

↑⌐_____⌐ 11 places

Large number, so the exponent is positive.
$100,000,000,000 = 1.0 \times 10^{11} = 10^{11}$

11. 281 million = 281,000,000

2.81,000,000.

↑⌐____⌐ 8 places

Large number, so the exponent is positive.
281 million $= 2.81 \times 10^8$

13. $\dfrac{67}{1,000,000,000} = 0.000000067$

0.00000006.7

⌐_____⌐↑ 8 places

Small number, so the exponent is negative.
$$\frac{67}{1,000,000,000} = 6.7 \times 10^{-8}$$

15. 8.74×10^7

Positive exponent, so the answer is a large number.
8.7400000.

⌐_____⌐↑ 7 places

$8.74 \times 10^7 = 87,400,000$

17. 5.704×10^{-8}

Negative exponent, so the answer is a small number.
0.00000005.704

↑⌐_____⌐ 8 places

$5.704 \times 10^{-8} = 0.00000005704$

19. $10^7 = 1 \times 10^7$

Positive exponent, so the answer is a large number.
1.0000000.

⌐_____⌐↑ 7 places

$10^7 = 10,000,000$

21. $10^{-5} = 1 \times 10^{-5}$

Negative exponent, so the answer is a small number.
0.00001.

↑⌐___⌐ 5 places

$10^{-5} = 0.00001$

23. $(3 \times 10^4)(2 \times 10^5) = (3 \cdot 2) \times (10^4 \cdot 10^5)$
$$= 6 \times 10^9$$

25. $(5.2 \times 10^5)(6.5 \times 10^{-2}) = (5.2 \cdot 6.5) \times (10^5 \cdot 10^{-2})$
$$= 33.8 \times 10^3$$

The answer at this stage is 33.8×10^3 but this is not scientific notation since 33.8 is not a number between 1 and 10. We convert 33.8 to scientific notation and simplify.

$33.8 \times 10^3 = (3.38 \times 10^1) \times 10^3 = 3.38 \times (10^1 \times 10^3) = 3.38 \times 10^4$

The answer is 3.38×10^4.

27. $(9.9 \times 10^{-6})(8.23 \times 10^{-8}) = (9.9 \cdot 8.23) \times (10^{-6} \cdot 10^{-8})$
$$= 81.477 \times 10^{-14}$$

The answer at this stage is 81.477×10^{-14}. We convert 81.477 to scientific notation and simplify.

$81.477 \times 10^{-14} = (8.1477 \times 10^1) \times 10^{-14} = 8.1477 \times (10^1 \times 10^{-14}) = 8.1477 \times 10^{-13}$.

The answer is 8.1477×10^{-13}.

29. $\dfrac{8.5 \times 10^8}{3.4 \times 10^{-5}} = \dfrac{8.5}{3.4} \times \dfrac{10^8}{10^{-5}}$

$\qquad\qquad\quad = 2.5 \times 10^{8-(-5)}$

$\qquad\qquad\quad = 2.5 \times 10^{13}$

31. $(3.0 \times 10^6) \div (6.0 \times 10^9) = \dfrac{3.0 \times 10^6}{6.0 \times 10^9}$

$\qquad\qquad\qquad\qquad\qquad = \dfrac{3.0}{6.0} \times \dfrac{10^6}{10^9}$

$\qquad\qquad\qquad\qquad\qquad = 0.5 \times 10^{6-9}$

$\qquad\qquad\qquad\qquad\qquad = 0.5 \times 10^{-3}$

The answer at this stage is 0.5×10^{-3}. We convert 0.5 to scientific notation and simplify.

$$0.5 \times 10^{-3} = (5.0 \times 10^{-1}) \times 10^{-3} =$$
$$5.0 \times (10^{-1} \times 10^{-3}) = 5.0 \times 10^{-4}$$

33. $\dfrac{7.5 \times 10^{-9}}{2.5 \times 10^{12}} = \dfrac{7.5}{2.5} \times \dfrac{10^{-9}}{10^{12}}$

$\qquad\qquad\quad = 3.0 \times 10^{-9-12}$

$\qquad\qquad\quad = 3.0 \times 10^{-21}$

35. There are 60 seconds in one minute and 60 minutes in one hour, so there are 60(60), or 3600 seconds in one hour. There are 24 hours in one day and 365 days in one year, so there are 3600(24)(365), or 31,536,000 seconds in one year.

$\qquad 4,200,000 \times 31,536,000$

$= (4.2 \times 10^6) \times (3.1536 \times 10^7)$

$= (4.2 \times 3.1536) \times (10^6 \times 10^7)$

$\approx 13.25 \times 10^{13}$

$\approx (1.325 \times 10) \times 10^{13}$

$\approx 1.325 \times (10 \times 10^{13})$

$\approx 1.325 \times 10^{14}$

About 1.325×10^{14} cubic feet of water is discharged from the Amazon River in 1 yr.

37. $\dfrac{1.908 \times 10^{24}}{6 \times 10^{21}} = \dfrac{1.908}{6} \times \dfrac{10^{24}}{10^{21}}$

$\qquad\qquad\quad = 0.318 \times 10^3$

$\qquad\qquad\quad = (3.18 \times 10^{-1}) \times 10^3$

$\qquad\qquad\quad = 3.18 \times (10^{-1} \times 10^3)$

$\qquad\qquad\quad = 3.18 \times 10^2$

The mass of Jupiter is 3.18×10^2 times the mass of Earth.

39. 10 billion trillion $= 1 \times 10 \times 10^9 \times 10^{12}$

$\qquad\qquad\qquad\qquad = 1 \times 10^{22}$

There are 1×10^{22} stars in the known universe.

41. We divide the mass of the sun by the mass of earth.

$\dfrac{1.998 \times 10^{27}}{6 \times 10^{21}} = 0.333 \times 10^6$

$\qquad\qquad\quad = (3.33 \times 10^{-1}) \times 10^6$

$\qquad\qquad\quad = 3.33 \times 10^5$

The mass of the sun is 3.33×10^5 times the mass of Earth.

43. First we divide the distance from the earth to the moon by 3 days to find the number of miles per day the space vehicle travels. Note that $240,000 = 2.4 \times 10^5$.

$$\dfrac{2.4 \times 10^5}{3} = 0.8 \times 10^5 = 8 \times 10^4$$

The space vehicle travels 8×10^4 miles per day. Now divide the distance from the earth to Mars by 8×10^4 to find how long it will take the space vehicle to reach Mars. Note that $35,000,000 = 3.5 \times 10^7$.

$$\dfrac{3.5 \times 10^7}{8 \times 10^4} = 0.4375 \times 10^3 = 4.375 \times 10^2$$

It takes 4.375×10^2 days for the space vehicle to travel from the earth to Mars.

45. Discussion and Writing Exercise

47. $2x - 4 - 5x + 8 = x - 3$

$\quad -3x + 4 = x - 3$ $\qquad$ Collecting like terms

$\quad -3x + 4 - 4 = x - 3 - 4$ $\qquad$ Subtracting 4

$\quad -3x = x - 7$

$\quad -3x - x = x - 7 - x$ $\qquad$ Subtracting x

$\quad -4x = -7$

$\quad \dfrac{-4x}{-4} = \dfrac{-7}{-4}$ $\qquad$ Dividing by -4

$\quad x = \dfrac{7}{4}$, or 1.75

The solution is $\dfrac{7}{4}$, or 1.75.

49. $8(2x + 3) - 2(x - 5) = 10$

$16x + 24 - 2x + 10 = 10$ $\qquad$ Removing parentheses

$14x + 34 = 10$ $\qquad$ Collecting like terms

$14x + 34 - 34 = 10 - 34$ $\qquad$ Subtracting 34

$14x = -24$

$\dfrac{14x}{14} = \dfrac{-24}{14}$ $\qquad$ Dividing by 14

$x = -\dfrac{12}{7}$, or $\qquad$ Simplifying

$x = -1.\overline{714285}$

The solution is $-\dfrac{12}{7}$, or $-1.\overline{714285}$.

51. $y = x - 5$

The equation is equivalent to $y = x + (-5)$. The y-intercept is $(0, -5)$. We find two other points.

When $x = 2$, $y = 2 - 5 = -3$.

When $x = 4$, $y = 4 - 5 = -1$.

x	y
0	−5
2	−3
4	−1

Plot these points, draw the line they determine, and label the graph $y = x - 5$.

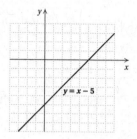

53. Discussion and Writing Exercise

55. $\dfrac{(5.2 \times 10^6)(6.1 \times 10^{-11})}{1.28 \times 10^{-3}} = \dfrac{(5.2 \cdot 6.1)}{1.28} \times \dfrac{(10^6 \cdot 10^{-11})}{10^{-3}}$

$\qquad = 24.78125 \times 10^{-2}$

$\qquad = (2.478125 \times 10^1) \times 10^{-2}$

$\qquad = 2.478125 \times 10^{-1}$

57. Observe that $\dfrac{6.4 \times 10^8}{1.28 \times 10^4} = \dfrac{6.4}{1.28} \times \dfrac{10^8}{10^4} = 5 \times 10^4$.

Thus, 6.4×10^8 is a multiple of 1.28×10^4, so the LCM is 6.4×10^8.

Chapter 10 Review Exercises

1. $(-4x + 9) + (7x - 15) = (-4 + 7)x + (9 - 15) = 3x - 6$

2. $\quad (7x^4 - 5x^3 + 3x - 5) + (x^3 - 4x + 2)$

$= 7x^4 + (-5 + 1)x^3 + (3 - 4)x + (-5 + 2)$

$= 7x^4 - 4x^3 - x - 3$

3. $\quad (9a^5 + 8a^3 + 4a + 7) - (a^5 - 4a^3 + a^2 - 2)$

$= 9a^5 + 8a^3 + 4a + 7 - a^5 + 4a^3 - a^2 + 2$

$= 8a^5 + 12a^3 - a^2 + 4a + 9$

4. $\quad (8a^3b^3 + 9a^2b^3) - (3a^3b^3 - 2a^2b^3 + 7)$

$= 8a^3b^3 + 9a^2b^3 - 3a^3b^3 + 2a^2b^3 - 7$

$= 5a^3b^3 + 11a^2b^3 - 7$

5. $-(12x^3 - 4x^2 + 9x - 3); \; -12x^3 + 4x^2 - 9x + 3$

6. $(-59)^0 = 1 \qquad (b^0 = 1 \text{ for any nonzero number } b.)$

7. $\quad 5t^3 + t = 5(-2)^3 + (-2)$

$= 5(-8) + (-2)$

$= -40 + (-2)$

$= -42$

8. $\quad -16t^2 + 200 = -16 \cdot 3^2 + 200$

$= -16 \cdot 9 + 200$

$= -144 + 200$

$= 56 \text{ ft}$

9. $(5x^3)(6x^4) = (5 \cdot 6)(x^3 \cdot x^4) = 30x^7$

10. $3x(6x^3 - 4x - 1) = 3x \cdot 6x^3 + 3x(-4x) + 3x(-1)$

$= 18x^4 - 12x^2 - 3x$

11. $2a^4b(7a^3b^3 + 5a^2b^3) = 2a^4b \cdot 7a^3b^3 + 2a^4b \cdot 5a^2b^3$

$= 14a^7b^4 + 10a^6b^4$

12. $(x - 7)(x + 9) = x \cdot x + 9 \cdot x - 7 \cdot x - 7 \cdot 9$

$= x^2 + 9x - 7x - 63$

$= x^2 + 2x - 63$

13. $(3x - 1)(5x - 2) = 3x \cdot 5x - 2 \cdot 3x - 1 \cdot 5x - 1(-2)$

$= 15x^2 - 6x - 5x + 2$

$= 15x^2 - 11x + 2$

14. $\quad (a^2 - 1)(a^2 + 2a - 1)$

$= a^2(a^2 + 2a - 1) - 1 \cdot (a^2 + 2a - 1)$

$= a^2 \cdot a^2 + a^2 \cdot 2a + a^2(-1) - 1 \cdot a^2 - 1 \cdot 2a - 1(-1)$

$= a^4 + 2a^3 - a^2 - a^2 - 2a + 1$

$= a^4 + 2a^3 - 2a^2 - 2a + 1$

15. Factor $45x^3 - 10x$.

The prime factorization of $45x^3$ is $3 \cdot 3 \cdot 5 \cdot x \cdot x \cdot x$.

The prime factorization of $10x$ is $2 \cdot 5 \cdot x$.

The largest common factor is $5 \cdot x$, or $5x$.

$45x^3 - 10x = 5x \cdot 9x^2 - 5x \cdot 2$

$= 5x(9x^2 - 2)$

16. Factor $7a - 35b - 49ac$.

The prime factorization of $7a$ is $7 \cdot a$.

The prime factorization of $35b$ is $5 \cdot 7 \cdot b$.

The prime factorization of $49ac$ is $7 \cdot 7 \cdot a \cdot c$.

The largest common factor is 7.

$7a - 35b - 49ac = 7 \cdot a - 7 \cdot 5b - 7 \cdot 7ac$

$= 7(a - 5b - 7ac)$

17. Factor $6x^3y - 9x^2y^5$.

The prime factorization of $6x^3y$ is $2 \cdot 3 \cdot x \cdot x \cdot x \cdot y$.

The prime factorization of $9x^2y^5$ is $3 \cdot 3 \cdot x \cdot x \cdot y \cdot y \cdot y \cdot y \cdot y$.

The largest common factor is $3 \cdot x \cdot x \cdot y$, or $3x^2y$.

$6x^3y - 9x^2y^5 = 3x^2y \cdot 2x - 3x^2y \cdot 3y^4$

$= 3x^2y(2x - 3y^4)$

18. $12^{-2} = \dfrac{1}{12^2} = \dfrac{1}{12 \cdot 12} = \dfrac{1}{144}$

19. $8a^{-7} = 8\left(\dfrac{1}{a^7}\right) = \dfrac{8}{a^7}$

20. $\dfrac{x^{-3}}{y^5z^{-6}} = \dfrac{z^6}{x^3y^5}$

21. $\left(\dfrac{4}{5}\right)^{-2} = \left(\dfrac{5}{4}\right)^2 = \dfrac{5}{4} \cdot \dfrac{5}{4} = \dfrac{25}{16}$

22. $\dfrac{1}{x^7} = x^{-7}$

23. $x^{-5} \cdot x^{-12} = x^{-5 + (-12)} = x^{-17} = \dfrac{1}{x^{17}}$

24. $(-7x^3y^{-5})(-2x^4y^{-2}) = -7 \cdot (-2) \cdot x^3 \cdot x^4 \cdot y^{-5} \cdot y^{-2}$

$\qquad = 14x^{3+4}y^{-5+(-2)}$

$\qquad = 14x^7y^{-7}$

$\qquad = \dfrac{14x^7}{y^7}$

25. $4.\underset{\longmapsto\ 7\ \text{places}}{2,700,000.}$

Large number, so the exponent is positive.

$42,700,000 = 4.27 \times 10^7$

26. $0.\underset{4\ \text{places}}{0001.924}$

Small number, so the exponent is negative.

$0.0001924 = 1.924 \times 10^{-4}$

27. $(5.1 \times 10^6)(2.3 \times 10^4) = (5.1 \cdot 2.3) \times (10^6 \cdot 10^4)$

$\qquad = 11.73 \times 10^{10}$

$\qquad = (1.173 \times 10) \times 10^{10}$

$\qquad = 1.173 \times (10 \cdot 10^{10})$

$\qquad = 1.173 \times 10^{11}$

28. $\dfrac{300,000,000}{1,200,000,000} = \dfrac{3 \times 10^8}{1.2 \times 10^9} = 2.5 \times 10^{-1}$

The wavelength is 2.5×10^{-1} m.

29. $(0.25)(2.5 \times 10^{-1}) = (2.5 \times 10^{-1})(2.5 \times 10^{-1})$

$\qquad = (2.5 \cdot 25) \times (10^{-1} \cdot 10^{-1})$

$\qquad = 6.25 \times 10^{-2}$

The length of the antenna should be 6.25×10^{-2} m.

30. *Discussion and Writing Exercise.* Adi is probably adding coefficients and multiplying exponents instead of the other way around.

31. *Discussion and Writing Exercise.* Because x^{-2} is $\dfrac{1}{x^2}$ and because x^2 is never negative, it follows that x^{-2} is never negative.

32. $(2349x^7 - 357x^2)(493x^{10} + 597x^5)$

$= 2349x^7 \cdot 493x^{10} + 2349x^7 \cdot 597x^5 - 357x^2 \cdot 493x^{10} -$

$\qquad 357x^2 \cdot 597x^5$

$= 1,158,057x^{17} + 1,402,353x^{12} - 176,001x^{12} - 213,129x^7$

$= 1,158,057x^{17} + 1,226,352x^{12} - 213,129x^7$

33. $-3x^5 \cdot 3x^3 - x^6(2x)^2 + (3x^4)^2 + (2x^4)^2 - 40x^2(x^3)^2$

$= -3x^5 \cdot 3x^3 - x^6 \cdot 2x \cdot 2x + 3x^4 \cdot 3x^4 + 2x^4 \cdot 2x^4 -$

$\qquad 40x^2 \cdot x^3 \cdot x^3$

$= -9x^8 - 4x^8 + 9x^8 + 4x^8 - 40x^8$

$= -40x^8$

34. Factor $39a^3b^7c^6 - 130a^2b^5c^8 + 52a^4b^6c^5$.

The prime factorization of $39a^3b^7c^6$ is

$3 \cdot 13 \cdot a \cdot a \cdot a \cdot b \cdot b \cdot b \cdot b \cdot b \cdot b \cdot b \cdot c \cdot c \cdot c \cdot c \cdot c \cdot c.$

The prime factorization of $130a^2b^5c^8$ is

$2 \cdot 5 \cdot 13 \cdot a \cdot a \cdot b \cdot b \cdot b \cdot b \cdot b \cdot c \cdot c \cdot c \cdot c \cdot c \cdot c \cdot c \cdot c.$

The prime factorization of $52a^4b^6c^5$ is

$2 \cdot 2 \cdot 13 \cdot a \cdot a \cdot a \cdot a \cdot b \cdot b \cdot b \cdot b \cdot b \cdot b \cdot c \cdot c \cdot c \cdot c \cdot c.$

The largest common factor is $13 \cdot a \cdot a \cdot b \cdot b \cdot b \cdot b \cdot b \cdot c \cdot c \cdot c \cdot c \cdot c$, or $13a^2b^5c^5$.

$39a^3b^7c^6 - 130a^2b^5c^8 + 52a^4b^6c^5$

$= 13a^2b^5c^5 \cdot 3ab^2c - 13a^2b^5c^5 \cdot 10c^3 + 13a^2b^5c^5 \cdot 4a^2b$

$= 13a^2b^5c^5(3ab^2c - 10c^3 + 4a^2b)$

35. Factor $w^5x^6y^4z^5 - w^7x^3y^7z^3 + w^6x^2y^5z^6 - w^6x^7y^3z^4$.

The variables w, x, y, and z appear in every term of the polynomial. Thus the largest common factor contains each variable raised to the smallest power that appears in the polynomial. This means that the largest common factor is $w^5x^2y^3z^3$.

$w^5x^6y^4z^5 - w^7x^3y^7z^3 + w^6x^2y^5z^6 - w^6x^7y^3z^4$

$= w^5x^2y^3z^3(x^4yz^2 - w^2xy^4 + wy^2z^3 - wx^5z)$

36. Factor $10a^4b^{-5} + 12a^7b^{-3}$.

First we consider the prime factorizations of the coefficients.

$10 = 2 \cdot 5$

$12 = 2 \cdot 2 \cdot 3$

The largest factor common to the coefficients is 2. Now consider the variables, a and b. Both appear in each term, so the largest common factor contains each variable raised to the smallest power that appears in the polynomial. Thus we use a^4 and b^{-5}. The largest common factor is $2a^4b^{-5}$.

$10a^4b^{-5} + 12a^7b^{-3} = 2a^4b^{-5} \cdot 5 + 2a^4b^{-5} \cdot 6a^3b^2$

$\qquad = 2a^4b^{-5}(5 + 6a^3b^2)$

This factorization can be checked by multiplying.

Chapter 10 Test

1. $\quad (12a^3 - 9a^2 + 8) + (6a^3 + 4a^2 - a)$

$= (12 + 6)a^3 + (-9 + 4)a^2 - a + 8$

$= 18a^3 - 5a^2 - a + 8$

2. $-(-9a^4 + 7b^2 - ab + 3); \ 9a^4 - 7b^2 + ab - 3$

3. $\quad (12x^4 + 7x^2 - 6) - (9x^4 + 8x^2 + 5)$

$= 12x^4 + 7x^2 - 6 - 9x^4 - 8x^2 - 5$

$= 3x^4 - x^2 - 11$

4. $19^1 = 19 \qquad (b^1 = b$ for any number $b.)$

5. $\quad (3x - 7)^0 = (3 \cdot 2 - 7)^0$

$\qquad = (6 - 7)^0$

$\qquad = (-1)^0$

$\qquad = 1 \quad (b^0 = 1$ for any nonzero number $b.)$

6. $-4.9t^2 + 15t + 2 = -4.9 \cdot 2^2 + 15 \cdot 2 + 2$

$\qquad = -4.9 \cdot 4 + 15 \cdot 2 + 2$

$\qquad = -19.6 + 30 + 2$

$\qquad = 12.4$ m

7. $(-5x^4y^3)(2x^2y^5) = (-5 \cdot 2)(x^4 \cdot x^2)(y^3 \cdot y^5)$
$$= -10x^{4+2}y^{3+5}$$
$$= -10x^6y^8$$

8. $2a(5a^2 - 4a + 3) = 2a \cdot 5a^2 + 2a(-4a) + 2a \cdot 3$
$$= 10a^3 - 8a^2 + 6a$$

9. $(x - 5)(x + 9) = x \cdot x + 9 \cdot x - 5 \cdot x - 5 \cdot 9$
$$= x^2 + 9x - 5x - 45$$
$$= x^2 + 4x - 45$$

10. $(2a + 1)(a^2 - 3a + 2)$
$$= 2a(a^2 - 3a + 2) + 1 \cdot (a^2 - 3a + 2)$$
$$= 2a \cdot a^2 + 2a(-3a) + 2a \cdot 2 + 1 \cdot a^2 + 1 \cdot (-3a) + 1 \cdot 2$$
$$= 2a^3 - 6a^2 + 4a + a^2 - 3a + 2$$
$$= 2a^3 - 5a^2 + a + 2$$

11. Factor $35x^6 - 25x^3 + 15x^2$.

The prime factorization of $35x^6$ is $5 \cdot 7 \cdot x \cdot x \cdot x \cdot x \cdot x \cdot x$.

The prime factorization of $25x^3$ is $5 \cdot 5 \cdot x \cdot x \cdot x$.

The prime factorization of $15x^2$ is $3 \cdot 5 \cdot x \cdot x$.

The largest common factor is $5 \cdot x \cdot x$, or $5x^2$.

$35x^6 - 25x^3 + 15x^2 = 5x^2 \cdot 7x^4 - 5x^2 \cdot 5x + 5x^2 \cdot 3$
$$= 5x^2(7x^4 - 5x + 3)$$

12. Factor $6ab - 9bc + 12ac$.

The prime factorization of $6ab$ is $2 \cdot 3 \cdot a \cdot b$.

The prime factorization of $9bc$ is $3 \cdot 3 \cdot b \cdot c$.

The prime factorization of $12ac$ is $2 \cdot 2 \cdot 3 \cdot a \cdot c$.

The largest common factor is 3.

$6ab - 9bc + 12ac = 3 \cdot 2ab - 3 \cdot 3bc + 3 \cdot 4ac$
$$= 3(2ab - 3bc + 4ac)$$

13. $5^{-3} = \dfrac{1}{5^3} = \dfrac{1}{5 \cdot 5 \cdot 5} = \dfrac{1}{125}$

14. $\dfrac{5a^{-3}}{b^{-2}} = \dfrac{5b^2}{a^3}$

15. $\left(\dfrac{3}{5}\right)^{-3} = \left(\dfrac{5}{3}\right)^3 = \dfrac{5}{3} \cdot \dfrac{5}{3} \cdot \dfrac{5}{3} = \dfrac{125}{27}$

16. $x^{-7} \cdot x^{-9} = x^{-7+(-9)} = x^{-16} = \dfrac{1}{x^{16}}$

17. $(3a^{-7}b^9)(-2a^{10}b^{-12}) = 3(-2) \cdot a^{-7} \cdot a^{10} \cdot b^9 \cdot b^{-12}$
$$= -6a^{-7+10}b^{9+(-12)}$$
$$= -6a^3b^{-3}$$
$$= \dfrac{-6a^3}{b^3}$$

18. $0.0004.7$

$\underline{\quad\uparrow}$ 4 places

Small number, so the exponent is negative.

$0.00047 = 4.7 \times 10^{-4}$

19. $8.250,000.$

$\underline{\uparrow\qquad\quad}$ 6 places

Large number, so the exponent is positive.

$8,250,000 = 8.25 \times 10^6$

20. $(3.2 \times 10^{-8})(5.7 \times 10^{-9}) = (3.2 \cdot 5.7) \times (10^{-8} \cdot 10^{-9})$
$$= 18.24 \times 10^{-17}$$
$$= (1.824 \times 10) \times 10^{-17}$$
$$= 1.824 \times 10 \cdot 10^{-17}$$
$$= 1.824 \times 10^{-16}$$

21. $0.041h - 0.018A - 2.69$
$$= 0.041(150) - 0.018(30) - 2.69$$
$$= 6.15 - 0.54 - 2.69$$
$$= 2.92 \text{ L}$$

22. $12a^6(2a^3 - 6a)^{-2}$
$$= \dfrac{12a^6}{(2a^3 - 6a)^2}$$
$$= \dfrac{12a^6}{(2a^3 - 6a)(2a^3 - 6a)}$$
$$= \dfrac{12a^6}{2a^3 \cdot 2a^3 + 2a^3(-6a) - 6a \cdot 2a^3 - 6a(-6a)}$$
$$= \dfrac{12a^6}{4a^6 - 12a^4 - 12a^4 + 36a^2}$$
$$= \dfrac{12a^6}{4a^6 - 24a^4 + 36a^2}$$
$$= \dfrac{3 \cdot 4 \cdot a^2 \cdot a^4}{4a^2(a^4 - 6a^2 + 9)}$$
$$= \dfrac{4a^2}{4a^2} \cdot \dfrac{3 \cdot a^4}{a^4 - 6a^2 + 9}$$
$$= \dfrac{3a^4}{a^4 - 6a^2 + 9}$$

Cumulative Review Chapters 1 - 10

1. $1.5 \text{ million} = 1.5 \times 1 \text{ million}$
$$= 1.5 \times 1,000,000$$
$$= 1,500,000$$

2. $1312 \text{ ft} = 1312 \,\cancel{\text{ft}} \times \dfrac{1 \text{ yd}}{3 \,\cancel{\text{ft}}}$
$$= \dfrac{1312}{3} \times 1 \text{ yd}$$
$$= 437\dfrac{1}{3} \text{ yd}$$

$1312 \text{ ft} = 1312 \times 1 \text{ ft}$
$$\approx 1312 \times 0.305 \text{ m}$$
$$\approx 400 \text{ m}$$

3. The lowest point on the graph represents 236 eggs per person. It corresponds to 1993 and to 1995.

4. The highest point on the graph represents 258 eggs per person. It corresponds to 2000.

5. Mean:
$$\frac{236 + 239 + 236 + 238 + 240 + 245 + 255 + 258}{8}$$
$$= \frac{1947}{8} = 243.375$$

Median: We first arrange the numbers from smallest to largest.

$$236, \; 236, \; 238, \; 239, \; 240, \; 245, \; 255, \; 258$$

The median is the average of the two middle numbers, 239 and 240.

$$\frac{239 + 240}{2} = \frac{479}{2} = 239.5$$

Mode: The number 236 occurs most often. It is the mode.

6. $\dfrac{240 + 245 + 255 + 258}{4} = \dfrac{998}{4} = 249.5$

7. $\dfrac{236 + 239 + 236 + 238}{4} = \dfrac{949}{4} = 237.25$

The average egg consumption during the years 1993 to 1996 is $249.5 - 237.25$, or 12.25, eggs per person lower than the consumption during the years 1997 to 2000.

8. *Familiarize.* First we subtract to find the amount of increase.

$$\begin{array}{r} 2\,5\,8 \\ -\,2\,3\,8 \\ \hline 2\,0 \end{array}$$

Let $p =$ the percent of increase.

Translate.

20 is $\underbrace{\text{what percent}}$ of 238?

$20 = \quad p \quad \times \; 238$

Solve.

$$20 = p \times 238$$
$$\frac{20}{238} = \frac{p \times 238}{238}$$
$$0.084 \approx p$$
$$8.4\% \approx p$$

Check. 8.4% of $238 = 0.084 \times 238 = 19.992 \approx 20$, so the answer checks. (Remember that we rounded the percent.)

State. Egg consumption increased about 8.4% from 1996 to 2000.

9. $\dfrac{\text{Number of ways to select a left-handed student}}{\text{Number of ways to select a student}}$

$= \dfrac{3}{20}$, or 0.15

10. $x + 22° + 40° = 180°$
$x + 62° = 180°$
$x = 180° - 62°$
$x = 118°$

11.
$$\begin{array}{r} {\scriptstyle 2\;1\;1} \\ 4\,9\,0\,3 \\ 5\,2\,7\,8 \\ 6\,3\,9\,1 \\ +\;4\,5\,1\,3 \\ \hline 2\,1,0\,8\,5 \end{array}$$

12.
$$5\frac{4}{9} = 5\frac{4}{9}$$
$$+3\boxed{\frac{1}{3}\cdot\frac{3}{3}} = +3\frac{3}{9}$$
$$\overline{8\frac{7}{9}}$$

13. $-29 + 53$

The absolute values are 29 and 53. The difference is 24. The positive number has the larger absolute value, so the answer is positive.

$-29 + 53 = 24$

14. $-543 + (-219)$

Add the absolute values: $543 + 219 = 762$

Make the answer negative: $-543 + (-219) = -762$

15. $-34.56 + 2.783 + 0.433 + (-13.02)$
$= -31.777 + 0.433 + (-13.02)$
$= -31.344 + (-13.02)$
$= -44.364$

16. $(4x^5 + 7x^4 - 3x^2 + 9) + (6x^5 - 8x^4 + 2x^3 - 7)$
$= (4 + 6)x^5 + (7 - 8)x^4 + 2x^3 - 3x^2 + (9 - 7)$
$= 10x^5 - x^4 + 2x^3 - 3x^2 + 2$

17.
$$\begin{array}{r} 6\,7\,4 \\ -\,4\,3\,1 \\ \hline 2\,4\,3 \end{array}$$

18. $-4x - 13x = (-4 - 13)x = -17x$

19. $\dfrac{2}{5} - \dfrac{7}{8} = \dfrac{2}{5}\cdot\dfrac{8}{8} - \dfrac{7}{8}\cdot\dfrac{5}{5}$
$= \dfrac{16}{40} - \dfrac{35}{40}$
$= \dfrac{16 - 35}{40}$
$= \dfrac{-19}{40}$

20.
$$4\boxed{\frac{1}{3}\cdot\frac{8}{8}} = 4\frac{8}{24} = 3\frac{32}{24}$$
$$-1\boxed{\frac{5}{8}\cdot\frac{3}{3}} = -1\frac{15}{24} = -1\frac{15}{24}$$
$$\overline{2\frac{17}{24}}$$

21.
$$\begin{array}{r} {\scriptstyle 1\;9\;9\;\;9\;9\,10} \\ 2\,0.\,0\,0\,0\,\cancel{0} \\ -\;\;\;0.\,0\,0\,2\,7 \\ \hline 1\,9.\,9\,9\,7\,3 \end{array}$$

22. $(7x^3 + 2x^2 - x) - (5x^3 - 3x^2 - 8x)$
$= 7x^3 + 2x^2 - x - 5x^3 + 3x^2 + 8x$
$= 2x^3 + 5x^2 + 7x$

23. $(9a^2b + 3ab) - (13a^2b - 4ab)$
$= 9a^2b + 3ab - 13a^2b + 4ab$
$= -4a^2b + 7ab$

24.
$$\begin{array}{r} 297 \\ \times\ 16 \\ \hline 1782 \\ 2970 \\ \hline 4752 \end{array}$$

25. $349 \cdot (-213)$

First we find $349 \cdot 213$.
$$\begin{array}{r} 349 \\ \times\ 213 \\ \hline 1047 \\ 3490 \\ 69800 \\ \hline 74{,}337 \end{array}$$

The product of a positive number and a negative number is negative, so $349 \cdot (-213) = -74{,}337$.

26. $2\frac{3}{4} \cdot 1\frac{2}{3} = \frac{11}{4} \cdot \frac{5}{3} = \frac{11 \cdot 5}{4 \cdot 3} = \frac{55}{12} = 4\frac{7}{12}$

27. $-\frac{9}{7} \cdot \frac{14}{15} = -\frac{9 \cdot 14}{7 \cdot 15} = -\frac{3 \cdot 3 \cdot 2 \cdot 7}{7 \cdot 3 \cdot 5} = \frac{3 \cdot 7}{3 \cdot 7} \cdot \left(-\frac{3 \cdot 2}{5}\right) =$
$-\frac{3 \cdot 2}{5} = -\frac{6}{5}$

28. $12 \cdot \frac{5}{6} = \frac{12 \cdot 5}{6} = \frac{2 \cdot 6 \cdot 5}{6 \cdot 1} = \frac{6}{6} \cdot \frac{2 \cdot 5}{1} = \frac{2 \cdot 5}{1} = 10$

29.
$$\begin{array}{r} 34.09 \quad \text{(2 decimal places)} \\ \times\ 7.6 \quad \text{(1 decimal place)} \\ \hline 20454 \\ 238630 \\ \hline 259.084 \quad \text{(3 decimal places)} \end{array}$$

30. $3(8x - 5) = 3 \cdot 8x + 3(-5) = 24x - 15$

31. $(9a^3b^2)(3a^5b) = 9 \cdot 3 \cdot a^3 \cdot a^5 \cdot b^2 \cdot b$
$= 27a^{3+5}b^{2+1}$
$= 27a^8b^3$

32. $7x^2(3x^3 - 2x + 8) = 7x^2 \cdot 3x^3 + 7x^2(-2x) + 7x^2 \cdot 8$
$= 21x^5 - 14x^3 + 56x^2$

33. $(x + 2)(x - 7) = x \cdot x + x(-7) + 2 \cdot x + 2(-7)$
$= x^2 - 7x + 2x - 14$
$= x^2 - 5x - 14$

34. $(a + 3)(a^2 - 5a + 4)$
$= a(a^2 - 5a + 4) + 3(a^2 - 5a + 4)$
$= a \cdot a^2 + a(-5a) + a \cdot 4 + 3 \cdot a^2 + 3(-5a) + 3 \cdot 4$
$= a^3 - 5a^2 + 4a + 3a^2 - 15a + 12$
$= a^3 - 2a^2 - 11a + 12$

35.
$$\begin{array}{r} 573 \\ 6\overline{)3438} \\ 3000 \\ \hline 438 \\ 420 \\ \hline 18 \\ 18 \\ \hline 0 \end{array}$$

The answer is 573.

36.
$$\begin{array}{r} 56 \\ 34\overline{)1914} \\ 1700 \\ \hline 214 \\ 204 \\ \hline 10 \end{array}$$

The answer is 56 R 10.

37. $\frac{4}{5} \div \left(-\frac{8}{15}\right) = \frac{4}{5} \cdot \left(-\frac{15}{8}\right) = -\frac{4 \cdot 15}{5 \cdot 8} = -\frac{4 \cdot 3 \cdot 5}{5 \cdot 2 \cdot 4} =$
$\frac{4 \cdot 5}{4 \cdot 5} \cdot \left(-\frac{3}{2}\right) = -\frac{3}{2}$

38. $-2\frac{1}{3} \div (-30) = -\frac{7}{3} \div (-30) = -\frac{7}{3} \cdot \left(-\frac{1}{30}\right) = \frac{7}{90}$

39.
$$\begin{array}{r} 39. \\ 2.7_\wedge\overline{)105.3_\wedge} \\ 810 \\ \hline 243 \\ 243 \\ \hline 0 \end{array}$$

The answer is 39.

40. $56\frac{10}{34} = 56\frac{5}{17}$

41. $10 \div 2 \times 20 - 5^2 = 10 \div 2 \times 20 - 25$
$= 5 \times 20 - 25$
$= 100 - 25$
$= 75$

42. $\frac{|3^2 - 5^2|}{2 - 2 \cdot 5} = \frac{|9 - 25|}{2 - 10} = \frac{|-16|}{-8} = \frac{16}{-8} = -2$

43. $\underbrace{14 \cdot 14 \cdot 14}_{3 \text{ factors}} = 14^3$

44. $6\,8,\boxed{4}\,8\,9$
$\uparrow$

The digit 8 is in the thousands place. Consider the next digit to the right. Since the digit, 4, is 4 or lower round down, meaning that 8 thousands stay as 8 thousands. Then change all digits to the right of the thousands digit to zeros.

The answer is 68,000.

45. Round
$21.8\,3\,\boxed{8}\,3 \ldots$ to the nearest hundredth.
Thousandths digit is 5 or higher.
21.84 Round up.

46. The sum of the digits, $1 + 3 + 6 + 8$, or 18, is divisible by 3, so 1368 is divisible by 3.

47. We find as many two-factor factorizations as we can.
$15 = 1 \cdot 15$
$15 = 3 \cdot 5$

The factors of 15 are 1, 3, 5, and 15.

48. $15 = 3 \cdot 5$

$35 = 5 \cdot 7$

The LCM is $3 \cdot 5 \cdot 7$, or 105.

49. $\dfrac{24}{33} = \dfrac{3 \cdot 8}{3 \cdot 11} = \dfrac{3}{3} \cdot \dfrac{8}{11} = \dfrac{8}{11}$

50. First we consider $\dfrac{18}{5}$. To convert $\dfrac{18}{5}$ to a mixed numeral, we divide.

$$5 \overline{\smash{\big)}\, 18} $$

$$\begin{array}{r} 3 \\ 5\,\overline{\smash{\big)}\,1\,8} \\ \underline{1\,5} \\ 3 \end{array}$$

$\dfrac{18}{5} = 3\dfrac{3}{5}$, so $-\dfrac{18}{5} = -3\dfrac{3}{5}$.

51. -17 is to the right of -29 on the number line, so $-17 > -29$.

52. $\dfrac{4}{7} = \dfrac{4}{7} \cdot \dfrac{5}{5} = \dfrac{20}{35}$

$\dfrac{3}{5} = \dfrac{3}{5} \cdot \dfrac{7}{7} = \dfrac{21}{35}$

Since $20 < 21$, it follows that $\dfrac{20}{35} < \dfrac{21}{35}$, so $\dfrac{4}{7} < \dfrac{3}{5}$.

53. To compare two positive numbers in decimal notation, start at the left and compare corresponding digits moving from left to right. When two digits differ, the number with the larger digit is the larger of the two numbers.

1.001

↑   Different; 1 is larger than 0.

0.9976

Thus, 1.001 is larger.

54. $\dfrac{a^2 - b}{3} = \dfrac{(-9)^2 - (-6)}{3} = \dfrac{81 - (-6)}{3} = \dfrac{81 + 6}{3} = \dfrac{87}{3} =$

$\dfrac{3 \cdot 29}{3 \cdot 1} = 29$

55. Factor $40 - 5t$.

The prime factorization of 40 is $2 \cdot 2 \cdot 2 \cdot 5$.

The prime factorization of $5t$ is $5 \cdot t$.

The largest common factor is 5.

$40 - 5t = 5 \cdot 8 - 5 \cdot t = 5(8 - t)$

56. Factor $18a^3 - 15a^2 + 6a$.

The prime factorization of $18a^3$ is $2 \cdot 3 \cdot 3 \cdot a \cdot a \cdot a$.

The prime factorization of $15a^2$ is $3 \cdot 5 \cdot a \cdot a$.

The prime factorization of $6a$ is $2 \cdot 3 \cdot a$.

The largest common factor is $3 \cdot a$, or $3a$.

$18a^3 - 15a^2 + 6a = 3a \cdot 6a^2 - 3a \cdot 5a + 3a \cdot 2$

$\qquad\qquad = 3a(6a^2 - 5a + 2)$

57. The rectangle is divided into 5 equal parts. The unit is $\dfrac{1}{5}$. The denominator is 5. We have 3 parts shaded. This tells us that the numerator is 3. Thus, $\dfrac{3}{5}$ is shaded.

58. $\dfrac{429}{10{,}000}$ 0.0429.

4 zeros   Move 4 places.

$\dfrac{429}{10{,}000} = 0.0429$

59. $-\dfrac{13}{25} = -\dfrac{13}{25} \cdot \dfrac{4}{4} = -\dfrac{52}{100} = -0.52$

60. $\dfrac{8}{9} = 8 \div 9$

$$\begin{array}{r} 0.\,8\,8 \\ 9\,\overline{\smash{\big)}\,8.\,0\,0} \\ \underline{7\,2} \\ 8\,0 \\ \underline{7\,2} \\ 8 \end{array}$$

Since 8 keeps reappearing as a remainder, the digits repeat and $\dfrac{8}{9} = 0.888\ldots$, or $0.\overline{8}$.

61. 7%

a) Replace the percent symbol with $\times 0.01$.

7×0.01

b) Move the decimal point two places to the left.

0.07.

Thus, $7\% = 0.07$.

62. 6.<u>71</u>   6.71. $\dfrac{671}{1\underline{00}}$

2 places   Move 2 places.   2 zeros

$6.71 = \dfrac{671}{100}$

63. First we consider $7\dfrac{1}{4}$.

$7\dfrac{1}{4} = \dfrac{29}{4}$ $(7 \cdot 4 = 28$ and $28 + 1 = 29)$

$7\dfrac{1}{4} = \dfrac{29}{4}$, so $-7\dfrac{1}{4} = -\dfrac{29}{4}$.

64. $40\% = \dfrac{40}{100}$ Definition of percent

$= \dfrac{2 \cdot 20}{5 \cdot 20}$

$= \dfrac{2}{5} \cdot \dfrac{20}{20}$

$= \dfrac{2}{5}$

65. $\dfrac{17}{20} = \dfrac{17}{20} \cdot \dfrac{5}{5} = \dfrac{85}{100} = 85\%$

66. 1.5

a) Move the decimal point two places to the right.

1.50.

b) Write a percent symbol: 150%

Thus, $1.5 = 150\%$.

67. $9.389 + 4.2105 \approx 9.4 + 4.2 = 13.6$

68.
$$234 + y = 789$$
$$234 + y - 234 = 789 - 234$$
$$y = 555$$

The number 555 checks. It is the solution.

69.
$$3.9a = 249.6$$
$$\frac{3.9a}{3.9} = \frac{249.6}{3.9}$$
$$a = 64$$

The number 64 checks. It is the solution.

70.
$$\frac{2}{3} \cdot t = \frac{5}{6}$$
$$t = \frac{5}{6} \div \frac{2}{3} \qquad \text{Dividing both sides by } \frac{2}{3}$$
$$t = \frac{5}{6} \cdot \frac{3}{2} = \frac{5 \cdot 3}{6 \cdot 2}$$
$$= \frac{5 \cdot 3}{2 \cdot 3 \cdot 2} = \frac{3}{3} \cdot \frac{5}{2 \cdot 2}$$
$$= \frac{5}{4}$$

The number $\frac{5}{4}$ checks. It is the solution.

71.
$$\frac{8}{17} = \frac{36}{x}$$
$$8 \cdot x = 17 \cdot 36 \qquad \text{Equating cross products}$$
$$\frac{8 \cdot x}{8} = \frac{17 \cdot 36}{8}$$
$$x = \frac{17 \cdot 4 \cdot 9}{2 \cdot 4} = \frac{4}{4} \cdot \frac{17 \cdot 9}{2}$$
$$x = \frac{153}{2}$$
$$x = 76.5, \text{ or } 76\frac{1}{2}$$

72.
$$7x - 9 = 26$$
$$7x - 9 + 9 = 26 + 9$$
$$7x = 35$$
$$\frac{7x}{7} = \frac{35}{7}$$
$$x = 5$$

The number 5 checks. It is the solution.

73.
$$-2(x - 5) = 3x + 12$$
$$-2x + 10 = 3x + 12$$
$$-2x + 10 + 2x = 3x + 12 + 2x$$
$$10 = 5x + 12$$
$$10 - 12 = 5x + 12 - 12$$
$$-2 = 5x$$
$$\frac{-2}{5} = \frac{5x}{5}$$
$$-\frac{2}{5} = x$$

The number $-\frac{2}{5}$ checks. It is the solution.

74. $\dfrac{\$20 + \$30 + \$25 + \$20}{4} = \dfrac{\$95}{4} = \$23.75$

75. *Familiarize*. Let $m =$ the number of minutes it takes to wrap 8710 candy bars.

Translate.

Number of bars per minute	times	Number of minutes	is	Number of bars wrapped
134	×	m	=	8710

Solve.
$$134 \times m = 8710$$
$$\frac{134 \times m}{134} = \frac{8710}{134}$$
$$m = 65$$

Check. $134 \cdot 65 = 8710$, so the answer checks.

State. It takes 65 min to wrap 8710 candy bars.

76. *Familiarize*. Let $f =$ the number of yards of fabric remaining on the bolt.

Translate.

Original amount	minus	Amount purchased	is	Amount remaining
$8\frac{1}{4}$	−	$3\frac{5}{8}$	=	f

Solve. We carry out the subtraction.
$$8\,\boxed{\frac{1}{4} \cdot \frac{2}{2}} = 8\frac{2}{8} = 7\frac{10}{8}$$
$$-3\frac{5}{8} = -3\frac{5}{8} = -3\;\frac{5}{8}$$
$$\rule{3cm}{0.4pt}$$
$$4\frac{5}{8}$$

Thus, $f = 4\frac{5}{8}$.

Check. $3\frac{5}{8} + 4\frac{5}{8} = 7\frac{10}{8} = 8\frac{2}{8} = 8\frac{1}{4}$, so the answer checks.

State. $4\frac{5}{8}$ yd of fabric remains on the bolt.

77. *Familiarize*. Let $t =$ the length of the trip, in miles.

Translate.

Starting mileage	plus	Miles driven	is	Ending mileage
27,428.6	+	t	=	27,914.5

Solve.
$$27,428.6 + t = 27,914.5$$
$$27,428.6 + t - 27,428.6 = 27,914.5 - 27,428.6$$
$$t = 485.9$$

Check. $27,428.6 + 485.9 = 27,914.5$, so the answer checks.

State. The trip was 485.9 mi long.

78. _Familiarize_. Let t = the amount that remains after the taxes are paid.

Translate.

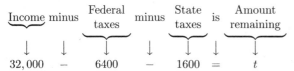

Income	minus	Federal taxes	minus	State taxes	is	Amount remaining
↓	↓	↓	↓	↓	↓	↓
32,000	−	6400	−	1600	=	t

Solve. We carry out the calculations on the left side of the equation.

$$32,000 - 6400 - 1600 = t$$
$$25,600 - 1600 = t$$
$$24,000 = t$$

Check. The total taxes paid were $6400 + $1600, or $8000, and $32,000 − $8000 = $24,000 so the answer checks.

State. $24,000 remains after the taxes are paid.

79. _Familiarize_. Let p = the amount Shannon was paid.

Translate.

Daily pay	times	Number of days	is	Amount paid
↓	↓	↓	↓	↓
85	×	7	=	p

Solve. We carry out the multiplication.

$$85 \times 7 = p$$
$$595 = p$$

Check. We can repeat the calculation. The answer checks.

State. Shannon was paid $595.

80. _Familiarize_. Let d = the distance the child would walk in $\frac{1}{2}$ hr, in kilometers.

Translate.

Speed	times	Time	is	Distance
↓	↓	↓	↓	↓
$\frac{3}{5}$	×	$\frac{1}{2}$	=	d

Solve. We carry out the multiplication.

$$\frac{3}{5} \times \frac{1}{2} = d$$
$$\frac{3}{10} = d$$

Check. We can repeat the calculation. The answer checks.

State. The child would walk $\frac{3}{10}$ km in $\frac{1}{2}$ hr.

81. _Familiarize_. Let s = the cost of each dress.

Translate.

Cost of each dress	times	Number of dresses	is	Total cost
↓	↓	↓	↓	↓
s	×	8	=	679.68

Solve.

$$s \times 8 = 679.68$$
$$\frac{s \times 8}{8} = \frac{679.68}{8}$$
$$s = 84.96$$

Check. $8 \cdot \$84.96 = \$679.68$, so the answer checks.

State. Each dress cost $84.96.

82. _Familiarize_. Let p = the number of gallons of paint needed to cover 3250 ft².

Translate. We translate to a proportion.

$$\text{Gallons} \rightarrow \frac{8}{2000} = \frac{p}{3250} \leftarrow \text{Gallons}$$
$$\text{Area covered} \rightarrow \phantom{\frac{8}{2000}} \phantom{\frac{p}{3250}} \leftarrow \text{Area covered}$$

Solve. We equate cross products.

$$\frac{8}{2000} = \frac{p}{3250}$$
$$8 \cdot 3250 = 2000 \cdot p$$
$$\frac{8 \cdot 3250}{2000} = \frac{2000 \cdot p}{2000}$$
$$13 = p$$

Check. We can substitute in the proportion and check the cross products.

$$\frac{8}{2000} = \frac{13}{3250}; \; 8 \cdot 3250 = 26,000; \; 2000 \cdot 13 = 26,000$$

The cross products are the same so the answer checks.

State. 13 gal of paint is needed to cover 3250 ft².

83. $\dfrac{\$3.06}{18 \text{ oz}} = \dfrac{306\cancel{c}}{18 \text{ oz}} = 17\cancel{c}/\text{oz}$

84. $I = P \cdot r \cdot t$

$$= \$4000 \times 8\% \times \frac{3}{4}$$
$$= \$4000 \times 0.08 \times \frac{3}{4}$$
$$= \$240$$

85. Commission = Commission rate × Sales

$$5800 = r \times 84,000$$

We divide both sides of the equation by 84,000 to find r.

$$\frac{5880}{84,000} = \frac{r \times 84,000}{84,000}$$
$$0.07 = r$$
$$7\% = r$$

The commission rate is 7%.

86. _Familiarize_. Let p = the population after a year.

Translate.

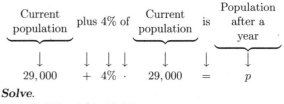

Current population	plus 4% of			Current population	is	Population after a year
↓	↓	↓	↓	↓	↓	↓
29,000	+	4%	·	29,000	=	p

Solve.

$$29,000 + 0.04 \cdot 29,000 = p$$
$$29,000 + 1160 = p$$
$$30,160 = p$$

Check. The new population will be 104% of the original population. Since 104% of $29,000 = 1.04 \cdot 29,000 = 30,160$, the answer checks.

State. After a year the population will be 30,160.

87. Familiarize. Let m = the number of miles Luis drove. We will express 15¢ as $0.15.

Translate.

Daily charge	plus	Rate per mile	times	Number of miles	is	Total cost
↓	↓	↓	↓	↓	↓	↓
35	+	0.15	·	m	=	68

Solve.

$$35 + 0.15 \cdot m = 68$$
$$35 + 0.15 \cdot m - 35 = 68 - 35$$
$$0.15 \cdot m = 33$$
$$\frac{0.15 \cdot m}{0.15} = \frac{33}{0.15}$$
$$m = 220$$

Check. The cost of driving 220 mi at $0.15 per mile is $220(\$0.15) = \$33$, and $\$35 + \$33 = \$68$, the total cost of the rental. The answer checks.

State. Luis drove 220 mi.

88. Familiarize. Let d = the dosage of Phenytoin, in mg, recommended for a child who weighs 32 kg.

Translate. We translate to a proportion.

$$\text{Weight} \rightarrow \frac{24}{42} = \frac{32}{d} \leftarrow \text{Weight} \atop \leftarrow \text{Dosage}$$

Solve.

$$\frac{24}{42} = \frac{32}{d}$$
$$24 \cdot d = 42 \cdot 32$$
$$\frac{24 \cdot d}{24} = \frac{42 \cdot 32}{24}$$
$$d = 56$$

Check. We substitute in the proportion and check the cross products.

$$\frac{24}{42} = \frac{32}{56}; \ 24 \cdot 56 = 1344; \ 42 \cdot 32 = 1344$$

The cross products are the same.

State. The recommended dosage of Phenytoin for a child who weighs 32 kg is 56 mg.

89. Familiarize. First we will find the volume V of the water. Then we will find its weight, w.

Translate and Solve.

$$V = l \cdot w \cdot h$$
$$V = 60 \text{ ft} \cdot 25 \text{ ft} \cdot 1 \text{ ft}$$
$$V = 1500 \text{ ft}^3$$

Weight of water	is	Volume of water	times	Weight per cubic foot
↓	↓	↓	↓	↓
w	=	1500	×	$62\frac{1}{2}$

We carry out the multiplication.

$$w = 1500 \times 62\frac{1}{2} = 1500 \times \frac{125}{2}$$
$$= \frac{1500 \times 125}{2} = \frac{\not{2} \times 750 \times 125}{\not{2} \times 1}$$
$$= 93,750$$

Check. We repeat the calculations. The answer checks.

State. The water weighs 93,750 lb.

90. $18^2 = 18 \cdot 18 = 324$

91. $37^0 = 1$ ($b^0 = 1$ for any nonzero number b.)

92. $\sqrt{121} = 11$

The square root of 121 is 11 because $11^2 = 121$.

93. $4^{-3} = \frac{1}{4^3} = \frac{1}{4 \cdot 4 \cdot 4} = \frac{1}{64}$

94. $\left(\frac{5}{4}\right)^{-2} = \left(\frac{4}{5}\right)^2 = \frac{4}{5} \cdot \frac{4}{5} = \frac{16}{25}$

95. $4.357,000.$

↑_____⌋ 6 places

Large number, so the exponent is positive.

$4,357,000 = 4.357 \times 10^6$

96. $(6.2 \times 10^7)(4.3 \times 10^{-23}) = (6.2 \cdot 4.3) \times (10^7 \cdot 10^{-23})$
$$= 26.66 \times 10^{-16}$$
$$= (2.666 \times 10) \times 10^{-16}$$
$$= 2.666 \times (10 \cdot 10^{-16})$$
$$= 2.666 \times 10^{-15}$$

97. To plot $(-5, 2)$, we locate -5 on the horizontal axis. From there go up 2 units and make a dot.

To plot $(4, 0)$, we locate 4 on the horizontal axis. Since the second coordinate is 0, we do not go up or down from that point. Make a dot there.

To plot $(3, -4)$, locate 3 on the horizontal axis. From there go down 4 units and make a dot.

To plot $(0, 2)$, do not move right or left from the origin on the horizontal axis (since the first coordinate is 0). From the origin, go up 2 units and make a dot.

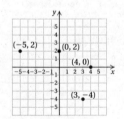

98. Graph $y = -\frac{1}{3}x$

Make a table of solutions, plot the points, and draw and label the line. Choosing x-values that are multiples of 3 avoids fraction values for y.

When x is -3, $y = -\frac{1}{3}(-3) = 1$.

When x is 0, $y = -\frac{1}{3} \cdot 0 = 0$.

When x is 3, $y = -\frac{1}{3} \cdot 3 = -1$.

x	y	(x, y)
-3	1	$(-3, 1)$
0	0	$(0, 0)$
3	-1	$(3, -1)$

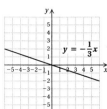

99.
$$\frac{24}{x} = \frac{15}{10}$$
$$24 \cdot 10 = 15 \cdot x$$
$$\frac{24 \cdot 10}{15} = x$$
$$16 = x$$

$$\frac{21}{y} = \frac{15}{10}$$
$$21 \cdot 10 = 15 \cdot y$$
$$\frac{21 \cdot 10}{15} = y$$
$$14 = y$$

100.
$$\frac{1}{3} \text{ yd} = \frac{1}{3} \times 1 \text{ yd}$$
$$= \frac{1}{3} \times 36 \text{ in.}$$
$$= \frac{36}{3} \text{ in.}$$
$$= 12 \text{ in.}$$

101. 3917 mm = _____ cm

Think: To go from mm to cm in the table is a move of 1 place to the left. Thus, we move the decimal point 1 place to the left.

3917   391.7.

$\uparrow\!\!\lrcorner$

3917 mm = 391.7 cm

102. 5.8 km = _____ m

Think: To go from km to m in the table is a move of 3 places to the right. Thus, we move the decimal point 3 places to the right.

5.8   5.800.

$\llcorner\!\!\longrightarrow\uparrow$

5.8 km = 5800 m

103. 60,000 g = _____ kg

Think: To go from g to kg in the table is a move of 3 places to the left. Thus, we move the decimal point 3 places to the left.

60,000   60.000.

$\uparrow\,\llcorner\!\!\lrcorner$

60,000 g = 60 kg

104. 10 lb = 10×1 lb
$$= 10 \times 16 \text{ oz}$$
$$= 160 \text{ oz}$$

105. 2.3 g = _____ mg

Think: To go from g to mg in the table is a move of 3 places to the right. Thus, we move the decimal point 3 places to the right.

2.3   2.300.

$\llcorner\!\!\longrightarrow\uparrow$

2.3 g = 2300 mg

106. 8190 mL = 8190×1 mL
$$= 8190 \times 0.001 \text{ L}$$
$$= 8.19 \text{ L}$$

107. 28 qt = $28 \,\cancel{\text{qt}} \times \dfrac{1 \text{ gal}}{4 \,\cancel{\text{qt}}}$
$$= \frac{28}{4} \times 1 \text{ gal}$$
$$= 7 \text{ gal}$$

108. Using a circle with 100 equally-spaced tick marks we first draw a line from the center to any tick mark. From that tick mark we count off 3 tick marks to graph 3% and label the wedge none and 3%. We continue in this manner with the other number of salads eaten per week. Finally we title the graph "Number of Salads Per Week." The graph is found on page A-31 in the text.

109. We will make a vertical bar graph. On the horizontal scale in four equally-spaced intervals indicate the number of salads eaten per week. Label this scale "Salads per Week." Then make ten equally-spaced tick marks on the vertical scale and label them by 10's. Label this scale "Percent." Finally draw vertical bars above the numbers of salads eaten to show the percents.

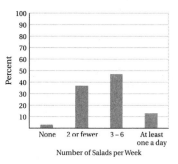

110. $P = 2 \cdot l + 2 \cdot w$
$$P = 2 \cdot 24 \text{ in.} + 2 \cdot 20 \text{ in.}$$
$$P = 48 \text{ in.} + 40 \text{ in.}$$
$$P = 88 \text{ in.}$$

111. $A = b \cdot h$
$$A = 15.4 \text{ cm} \cdot 4 \text{ cm}$$
$$A = 61.6 \text{ cm}^2$$

112. $A = \frac{1}{2} \cdot b \cdot h$

$A = \frac{1}{2} \cdot 10 \text{ in.} \cdot 5 \text{ in.}$

$A = 25 \text{ in}^2$

113. $A = \frac{1}{2} \cdot h \cdot (a + b)$

$A = \frac{1}{2} \cdot 8.3 \text{ yd} \cdot (10.8 \text{ yd} + 20.2 \text{ yd})$

$A = \frac{8.3 \cdot 31}{2} \text{ yd}^2$

$A = 128.65 \text{ yd}^2$

114. The area is the sum of the areas of a rectangle that measures 10.3 m by 2.5 m and a semicircle with diameter 10.3 m. First we find the area of the rectangle.

$A = l \cdot w$

$A = 10.3 \text{ m} \cdot 2.5 \text{ m} = 25.75 \text{ m}^2$

Next we find one-half the area of a circle with diameter 10.3 m.

$r = \frac{d}{2} = \frac{10.3 \text{ m}}{2} = 5.15 \text{ m}$

$A = \frac{1}{2} \cdot \pi \cdot r \cdot r$

$A \approx \frac{1}{2} \cdot 3.14 \cdot 5.15 \text{ m} \cdot 5.15 \text{ m} = 41.640325 \text{ m}^2$

Finally, we add to find the total area.

$25.75 \text{ m}^2 + 41.640325 \text{ m}^2 = 67.390325 \text{ m}^2$

115. $d = 2 \cdot r = 2 \cdot 10.4 \text{ in.} = 20.8 \text{ in.}$

$C = 2 \cdot \pi \cdot r$

$C \approx 2 \cdot 3.14 \cdot 10.4 \text{ in.} = 65.312 \text{ in.}$

$A = \pi \cdot r \cdot r$

$A \approx 3.14 \cdot 10.4 \text{ in.} \cdot 10.4 \text{ in.} = 339.6224 \text{ in}^2$

116. $V = l \cdot w \cdot h$

$V = 10 \text{ m} \cdot 2.3 \text{ m} \cdot 2.3 \text{ m}$

$V = 23 \cdot 2.3 \text{ m}^3$

$V = 52.9 \text{ m}^3$

117. $V = Bh = \pi \cdot r^2 \cdot h$

$V \approx 3.14 \cdot 4 \text{ ft} \cdot 4 \text{ ft} \cdot 16 \text{ ft}$

$V = 803.84 \text{ ft}^3$

118. $V = \frac{4}{3} \cdot \pi \cdot r^3$

$V \approx \frac{4}{3} \cdot 3.14 \cdot (4 \text{ mi})^3$

$V = \frac{4 \cdot 3.14 \cdot 64 \text{ mi}^3}{3}$

$V = 267.94\overline{6} \text{ mi}^3$

119. $a^2 + b^2 = c^2$

$a^2 + 6^2 = 11^2$

$a^2 + 36 = 121$

$a^2 = 121 - 36$

$a^2 = 85$

$a = \sqrt{85} \text{ ft} \qquad \text{Exact answer}$

$a \approx 9.220 \text{ ft} \qquad \text{Approximation}$